Asteroids

Viorel Badescu
Editor

Asteroids

Prospective Energy and Material Resources

Springer

Editor
Viorel Badescu
Polytechnic University of Bucharest
Candida Oancea Institute
Bucharest
Romania

ISBN 978-3-642-39243-6 ISBN 978-3-642-39244-3 (eBook)
DOI 10.1007/978-3-642-39244-3
Springer Heidelberg New York Dordrecht London

Library of Congress Control Number: 2013941374

Printed on acid-free paper

Springer is part of Springer Science+Business Media (www.springer.com)

Motto

And the fox said to the little prince: men have forgotten this truth, but you must not forget it. You become responsible, forever, for what you have tamed.

Antoine de Saint-Exupery, The Little Prince

Foreword: Asteroids in the Service of Humanity

There are at least three compelling reasons for the human race to initiate a major programme to explore and better understand the 'minor planets' of the Solar System:

(1) Enhancing scientific knowledge: Asteroids represent material that failed to get incorporated into planets when the Solar System formed. As such they constitute a metaphorical 'gold mine' of scientific information relating to the properties of the early Sun, astrophysical processes in the protoplanetary disk, and the early stages of planetesimal formation and evolution (US National Research Council, 2011). The fact that many asteroids (i.e. the parent bodies of chondritic meteorites) are undifferentiated, and therefore largely unaltered since the solar system formed, is especially important in this respect. Moreover, the differentiated asteroids (i.e. the sources of metallic and achondritic meteorites) are also of great scientific interest, as their study will shed light on our understanding of the earliest stages of planetary differentiation into cores, mantles and crusts. These scientific objectives can only be fully addressed in the context of an ambitious programme of space exploration able to conduct *in situ* investigations of, and sample collection and return to Earth from, a large number of different types of asteroid.

(2) Mitigating the impact hazard: Some asteroids, especially those near-Earth asteroids (NEAs) on Earth-crossing orbits, represent an impact hazard for our planet. Based on data from NASA's Wide-Field Infrared Survey Explorer (WISE) spacecraft, more than 20,000 NEA's larger than 100m across are thought to exist, and there are probably over a million smaller objects (Mainzer et al, 2011). Although the probability of a civilisation-destroying impact is low (no more than one is expected every 100,000 years), as the Tunguska (1908) and Chelyabinsk (2013) impacts remind us, the Earth is hit by asteroidal (and/or cometary) debris capable of causing significant damage and loss of life much more frequently. In order to better mitigate this threat we need to increase our knowledge of the numbers and orbits of NEAs (especially for the smaller sizes where current catalogues are incomplete), and start developing techniques to disrupt or deflect any objects

found on Earth collision trajectories. Achieving that latter capability requires both a spacefaring capability able to visit asteroids at will (and possibly at short notice) in order to place explosive or propulsive devices on their surfaces, *and* a detailed understanding of the nature of asteroidal surfaces so that we can predict how they will react to such interventions.

(3) Utilizing extraterrestrial resources: Asteroids represent a significant potential resource of raw materials, both in support of continued space exploration activities and for the wider global economy (e.g. Martin, 1985; Hartmann, 1986; Lewis et al., 1993; see also other chapters in this book). Many NEAs are relatively easy to reach in energy terms and have very low surface gravities, which would minimise the cost of transferring materials extracted from them to the vicinity of the Earth. Moreover, for many of these objects nature has already performed significant refining, or at least beneficiation, for us. For example, metallic asteroids (which constitute a few percent of the NEA population) consist of essentially pure nickel-iron alloy, and although Earth has significant reserves of both these elements they may still be very useful in the context of future space development. Perhaps of greater interest is the fact that metallic asteroids also contain about one hundred parts per million of gold and platinum group elements (PGEs), which are of sufficiently high value (for example as industrial catalysts) that they may be worth importing directly to Earth (e.g. Kargel, 1994). At today's prices for these elements ($20,000 to $50,000 per kilogram) it follows that a single small metallic asteroid about 200 metres across could be worth of the order of $100 billion dollars. Thus, in addition to being metaphorical scientific gold mines, some asteroids may prove to be *literal* gold mines as well! Moreover, although essentially rocky objects, ordinary chondritic asteroids (which probably account for the majority of NEAs) themselves consist of several percent Ni-Fe metal, which similarly contains hundred ppm-levels of PGEs. In addition, carbonaceous chondrites (which make up perhaps 10-15% of NEAs) are relatively rich in volatiles, which could be of great value to a future space economy by providing water, hydrogen, and oxygen for future space missions without the need to haul these materials out of Earth's gravity. Last but not least, there are also strong environmental arguments for mining even relatively common materials (such as iron, nickel, copper, and the increasingly important rare earth elements) from asteroids as an alternative to invasive strip-mining on Earth – asteroids do not have indigenous ecosystems that may be disrupted by mining activities whereas our planet does (see the discussion by Hartmann, 1986). For all these reasons, developing the capability of extracting useful resources from asteroids, and from other extraterrestrial sources, can be seen as an important investment in the future of the world economy (e.g. Crawford, 1995).

This book focusses mainly on the latter of these motivations, but clearly there are strong synergies between all three. For example, before either impact mitigation or economic utilization can be implemented it will be necessary to learn a great deal more about the nature and compositions of NEAs, both in general and for specific objects of interest. This implies initiating a major programme of scientific investigation, using both astronomical remote-sensing and *in situ* spacecraft investigations.

Moreover, the potential synergies also act the other way. As Martin Elvis (2012, see also his chapter in this book) has convincingly argued, the human and robotic missions needed to fully explore the Solar System, as well as the next generation of space telescopes required to advance our knowledge of the wider universe, will be so expensive that they may be unaffordable unless additional sources of funding can be identified. Leveraging some of the economic wealth locked up in NEAs (for example in the PGEs) may be one way to help finance ambitious future space exploration activities. And, of course, it goes without saying that the kind of infrastructural investments that will be required to extract the economic wealth of asteroids will be essentially the same as those required to destroy or deflect Earth-impacting asteroids should this ever prove to be necessary.

All these activities would benefit from greater international cooperation in space exploration by the World's space agencies, and the recognition that asteroids are important targets for human and robotic exploration. In this respect it is heartening that in September 2011 twelve of the world's space agencies came together, under the auspices of the International Space Exploration Coordination Group (ISECG), to produce a Global Exploration Roadmap for the human and robotic exploration of the inner Solar System (ISECG, 2011). In addition to missions to the Moon and Mars, which are important exploration targets in their own right, the Roadmap includes a strong focus on NEAs. Implementation of the Global Exploration Roadmap would both contribute to, and benefit from, the economic utililization of asteroidal resources.

Much also depends on the extent to which the cost of access to space can be reduced by new launch service providers such as SpaceX, or by new generations of vehicles such as the Skylon re-useable space-plane concept (e.g. Hempsell, 2010). The more the cost of accessing space can be reduced the greater will be the incentives for entrepreneurs to invest, and the sooner the economic utilization of asteroidal resources is likely to begin. By providing raw materials for *in situ* resource utilization (ISRU) for future space activities, as well as possible additional financing arising from the sale of valuable raw materials, asteroid mining may greatly facilitate ambitious space exploration activities such as envisaged by the Global Exploration Roadmap. Ultimately, it is not too much to hope that these activities will result in the creation of an economic and industrial infrastructure in the inner Solar System from which all humanity will benefit.

For all of these reasons, a book reviewing opportunities for the exploration and utilisation of asteroids is especially timely. Professor Viorel Badescu is therefore to be congratulated for compiling this volume, which I am sure will make a lasting contribution to the field.

Ian A. Crawford
Department of Earth and Planetary Sciences,
Birkbeck College, University of London

References

1. Crawford, I.A.: Space development: social and political implications. Space Policy 11, 219–225 (1995)
2. Elvis, M.: Let's mine asteroids —for science and profit. Nature 485, 549 (2012)
3. Hartmann, W.K.: Space exploration and environmental issues. In: Hargrove, E.C. (ed.) Beyond Spaceship Earth: Environmental Ethics and the Solar System, pp. 119–139. Sierra Club Books, San Francisco (1986)
4. Hempsell, M.: An Analysis of the Skylon Infrastructure system. Journal of the British Interplanetary Society 63, 129–135 (2010)
5. ISECG, The Global Exploration Roadmap (2011), `http://www.nasa.gov/pdf/591067main_GER_2011_small_single.pdf`
6. Kargel, J.S.: Metalliferous asteroids as potential sources of precious metals. Journal of Geophysical Research 99, 21129–21141 (1994)
7. Lewis, J.S., Matthews, M.S., Guerrieri, M.L. (eds.): Resources of Near-Earth Space, p. 977. University of Arizona Press, Tucson (1993)
8. Mainzer, A., et al.: NEOWISE Observations of Near-Earth Objects: Preliminary Results. Astrophysical Journal 743, 156 (2011)
9. Martin, A.R.: Space Resources and the Limits to Growth. Journal of the British Interplanetary Society 38, 243–252 (1985)
10. US National Research Council, Vision and voyages for planetary science in the decade 2013-2022. National Research Council. National Academies Press, Washington, DC (2011)

Preface

This is a third volume of a Springer book series making an inventory of the material and energy resources of our Solar system. The first two books, referring to resources existent on Mars and Moon, were published in 2009 and 2012, respectively.

This book presents a present-day perspective on the asteroids' energy and material resources for prospective human use. One investigates the advantages and limitations of various systems thought-out for future mankind utilization. The book collects together recent proposals and innovative options and solutions. It is a good starting point for researchers involved in current and impending asteroid-related activities.

The book is structured along logical lines of progressive thought and may be conceptually divided into eight sections.

The first section deals with *asteroid detection* and contains three chapters. After the introductory Chap. 1, showing what we have learned from twenty years of space exploration, Chap. 2 refers to Trojan asteroids in the inner Solar System and Chap. 3 describes the orbital and dynamical characteristics of small bodies in the region of inner planets.

The second section of the book deals with *asteroid characterization* and contains four chapters. Chapter 4 enters the details of prospecting the asteroid resources while Chap. 5 describes asteroid models for target selection and mission planning. Chapter 6 presents the first steps towards the utilization of asteroid resources and Chap. 7 shows a system perspective concerning human missions to Near Earth Objects (NEOs).

The third section of the book, dealing with the *design, building and operation of robotic miners,* consists of three chapters. Chapter 8 shows how robots are designed for gravity-independent locomotion while Chap. 9 describes bio-inspired landing approaches and their potential use on extraterrestrial bodies. Also, Chap. 10 presents the electric power system options for robotic miners.

The fourth section of the book, referring to *technology and in-situ resources utilization (ISRU),* consists of seven chapters. Chapter 11 describes the granular physics of the rubble-pile NEOs while Chap. 12 presents the anchoring and sample acquisition approaches for asteroid In situ Resource Utilization. Chapters 13 and 14 refer to the closed-cycle pneumatics for asteroid regolith mining and to the extraction of the asteroidal mass for robotic construction, respectively. The curing of the construction composite materials on asteroids is presented in Chap. 15 while in Chap. 16 the architecture of an asteroid-mining spacecraft is described. Chapter 17 refers to prospecting, orbit modification, mining and habitation of Near Earth Asteroids (NEAs).

The fifth section of the book, referring *to transportation between Earth, low Earth orbit and asteroids,* consists of five chapters. Chapter 18 deals with available asteroid resources in the Earth' neighborhood. Chapters 19 and 20 refer to asteroid capture and how the asteroid trajectory may be changed, respectively. The opportunities for asteroid retrieval missions are presented in Chap. 21 while Chap. 22 focuses on shaped metal Earth-delivery systems.

The sixth section of the book, dealing with *less common asteroid utilizations,* consists of five chapters. Chapters 23 and 24 refer to artificial gravitation on asteroids and on how to make the asteroids habitable, respectively. The usage of asteroid resources for space-based geo-engineering is treated in Chap. 25 while Chap. 26 shows how the asteroids may be used for launching and/or changing the trajectory and acceleration of space ships. Chapter 27 describes the observation of asteroids for searching extraterrestrial artifacts.

The seventh section of the book, dealing with the *management of the space environment*, consists of Chap. 28, with commercial concerns and considerations.

Finally, the eighth section of the book refers to *laws and regulation* and consists of Chap. 29 with legal considerations on asteroid exploitation and deflection.

Most of the chapters are interdisciplinary by nature and some of them might be equally well included into more than one section. For example, Chaps. 12, 13, 16 and 17 may be included into the third section, while Chap. 14 could be associated with the third or fifth sections.

More details about the twenty-nine chapters of the book are given below.

Chapter 1, by Ivano Bertini, shows that asteroids are the remnant debris from the formation of the inner solar system. Their study provides therefore important clues on the original materials and the mechanisms which formed the terrestrial planets. These small bodies can be object of a strong interaction with Earth's biosphere both as possible carriers of water and organics to our planet in the past and as potential impact hazards. They also represent an extraordinary source of minerals which can be exploited for the increasing necessities of our civilization and for the future exploration and colonization of the solar system. In the last twenty years dedicated space missions revealed asteroids as intriguing complex worlds, combining the advantages of observing in a wider range of wavelengths and geometries with respect to Earth to the high resolution possible only with close encounters. Spacecraft data have so far complemented, improved, and in many cases revolutionized, the theories and findings derived from ground-based data. This chapter contains a summary of the main scientific results obtained in these exciting years of asteroid space exploration, starting from the encounter of the NASA Galileo mission with asteroid (951) Gaspra in 1991 and ending with the visit of asteroid (4179) Toutatis by the Chinese Cheng'E 2 spacecraft in 2012. Future missions already approved and under development by the various space agencies are also outlined at the end of this work.

Chapter 2, by Michael Todd, shows that most of the asteroids in our Solar System are located between the orbits of Mars and Jupiter; however there are many less populated regions where asteroids are found. One special case is the Trojan asteroids, which occupy the so-called stable Lagrangian regions and orbit the Sun in a 1:1 mean motion resonance with a planet. Very few Trojan asteroids have been discovered in the inner Solar System in the orbits of Earth and Mars. Modelling and simulations indicate that there should be more waiting to be discovered. The regions in which Trojans for Earth and Mars are most likely to exist have been identified. The probability distributions for Earth and Mars show that the longitudes where Trojans are most likely to exist are consistent with classical Lagrangian points, but that they are much more likely to be inclined orbits. The small number of predicted Trojan asteroids means the chance of discovery is small. The optimal method of searching is to concentrate on the regions of highest probability near the Lagrangian points. In these regions the amount of sky area to be searched is still very high, about 3500 deg^2 for Earth Trojans and up to 17000 deg^2 for Mars Trojans. A survey of the whole area would be time-consuming, but could be supplemented by examining the data from other observing programmes.

Chapter 3, by Bojan Novaković, shows that for planetary scientists, asteroids are real treasures, because they are very useful in a wide variety of topics. Moreover, in the last years it becomes evident that an asteroid mining mission is feasible, making asteroids valuable economic resources as well. Those objects located in the region of terrestrial planets are of particular interest because of their proximity to the Earth. Three groups of asteroids approach inner planets. These are Inner Earth Objects (IEOs), Near-Earth Asteroids (NEAs), and Mars Crossers (MCs). In this chapter the author has reviewed basic gravitational and non-gravitational mechanisms at work, such are resonances and Yarkovsky effect, which control the dynamic of an asteroid located in this region. Both, short- and long-term dynamical characteristics have been discussed. Furthermore, the advantages (or disadvantages) of different orbital sub-classes with respect to other classes, from astro-dynamical point of view, have been analyzed. The populations of asteroids in the region of terrestrial planets have a relatively short dynamical lifetime. Thus, besides the objects present currently in this region, of grate importance are also asteroids located in the possible source regions, which continuously supply new objects to the region of terrestrial planets. A reasonable hypothesis is that these source regions are related to the most prominent gaps in the main asteroid belt, in the distribution of orbital semi-major axis. This is because these gaps (known as Kirkwood gaps) correspond to the most powerful mean-motion resonances, and it is well known that asteroids become planet crossers by increasing their orbital eccentricity under the action of a variety of resonant phenomena. The most plausible mechanisms of transport from the main-asteroid belt to the region of terrestrial planets, as well as contribution of different resonances, have been reviewed here.

Chapter 4, by Martin Elvis, shows that prospecting the asteroids for valuable resources - ore - has three stages: (1) Discovery, including orbit determination; (2) Remote, telescopic, characterization; and (3) Local characterization using robotic spacecraft. The first two of these phases uses exclusively astronomical techniques: imaging, astrometry, photometry, and spectroscopy. The implementation of the third phase depends heavily on how good the results of the first two phases are, and will make use of space astronomy, and planetary sciences, techniques. In this review a brief outline of the prospecting problem is given. Then each prospecting phase is reviewed, taking care to explain astronomical terms, but retaining them so that the reader should be able to read the original research literature. Brief consideration is given to timescales and costs for implementing an adequate program of near-Earth object characterization on a decade timescale, in order to enable asteroid mining to begin.

Chapter 5, by Mikko Kaasalainen and Josef Durech, gives an overview of the pre-mission sources of information about asteroids. The data available to us are mostly Earth-based, from ground or satellite observatories. With sufficient remote-sensing data, we can construct models of individual asteroids by solving the corresponding inverse problems. The most extensive source for modeling is photometry, available for hundreds of thousands of asteroids via all-sky surveys, while the most detailed models are obtained with radar for near-Earth asteroids. Other data sources are most effectively used together with photometry, such as adaptive optics or interferometry for main-belt asteroids. The authors discuss the use of these data sources and stellar occultations, thermal infrared data, and spectroscopy for determining or estimating the shapes, spin states, sizes, mass/density, structure, and composition of asteroids. These sources can be used for a top-to-bottom scenario for selecting interesting targets and obtaining increasingly detailed ground-based information about them, from lower to higher resolution. This can be followed by in situ data from flybys and radio tomography prior to surface operations.

Chapter 6, by Mikael Granvik, Robert Jedicke, Bryce Bolin, Monique Chyba, Geoff Patterson and Gautier Picot, shows that it has recently been predicted that the Earth is surrounded by a cloud of very small, meter- and sub-meter-sized asteroids that have been temporarily captured from the near-Earth-object population into Earth-centered orbits. The >2.1-meter-diameter asteroid 2006 RH_{120} is the first natural object known to have been temporarily captured by the Earth and it agrees virtually perfectly with the predicted characteristics of these objects. We expect that within a decade the ever-improving survey systems will allow these objects to be discovered regularly. Temporarily-captured Earth orbiters (TCOs) form a natural target for any asteroid mission, because they are energetically very easy, and thus inexpensive, to get to. The authors review our current knowledge of the characteristics of the TCO population and their detectability, and argue that sample- or asteroid-return missions to TCOs would benefit entities interested in the utilization of asteroid resources. The authors discuss the feasibility of a TCO rendezvous mission in terms of possible transfer orbits and highlight some of the other remaining challenges. Discovering a target object with enough lead time is arguably a major challenge for a space mission to a TCO.

Chapter 7, by Marco Cenzon and Dragoş Alexandru Păun, shows that Near Earth Objects have attracted humanity's fascination for centuries but it is only in the last two decades that our understanding of these objects has shed light on their potentially destructive capabilities as well as their potential to fuel humanity's needs both on Earth as well as beyond low Earth orbit. The purpose of the present chapter is that of providing overview on the state of the art regarding human missions to NEOs, the fundamental challenges that such missions would face and what would the benefits be. Furthermore, the authors propose a flexible architecture aimed at serving both as a human NEO study platform and as an industrial NEO utilization/exploitation platform.

Chapter 8, by Marco Chacin, shows that asteroids' physical characteristics provide a very hostile environment distinguished by the absence of gravity. In recent years, the scientific community has had an increased interest in exploring the asteroids of the solar system. The in-situ study of asteroids can lead to important scientific findings in the effort to map the main asteroid belt. However, relatively little attention from planetary scientists and planetary robotics engineers has been focused on surface mobility on asteroids due to the challenging gravity conditions. As a result, there exists some risk that premature conclusions about the feasibility of stable mobility on asteroid surfaces may be drawn without thorough consideration of all possible alternatives. In spite of these difficulties it is clear that in order to maximize the scientific return from any given mission on an asteroid's surface, future missions must have the ability to conduct stable mobility and accurate positioning on the rough terrain. Strategies would require a closer look at stability control against the forces interacting between bodies in a microgravity environment. In this chapter, various approaches to gravity-independent locomotion on weak gravity surfaces of asteroids are discussed along with related technologies that could improve the operational performance and efficiency of robots capable of position-based controlled motion. Challenges are also described that affect planning and control of surface exploration robots that use hopping and rolling mechanisms and/or articulated limbs for the ground contact. An in-depth representative example of an asteroid mobility solution and control approach is provided. The control approach considers reaction and friction forces with the asteroid surface and is demonstrated using a prototype robot and laboratory testbed that emulates microgravity. This example considers issues that most solutions must address related to the microgravity environment and its effect on dynamics of robotic systems on asteroids.

Chapter 9, by Thibaut Raharijaona, Guillaume Sabiron, Stephane Viollet, Nicolas Franceschini and Franck Ruffier, states that automatic landing on extraterrestrial bodies is still a challenging and hazardous task. The authors propose a new type of autopilot designed to solve landing problems, which is based on neurophysiological, behavioral, and biorobotic findings on flying insects. Flying insects excel in optic flow sensing techniques and cope with highly parallel data at a low energy and computational cost using lightweight dedicated motion processing circuits. In the first part of this chapter, the authors present a biomimetic approach in the context of a lunar landing scenario, assuming a 2-degree-of-freedom spacecraft approaching the moon, which is simulated with the PANGU software. The autopilot proposed by the authors relies only on optic flow (OF) and inertial measurements, and aims at regulating the OF generated during the landing approach, by means of a feedback control system whose sensor is an OF sensor. The authors put forward an estimation method based on a two-sensor setup to accurately estimate the orientation of the lander's velocity vector, which is mandatory to control the lander's pitch in a near optimal way with respect to the fuel consumption. In the second part, the authors present a lightweight Visual Motion Sensor (VMS) which draws on the results of neurophysiological studies on the insect visual system. The VMS was able to perform local 1-D angular speed measurements in the range 1.5^o/s - 25°/s. The sensor was mounted on an 80 kg unmanned helicopter and test-flown outdoors over various fields. The OF measured onboard was shown to match the ground-truth optic flow despite the dramatic disturbances and vibrations experienced by the sensor.

Chapter 10, by Simon Fraser, shows that one of the most important technological challenges in designing mission elements for surface operations on asteroids is power system design. A safe and in terms of available energy and output power sufficiently dimensioned power system is essential to fulfill the goals of each mission element operated on the surface of asteroids. Power system design for asteroid surface applications therefore requires a very careful evaluation of the possibilities and limitations of each individual power system element in order to find a single element - or a hybrid system of multiple power system elements - satisfying the energy storage and power generation requirements of a specific asteroid surface exploration and/or exploitation system in the best possible way; and this not only with respect to technical performance data, but also with respect to the economical framework of a commercial asteroid exploitation mission. A wide range of different nuclear and non-nuclear power system technologies are available or are currently developed for future applications in space exploration and exploitation. This chapter provides a high-level discussion about the most relevant power system options for mobile and robotic miners designed to collect rubble on the surface of asteroids.

Chapter 11, by Karen Daniels, shows that most Near Earth Objects (NEOs) are composed of fractured rock, sometimes highly fractured and porous, and they have come to be known as rubble piles. The constituent particles, ranging from millimeters up to tens of meters, are weakly held together by a combination of both gravitational and van der Waals forces, which can be of comparable strength. Future missions to these rubble NEOs, whether human or robotic, will need to operate in such a way that they can safely and successfully interact with a fragile object. Of key importance is the ability to predict and control the circumstances under which the NEO material will remain intact or become unstable during activities such as digging, sample-collection, anchoring, or lift-off. While a comprehensive theory of these granular materials remains elusive, extensive laboratory experiments and computer simulations on these materials have lead towards an improved theoretical understanding of their dynamics. This chapter reviews the current state of knowledge of granular materials, and provides guidance about how to apply this knowledge to rubble-pile asteroids.

Chapter 12, by Kris Zacny, Philip Chu, Gale Paulsen, Magnus Hedlund, Bolek Mellerowicz, Stephen Indyk, Justin Spring, Aaron Parness, Don Wegel, Robert Mueller and David Levitt shows that Near Earth Objects (NEOs) are of interest for two reasons: for scientific study, and as a source of space resources. So far, all missions to NEOs have been motivated by scientific exploration. However, given recent advancements in space propulsion and communication technologies, mining NEOs for resources is becoming ever more feasible. A significant portion of NEOs' value is derived from their location; the resources contained in NEOs do not need to be lifted at a great expense from the surface of the Earth in order to be utilized in space. Instead, resources can be mined, processed, and used in space. To help represent this notion, a new term was coined: In Situ Resource Utilization (ISRU). ISRU facilitates planetary exploration by drawing needed resources, such as water, from the local environment. For decades, NASA and other space agencies have focused significant resources on developing and flying missions for *in situ* exploration of the Moon and Mars. These missions included the Apollo and Luna programs that explored the Moon and brought samples back, as well as the Viking, Mars Exploration Rovers, and Mars Science Laboratory Rover missions, to name a few, that helped explore the surface and near subsurface of Mars. Comparatively very little has been committed to *in situ* exploration or advancing technologies for *in situ* exploration of NEOs. The goal of this chapter is to present information pertaining to technologies that have either been developed or are being developed for *in situ* exploration of NEOs. In particular, the chapter describes various methods of anchoring to a small body (a prerequisite for sampling and mining missions to large asteroids) as well as sample acquisition technologies and large scale mining options. These technologies are critical to enabling the exploration and utilization of asteroids by NASA, other space agencies, and private companies.

Chapter 13, by Leonhard Bernold, states that it is surmised that colliding Asteroids once delivered us valued resources from deep space causing giant explosions shattering the Earth's crust. However, recent scientific interest and visits by spacecrafts gives rise to the expectation that humankind will be able to reap their bounty without "deep impacts." While Hollywood offers visions of gigantic human-controlled miners crushing mountains of basaltic rock on Asteroids, the risks and economic reality of our terrestrial gravity well asks for a mining technology that is lightweight and remote-controlled. This chapter presents the design and results of testing an experimental miner for very dense Asteroid regolith utilizing pneumatic conveying technology to excavate and transport. Taking synergistically advantage of several similitudes the terrestrial prototype excavated regolith simulants from an intake nozzle to a separator via a piping system. Two different nozzle designs were successfully tested.

Chapter 14, by Narayanan Komerath, Thilini Rangedera and Scott Bennett, deals withy the conceptual design of a solar-powered robotic craft to land on, attach to, and extract materials from, a typical Near Earth Object (NEO) on the scale required to construct habitat modules in orbit, or industrial processing for minerals. A solar-powered trajectory to a candidate NEO is used to estimate requirements. A reconfigurable solar sail/collector is the primary propulsion and power source for the craft. Following a journey of nearly 5 years, the craft will use a pulsed plasmajet torque-hammer concept to attach to the NEO. The basic cutting tool element is a solar-powered Neodymium fiber laser beam sheathed in a plasma jet, expanded through a truncated aerospike nozzle. Two telescoping, rotating arms carrying a total of 60 such nozzles at the ends of "fingers" enable the craft to dig and "float" out NEO material at a rate adequate to build a 50m diameter, 50m-long, 2m thick, walled cylinder within 19 days. The system is also amenable to applications requiring excavation of a large mass of near-surface material for resource processing. The present design with no return to Earth appears to close with a total payload to LEO of 37,500 kg, with a total mass of 30,000 kg including the sail/collector at earth escape. The primary consumables on the system are the plasma gas for cutting and maneuvering, and electrodes of the plasma cutters, sufficient to cut and release 1.4 million kilograms of NEO material. Options to collect the mass and return it to Earth orbit are discussed, centered on solar sail and mass driver propulsion.

Chapter 15, by Alexey Kondyurin, describes for the first time a polymerisation technique on asteroids and is backed up with the experimental results obtained in simulated space chambers as well as flight experiments of composite curing in near space environment. These results for the first time consider a number of asteroids that would be suitable for the polymerisation technique. As a consequence it will be possible to create new materials for building constructions in space for exploitation of asteroids. The chapter reviews in details space factors and their impact on polymerisation processes including high vacuum, cosmic rays, temperature changes and orbits of asteroids. It provides experimental information how to utilise the environmental condition in space.

Chapter 16, by Haym Benaroya, suggests a spacecraft concept that has the possibility of operating on a variety of asteroid types, shapes and sizes.

Mining activities on asteroids will be very challenging. The proposed craft – a deployable tetrahedral craft – would provide the option of standardization; it has not been seen in the published literature.

Chapter 17, by Werner Grandl and Akos Bazso, presents conceptual research on technological, commercial and architectural development related to Near Earth Asteroids (NEAs). In a first step NEAs, which are expected to contain resources like nickel-iron, platinum group metals or rare-earth elements, will be prospected by robotic probes. A number of asteroids with a minimum density of 2 g/cm^3 and a diameter of 100 to 500 m will be selected for mining. The authors propose to modify the orbits of those celestial bodies by orbital maneuvers from solar orbits to Earth orbits. To move the NEAs by remote control, unmanned *space tugs* with advanced propulsion types were designed. A pair of space tugs is docked to each asteroid using drilling anchors. The rocket engines of the tugs then apply the thrust forces for the maneuvers. Once stabilized in Earth orbit beyond the Moon, the mining process is started along the major axis of the asteroid. A *manned space station* will be connected to the asteroid, carrying digging, conveying and processing machinery and storage modules. The *Active Mining Head* initially digs a *main central tunnel* of 8 m diameter to the center of the asteroid. Then it excavates a spherical or cylindrical cave up to 50% of the NEA's volume. While the mass of the asteroid decreases constantly, its orbit is stabilized by the two *space tugs*. The processed material is transported by *unmanned cargo ships* to Low Earth Orbit or to the Lagrange Points of the Earth-Moon System for further industrial use. In the last phase of mining the inner surface of the cave is sintered by a robotic laser device. After the end of the mining process a *rotating human habitat* may be built inside the cave. The remaining shell of the asteroid will provide shelter against cosmic rays, solar flares and micrometeorites. The various materials produced by asteroid mining can be used for construction. Oxygen, hydrogen and carbon can be extracted from C-Type and similar NEAs. Natural sunlight can be collected outside the asteroid's shell by parabolic mirrors and beamed into the cave through the central tunnel. In a future scenario we can imagine dozens of *asteroid colonies* orbiting the planet Earth each with several thousand inhabitants.

Chapter 18, by Joan-Pau Sanchez and Colin R. McInnes, shows that exploiting the resources of near-Earth space has long been suggested as a means of lowering the costs of future space endeavours. Asteroids and comets, in particular, are generally agreed to be ideal resources, both in terms of their accessibility and their potential wealth. The intense survey efforts of the past decades have led to a growing catalogue of accessible near-Earth objects, but also to the realisation of the potential for exploitation and science of the myriad of objects that the Earth encounters along its orbit. The questions that now arise are how much near-Earth asteroid material is known, how much remains to be discovered, and, more importantly, how much can be easily accessed for future asteroid exploitation missions, such as those recently proposed by Planetary Resources Inc. This chapter demonstrates that a substantial quantity of resources can indeed be accessed at relatively low energy; on the order of 10^{14} kg of material could potentially be harvested at an energy cost lower than that required to access the resources of the Moon. More importantly, asteroid resources can be accessed across a wide spectrum of energies, and thus, it is show that current technology could be adapted to return to the Earth's neighbourhood objects from 10 to 30 meters diameter for scientific exploration and resource utilisation purposes. While the exploitation of the resources of near-Earth space has been considered for some time, it is here demonstrated that emerging technologies can now enable these resources to be accessed at modest cost.

Chapter 19, by Didier Massonnet, shows that the artificial capture of an asteroid from the near-earth objects family into an earth-bound orbital position is a very fascinating prospect. The main two questions raised by this issue are: 1) How to capture the asteroid and what are the associated orders of magnitude in terms of cost delay and the availability of technologies and 2) Why to capture an asteroid. The additional question of the benefit versus risks may also be addressed. Regarding the first question, the author reviews the main possibilities of propulsion and conclude that the only possibility that remains within a reasonable timescale is to use a mechanical throwing device using the material of the asteroid itself. The capture itself takes advantage of the L1/L2 Sun-Earth Lagrange points but still requires more ΔV than most scenarios of asteroid deviation. The second question is less straightforward. Asteroids can be used for mining materials uncommon or difficult to extract on earth, which supposes bringing back to Earth a very small fraction of the asteroid mass. They can also be used for mining materials to be used in space, with the advantage of not having to propel these materials into orbit: this is the case of liquid oxygen with chemical propulsion of large rockets or of bulk mass for the radiation protection of space crews. The extracted matter could then constitute a few percent of the asteroid mass. An even higher percentage of the mass can also be used in some schemes of nuclear space propulsion. Finally, the asteroid itself can be used as a shield against an earth-impact-bound asteroid, possibly much larger. In this case the whole mass of the asteroid is used and it has to be piloted into a collision trajectory with the incoming threat. The author discusses the likeliness and difficulty associated with each use, closely related to the complexity of the facility that has to be added to the propulsion device in each case.

Chapter 20, by Alexander Bolonkin, shows that for delivery asteroids to Earth we need in methods for changing the asteroid trajectory and theory for an estimation or computation the impulse which produces these methods. The author develops some methods of this computation. There are: impact of the space apparatus to asteroid, explosion the conventional explosive having form of plate and ball on asteroid surface, explosion the small nuclear bomb on the asteroids surface, entry asteroid to Earth atmosphere, braking of asteroid by parachute. Offered method may be also used for braking of apparatus reentering in the Earth from a space flight. The offered theory also may be used for protection the Earth from impact of a big asteroid.

Chapter 21, by Daniel García Yárnoz, Joan-Pau Sanchez and Colin R. McInnes, shows that asteroids and comets are of strategic importance for science in an effort to uncover the formation, evolution and composition of the Solar System. Near-Earth Objects (NEOs) are of particular interest because of their accessibility from Earth, but also because of their speculated wealth of material resources. The exploitation of these resources has long been discussed as a means to lower the cost of future space endeavours. In this chapter, the authors analyze the possibility of retrieving entire objects from accessible heliocentric orbits and moving them into the Earth's neighbourhood. The asteroid retrieval transfers are sought from the continuum of low energy transfers enabled by the dynamics of invariant manifolds; specifically, the retrieval transfers target planar, vertical Lyapunov and halo orbit families associated with the collinear equilibrium points of the Sun-Earth Circular Restricted Three Body problem. The judicious use of these dynamical features provides the best opportunity to find extremely low energy transfers for asteroidal material. With the objective to minimise transfer costs, a global search of impulsive transfers connecting the unperturbed asteroid's orbit with the stable manifold phase of the transfer is performed. A catalogue of asteroid retrieval opportunities of currently known NEOs is presented here. Despite the highly incomplete census of very small asteroids, the catalogue can already be populated with 12 different objects retrievable with less than 500 m/s of Δv. All, but one, of these objects have an expected size in the range that can be met by current propulsion technologies. Moreover, the methodology proposed represents a robust search for future retrieval candidates that can be automatically applied to a growing survey of NEOs.

Chapter 22, by Richard B. Cathcart, Alexander Bolonkin, Viorel Badescu and Dorin Stanciu, states that some asteroids are assumed to either contain some valuable mineral resources or consist of the only immutable, rust-free forms of iron in the Universe. These resources may become in the future a useful and ecologically sound supply meeting our iron mineral demands. In order to use these valuable iron resources, some actions should be engaged. The first is the extraterrestrial mining by which one has to break into some relatively small pieces the targeted iron-rich asteroid (when necessary). In some cases, these iron pieces may be further reshaped into a quite aerodynamic shape and even be foamed to be prepared for the next action. This action aims to send these pieces firstly on the near-Earth elliptic orbit, and after that on to the Earth-surface. In this chapter, two delivery methods are proposed. One of them is based on remotely-controlled, aerodynamic, shaped gliders resembling the USA's 1960–1965 ASV-3 ASSET vehicles, the first vehicles to return data from a lifting re-entry craft entering the atmosphere from near orbital speeds. The chosen vehicle/cargo is a much simpler version of the 1970s Space Shuttle design named the "Mega-ASSET" machine. Within this method, the asteroid fashioning is obtained through an asteroid mining process occurring in extraterrestrial space. Asteroid mining may initiate a new phase of the Earth's Anthropocene, delinking Homo sapiens' increasingly bulky infrastructure from (possibly all) Earth-crust mining. The idea and basis of the second method comes from the fact that in recent years the materials industry has produced high-temperature carbon fiber and whiskers. The authors examined, and proposed, the use of high temperature tolerant AB parachute for atmospheric braking. Though it is not large, a light AB parachute could decrease asteroid speed from 11 km/s to 50 m/s and a friction-induced heat flow by tens times. The parachute surface is opened behind the incoming object so that it can emit heat radiation efficiently to the Earth-atmosphere. The temperature of an AB parachute may become about 1000–1300°C. Carbon fiber is able to keep its functionality up to a temperature of 1500–2000°C. There is no conceivable problem to manufacture the AB parachute from carbon fiber. The second method consists of capturing unshaped asteroids by using an apparatus formed by a capture net, a long cable, a mechanical energy accumulator, and an AB carbon fiber parachute. The apparatus is launched into outer space by a rocket and intercepts the targeted asteroid on its elliptic orbit having in focus the Earth. After interception, the apparatus retards the asteroid's inbound flight through a kinetic process, so that the whole ensemble quits the Earth elliptic trajectory and enters the upper atmosphere. At this time, the AB carbon fiber parachute deploys from the asteroid's backside, retarding the whole ensemble (asteroid and delivery apparatus) and assures its landing at desired and assigned place inside the Earth-biosphere. Within this method the final mining process of delivered asteroid occurs in the Earth-biosphere. A combination of the two methods is also possible. Some small, natural whole or artificial asteroids fashioned to resemble the "Mega-ASSET" machine may best be delivered via carbon fiber AB parachutes.

Chapter 23, by Alexander Bolonkin, offers and researches the conditions which allow people and vehicles to levitate on asteroids using the electrostatic repulsive force. The author shows that by using small electrically charged balls or electret, people can create the artificial electrostatic gravity. The author has computed projects and discusses the problems which can appear in the practical development of this method. It is also shown how this method may be used for creating artificial gravity (attraction force) on asteroids.

Chapter 24, by Alexander Bolonkin, states that on an asteroid without atmosphere, sustaining human life is very difficult, especially during short sunlit period when low temperature prevails. To counter these environmental stresses, the author offers an innovative artificial "Evergreen" dome, an inflated hemisphere with interiors continuously providing a climate like that of Florida, Italy and Spain. The "Evergreen" dome theory is developed, substantiated by computations that show it is possible for current technology to construct and heat large enclosed volumes inexpensively. Specifically, a satisfactory result is reached by using high altitude magnetically supported sunlight reflectors and a special double thin film as an enclosing skin, which concentrates solar energy inside the dome while, at the same time, markedly decreasing the heat loss to exterior space. Offered design may be employed for settlements on asteroids. The author also methodically researched a revolutionary macro-engineering idea for a closed-loop freshwater irrigation and in this chapter it is unveiled in some useful detail. One offers to cover a given site by a thin, enclosure film (with controlled heat conductivity and clarity) located at an altitude of 50 – 300 m. The film is supported, at its working altitude, by small additional induced air over-pressuring, and anchored to the ground by thin cables. The author shows that this closed dome allows full control of the weather within at a given asteroid surface region. This is, today, a realistic and cheap method of evaporation-economical irrigation and virtual weather control on asteroids.

Chapter 25, by Russell Bewick, Joan-Pau Sanchez and Colin R. McInnes, focuses on the use of captured near-Earth asteroids as an integral part of three, large-scale, space-based geoengineering methods. These methods are; an unstable dust cloud at the Sun-Earth first Lagrange point, a gravitationally anchored dust cloud at the same position and a dust ring around Earth. Geoengineering, otherwise known as climate engineering, is the deliberate modification of the Earth's climate to offset the effects of an increase in mean global temperature. It has been shown that an increase in global temperatures by 2°C can be offset by a 1.7% decrease in solar insolation. The three methods discussed in this chapter will aim to achieve this target. The unstable dust cloud uses the linearised dynamics of the circular restricted three-body problem to create a steady state dust cloud that is used, in conjunction with a model of the incident solar radiation, to determine the mass of dust required per year to achieve the insolation reduction target. In contrast, the gravitationally anchored dust cloud uses four-body dynamics to determine a zero-velocity curve that bounds a region around the first Lagrange position within which dust ejected from the surface of an asteroid will not escape. A list of known near-Earth asteroids is used to determine the maximum insolation reduction achievable for the given asteroids mass. Finally, the Earth ring uses the perturbations of solar radiation pressure and the J_2 effect to find stable, elliptical, heliotropic orbits around Earth. These orbits are then used to generate a model of a ring with a dust grain size distribution, from which the mass of dust and insolation reduction can be calculated. In summary of the mass of asteroid material required, the unstable dust cloud requires 1.87-7.60x10^{10}kg/yr, whilst the Earth ring requires 2x10^{12}kg of material. Finally, it is demonstrated that this mass of material can be captured from the populations of near-Earth asteroids below the velocity required to exploit lunar resources.

Chapter 26, by Alexander Bolonkin, shows that currently, rockets are used to change the trajectory of space ships and probes. Sometimes space probes use the gravity field of a planet. However, there are only nine planets in the solar system, all separated by great distances. There are tens of millions of asteroids in outer space. This chapter contains three parts. In the first part the author offers and researches the space elevator for rotated asteroid having the diameter more 20 - 100m. This elevator allows landing and starting (land/launch), braking and accelerating of the space ship without spending of fuel in many cases. The space ships use the rotational energy of asteroid. The chapter brings together research on the space elevator and a new transportation system for it. This transportation system uses mechanical energy transfer and requires only minimal energy so that it provides a "Free Trip" into space. It uses the rotational energy of asteroids. In the second part of this chapter the author offers a revolutionary cable method for changing the trajectory of space probes. This method uses the kinetic or rotational energy of asteroids. In third part of this chapter the author offers an electrostatic method for changing the trajectory of space probes by asteroids. The method uses electrostatic force and the kinetic or rotational energy of asteroids. These methods allows to increase (to decrease) ship (probe) speed up to 1000 m/s (or more) and to achieve any new direction in outer space.

Chapter 27, by Csaba Kecskes, discusses shortly the problem of "mankind is not unique - where are the extraterrestrials". An evolutionary model of the development of technical civilisations is presented in which the civilisations evolve in such a way that Earth-like planets (and, eventually, Sun-like stars) are becoming unimportant for them. A testing method (searching the traces of extraterrestrial visitations in the asteroid belt) is suggested. NASA's Lunar Reconnaissance Orbiter mission is presented as an example of searching artificial objects on the Moon. The problem of high resolution photography from spacecrafts is discussed. A multiple asteroid flyby mission is suggested as an economical method for surveying many objects in the asteroid belt. The once planned Soviet Vesta mission is presented as an example.

Chapter 28, by Mike H. Ryan and Ida Kutschera, describes some of the unique opportunities that could develop from proposed asteroid mining ventures. The use of extra-terrestrial resources is examined from a business perspective and numerous challenges enumerated along with possible solutions and alternatives. Several historical analogs are examined and their implications for asteroid mining developed. The roles that human explorers and miners might play within such undertakings are discussed in conjunction with robotic approaches. The Chapter looks in detail at issues such as self-sufficiency, command and control, property rights, risk and the types of business models that might develop. The Chapter concludes with an assessment of the probability of such ventures occurring in the near future.

Chapter 29, by Virgiliu Pop, shows that the asteroids are mines in both meanings of the word; they are a likely source of minerals, as well as a danger to human civilization. Efforts are in place to deal with asteroids in both these meanings to mine them for their riches, and to de-mine the sky. These activities raise important legal questions: how is this wealth going to be appropriated and shared? Who, after all, really own the asteroids? What is the place and significance of property rights in the context of space activities? Is planetary defense a right, or an obligation? Who is entitled - or obliged - to defend the Earth from the PHO menace? Are all deflection technologies legal? Can nuclear explosions be used not only for deflection, but also for exploiting the mineral riches of asteroids? Could asteroids be used for waging war? It is hereby offered that property rights produce the best answer when included in these equations; private enterprise, if allowed to make a profit from exploiting the asteroids, would survey and track them better than a voluntary or a government program; private enterprise would devise means of exploiting and moving the asteroids that would not only extract their riches, but would also be able to alter their orbits. Mining the sky would, then, mean also de-mining it but in order for this to happen, humans ought to be able to say the word "mine" in its third meaning - that of property - when it comes to asteroids.

The book allows the reader to acquire a clear understanding of the scientific fundamentals behind specific technologies to be used on asteroids in the future. The principal audience consists of researchers (engineers, physicists) involved or interested in space exploration in general and in asteroid exploration in special. Also, the book may be useful for industry developers interested in joining national or international space programs. Finally, it may be used for undergraduate, postgraduate and doctoral teaching in faculties of engineering and natural sciences.

The Editor

Acknowledgments

A critical part of writing any book is the review process, and the authors and editor are very much obliged to the following researchers who patiently helped them read through subsequent chapters and who made valuable suggestions: Prof. Haim Baruh (Rutgers University, Piscataway, New Jersey, US), Dr. William R. Bigler, Jr. (Louisiana State University, Shreveport, LA, US), Prof. Haym Benaroya (Rutgers University, Piscataway, New Jersey, USA), Prof. Leonhard Bernold (University of New South Wales, Sydney, Australia), Dr. William Bottke (NASA/NLSI Center for Lunar Origin and Evolution, Southwest Research Institute, CO, US), Dr. John Brophy (Jet Propulsion Laboratories, US), Dr. Elisabet Canalias (CNES, France), Dr. Benoit Carry (Institut de Mecanique Celeste, Paris, France), Dr. Robert Cassanova (NASA Institute of Advanced Concepts, US), Dr. Paul Chodas (NASA/ Jet Propulsion Laboratories, US), Dr. Wojciech Chrzanowski (University of Sydney, Australia), Dr. Camilla Colombo (University of Southampton, UK), Prof. David Coward (University of Western Australia), Dr. Marco Delbo (Observatorie de la Cote d'Azur, Nice, France), Dr. Christos Efthimiopoulos (Research Center for Astronomy, Academy of Athens, Greece), Dr. Martin Elvis (Harvard-Smithsonian Center for Astrophysics, Cambridge, MA, US), Dr. Martyn Fogg (University of London, UK), Dr. Robert A. Freitas (Institute for Molecular Manufacturing, Palo Alto, CA, US), Dr. Leslie Gertsch (Missouri University of Science and Technology, US), Prof. Giovanni Gronchi (Universita di Pisa, Italy), Dr. David Gucik-Derigny (University of Bordeaux, France), Prof. Viktor Hacker (TU Graz, Austria), Dr. Henrik Hargitai (Eotvos Lorand University, Budapest, Hungary), Dr. Daniel Hernandez (Devil-Hop Consultancy, France), Dr. Marvin Herndon (Transdyne Corporation, San Diego, CA, US), Dr. Genya Ishigami (Space Exploration Robotics Laboratory, JAXA/ISAS, Japan), Dr. Mark Krinker (City College of Technology, CUNY, New York, US), Prof. Monica Lazzarin (Università di Padova, Italy), Prof. John S. Lewis (University of Arizona, AZ, US), Prof. Josep Masdemont (Universitat Politècnica de Catalunya, Spain), Dr. John-Jairo Martinez-Molina (GIPSA, Grenoble-INP, Saint Martin D'Hères, France), Dr. Andres Mora (School of Earth and Space Exploration, Arizona State University, US), Dr. Naomi Murdoch (Observatoire de la Côte d'Azur, Nice, France), Dr. Galina Nechitailo (Institute of Biochemical Physics, Russian Academy of Science, Moscow, Russia), Prof. Shmuel Neumann (Michlala Jerusalem College, Israel), Dr. Jerome Pearson (Star Technology and Research, Inc., Mount Pleasant, SC, US),

Dr. Emanuele Pensavalle (Aviospace S.r.l., Torino, Italy), Prof. Oleg Pensky (National Research Perm University, Russia), Prof. Derek C. Richardson (University Maryland, MD, US), Dr. Holger Sierks (Max Planck Institute for Solar System Research, Katlenburg-Lindau, Germany), Dr. W. Scott Sherman (Texas A&M University, Corpus Christi, TX, US), Dr. Jean Souchay (Observatoire de Paris, France), Dr. Christopher Stott (International Institute of Space Commerce, Isle of Man, UK), Prof. Curtis J. Struck (Iowa State University, US), Dr Cristina Thomas (GSFC, NASA, US), Prof. Harry Varvoglis (Aristotle University of Thessaloniki, Greece), Dr. Lotta Viikari (University of Lapland, Finland), Dr. Kris Zacny (Honeybee Robotics Spacecraft Mechanisms Corporation, Pasadena, US), Prof. Marjan Zadnik (Curtin University, Australia).

The editor, furthermore, owes a debt of gratitude to all authors. Collaborating with these stimulating colleagues has been a privilege and a very satisfying experience.

Contents

List of Contributors

Viorel Badescu
Candida Oancea Institute
Polytechnic University of Bucharest
Spl. Independentei 313,
Bucharest 060042,
Romania
badescu@theta.termo.pub.ro

Akos Bazso
Institute of Astronomy,
University of Vienna
Tuerkenschanzstrasse 17,
A-1180 Vienna,
Austria
akos.bazso@univie.ac.at

Haym Benaroya
Department of Mechanical and
Aerospace Engineering
Rutgers University,
98 Brett Road,
Piscataway, New Jersey 08854-8058
USA
benaroya@rci.rutgers.edu

Scott Bennett
Daniel Guggenheim School of
Aerospace Engineering
Georgia Institute of
Technology, Atlanta, GA,
30332-0150, USA
dbennett9@gatech.edu

Leonhard E. Bernold
School of Civil and
Environmental Engineering
University of New South Wales,
Sydney,
Australia
leonhard.bernold@gmail.com

Ivano Bertini
Center of Studies and Activities
for Space (CISAS) 'G. Colombo'
University of Padova,
Via Venezia 15,
35131 Padova,
Italy
ivano.bertini@unipd.it

Russell Bewick
Advanced Space Concepts
Laboratory
Department of Mechanical
Engineering
University of Strathclyde -
Glasgow G1 1XJ - UK
rusbewick@hotmail.com

Bryce Bolin
University of Hawaii,
Institute for Astronomy
2680 Woodlawn Dr.,
Honolulu, HI,
96822, USA
bolin@ifa.hawaii.edu

Alexander A. Bolonkin
C&R, 1310 Avenue R, #6-F,
Brooklyn, NY 11229, USA
aBolonkin@juno.com,
aBolonkin@gmail.com

Richard B. Cathcart
Geographos, 1300 West Olive Avenue, Suite M
Burbank, California 91506-2225, USA
rbcathcart@gmail.com

Marco Cenzon
Via M. Stelvio 16, Caldogno (VI),
36030, Italy
marco.cenzon@gmail.com

Marco Chacin
Singularity University
NASA Research Park
Moffett Field,
CA 94035
mchacin@ieee.org

Phil Chu
Honeybee Robotics Spacecraft Mechanisms Corporation
398 W Washington Blvd., Suite 200,
Pasadena, CA 91103, USA
chu@honeybeerobotics.com

Monique Chyba
Department of Mathematics,
University of Hawaii
2565 McCarthy Mall, Honolulu, HI,
96822, USA
chyba@hawaii.edu

Karen E. Daniels
Department of Physics
North Carolina State University,
Raleigh, NC, USA
kdaniel@ncsu.edu

Josef Ďurech
Charles University in Prague
Faculty of Mathematics and Physics
Astronomical Institute,
Czech Republic
durech@sirrah.troja.mff.cuni.cz

Martin Elvis
Harvard-Smithsonian Center for Astrophysics
60 Garden St.,
Cambridge MA 02138
USA
elvis@cfa.harvard.edu

Nicolas Franceschini
Aix-Marseille University, CNRS
Institute of Movement Science,
Biorobotics Dept.
UMR7287, 13288, Marseille,
France
nicolas.franceschini@univ-amu.fr

Simon D. Fraser
Engineering consultant
Inge-Morath-Strasse 45d,
8045 Graz,
Austria
simon.fraser@gmx.at

Werner Grandl
Dr. Billrothstrasse 6,
A-3430 Tulln,
Austria
archigran@gmx.at

Mikael Granvik
Department of Physics,
P.O. Box 64
00014 University of Helsinki,
Finland
mgranvik@iki.fi

Magnus Hedlund
Honeybee Robotics Spacecraft Mechanisms Corporation
398 W Washington Blvd.,
Suite 200, Pasadena,
CA 91103, USA
hedlund@honeybeerobotics.com

Stephen Indyk
Honeybee Robotics Spacecraft Mechanisms Corporation
398 W Washington Blvd.,
Suite 200, Pasadena,
CA 91103, USA
indyk@honeybeerobotics.com

Robert Jedicke
University of Hawaii, Institute for Astronomy
2680 Woodlawn Dr., Honolulu,
HI, 96813, USA
jedicke@ifa.hawaii.edu

Mikko Kaasalainen
Department of Mathematics,
Tampere University of Technology
PO Box 553, 33101 Tampere,
Finland
mikko.kaasalainen@tut.fi

Csaba Kecskes
1025 Budapest,
Felsozoldmali ut 60,
Hungary
csaba56@freemail.hu

Narayanan Komerath
Daniel Guggenheim School of Aerospace Engineering
Georgia Institute of Technology,
Atlanta, GA, 30332-0150, USA
narayanan.komerath@aerospace.gatech.edu;
komerath@gatech.edu

Alexey Kondyurin
School of Physics, A28,
University of Sydney, N.S.W. 2006,
Australia
kond@lycos.com

Ida Kutschera
Rubel School of Business
Bellarmine University, Louisville,
KY 40205, USA
Ikutschera@bellarmine.edu

David Levitt
Cadtrak Engineering
San Anselmo, CA 94960,
USA
dlevitt@cadtrak.com

Didier Massonnet
CNES Centre National d'Etudes Spatiales
18 Avenue E. Belin, 31401,
Toulouse cedex 09,
France
Didier.Massonnet@cnes.fr

Colin McInnes
Advanced Space Concepts Laboratory
Department of Mechanical Engineering
University of Strathclyde -
Glasgow G1 1XJ - UK
colin.mcinnes@strath.ac.uk

Bolek Mellerowicz
Honeybee Robotics Spacecraft Mechanisms Corporation
398 W Washington Blvd.,
Suite 200, Pasadena,
CA 91103, USA
mellerowicz@honeybeerobotics.com

Robert Mueller
Surface Systems Office, NE-S
Engineering & Technology Directorate
NASA, Kennedy Space Center, FL, USA
rob.mueller@nasa.gov

Bojan Novaković
Department of Astronomy, Faculty of Mathematics
University of Belgrade, Studentski trg 16,
11000 Belgrade, Serbia
bojan@matf.bg.ac.rs

Aaron Parness
NASA Jet Propulsion Laboratory
Robotic Hardware Systems Group
4800 Oak Grove Dr Pasadena, CA 91109, USA
Aaron.Parness@jpl.nasa.gov

Geoff Patterson
Department of Mathematics, University of Hawaii
2565 McCarthy Mall, Honolulu, HI, 96822, USA
gpatters@hawaii.edu

Gale Paulsen
Honeybee Robotics Spacecraft Mechanisms Corporation
398 W Washington Blvd., Suite 200, Pasadena, CA 91103, USA
paulsen@honeybeerobotics.com

Dragoş Alexandru Păun
Corso Galileo Ferraris 149, Torino, 10128, Italy
alexandru.paun@yahoo.co.uk

Gautier Picot
Department of Mathematics, University of Hawaii
2565 McCarthy Mall, Honolulu, HI, 96822, USA
gpicot@hawaii.edu

Virgiliu Pop
Romanian Space Agency
Mendeleev 21-25, Bucharest, Romania
virgiliu.pop@rosa.ro

Thibaut Raharijaona
Aix-Marseille University, CNRS
Institute of Movement Science, Biorobotics Dept.
UMR7287, 13288, Marseille, France
thibaut.raharijaona@univ-amu.fr

Thilini Rangedera (Schlesinger)
Daniel Guggenheim School of Aerospace Engineering
Georgia Institute of Technology, Atlanta, GA, 30332-0150, USA
thilini.schlesinger@gmail.com

Franck Ruffier
Aix-Marseille University, CNRS
Institute of Movement Science, Biorobotics Dept.
UMR7287, 13288, Marseille, France
franck.ruffier@univ-amu.fr

Mike H. Ryan
Rubel School of Business
Bellarmine University, Louisville, KY 40205, USA
mryan@bellarmine.edu

Guillaume Sabiron
French Aerospace Lab
ONERA, Systems Control and Flight Dynamics -DCSD
31055 Toulouse,
France
guillaume.sabiron@onera.fr

Joan Pau Sánchez
Advanced Space Concepts Laboratory
Department of Mechanical Engineering
University of Strathclyde -
Glasgow G1 1XJ -
UK
jpau.sanchez@strath.ac.uk
jpau.sanchez@upc.edu

Justin Spring
Honeybee Robotics Spacecraft Mechanisms Corporation
398 W Washington Blvd.,
Suite 200, Pasadena,
CA 91103,
USA
spring@honeybeerobotics.com

Dorin Stanciu
Polytechnic University of Bucharest
Faculty of Mechanical Engineering
Spl. Independentei 313,
Bucharest 060042,
Romania
sdorin_ro@yahoo.com

Michael Todd
Department of Imaging and Applied Physics
Bldg 301, Curtin University, Kent St,
Bentley, WA 6102, Australia
michael.todd@icrar.org

Stephane Viollet
Aix-Marseille University, CNRS,
Institute of Movement Science,
Biorobotics Dept.
UMR7287, 13288, Marseille,
France
stephane.viollet@univ-amu.fr

Don Wegel
NASA Goddard Space Flight Center
Greenbelt, MD 20771, USA
donald.c.wegel@nasa.gov

Daniel Garcia Yarnoz
Advanced Space Concepts Laboratory
Department of Mechanical Engineering
University of Strathclyde -
Glasgow G1 1XJ - UK
daniel.garcia-yarnoz@strath.ac.uk

Kris Zacny
Honeybee Robotics Spacecraft Mechanisms Corporation,
398 W Washington Blvd.,
Suite 200, Pasadena,
CA 91103, USA
zacny@honeybeerobotics.com

Chapter 1
Asteroids Close-Up: What We Have Learned from Twenty Years of Space Exploration

Ivano Bertini

University of Padova, Italy

1.1 Introduction

Asteroids are objects of fundamental scientific importance for several reasons. They are the remnant debris from the formation of the inner solar system. They offer therefore the unique opportunity to study the original materials and the mechanisms which formed the terrestrial planets. Moreover, they can be the objects of a strong interaction with Earth's biosphere. Asteroids may have played a role in bringing water and organic substances on the Earth, influencing the formation of life. With catastrophic impacts they may have changed in the past the evolutionary path of life forms and may still constitute a serious threat to the human presence on the planet. Finally, asteroids represent an extraordinary source of minerals which can be exploited for the increasing necessities of our civilization and for the future exploration and colonization of the solar system.

In the last twenty years asteroids passed from being distant point-like light sources in our telescopes to revealing themselves as intriguing complex worlds thanks to the vast amount of scientific results coming from dedicated space missions. The in-situ exploration combines the advantages of observing in a wide range of wavelengths, overcoming the limits imposed by Earth's atmosphere, to the high resolution possible only with close encounters. Moreover, spacecraft observations cover a range of observational geometries often unattainable from Earth. Spacecraft data have so far complemented, improved, and in many cases revolutionized, the theories and findings derived from ground-based asteroid data. Finally, space missions have the notable scientific expansion of triggering extensive ground- and space-based observational campaigns to characterize the targets before and during the in-situ operations of the spacecrafts. This is necessary both for mission planning purposes and for connecting the ground truth to the in-situ results.

Several scientific subjects can be addressed in a better, or exclusive, way from in-situ instruments in comparison with observations from ground-based or Earth-orbiting telescopes. Spacecraft exploration allowed the birth of asteroid geology, giving hints on the origin and evolutionary processes of the objects. Crater counting allows the measure of surfaces age and tells us about the collisional evolution of asteroids. Deriving the craters depth-to-diameter ratio provides information on the nature of these bodies and the depth of the regolith covering the surfaces. The detection of existing boulders on the surface is a direct evidence of regolith presence on asteroids. As retained ejecta fragments, these rocks give information on the impact cratering processes on bodies with low gravity (Lee et al. 1996). The derivation of a precise shape model can give hints on a possible effect acting on the rotation rate, and therefore shaping, of small bodies caused by the anisotropic emission of thermal photons, called YORP effect. Resolved spectrophotometry, investigating the variegation of the surface in terms of albedo and color give hints on dynamical and space weathering aging processes on asteroid surfaces (Schröder et al. 2010). Resolved ultraviolet, visible, and infrared spectroscopy provide information on the distribution of minerals across the surface. Resolved observations of the thermal emission tell us about the variegation of the subsurface in terms of texture and composition (Tosi et al. 2010). The measure of a global magnetic field places significant constraints on the composition, origin, and thermal evolution of the asteroids and their parent bodies (Blanco-Cano et al. 2003; Auster et al. 2010). Finally, spacecraft encounters allow very effective searches and discoveries of asteroid companions. The measure of the orbit of small satellites permits the mass of the primary to be measured and hence its bulk density, once the volume is known. This give hints on the physical composition of the object and its internal structure. The study of the connected systems provides also clues on the collisional events which occurred during the early stages of the formation of the inner solar system (Merline et al. 2002). Asteroid masses can also be obtained with high precision from the measure of the perturbations the objects are exerting on the spacecraft's trajectory, in case of large masses and/or extremely close flybys, or during orbital manoeuvres.

This chapter contains a summary of the main scientific results obtained in the last twenty years of asteroid space exploration, starting from the encounter of the NASA Galileo mission with asteroid (951) Gaspra in 1991 and ending with the visit of asteroid (4179) Toutatis by the chinese Cheng'E 2 spacecraft in 2012. Future missions already approved and under development by the various space agencies are also outlined at the end of this work.

1.2 The Scientific Results from 20 Years of In-Situ Asteroid Space Exploration

Eight space missions were devoted so far to the in-situ exploration of eleven asteroids. Although the most common mission type was a fast flyby, often en route to another primary target, two orbiting, and one orbiting and sample return mission were also performed. A summary of the past space missions to asteroids and achieved scientific results is portrayed in the following sub-sections. This summary represents an up-to-date portrait of our knowledge about asteroids coming from in-situ studies.

1.2.1 The First Encounter with an Asteroid: Galileo at Gaspra

En route to explore the Jupiter system, the NASA Galileo spacecraft flew-by the S-type main belt asteroid (951) Gaspra at a distance of 1600 km and a relative speed of 8 km s^{-1}, obtaining the first close-up view of an asteroid in history on 21 October 1991 (Belton et al. 1992b).

During the encounter, the Solid State Imaging Camera (SSI) instrument (Belton et al. 1992a) took several images of the asteroid with the highest resolution of 54 m px^{-1} (Chapman et al. 1996b). Only the 80% of the surface was revealed by SSI images due to the geometry of the flyby. Gaspra was found to be an highly irregular body which shape can be best fitted by a triaxial ellipsoid having dimensions of 18.2×10.5×8.8 km^3 (Veverka et al. 1994). An image of Gaspra taken by Galileo is shown in Fig. 1.1. The shape is dominated by large flat to slightly concave regions, termed facets, of controversial origin. They may be the remnants of impact scars or being formed as a parent body broke up along pre-existing fractures (Stooke 1996, 1997). Gaspra has a surface battered with craters. Crater counting yielded a very steep differential population beyond the theoretical limit for collisional equilibrium, a situation in which as many craters are being destroyed as are being produced. This result pushed to a reassessment of previous asteroid collisional models (Chapman et al. 1996b). Estimates of the cratering age of Gaspra's surface gave very young values ranging from 20 to 300 Myr (Veverka et al. 1994). Several linear features were found, including sets of grooves and linear

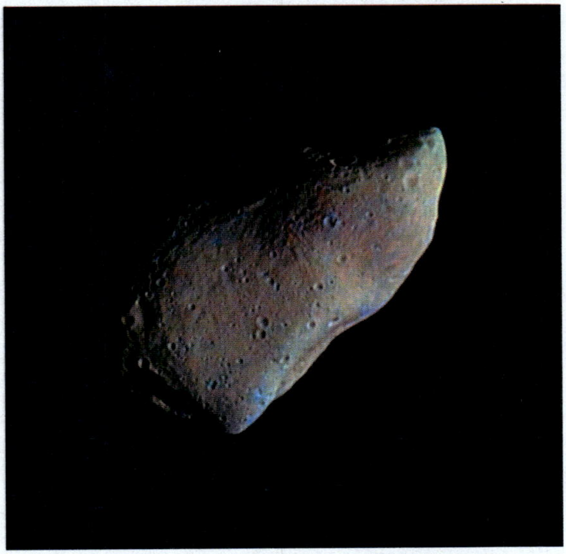

Fig. 1.1 Mosaic picture of Gaspra taken by the Galileo spacecraft from a distance of 5300 km ten minutes before closest approach on 29 October 1991 with a resolution of ~ 54 m px^{-1}. The portion illuminated in this view is about 18 km from lower left to upper right. Courtesy NASA/JPL/USGS.

depressions. The analysis of the morphological features pointed towards an interpretation of the origin of Gaspra by a catastrophic collision involving a previously larger body. Another conclusion derived was that the asteroid is probably a single coherent monolithic body rather than a contact binary system or a rubble pile (Veverka et al. 1994; Stooke 1996).

SSI images were combined with the Near Infrared Mapping Spectrometer (NIMS) instrument (Carlson et al. 1992) data to derive the spectral behavior of the asteroid up to 5.2 μm. This allowed the investigation of the surface composition and of the distribution of the mineral phases on the surface (Granaham 2011). Modest color differences (5%) were observed across the surface (Carr et al. 1994; Helfstein et al. 1994). Two 'extreme' units were identified. The first one has a slightly higher albedo, bluer color, and deeper 1 μm and 2 μm silicate absorption bands than the second one. The average Gaspra's surface is closer to the second unit, with an average geometric albedo of 0.23 at 0.56 μm (Veverka et al. 1994; Chapman 1996; Granaham 2011). There is also a clear correlation between color and elevation. The bluer unit is located preferentially on ridges and appears to be correlated with several small fresh craters located on the ridges. The redder unit is preferentially located on the lower, flatter facets. These correlations led to conclude that there is a space weathering process that changes the color with time, converting the first unit materials into the second ones, and that there is a tendency for the older material to migrate downhill from the ridges and fill in the lower planar regions (Belton et al 1992; Helfenstein et al 1994; Carr et al. 1994). The latter fact requires the presence of particulate regolith, also suggested by the results on thermal inertia modeling (Weissman et al 1992). A nominal relative abundance of olivine (~90%) with respect to orthopyroxene (~10%) was derived modeling the ratio of the 1 μm and 2 μm bands, consistently with the spectra of meteorites with monomineralic olivine (Granaham 2011). All of these meteorites have been subjected to igneous processes on their parent body asteroids. Hence, the spectra of Gaspra indicate that it has been subjected to igneous differentiation.

A very interesting result came also from the onboard magnetometer (Kivelson et al. 1992). Two large interplanetary magnetic field rotations with little changes in the field magnitude were recorded one minute before and two minutes after the closest approach, respectively. The time and the geometry of the field changes were interpreted as being consistent with the magnetic signature of an obstacle that is removing momentum from the outward-flowing solar wind (Kivelson et al.1993). Considering the size of the disturbed region, it was derived that Gaspra does not possess a magnetic field strong enough to build a magnetosphere impenetrable to the solar wind. The magnetic interaction would than result in a whistler wake downstream the asteroid, in contrast to the bow wave formed in front of planets (Kivelson et al. 1993; Baumgartel et al. 1994; Wang et al. 1995). The inferred magnetic moment was estimated to be [6×10^{12}-2×10^{14}] A m^2, more than eight orders of magnitude smaller than the value for Earth, and the resulting specific moment is [0.001-0.03] A m^2 kg^{-1}, in the range observed for iron meteorites and highly magnetized chondrites. The conclusion was that Gaspra might then be a fragment of a differentiated parent body. Alternatively, Gaspra or its parent body may have cooled in a magnetic field capable of aligning ferromagnetic material that was present in the early solar system (Kivelson et al. 1993). This interpretation was later challenged by Blanco-Cano et al. (2003) who considered the perturbation observed near Gaspra not generated by the interaction of the solar wind with a magnetized asteroid. The authors found that the signatures were linearly polarized, resembling the magnetic discontinuities commonly found in the solar wind, in contrast to the nearly circularly polarized signatures of their theoretical simulations of the solar wind-magnetized asteroid interaction.

Finally, no evidence for satellites down to the size limit of 27 m in radius, considering Gaspra-like photometric properties, was found in SSI images. The search covered the region from about 100 to 200 m above Gaspra's surface out to about 10 asteroid radii, only a tiny fraction of the region of gravitational influence of the asteroid, called Hill sphere (Belton et al. 1992b).

1.2.2 Asteroids DO Have Satellites: Galileo at Ida

On 28 August 1993 Galileo performed its second flyby with a main belt S-type asteroid: (243) Ida, which belongs to the Koronis family. The closest distance reached was 2400 km at a relative speed of 12.4 km s^{-1}. The same instruments operating during the flyby of Gaspra were switched on (Belton et al. 1994).

SSI imaged ~95% of the surface with a maximum resolution of 25 m px^{-1}. Ida resulted to be an irregular elongated body with dimensions along the principle axis of momentum of 60×25×19 km^3 (Belton et al. 1996). An image of Ida taken by Galileo is showed in Fig. 1.2. The surface was found to be rich in geological features. Craters as large as 8 km without complex morphologies were identified. The high crater density and size-frequency distribution indicated a surface in equilibrium with saturated cratering. A minimum model crater age for Ida was estimated to be 1-2 Gyr (Belton et al. 1994; Chapman et al. 1996a). Several relief features were interpreted as boulders, providing a direct evidence for regolith retention on the asteroid. A deep regolith layer was inferred covering the surface (Lee et al. 1996; Sullivan et al. 1996).

Galileo's photometry combined with near opposition ground-based data yielded a geometric albedo of 0.21 at 0.56 μm. Two primary color units were identified on the surface, similarly to Gaspra: the first one exhibits a relatively lower albedo, shallower 1μm band and a steeper spectral slope with respect to the second. Average photometric properties are similar to the first unit. The first unit is the ubiquitous background of the surface while the second one is correlated with small craters and possible ejecta from big craters. The most natural interpretation of the difference implies effects of space weathering rather than intrinsic compositional heterogeneity (Helfenstein et al. 1996; Veverka et al. 1996a). The combination of SSI and NIMS data suggests the silicate composition of the surface is 65% olivine and 35% orthopyroxene (Granahan 2002).

The onboard magnetometer found magnetic signatures close to the closest approach, as was previously detected at Gaspra. A similar interpretation of the results was provided (Wang et al. 1995; Kivelson et al. 1995; Blanco-Cano et al. 2003).

The most striking event during the flyby was the serendipitous discovery of a small round moon, named Dactyl, of approximate dimensions of $1.6 \times 1.4 \times 1.2 \ km^3$. An image of Dactyl taken by Galileo is shown in Fig. 1.2.

Fig. 1.2 Composite color picture taken by the Galileo spacecraft about 14 minutes before its closest approach to Ida on 28 August 1993. The distance from Galileo was ~10500 km, yielding a resolution of ~100 m px^{-1}. The moon Dactyl is visible to the right of the asteroid. In the upper right corner a monochromatic image of Dactyl, taken from a distance of ~3900 km, 4 minutes before closest approach, has been added. The resolution of this image is ~39 m px^{-1}. Courtesy NASA/JPL.

This was the first and definitive proof that asteroids have satellites and binary systems exist. Dactyl has its longest axis pointed towards Ida and the shortest perpendicular to the orbital plane. Well defined craters were found on its surface, the crater population resulting in equilibrium with saturation (Chapman et al. 1995). There were no evidences of grooves, ridges, sharp edges, or other geological features at the resolution of the SSI imaging system (Chapman et al. 1995; Veverka et al. 1996b). The moon was found having a geometric albedo of 0.20 at 0.56 μm, similar to Ida's one and a consistent phase function (Chapman et al. 1995; Helfenstein et al. 1996). The spectral differences between Ida and its moon are minor. Dactyl is slightly less red in color and with a deeper 1 μm absorption band. Combining all the results, the most likely explanation for the small photometric differences between Ida and Dactyl is a slightly different composition between the two objects (Chapman et al. 1995; Veverka et al. 1996b). Almost certainly Dactyl was formed either from Ida via impact or at the same time as Ida during the catastrophic collision that formed the Koronis family (Chapman et al. 1995; Giblin et al. 1998).

The presence of Dactyl and constraints on the range of its possible orbits around Ida, combined with Ida's volume estimate, allowed for the first time in history the precise estimation of an asteroid density. Ida was found having a density of 2.6±0.5 g cm^{-3}, being consistent with a bulk condritic composition and excluding therefore, unless the bulk porosity is exceptionally high, a larger content of nickel and iron (Belton et al. 1995, 1996). Moreover, Ida's spin axis was found to be aligned with the principal axis of inertia, consistently with a homogeneous density distribution and ruling out extreme density asymmetries (Thomas et al. 1994; Belton et al. 1996).

1.2.3 The First C-type Asteroid: NEAR at Mathilde

On its way to the mission main target, the near Earth asteroid (433) Eros, the NASA Near Earth Asteroid Rendezvous (NEAR) spacecraft passed within 1212 kilometers of the main belt C-type asteroid (253) Mathilde at a speed of 9.93 km s^{-1} on 27 June 1997 (Veverka et al. 1997b).

To conserve power, only the Multispectral Imager (MSI) instrument (Veverka et al. 1997a) was turned on for the encounter. The images were taken with a highest resolution of 160 m px^{-1}. Because of Mathilde's slow rotation, only about 60% of its surface was seen during the flyby (Veverka et al. 1997b). Its irregular shape was modeled with a triaxial ellipsoid having dimensions of 66×48×44 km^3 (Veverka et al. 1999). A picture of Mathilde taken by NEAR is shown in Fig. 1.3. The most remarkable morphological features detected were five big craters with diameters between 19 and 33.4 km, comparable to the mean radius of the body. No evidences of ejecta were found on the surface. It was immediately considered quite exceptional that Mathilde survived many major catastrophic events without being completely destroyed. A possible explanation could come from a weak, low-density bulk material and/or a rubble pile internal structure (Chapman et al. 1999). In addition to this, the observed morphology may be explained in terms of crater formation by oblique impacts (Cheng and Barnouin-Jha 1999) or by compaction rather than excavation and ejection of the surface material (Housen et al. 1999). However, an identified 20-km long scarp on the surface, together with polygonal strength-controlled craters, indicate that Mathilde is not completely strengthless (Thomas et al. 1999). The presence of the long scarp implies that if Mathilde is a rubble-pile at least one of its component bodies appears coherent over scales of 20 km. Therefore, a rubble-pile Mathilde cannot be formed entirely from small bits of rubble (Cheng 2004). The surface density of craters was found to be close to equilibrium saturation. Estimates of the collisional lifetime of the asteroid placed it approximately to 4 Gyr, essentially indistinguishable from the age of the solar system. However, uncertainties in the number of small impactors could allow an age as young as 2 Gyr (Davis 1999). The depth-to-diameter ratio (d/D) for the craters was found to range between 0.12 and 0.25, comparable to values ~0.2 coming from fresh lunar craters (Thomas et al. 1999).

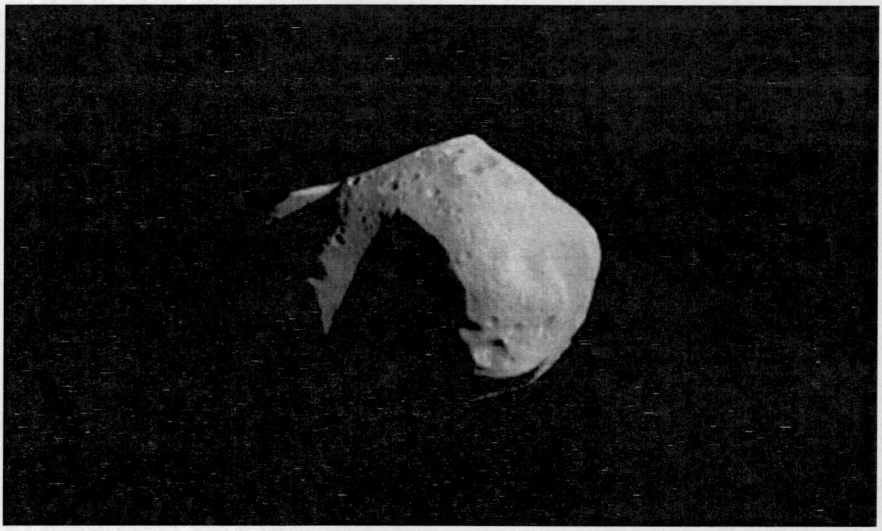

Fig. 1.3 Image mosaic of Mathilde by the NEAR spacecraft taken from a distance of 2400 km on 27 June 1997. The part of the asteroid shown is about 59×47 km^2 across. The resolution of the image is 380 m px^{-1}. Courtesy NASA/JPL/JHUAPL.

The measure of the perturbations exerted on the trajectory of the spacecraft (Yeomans et al. 1997) together with the best volume estimate (Thomas et al. 1999) yielded a mean density of 1.3 ± 0.2 g cm^{-3}. Such a low value is less than half of the value measured for the most likely analog materials, carbonaceous chondrite meteorites containing high percentage of water and organic compounds (CM), indicating a porous, low-density structure for Mathilde's interior. The porosity of the asteroid must be up to 50%. It is therefore possible that either Mathilde is a rubble-pile asteroid, the interior having been pulverized by a long history of impacts, or it is made of primitive, unprocessed materials, with primordial low density (Veverka et al. 1997b; Davis 1999; Chang and Barnouin-Jha 1999).

Combining disk-resolved images and ground-based low-phase angle data, a geometric albedo of 0.047 at 0.55 μm, essentially in the middle of the 0.03-0.06 range for C-type asteroids, was derived. The asteroid was also found remarkably homogeneous in reflectance across the surface. The lack of color and normal reflectance variation on the asteroid suggests uniform regolith texture, an interior similar to the surface, and compositional homogeneity to the depth of crater penetration (Clark et al. 1999).

Finally, an extensive search for bound companions, which may have accounted for the Mathilde very slow rotation period of 418 h was performed. In the images covering the whole Hill sphere, no satellite larger than 10 km in radius, assuming the same albedo as Mathilde, was found. The post-flyby coverage was limited to about 20 radii around Mathilde: no satellite down to a limiting size of 40 m was detected (Veverka et al. 1999). Another possible explanation of the long rotation period of the asteroid could come from an impact of the porous asteroid with a 3-km size projectile, being this the size of the largest impactor needed to form the largest craters on Mathilde, which could have despun the asteroid if it was initially a more rapid rotator (Davis 1999).

1.2.4 The First Encounter with a Near Earth Asteroid: Deep Space 1 at Braille

The NASA Deep Space 1 mission was designed as a low-cost attempt to test new technologies for future space flight projects. As part of the validation of these technologies, the spacecraft flew-by the small Q-type near Earth asteroid (9969) Braille on 29 July 1999 with a relative velocity of 15.5 km s^{-1}, at the closest distance of 28 km (Buratti et al. 2004; Richter et al. 2001).

During the encounter, the Miniature Integrated Camera and Imaging Spectrometer (MICAS) instrument (Soderblom et al. 2001) took two medium-resolution images and three infrared spectra of the object from ~13000 km. The two images were combined with data from a coordinated ground-based photometric campaign to constrain the size, shape, and albedo of the object. Using a simple three-dimensional shape model made by three spheres in contact which sizes and relative positions were determined by fitting the observations, Braille was estimated to have a size of $2.1 \times 1.0 \times 1.0$ km^3 (Oberst et al. 2001). The spectra obtained in the [1.25-2.6] μm range showed a 10% absorption band centred at 2 μm and a reflectance peak at 1.6 μm. The analysis of these features suggests that the composition of Braille is roughly equal parts pyroxene and olivine. The spectra also suggested that the asteroid is closely related to the ordinary chondrites, the most common type of terrestrial meteorite. The geometric albedo, derived from a combination with ground-based spectra, is unusually high in the visible (0.34), which is also consistent with its placement within the rarer classes of stony asteroids, and which suggests it has a relatively fresh, unweathered surface, perhaps due to a recent collision (Buratti et al. 2001).

The most important discovery at Braille came from the magnetometers, two ion engine diagnostic sensors able to resolve magnetic fields with a resolution of only 0.04 nT. Once the data were properly cleaned from the magnetic disturbance caused by the spacecraft itself and the propulsive ion beam, a flux bump at the time of closest approach was found. Considering the conditions for the generation of asteroids magnetosphere, it was concluded that Braille cannot build up a magnetosphere itself. The measured field was therefore interpreted as the unperturbed dipole field of the asteroid, with a dipole magnetic moment of 2.1×10^{11} A m^2. Consequently, Deep Space 1 was the first spacecraft which measured the magnetic field of an asteroid directly (Richter et al. 2001).

1.2.5 The First Asteroid Orbiting Mission: NEAR at Eros

The NASA NEAR spacecraft successfully entered in orbit around the S-type near Earth asteroid (433) Eros on 14 February 2000 to carry out a global survey of the surface properties and internal structure of the asteroid (Veverka et al. 2000). The mission ended with a final descent on the surface on 12 February 2001, which provided the most detailed ever look at the surface of an asteroid with the amazing resolution of 1cm px^{-1} (Veverka et al. 2001b).

Eros was found to be a very elongated curved asteroid in MSI images with dimensions of 34×11×11 km^3. Color images of Eros taken by NEAR are shown in Fig. 1.4. Most of the surface of the asteroid is saturated with craters smaller than 1 km in diameter with depth-to-diameter ratio from 0.12 to 0.16, indicating old degraded structures. The largest crater is ~6 km across. Surface linear features, both grooves and ridges, are prominent; some probably exploit planes of weakness produced by collisions on Eros and/or its parent body. Ejecta boulders are abundant but not uniformly distributed over the surface which results to be covered by fine regolith particles (Veverka et al. 2000, 2001a). Boulders were found originating from a young large crater (Thomas et al. 2001). A detailed analysis of the morphological features suggested the asteroid is a largely coherent but fractured body (Prockter et al. 2002).

Fig. 1.4 Color images of Eros acquired by NEAR on 12 February 2000, at a distance of 1800 km, during the final approach imaging sequence prior to orbit insertion. Courtesy NASA/JPL/JHUAPL.

Albedo, which average value was found to be 0.25 at 0.55 μm, variations are restricted to the inner walls of certain craters and may be related to downslope movement of regolith. (Veverka et al. 2000).

The NEAR Infrared Spectrometer (NIS) instrument (Veverka et al. 1997a) spectra exhibit the silicate 1 μm and 2 μm absorption bands and result consistent with an ordinary primitive chondritic composition, confirmed also by MSI color data (Veverka et al. 2000). This consistency, with the exception of a strong depletion in sulfur, was also confirmed by the X-ray/Gamma-ray Spectrometer (XGRS) instrument (Goldsten et al. 1997) data which pointed towards an undifferentiated body of chondritic origin (Trombka et al. 2000). Although Eros shows no evidence of mineralogical heterogeneity, modest spectral variations correlate with morphologically and geographically distinct areas of the asteroid. Eros bright-to-dark spectral ratios and the low abundance of sulfur found at the surface are largely consistent with laboratory space weathering experiment results and modeling of space weathering effects on chondritic materials (Izenberg et al. 2003; Foley et al. 2006; Loeffler et al. 2008, Lim and Nittler 2008).

Combining spacecraft tracking data for the mass determination and the measured volume, Eros resulted having a mean density of 2.67±0.03 g cm^{-3}. Considering a chondritic composition, this result implies internal porosities ranging from about ~20% to 30%.A rubble pile structure is anyway ruled out (Veverka et al. 2000; Yeomans et al. 2000).

The Magnetometer (MAG) investigation aboard (Acuña et al. 1997) obtained extensive magnetic field observations throughout the Eros environment, from distances in excess of 100000 km to those conducted after landing on 12 February 2001. Data indicate the apparent absence of a global scale magnetization of this asteroid (global magnetization lower than 0.005 A m^{-1}, specific moment lower than 1.9×10^{-6} A m^2 kg^{-1}), orders of magnitude less than the intense magnetization attributed to S-class asteroids Gaspra and Braille. The extremely low magnetization state of Eros places this object significantly below the levels generally associated with low total iron-low metal (LL) chondrites and undifferentiated primitive bodies (Acuña et al. 2002). This result raised the need of revising the meteorite magnetism laboratory record (Wasilewski et al. 2002).

An intensive search for satellites out to 100 asteroid radii was performed during the approach. No satellite larger than 20 m was found, assuming the same albedo as the asteroid (Veverka et al. 2000).

1.2.6 An Engineering Test with Important Scientific Results: Stardust at Annefrank

En route to collect for the first time in history the dust of a comet, 81P/Wild2, and bringing it back to Earth, the NASA Stardust spacecraft flew-by the main belt S-type asteroid (5535) Annefrank on 2 November 2002. The closest distance to the asteroid was ~3100 km and the relative velocity 7.4 km s^{-1}.

During the encounter, images were taken with the Stardust Imaging Camera (Newburn et al. 2003b) at a maximum resolution of 185 m px^{-1}. Less than 40% of the entire surface was observed, significantly constraining the size and shape determination. An image of Annefrank taken by Stardust is shown in Fig. 1.5. The object was found to be highly irregular in shape, giving the appearance of a contact binary. The overall impression was that the asteroid is similar to a triangular prism, with few rounded bodies in contact with the base of the prism. A fit of a simple ellipsoidal shape model to the images gave dimensions of $3.3\times2.5\times1.7$ km^3. Few craters half-km-sized were seen and surface brightness variation appeared to follow the highly irregular topography rather than being dominated by albedo variations. It was concluded that Annefrank possibly represents a significant fraction of asteroids that are fragments of larger bodies and have accreted smaller bodies through contact, possibly even fragments of itself ejected during impacts and re-accreted at low speed (Duxbury et al. 2004).

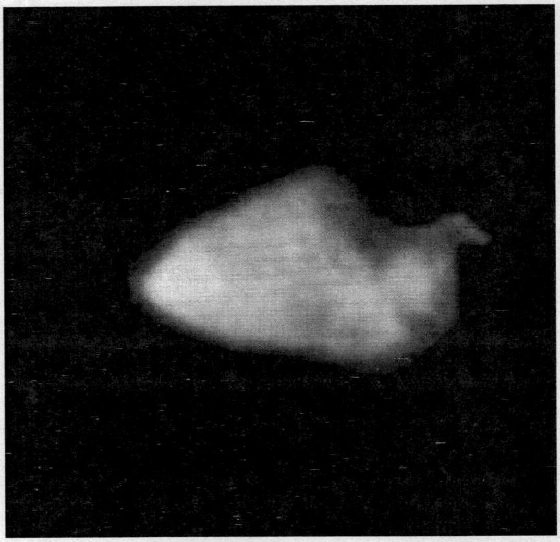

Fig. 1.5 Image of Annefrank taken by the Stardust spacecraft on 2 November 2002 with a resolution of ~185 m px^{-1}. The straight edge in the right side of the image is an artifact of processing. Courtesy NASA/JPL-Caltech.

A broadband (0.47-0.94 μm) geometric albedo of 0.24 was derived (Newburn et al. 2003a). Combining the spacecraft and ground-based low-phase angle data it was found that the asteroid appears to be on the bright side of S-type asteroids, suggesting that it may be a recent collisional fragment with a relative immature surface which has had relatively little time to be affected by space weathering (Hillier et al. 2011).

1.2.7 The First Sample Return Mission: Hayabusa at Itokawa

During the interval from September through early December 2005, the JAXA Hayabusa spacecraft, the first sample return mission to an asteroid in history, was in close proximity to the small S-type near Earth asteroid (25143) Itokawa, orbiting at an altitude of ~7 km from the surface. A microrover was released, but its landing onto the surface was unsuccessful. On 19 and 25 November 2005 two touchdowns, 30-minutes stays on the surface, and liftoffs were performed. The first touchdown allowed collecting some surface material while the second one suffered operational problems (Fujiwara et al. 2006).

The Asteroid Multiband Imaging Camera (AMICA) instrument (Nakamura et al. 2001) imaged the entire surface of Itokawa with a resolution of 70 cm px^{-1} (Saito et al. 2006). The shape of Itokawa was found resembling a sea otter. It appears to be composed of two rather rounded parts called 'head' and 'body', connected by a depressed neck-zone. The best fit ellipsoidal model has dimensions of $535 \times 294 \times 209$ m^3 (Demura et al. 2006). A picture of the asteroid taken by Hayabusa is shown in Fig. 1.6. The surface is divided into a rough terrain, mostly consisting of numerous boulders, and a smooth terrain. The smooth region is found to be composed of fragmental debris with grain sizes of cm to mm scales from close-up images (Fujiwara et al. 2006; Yano et al. 2006). Several boulders were also observed. There are no long linear structures extending nearly the entire length of the asteroid, and this, coupled with the existence of some faint local facets with scales of at most several tens of meters, suggests that Itokawa is not a single consolidated, coherent body but rather an aggregate of rubble with sizes ranging up to about 50 m. (Fujiwara et al. 2006; Saito et al. 2006). Crater counting gave a large range of possible ages for the surface, spanning from 75 Myr up to 1 Gyr, depending on the scaling law used in the calculation (Michel et al. 2009).

The mass of Itokawa was estimated from the Hayabusa's tracking and navigation data. Combining the mass measure with the best volume estimate yielded a bulk density of 1.95 ± 0.14 g cm^{-3}. Assuming that LL ordinary chondrites, having a density of 3.2 g cm^{-3}, are analogs for Itokawa's composition, the macroporosity of the asteroid is estimated to be ~40% (Fujiwara et al. 2006; Abe S. et al. 2006). Itokawa is considered to be the first clearly observed rubble-pile body due to its low bulk density, high porosity, boulder-rich appearance, and shape (Fujiwara et al. 2006).

There is no substantial difference in mineralogical composition over the whole asteroid surface in spite of the bifurcated appearance (Fujiwara et al. 2006). A homogeneous composition for Itokawa's surface was also suggested from the results of the X-ray Spectrometer (XRS) instrument (Okada et al. 1999) during the first touchdown (Okada et al. 2006). The Near Infrared Spectrometer (NIRS) data found instead a variation of more than 10% in albedo, color, and absorption band depth in the surface reflectance for wavelengths up to 2 μm, possibly resulting from a combination of different degrees of space weathering and different grain sizes. The spectral shape over the 1 μm absorption band indicates that the surface of this body has an olivine-rich mineral assemblage potentially similar to that of thermally metamorphosed and aqueous altered LL (LL5 or LL6) chondrites (Abe M. et al. 2006). The similarity with LL chondrites is also consistent with XRS results during the first touchdown (Okada et al. 2006).

Fig. 1.6 Itokawa as seen from the Hayabusa spacecraft at the distance of ~8 km on 29 September 2005. Image credit: JAXA.

A search for satellites in AMICA images covering the entire Hill sphere of the asteroid was performed. No satellites dawn to the size of 1 m were found (Fuse et al. 2008).

A new chapter of space science was opened when the surface samples were successfully brought back to Earth for laboratory analysis on 13 June 2010. Synchrotron-radiation x-ray diffraction and transmission and scanning electron microscope analyses indicate that the mineralogy and mineral chemistry of the Itokawa dust particles are identical to those of thermally metamorphosed LL chondrites, consistent with spectroscopic observations made from Earth and by the Hayabusa spacecraft. These results directly demonstrate and give the definitive proof that ordinary chondrites come from the abundant S-type asteroids. Mineral chemistry indicates that the majority of regolith surface particles suffered long-term thermal annealing and subsequent impact shock, suggesting that Itokawa is an asteroid made of reassembled pieces of the interior portions of a once larger asteroid (Nakamura et al. 2011; Yurimoto et al. 2011; Ebihara et al. 2011). Evidences of space weathering effects were also found in the captured sample. This result helps interpreting the observational differences between the spectra of ordinary chondrites meteorites and of the S-type asteroids exactly in terms of space weathering (Noguchi et al. 2011; Nagao et al. 2011).

1.2.8 The First E-type Asteroid: Rosetta at Steins

On 5 September 2008 the ESA Rosetta spacecraft, aimed at reaching and escorting the comet 67P/Churyumov-Gerasimenko, passed near the E-type main belt asteroid (2867) Steins. The closest distance reached was ~800 km with a relative speed of 8.6 km s^{-1} (Accomazzo et al. 2010). Fourteen scientific instruments were switched on during the flyby, providing a detailed characterization of the target (Schulz 2010).

Fig. 1.7 Steins as seen from the Rosetta spacecraft at a distance of 800 km on 5 September 2008. Image credit: ESA ©2008 MPS for OSIRIS Team MPS/UPD/LAM/IAA/RSSD/INTA/UPM/DASP/IDA.

The Optical, Spectroscopic, and Infrared Remote Imaging System (OSIRIS) two-camera instrument (Keller et al. 2007) imaged ~60% of the surface with a maximum resolution of 80 m px^{-1}. The object resulted having a shape resembling that of a brilliant cut diamond which is best approximated by an oblate spheroid rotating about its shortest axis with mean equatorial and polar radii of 3.1 and 2.2 km, respectively. An image of Steins from Rosetta can be seen in Fig. 1.7. The morphology of Steins is dominated by linear faults and a large 2.1-km-diameter crater near its south pole. The presence of an equatorial bulge and a relatively smooth, roughly rotationally symmetric northern hemisphere led to the interpretation of being Steins' shape attributed to spin-up by YORP effect (Keller et al. 2010). Hydro-dynamical simulations of the large crater formation concluded that the effect of the impact was transforming Steins into a rubble pile, consistently with the interpretation of subsequent YORP reshaping (Jutzi et al. 2010). Crater counting revealed an age for the surface ranging from few hundred Myr to more than 1Gyr, depending on the adopted scaling law and asteroid physical parameters (Marchi et al. 2010).

OSIRIS disk-integrated geometric albedo at 0.632 μm was found to be 0.40, consistently with the high albedos of other E-type asteroids. Spectrophotometric data showed Steins is slightly reddish. The visible spectrum exhibits a steep drop below 0.4 μm, typical of low-iron content minerals (Keller et al. 2010). No surface color variegation larger than 4% was detected. This uniformity suggests that the asteroid is compositionally homogeneous and its regolith does not display signs of space weathering (Keller et al. 2010; Leyrat et al. 2010). The Visible and Infrared Thermal Imaging Spectrometer (VIRTIS) instrument (Coradini et al. 2007) data confirmed the spectral homogeneity of the surface up to 4 μm and found a new absorption band centered at approximately 0.8 μm, possibly indicative of the presence of sulfide minerals (Tosi et al. 2010). VIRTIS thermal mapping suggested a low-porosity surface and the absence of a thick regolith layer (Leyrat et al. 2011). The results from the Microwave Instrument for the Rosetta Orbiter (MIRO), i.e. a submillimeter and millimeter radiometer and spectrometer (Gulkis et al. 2007), were consistent with this interpretation, supporting a high thermal inertial surface, characteristics of rock-dominated regolith rather than the powdered-regolith surface of the Moon (Gulkis et al. 2010). The Ultraviolet Imaging Spectrometer (ALICE) instrument (Stern et al. 2007) measured the first ever far-ultraviolet reflectivity spectrum of an asteroid, in the [0.085-0.200] μm range. Steins was found to have a low ultraviolet albedo of 0.04 with no evidences of albedo variation across the surface. A broad absorption feature centered at 0.165 μm was detected. The shape of the feature implies a very low abundance of iron ions in the surface minerals. ALICE was also used for the search of a possible exosphere of atoms sputtered from the surface, mainly hydrogen and oxygen. The lack of positive detection set an upper limit of 1.5×10^{9} cm^{-2} on Steins exospheric oxygen abundance (A'Hearn et al. 2010).

The Rosetta Plasma Consortium (RCP) magnetometer on the orbiter (Glassmeier et al. 2007) and the Rosetta Lander Magnetometer and Plasma Monitor (ROMAP) on the lander (Auster et al. 2007) were activated during the flyby to search for possible magnetic signatures of the asteroid. There were no univocal detections. It was therefore derived that Steins may have a magnetic field only lower than the resolution limit of 1 nT. This implies that the magnetic specific moment of the asteroid is less than 10^{3} A m^{2} kg^{-1}, close to the low values characteristics of aubrites (Auster et al. 2010).

1.2.9 The First Visit to an Intact Planetesimal: Rosetta at Lutetia

On 10 July 2010, during its second passage through the main belt, Rosetta flew-by the compositionally puzzling asteroid (21) Lutetia at a distance of ~3170 km with a relative velocity of 15 km s^{-1} (Schulz et al. 2012).

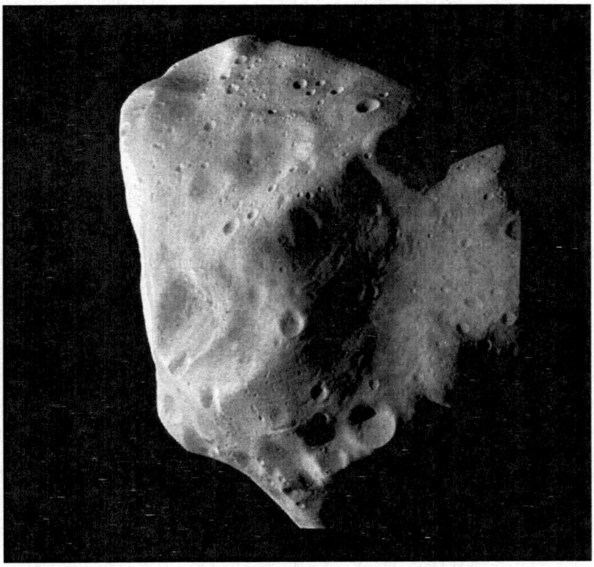

Fig. 1.8 Lutetia as seen from the Rosetta spacecraft at closest approach on 10 July 2010. Image credit: ESA 2010 MPS for OSIRIS Team MPS/UPD/LAM/IAA/RSSD/INTA/UPM/DASP/IDA.

OSIRIS data covered more than 50% of the asteroid surface, mostly of the northern hemisphere. The images revealed an irregular world with overall dimensions 121×101×75 km^3 along the principal axis of inertia. An image of Lutetia taken from Rosetta is shown in Fig. 1.8. The surface was found battered with craters, the largest one, named Massilia, has a diameter of ~55 km. Numerous linear features, presence of regolith, and ejecta boulders were also detected (Sierks et al. 2011a; Küppers et al. 2012; Vincent et al. 2012). The body was found having undergone a complex geological history reflected in five distinct main geological units on its surface, defined in terms of crater density, overlapping and cross-cutting relationships, and presence of linear features (Massironi et al. 2012; Thomas et al. 2012). Crater counting revealed an age of 3.6 Gyr for the oldest region and of tens to hundreds of Myr for the youngest one (Marchi et al 2012a) . The Lutetia's disk integrated geometric albedo was found to be 0.19 at 0.55 μm. Albedo variegation is seen across the surface (Sierks et al. 2011a; Magrin et al. 2012). The presence of a fine lunar-like regolith with low thermal inertia in the upper 1-3 cm of the surface, overlaying a layer of rapidly increasing density and thermal conductivity, is also required to explain MIRO observations in both hemispheres (Gulkis et al. 2012). Low-inertia regolith was also inferred from VIRTIS spectra (Coradini et al. 2011). VIRTIS data confirmed also the absence of OH hydration and other absorption features up to a wavelength of 3.5 μm at northern latitudes (Coradini et al. 2011; Tosi et al. 2012).

The best shape model derived from OSIRIS and ground-based data (Sierks et al. 2011a), combined with the mass determination from the deviation of the spacecraft trajectory during the flyby caused by the asteroid (Pätzold et al. 2011), yielded a high density measure of 3.4 ± 0.3 g cm^{-3} for Lutetia. This value, exceeding that of most known chondritic meteorites, implies that Lutetia has likely a low bulk macroporosity or even a partially differentiated interior or at least large metal-rich regions. By these arguments, a rubble-pile structure is ruled out (Sierks et al. 2011a; Weiss et al. 2012).

Seven instruments tried to detect an exosphere with a coordinated observational campaign. The lack of positive detections allowed only the determination of upper limits for the production rates of water (4.3×10^{23} molecules s^{-1}) and CO (1.7×10^{25} molecules s^{-1}). The upper limit for a water exosphere density was set to 3.5×10^{3} cm^{-3} at the flyby distance of 3160 km (Morse et al. 2012; Altwegg et al. 2012).

The magnetometers did not detect any conclusive signature of the asteroid magnetic field, possibly because of the large flyby distance, below the detection limit. Consequently, only upper limits for the global magnetic properties of Lutetia could be derived. The dipole magnetic moment, the global magnetization, and the specific moment would be lower than 1.0×10^{12} A m^{2}, 2.1×10^{3} A m^{-1}, and 5.9×10^{-7} A m^{2} kg^{-1}, respectively (Richter et al. 2012).

No satellite larger than ~160 m was found inside the Hill sphere. The size limit was extended to ~30 m within distances from the asteroid lower than 20 radii (Bertini et al. 2012).

In summary, it can be concluded that the asteroid has a condritic surface, possibly composed of mixtures of different types of materials: carbonaceous and enstatite condrites, being the juxtaposition of these materials a possible consequence of the large impact that created the present surface (Coradini et al. 2011; Barucci et al. 2012). Lutetia's geologically complex surface, ancient surface age, high density, and hydro-dynamical modeling of the craters suggest that the asteroid is most likely a primordial planetesimal which survived intact the age of the solar system, the first one visited and investigated through a close encounter (Sierks et al. 2011a; Cremonese et al. 2012).

1.2.10 The First Visit to a Protoplanet: Dawn at Vesta

On 16 July 2011 the NASA Dawn spacecraft, aimed at the exploration of the first and third largest main belt asteroids, (1) Ceres and (4) Vesta, entered into orbit around Vesta. This asteroid is an almost spherical body with a radius of 530 km. It is believed to be a remnant intact differentiated protoplanet from the earliest epoch of the solar system formation, based on the analysis of howardite-eucrite-diogenite (HED) meteorites which are known to originate from it (Russell et al. 2012). Moreover, Vesta is the parent body of the so-called Vestoids asteroids (Binzel and Xu 1993; Marzari et al. 1996), indicating it underwent a complex collisional history.

Dawn arrived in vestan southern summer and mapped ~ 80% of the surface from an altitude of ~2700 km with the Framing Camera (FC) instrument (Sierks et al. 2011b) and the Visible and Infrared Spectrometer (VIR) instrument (De Sanctis et al. 2011) with resolution of ~260 m px^{-1} and ~700 m px^{-1}, respectively (Jaumann et al. 2012). The volume estimation yielded a best fit ellipsoid model of 286×279×223 km^3 (Russell et al. 2012). An image of Vesta as seen from Dawn is shown in Fig. 1.9. FC images revealed, confirming previous Hubble Space Telescope data, a giant impact basin, named Rheasilva, in the south-polar region (Schenk et al. 2012). This impact, whose age was estimated to be about 1.0 Gyr from crater counting (Marchi et al. 2012b), is consistent with the production of HEDs and the Vestoids (Schenk et al. 2012). The Rheasilva event resulted in a strong dichotomy between the northern and southern emispheres, reflected in surface albedo and crater densities (Russell et al. 2012). An older basin, named Veneneia and underlying Rheasilva, was also found in FC images. Its age was estimated to be ~2.0 Gyr and it could have provided an earlier additional source of HEDs (Schenk et al. 2012). Vesta's geology displays morphological features characteristic of the Moon and terrestrial planets (presence of an igneous crust) as well as those of other asteroids (e.g. the presence of ponds), underlining Vesta's unique role as a transitional solar system body. No unambiguous volcanic deposits have been identified, although they might have been expected from the analysis of HEDs. The present lack of volcanic relicts on Vesta suggests that such features were only produced during the short period of rapid cooling of the asteroid interior within the first 100 Myr after formation and have been eroded and gardened by impacts. Volcanic materials should be deeply buried by impact ejecta from the Rheasilva and other large basins whose ejecta must cover the surface (Jaumann et al. 2012).

VIR spectral mapping revealed a diverse surface with considerable and local variations consistent with the mineralogy of HEDs being physical mixtures of eucrites (crustal basalts) and diogenites (ultramafic cumulates) formed by impact processes. Spectrally distinct regions include the Rheasilva basin, which displays a higher diogenetic pyroxene-rich component signature of a deeper crust exposed after excavation by the large impact which formed the basin, and equatorial regions which show a higher eucritic component, signature of an upper crust. Evidence for mineralogical stratigraphic layering is observed on crater walls and in ejecta. Overall, the mineralogy indicates a complex magmatic evolution that led to a differentiated crust and mantle. The record of early magmatic processes is reflected in the largest color and albedo (0.10-0.67) variation across the surface

Fig. 1.9 Mosaic image of Vesta as seen from the Dawn spacecraft. Courtesy NASA/JPL-Caltech/UCAL/MPS/DLR/IDA.

observed so far in asteroids (De Sanctis et al. 2012a; Reddy et al. 2012). VIR also detected a widespread 2.8 μm OH absorption band on the surface of the asteroid. This feature is distributed across the surface and shows areas enriched and depleted in hydrated materials. The origin of vestan OH provides new insights on the presence of hydrous material in the main belt and may offer new scenarios on the delivery of water in the inner solar system (De Sanctis et al. 2012b). For a recent review on the theories about the origin of water on Earth the reader is addressed to Bertini (2011).

Combining the knowledge of the mass from the spacecraft orbital data and the best volume estimate, a density of 3.456 g cm^{-3} with an uncertainty of 1% was measured. This value is comparable to the bulk density of HEDs. The measure of the J2 gravitational moment confirmed the presence of a high-density central concentration. From these results a porosity of 5-6% could be estimated in the mantle and the crust of the asteroid (Russell et al. 2012).

In summary, this preliminary but exhaustive analysis of Vesta exploration confirmed that the asteroid is a surviving protoplanet which properties are as inferred from HED meteorites. The object appears to have accreted early and differentiated, forming an iron core that may have sustained a magnetic dynamo (Russell et al. 2012).

1.2.11 The First Chinese Mission: Cheng'E 2 at Toutatis

This summary of scientific results from past space missions to asteroids ends with the flyby of the second chinese lunar probe Cheng'E 2 with the S-type near Earth asteroid (4179) Toutatis.

At the time of writing no scientific publication has been released and the few news about the event are coming from astronomy dedicated web forums. The encounter took place on 13 December 2012 at the closest distance of only 3.2 km and with a relative speed of 10.73 km s^{-1}. In the images obtained with the star observation cameras onboard the spacecraft, Toutatis resulted to be an highly irregular object, probably made by two distinct lobes, as already discovered in previous observations from the Goldstone's Solar System Radar (USA).

1.3 The Future of Asteroids Space Exploration

Several space missions are already in construction or planned to deepen our knowledge of the asteroids in the near future.

The NASA DAWN spacecraft, after leaving Vesta, is now heading towards the largest main belt asteroid, the protoplanet (1) Ceres which will be reached in 2015 (Russell et al. 2007).

NASA has also approved the OSIRIS-REx mission, designed to approach and study the near Earth asteroid 101955 (1999 RQ36), which is both the most accessible primitive carbonaceous asteroid and one of the most potentially hazardous asteroids known. OSIRIS-Rex will be launched in 2016, and will orbit its target in 2020. The spacecraft will collect a sample form the surface of the asteroid and return it to Earth (Lauretta and OSIRIS-REx Team 2012).

JAXA is planning the successor of Hayabusa: Hayabusa2. The mission is planned to take off in 2014, reach its target, the C-type near Earth asteroid 162173 (1999 JU3) in 2018, orbit around it, create a small crater with a collision device, collect unaltered samples from it, and return the samples back to Earth in 2020 (Takagi et al. 2011).

Finally, ESA is developing another sample and return mission, named Marco Polo-R (Barucci et al. 2009), which target has been recently chosen: the primitive near Earth asteroid 2008 EV5.

1.4 Conclusions

The scientific results of past space mission to asteroids have been described in detail in the previous sections. In-situ studies have been a test bench for new technologies and for validating theories based on ground-based observations. Space missions to asteroids provided an accurate description of the objects visited by the spacecrafts, but their importance goes behind that. Their results often provided scientific proofs applicable to a broader context. For example, spacecraft data proved that asteroids possess a magnetic field, telling us that ferromagnetic materials were indeed present in the forming solar nebula. Asteroids have satellites, as also proved later to a larger extent with the use of radar data. This means the collisional environment of these objects was still active in recent times. Collisions have shaped the surface of all asteroids visited so far and low-velocity accretion of smaller bodies, leading to the formation of larger rubble-pile structures, has been proved to exist. It was demonstrated that thermal forces act effectively on small asteroids, influencing their rotational state and shape. Large asteroids can survive the age of the solar system as intact planetesimals or differentiated protoplanets. Definitive evidence of the relationship between the massive protoplanet Vesta and its family and meteorites has been given. The presence of hydrous material in the main belt has been proved, offering new scenarios on the delivery of water to the inner solar system. Laboratory analysis of material brought back from Itokawa proved that ordinary chondrites are coming from S-type asteroids. It is not a case that all future missions to asteroids already planned have the return of early solar system material for laboratory studies as key scientific aim. These pristine samples coming from easily accessible carbonaceous asteroids will further deepen our knowledge on the processes occurring in the early solar system and accompanying the planets formation, bringing light on the nature and origin of organics in primitive asteroids and their relationship with the molecules necessary for life. What was obtained so far constitutes undoubtedly an extraordinary scientific success, but it is only the first step towards a future where robots and human beings will investigate in detail our past and possibly get a new source of useful materials for our future visiting the asteroids.

References

Abe, M., Takagi, Y., Kitazato, K., Abe, S., Hiroi, T., Vilas, F., Clark, B.E., Abell, P.A., Lederer, S.M., Jarvis, K.S., Nimura, T., Ueda, Y., Fujiwara, A.: Near-Infrared Spectral Results of Asteroid Itokawa from the Hayabusa Spacecraft. Science 312, 1334–1338 (2006)

Abe, S., Mukai, T., Hirata, N., Barnouin-Jha, O.S., Cheng, A.F., Demura, H., Gaskell, R.W., Hashimoto, T., Hiraoka, K., Honda, T., Kubota, T., Matsuoka, M., Mizuno, T., Nakamura, R., Scheeres, D.J., Yoshikawa, M.: Mass and Local Topography Measurements of Itokawa by Hayabusa. Science 312, 1344–1349 (2006)

Accomazzo, A., Wirth, K.R., Lodiot, S., Küppers, M., Schwehm, G.: The flyby of Rosetta at asteroid Steins – mission and science operations. Planetary and Space Science 58, 1058–1065 (2010)

Acuña, M.H., Russell, C.T., Zanetti, L.J., Anderson, B.J.: The NEAR magnetic field investigation: Science objectives at asteroid Eros 433 and experimental approach. Journal of Geophysical Research 102, 23751–23760 (1997)

Acuña, M.H., Anderson, B.J., Russell, C.T., Wasilewski, P., Kletetshka, G., Zanetti, L., Omidi, N.: NEAR Magnetic Field Observations at 433 Eros: First Measurements from the Surface of an Asteroid. Icarus 155, 220–228 (2002)

A'Hearn, M.F., Feaga, L.M., Bertaux, J.L., Feldman, P.D., Parker, J.W., Slater, D.C., Steffl, A.J., Stern, S.A., Throop, H., Versteeg, M., Weaver, H.A., Keller, H.U.: The farultraviolet albedo of Steins measured with Rosetta-Alice. Planetary and Space Science 58, 1088–1096 (2010)

Altwegg, K., Balsiger, H., Calmonte, U., Hässig, M., Hofer, L., Jäckel, A., Schläppi, B., Wurz, P., Berthelier, J.J., De Keyser, J., Fiethe, B., Fuselier, S., Mall, U., Rème, H., Rubin, M.: In situ mass spectrometry during the Lutetia flyby. Planetary and Space Science 66, 173–178 (2012)

Auster, H.U., Apathy, I., Berghofer, G., Remizov, A., Roll, R., Fornacon, K.H., Glassmeier, K.H., Haerendel, G., Hejja, I., Kührt, E., Magnes, W., Moehlmann, D., Motschmann, U., Richter, I., Rosenbauer, H., Russell, C.T., Rustenbach, J., Sauer, K., Schwingenschuh, K., Szemerey, I., Waesch, R.: ROMAP: Rosetta Magnetometer and Plasma Monitor. Space Science Reviews 128, 221–240 (2007)

Auster, H.U., Richter, I., Glassmeier, K.H., Berghofer, G., Carr, C.M., Motschmann, U.: Magnetic field investigations during ROSETTA's 2867 Steins flyby. Planetary and Space Science 58, 1124–1128 (2010)

Barucci, M.A., Yoshikawa, M., Michel, P., Kawagushi, J., Yano, H., Brucato, J.R., Franchi, I.A., Dotto, E., Fulchignoni, M., Ulamec, S.: MARCO POLO: near earth object sample return mission. Experimental Astronomy 23, 785–808 (2009)

Barucci, M.A., Belskaya, I.N., Fornasier, S., Fulchignoni, M., Clark, B.E., Coradini, A., Capaccioni, F., Dotto, E., Birlan, M., Leyrat, C., Sierks, H., Thomas, N., Vincent, J.B.: Overview of Lutetia's surface composition. Planetary and Space Science 66, 23–30 (2012)

Baumgartel, K., Sauer, K., Bogdanov, A.: A Magnetohydrodynamic Model of Solar Wind Interaction with Asteroid Gaspra. Science 263, 653–655 (1994)

Belton, M.J.S., Klaasen, K.P., Clary, M.C., Anderson, J.L., Anger, C.D., Carr, M.H., Chapman, C.R., Davies, M.E., Greeley, R., Anderson, D.: The Galileo Solid State Imaging experiment. Space Science Reviews 60, 413–455 (1992a)

Belton, M.J.S., Veverka, J., Thomas, P., Helfenstein, P., Simonelli, D., Chapman, C., Davies, M.E., Greeley, R., Greenberg, R., Head, J., Murchie, S., Klaasen, K., Johnson, T.V., McEwen, A., Morrison, D., Neukum, G., Fanale, F., Anger, C., Carr, M., Pilcher, M.: Galileo Encounter with 951 Gaspra: First Pictures of an Asteroid. Science 257, 1647–1652 (1992b)

Belton, M.J.S., Chapman, C.R., Veverka, J., Klaasen, K.P., Harch, A., Greeley, R., Greenberg, R., Head III, J.W., McEwen, A., Morrison, D., Thomas, P.C., Davies, M.E., Carr, M.H., Neukum, G., Fanale, F.P., Davis, D.R., Anger, C., Gierasch, P.J., Ingersoll, A.P., Pilcher, C.B.: First Images of Asteroid 243 Ida. Science 265, 1543–1547 (1994)

Belton, M.J.S., Chapman, C.R., Thomas, P.C., Davies, M.E., Greenberg, R., Klaasen, K., Byrnes, D., D'Amario, L., Synnott, S., Johnson, T.V., McEwen, A., Merline, W.J., Davis, D.R., Petit, J.-M., Storrs, A., Veverka, J., Zellner, B.: Bulk density of asteroid 243 Ida from the orbit of its satellite Dactyl. Nature 374, 785–788 (1995)

Belton, M.J.S., Chapman, C.R., Klaasen, K.P., Harch, A.P., Thomas, P.C., Veverka, J., McEwen, A.S., Pappalardo, R.T.: Galileo's Encounter with 243 Ida: an Overview of the Imaging Experiment. Icarus 120, 1–19 (1996)

Bertini, I.: Main Belt Comets: A new class of small bodies in the solar system. Planetary and Space Science 59, 365–377 (2011)

Bertini, I., Sabolo, W., Gutierrez, P.J., Marzari, F., Snodgrass, C., Tubiana, C., Moissl, R., Pajola, M., Lowry, S.C., Barbieri, C., Ferri, F., Davidsson, B., Sierks, H., The OSIRIS Team: Search for satellites near (21) Lutetia using OSIRIS/Rosetta images. Planetary and Space Science 66, 64–70 (2012)

Binzel, R.P., Xu, S.: Chips off of asteroid 4 Vesta - Evidence for the parent body of basaltic achondrite meteorites. Science 260, 186–191 (1993)

Blanco-Cano, X., Omidi, N., Russell, C.T.: Hybrid simulations of solar wind interaction with magnetized asteroids: Comparison with Galileo observations near Gaspra and Ida. Journal of Geophysical Research 108, A5 (2003)

Buratti, B.J., Britt, D.T., Soderblom, L.A., Hicks, M.D., Boice, D.C., Brown, R.H., Meier, R., Nelson, R.M., Oberst, J., Owen, T.C., Rivkin, A.S., Sandel, B.R., Stern, S.A., Thomas, N., Yelle, R.V.: 9969 Braille: Deep Space 1 infrared spectroscopy, geometric albedo, and classification. Icarus 167, 129–135 (2004)

Carlson, R.W., Weissman, P.R., Smythe, W.D., Mahoney, J.C.: Near-Infrared Mapping Spectrometer experiment on Galileo. Space Science Reviews 60, 457–502 (1992)

Carr, M.H., Kirk, R.L., McEwen, A., Veverka, J., Thomas, P., Head, J.W., Murchie, S.: The geology of Gaspra. Icarus 107, 61–71 (1994)

Chapman, C.R., Veverka, J., Thomas, P.C., Klaasen, K., Belton, M.J.S., Harch, A., McEwen, A., Johnson, T.V., Helfenstein, P., Davies, M.E., Merline, W.J., Denk, T.: Discovery and physical properties of Dactyl, a satellite of asteroid 243 Ida. Nature 374, 783–785 (1995)

Chapman, C.R.: S-Type Asteroids, Ordinary Chondrites, and Space Weathering: The Evidence from Galileo's Flybys of Gaspra and Ida. Meteoritics & Planetary Science 31, 699–725 (1996)

Chapman, C.R., Ryan, E.V., Merline, W.J., Neukum, G., Wagner, R., Thomas, P.C., Veverka, J., Sullivan, R.J.: Cratering on Ida. Icarus 120, 77–86 (1996a)

Chapman, C.R., Veverka, J., Belton, M.J.S., Neukum, G., Morrison, D.: Cratering on Gaspra. Icarus 120, 231–245 (1996b)

Chapman, C.R., Merline, W.J., Thomas, P.: Cratering on Mathilde. Icarus 140, 28–33 (1999)

Cheng, A.F., Barnouin-Jha, O.S.: Giant Craters on Mathilde. Icarus 140, 34–48 (1999)

Cheng, A.F.: Implications of the NEAR mission for internal structure of Mathilde and Eros. Advances in Space Research 33, 1558–1563 (2004)

Clark, B.E., Veverka, J., Helfenstein, P., Thomas, P.C., Bell, J.F., Harch, A., Robinson, M.S., Murchie, S.L., McFadden, L.A., Chapman, C.R.: NEAR Photometry of Asteroid 253 Mathilde. Icarus 140, 53–65 (1999)

Coradini, A., Capaccioni, F., Drossart, P., Arnold, G., Ammannito, E., Angrilli, F., Barucci, M.A., Bellucci, G., Benkhoff, J., Bianchini, G., Bibring, J.P., Blecka, M., Bockelee-Morvan, D., Capria, M.T., Carlson, R., Carsenty, U., Cerroni, P., Colangeli, L., Combes, M., Combi, M., Crovisier, J., De Sanctis, M.C., Encrenaz, E.T., Erard, S., Federico, C., Filacchione, G., Fink, U., Fonti, S., Formisano, V., Ip, W.H., Jaumann, R., Kuehrt, E., Langevin, Y., Magni, G., McCord, T., Mennella, V., Mottola, S., Neukum, G., Palumbo, P., Piccioni, G., Rauer, H., Saggin, B., Schmitt, B., Tiphene, D., Tozzi, G.: Virtis: An Imaging Spectrometer for the Rosetta Mission. Space Science Reviews 128, 529–559 (2007)

Coradini, A., Capaccioni, F., Erard, S., Arnold, G., De Sanctis, M.C., Filacchione, G., Tosi, F., Barucci, M.A., Capria, M.T., Ammannito, E., Grassi, D., Piccioni, G., Giuppi, S., Bellucci, G., Benkhoff, J., Bibring, J.P., Blanco, A., Blecka, M., Bockelee-Morvan, D., Carraro, F., Carlson, R., Carsenty, U., Cerroni, P., Colangeli, L., Combes, M., Combi, M., Crovisier, J., Drossart, P., Encrenaz, E.T., Federico, C., Fink, U., Fonti, S., Giacomini, L., Ip, W.H., Jaumann, R., Kuehrt, E., Langevin, Y., Magni, G., McCord, T., Mennella, V., Mottola, S., Neukum, G., Orofino, V., Palumbo, P., Schade, U., Schmitt, B., Taylor, F., Tiphene, D., Tozzi, G.: The Surface Composition and Temperature of Asteroid 21 Lutetia As Observed by Rosetta/VIRTIS. Science 334, 492–494 (2011)

Cremonese, G., Martellato, E., Marzari, F., Kuhrt, E., Scholten, F., Preusker, F., Wünnemann, K., Borin, P., Massironi, M., Simioni, E., Ip, W.: OSIRIS Team Hydrocode simulations of the largest crater on asteroid Lutetia. Planetary and Space Science 66, 147–154 (2012)

Davis, D.R.: The Collisional History of Asteroid 253 Mathilde. Icarus 140, 49–52 (1999)

Demura, H., Kobayashi, S., Nemoto, E., Matsumoto, N., Furuya, M., Yukishita, A., Muranaka, N., Morita, H., Shirakawa, K., Maruya, M., Ohyama, H., Uo, M., Kubota, T., Hashimoto, T., Kawaguchi, J., Fujiwara, A., Saito, J., Sasaki, S., Miyamoto, H., Hirata, N.: Pole and Global Shape of 25143 Itokawa. Science 312, 1347–1349 (2006)

De Sanctis, M.C., Coradini, A., Ammannito, E., Filacchione, G., Capria, M.T., Fonte, S., Magni, G., Barbis, A., Bini, A., Dami, M., Ficai-Veltroni, I., Preti, G.: The VIR Spectrometer. Space Science Reviews 163, 329–369 (2011)

De Sanctis, M.C., Ammannito, E., Capria, M.T., Tosi, F., Capaccioni, F., Zambon, F., Carraro, F., Fonte, S., Frigeri, A., Jaumann, R., Magni, G., Marchi, S., McCord, T.B., McFadden, L.A., McSween, H.Y., Mittlefehldt, D.W., Nathues, A., Palomba, E., Pieters, C.M., Raymond, C.A., Russell, C.T., Toplis, M.J., Turrini, D.: Spectros-copic Characterization of Mineralogy and Its Diversity Across Vesta. Science, 697–700 (2012a)

De Sanctis, M.C., Combe, J.P., Ammannito, E., Palomba, E., Longobardo, A., McCord, T.B., Marchi, S., Capaccioni, F., Capria, M.T., Mittlefehldt, D.W., Pieters, C.M., Sunshine, J., Tosi, F., Zambon, F., Carraro, F., Fonte, S., Frigeri, A., Magni, G., Raymond, C.A., Russell, C.T., Turrini, D.: Detection of Widespread Hydrated Materials on Vesta by the VIR Imaging Spectrometer on board the Dawn Mission. The Astrophysical Journal Letters 758, L36 (2012b)

Duxbury, T.C., Newburn, R.L., Acton, C.H., Carranza, E., McElrath, T.P., Ryan, R.E., Synnott, S.P., You, T.H., Brownlee, D.E., Cheuvront, A.R., Adams, W.R., Toro-Allen, S.L., Freund, S., Gilliland, K.V., Irish, K.J., Love, C.R., McAllister, J.G., Mumaw, S.J., Oliver, T.H., Perkins, D.E.: Asteroid 5535 Annefrank size, shape, and orientation: Stardust first results. Journal of Geophysical Research 109, E02002 (2004)

Ebihara, M., Sekimoto, S., Shirai, N., Hamajima, Y., Yamamoto, M., Kumagai, K., Oura, Y., Ireland, T.R., Kitajima, F., Nagao, K., Nakamura, T., Naraoka, H., Noguchi, T., Okazaki, R., Tsuchiyama, A., Uesugi, M., Yurimoto, H., Zolensky, M.E., Abe, M., Fujimura, A., Mukai, T., Yada, Y.: Neutron Activation Analysis of a Particle Returned from Asteroid Itokawa. Science 333, 1119–1121 (2011)

Foley, C.N., Nittler, L.R., McCoy, T.J., Lim, L.F., Brown, M.R.M., Starr, R.D., Trombka, J.I.: Minor element evidence that Asteroid 433 Eros is a space-weathered ordinary chondrite parent body. Icarus 184, 338–343 (2006)

Fujiwara, A., Kawaguchi, J., Yeomans, D.K., Abe, M., Mukai, T., Okada, T., Saito, J., Yano, H., Yoshikawa, M., Scheeres, D.J., Barnouin-Jha, O., Cheng, A.F., Demura, H., Gaskell, R.W., Hirata, N., Ikeda, H., Kominato, T., Miyamoto, H., Nakamura, A.M., Nakamura, R., Sasaki, S., Uesugi, K.: The Rubble-Pile Asteroid Itokawa as Observed by Hayabusa. Science 312, 1330–1334 (2006)

Fuse, T., Yoshida, F., Tholen, D., Ishiguro, M., Saito, J.: Searching satellites of asteroid Itokawa by imaging observation with Hayabusa spacecraft. Earth, Planets and Space 60, 33–37 (2008)

Giblin, I., Petit, J.M., Farinella, P.: Impact Ejecta Rotational Bursting as a Mechanism for Producing Stable Ida-Dactyl Systems. Icarus 132, 43–52 (1998)

Glassmeier, K.H., Richter, I., Diedrich, A., Musmann, G., Auster, U., Motschmann, U., Balogh, A., Carr, C., Cupido, E., Coates, A., Rother, M., Schwingenschuh, K., Szegö, K., Tsurutani, B.: RPC-MAG The Fluxgate Magnetometer in the ROSETTA Plasma Consortium. Space Science Reviews 128, 649–670 (2007)

Goldsten, J.O., McNutt, R.L.J., Gold, R.E., Gary, S.A., Fiore, E., Schneider, S.E., Hayes, J.R., Trombka, J.I., Floyd, S.R., Boynton, W.V., Bailey, S., Brueckner, J., Squyres, S.W., Evans, L.G., Clark, P.E., Starr, R.: The X-ray/Gamma-ray Spectrometer on the Near Earth Asteroid Rendezvous Mission. Space Science Reviews 82, 169–216 (1997)

Granahan, J.C.: A compositional study of asteroid 243 Ida and Dactyl from Galileo NIMS and SSI observations. Journal of Geophysical Research 107, 20 (2002)

Granahan, J.C.: Spatially resolved spectral observations of Asteroid 951 Gaspra. Icarus 213, 265–272 (2011)

Gulkis, S., Frerking, M., Crovisier, J., Beaudin, G., Hartogh, P., Encrenaz, P., Koch, T., Kahn, C., Salinas, Y., Nowicki, R., Irigoyen, R., Janssen, M., Stek, P., Hofstadter, M., Allen, M., Backus, C., Kamp, L., Jarchow, C., Steinmetz, E., Deschamps, A., Krieg, J., Gheudin, M., Bockelée-Morvan, D., Biver, N., Encrenaz, T., Despois, D., Ip, W., Lellouch, E., Mann, I., Muhleman, D., Rauer, H., Schloerb, P., Spilker, T.: MIRO: Microwave Instrument for Rosetta Orbiter. Space Science Reviews 128, 561–597 (2007)

Gulkis, S., Keihm, S., Kamp, L., Backus, C., Janssen, M., Lee, S., Davidsson, B., Beaudin, G., Biver, N., Bockelée-Morvan, D., Crovisier, J., Encrenaz, P., Encrenaz, T., Hartogh, P., Hofstadter, M., Ip, W., Lellouch, E., Mann, I., Schloerb, P., Spilker, T., Frerking, M.: Millimeter and submillimeter measurements of asteroid (2867) Steins during the Rosetta fly-by. Planetary and Space Science 58, 1077–1087 (2010)

Gulkis, S., Keihm, S., Kamp, L., Lee, S., Hartogh, P., Crovisier, J., Lellouch, E., Encrenaz, P., Bockelee-Morvan, D., Hofstadter, M., Beaudin, G., Janssen, M., Weissman, P., von Allmen, P.A., Encrenaz, T., Backus, C.R., Ip, W.-H., Schloerb, P.F., Biver, N., Spilker, T., Mann, I.: Continuum and spectroscopic observations of asteroid (21) Lutetia at millimeter and submillimeter wavelengths with the MIRO instrument on the Rosetta spacecraft. Planetary and Space Science 66, 31–42 (2012)

Helfenstein, P., Veverka, J., Thomas, P.C., Simonelli, D.P., Lee, P., Klaasen, K., Johnson, T.V., Breneman, H., Head, J.W., Murchie, S.: Galileo photometry of Asteroid 951 Gaspra. Icarus 107, 37 (1994)

Helfenstein, P., Veverka, J., Thomas, P.C., Simonelli, D.P., Klaasen, K., Johnson, T.V., Fanale, F., Granahan, J., McEwen, A.S., Belton, M., Chapman, C.: Galileo Photometry of Asteroid 243 Ida. Icarus 120, 48–65 (1996)

Hillier, J.K., Bauer, J.M., Buratti, B.J.: Photometric modeling of Asteroid 5535 Annefrank from Stardust observations. Icarus 211, 546–552 (2011)

Housen, K.R., Holsapple, K.A., Voss, M.E.: Compaction as the origin of the unusual craters on the asteroid Mathilde. Nature 402, 155–157 (1999)

Izenberg, N.R., Murchie, S.L., Bell, J.F., McFadden, L.A., Wellnitz, D.D., Clark, B.E., Gaffey, M.J.: Spectral properties and geologic processes on Eros from combined NEAR NIS and MSI data sets. Meteoritics & Planetary Science 38, 1053–1077 (2003)

Jaumann, R., Williams, D.A., Buczkowski, D.L., Yingst, R.A., Preusker, F., Hiesinger, H., Schmedemann, N., Kneissl, T., Vincent, J.B., Blewett, D.T., Buratti, B.J., Carsenty, U., Denevi, B.W., De Sanctis, M.C., Garry, W.B., Keller, H.U., Kersten, E., Krohn, K., Li, J.Y., Marchi, S., Matz, K.D., McCord, T.B., McSween, H.Y., Mest, S.C., Mittlefehldt, D.W., Mottola, S., Nathues, A., Neukum, G., O'Brien, D.P., Pieters, C.M., Prettyman, T.H., Raymond, C.A., Roatsch, T., Russell, C.T., Schenk, P., Schmidt, B.E., Scholten, F., Stephan, K., Sykes, M.V., Tricarico, P., Wagner, R., Zuber, M.T., Sierks, H.: Vesta's Shape and Morphology. Science 336, 687–690 (2012)

Jutzi, M., Michel, P., Benz, W.: A large crater as a probe of the internal structure of the E-type asteroid Steins. Astronomy and Astrophysics 509, L2 (2010)

Keller, H.U., Barbieri, C., Lamy, P., Rickman, H., Rodrigo, R., Wenzel, K.P., Sierks, H., A'Hearn, M.F., Angrilli, F., Angulo, M., Bailey, M.E., Barthol, P., Barucci, M.A., Bertaux, J.L., Bianchini, G., Boit, J.L., Brown, V., Burns, J.A., Büttner, I., Castro, J.M., Cremonese, G., Curdt, W., Da Deppo, V., Debei, S., De Cecco, M., Dohlen, K., Fornasier, S., Fulle, M., Germerott, D., Gliem, F., Guizzo, G.P., Hviid, S.F., Ip, W.H., Jorda, L., Koschny, D., Kramm, J.R., Kührt, E., Küppers, M., Lara, L.M., Llebaria, A., López, A., López-Jimenez, A., López-Moreno, J., Meller, R., Michalik, H., Michelena, M.D., Müller, R., Naletto, G., Origné, A., Parzianello, G., Pertile, M., Quintana, C., Ragaz-zoni, R., Ramous, P., Reiche, K.U., Reina, M., Rodríguez, J., Rousset, G., Sabau, L., Sanz, A., Sivan, J.P., Stöckner, K., Tabero, J., Telljohann, U., Thomas, N., Timon, V., Tomasch, G., Wittrock, T., Zaccariotto, M.: OSIRIS The Scientific Camera System Onboard Rosetta. Space Science Reviews 128, 433–506 (2007)

Keller, H.U., Barbieri, C., Koschny, D., Lamy, D., Rickman, H., Rodrigo, R., Sierks, H., A'Hearn, M.F., Angrilli, F., Barucci, M.A., Bertaux, J.L., Cremonese, G., Da Deppo, V., Davidsson, B., De Cecco, M., Debei, S., Fornasier, S., Fulle, M., Groussin, O., Gutierrez, P.J., Hviid, S.F., Ip, W.H., Jorda, L., Knollenberg, J., Kramm, J.R., Kührt, E., Küppers, M., Lara, L.M., Lazzarin, M., Lopez Moreno, J., Marzari, F., Michalik, H., Naletto, G., Sabau, L., Thomas, N., Wenzel, K.P., Bertini, I., Besse, S., Ferri, F., Kaasalainen, M., Lowry, S., Marchi, S., Mottola, S., Sabolo, W., Schröder, S.E., Spjuth, S., Vernazza, P.: E-Type Asteroid (2867) Steins as Imaged by OSIRIS on Board Rosetta. Science 327, 190–193 (2010)

Kivelson, M.G., Khurana, K.K., Means, J.D., Russell, C.T., Snare, R.C.: The Galileo magnetic field investigation. Space Science Reviews 60, 357–383 (1992)

Kivelson, M.G., Bargatze, L.F., Khurana, K.K., Southwood, D.J., Walker, R.J., Coleman, P.J.: Magnetic Field Signatures Near Galileo's Closest Approach to Gaspra. Science 261, 331–334 (1993)

Kivelson, M.G., Wang, Z., Joy, S., Khurana, K.K., Polanskey, C., Southwood, D.J., Walker, R.J.: Solar wind interaction with small bodies. 2: What can Galileo's detection of magnetic rotations tell us about Gaspra and Ida. Advances in Space Research 16, 59–68 (1995)

Küppers, M., Moissl, R., Vincent, J.B., Besse, S., Hviid, S.F., Carry, B., Grieger, B., Sierks, H., Keller, H.U., Marchi, S., The OSIRIS Team: Boulders on Lutetia. Planetary and Space Science 66, 71–78 (2012)

Lauretta, D.S., OSIRIS-REx Team: An Overview of the OSIRIS-REx Asteroid Sample Return Mission (2012). In: Asteroids, Comets, Meteors 2012. LPI contribution 1659 (2012)

Lee, P., Veverka, J., Thomas, P.C., Helfenstein, P., Belton, M.J.S., Chapman, C.R., Greeley, R., Pappalardo, R.T., Sullivan, R., Head, J.W.: Ejecta Blocks on 243 Ida and on Other Asteroids. Icarus 120, 87–105 (1996)

Leyrat, C., Fornasier, S., Barucci, A., Magrin, S., Lazzarin, M., Fulchignoni, M., Jorda, L., Belskaya, I., Marchi, S., Barbieri, C., Keller, H.U., Sierks, H., Hviid, S.: Search for Steins' surface inhomogeneities from OSIRIS Rosetta images. Planetary and Space Science 58, 1097–1106 (2010)

Leyrat, C., Coradini, A., Erard, S., Capaccioni, F., Capria, M.T., Drossart, P., De Sanctis, M.C., Tosi, F., Virtis Team: Thermal properties of the asteroid (2867) Steins as observed by VIRTIS/Rosetta. Astronomy and Astrophysics 531, A168 (2011)

Lim, L.F., Nittler, L.R.: Elemental composition of 433 Eros: New calibration of the NEAR-Shoemaker XRS data. Icarus 200, 129–146 (2009)

Loeffler, M.J., Dukes, C.A., Chang, W.Y., McFadden, L.A., Baragiola, R.A.: Laboratory simulations of sulfur depletion at Eros. Icarus 193, 622–629 (2008)

Magrin, S., La Forgia, F., Pajola, M., Lazzarin, M., Massironi, M., Ferri, F., Da Deppo, V., Barbieri, C., Sierks, H., The OSIRIS Team: (21) Lutetia spectrophotometry from Rosetta-OSIRIS images and comparison to ground-based observations. Planetary and Space Sceince 66, 43–53 (2012)

Marchi, S., Barbieri, C., Küppers, M., Marzari, F., Davidsson, B., Keller, H., Besse, S., Lamy, P., Mottola, S., Massironi, M., Cremonese, G.: The cratering history of asteroid (2867) Steins. Planetary and Space Science 58, 1116–1123 (2010)

Marchi, S., Massironi, M., Vincent, J.B., Morbidelli, A., Mottola, S., Marzari, F., Küppers, M., Besse, S., Thomas, N., Barbieri, C., Naletto, G., Sierks, H.: The cratering history of asteroid (21) Lutetia. Planetary and Space Science 66, 87–95 (2012a)

Marchi, S., McSween, H.Y., O'Brien, D.P., Schenk, P., De Sanctis, M.C., Gaskell, R., Jaumann, R., Mottola, S., Preusker, F., Raymond, C.A., Roatsch, T., Russell, C.T.: The Violent Collisional History of Asteroid 4 Vesta. Science 336, 690–694 (2012b)

Marzari, F., Cellino, A., Davis, D.R., Farinella, P., Zappala, V., Vanzani, V.: Origin and evolution of the Vesta asteroid family. Astronomy and Astrophysics 316, 248–262 (1996)

Massironi, M., Marchi, S., Pajola, M., Snodgrass, C., Thomas, N., Tubiana, C., Vincent, J.B., Cremonese, G., Da Deppo, V., Ferri, F., Magrin, S., Sierks, H., Barbieri, C., Lamy, P., Rickman, H., Rodrigo, R., Koschny, D., The OSIRIS Team: Geological map and stratigraphy of asteroid 21 Lutetia. Planetary and Space Science 66, 125–136 (2012)

Merline, W.J., Weidenschilling, S.J., Durda, D.D., Margot, J.L., Pravec, P., Storrs, A.D.: Asteroids Do Have Satellites. In: Bottke, W.F., Cellino, A., Paolicchi, P., Binzel, R. (eds.) Asteroids III, pp. 289–312. University of Arizona Press, Tucson (2002)

Michel, P., O'Brien, D.P., Abe, S., Hirata, N.: Itokawa's cratering record as observed by Hayabusa: Implications for its age and collisional history. Icarus 200, 503–513 (2009)

Morse, A.D., Altwegg, K., Andrews, D.J., Auster, H.U., Carr, C.M., Galand, M., Goesmann, F., Gulkis, S., Lee, S., Richter, I., Sheridan, S., Stern, S.A., A'Hearn, M.F., Feldman, P., Parker, J., Retherford, K.D., Weaver, H.A., Wright, I.P.: The Rosetta campaign to detect an exosphere at Lutetia. Planetary and Space Science 66, 165–172 (2012)

Nagao, K., Okazaki, R., Nakamura, T., Miura, Y.N., Osawa, T., Bajo, K., Matsuda, S., Ebihara, M., Ireland, T.R., Kitajima, F., Naraoka, H., Noguchi, T., Tsuchiyama, A., Yurimoto, H., Zolensky, M.E., Uesugi, M., Shirai, K., Abe, M., Yada, T., Ishibashi, Y., Fujimura, A., Mukai, T., Ueno, M., Okada, T., Yoshikawa, M., Kawaguchi, J.: Irradiation History of Itokawa Regolith Material Deduced from Noble Gases in the Hayabusa Samples. Science 333, 1128–1131 (2011)

Nakamura, T., Nakamura, A.M., Saito, J., Sasaki, S., Nakamura, R., Demura, H., Akiyama, H., Tholen, D., AMICA Team: Multiband imaging camera and its sciences for the Japanese near-earth asteroid mission MUSES-C. Earth, Planets and Space 53, 1047–1063 (2001)

Nakamura, T., Noguchi, T., Tanaka, M., Zolensky, M.E., Kimura, M., Tsuchiyama, A., Nakato, A., Ogami, T., Ishida, H., Uesugi, M., Yada, T., Shirai, K., Fujimura, A., Okazaki, R., Sandford, S., Ishibashi, Y., Abe, M., Okada, T., Ueno, M., Mukai, T., Yoshikawa, M., Kawaguchi, J.: Itokawa Dust Particles: A Direct Link Between S-Type Asteroids and Ordinary Chondrites. Science 333, 1113–1116 (2011)

Newburn, R.L., Duxbury, T.C., Hanner, M., Semenov, B.V., Hirst, E.E., Bhat, R.S., Bhaskaran, S., Wang, T.C.M., Tsou, P., Brownlee, D.E., Cheuvront, A.R., Gingerich, D.E., Bol-lendonk, G.R., Vellinga, J.M., Parham, K.A., Mumaw, S.J.: Phase curve and albedo of asteroid 5535 Annefrank. Journal of Geophysical Research 108, 5117 (2003a)

Newburn, R.L., Bhaskaran, S., Duxbury, T.C., Fraschetti, G., Radey, T., Schwochert, M.: Stardust Imaging Camera. Journal of Geophysical Research 108, 8116 (2003b)

Noguchi, T., Nakamura, T., Kimura, M., Zolensky, M.E., Tanaka, M., Hashimoto, T., Konno, M., Nakato, A., Ogami, T., Fujimura, A., Abe, M., Yada, T., Mukai, T., Ueno, M., Okada, T., Shirai, K., Ishibashi, Y., Okazaki, R.: Incipient Space Weathering Observed on the Surface of Itokawa Dust Particles. Science 333, 1121–1125 (2011)

Oberst, J., Mottola, S., Di Martino, M., Hicks, M., Buratti, B., Soderblom, L., Thomas, N.: A Model for Rotation and Shape of Asteroid 9969 Braille from Ground Based Observations and Images Obtained during the Deep Space 1 (DS1) Flyby. Icarus 153, 16–23 (2001)

Okada, T., Kato, M., Fujimura, A., Tsunemi, H., Kitamoto, S.: X-ray Fluorescence Spectrometry with the SELENE Orbiter. Advances in Space Research 23, 1833–1836 (1999)

Okada, T., Shirai, K., Yamamoto, Y., Arai, T., Ogawa, K., Hosono, K., Kato, M.: X-ray Fluorescence Spectrometry of Asteroid Itokawa by Hayabusa. Science 312, 1338–1341 (2006)

Pätzold, M., Andert, T.P., Asmar, S.W., Anderson, J.D., Barriot, J.P., Bird, M.K., Häusler, B., Hahn, M., Tellmann, S., Sierks, H., Lamy, P., Weiss, B.P.: Asteroid 21 Lutetia: Low Mass, High Density. Science 334, 491–492 (2011)

Prockter, L., Thomas, P., Robinson, M., Joseph, J., Milne, A., Bussey, B., Veverka, J., Cheng, A.: Surface Expressions of Structural Features on Eros. Icarus 155, 75–93 (2002)

Reddy, V., Nathues, A., Le Corre, L., Sierks, H., Li, J.Y., Gaskell, R., McCoy, T., Beck, A.W., Schröder, S.E., Pieters, C.M., Becker, K.J., Buratti, B.J., Denevi, B., Blewett, D.T., Christensen, U., Gaffey, M.J., Gutierrez-Marques, P., Hicks, M., Keller, H.U., Maue, T., Mottola, S., McFadden, L.A., McSween, H.Y., Mittlefehldt, D., O'Brien, D.P., Raymond, C., Russell, C.: Color and Albedo Heterogeneity of Vesta from Dawn. Science 336, 700–704 (2012)

Richter, I., Brinza, D.E., Cassel, M., Glassmeier, K.H., Kuhnke, F., Musmann, G., Othmer, C., Schwingenschuh, K., Tsurutani, B.T.: First direct magnetic field measurements of an asteroidal magnetic field: DS1 at Braille. Geophysical Research Letters 28, 1913–1916 (2001)

Richter, I., Auster, H.U., Glassmeier, K.H., Koenders, C., Carr, C.M., Motschmann, U., Müller, J., McKenna-Lawlor, S.: Magnetic field measurements during the ROSETTA flyby at asteroid (21)Lutetia. Planetary and Space Science 66, 155–164 (2012)

Russell, C.T., Capaccioni, F., Coradini, A., De Sanctis, M.C., Feldman, W.C., Jaumann, R., Keller, H.U., McCord, T.B., McFadden, L.A., Mottola, S., Pieters, C.M., Prettyman, T.H., Raymond, C.A., Sykes, M.V., Smith, D.E., Zuber, M.T.: Dawn Mission to Vesta and Ceres. Symbiosis Between Terrestrial Observations and Robotic Exploration. Earth, Moon, and Planets 101, 65–91 (2007)

Russell, C.T., Raymond, C.A., Coradini, A., McSween, H.Y., Zuber, M.T., Nathues, A., De Sanctis, M.C., Jaumann, R., Konopliv, A.S., Preusker, F., Asmar, S.W., Park, R.S., Gaskell, R., Keller, H.U., Mottola, S., Roatsch, T., Scully, J.E.C., Smith, D.E., Tricarico, P., Toplis, M.J., Christensen, U.R., Feldman, W.C., Lawrence, D.J., McCoy, T.J., Prettyman, T.H., Reedy, R.C., Sykes, M.E., Titus, T.N.: Dawn at Vesta: Testing the Protoplanetary Paradigm. Science 336, 684–686 (2012)

Saito, J., Miyamoto, H., Nakamura, R., Ishiguro, M., Michikami, T., Nakamura, A.M., Demura, H., Sasaki, S., Hirata, N., Honda, C., Yamamoto, A., Yokota, Y., Fuse, T., Yoshida, F., Tholen, D.J., Gaskell, R.W., Hashimoto, T., Kubota, T., Higuchi, Y., Nakamura, T., Smith, P., Hiraoka, K., Honda, T., Kobayashi, S., Furuya, M., Matsumoto, N., Nemoto, E., Yukishita, A., Kitazato, K., Dermawan, B., Sogame, A., Terazono, J., Shinohara, C., Akiyama, H.: Detailed Images of Asteroid 25143 Itokawa from Hayabusa. Science 312, 1341–1344 (2006)

Schenk, P., O'Brien, D.P., Marchi, S., Gaskell, R., Preusker, F., Roatsch, T., Jaumann, R., Buczkowski, D., McCord, T., McSween, H.Y., Williams, D., Yingst, A., Raymond, C., Russell, C.: The Geologically Recent Giant Impact Basins at Vesta's South Pole. Science 336, 694–697 (2012)

Schröder, S.E., Keller, H.U., Gutierrez, P.J., Hviid, S.F., Kramm, R., Sabolo, W., Sierks, H.: Evidence for surface variegation in Rosetta OSIRIS images of asteroid 2867 Steins. Planetary and Space Science 58, 1107–1115 (2010)

Schulz, R.: The Rosetta mission and its fly-by at asteroid 2867 Steins. Planetary and Space Science 58, 1057 (2010)

Schulz, R., Sierks, H., Küppers, M., Accomazzo, A.: Rosetta fly-by at asteroid (21) Lutetia: An overview. Planetary and Space Science 66, 2–8 (2012)

Sierks, H., Lamy, P., Barbieri, C., Koschny, D., Rickman, H., Rodrigo, R., A'Hearn, M.F., Angrilli, F., Barucci, M.A., Bertaux, J.L., Bertini, I., Besse, S., Carry, B., Cremonese, G., Da Deppo, V., Davidsson, B., Debei, S., De Cecco, M., De Leon, J., Ferri, F., Fornasier, S., Fulle, M., Hviid, S.F., Gaskell, R.W., Groussin, O., Gutierrez, P.J., Ip, W., Jorda, L., Kaasalainen, M., Keller, H.U., Knollenberg, J., Kramm, R., Kührt, E., Küppers, M., Lara, L.M., Lazzarin, M., Leyrat, C., Moreno, J.J.L., Magrin, S., Marchi, S., Marzari, F., Massironi, M., Michalik, H., Moissl, R., Naletto, G., Preusker, F., Sabau, L., Sabolo, W., Scholten, F., Snodgrass, C., Thomas, N., Tubiana, C., Vernazza, P., Vincent, J.B., Wenzel, K.P., Andert, T., Pätzold, M., Weiss, B.P.: Images of Asteroid 21 Lutetia: A Remnant Planetesimal from the Early Solar System. Science 334, 487–490 (2011a)

Sierks, H., Keller, H.U., Jaumann, R., Michalik, H., Behnke, T., Bubenhagen, F., Büttner, I., Carsenty, U., Christensen, U., Enge, R., Fiethe, B., Gutiérrez Marqués, P., Hartwig, H., Krüger, H., Kühne, W., Maue, T., Mottola, S., Nathues, A., Reiche, K.U., Richards, M.L., Roatsch, T., Schröder, S.E., Szemerey, I., Tschentscher, M.: The Dawn Framing Camera. Space Science Reviews 163, 263–327 (2011b)

Soderblom, L.A., Boice, D.C., Britt, D.T., Brown, R.H., Buratti, B.J., Hicks, M.D., Nelson, R.M., Oberst, J., Sandel, B.R., Stern, S.A., Thomas, N., Yelle, R.V.: Observations of Comet 19P/Borrelly from the Miniature Integrated Camera and Spectrometer (MICAS) aboard Deep Space 1 (DS1). Bulletin of the American Astronomical Society 33, 1087 (2001)

Stern, S.A., Slater, D.C., Scherrer, J., Stone, J., Versteeg, M., A'Hearn, M.F., Bertaux, J.L., Feldman, P.D., Festou, M.C., Parker, J.W., Siegmund, O.H.W.: Alice: The rosetta Ultraviolet Imaging Spectrograph. Space Science Reviews 128, 507–527 (2007)

Stooke, P.J.: Linear Features on Asteroid 951 Gaspra. Earth, Moon and Planets 74, 131–149 (1996)

Stooke, P.J.: The Surface of Asteroid 951 Gaspra. Earth, Moon and Planets 75, 53–75 (1997)

Sullivan, R., Greeley, R., Pappalardo, R., Asphaug, E., Moore, J.M., Morrison, D., Belton, M.J.S., Carr, M., Chapman, C.R., Geissler, P., Greenberg, R., Granahan, J., Head, J., Kirk, R., McEwen, A., Lee, P., Thomas, P.C., Veverka, J.: Geology of 243 Ida. Icarus 120, 119–139 (1996)

Takagi, Y., Yoshikawa, M., Abe, M., Tachibana, S., Okada, T., Kitazato, K., Nakamura, R., Hirata, N., Yano, H., Demura, H., Nakazawa, S., Iijima, Y., Shirai, K., Hayakawa, M., Hayabusa 2 Project Team: Hayabusa2, C-type Asteroid Sample Return Mission. In: American Geophysical Union, Fall Meeting (2011)

Thomas, P.C., Veverka, J., Simonelli, D., Helfenstein, P., Carcich, B., Belton, M.J.S., Davies, M.E., Chapman, C.: The shape of Gaspra. Icarus 107, 23–36 (1994)

Thomas, P.C., Veverka, J., Bell, J.F., Clark, B.E., Carcich, B., Joseph, J., Robinson, M., McFadden, L.A., Malin, M.C., Chapman, C.R., Merline, W., Murchie, S.: Mathilde: Size, Shape, and Geology. Icarus 140, 17–27 (1999)

Thomas, P.C., Veverka, J., Robinson, M.S., Murchie, S.: Shoemaker crater as the source of most ejecta blocks on the asteroid 433 Eros. Nature 413, 394–396 (2001)

Thomas, N., Barbieri, C., Keller, H.U., Lamy, P., Rickman, H., Rodrigo, R., Sierks, H., Wenzel, K.P., Cremonese, G., Jorda, L., Küppers, M., Marchi, S., Marzari, F., Massironi, M., Preusker, F., Scholten, F., Stephan, K., Barucci, M.A., Besse, S., El-Maarry, M.R., Fornasier, S., Groussin, O., Hviid, S.F., Koschny, D., Kührt, E., Martellato, E., Moissl, R., Snodgrass, C., Tubiana, C., Vincent, J.B.: The geomorphology of (21) Lutetia: Results from the OSIRIS imaging system onboard ESA's Rosetta spacecraft. Planetary and Space Science 66, 96–124 (2012)

Tosi, F., Coradini, A., Capaccioni, F., Filacchione, G., Grassi, D., de Sanctis, M.C., Capria, M.T., Barucci, M.A., Fulchignoni, M., Mottola, S., Erard, S., Dotto, E., Baldetti, C., VIRTIS Team: The light curve of asteroid 2867 Steins measured by VIRTIS-M during the Rosetta fly-by. Planetary and Space Science 66, 1066–1076 (2010)

Tosi, F., Capaccioni, F., Coradini, A., Erard, S., Filacchione, G., De Sanctis, M.C., Capria, M.T., Giuppi, S., Carraro, F., VIRTIS Team: The light curve of asteroid 21 Lutetia measured by VIRTIS-M during the Rosetta fly-by. Planetary and Space Science 66, 9–22 (2012)

Trombka, J.I., Squyres, S.W., Brückner, J., Boynton, W.V., Reedy, R.C., McCoy, T.J., Gorenstein, P., Evans, L.G., Arnold, J.R., Starr, R.D., Nittler, L.R., Murphy, M.E., Mikhee-va, I., McNutt, R.L., McClanahan, T.P., McCartney, E., Goldsten, J.O., Gold, R.E., Floyd, S.R., Clark, P.E., Burbine, T.H., Bhangoo, J.S., Bailey, S.H., Petaev, M.: The Elemental Composition of Asteroid 433 Eros: Results of the NEAR-Shoemaker X-ray Spectrometer. Science 289, 2101–2105 (2000)

Veverka, J., Belton, M., Klaasen, K., Chapman, C.: Galileo's Encounter with 951 Gaspra: Overview. Icarus 107, 2–17 (1994)

Veverka, J., Helfenstein, P., Lee, P., Thomas, P., McEwen, A., Belton, M., Klaasen, K., Johnson, T.V., Granahan, J., Fanale, F., Geissler, P., Head, J.W.: Ida and Dactyl: Spectral Reflectance and Color Variations. Icarus 120, 66–76 (1996a)

Veverka, J., Thomas, P.C., Helfenstein, P., Lee, P., Harch, A., Calvo, S., Chapman, C., Belton, M.J.S., Klaasen, K., Johnson, T.V., Davies, M.: Dactyl: Galileo Observations of Ida's Satellite. Icarus 120, 200–211 (1996b)

Veverka, J., Bell, J.F., Thomas, P., Harch, A., Murchie, S., Hawkins, S.E., Warren, J.W., Darlington, H., Peacock, K., Chapman, C.R., McFadden, L.A., Malin, M.C., Robinson, M.S.: An overview of the NEAR multispectral imager-near-infrared spectrometer investigation. Journal of Geophysical Research 102, 23709–23728 (1997a)

Veverka, J., Thomas, P., Harch, A., Clark, B., Bell, J.F., Carcich, B., Joseph, J., Chapman, C., Merline, W., Robinson, M., Malin, M., McFadden, L.A., Murchie, S., Hawkins, S.E., Farquhar, R., Izenberg, N., Cheng, A.: NEAR's Flyby of 253 Mathilde: Images of a C Asteroid. Science 278, 2109–2114 (1997b)

Veverka, J., Thomas, P., Harch, A., Clark, B., Bell, J.F., Carcich, B., Joseph, J., Murchie, S., Izenberg, N., Chapman, C., Merline, W., Malin, M., McFadden, L., Robinson, M.: NEAR Encounter with Asteroid 253 Mathilde: Overview. Icarus 140, 3–16 (1999)

Veverka, J., Robinson, M., Thomas, P., Murchie, S., Bell, J.F., Izenberg, N., Chapman, C., Harch, A., Bell, M., Carcich, B., Cheng, A., Clark, B., Domingue, D., Dunham, D., Farquhar, R., Gaffey, M.J., Hawkins, E., Joseph, J., Kirk, R., Li, H., Lucey, P., Malin, M., Martin, P., McFadden, L., Merline, W.J., Miller, J.K., Owen, W.M., Peterson, C., Prockter, L., Warren, J., Wellnitz, D., Williams, B.G., Yeomans, D.K.: NEAR at Eros: Imaging and Spectral Results. Science 289, 2088–2097 (2000)

Veverka, J., Thomas, P.C., Robinson, M., Murchie, S., Chapman, C., Bell, M., Harch, A., Merline, W.J., Bell, J.F., Bussey, B., Carcich, B., Cheng, A., Clark, B., Domingue, D., Dunham, D., Farquhar, R., Gaffey, M.J., Hawkins, E., Izenberg, N., Joseph, J., Kirk, R., Li, H., Lucey, P., Malin, M., McFadden, L., Miller, J.K., Owen, W.M., Peterson, C., Prockter, L., Warren, J., Wellnitz, D., Williams, B.G., Yeomans, D.K.: Imaging of Small-Scale Features on 433 Eros from NEAR: Evidence for a Complex Regolith. Science 292, 484–488 (2001a)

Veverka, J., Farquhar, B., Robinson, M., Thomas, P., Murchie, S., Harch, A., Antreasian, P.G., Chesley, S.R., Miller, J.K., Owen, W.M., Williams, B.G., Yeomans, D., Dunham, D., Heyler, G., Holdridge, M., Nelson, R.L., Whittenburg, K.E., Ray, J.C., Carcich, B., Cheng, A., Chapman, C., Bell, J.F., Bell, M., Bussey, B., Clark, B., Domingue, D., Gaffey, M.J., Hawkins, E., Izenberg, N., Joseph, J., Kirk, R., Lucey, P., Malin, M., McFadden, L., Merline, W.J., Peterson, C., Prockter, L., Warren, J., Wellnitz, D.: The landing of the NEAR-Shoemaker spacecraft on asteroid 433 Eros. Nature 413, 390–393 (2001b)

Vincent, J.B., Besse, S., Marchi, S., Sierks, H., Massironi, M., The OSIRIS Team: Physical properties of craters on asteroid (21) Lutetia. Planetary and Space Science 66, 79–86 (2012)

Wang, Z., Kivelson, M.G., Joy, S., Khurana, K.K., Polanskey, C., Southwood, D.J., Walker, R.J.: Solar wind interaction with small bodies: 1. Whistler wing signatures near Galileo's closest approach to Gaspra and Ida. Advances in Space Research 16, 47–57 (1995)

Wasilewski, P., Acuna, M.H., Kletetschka, G.: 433 Eros: Problems with the meteorite magnetism record in attempting an asteroid match. Meteoritics & Planetary Science 37, 937–950 (2002)

Weiss, B.P., Elkins-Tanton, L.T., Barucci, M.A., Sierks, H., Snodgrass, C., Vincent, J.B., Marchi, S., Weissman, P.R., Pätzold, M., Richter, I., Fulchignoni, M., Binzel, R.P., Schulz, R.: Possible evidence for partial differentiation of asteroid Lutetia from Rosetta. Planetary and Space Science 66, 137–146 (2012)

Weissman, P.R., Carlson, R.W., Smythe, W.D., Byrne, L.C., Ocampo, A.C., Kieffer, H.H., Soderblom, L.A., Fanale, F.P., Granahan, J.C., McCord, T.B.: Thermal Modelling of Asteroid 951 Gaspra. Bulletin of the American Astronomical Society 24, 933 (1992)

Yano, H., Kubota, T., Miyamoto, H., Okada, T., Scheeres, D., Takagi, Y., Yoshida, K., Abe, M., Abe, S., Barnouin-Jha, O., Fujiwara, A., Hasegawa, S., Hashimoto, T., Ishiguro, M., Kato, M., Kawaguchi, J., Mukai, T., Saito, J., Sasaki, S., Yoshikawa, M.: Touchdown of the Hayabusa Spacecraft at the Muses Sea on Itokawa. Science 312, 1350–1353 (2006)

Yeomans, D.K., Barriot, J.P., Dunham, D.W., Farquhar, R.W., Giorgini, J.D., Helfrich, C.E., Konopliv, A.S., McAdams, J.V., Miller, J.K., Owen, W.M.J., Scheeres, D.J., Synnott, S.P., Williams, B.G.: Estimating the Mass of Asteroid 253 Mathilde from Tracking Data During the NEAR Flyby. Science 278, 2106–2109 (1997)

Yeomans, D.K., Antreasian, P.G., Barriot, J.P., Chesley, S.R., Dunham, D.W., Farquhar, R.W., Giorgini, J.D., Helfrich, C.E., Konopliv, A.S., McAdams, J.V., Miller, J.K., Owen, W.M., Scheeres, D.J., Thomas, P.C., Veverka, J., Williams, B.G.: Radio Science Results During the NEAR-Shoemaker Spacecraft Rendezvous with Eros. Science 289, 2085–2088 (2000)

Yurimoto, H., Abe, K., Abe, M., Ebihara, M., Fujimura, A., Hashiguchi, M., Hashizume, K., Ireland, T.R., Itoh, S., Katayama, J., Kato, C., Kawaguchi, J., Kawasaki, N., Kitajima, F., Kobayashi, S., Meike, T., Mukai, T., Nagao, K., Nakamura, T., Naraoka, H., Noguchi, T., Okazaki, R., Park, C., Sakamoto, N., Seto, Y., Takei, M., Tsuchiyama, A., Uesugi, M., Wakaki, S., Yada, T., Yamamoto, K., Yoshikawa, M., Zolensky, E.: Oxygen Isotopic Compositions of Asteroidal Materials Returned from Itokawa by the Hayabusa Mission. Science 333, 1116–1119 (2011)

Chapter 2
Trojan Asteroids in the Inner Solar System

Michael Todd

Curtin University, Bentley, WA, Australia

Most of the asteroids in our Solar System are located between the orbits of Mars and Jupiter, the so-called Main Belt asteroids. However there are many less-populated regions where asteroids are also found. One such example is the Trojan asteroids, which are a particularly special case since they share the orbit of a planet around the Sun.

In 1772, the Italian-French mathematician Joseph-Louis Lagrange provided a solution to the restricted three-body problem (Lagrange 1772), describing five positions in the orbit of one body around another. These are now known as the Lagrangian points.

Any system with two bodies will orbit around their common centre of mass. When this orbit is near-circular, and a third body of negligible mass orbits these two larger bodies, if it is positioned at any of these five Lagrangian points (Fig. 2.1) then it experiences zero net force as it follows the circular orbit of its host bodies. At these points the third body can exist in a stable orbit in a 1:1 mean motion resonance with the planet.

The Solar System is more complex than a three-body model as it has many more objects for which gravitational interactions must be considered. As a result Trojan asteroids are typically located in stable regions around the L4 and L5 Lagrangian points rather than exactly at these classical locations. Outside these regions the orbits tend to destabilise quickly.

The first asteroid discovered at a planetary Lagrangian point was 588 Achilles, in the orbit of Jupiter, in 1906 by the German astronomer Max Wolf (Nicholson 1961). Objects in the stable zones near the equilateral L4 and L5 Lagrangian points are known as "Trojans". There are currently about 600,000 known asteroids in the Solar System. Almost 10% of these are Jupiter Trojans (IAU Minor Planet Center 2012).

The Trojan asteroid population in the inner Solar System is much smaller, with only a handful of known Trojans, most of which are Mars Trojans. The first of these to be discovered (serendipitously), in 1990, was 5261 Eureka (Bowell et al. 1990). Two additional Mars Trojans have since been discovered. Simulations (Tabachnik and Evans 1999, 2000a,b) suggest that there could exist as many as 50 asteroids larger than 1km diameter in the Trojan regions of Mars' orbit around the Sun.

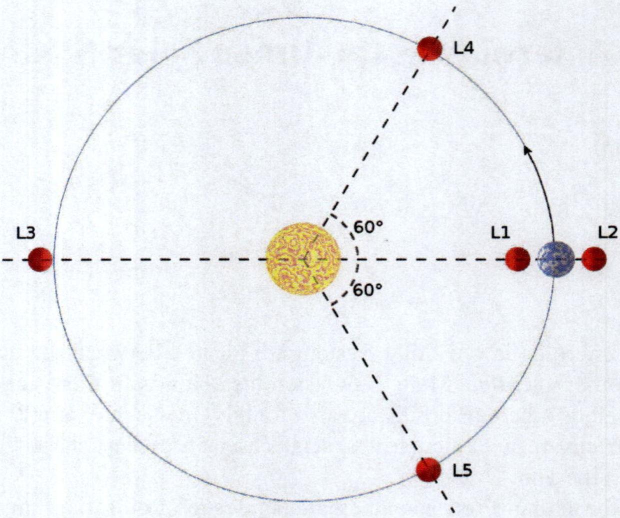

Fig. 2.1 The five stable Lagrangian points

There have been several attempts at searching for Earth Trojans (Dunbar and Helin 1983; Whiteley and Tholen 1998; Connors et al. 2000) however it wasn't until 2010 when the unassumingly-named 2010 TK7 (Fig. 2.2) was discovered by NASA's Wide-field Infrared Survey Explorer (WISE) satellite (Connors et al. 2011). With a diameter of 300m this is probably the largest asteroid in Earth's Trojan regions but there may yet be a small number of undiscovered Earth Trojans with smaller diameters (Morais and Morbidelli 2002).

The semi-major axis of a Trojan asteroid must necessarily be similar to the planet, so that the period of revolution about the Sun is similar to the planet. The major planets of the Solar System have almost-circular orbits, and so to remain in the Trojan region the orbit must also be almost circular, that is they must have a low eccentricity.

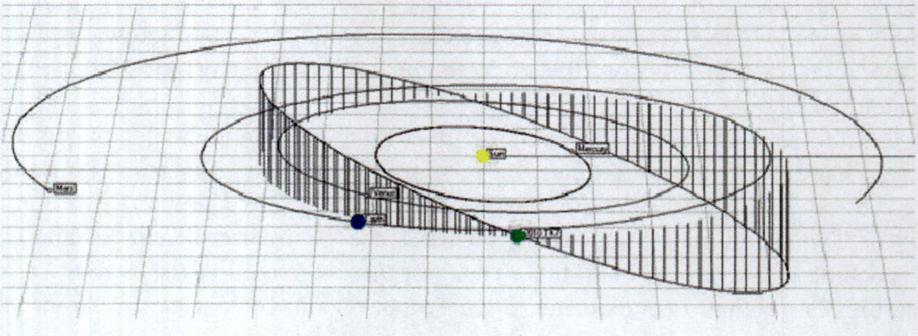

Fig. 2.2 Orbit of Earth Trojan 2010 TK7 (Jablonski 2003)

A peculiarity of the inner Solar System Trojans is that, instead of orbiting the Sun in the plane of the ecliptic – orbiting in the same plane as the planets – the stable orbits are inclined with respect to the plane of the ecliptic. Modelling and simulations to find stable inclinations for Trojan asteroid orbits for the Trojans of Earth (Morais and Morbidelli 2002). and Mars (Scholl et al. 2005) revealed that the most stable orbits were those that were moderately inclined (between about 10° to 40°) to the plane of the ecliptic. These models also found that orbits which had small inclinations were found to be unstable, resulting in the asteroids drifting into new orbits.

Regions in which Trojans for Earth and Mars are most likely to exist have been identified (Todd et al. 2012b,c). The probability distributions for Earth (Fig. 2.3) and Mars (Fig. 2.4) both show that the longitudes where Trojans are most likely to exist are consistent with classical Lagrangian points, but that they are much more likely to be inclined orbits than to lie in the plane of the ecliptic. For each plot the zero longitude point corresponds to the position of the planet, so that the longitude is the heliocentric longitude relative to the planet.

Superimposing the orbit inclinations and longitudes of the known Earth and Mars Trojans shows that their locations are consistent with this distribution model. Predictions about the future position of the Earth Trojan 2010 TK7 (Connors et al. 2011) suggest that its longitude will drift back and forth around the classical Lagrangian point, meaning that it will stay within the higher probability region.

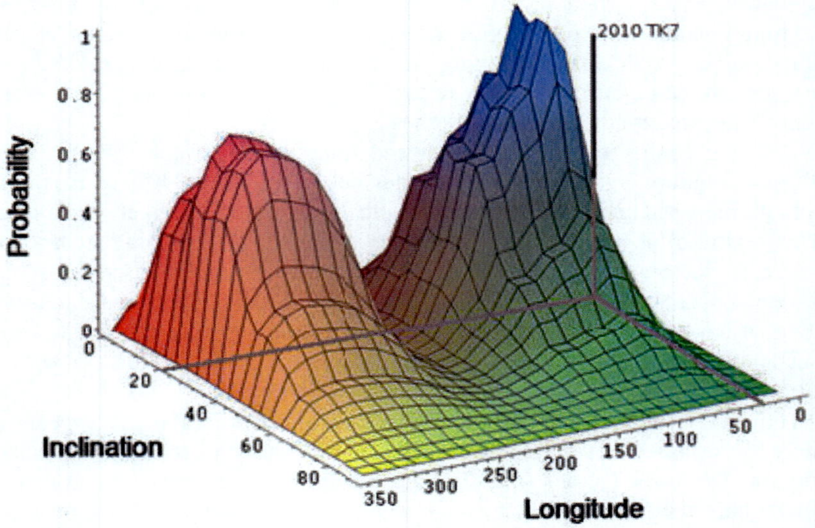

Fig. 2.3 Earth Trojans probability distribution showing location of Earth Trojan 2010 TK7 (adapted from Todd et al. 2012b)

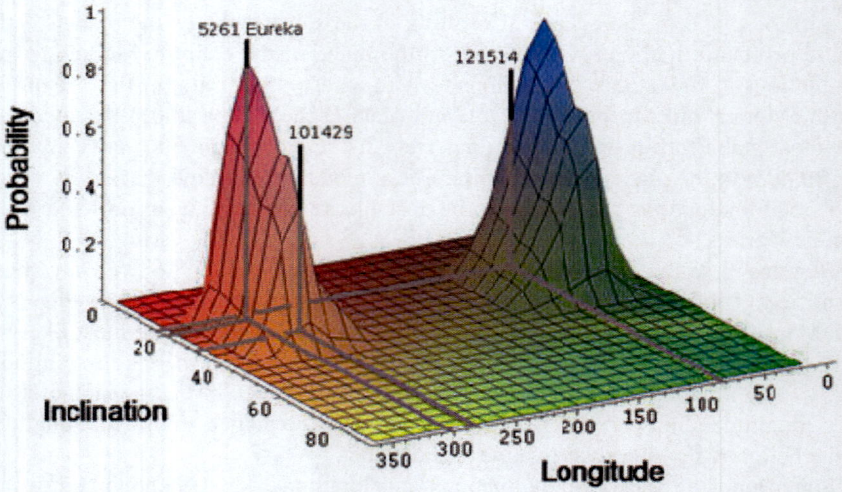

Fig. 2.4 Mars Trojans probability distribution showing locations of known Mars Trojans (adapted from Todd et al. 2012c)

For a known Trojan the exact position can easily be determined for any given date. However these probability distributions can be used to identify the regions of sky through which the Trojans move during their orbits around the Sun. The same regions can be used when searching for additional Trojans. By restricting the search area to the higher probability regions provides a starting point for a search, and limits the amount of sky to be searched.

Within the limits of 45° inclination and longitudes between 30° - 130° for Earth's L4 region (Fig. 2.5), and longitudes between 240° - 340° for Earth's L5 region, define a region of sky where any Earth Trojans are most likely to be found.

This region of sky is still very large, about 3500 deg². Searching such a large sky area is not practical with any currently available ground-based telescopes. When one considers the small predicted number of Earth Trojans and the amount of sky that would need to be searched, it is perhaps unsurprising that it was a careful analysis of data from a space-based telescope that led to the discovery of the first Earth Trojan.

When one then considers the relative rarity of Mars Trojans it appears extremely fortuitous that even three have been discovered. Even though the probability regions for Mars Trojans (Fig. 2.4) are much more narrowly defined than for Earth Trojans, the resulting sky area is much larger due mainly to the larger size of Mars' orbit (compared to Earth) and its relative proximity. At opposition (when on the opposite side of the sky from the Sun) the maximum sky area encompassed within the limits of 35° inclination and longitudes between 40° - 90° for Mars' L4 region (Fig. 2.6), and longitudes between 270° - 320° for Mars' L5 region, varies between 11000 – 17000 deg² due to the eccentricity of Mars' orbit around the Sun.

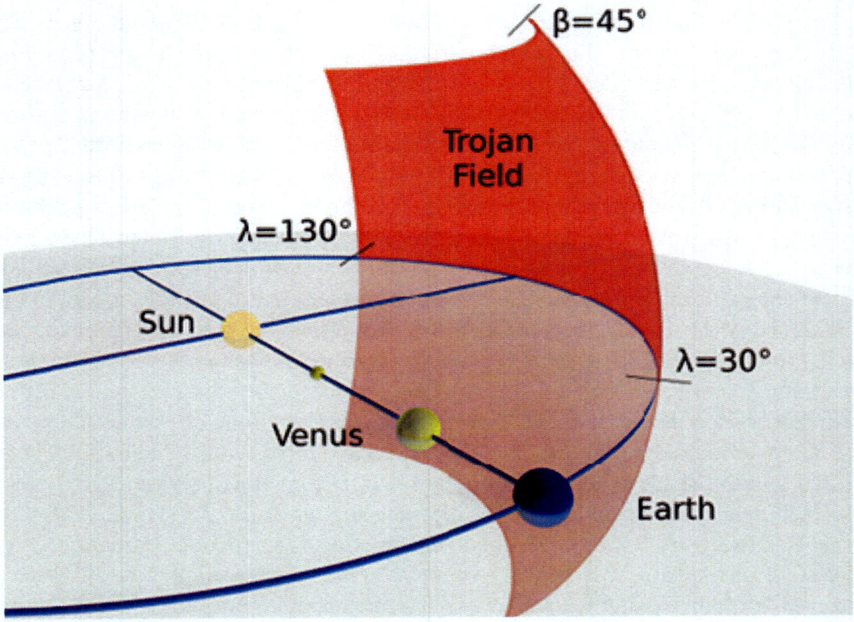

Fig. 2.5 Earth Trojan (L4) region

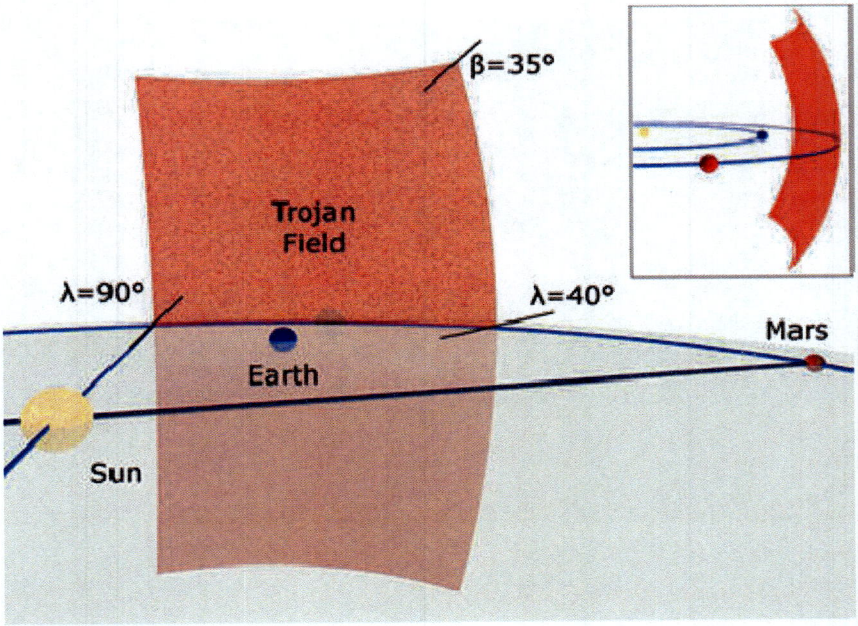

Fig. 2.6 Mars Trojan (L4) region

The Earth Trojan asteroid 2010 TK7 shares Earth's orbit around the Sun. It lies in the Earth's L4 Lagrangian region, ahead of Earth in its orbit. Its orbit is somewhat elliptical with an eccentricity $e = 0.19$, and is inclined at 20.9° to the plane of the ecliptic. Plotting its relative position in Fig. 2.3 shows that it resides in a region that was predicted to have a higher probability of containing an Earth Trojan. Its current apparent position on the sky is close to the nearest edge of the region shown in Fig. 2.5. A plot of its orbit (Fig. 2.2) shows it ahead of Earth's position. The height of its orbit above or below the ecliptic plane is indicated by vertical lines.

The composition of the inner Solar System Trojans is not known for certain but is assumed to be silicaceous (stony), and to have a relatively high reflectivity, similar to other inner Solar System asteroids. Using this assumption the brightness (magnitude) can be calculated for any size object at any distance and position relative to the Earth and the Sun.

The apparent magnitude (V-band) of an Earth Trojan with a diameter of 1km, located within the identified Trojan region (Fig. 2.5), varies between V = 17.9 to V = 19.5 depending on its location within the region (Fig. 2.7). These values assume an albedo of 0.20 and without considering atmospheric extinction. Any objects in this region will appear brightest at the nearest point. The brightness decreases with increasing distance, such that the apparent brightness is least at the farthest points in the region.

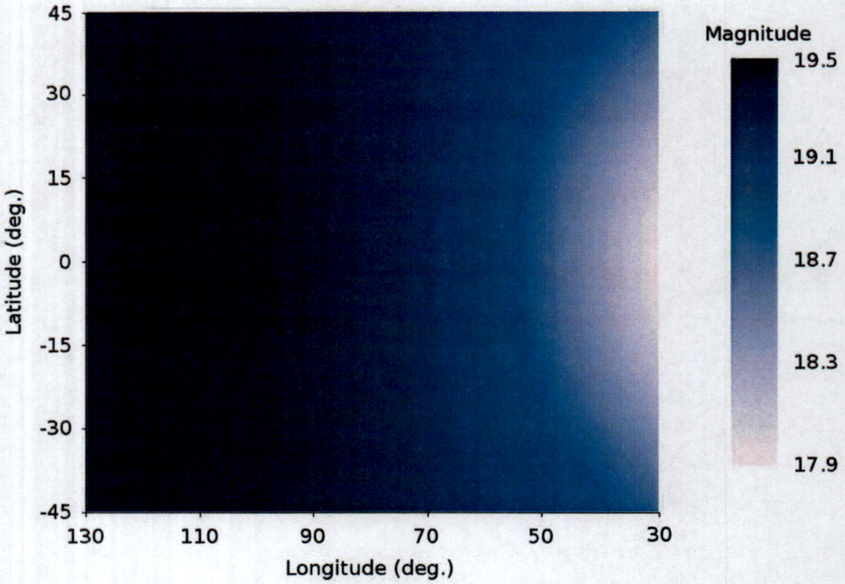

Fig. 2.7 Earth Trojans variation in apparent magnitude across field

Similarly, the apparent brightness of Trojans in Mars' orbit varies with distance from Earth. These are brightest at opposition, with some small variation depending whether Mars is near perihelion or aphelion (Fig. 2.8). Near the centre of the image, at latitude 0 and at opposition, the object would be closest to Earth and appear brightest.

In comparing the magnitude plots (Fig. 2.7 and 2.8) one will note the difference in coordinate systems. In both figures the vertical axis represents heliocentric latitude, the position above or below the plane of the ecliptic. This is a convenient method of presenting this component, and the apparent height above or below the ecliptic can be deduced from this geometry.

In Fig. 2.7 for the Earth Trojans the horizontal axis represents the heliocentric longitude. In this frame, with Earth defined as the origin for longitude, it is convenient to express the position in Earth's orbit as a coordinate relative to Earth's position.

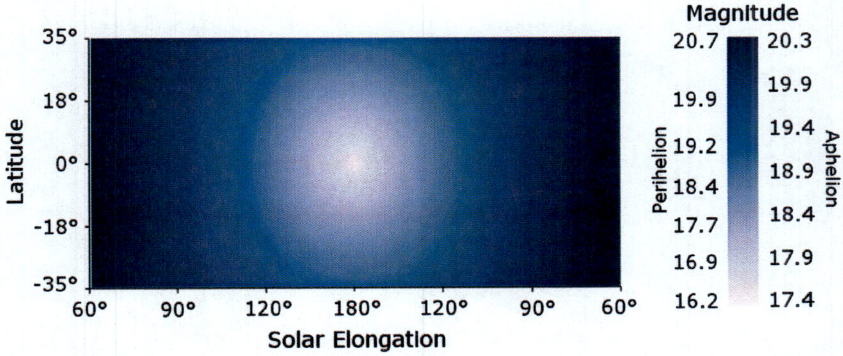

Fig. 2.8 Mars Trojans variation in apparent magnitude across field

In Fig. 2.8 for the Mars Trojans the horizontal axis represents solar elongation, the angle between the direction of the Sun and the direction of the object. Mars Trojans are (by definition) co-orbital with Mars, and therefore orbit the Sun with a different period than Earth and so are constantly changing position relative to Earth. The elongation of a specific object can be calculated for any point in time. Rather than express the coordinate as longitude, it is more convenient to express the position in terms of elongation.

The Trojan regions can be illustrated by simulating numbers of objects in the defined space. Using the orbit parameters derived from the distribution model enables the creation of an artificial population within the region. Plotting this as a snapshot in time then illustrates the region around Earth's orbit (Fig. 2.9) and Mars' orbit (Fig. 2.10) where Trojans could exist.

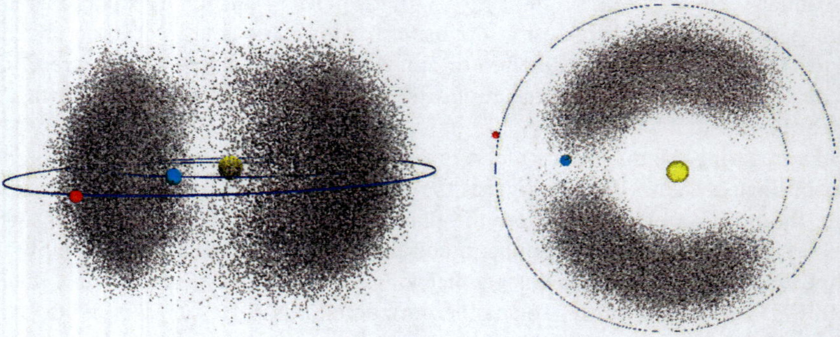

Fig. 2.9 Simulated Earth Trojans

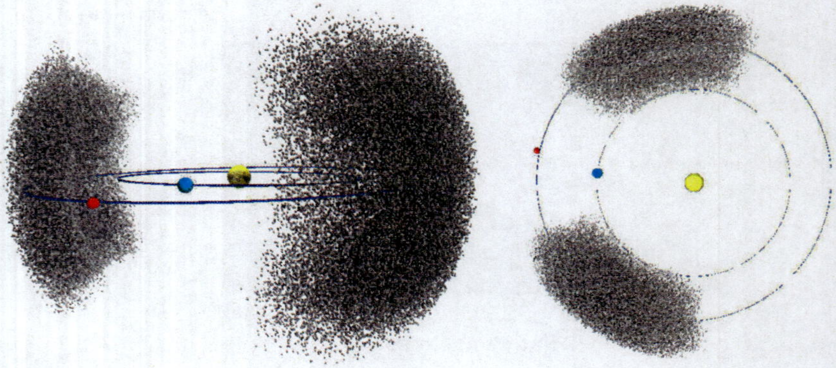

Fig. 2.10 Simulated Mars Trojans

These images (Fig. 2.9 and 2.10) reflect the distributions (Figs. 2.3 and 2.4) and show the tighter constraints on the Mars Trojans. The smearing observed in the Earth Trojan distribution is strongly influenced by the planet Venus. Conversely the constriction in the Mars Trojan distribution is influenced by both Earth and Jupiter.

Very few Trojan asteroids have been discovered in the inner Solar System, and the various simulations (Tabachnik and Evans 1999, 2000a,b; Morais and Morbidelli 2002) have predicted low numbers, which in itself is not especially surprising since the vast majority of all known asteroids are located in the Main Belt between Mars and Jupiter. The number of inner Solar System Trojans can literally be counted on one's fingers. The exact number of Trojans of any significant size is an estimate at best given the enormous amount of sky that would have to be searched in detail to have any chance of discovering additional Trojans.

It is possible, though generally considered unlikely, that those discovered so far are all that exist. The sole known Earth Trojan was discovered by a careful examination of data from NASA's WISE satellite. Future missions will likely be planned to launch telescopes into space for searching for Near-Earth Asteroids or conducting surveys of the Solar System, our Galaxy or the local Universe. Whether the mission is specific to Solar System studies or not, it is probable that the collected data will be analysed to search for evidence of things such as inner Solar System Trojans.

The only truly certain means of finding such small objects in such a large area of sky is by surveying the entire sky, which requires a mission dedicated to exactly that. The optimal search method for finding such small objects would be to survey those parts of the sky which contain stable orbits in order to maximise the chance of discovery (Todd et al. 2012a). Such small chances of positive discovery in an all-sky search makes it more likely that a Trojan search would be piggybacked onto another mission, where the mission data could be searched for these objects.

References

Bowell, E., Holt, H.E., Levy, D.H., Innanen, K.A., Mikkola, S., Shoemaker, E.M.: 1990 MB: the first Mars Trojan. Bulletin of the Astronomical Society 22, 1357 (1990)

Connors, M., Veillet, C., Wiegert, P., Innanen, K., Mikkola, S.: Initial Results of a Survey of Earth's L4 Point for Possible Earth Trojan Asteroids. Bulletin of the American Astronomical Society 32, 1019 (2000)

Connors, M., Wiegert, P., Veillet, C.: Earth's Trojan asteroid. Nature 475, 481–483 (2011)

Dunbar, R.S., Helin, E.F.: Estimation of an Upper Limit on the Earth Trojan Asteroid Population from Schmidt Survey Plates. Bulletin of the Astronomical Society 15, 830 (1983)

IAU Minor Planet Center, Trojan Minor Planets (2012), `http://www.minorplanetcenter.net/iau/lists/Trojans.html`

Jablonski, M.: euphOrbit. Version 1.0. software (2003)

Lagrange, J.-L.: Essai sur le Problème des Trois Corps (Essay on the Three Body Problem). Prix de l'Acadèmie Royale des Sciences de Paris, tome IX. 6 (1772)

Morais, M.H.M., Morbidelli, A.: The Population of Near-Earth Asteroids in Coorbital Motion with the Earth. Icarus 160, 1–9 (2002)

Nicholson, S.B.: The Trojan Asteroids. Astronomical Society of the Pacific Leaflets 8, 239 (1961)

Scholl, H., Marzari, F., Tricarico, P.: Dynamics of Mars Trojans. Icarus 175, 397–408 (2005)

Tabachnik, S., Evans, N.W.: Cartography for Martian Trojans. The Astrophysical Journal 517, L63–L66 (1999)

Tabachnik, S., Evans, N.W.: Asteroids in the inner Solar system - I. Existence. Monthly Notices of the Royal Astronomical Society 319, 63–79 (2000a)

Tabachnik, S., Evans, N.W.: Asteroids in the inner Solar system - II. Observable properties. Monthly Notices of the Royal Astronomical Society 319, 80–94 (2000b)

Todd, M., Coward, D.M., Zadnik, M.G.: Search strategies for Trojan asteroids in the inner Solar System. Planetary and Space Science 73, 39–43 (2012a)

Todd, M., Tanga, P., Coward, D.M., Zadnik, M.G.: An optimal Earth Trojan asteroid search strategy. Monthly Notices of the Royal Astronomical Society: Letters 420, L28–L32 (2012b)

Todd, M., Tanga, P., Coward, D.M., Zadnik, M.G.: An optimal Mars Trojan asteroid search strategy. Monthly Notices of the Royal Astronomical Society 424, 372–376 (2012c)

Whiteley, R.J., Tholen, D.J.: A CCD Search for Lagrangian Asteroids of the Earth-Sun System. Icarus 136, 154–167 (1998)

Chapter 3
Orbital and Dynamical Characteristics of Small Bodies in the Region of Inner Planets

Bojan Novaković

University of Belgrade, Serbia

3.1 Introduction

In addition to the planets, there are numerous smaller objects orbiting around the Sun as well. These objects exhibit a wide range of sizes, from dust grains to dwarf planets. The minor objects, larger than about 1 meter in diameter, orbiting the Sun interior to Saturn's orbit are called asteroids. Asteroids occupy a wide variety of orbits, but most of them are located in the so-called main asteroid belt, between the orbits of Mars and Jupiter. Apart from this, there are also other large groups of asteroids such as near-Earth asteroids, Hilda group and Jupiter's Trojans. Moreover, several thousands of planet-crossing asteroids, with perihelia inside the orbits of the inner planets, are known as well.

The majority of asteroids reside on stable orbits, but a significant fraction of them is located in dynamically unstable (i.e. chaotic) regions. To understand their dynamical characteristics, and to distinguish between the stable and unstable orbits, is often not straightforward. To achieve this goal different tools and techniques must be used.

The best characterized is the population of near-Earth asteroids (NEAs), the objects that occasionally come close to the Earth. This is because some of these objects are easier to reach than the Moon, and a typical near-Earth asteroid is much easier to reach by a space mission than a typical main-belt asteroid.

In this chapter we will focus on the near-Earth asteroids, but some other groups of asteroids will be discussed as well, in particular those that may evolve from their current locations and to become typical NEAs. The chapter is organized as follows. In Sect. 3.2, the asteroids nomenclature and orbital elements are explained. Sections 3.3 and 3.4 are devoted to the two most important phenomena that shape the orbits of asteroids, namely i) orbital resonances, and ii) Yarkovsky and YORP thermal forces. In Sect. 3.5, the near-Earth asteroids are described, while Sect. 3.6 is devoted to the asteroids orbiting inside the Earth's orbit. Long-term dynamical features of near-Earth asteroids are presented in Sect. 3.7. Mechanisms of transport from the main-asteroid belt to the near-Earth space as well as main source regions are discussed in Sect. 3.8. Finally, in Sect. 3.9 we describe the populations of Mars-crosser and Hungaria asteroids.

3.2 Asteroid Orbital Elements and Nomenclature

3.2.1 *Nomenclature*

After an asteroid is discovered, but before its orbit is well determined, it gets a provisional designation. The standard designation is composed of the year of discovery, followed by two letters and, optionally, a number. Taken all together these elements refer to the date of discovery. The first four digits indicate the year, then a letter to denote the half-month of discovery within that year (A for January 1-15, etc., I is omitted), and then again a letter to indicate the order of discovery within the corresponding half-month (A for 1st, Z for 25th, I is again omitted). Finally, at the end of the name there is the number of cycles of the second letter. For instance, asteroid 2005YF127 was (127 x 25) + 6 = 3181 object discovered during 2005 December 16-31. Later, when the orbit of an asteroid becomes well determined it receives a catalog number in chronological order. As of May 2012, more than 300 000 asteroids have been numbered, and this number continuously increases.

3.2.2 *Orbital Elements*

Orbital elements are a set of independent parameters that describe the orbital motion of a body and are sufficient to predict its position at any given time. In the case of Keplerian elliptical orbits, there are six orbital elements (see Fig. 3.1).

The first two elements, *semi-major axis* (a) and *eccentricity* (e), define the shape and size of the elliptical trajectory. An ellipse is the set of points in a plane whose sum of the distances from two fixed points of the same plane, named foci, is constant. In Keplerian elliptical orbits, the primary body (the Sun for our purposes) is located in one of the two foci. If we indicate as $2a$ the (constant) sum of the distances of any point of the ellipse from the two foci, the a parameter is called semi-major axis, and the distance between the two foci is equal to $2ea$, where e is the eccentricity. An ellipse reduces to a circle when the two foci are coincident, so that $e=0$. The upper limit for the value of e is 1. Roughly speaking, larger ellipses correspond to larger values of a, while e determines the elongation (deformation from a circle) of the ellipse.

Given an elliptical orbit, the point closest to the primary body is called the pericenter (or alternatively perihelion if the primary body is the Sun), and its distance q from the primary body is equal to $a(1-e)$; the farthest point is called the apocenter (or, respectively, aphelion), and its distance Q is equal to $a(1+e)$.

The next two elements define the orientation of the orbital plane in which the ellipse is embedded. These are the *inclination* (i) and the *longitude of the ascending node* (Ω). The inclination is the angle between the plane of the orbit and a reference plane (usually chosen to be the Ecliptic plane in solar system studies). The intersection between the orbit plane and the reference plane defines a line, called the *line of the nodes*, where the nodes are the two points corresponding to the interception of the elliptic orbit with the reference plane. The ascending node is the point in which the orbiting body crosses the reference plane moving upward. The longitude of the ascending node describes the position of the ascending node in the reference plane, relative to a fixed direction, which is usually chosen to be the Earth's vernal point.

The last two orbital elements are the *argument of pericentre* (ω) and the *mean anomaly* (*M*). The argument of pericentre (or argument of perihelion when the Sun is the primary body) defines the orientation of the elliptical orbit in its plane. In particular, ω is defined as the angle measured in the orbital plane from the ascending node to the direction of the line of the apsides (defined as the line connecting the two foci of the ellipse). The mean anomaly defines the position of the orbiting body along the ellipse at a specific time (the "epoch"). This is an orbital element that changes linearly with time. It can be defined as $M = n(t\text{-}t_0)$ where *n* is the *mean motion* and t_0 is a time of passage through the perihelion.

It should be noted that in the definition of the orbital elements above, in the case when the inclination is zero, ω and M are not defined, because the position of the ascending node is not determined. In addition, *M* is not defined also when the eccentricity is zero, because the position of the pericenter is not determined. It is convenient, therefore, to introduce the *longitude of perihelion* $\varpi = \omega+\Omega$, and the *mean longitude* $\lambda = M+\omega+\Omega$. The first angle is well defined when i=0, while the second one is well defined when i=0 and/or e=0. It is evident that the set of orbital elements a, e, i, ϖ, Ω, λ unequivocally defines the position and velocity of the body.

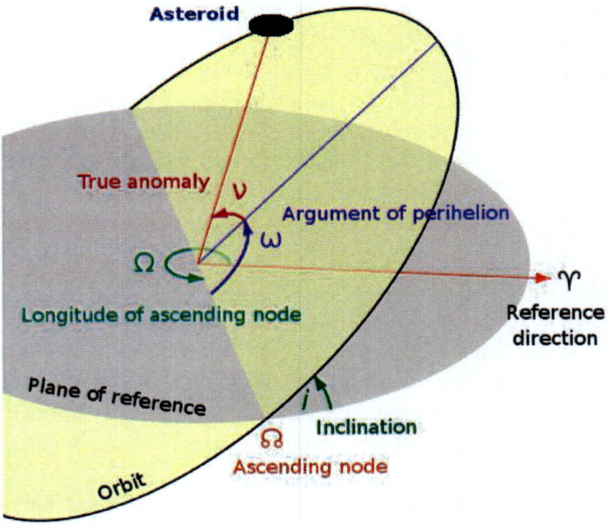

Fig. 3.1 Schematic definition of asteroid's orbital elements

There are two types of orbital elements that we will use through this chapter, the osculating and the proper ones. The osculating elements of an object in a space, at a given time, describe the Keplerian orbit that this object would pursue by its instantaneous position and velocity in space at a certain time, if all perturbations besides the gravitational force of the Sun where no longer present after that time. This kind of elements can be derived directly from the observations, and its most common application is to predict very accurately position of a body in a near future (or past), e.g. to produce ephemeredes. On the other hand, the osculating elements of an asteroid continuously change with a time due to the gravitational perturbations of other objects, first of all the major planets. These perturbations change the mutual distances among the asteroids in the space of osculating semi-major axis, eccentricity and inclination on the timescale of the orbital precessions. Because of these reasons, these elements are not suitable to study long-term characteristics of orbits.

To study dynamical characteristics of the asteroids over long time scales another type of elements has been developed, the so-called proper orbital elements. The proper elements are derived from the instantaneous osculating elements by removing the short and long-term periodic perturbations. By definition they represent integrals of motion and, thus, are supposed to be a sort of average characteristics of motion that are constant for a very long time. It is, however, well known that a full **N**-body problem is non integrable, and that therefore it does not have such integrals. Consequently, the proper elements can only be obtained as quasi integrals of motion, that is, more or less good approximations of the real dynamics, or as true integrals of motion, but in a significantly simplified model.

The proper elements for the majority of asteroids of low to moderate eccentricities and/or inclinations are computed by means of analytical theories based on the series development of the perturbing Hamiltonian. These theories are extremely complex and cannot be pursued too far. They require handling of complicated, cumbersome relations and are subject to problems of the convergence of solutions. The most advanced theory of the kind, developed so far, is the theory of Milani and Knežević (1990, 1994), based on the Lie series canonical transformations (Yuasa 1973). It takes into account terms in the expansion of the perturbing Hamiltonian up to the second order in perturbing mass and up to degree four in eccentricity and inclination. Once developed, the procedure to compute proper elements by means of the analytical theory is very efficient and suitable for the computation of large catalogs of proper elements for hundreds of thousands of asteroids. However, the analytical proper elements are known to be of limited accuracy. Moreover, they are usually supplied without error estimates, despite existing possibility to perform such calculations based upon the size of neglected terms.

More recently Knežević and Milani (2000) have developed a new method for computation of the so-called synthetic proper elements of asteroids, which consists of a set of purely numerical procedures, collectively called the synthetic theory. The procedure includes: (i) numerical integration of asteroid orbits in the framework of a realistic dynamical model; (ii) on-line digital filtering of the short periodic perturbations to compute the mean elements and the proper semi-major axis; (iii) Fourier analysis of the output to remove main forced terms and extract proper eccentricity, proper inclination, and the corresponding fundamental frequencies; (iv) check of the accuracy of the results by means of running box tests. The accuracy of the synthetic proper elements is better by a factor of more than 3 on the average with respect to the results derived by means of the above mentioned most advanced version of the analytical theory (Milani and Knežević 1994).

Proper elements for planet-crossing asteroids have been developed as well. The application of a perturbative method based on an averaging principle to solve the equations of motion of an asteroid under the perturbation of the planets fails whenever the perturbation has a crossing singularity along the domain of the solution. In order to avoid this problem, Gronchi and Milani (1999, 2001) introduced a generalized averaging principle. Their solution also provides the proper frequencies and the encounter circumstances with each planet. The latter information allows predicting the occurrence of node crossings and is particularly interesting in the case of Earth crossing asteroids, since it provides a way to evaluate potential Earth impactors. The stability of proper elements for planet-crossing asteroids is guaranteed only over a short time scale, either of the order of the period of a complete oscillation of ω, or until the next very close approach to a planet. Nevertheless, they are a very important tool to study the dynamical behavior of this class of objects.

In addition, several specially adapted theories exist for dynamically specific populations, as for the high eccentricity and/or inclination asteroids (Lemaitre and Morbidelli 1994), Trojans (Milani 1993; Beauge and Roig 2001), Hildas (Schubart 1982), etc.

Proper elements (and many other information) for main-belt and near-Earth asteroids are available at Asteroids Dynamic Site [AstDys: http://hamilton.dm.unipi.it/astdys/] and Near Earth Objects Dynamic Site [NEODys: http://newton.dm.unipi.it/neodys/], respectively.

3.3 Orbital Resonances

An orbital resonance is a gravitational phenomenon that implies commensurability between two or more frequencies of the motion of the bodies that are orbiting around the same central body. There are numerous examples of orbital resonances in the solar system. Resonances can be a source of both instability and long-term stability. Thus, resonant dynamics plays a critical role in understanding the distribution, transport and dynamical lifetimes of the small bodies. There are three general types of resonance phenomena in the Solar system involving orbital motions: mean motion, secular and spin-orbit resonances. Here we will only discuss the first two types. For more details see e.g. Malhotra (1998) and Morbidelli (2002).

Mean motion resonances (MMRs), occur when the orbital periods of an asteroid and a planet are close to a ratio of small integers, i.e. when $kn - k_j n_j \approx 0$, where k and k_j are positive integers, n_j is a mean motion of a j-th planet and n is the mean motion of an asteroid.

Special type of MMRs are three-body resonances that involve mean motion of an asteroid and two perturbing planets, i.e they occur when $kn + k_i n_i + k_j n_j \approx 0$, where k, k_i and k_j are integers, and n, n_i and n_j are mean motions of an asteroid, i-th and j-th planet respectively (Nesvorny and Morbidelli 1998a,b).

Secular resonances (SRs) are consequences of commensurabilities among the precession rates of the orbit of an asteroid and those of major planets (Knežević et al. 1991). These resonances concern slow angles like the argument of perihelion or the longitude of nodes. The strongest are linear SRs which occur when a secular frequency of an asteroid's secular angle is equal to the corresponding frequency of

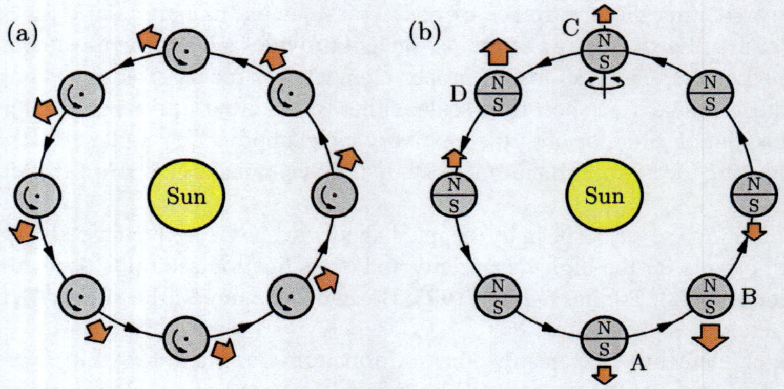

Fig. 3.2 A schematic representation of the two components of the Yarkovsky effect: (i) the diurnal component (left), and (ii) the seasonal component (right). A circular orbit and optimum values of the obliquity are assumed for simplicity, $\gamma=0^\circ$ on the left figure, and $\gamma=90^\circ$ on the right figure. Sunlight always heats the body on the nearside (noon), but due to finite thermal inertia, the point on the surface of the asteroid where the maximum temperature occurs, and hence the direction of the maximum recoil force due to thermal radiation, is displaced from the solar direction. In the diurnal variant (left), the body's rotation forces the maximum emissivity to be skewed toward the *afternoon* side on the body; thus the recoil force is always directed along the arrows. A net positive along-track force makes the body systematically accelerated; thus spiraling outward from the Sun. The effect would have an opposite direction if the body had a retrograde rotation (i.e., with obliquity $\gamma=180^\circ$) In the seasonal variant (right), thermal relaxation occurs on the timescale comparable to the orbital period of the given body. The seasonal force is directed along the spin axis and is due to north/south temperature difference on the body. The net, orbit-averaged, along-track force is always negative and the seasonal variant of the Yarkovsky effect makes the orbital semi-major axis, of an asteroid, to constantly decrease. For extreme values of the obliquity, $\gamma=0^\circ$ or $\gamma=180^\circ$, the seasonal component is zero because of symmetry between the north and south hemispheres.

a planet. An example is $\nu_6 = g\text{-}g_6$ secular resonance that involves frequencies of perihelions of an asteroid and the Saturn (here 6 stands for the 6th planet, namely Saturn itself).

Kozai resonance is a kind of secular resonance which occurs when the precession rate g of the longitude of perihelion ϖ of an asteroid is equal to the precession rate s of its longitude of the ascending node Ω (Kozai 1962). Therefore, this resonance dose not involve precession rates of the planetary orbits, as in the case of typical secular resonances. It is characterized by a libration of ω around 90° or 270° due to the perturbation by Jupiter, and around 180° due to the perturbation by the Earth and Venus (Michel and Thomas 1996). Also, it results in large coupled oscillations of the eccentricity e and the inclination i, as a consequence of the relation $(1\text{-}e^2)^{1/2}\cos i = const.$, that holds in the case of a Kozai resonance.

3.4 Thermal Phenomenon (Yarkovsky and YORP Effect)

Apart from gravitational perturbations, dynamical stability of small objects is strongly influenced by non-gravitational perturbations. In the case of asteroids, the most relevant is the Yarkovsky effect (Figs. 3.2 and 3.3). This effect is a consequence of a small but still relevant force that affects the orbital motion of asteroids smaller than about 30 kilometers in diameter. The Yarkovsky effect is the result of partial absorption of solar radiation at the surface of an asteroid and its anisotropic re-emission in the infrared band.

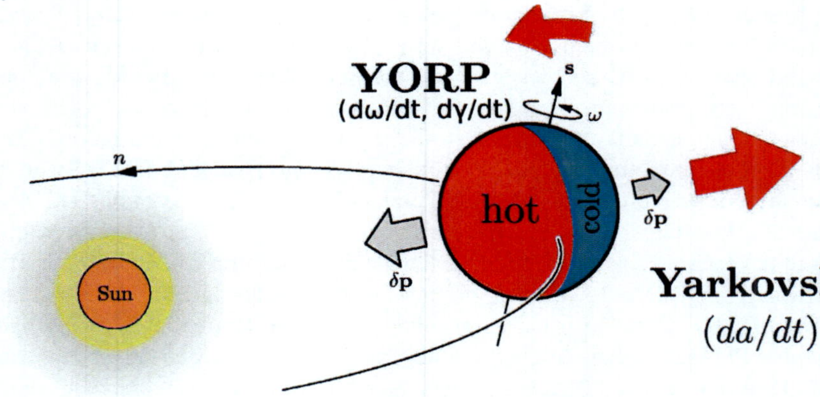

Fig. 3.3 An illustration of the Yarkovsky/YORP effect principle. As an asteroid absorbs the solar radiation, its side facing the Sun becomes hotter than the opposite one. The infrared emission from the surface is in that case anisotropic, giving rise to the Yarkovsky force that affects the orbital motion of the asteroid, and the YORP torque that modifies the spin state (Brož 2006).

The thermal radiation from the hottest part of the surface carries away more linear momentum than from the coldest part and this imbalance results in a recoil force. It mainly acts on the semi-major axis and its magnitude is size dependent. For asteroids it scales as 1/D, where D is the body's diameter. The resulting drift speed in the semi-major axis (da/dt) depends also on several other physical and dynamical parameters, such as thermal inertia, rotational period (P), spin obliquity (γ), and orbital geometry. This recoil acceleration is much weaker than solar and planetary gravitational forces, but it can produce substantial orbital changes over timescales ranging from millions to billions of years.

The reflection and re-emission of sunlight from an asteroid surface also produces a net thermal torque on asteroids with irregular shape. This effect is known as the Yarkovsky–O'Keefe–Radzievskii–Paddack (YORP). Over time, this torque can affect the spin rate and obliquity of small asteroids. YORP is important because it controls not only the long-term evolution of asteroid spin vectors but also the magnitude and direction of the Yarkovsky drift in the semi-major axis (see e.g. Rubincam 2000; Bottke et al. 2006).

3.5 Near Earth Asteroids

Near Earth Asteroids (NEAs) are those with perihelion distance $q \leq 1.3$ AU and aphelion distance $Q \geq 0.983$ AU. This population has been conventionally divided in three sub-populations: Atens, Apollos and Amors, each one named after the first discovered asteroid from the corresponding sub-population. Some NEAs are of high interest because they can be explored with lower mission velocity even than the Moon, due to their combination of low velocity with respect to Earth (ΔV) and small gravity, so they may present interesting scientific opportunities both for direct geochemical and astronomical investigation, and as potentially economical sources of extraterrestrial materials for human exploitation. This makes them the attractive targets for future explorations (Xu et al. 2007). In addition, many NEAs are considered to be Potentially Hazardous Asteroids (PHAs). Potentially Hazardous Asteroids are currently defined based on parameters that measure the asteroid's potential to make threatening close approaches to the Earth. Specifically, all asteroids with an Earth Minimum Orbit Intersection Distance (MOID) of 0.05 AU or less and an absolute magnitude (H) of 22.0, or less, are considered PHAs. In other words, asteroids that can not get any closer to the Earth than 0.05 AU (roughly 20 times further than Moon) or are smaller than about 150 m in diameter (*i.e.* H = 22.0 mag with an assumed albedo of 0.13) are not considered PHAs.

As of May 2012, 8,880 near-Earth asteroids are known, ranging in size from 1 meter up to about 32 kilometers. A total number of NEAs with $H < 18$ (i.e. roughly larger than 1 km in size) and $a < 7.4$ AU, was estimated by Bottke et al. (2002), to be 960±120, in an agreement with recent findings by NEOWISE (Mainzer et al. 2011) who found 981±19 NEAs with D >1 km. Mainzer et al. (2011) also estimated a total number of NEAs larger than 100 m in diameter to be 20,500±3,000, a somewhat less than in previous works (Rabinowitz et al. 2000; Harris 2008).

3.5.1 Aten Asteroids

Aten asteroids are those with an orbital semi-major axis $a < 1$ AU and aphelion distance $Q \geq 0.983$ AU. This is the smallest of the three NEO sub-populations. Since (2062) Aten was discovered in 1976 (Helin and Shoemaker 1977), the first asteroid that belongs to this class, more than 700 additional members have been found. Mainzer et al. (2012) estimated total number of Atens to be 1,600±760 larger than 100 m, and 42±31 larger than 1 km. Most of the time these objects spend orbiting interior to the Earth's orbit, but they occasionally cross the orbit of the Earth and may potentially collide with it.

3.5.2 Apollo Asteroids

Apollos are asteroids with $a > 1$ AU and perihelion distance $q \leq 1.017$ AU, where 1.017 AU is the aphelion distance of the Earth. The first member, asteroid (1862) Apollo, was discovered by Karl Wilhelm Reinmuth in 1932. Today, about 4,500

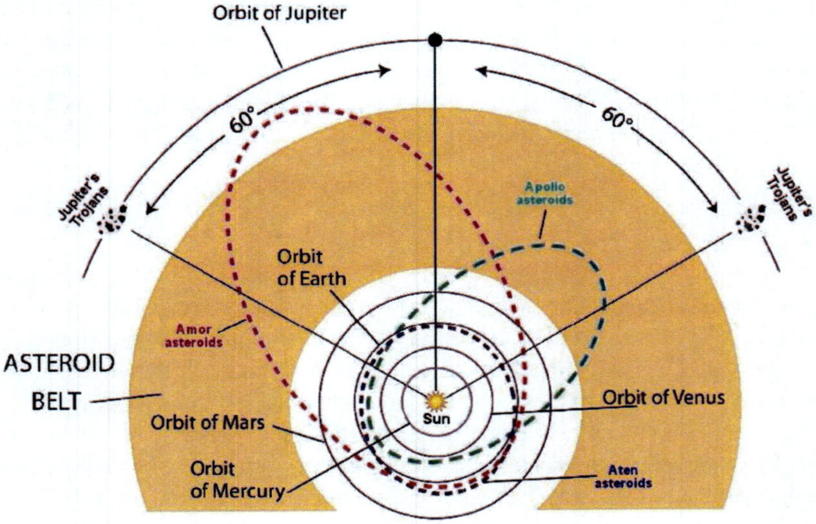

Fig. 3.4 The paths of three near-Earth asteroids, that represent typical orbits of Amor, Apollo and Aten asteroids. Amor crosses the orbit of Mars, and almost reaches the Earth's orbit. Apollo crosses the orbits of Mars, Earth and Venus, while Aten is always fairly close to the Earth's orbit.

Apollos are known. It is estimated that about 62% of the total number of NEAs are Apollos (Bottke et al. 2002). According to Mainzer et al. (2012) there are 462±110 Apollos larger than 1 km and 11,200±2,900 larger than 100 m in diameter.

The largest member is (1866) Sisyphus, with a diameter of about 10 km. The list of well known members include (3200) Phaethon, an asteroid believed to be a parent body of the Geminids meteor shower (Jewitt and Li 2010) which itself may be a fragment of asteroid (2) Pallas (de Leon et al. 2010), (4179) Toutatis, a famous example of a Potentially Hazardous Asteroid, (6489) Golevka, an asteroid on which the Yarkovsky effect was measured for the first time (Chesley et al. 2003), (25143) Itokawa, the first asteroid to be a target of a sample return mission, the Japanese space probe Hayabusa, and (54509) YORP whose determination of the rotation rate provided the first observational evidence of the YORP effect (Taylor et al. 2007), hence the name of the asteroid.

3.5.3 Amor Asteroids

Amor asteroids are defined by $1.017 \leq q \leq 1.3$ AU (Fig. 3.4). These objects may come close to the Earth but do not cross its orbit and, thus, cannot currently collide with it (Fig. 3.5). The first discovered NEA, (433) Eros, belongs to this class. It was discovered in 1898 by Carl Gustav Witt at the Urania Sternwarte (Berlin, Germany), and independently by Auguste Charlois at the Observatoire de Nice (France). With a diameter of about 17 km, (433) Eros is the second-largest NEA known. The largest known NEA, (1036) Ganymed, is also an Amor asteroid, with a diameter estimated to be about 32 km. Amors are the second most-populous group among NEAs with an estimated total of 320±90 members larger than 1 km and 7,700±3,200 larger than 100 m (Mainzer et al. 2012). So far, about 3,800 Amors have been discovered.

3.5.4 Earth Co-Orbital Asteroids (ECOAs)

Earth co-orbital asteroids are those that orbit at the same, or very similar, distance from the Sun as the Earth, i.e. they are in a 1/1 mean motion resonance with our planet. Each of the ECOAs belongs to one of three NEAs sub-populations mentioned above, however, as these objects have some specific characteristics, they are often analyzed separately. There are several classes of co-orbital objects, depending on their point of libration. The most common and best-known class are Trojans (tadpole orbits), which librate around one of the two stable Lagrangian points, L_4 and L_5, 60° ahead of and behind the planet, respectively. Another class is the horseshoe orbit, in which objects librate around 180° from the larger body. Objects librating around 0° are called quasi-satellites (see Figs. 3.5 and 3.6).

Earth co-orbital asteroids present advantages as potential targets for future asteroid rendezvous missions. Their prolonged proximity to Earth facilitates communication, while their Earth-like orbits mean a steady flux of solar power and no significant periodic heating and cooling of the spacecraft throughout the course of the mission. Theoretical studies show that low-inclination co-orbital orbits are more stable than high-inclination orbits. As inclination is the most significant indicator of low delta-v rendezvous orbits, there is the potential for a large population of easily accessible asteroids, with favorable engineering requirements. While rendezvous orbits to co-orbital objects do not necessarily have a low delta-v, energy requirements to reach some of these objects are significantly less than those in previous rendezvous missions. The best candidates seem to be Earth's Trojans that present very low-energy requirements (Stacey and Connors 2009). Morais and Morbidelli (2002) predict that the number of the ECOAs with absolute magnitudes H<18 and H<22 is 0.65±0.12 and 16.3±3.0, respectively.

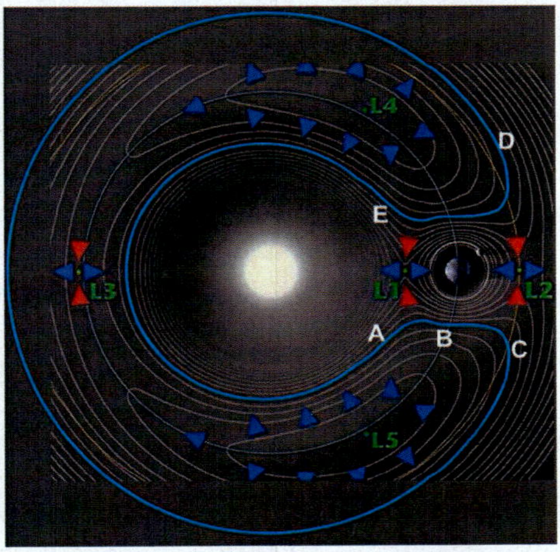

Fig. 3.5 Diagram indicating the Lagrangian points and the regions where the stable co-orbital trajectories can happen. The horseshoe and tadpole orbits are viewed in the rotating frame. Horseshoe orbits (light blue) encompass the L4, L3 and L5 points. Tadpole orbits oscillate about the L4 or L5 Lagrangian points (Trojan asteroids).

Trojan Asteroids

Trojans are objects that share an orbit with a planet, but do not collide with it because they orbit around one of two Lagrangian points L_4 and L_5, which lie approximately 60° ahead and 60° behind the planet, respectively. The giant planets

Jupiter and Neptune are known to host Trojan asteroids, and also Mars is hosting several asteroids of this type. The most famous and well known are Jupiter's Trojans.

Several theoretical studies were performed to establish the stability of the objects near Lagrange points in simplified models (see e.g. Bien and Schubart 1984; Efthymiopoulos 2005; Lhotka et al. 2008; Erdi et al. 2009, and references therein). Numerical studies undertaken to find an extension of the stability regions around the equilibrium points for the planets are even more numerous (e.g. Mikkola and Innanen 1992; Tabachnik and Evans 2000; Nesvorný and Dones 2002; Dvorak and Schwarz 2005; Robutel et al. 2005). Although many studies have predicted that Earth's Trojans may exist, their existence was confirmed just recently, when the first such object, the asteroid 2010TK7, has been discovered (Connors et al. 2011). For more details about Trojans see Chap. 2 by Michael Todd.

Horseshoe Companions to the Earth

Horseshoe-type orbits appear to be less prevalent than tadpoles orbits (Trojans), probably due to their different stability characteristics. An often-cited example of objects horseshoeing with each other are the Saturnian satellites Janus and Epimetheus. There are currently 3 asteroids (54509 YORP, 2002 AA29, 2010 SO16) known to follow horseshoe trajectories with respect to the Earth (Brasser et al. 2004; Christou and Asher 2011). In addition, asteroid 2001 GO2 may be of this class as well. However, confirmation of its status as the fourth Earth horseshoe will have to await further refinement of its orbit (Brasser et al. 2004).

The libration periods and lifetimes of these objects range from several thousand to a few hundred years. The orbital eccentricities of (54509) YORP and 2001 GO2 allow close encounters with the Earth. Such encounters, however, do not necessarily eject the asteroids from the co-orbital resonance, but instead they initiate their transition into another mode of libration or circulation.

Unlike Trojans, horseshoe co-orbitals are not generally considered to be stable over long time-scales (Dermott and Murray 1981). However, recent results of Ćuk et al. (2012), who numerically investigated long-term stability of Earth's and Venus's horseshoe co-orbitals, show that, contrary to analytical estimates, many objects of this type could be generally long lived (and potentially stable) for systems with primary-to-secondary mass ratio larger than about 1200. Horseshoe orbits at smaller mass ratios are unstable because they approach within 5 Hill radii of the corresponding planet. On the other hand, tadpole orbits are more robust and can remain stable even in the cases when approaching within 4 Hill radii of the planet.

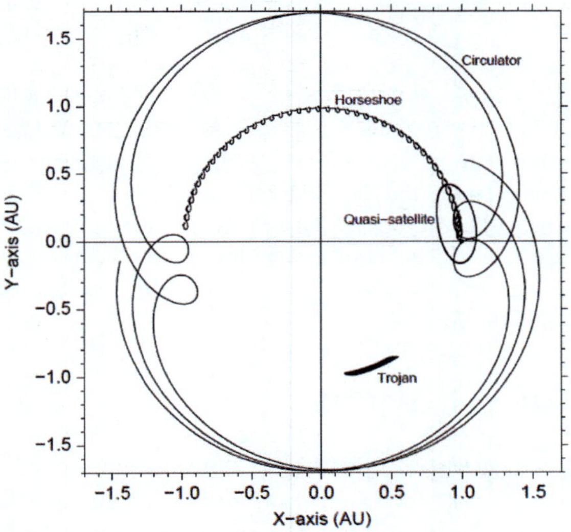

Fig. 3.6 Orbits of several asteroids viewed in Earth's co-rotating plane. The horseshoe (2002 AA29), the quasi-satellite (2004 GU9), and the Trojan (a synthetic Trojan) are Earth co-orbitals, while the circulator (Itokawa) is not. The circulator and the quasi-satellite are presented for a seven-year arc, while the horseshoe and the Trojan are presented for a fifty year arc in order to fully demonstrate their motions. The horseshoe remains close to the Earth for approximately ten years of the fifty years shown in this figure. The quasi-satellite remains close to the Earth for the extent of its motion. All three co-orbitals remain at a relatively constant distance from the Sun. The circulator does not have either of these advantages. Adopted from Stacey and Connors (2009).

3.5.5 Natural Earth Satellites (NESs)

In addition to the ECOAs, there is another population of interesting Earth-approaching asteroids. These move on extremely Earth-like orbits and their motion is dominated by close approaches with the Earth, although they are not in a 1/1 MMR with the Earth. Still, every synodic period they approach our planet with a small relative velocity. Thus, such objects can be potentially captured by Earth's gravity (Brasser and Wiegert 2008). Following Granvik et al. (2012) we will call these objects Natural Earth Satellites (NESs). The Earth's quasi-satellites (which are Earth's co-orbitals) have some common characteristics with the NESs. The essential difference between two populations is that the orbits of NESs depend critically on the gravity of the Earth-Moon system (EMS), while the orbits of quasi-satellites would only slightly change if the EMS suddenly disappeared, because they are simply orbiting the Sun on Earth-like orbits in the vicinity of the Earth, but without a strong gravitational interaction with it.

The first discovered NES was 1991 VG, which became a temporary satellite of the Earth in 1991 (Tancredi 1997). It was captured by Earth's gravitation, but most of the open loop it made around the planet was outside Earth's Hill sphere. Some early observations pointed towards the possibility that this particular object was of artificial origin, but newer observations render this a rather unlikely scenario. 2006 RH120 (also known as J002 E3) is the second known object to orbit the Earth, discovered in 2002. It came from the heliocentric orbit through the L_1 Lagrange point, spent a year inside the Earth's Hill sphere orbiting it six times on an open-loop orbit, and then left the Earth-Moon system.

According to the results of Granvik et al. (2012) at any given time there should be at least one NES of 1-meter diameter orbiting the Earth. An early identification of such objects creates an opportunity for a low-cost low-delta-v meteoroid return mission (Elvis et al. 2011).

3.6 Inner Earth Objects (IEOs)

By definition, asteroids whose orbits are wholly interior to the Earth's orbit are usually called Inner Earth Objects (IEOs) (Michel et al. 2000a). To this population belongs any small object with aphelion $Q < 0.983$ AU, where the latter number is the Earth's perihelion distance. According to the type of their orbits, the IEOs are further characterized as Vulcanoids ($Q < 0.307$ AU), Vatiras ($0.307 < Q < 0.718$ AU), or as Atiras ($0.718 < Q < 0.983$ AU). The IEOs are noticeably less populous than the NEAs, and a complete population consists of very limited number of objects. According to estimation of Bottke et al. (2002), there are only about 2% as many IEOs as NEOs at any H magnitude. A recent study performed by Greenstreet et al. (2012a), based on a *steady-state* orbital distribution model, found that among the asteroids with perihelion distance $q \leq 1.3$ AU only 1.38 ± 0.04 and 0.22 ± 0.03 % are Atiras and Vatiras, respectively. The latter authors also set the upper limit for the fraction of Vulcanoids to be 0.006 %.

The Vulcanoids are (still) hypothetical objects orbiting the Sun at distances interior to Mercury's orbit, i.e. $Q < 0.307$ AU. Possible existence of this population was first proposed by Weidenschilling (1978), as a potential explanation of contradictions in the cratering history of Mercury. An inner border of the population, in terms of orbital semi-major axis, is estimated to be at about 0.09 AU. Inside this limit, asteroids can not survive because of the evaporation, and the Yarkovsky effect (Vokrouhlický et al. 2000). Evans and Tabachnik (1999, 2000) conducted numerical simulations in order to study the long-term stability of Vulcanoids. They found that there is a belt ranging from ~0.1 to ~0.2 AU, where dynamically stable orbits exist. Thus, minor bodies may survive for a long period of time (comparable to the age of the Solar System) on these orbits, provided they are larger than about 0.1 km in diameter. According to Evans and Tabachnik (1999, 2000) objects beyond 0.21 AU are dynamically unstable and are excited into Mercury-crossing orbits on 100 Myr timescales. It is, however, important to note that these authors did not take into account the Yarkovsky effect that may significantly change the dynamical picture in this region.

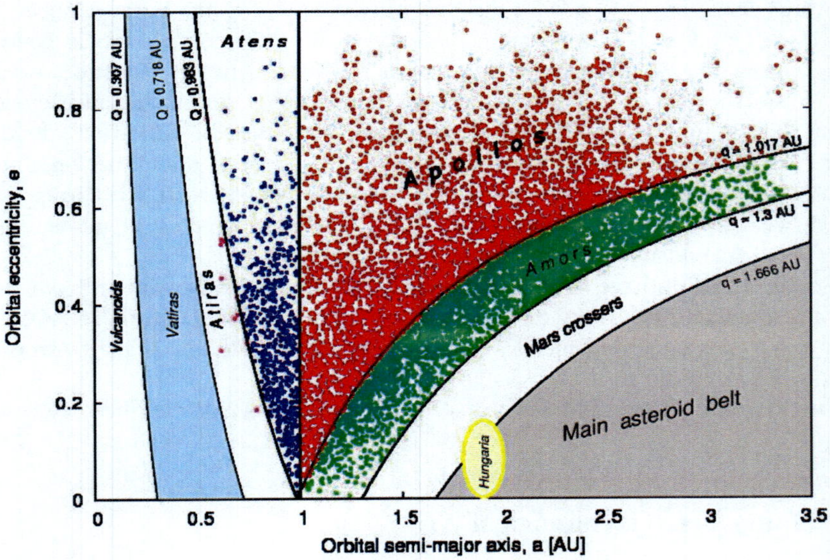

Fig. 3.7 IEOs and NEAs class distinctions in the semi-major axis vs eccentricity plane. We restrict the NEA population to orbits with $q < 1.3$ AU. Three NEAs classes we discriminate in this paper are Amors ($1.017 < q < 1.3$ AU), Apollos ($a > 1.0$ AU, $q < 1.017$ AU), and Atens ($a < 1.0$ AU, $Q > 0.983$ AU), while three IEOs classes are Atiras ($0.718 < Q < 0.983$ AU), Vatiras ($0.307 < Q < 0.718$ AU), and Vulcanoids ($Q < 0.307$ AU). A reader should be aware that objects could switch from one class to another on a relatively short time scale (see text for additional explanations). A complete list of asteroids that belong to each of these populations can be found at the Minor Planet Center web page: *http://www.minorplanetcenter.net/iau/lists/MPLists.html.*

Searching for Vulcanoids is exceedingly challenging due to their proximity to the Sun. Several searches for Vulcanoids have been performed so far (Leake et al. 1987; Campins et al. 1996; Durda et al. 2000; Schumacher and Gay 2001; Zhao et al. 2009), but none have discovered any object. Although these surveys did not find any Vulcanoid, the obtained results seem to rule out possibility that objects larger than about 10 km in diameter exist at this location (Durda et al. 2000; Zhao et al. 2009). These findings and the fact that km-size objects are unlikely to survive in that region due to the Yarkovsky effect (Vokrouhlický et al. 2000), put hypothesis about present-day existence of Vulcanoids in question.

According to the definition adopted here, Vatiras are asteroids with aphelion distance between 0.307 and 0.718 AU, but, as in the case of Vulcanoids, these are still hypothetical objects. Orbits of such asteroids would be wholly interior to the Venus's orbit, but they may be Mercury-crossers if the perihelion distance is below 0.307 AU.

Atiras are asteroids that orbit the Sun wholly interior to the Earth's trajectory ($0.718 < Q < 0.983$ AU). They are named after their prototype, i.e. first discovered object from the population, namely asteroid 163693 Atira (Greenstreet et al. 2012a). As this is a relatively new class of asteroids there are some inconsistencies about their definition used by different authors. Note that the definition adopted here clearly distinct Atiras from Atens (see Fig. 3.7), making the former ones non a sub-class of the latter ones. On the other hand, although we classified them as IEOs (what they certainly are), Atiras may also be considered as NEAs, because they may come relatively close to the Earth as well.

So far, 11 Atiras have been discovered (as of July 5th, 2012), with only two being already numbered. All of them have a perihelion distance below 0.718 AU, i.e. they are Venus-crossing objects. The largest member, asteroid 163693 Atira, is about 2 km in diameter.

Despite orbiting wholly interior to the Earth's orbit, Atiras may become a so-called PHAs. Two Atiras are currently listed as PHAs: 2004 JG6 and 2008 UL90.

3.7 Long Term Dynamics of NEAs

Numerical integrations have shown that the typical lifetime of NEAs is about 10 Myr. They are eliminated by collision with the Sun, ejection from the Solar System, or collision with the planets. Understanding the long-term dynamical characteristics of these objects is another task that requires modeling the location and the strength of resonances as well as the effects of close encounters with the planets.

The dynamics of the bodies in the region of the terrestrial planets is strongly influenced by close encounters with these planets. Each encounter gives rise to an impulse velocity to the body's trajectory, causing its orbital semi-major axis to change instantly by a quantity that depends on the geometry of the encounter and on the mass of the planet. The change in the semi-major axis is correlated with the change in eccentricity (and inclination) due to the quasi conservation of the so-called *Tisserand* parameter T (i.e. the pseudo-energy of the Jacobi integral that must be conserved in the restricted circular 3-body problem), defined as $T = a_{\mathrm{pla}}/a + 2[a/a_{pla}\,(1-\mathrm{e}^2)]^{1/2} \cos(i)$ relative to the encountered planet with semi-major axis a_{pla} (Öpik 1976). An encounter with Jupiter can easily eject the body from the Solar system, while this is practically impossible in encounters with the terrestrial planets.

Under the sole effect of close encounters with a unique planet, and neglecting the effects on the inclination, a body would perform a random walk on a curve in the (a, e) plane defined by $T = const.$ These curves are transverse to all MMRs and to most secular resonances, so that the body can be extracted from one resonance and be transported into another one. Resonances, on the other hand, change the eccentricity and/or the inclination of the asteroids, keeping the semi-major axis constant (apart from the short periodic oscillations). Thus, the real dynamic in the NEO region is the result of a complicated interplay between resonances and close encounters (Michel et al. 1996), and is therefore eminently chaotic.

MMRs present in the region of terrestrial planets are not so powerful as those located in the main-asteroid belt. This is because nearby terrestrial planets are significantly less massive than Jupiter, while Jupiter itself is not so close to this region. However, although they are relatively weak, MMRs located inside the orbit of Mars are very dense and often overlap with each other. This tends to extend the chaotic zones and increase the efficiency of these resonances, making them very important for the dynamics of NEAs.

Several studies of NEOs dynamics show that the secular resonances may play an important role in the orbital evolutions of small bodies in this region as well. SRs can either provide a protection mechanism from close approaches or a transport mechanism from one region of the phase space to another. The locations of linear secular resonances are obtained by means of a semi-analytical theory. It turns out that several SRs are present in the region of terrestrial planets, and some can even overlap (see Figs. 3.8 and 3.9).

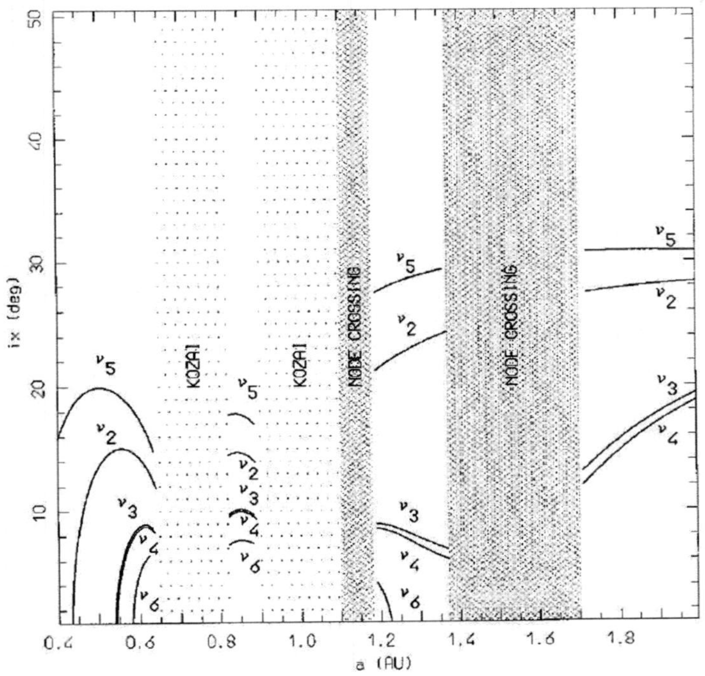

Fig. 3.8 The location in the (a (AU), i_x (degrees)) plane of the secular resonances involving the precession rates of the perihelion longitudes of a small body and the planets for $e_x = 0.1$. We follow the notations of Williams (1969) in naming the resonances. The region marked KOZAI corresponds to the regions of libration of the perihelion argument of the small body around 0° or 180°. In the regions marked NODE-CROSSING, the calculation of the free frequencies was not performed. Adopted from Michel and Froeschlé (1997).

Finally, the Kozai resonance also plays an important role for the dynamics of NEAs. Although it can act as a destabilizing factor, this resonance may also provide a protection mechanism that prevents an object to have close approaches with the planets. For example, for an orbital semi-major axis larger than that of the Earth, close encounters with the Earth can occur only near perihelion, but, since ω librates around 90° or 270°, and the inclination is large, perihelion always lie well outside the planetary orbital plane; thus preventing close approaches between the asteroid and the planet. The location of Kozai resonances is shown in Figs. 3.8 and 3.9.

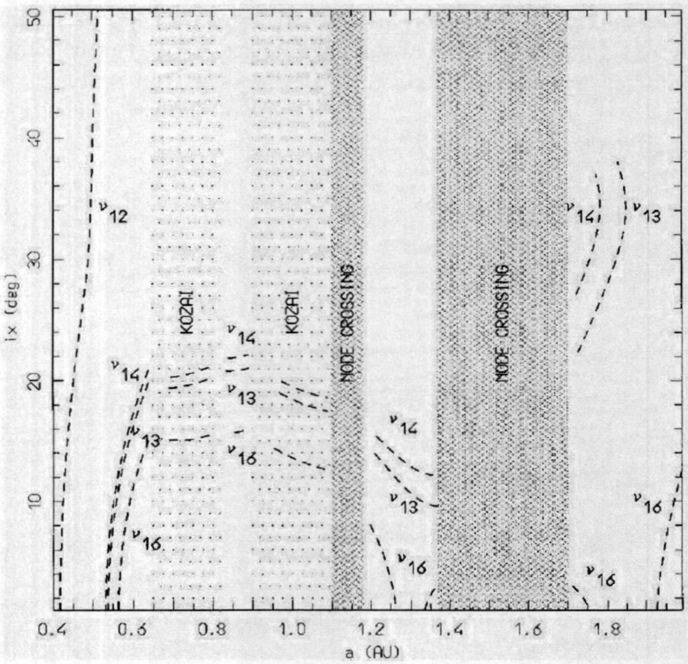

Fig. 3.9 The same as Fig. 8, but for the secular resonances involving the precession rates of the nodal longitudes of a small body and the planets. Adopted from Michel and Froeschlé (1997).

A dynamical classification of NEAs was performed by Milani et al. (1989) within the SPACEGUARD project (Milani 1989). This classification was based on the numerical integrations of orbits of 410 NEAs known at that time over time-scale of 200,000 yr. It used as criteria the main characteristics of planet-crossing dynamics: node crossings and close approaches to the Earth, MMRs with the Earth, value of q (below/above 1 AU), MMRs and approaches with Jupiter. According to this classification NEAs were divided into six (seven including comets) orbital classes; each one named after its best-known member. The decision algorithm used by Milani et al. (1989) is summarized in the diagram shown in Fig. 3.10, and the obtained classes are summarized in what follows.

Geographos class: The Earth-crossing orbits for which the evolution of a is dominated by close approaches to the Earth (and in some cases to Venus).

Toro class: Objects that are sometimes locked in a MMR, and this resonance is effective to avoid close approaches to the Earth.

Kozai class: Objects having perihelion distance below 1 AU at least for part of the time during a period of Ω, but being not Earth-crossers due to secular protection mechanisms.

Alinda class: Asteroids that are not presently Earth-crossing, but strongly perturbed by a MMR with Jupiter that causes large variations in eccentricity.

Eros class: Objects with a perihelion distance above 1 AU, which are not in any major MMR with Jupiter.

Oljato class: These are objects which are not in a resonance with Jupiter (at least not for a significant span of time), do not undergo orbital changes dominated by close approaches to the terrestrial planets, and are neither Jupiter crossing nor Jupiter approaching.

In addition to these six classes, there are also comet-like objects, i.e. those having an asteroid appearance (without any detectable cometary activity) but moving along typically cometary orbits.

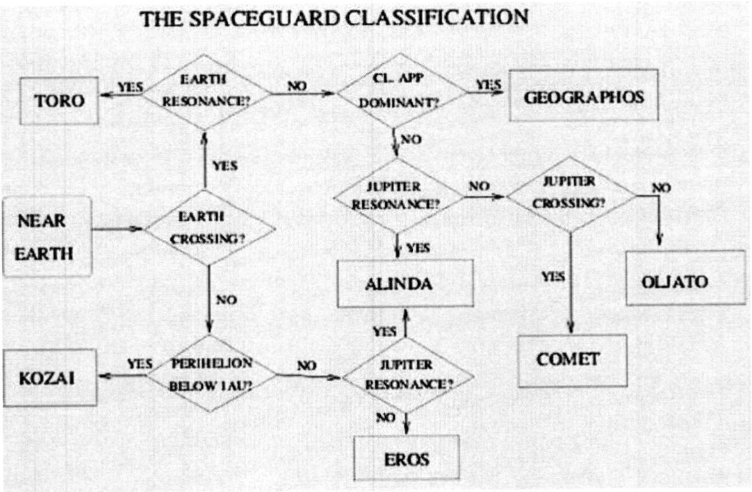

Fig. 3.10 A scheme of the SPACEGUARD NEAs' classification algorithm based on their dynamical characteristics (Milani et al. 1989)

In conclusion, we would like to note that the dynamical classification of Milani et al. (1989) is a useful tool to understand the dynamics of the NEAs. However, it must be taken into account that all these orbits are strongly chaotic, and that all the orbital elements can undergo large relative changes, as a result of either deep close approaches or the action of some resonance. Thus transitions can occur (and often do), between different dynamical classes. The Toro and Alinda classes are especially unstable, with typical residence times in these classes being less than 100,000 years. For example, transitions can occur between the Alinda, comet-like and Oljato classes.

Thus, if one investigates a long-term dynamics of NEAs, then problems do arise when using any classification based on current orbital characteristics. Due to the strong chaoticity of the motion, the objects may change dynamical class after a certain time. This phenomenon is named mixing. Due to the mixing occurring as a function of time, it is obvious that any inference derived from any classification valid at a certain epoch cannot be safely extrapolated to subsequent epochs. To investigate NEAs behavior over longer time scales some different statistical approach must be developed (Freistetter 2009).

3.8 Origin of NEAs

The short lifetime of NEAs means that this population is not primordial, but it is kept in a sort of steady state by a constant supply of new objects from some source regions. It has long been debated in the past whether the NEAs are mostly of asteroidal or cometary origin. Today, with an improved knowledge about resonant dynamics, it is clear that the asteroid belt is able to efficiently supply most NEAs. In particular, main belt asteroids can become planet crossers by increasing their orbital eccentricity under the action of a variety of resonant phenomena.

It is a reasonable hypothesis that the origin of NEAs is related to the most prominent gaps (as the so-called Kirkwood gaps, corresponding to the most important MMR with Jupiter) in the distribution of orbital semi-major axis in the main asteroid belt (Fig. 3.11). An interplay among collisional processes, Yarkovsky and YORP forces, may steadily bring new bodies into the gaps. The fast action of the most powerful secular and mean motion resonances keeps these gaps clear, compelling the asteroids to leave the main belt and to reach the near-Earth region. However, the concentration of Mars-crossing asteroids between 2.1 and 2.5 AU suggests that many bodies escape also from the portion of the main belt that is located in this semi-major axis range, despite the absence of any evident gap. It was shown that this is due to the action of numerous weak resonances, whose destabilizing effects manifest themselves on timescales that are long enough to allow their complete replenishment (Nesvorný et al. 2002). In this respect, it is suitable to distinguish between “powerful resonances” and “diffusive resonances”, where the former ones are characterized by the existence of associated gaps.

Some particular zones in the main belt provide NEAs via both powerful and diffusive resonances. In this respect, the most powerful resonances are the ν_6

secular resonance located at the inner edge of the asteroid belt, and the mean motion resonances with Jupiter 3/1, 5/2, and 2/1 located at 2.5, 2.8, and 3.27 AU respectively (Morbidelli et al. 2002). On the other hand, diffusive resonances are very numerous so that they cannot be effectively enumerated. Therefore, only their generic dynamical effects will be mentioned here. For more details the reader can refer to Nesvorný et al. (2002).

3.8.1 ν_6 Resonance

Let us first recall that the ν_6 secular resonance occurs when the precession frequency of the asteroid's longitude of perihelion is equal to the mean precession frequency of Saturn's longitude of perihelion. As shown in Fig. 3.12, the ν_6 resonance marks the inner edge of the main belt. The effect of this resonance rapidly decays with the distance from the shown curve. Asteroids located very close to the resonance (closer than 0.04 AU) exhibit a regular but large increase of the eccentricity (Morbidelli et al. 2002). This increase in eccentricity is large enough for these bodies to reach planet-crossing orbits. The median time required to become Earth-crosser, starting from a quasi-circular orbit, is about 0.5 Myr. Accounting for their subsequent evolution in the near-Earth region, the median lifetime of bodies initially in the ν_6 resonance is about 2 Myr, with the typical end states being collision with the Sun (80% of the cases) and ejection onto hyperbolic orbit via a close encounter with Jupiter (12%) (Gladman et al. 1997). The mean time spent in the NEO region is 6.5 Myr, longer than the median time because ν_6 bodies often reach semi-major axis $a < 2$ AU orbits where they often reside for tens of Myr (Bottke et al. 2002). In the border region (between 0.04 and 0.08 AU from the resonance), the effect of the ν_6 resonance is less powerful, but is still able to force the asteroids to cross the orbit of Mars at the top of the secular oscillation cycle of their eccentricity. To enter the NEA space, these asteroids must evolve under the effect of Martian encounters, and the required time decreases sharply when approaching the resonance (Morbidelli and Gladman, 1998; Migliorini et al. 1998).

3.8.2 3/1 Resonance

The 3/1 MMR with Jupiter is located at about 2.5 AU, where the orbital period of the asteroid is one third of that of the giant planet. The resonance width is an increasing function of the eccentricity (about 0.02 AU at $e = 0.1$ and 0.04 AU at $e = 0.2$), while it does not vary appreciably with the inclination. Inside the resonance, one can distinguish two regions: a narrow central region where the asteroid eccentricity has regular oscillations that make them to periodically cross the orbit of Mars, and a larger border region where the evolution of the eccentricity is strongly chaotic and unbounded, so that the bodies can rapidly reach Earth-crossing orbits. Under the effect of Martian encounters, bodies in the central region of the resonance can easily be transported to the border region and be rapidly pushed towards near-Earth space (Morbidelli et al. 2002). For a population initially uniformly distributed inside the 3/1 resonance, the median time required to cross the orbit of the Earth is about 1 Myr, while the median lifetime is approximately 2 Myr. The typical end states of this population are the collision with the Sun (70%) and the ejection on hyperbolic orbit (28%) (Gladman et al. 1997). The mean time spent in the near-Earth region is 2.2 Myr (Bottke et al. 2002).

3.8.3 5/2 Resonance

The 5/2 mean-motion resonance with Jupiter is located at about 2.8 AU. The rapid and chaotic eccentricity evolution observed in the border region of the 3/1 resonance in this case extend to the entire resonance (Moons and Morbidelli 1995). As a consequence, this resonance is the one that increases the orbital eccentricity on the shortest timescale. The median time required to reach Earth-crossing orbit, for objects initially located insude the 5/2 resonance, is about 0.3 Myr, and the median lifetime is 0.5 Myr. Because the 5/2 resonance is closer to Jupiter than the 3/1, the ejection on hyperbolic orbit is the most typical end state (92%), while the collision with the Sun accounts only for 8% of the losses (Gladman et al. 1997). The mean time spent in the near-Earth region is 0.4 Myr.

3.8.4 2/1 Resonance

The 2/1 resonance is located at 3.27 AU, but despite the fact that it is associated with a deep gap in the asteroid distribution, there are no mechanisms capable of destabilizing the resonant asteroid motion on the short timescales typical of the other resonances. Actually, the dynamical structure of this resonance is very complex (Nesvorný and Ferraz-Mello 1997; Moons et al. 1998). At the center of the resonance and at moderate eccentricity, there are large regions where the dynamical lifetime is of the order of the age of the Solar system. Some asteroids are presently located in these regions, but it is still not completely understood why their number is so small (Nesvorný and Ferraz-Mello 1997). The regions close to the borders of the resonance are unstable, but several million years are required before an Earth-crossing orbit can be reached (Moons et al. 1998).

Once in NEA space, the dynamical lifetime of asteroids coming from the 2/1 resonance is only on the order of 0.1 Myr, because the bodies are rapidly ejected by Jupiter onto hyperbolic orbit.

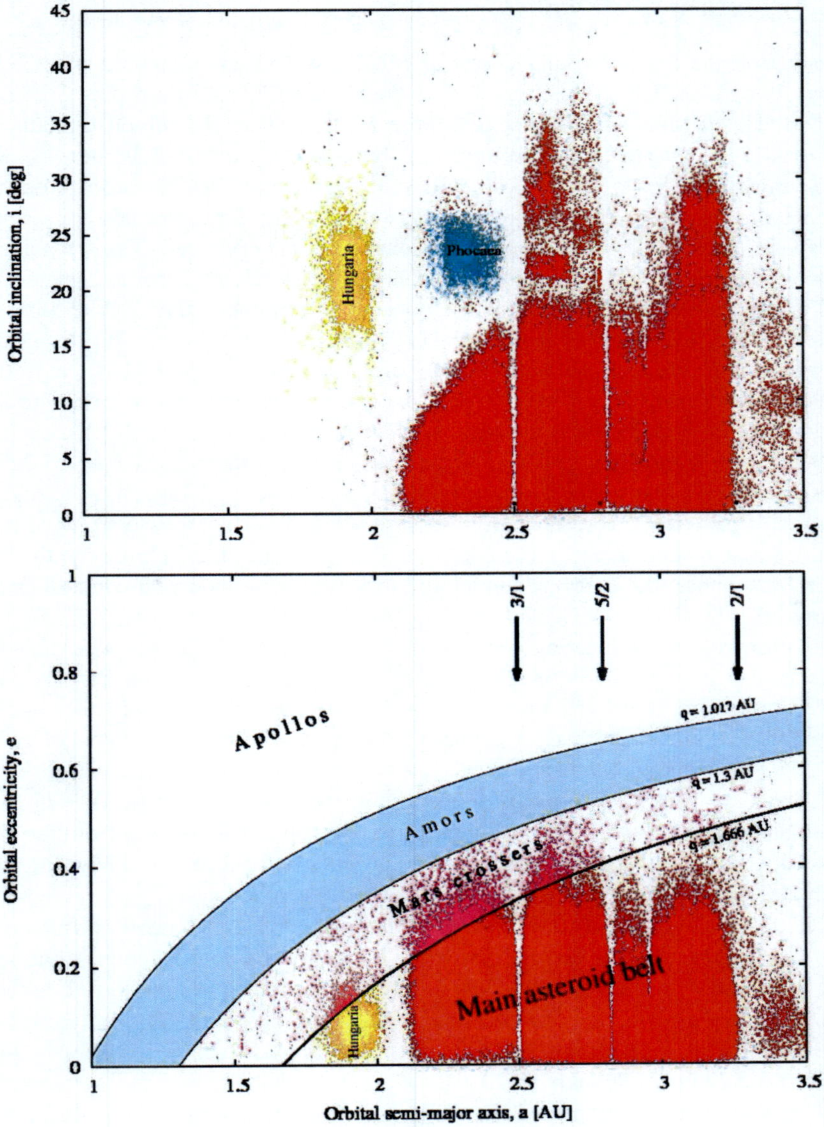

Fig. 3.11 Orbital distribution of Mars-crossers (purple), Hungaria (yellow), Phocaea (light blue) and Main belt (red) asteroids. Vertical arrows denote the positions of the 3/1, 5/2, and 2/1 mean motion resonances with Jupiter.

3.8.5 Diffusive Resonances

Apart from the few wide and powerful MMRs with Jupiter that we mentioned above, the main belt is densely crossed by numerous thin resonances: such as higher-order MMRs with Jupiter (i.e. those MMRs where the orbital frequencies are in a ratio of large integer numbers), 3-body resonances with Jupiter and Saturn, and MMRs with Mars (Morbidelli and Nesvorný 1999). Because of the influence of these resonances, many (possibly most) main-belt asteroids are chaotic (Nesvorný et al. 2002). The effect of this chaoticity is very weak. The semi-major axis is bounded within the narrow resonant region, while the proper eccentricity and inclination slowly change with time, in a chaotic diffusion-like process (Knežević et al. 2002; Novaković et al. 2010). Thus, the time needed to reach a Mars-crossing orbit (or a Jupiter-crossing orbit) ranges from several 10^7 yr to billions of years, depending on the resonance involved and starting eccentricity (Murray and Holman 1997). Integrating real objects in the inner belt ($2 < a < 2.5$ AU) for 100 Myr, Morbidelli and Nesvorný (1999) estimated that chaotic diffusion transports about two asteroids larger than 5 km into the Mars-crossing region every million years. To reach Earth-crossing orbit, the Mars-crossers (MCs) randomly walk in semi-major axis under the effect of Martian encounters until they enter a resonance that is strong enough to further decrease their perihelion distance below 1.3 AU. For the MCs with semi major axis between 2.06 and 2.8 AU, and inclination below the location of the ν_6 secular resonance (this subpopulation of MCs is also known as Intermediate Mars-crossers (IMCs)), the median time required to become the Earth-crossers is about 60 Myr; thus, roughly two bodies larger than 5 km in diameter become NEAs every million years (Michel et al. 2000b), consistent with the supply rate from the main belt estimated by Morbidelli and Nesvorný (1999). The mean time spent in the near-Earth space is 3.75 Myr (Bottke et al. 2002), while the median time to become Earth-crossers from the two groups of high-inclination MCs (Hungaria and Phocaea) exceeds 100 Myr (Michel et al. 2000b).

The lack of Mars-crossers at $a > 2.8$ AU is not because the chaotic diffusion in the outer asteroid belt is less inefficient. It is simply the consequence of the fact that the dynamical lifetime of bodies in the Mars-crossing region decreases with increasing semi-major axis toward the Jupiter-crossing limit. The outer belt is densely crossed by high-order MMRs with Jupiter and 3-body resonances with Jupiter and Saturn (Novaković 2010), so an important escape rate towards the NEA region should be expected. Bottke et al. (2002) numerically integrated about 2,000 main-belt asteroids, with $2.8 < a < 3.5$ AU, $i < 15°$ and $q < 2.6$ AU, for 100 Myr. They found that almost 20% of these asteroids entered the near-Earth region. According to Bottke et al. results, in a steady-state scenario this population could provide about 600 new NEAs with $H < 18$ mag per million years, but the mean time that these objects spend in the NEAs region is only about 0.15 Myr.

3.8.6 Asteroid Families as Source of NEAs

Dynamical families are believed to be the products of the disruptions of a number of single parent bodies. In the main asteroid belt, several tens of families have been identified so far (Zappalà et al. 1995; Nesvorný et al. 2005; Novaković et al. 2011). Moreover, families have been identified among the Trojans (Milani 1993; Brož and Rozehnal 2011), and are suspected to exist also in the Trans-neptunian region (Brown et al. 2007). Similar groups are expected to exist among NEAs as well, but we are still waiting for this to be confirmed (Schunová et al. 2012).

The dynamical lifetime of the objects located inside the resonances that are identified to be the main source of NEAs is very short, in some cases only a few Myr. Thus, it was obvious that in order to keep population of NEAs in a steady-state regime new objects must be constantly injected into the resonances. Starting from the early 1990s, when the first reliable systematic identification of families was done, it was realized that some families are located just on the borders of some of the most powerful resonances. This was interpreted as a good indication that, during the family forming events, many fragments are injected into these resonances, supplying in this way enough asteroids to explain the flux towards the region of terrestrial planets.

Following this approach, Zappalà et al. (1998) concluded that the formation of some of the most populous families which are located in the main belt very close to the borders of some of the powerful MMR with Jupiter, should have produced transient episodes of intense craterization of the terrestrial planets ("asteroid showers") whose duration depended in each case upon the involved MMR, and on the number of fragments injected into it. These conclusions were initially in partial disagreement with the predictions of modeling of family forming collisions, since some results of numerical modeling predicted that the ejection velocities of the fragments in family-forming phenomena could have been too low to inject many objects into nearby MMRs.

Later, however, the recognition of the importance of the Yarkovsky drift in orbital semi-major axis leads to conclude that objects up to some km in diameter produced in family forming events should be expected to reach in any case nearby resonances, although over timescales longer than those predicted by immediate injection hypothesis. In the case of large NEOs, with sizes of the order of 10 km, the Yarkovsky effect becomes much slower and significantly less relevant. Therefore, the existence of these objects must be found in other diffusive processes (Zappalà et al. 2002).

3.8.7 Cometary Contribution

As discussed earlier, both asteroids and comets are believed to contribute to the population of discovered near-Earth objects. However, the question remains: How many of these objects are asteroids and how many are comets? Observationally, any object that does not display a detectable cometary activity (i.e. coma) is catalogued as an asteroid, regardless of its genesis. From a mitigation or space resource utilization point of view, there may be a substantial difference in response or interest, as we presently presume that cometary bodies are more likely to have a high content of water and other volatiles.

Comets can be divided into two groups: those coming from the Trans-neptunian region (or, more likely, the scattered disk; Levison and Duncan 1997) and those coming from the Oort cloud (e.g. Weissman et al. 2002). The first group includes the Jupiter-family comets (JFCs), while the second group includes the long periodic and Halley-type comets. In addition, some of the near-Earth objects with comet-like properties may come from the population of Jupiter's Trojans as well, though it is believed that their contribution is small compared to those coming from the Trans-neptunian region and Oort cloud (Levison and Duncan 1997). Finally, after discovery of the so-called main-belt comets in the recent years (Hsieh and Jewitt 2006; Hsieh et al. 2012), it becomes evident that some of the comets from the near-Earth region may even originate in the main asteroid belt.

Measured physical properties of cometary nuclei indicate low albedos, typically reflecting less than 7% of the light. As this value is substantially below the average of all near-Earth objects, possible comets in the near-Earth space are less likely to be discovered. Given the available data on search statistics and near-Earth objects physical properties (taxonomic distributions and their corresponding albedos), "bias-corrected" models for this population were produced by Bottke et al. (2002) and Stuart and Binzel (2004). According to the results of these analyzes, 30% of all near-Earth objects reside in highly elliptical orbits that are strongly perturbed by Jupiter, similar to the identical orbital characteristics of many short-period comets. Half of this object subset have low albedo, which may be diagnostic of a cometary origin. It seems therefore that about 15% of near-Earth objects are likely extinct (i.e., having expelled most of their volatiles and having little left to produce a tail or coma) or dormant (i.e., inactive, asteroid-looking objects with low albedos, but still having considerable amounts of volatiles just under the crust) comets. According to Whitman et al. (2006) there are about 75 dormant JFCs (with $H < 18$ mag) in the near-Earth region.

3.8.8 Production of NEAs on Retrograde Orbits

So far, there are two known NEAs orbiting the Sun "backward", i.e. on retrograde orbits. These objects were thought to be of cometary origin. However, recently, Greenstreet et al. (2012b) re-analyzed transfer from the main-asteroid belt to the near-Earth region using the numerical integrations, and found that objects on such orbits could be produced from asteroid belt source regions as well. According to their model, about 0.1% of the steady-state NEAs population is on retrograde orbits. These objects are found to come from 3/1 MMR with Jupiter.

3.9 Mars Crossers

Similarly as in Michel et al. (2000b) (see also Migliorini et al. 1998) we will use here the term Mars-crossers (MCs) to refer to asteroids that intersect the orbit of Mars, but which cannot come close to the Earth, i.e. those with $1.3 < q < 1.666$ AU. Thus, in this way we formally exclude Amor asteroids from the population of MCs. As we mentioned in Sect. 3.8.5, all MCs are unstable and may potentially evolve to Earth-crossing orbits.

The population of MCs can be classified based on their current orbital distribution. Such a classification is useful for distinguishing among different dynamical behaviors, lifetimes, and possible final states. Figures 3.11 and 3.12 show the orbital distribution of Mars-crossers with respect to current semi-major axis and inclination. In the following, we give a qualitative description of the different groups and then their exact limits in semi-major axis and inclination.

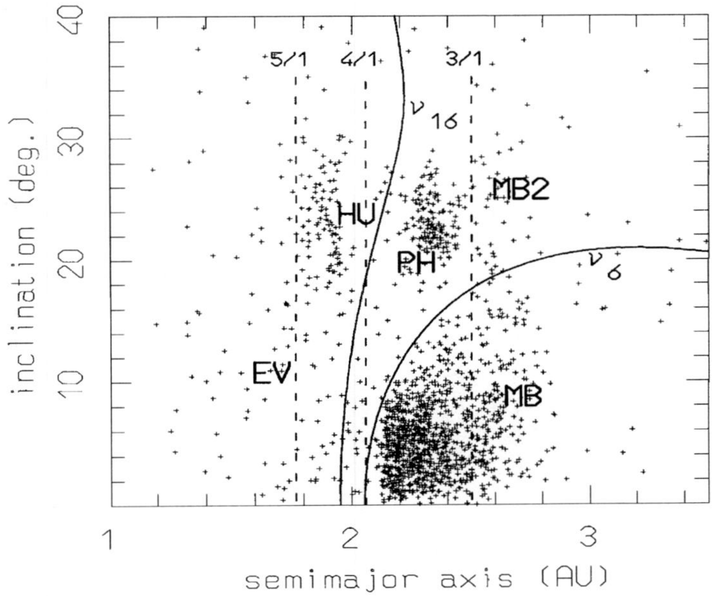

Fig. 3.12 Orbital distribution of Mars-crossers (inclination vs semi-major axis). Labels denote the different groups described in the text. The two curves denote the location of the ν_6 and ν_{16} secular resonances; the dashed lines correspond to the positions of the 5/1, 4/1, and 3/1 mean motion resonances with Jupiter. Adopted from Michel et al. (2000b).

Three main groups are relatively easy to identify (see Figs. 3.11 and 3.12). These are: main belt asteroids below the ν_6 resonance (MB); Hungarias (HU) and Phocaeas (PH). The fact that most objects from these groups are non-planet-crossing suggests that they might supply objects to the Mars-crossing region, "feeding" the MB, HU, and PH groups. This has already been confirmed by Migliorini et al. (1998). Among the remaining Mars-crossers, those on the left of the 4/1 resonance with Jupiter have orbital elements which differ from those of all non-planet crossing asteroidal populations. As was shown by Michel et al. (2000b), they have evolved relative to the orbit that they had when they first crossed the orbit of Mars. For this reason they are denoted as EV. Finally, the Mars-crossers on the right of the 3/1 and above the ν_6 resonances could be related to the non-planet-crossing asteroids located in this region: this group is denoted as MB2.

Michel et al. (2000b) defined the following limits in the semi-major axis and the inclination of each group:

- MB are Mars-crosser asteroids with a > 2.06 AU (location of the 4/1 MMR with Jupiter) and an inclination such that they are below the ν_6 secular resonance; this is a population of MCs often called the Intermediate Mars-crossers;
- HU have $1.77 < a < 2.06$ AU and $i > 15°$;
- PH have $2.1 < a < 2.5$ AU (between the 4/1 and 3/1 MMRs with Jupiter) and are situated above the ν_6 secular resonance;
- MB2 have a > 2.5 AU and are situated above the ν_6 secular resonance;
- EV are Mars-crosser objects having either current semi-major axis $a < 1.77$ AU or $1.77 < a < 2.06$ AU (i.e. between the 5/1 and 4/1 MMRs with Jupiter) and inclination $i < 15°$.

At present, there are more than 5,000 asteroids that satisfy the definition to be classified as MCs ($1.3 < q < 1.666$ AU, $a < 3.5$ AU). The MB group is the most numerous group of MCs. Numerical simulations performed by Michel et al. (2000b) have shown that 50% of the MB Mars-crossers become Earth-crossers within about 60 Myr, and that the contribution of this group dominates the production of ECs (Migliorini et al. 1998).

Mars-crossers belonging to the HU group become ECs on longer time scales, due to the reduced strength and frequency of encounters with Mars at high orbital inclinations. Within the integration time span of 100 Myr, 51.5% of the HUs never become ECs, but about 1.5% of the them reach the Aten region. Among the groups constituting the Mars-crosser population, the PH group is the most stable one, mainly because close approaches with Mars happen at low frequency and high relative velocities, decreasing thus the probability of great changes in semi-major axis. Thus, only 31% of these objects reach the Earth-crossing zone in 100 Myr.

The typical evolutions of MB2 asteroids are qualitatively similar to those of MB asteroids, but given their initial semi-major axis ($a > 2.5$ AU), the resonances transporting them are preferentially 3/1, 8/3, 7/3, and 5/2 with Jupiter. Evolving closer to the Jupiter-crossing region, their ejection outside Saturn's orbit as a result of a Jovian encounter is more probable than for the other groups. For this reason the MB2s do not contribute much to sustain the Earth-crossing population.

Orbital elements of EV asteroids differ from those of all non-planet-crossing asteroidal populations. Therefore they must have evolved relative to the orbit that they had when they first crossed the orbit of Mars. As an example, some asteroids belonging originally to MB or HU groups may temporarily become EVs (Michel et al. 2000b). Michel et al. simulations have also shown that EV Mars-crossing asteroids are mainly generated in two different ways. The first possibility applies to MB Mars-crossers. Asteroids from this group first become Earth-crossers, decrease their semi-major axis under the action of Earth encounters, and then temporarily decrease their eccentricity under the effect of some resonance. Secular resonances with the inner planets can be important in this phase (Michel and Froeschlé 1997, Michel 1997). Consequently, the perihelion distance is raised above the Earth-crossing limit, so that a previously Earth-crossing object may return to a purely Mars-crosser state. Very few MBs become EVs without first being Earth-crossers. Thus, the overall scenario for the origin of Mars-crossers and their subsequent evolution to Earth-crossing orbits is the one shown in Fig. 3.13.

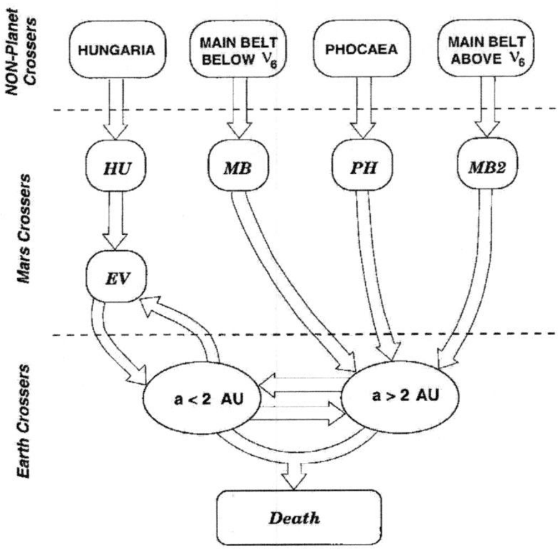

Fig. 3.13 Schematic view of the scenario of the origin of Mars-crossers and their subsequent evolution to Earth-crossing orbits. The HU, PH, MB, and MB2 Mars-crossers are sustained, respectively, by the non-planet-crossing asteroids of the Hungaria, Phocaea, and main-belt (below and above the ν_6 resonance) populations. In turn, they supply some fraction of the Earth-crossers and EV Mars-crossers. The arrows denote the main fluxes among the populations. (Reprinted from Michel et al. (2000b) with a permission of Elsevier).

3.9.1 Hungaria Asteroids

The inner edge of the asteroid main belt has a comparatively populated portion at high inclination and low to moderate eccentricity: the Hungaria region, named after the first discovered member, namely asteroid (434) Hungaria. Hungarias are a distinct population of asteroids located between 1.78 and 2.06 AU. Bounded by the ν_5 and ν_{16} secular resonances, the 4/1 mean motion resonance with Jupiter, and Mars-crossing orbital space (Gradie et al. 1979; Milani et al. 2010), its members have relatively high inclinations ($16° < i < 34°$) and eccentricities typically less than 0.18. The Hungarias are interior to the inner main belt (see Figs. 3.11 and 3.12).

As was shown by Milani et al. (2010) the Hungaria region has natural dynamical boundaries, where strong instabilities, arising from either secular resonances with Jupiter, Saturn and Mars, or close approaches to Mars, depopulate leaving large gaps in the distribution of asteroids. In terms of being a possible source of NEAs, the most important feature seems to be its inner boundary, which is entirely due to the instability resulting from deep Mars crossing. At present there are almost 900 MCs among the Hungaria asteroids. This sub-population is resupplied via the diffusive resonances that cross the region, but all these resonances are week; thus, effective only on a long-time scale.

With their, on average, high albedos (30 - 40%), Hungarias are among the smallest main belt objects that can be readily studied with modest-size instruments. Since they are also not generally subject to tidal encounters with the terrestrial planets, these asteroids provide a control set within the main belt to compare against the NEAs for such characteristics as rotation rate, distribution and binary population. Studies along those lines have shown some striking similarities between the NEAs and Hungarias (Warner et al. 2009). This suggests that forces other than tidal encounters (the most promising candidate being the YORP effect) were responsible for the binary and/or paired asteroids in the Hungaria group.

Asteroid (434) Hungaria has also been identified as the largest fragment of what is likely to be the region's sole asteroid family, created by a catastrophic collision about 0.2 - 0.5 Gyr ago (Warner et al. 2009; Milani et al. 2010). A majority of the approximately 7,000 bodies in the region are thought to be part of this family. There are also some (still weak) indications of possible sub-structures within the Hungaria family. These could belong to two types: sub-families and couples. A sub-family is a sub-group, of family asteroids more tightly packed than the surrounding members, which may be possibly produced by the subsequent breakup of a member of the original family, possibly at an epoch much more recent than the family formation event (Milani et al. 2010).

It is likely that, because of their dynamical stability, the objects trapped in Hungaria space have resided there since the planets assumed their final configuration. This makes the asteroids in the region prime targets for studies concerning the formation and evolution of the solar system in terms of orbital dynamics, space weathering, and Yarkovsky and YORP forces, among others.

Acknowledgements. I would like to express my gratitude to the reviewers G. Gronchi and C. Efthymiopoulos whose suggestions helped to improve this chapter. I also would like to thank to A. Cellino and Z. Knežević who read the manuscript and provided numerous useful suggestions. This work has been supported by the Ministry of Education and Science of Serbia under the Project 176011.

References

Beaugè, C., Roig, F.: A Semianalytical Model for the Motion of the Trojan Asteroids: Proper Elements and Families. Icarus 153, 391–415 (2001)

Bien, R., Schubart, J.: Trojan orbits in secular resonances. Celestial Mechanics 34, 425–434 (1984)

Bottke, W.F., Morbidelli, A., Jedicke, R., Petit, J.-M., Levison, H.F., Michel, P., Metcalfe, T.S.: Debiased Orbital and Absolute Magnitude Distribution of the Near-Earth Objects. Icarus 156, 399–433 (2002)

Bottke, W.F., Vokrouhlickỳ, D., Rubincam, D.P., Nesvornỳ, D.: The Yarkovsky and Yorp Effects: Implications for Asteroid Dynamics. Annual Review of Earth and Planetary Sciences 34, 157–191 (2006)

Brasser, R., Innanen, K.A., Connors, M., Veillet, C., Wiegert, P., Mikkola, S., Chodas, P.W.: Transient coorbital asteroids. Icarus 171, 102–109 (2004)

Brasser, R., Wiegert, P.: Asteroids on Earth-like orbits and their origin. Monthly Notices of the Royal Astronomical Society 386, 2031–2038 (2008)

Brož, M.: Yarkovsky Effect and the Dynamics of the Solar System. Ph.D. thesis, Charles University, Prague (2006)

Brož, M., Rozehnal, J.: Eurybates - the only asteroid family among Trojans. Monthly Notices of the Royal Astronomical Society 414, 565–574 (2011)

Brown, M.E., Barkume, K.M., Ragozzine, D., Schaller, E.L.: A collisional family of icy objects in the Kuiper belt. Nature 446, 294–296 (2007)

Campins, H., Davis, D.R., Weidenschilling, S.J., Magee, M.: Searching for vulcanoids. In: Rettig, T.W., Hahn, J.M. (eds.) Completing the Inventory of the Solar System. ASP Conf. Series, pp. 85–96 (1996)

Chesley, S.R., Ostro, S.J., Vokrouhlický, D., Capek, D., Giorgini, J.D., Nolan, M.C., Margot, J.L., Hine, A.A., Benner, L.A.M., Chamberlin, A.B.: Direct Detection of the Yarkovsky Effect by Radar Ranging to Asteroid 6489 Golevka. Science 302, 1739–1742 (2003)

Christou, A.A., Asher, D.J.: A long-lived horseshoe companion to the Earth. Monthly Notices of the Royal Astronomical Society 414, 2965–2969 (2011)

Connors, M., Wiegert, P., Veillet, C.: Earth's Trojan asteroid. Nature 475, 481–483 (2011)

Ćuk, M., Hamilton, D.P., Holman, M.J.: Long-Term Stability of Horseshoe Orbits. Monthly Notices of the Royal Astronomical Society 426, 3051–3056 (2012)

de Leòn, J., Campins, H., Tsiganis, K., Morbidelli, A., Licandro, J.: Origin of the near-Earth asteroid Phaethon and the Geminids meteor shower. Astronomy and Astrophysics 513, A26 (2010)

Dermott, S.F., Murray, C.D.: The dynamics of tadpole and horseshoe orbits. I - Theory. II - The coorbital satellites of Saturn. Icarus 48, 1–22 (1981)

Durda, D.D., Stern, S.A., Colwell, W.B., Parker, J.W., Levison, H.F., Hassler, D.M.: A New Observational Search for Vulcanoids in SOHO/LASCO Coronagraph Images. Icarus 148, 312–315 (2000)

Dvorak, R., Schwarz, R.: On the Stability Regions of the Trojan Asteroids. Celestial and Dynamical Astronomy 92, 19–28 (2005)

Dvorak, R., Schwarz, R., Sűli, Á., Kotoulas, T.: On the stability of the Neptune Trojans. Monthly Notices of the Royal Astronomical Society 382, 1324–1330 (2007)

Elvis, M., McDowell, J., Hoffman, J.A., Binzel, R.P.: Ultralow deltav objects and the human exploration of asteroids. Planetary and Space Science 59, 1408–1412 (2011)

Efthymiopoulos, C.: Formal Integrals and Nekhoroshev Stability in a Mapping Model for the Trojan Asteroids. Celestial Mechanics and Dynamical Astronomy 92, 29–52 (2005)

Érdi, B., Forgács-Dajka, E., Nagy, I., Rajnai, R.: A parametric study of stability and resonances around L4 in the elliptic restricted three body problem. Celestial Mechanics and Dynamical Astronomy 104, 145–158 (2009)

Evans, N.W., Tabachnik, S.: Possible long-lived asteroid belts in the inner Solar System. Nature 399, 41–43 (1999)

Evans, N.W., Tabachnik, S.: Asteroids in the inner Solar system - II. Observable properties. Monthly Notices of the Royal Astronomical Society 319, 80–94 (2000)

Freistetter, F.: Fuzzy characterization of near-earth-asteroids. Celestial Mechanics and Dynamical Astronomy 104, 93–102 (2009)

Gladman, B., Migliorini, F., Morbidelli, A., Zappalà, V., Michel, P., Cellino, A., Froeschlé, C., Levison, H., Bailey, M., Duncan, M.: Dynamical lifetimes of objects injected into asteroid belt resonances. Science 277, 197–201 (1997)

Gradie, J.C., Chapman, C.R., Williams, J.G.: Families of minor planets. In: Gehrels, T. (ed.) Asteroids, pp. 359–390. University of Arizona Press (1979)

Granvik, M., Vaubaillon, J., Jedicke, R.: The population of natural Earth satellites. Icarus 218, 262–277 (2012)

Greenstreet, S., Ngo, H., Gladman, B.: The orbital distribution of Near-Earth Objects inside Earth's orbit. Icarus 217, 355–366 (2012a)

Greenstreet, S., Gladman, B., Ngo, H., Granvik, M., Larson, S.: Production of Near-Earth Asteroids on Retrograde Orbits. The Astrophysical Journal 749, L39 (2012b)

Gronchi, G.F., Milani, A.: Averaging on Earth-Crossing Orbits. Celestial Mechanics and Dynamical Astronomy 71, 109–136 (1999)

Gronchi, G.F., Milani, A.: Proper Elements for Earth-Crossing Asteroids. Icarus 152, 58–69 (2001)

Harris, A.: What Spaceguard did? Nature 453, 1178–1179 (2008)

Helin, E.F., Shoemaker, E.M.: Discovery of Asteroid 1976 AA. Icarus 31, 415–419 (1977)

Hsieh, H.H., Jewitt, D.: A Population of Comets in the Main Asteroid Belt. Science 312, 561–563 (2006)

Hsieh, H.H., 41 colleagues: Discovery of Main-belt Comet P/2006 VW139 by Pan-STARRS1. The Astrophysical Journal 748, L15 (2012)

Jewitt, D., Li, J.: Activity in Geminid Parent (3200) Phaethon. The Astronomical Journal 140, 1519–1527 (2010)

Knežević, Z., Milani, A., Farinella, P., Froeschle, C., Froeschle, C.: Secular resonances from 2 to 50 AU. Icarus 93, 316–330 (1991)

Knežević, Z., Milani, A.: Synthetic Proper Elements for Outer Main Belt Asteroids. Celestial Mechanics and Dynamical Astronomy 78, 17–46 (2000)

Knežević, Z., Lemaitre, A., Milani, A.: Asteroid proper elements determination. In: Bottke, W.F., Cellino, A., Paolicchi, P., Binzel, R. (eds.) Asteroids III, pp. 603–612. University of Arizona Press, Tucson (2002)

Kozai, Y.: Secular perturbations of asteroids with high inclination and eccentricity. The Astronomical Journal 67, 591–591 (1962)

Leake, M.A., Chapman, C.R., Weidenschilling, S.J., Davis, D.R., Greenberg, R.: The chronology of Mercury's geological and geophysical evolution - The Vulcanoid hypothesis. Icarus 71, 350–375 (1987)

Lemaitre, A., Morbidelli, A.: Proper elements for highly inclined asteroidal orbits. Celestial Mechanics and Dynamical Astronomy 60, 29–56 (1994)

Levison, H.F., Duncan, M.J.: From the Kuiper belt to Jupiter-family comets: the spatial distribution of ecliptic comets. Icarus 127, 13–32 (1997)

Lhotka, C., Efthymiopoulos, C., Dvorak, R.: Nekhoroshev stability at L4 or L5 in the elliptic-restricted three-body problem - application to Trojan asteroids. Monthly Notices of the Royal Astronomical Society 384, 1165–1177 (2008)

Mainzer, A., 36 colleagues: NEOWISE Observations of Near-Earth Objects: Preliminary Results. The Astrophysical Journal 743, 156 (2011)

Mainzer, A., 12 colleagues: Characterizing Subpopulations within the near-Earth Objects with NEOWISE: Preliminary Results. The Astrophysical Journal 752, 110 (2012)

Malhotra, R.: Orbital Resonances and Chaos in the Solar System. In: Lazzaro, D., Vieira Martins, R., Ferraz-Mello, S., Fernandez, J., Beauge, C. (eds.) Solar System Formation and Evolution. ASP Conf. Series, vol. 149, pp. 37–63 (1998)

Michel, P.: Effects of linear secular resonances in the region of semimajor axes smaller than 2 AU. Icarus 129, 348–366 (1997)

Michel, P., Thomas, F.: The Kozai resonance for near-Earth asteroids with semimajor axes smaller than 2 AU. Astronomy and Astrophysics 307, 310–318 (1996)

Michel, P., Froeschlé, C.: The location of linear secular resonances for semimajor axes smaller than 2 AU. Icarus 128, 230–240 (1997)

Michel, P., Froeschlé, C., Farinella, P.: Dynamical evolution of NEAs: Close encounters, secular perturbations and resonances. Earth Moon Planets 72, 151–164 (1996)

Michel, P., Zappalà, V., Cellino, A., Tanga, P.: NOTE: Estimated Abundance of Atens and Asteroids Evolving on Orbits between Earth and Sun. Icarus 143, 421–424 (2000a)

Michel, P., Migliorini, F., Morbidelli, A., Zappalà, V.: The population of Mars crossers: Classification and dynamical evolution. Icarus 145, 332–347 (2000b)

Migliorini, F., Michel, P., Morbidelli, A., Nesvorný, D., Zappalà, V.: Origin of multikilometer Earth- and Mars-crossing asteroids: A quantitative simulation. Science 281, 2022–2024 (1998)

Mikkola, S., Innanen, K.: A numerical exploration of the evolution of Trojan-type asteroidal orbits. The Astronomical Journal 104, 1641–1649 (1992)

Milani, A.: Planet Crossing Asteroids and Parallel Computing: Project Spaceguard. Celestial Mechanics 45, 111–118 (1989)

Milani, A.: The Trojan asteroid belt: Proper elements, stability, chaos and families. Celestial Mechanics and Dynamical Astronomy 57, 59–94 (1993)

Milani, A., Carpino, M., Hahn, G., Nobili, A.M.: Dynamics of planet-crossing asteroids - Classes of orbital behavior. Icarus 78, 212–269 (1989)

Milani, A., Knežević, Z.: Secular perturbation theory and computation of asteroid proper elements. Celestial Mechanics and Dynamical Astronomy 49, 347–411 (1990)

Milani, A., Knežević, Z.: Asteroid proper elements and the dynamical structure of the asteroid main belt. Icarus 107, 219–254 (1994)

Milani, A., Knežević, Z., Novaković, B., Cellino, A.: Dynamics of the Hungaria asteroids. Icarus 207, 769–794 (2010)

Moons, M., Morbidelli, A.: Secular resonances in mean motion commensurabilities: The 4:1, 3:1, 5:2 and 7:3 cases. Icarus 114, 33–50 (1995)

Moons, M., Morbidelli, A., Migliorini, F.: Dynamical structure of the 2:1 commensurability and the origin of the resonant asteroids. Icarus 135, 458–468 (1998)

Morais, M.H.M., Morbidelli, A.: The Population of Near-Earth Asteroids in Coorbital Motion with the Earth. Icarus 160, 1–9 (2002)

Morbidelli, A.: Modern celestial mechanics: aspects of solar system dynamics. Taylor and Francis, London (2002) ISBN 0415279399

Morbidelli, A., Gladman, B.: Orbital and temporal distribution of meteorites originating in the asteroid belt. Meteoritics & Planet. Sci. 33, 999–1016 (1998)

Morbidelli, A., Nesvorný, D.: Numerous weak resonances drive asteroids towards terrestrial planets orbits. Icarus 139, 295–308 (1999)

Morbidelli, A., Bottke, W.F., Froeschlé, C., Michel, P.: Origin and Evolution of Near-Earth Objects. In: Bottke, W.F., Cellino, A., Paolicchi, P., Binzel, R. (eds.) Asteroids III, pp. 409–422. University of Arizona Press, Tucson (2002)

Murray, N., Holman, M.: Diffusive chaos in the outer asteroid belt. The Astronomical Journal 114, 1246–1252 (1997)

Nesvorný, D., Ferraz-Mello, S.: On the asteroidal population of the first-order Jovian resonances. Icarus 130, 247–258 (1997)

Nesvorný, D., Morbidelli, A.: Three-Body Mean Motion Resonances and the Chaotic Structure of the Asteroid Belt. The Astronomical Journal 116, 3029–3037 (1998a)

Nesvorný, D., Morbidelli, A.: An Analytic Model of Three-Body Mean Motion. Celestial Mechanics and Dynamical Astronomy 71, 243–271 (1998b)

Nesvorný, D., Dones, L.: How Long-Lived Are the Hypothetical Trojan Populations of Saturn, Uranus, and Neptune? Icarus 160, 271–288 (2002)

Nesvorný, D., Ferraz-Mello, S., Holman, M., Morbidelli, A.: Regular and chaotic dynamics in the mean motion resonances: Implications for the structure and evolution of the asteroid belt. In: Bottke, W.F., Cellino, A., Paolicchi, P., Binzel, R. (eds.) Asteroids III, pp. 379–394. University of Arizona Press, Tucson (2002)

Nesvorný, D., Jedicke, R., Whiteley, R.J., Ivezić, Ž.: Evidence for asteroid space weathering from the Sloan Digital Sky Survey. Icarus 173, 132–152 (2005)

Novaković, B., Tsiganis, K., Knežević, Z.: Chaotic transport and chronology of complex asteroid families. Monthly Notices of the Royal Astronomical Society 402, 1263–1272 (2010)

Novaković, B.: Portrait of Theobalda as a young asteroid family. Monthly Notices of the Royal Astronomical Society 407, 1477–1486 (2010)

Novaković, B., Cellino, A., Knežević, Z.: Families among high-inclination asteroids. Icarus 216, 69–81 (2011)

Öpik, E.J.: Interplanetary encounters - Close-range gravitational interactions. In: Kopal, Z., Cameron, A.G.W. (eds.) Developments in Solar System and Space Science, p. 155. Elsevier, Amsterdam (1976)

Rabinowitz, D., Helin, E., Lawrence, K., Pravdo, S.: A reduced estimate of the number of kilometre-sized near-Earth asteroids. Nature 403, 165–166 (2000)

Robutel, P., Gabern, F., Jorba, A.: The Observed Trojans and the Global Dynamics Around The Lagrangian Points of the Sun Jupiter System. Celestial Mechanics and Dynamical Astronomy 92, 53–69 (2005)

Rubincam, D.P.: Radiative Spin-up and Spin-down of Small Asteroids. Icarus 148, 2–11 (2000)

Schubart, J.: Three characteristic parameters of orbits of Hilda-type asteroids. Astronomy and Astrophysics 114, 200–204 (1982)

Schumacher, G., Gay, J.: An attempt to detect Vulcanoids with SOHO/LASCO images. I. Scale relativity and quantization of the solar system. Astronomy and Astrophysics 368, 1108–1114 (2001)

Schunová, E., Granvik, M., Jedicke, R., Gronchi, G., Wainscoat, R., Abe, S.: Searching for the first near-Earth object family. Icarus 220, 1050–1063 (2012)

Stacey, G.R., Connors, M.: Delta-v requirements for earth co-orbital rendezvous missions. Planetary and Space Science 57, 822–829 (2009)

Stuart, J.S., Binzel, R.P.: Bias-Corrected Population, Size Distribution, and Impact Hazard for the Near-Earth Objects. Icarus 170, 295–311 (2004)

Tabachnik, S.A., Evans, N.W.: Asteroids in the inner Solar system - I. Existence. Monthly Notices of the Royal Astronomical Society 319, 63–79 (2000)

Tancredi, G.: An Asteroid in a Earth-like Orbit. Celestial Mechanics and Dynamical Astronomy 69, 119–132 (1997)

Taylor, P.A., 11 colleagues: Spin Rate of Asteroid (54509) 2000 PH5 Increasing Due to the YORP Effect. Science 316, 274 (2007)

Vokrouhlický, D., Farinella, P., Bottke, W.F.: The Depletion of the Putative Vulcanoid Population via the Yarkovsky Effect. Icarus 148, 147–152 (2000)

Warner, B.D., Harris, A.W., Vokrouhlický, D., Nesvorný, D., Bottke, W.F.: Analysis of the Hungaria asteroid population. Icarus 204, 172–182 (2009)

Weidenschilling, S.J.: Iron/silicate fractionation and the origin of Mercury. Icarus 35, 99–111 (1978)

Weissman, P.R., Bottke, W.F., Levison, H.: Evolution of comets into asteroids. In: Bottke, W.F., Cellino, A., Paolicchi, P., Binzel, R. (eds.) Asteroids III, pp. 669–686. University of Arizona Press, Tucson (2002)

Whitman, K., Morbidelli, A., Jedicke, R.: The size frequency distribution of dormant Jupiter family comets. Icarus 183, 101–114 (2006)

Williams, J.G.: Secular Perturbations in the Solar System. Ph.D. Thesis, University of California, Los Angeles (1969)

Xu, R., Cui, P., Dong Qiao, D., Luan, E.: Design and optimization of trajectory to Near-Earth asteroid for sample return mission using gravity assists. Advances in Space Research 40, 200–225 (2007)

Yuasa, M.: Theory of Secular Perturbations of Asteroids Including Terms of Higher Orders and Higher Degrees. Publications of the Astronomical Society of Japan 25, 399 (1973)

Zappalà, V., Bendjoya, P., Cellino, A., Farinella, P., Froeschle, C.: Asteroid families: Search of a 12,487-asteroid sample using two different clustering techniques. Icarus 116, 291–314 (1995)

Zappalà, V., Cellino, A., Gladman, B.J., Manley, S., Migliorini, F.: NOTE: Asteroid Showers on Earth after Family Breakup Events. Icarus 134, 176–179 (1998)

Zappalà, V., Cellino, A., Dell'Oro, A.: A Search for the Collisional Parent Bodies of Large NEAs. Icarus 157, 280–296 (2002)

Zhao, H., Lu, H., Zhaori, G., Yao, J., Ma, Y.: The search for vulcanoids in the 2008 total solar eclipse. Science in China G: Physics and Astronomy 52, 1790–1793 (2009)

Chapter 4
Prospecting Asteroid Resources

Martin Elvis

Harvard-Smithsonian Center for Astrophysics, Cambridge, MA, USA

4.1 Introduction

Mining the asteroids has long been the stuff of science fiction, but is rapidly becoming an engineering reality [http://www.planetaryresources.com, http:// deepspaceindustries.com].The idea that asteroid mining could be a profitable industry in the near future gives the prospecting phase of mining a new urgency. Finding suitable asteroids to mine could well be the bottleneck to developing asteroid resources. Though the population of near-Earth objects (NEOs) is huge, with some 20,000 NEOs larger than 100m diameter (Mainzer et al. 2011b), and vastly smaller ones, there may be only a small number of NEOs that are initially profitable. Thorough prospecting could be needed to find these precious objects. This review identifies the state of the art for each stage of NEO prospecting, with an emphasis on remote telescopic techniques, and sets out options for upgrading our capabilities to the requisite industrial scale.

Our goal is to find ore. Here we take the definition that "*Ore is commercially profitable material*" (Sonter 1997), not merely a concentration of some minable resource. Ore in space could be precious metals, Helium-3, water, complex organics, or any other material that has commercial value. For the asteroids, all of the above except Helium-3 are abundantly available, but may not be profitable. Establishing whether some resource is potentially profitable, and thus qualifies as ore, depends on many factors. Prospecting is the first step, and is the process of establishing the presence of potentially valuable resources in a subset of the asteroid population. The costs of extracting those resources must then be determined in order to know if they are, in fact, ore. Only asteroids orbiting the Sun close to the Earth's orbit, the NEOs, are considered in this review because of the energetic favorability of their orbits, imposed by the unforgiving rocket equation (e.g. Elvis et al. 2011), which makes them the first targets for asteroid mining. If NEO mining becomes profitable, then the million times greater resources of the Main Belt (Sect. 4.4) will become available.

Much of this review depends on the techniques of astronomy. Because this is a new, interdisciplinary, field, attracting specialists from many areas, I deliberately explain terms that may appear elementary to astronomers. I also retain the professional astronomical usage, to make the original papers easier to read for those who want to dig deeper. Similarly, at the risk of overlapping in content with other chapters, I include some basic information about asteroids, as this may be unfamiliar to some astronomers. For a general introduction to NEOs, see Yeomans (2013). For a more technical overview of asteroid properties see Bottke et al. (2002a).

In this review Sect. 4.2 outlines the prospecting problem; Sect. 4.3 reviews asteroid composition from an ore-seeking point of view; Sect. 4.4 reviews the properties of NEOs; Sect. 4.5 examines NEO discovery survey techniques for discovering NEOs; Sect. 4.6 moves on to remote, telescopic, characterization; Sect. 4.7 looks, fairly briefly, at local characterization, both nearby an asteroid and in contact with its surface. Finally, Sect. 4.8 faces the implementation costs and timescales involved and emphasizes the need for new research. In concluding, in Sect. 4.9, I stress the commonality of means required to satisfy several motivations for going to the asteroids: exploration, hazards, and science, as well as mining. These all point towards an energetic asteroid prospecting program over the next decade.

4.2 Prospecting Overview

Prospecting asteroids falls naturally into three phases, based on the techniques that are used. (1) *Discovery* of a large enough sample of asteroids. (2) *Remote telescopic characterization* of sufficient asteroids to sufficient accuracy to find enough promising mining sites. (3) *Local characterization* of promising sites by visiting the asteroids, to prove enough viable mining targets. [Local characterization is often called *in situ* characterization. The local *use* of the resource is often termed *in situ* resource utilization (ISRU)]. Local characterization is divided into *"nearby"* and *"contact"* steps. Each stage requires different techniques, and becomes progressively more demanding, and so must be carried out for smaller and smaller numbers of asteroids.

While the mineral content of an asteroid is the most important property to determine, other properties are also critical to determining whether the rock can be mined at reasonable cost. For example, the granularity and cohesive strength of the asteroidal material, on all scales, will affect mining costs.

In a little more detail the prospecting phases are:

(1) *Discovery*. To count as "discovered" a NEO must have its orbit well enough determined that it can be located on its next pass near to the Earth (its next "apparition") with available telescopes. A few arcseconds is generally necessary accuracy when the orbit is extrapolated a year ahead [Astronomers use the hexadecimal units of degrees, arcminutes, arcseconds. 1 deg = 60 arcmin = 3600 arcSect. Engineers and planetary scientists tend to use radians. 1 arcsec = 4.85 microradians].

(2) *Remote Characterization.* Two primary types of characterization are needed: *spectroscopic* (to determine the surface mineral content) and *photometric* (to determine the rotation rate and shape). Photometry is generally done in the normal optical bands (0.4 – 0.9 microns wavelength), while spectroscopy provides optimal discrimination if it covers the redder optical to near-infrared bands (0.8 - 2.5 microns) or, even better, thermal infrared coverage (from 3.5 microns to 10 microns) – for reasons explained in Sect. 4.6. Other characterization tools are radar and interferometry, which directly determine asteroid sizes, and astrometry – accurate positional measurement - which can measure masses when the asteroid is a pair orbiting one another and, when combined with other data, via the Yarkovsky effect (Sect. 4.6.4).

(3) *Local Characterization.* Powerful additional prospecting tools can be employed if we can send a spacecraft close to an NEO (to kilometer scale distances, or to contact). For example, an accurate mass can be measured by orbiting the NEO and, coupled with imaging, a density. X-ray and gamma-ray spectrometers can be used to measure surface element composition.

4.2.1 The Scope of the Prospecting Problem

How long will it take to prospect the NEOs? Table 4.1 (Beeson et al. 2013) presents a quick summary of the flow of each phase of prospecting from discovery through remote and local characterization. The years needed to complete each phase at the current NEO discovery rate of ~1000/year (Sect. 4.4.2) and characterization rate of ~100/year (Sect. 4.6.2.1) are also given. Numbers are given for both 50 m and 100 m diameter NEOs. The first row begins with the predicted total NEO population (Mainzer et al. 2011b). In practice 30% of the 100 m or larger NEOs have been found, but discoveries are progressing at ~400/year.

For this table 4.1, I assumed that 10% of the discovered NEOs would have promising orbits that merit characterizing the objects better, and that 10% of these would remain promising targets that merit detailed assays. Obviously these can only be illustrative numbers at this early stage.

For large-scale extractive industries a time horizon of 10 years is not exceptional. The current discovery rate is then not unreasonable. However, current characterization rates to determine sizes and compositions are too slow by an order of magnitude, unless ore-bearing NEOs turn out to be very common. The final two columns of Table 4.1 list some ways to enhance the discovery and characterization rates, and the years then needed to complete the task.

Table 4.1 What we want to know, how long it will take, how we can do better

Prospecting Stage	Properties	Technique	# needed @100m @ 50m	Years to Complete	Enhancement Options
1. Discovery	Orbits, rough sizes	Wide-field ground-based optical surveys	20,000 80,000	30 40	Larger solid angle, collecting area ground-based surveys. Space-based thermal-IR survey(s)
2. Remote Characterization	Accurate Size[a]	Radar, Ground-based thermal-IR	2,000 8,000	20 80	Space-based thermal-IR survey
	Mineral Composition[a] Mass	Ground-based opt-Near-IR spectroscopy; comparison with meteorites Binaries Yarkovsky effect	2,000 8,000	20 80	Dedicated 10m class ground-based telescope with OH suppression, Space-based 0.5m opt-IR spectrometer
3. Local Characterization	Mass, density[b]	Close approach	200 800	> 100 > 100	Precision photometry, Precision astrometry, Swarms of small spacecraft
	Elemental composition[b]	Ground-based opt-Near-IR spectroscopy; comparison with meteorites	200 800	>100 >100	Nearby solar fluorescence X-ray spectroscopy Contact sampling

a. Only the 10% with low delta-v orbits
b. Only the 1% having low delta-v orbits, sufficient size and promising mineralogy

4.2.2 How Many Ore-Bearing NEOs?

If ore-bearing NEOS are common, then a complete reconnaissance would be unnecessary. How many NEOs must we characterize before we have compiled a sufficient inventory of highly promising mining targets? The answer will determine how complete our surveys of NEOs have to be.

We can quantify the number of ore-bearing NEOs (Elvis 2013) as

$$N_{ore} = P_{type} \cdot P_{rich} \cdot P_{low\text{-}\delta v} \cdot P_{eng} \cdot N\,(>D_{min}),$$

where P_{type} is the probability that an asteroid is of the resource bearing taxonomic type (DeMeo et al. 2009), P_{rich} is the probability that this type of asteroid is sufficiently rich in the resource, $P_{low\text{-}\delta v}$ is the probability that the asteroid is in an orbit with a sufficiently low delta-v, and $N(>D_{min})$ is the total number of asteroids larger than a threshold diameter, D_{min} , for profitability for the identified resource. P_{eng} is the probability that the engineering challenges of mining this asteroid can be overcome. Other factors can be added to this equation as the calculations become more refined, but this captures the essence of the problem.

We begin with the platinum group metals (PGMs: Pt, Rh, Os, Ir, Pd, Re, see Sect. 4.3.2) which Kargel (1994) has identified as a promising asteroidal ore type, because of their high value if returned to Earth (~US$50k/kg). Kargel (1994) discusses the fraction of NEOs that might be rich in PGMs. He estimates, from meteorite numbers, that P_{type} = 2-5% of NEOs are nickel-iron (Ni-Fe). (A much higher proportion of meteorites are Ni-Fe as they survive passage through the Earth's atmosphere much better than stony meteors.) Ni-Fe meteorites span 4 orders of magnitude (4 dex [1 dex = a factor 10 = 10 dB = one "order of magnitude".]) in iridium (Ir) richness [0.01 – 100 parts per million (ppm), Kargel 1994, Fig. 4.1]. Iridium is a useful indicator of overall PGM content. The richest dex has 10-100 ppm of Ir. The curve is steep, so the richness of the second 10% is well below that of the top 10%. This curve thus defines P_{rich} = 10%.

Perhaps $P_{low\text{-}\delta v}$ = 10% (δv < 5.5 km s^{-1}, vs. peak δv=6.6 km s^{-1}, Elvis et al. 2011). So the ore bearing Ni-Fe NEOs are approximately 4 x 10^{-4} of the total NEO population, or 1 in 2500. The situation is better than these odds make it seem, as all the terms except P_{rich} (and P_{eng}) can be determined remotely at modest cost. The 1 in 10 odds of finding a PGM-rich asteroid from the good candidates is much more acceptable.

The total number of ore-bearing asteroids depends on D_{min}, the minimum diameter NEO worth mining for PGMs. To estimate D_{min} requires considering ore value. At a density of 4,000 kg m^{-3}, a 100 m diameter asteroid would have a mass of 2.09 x10^6 metric tons (mt). At 10 ppm, the mass of platinum would be 20.9 mt. If this mass of platinum could be extracted and returned to Earth then, at present approximate prices of US$50 k kg^{-1}, they would be worth US$1.05 B. Factoring in the other PGMs increases the return by 60% (B.C. Crandall, private communication 2012) to US$1.7 B. Larger asteroids quickly become more valuable. An otherwise identical 150m dia. NEO could contain US$5.7 B of PGMs, for example.

However, smaller asteroids become unpromising just as rapidly. A 50 m diameter asteroid would contain one eighth of the platinum mass (7.8 mt), worth about US$220 M. Given that extraction is unlikely to be 100% efficient at first (Kargel 1994), and the as yet unknown, but likely substantial, mining and transport cost involved, D_{min} = 100 m dia. seems a reasonable starting point to take for profitability.

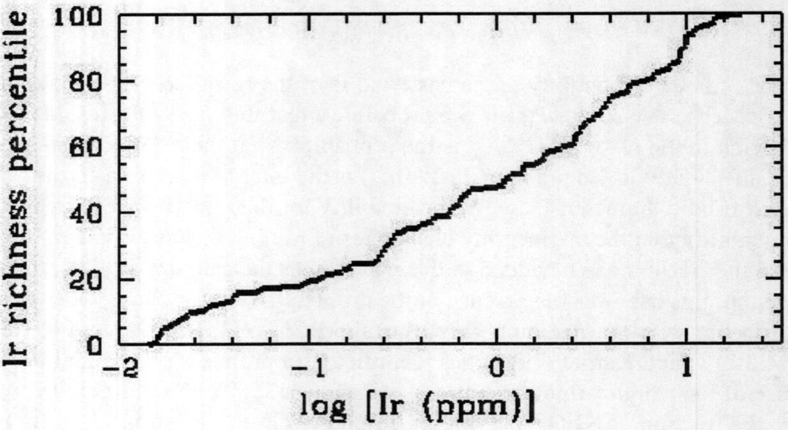

Fig. 4.1 Cumulative percentage of metalliferous type IIIAB meteorites with a given Iridium richness (ppm). Only the richest ~10% have a promising Ir richness > 10 ppm. (Data from table 2 of Scott and Buchwald 1973).

N(>100), the number of NEOs larger than 100 m diameter, is about 20,000 (Mainzer et al. 2011b). Combining these estimates, there are likely to be about eight NEOs that are good initial PGM mining targets. This is surprisingly small number and is certainly subject to large uncertainties. The total PGM value in these asteroids is US$14 B, so is still substantial. The most likely path to increasing the number is improved propulsion as $P_{low\text{-}\delta v}$ increases rapidly with δv.

If, instead, our goal is to mine water, then the larger population of carbonaceous asteroids is the target, for which P_{type} = 25% - 50%. Meteorites show that these also contain larger resource fractions than metallic asteroids, in the range 1% to 20% water. The distribution of water richness is not well determined. Taking the same P_{rich} = 10%, and the same $P_{low\text{-}\delta v}$=10%, good water targets will be (2.5-5) · 10^3, 1 in ~300, of the total population. This is a total of ~60 of the 20,000 NEOs with larger than 100 m diameter.

However, D_{min} is smaller for water-bearing asteroids. Water is about 1 dex less valuable per kg in LEO, than PGMs are on Earth, though its value doubles in GEO, and perhaps equals that of PGMs at the Earth-Moon L1 point. However, the much higher richness of water in the best targets more than compensates, so that smaller, ~15 m diameter, NEOs cross the nominal ~US$1.7 B threshold at 20%

water content. There may be about a million of these, for a total population of water-ore-bearing NEOs of ~5000 - ~10,000. This is a more promising number. For 1% water content NEOs D_{min} increases to 40 m.

The disadvantage, as we will see later, is such small NEOs are significantly harder to discover and characterize than 100 m diameter ones. Solutions are available though, and may be implemented for other reasons.

P_{eng} has not been considered here. Doing so is inherently complex and will greatly affect P_{ore}. Several considerations favor water over PGMs. The lower mass NEOs required for water-ore may have higher $P_{low-\delta v}$, increasing their numbers. Mining water will be less demanding on the miner spacecraft, especially energetically, reducing extraction costs. Returning ore to a high orbit rather than to the Earth's surface also reduces engineering complexity and energetic costs. Whether there is a sufficient market for water in high orbits is beyond the present discussion.

There are then a few tens to a few thousand NEOs that fulfill all the criteria for being ore-bearing, and thus potentially profitable. The small number of PGM-ore bearing NEOs argues for a virtually complete survey of 100 m diameter NEO orbits, size and compositions, while the larger number of water-ore-bearing NEOs argues for less complete surveys for smaller objects.

4.3 Asteroid Ore

Asteroids are flying mountains in space. On Earth only a small minority of mountains contain ore. We should expect that most asteroids are similarly uninteresting commercially. Extensive surveys will likely be needed to locate the small number of ore-bearing asteroids. Unfortunately, as we will see, many of the available diagnostics have limited ability to find resources of interest.

To know if sufficient ore is present we need several asteroid parameters: the *size* of the asteroid, its *density,* and hence *mass*, and the *concentration* of the target material. Asteroid sizes, densities and masses are considered in Sect. 4.4. Here I look at concentration diagnostics.

4.3.1 Astronomical versus Commercial Accuracy

We must keep in mind what level of confidence in our resource determination we need to establish. The accuracy of many astronomical observations is modest. Until recently, for example, the scale of the Universe (the Hubble Constant) was known only to a factor of 2, and is now determined to ~5% (Freedman et al. 2001). Most astronomical measurements of asteroids are similar to the Hubble constant in accuracy, and often to its earlier poor state. For example, asteroid diameters are mostly uncertain by a factor 2, so their volumes are uncertain by about an order of magnitude (see Sect. 4.4.3), which is far too large to allow a decent estimate of Net Present Value (NPV). Better evaluations of resource content are needed to make the business case for mounting a mining expedition to a promising asteroid.

An evaluation of NEO ore-value would have to be made using a low-end estimate. To have a 90% chance of not having a lower resource content than the value chosen means using an estimate two standard deviations (2σ) down from the best estimate. If the resource content of a NEO has a standard deviation of a factor two, then the 2σ low value is four times below the best estimate. In the PGM-rich 100 m diameter object discussed in Sect. 4.2.2, that means working to a low ore value of US$1.3 B, rather than a best estimate of US$5.3 B. The return on investment (ROI) is clearly affected strongly by this uncertainty.

In order to be useful for mining, the measurements need to be improved from these rough characterizations.

4.3.2 Composition: Resource Concentration

Asteroids and meteorites come in a profusion of compositions. Knowing which ones are likely to contain high concentrations of potential ore is far from straightforward.

This compositional profusion is due to the complex histories of asteroids. Current understanding is that the progenitors of the asteroids were small proto-planets, called "planetisimals" (10 km radius or larger). Radioactive heating (primarily from the decay of ^{26}Al to ^{26}Mg) melted the cores of the larger of these planetisimals, allowing gravity to differentiate their elements, with heavier ones, especially *siderophiles*, settling toward their cores. Siderophiles are "iron-loving" elements that dissolve easily in liquid iron. They include the PGMs. The details are complex, and models with sufficient detail to locate where, for example, PGMs end up, are not fully developed. An example of current models is the study by Yang et al. (2010), who find that Type IVB iron meteorites, which have some of the highest measured levels of platinum content (~30 ppm, Campbell & Humayun 2005), formed in a 70 km radius core within a 140 km radius planetisimal. This core was later split apart by collisions into the smaller asteroids we see today. This is still an active research topic. We do not know how many planetisimals of 140 km radius formed. As a result we do not know how much high PGM concentration rock formed in the early Solar System. Only 14 Type IVb meteorites are known, and their orbits before they impacted the Earth are unknown [Meteoritical Bulletin Database, see http://www.lpi.usra.edu/meteor/metbull.php]. Given the likely small fraction of PGM rich asteroids (2 – 5%), there's no guarantee that any of them will be NEOs in low delta-v orbits, and thus economically accessible.

The three main types of asteroid, based on meteorite compositions, are: stony, carbonaceous and metallic. Stony asteroids are chondritic, i.e. the rock contains small almost spherical lumps of different material. Chondrites can be metal rich (H) or metal poor (L). Carbonaceous asteroids contain high levels of complex organic molecules and ices, particularly water ice. Metallic asteroids are predominantly nickel-iron (Ni-Fe) and can contain high concentrations of heavy metals.

These three primary types map into 24 classification sub-types for asteroids (De Meo et al. 2009) and 34 for meteorites (Burbine et al. 2002). If we could match one with the other then prospecting would be relatively straightforward. Unfortunately, the links between the two systems are mostly quite uncertain. Burbine et al. (2002) list 27 suggested asteroid-meteorite associations. If these are correct then selecting, for example, S(IV) class asteroids would maximize the probability of finding ordinary chondrites (Gaffey et al. 1993). In addition, many of the clearly different asteroid sub-type assignments based on spectra are ambiguous as to composition. It is unclear, to take an important example for mining, whether an individual X-class asteroids, which all have rather featureless optical-near-IR spectra, belong to the E- M- or P-class sub-type, some of which are nickel-iron bodies. An albedo measurement can discriminate between the E, M and P categories (Thomas et al. 2011a). Only a high radar reflectivity can indicate a metallic asteroid surface, as it has in a few cases (Ostro et al. 1991), but this evidence is not completely convincing (Shepard et al. 2010).

Most meteorites are of unknown age. Fresh falls of meteorites allow minimally contaminated or chemically altered samples to be recovered. This is particularly important for volatiles. The Tagash Lake fall in 2000 was especially valuable because it was witnessed and fell on a frozen lake, helping to preserve the volatile content of the meteorites. A new method, that may produce many fresh falls, is to locate falls using weather radar. This method was proven in the Sutter's Mill fall of 2012 (Jenniskens et al. 2012).

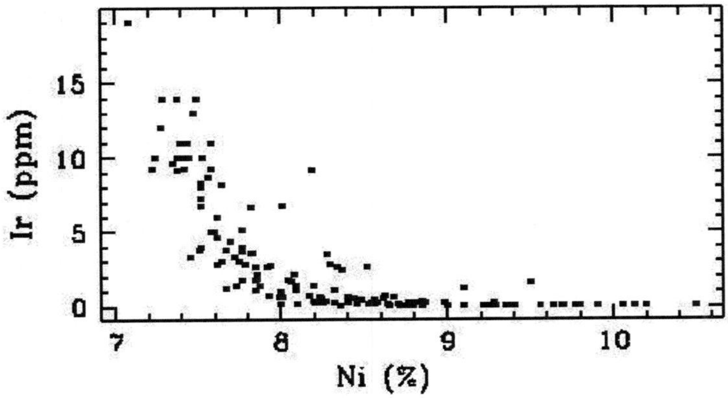

Fig. 4.2 Anticorrelation of Iridium (Ir) richness (ppm) with Nickel (Ni) content for 140 metaliferous type IIIAB meteorites. (Data from table 2 of Scott and Buchwald 1973).

Most spectroscopy is sensitive to silicates rather than to potentially valuable elements. From a mining perspective, then, it is important to know whether any silicates are a good tracer for elements of value. Among metallic meteorites (at least of types IIAB, IVA and IVB) the PGMs correlate with one another (Morgan et al. 1992). However, there is a large spread: about 1 dex for platinum (Pt) and rubidium (Ru), and 2-3 dex for rhenium (Re), osmium (Os), and iridium (Ir), in their abundances relative to palladium (Pd) (Petaev and Jacobsen 2004, McCoy et al. 2011). PGMs are anticorrelated with Nickel (Ni) abundance (Morgan et al. 1992, Fig. 4.2).

Meteorites measured with sufficient accuracy to detect PGMs are few. The 20,000, or so, meteorites collected by the Antarctic Search for Meteorites program [http://geology.cwru.edu/~ansmet/], have all been classified by the Smithsonian's National Museum of Natural History (NMNH). X-ray spectra generated with an electron probe identify elemental richness down to a few hundred ppm. PGMs have lower richness, however. As a result of the limited data there may well be useful correlations of abundant elements with PGMs that do not appear in the current data. If we are serious about asteroid mining, it would make sense to take a "human genome" approach to characterizing the nearly 45,000 meteorite samples in collections (Grady 2000) down to 10 ppm or thereabouts.

4.3.3 Mineral Diagnostics from Spectroscopy

An asteroid does not act as a perfect mirror, but reflects better at some wavelengths than others. The shape of the reflectivity curve depends on the surface minerals, and so can be used as a diagnostic. The reflectivity curve is often loosely called the asteroid spectrum, whereas it is really the observed spectrum divided by a solar spectrum. Nearby stars that are close solar analogs are usually used rather than the actual spectrum of the Sun.

Many asteroids and meteorites show two broad, shallow, spectral absorption bands at ~1 micron and ~2 microns. In meteorites the centers and band area ratios distinguish two types of silicates, olivine and pyroxene (Dunn et al. 2010). Slight changes in the wavelength of the centers of these features, and in the ratio of their strengths (their "Band Area Ratio", Thomas & Binzel 2010), indicate changes in composition. These features differ from one asteroid class to another, so these bands are used to link S-, Q- and V-type asteroid spectra to meteorite types, which then suggest a particular olivine to pyroxene ratio. The range of possible meteorite types for a given asteroid type, however, can be quite large (Gaffey et al. 1993, Thomas and Binzel 2010).

4.3.4 Water Diagnostics from Spectroscopy

The 3.1 micron hydration feature is a valuable diagnostic (Nelson et al. 1993, Vilas 1994). If water or other volatile materials are being sought, then this is a good signpost. It may be also be a good counter-indicator for a rich source of PGMs.

Observationally it is tricky, though. The feature lies at the short wavelength end of the L band (3.2 – 3.8 microns), whose width is set by atmospheric transmission, and the L-band background is strongly dominated by the 300 K black body emission of the Earth's atmosphere and the telescope (Sect. 4.4.4). Earth-based detection will be infeasible for all but the brightest NEOs.

An alternative to the 3.1 micron hydration feature is the feature at 0.7 micron, which lies well within the readily accessed optical band (Vilas 1994). These two features correlate well in the carbonaceous C-class Bus-DeMeo objects (Howell et al. 2011). The 0.7 micron feature is rare though in the X-class objects, despite the 3.1-micron feature being common. So the 0.7 micron feature appears to be a sufficient, but not necessary, diagnostic for hydration. The 0.7 micron feature is common in Main Belt asteroids, but is unfortunately rare or weak in most NEOs (Howell et al. 2011), and so is of limited utility for early mining operations.

4.3.5 Gardening and Space Weathering

Optical-near-infrared spectra can only tell us the mineral content of the asteroid surface, usually loose regolith. Most asteroids have likely undergone a collision that broke the asteroid up, reassembling it in a new random configuration. This "gardening" process creates a rubble pile asteroid and is thought to make the surface mineralogy representative of the whole body. Supporting this idea is the observation that pairs of asteroids, which likely result from a single asteroid breaking up, have similar colors (Moskovitz 2012). The gentler process of tidal stressing from near-Earth encounters can also refresh the surface (Binzel et al. 2010).

Some of the variety in asteroid sub-types (Binzel et al. 2012) seems to be due to space weathering (Hapke 2001), which alters the mineralogy of the surface layers. The asteroid surface material is irradiated by Solar ultraviolet photons, by high energy particles in the Solar wind, and by the high energy protons that form most of the Galactic cosmic rays. All of these forms of energetic particles can change the surface mineralogy compared with the bulk of the asteroid. One, possibly dominant process, is the coating of the silicate grains with a layer of nano-phase iron (Pieters et al. 2000); being opaque the iron removes the silicate spectral signatures. Micrometeorites also bombard the surface, as on the Moon. Outgassing of volatiles will occur near the surface if they are suddenly exposed to the space environment by this bombardment.

Space weathering could be quite a rapid process, by astronomical standards. A timescale of <1 million years (Myr) is likely (Vernazza et al. 2009), which is short compared to the ~10 Myr that an asteroid spends in a NEO orbit. However, this is by no means settled (Willman et al. 2012, Thomas et al. 2011b). Research on the changes induced by space weathering is underway, including laboratory research where meteorite samples are irradiated with lasers to induce the same effects. Once completed, this research should enable us to work back from the current surface mineralogy to that of the bulk material.

4.3.6 Direct Asteroid-Meteorite Comparisons

The best way to connect asteroids with meteorites mineralogically is to find small asteroids on their way to impacting the Earth. If found a few days to weeks before impact, we could characterize them telescopically in space to establish their spectral types, orbits and light curves. Then, when they are partially recovered as meteorites, detailed laboratory analysis can be compared directly with the telescopic spectra. With a sample of a few tens we could reliably tie NEO telescopic characterizations of a number of common asteroid types to their laboratory compositions.

Only one such "death plunge" asteroid has been found to date, 2008 TC3. This lone example has both space and laboratory measurements as, luckily, pieces of it landed in the northern Sudanese desert (Jenniskens et al. 2009) [Its meteoritic fragments were designated the Almahata Sitta fall, after a nearby railroad station. (almahata sitta means "station six" in Arabic.)]. 2008 TC3 had a diameter of about one meter. Similar sized asteroids impact the Earth about a dozen times a year (with significant uncertainty, being based on only 12 known events, Brown et al. 2002). Two thirds will fall in the oceans and at least half the remainder will fall where their meteorites cannot be recovered. A few per year may be retrievable. However, they could all be studied as they enter the atmosphere. Fast imaging and spectroscopy of their meteor trails could be carried out from aircraft, as is now done for meteor showers (Jenniskens 2007). The next generation of optical NEO surveys (Sect. 4.5) is likely to have the capability to find many of these incoming NEOs.

4.4 Near-Earth Objects (NEOs)

Millions of asteroids, with a total mass $5\text{x}10^{-4}$ Earth masses [1 Earth mass = $5.97\text{x}10^{24}$ kg (Cox 1999)], or ~$3\text{x}10^{18}$ metric tons (Petit et al. 2002), lie in the Main Belt, between the orbits of Mars and Jupiter, where Jupiter-induced and mutual perturbations among the Main Belt asteroids cause collisions and dynamical excitation (Weidenschilling 2000). Main Belt asteroids require significantly more energy to send spacecraft to (as measured as "delta-v", Sect. 4.4.1), and round trip journey times from Earth take a decade or more with current technology. Such a long timescales, even once a mining expedition is underway, have a bad effect on the profitability of any mining venture.

The much smaller population of Near-Earth objects (NEOs) have orbits that approach or cross the Earth's orbit at some point. The NEOs were scattered into these orbits from the Main Belt (Bottke et al. 2002b) by entering orbital resonances (i.e. orbits having integer orbital period ratios) with Mars or Jupiter (Greenstreet and Gladman 2012). Asteroids remain as NEOs for about 10 million years before being scattered again. NEOs are mostly asteroids, with a small admixture of dormant comets, although more of these are becoming recognized (Jewett 2012). Hence the term "NEO" rather than NEA (near-Earth asteroid) is usually used.

The dust from some small NEOs (and from comets) enters the atmosphere and burns up as meteors. Larger pieces make it to the ground as meteorites. Occasionally large pieces hit the Earth, such as the 20th century Tunguska (1908, e.g. Longo 2007), Rio Curaça (1930, Steel 1995) and the 2013 Chelyabinsk objects, which had diameters of a few tens of meters. This hazard makes NEOs worrysome, as well as being enticing as sites of possibly valuable resources.

4.4.1 NEO Types

NEOs are defined to lie within Mars' orbit (perihelion, $q < 1.3$ AU [AU = Astronomical Unit, the mean distance from the Sun to the Earth = 149 x 10^6 km (Cox 1999)]). The asteroid scientific literature sub-divides NEOs into types depending on their orbits [http://neo.jpl.nasa.gov/neo/ groups.html]:

- *Amors* ($1.017 < q < 1.3$ AU AND semi-major axis, a>1 AU) are Earth-approaching NEOs with a perihelion between Mars and Earth orbit.
- *Apollos* ($q < 1.017$ AND $a > 1$ AU) are NEOs that dip inward to cross the Earth's orbit, and so are Earth-crossing objects.
- *Atens* (aphelion, Q > 0.983 AND a < 1.0 AU) NEOs rise outward to be Earth-crossing objects.
- *Atiras* (Q < 0.983, AND $a < 1$ AU), NEOs that are always within the Earth's orbit.

While useful for astronomical observations, these distinctions are not of great utility to asteroid miners.

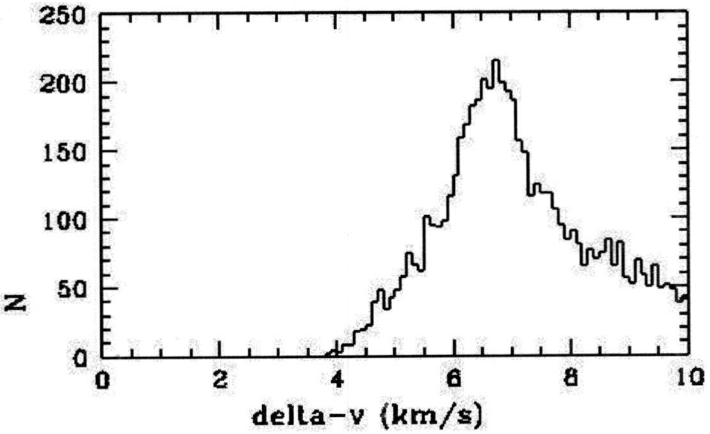

Fig. 4.3 Distribution of outbound delta-v from LEO to NEO (Elvis et al. 2011)

A more interesting parameter for mining, or any spacecraft visit, is delta-v, the total change in velocity a spacecraft must achieve to make a round trip to the NEO. Because of the rocket equation, which calculates the penalty a rocket must pay for having to carry its propellant with it, small differences in delta-v lead to large changes in the payload mass that can be delivered to the NEO. Reducing the outbound delta-v from 6 km s^{-1} to 4 km s^{-1} doubles, and can even quadruple, the payload that can be delivered to a NEO from LEO (Elvis et al. 2011). Hence, for example, a NEO on a circular 1.2 AU orbit is not of great interest for early mining operations (or for human exploration), as it has a high delta-v. The tail of small delta-v NEOs in Fig. 4.3 is thus of great importance. To get accurate delta-v estimates is computationally demanding but feasible. The Near-Earth Object Human Space Flight Accessible Targets Study (NHATS [http://neo.jpl.nasa.gov/nhats/]) calculated delta-v for several thousand NEOs. A preliminary estimate of the outbound delta-v can be calculated using minimum energy Hohman transfer orbits (Shoemaker and Helin 1978) [Near-Earth Asteroid Delta-V for Spacecraft Rendezvous: http://echo.jpl.nasa.gov/~lance/delta_v/delta_v.rendezvous.html].

Potentially Hazardous Objects (PHOs), which have a relatively high probability of striking Earth, tend to have low delta-v. There are 1370 PHOs known at present, about 15% of all NEOs, but 25% of the NEOs larger than 140 m diameter. PHO orbits must have an asteroid-Earth minimum orbit intersection distance, MOID < 0.05 AU (= 19.6 lunar distances = 7.45×10^6 km). PHOs must also be large enough to reach the Earth's surface where they would be highly damaging. Lacking accurate sizes their absolute visual magnitude, H is used as a proxy (Sect. 4.4.4). An NEO must have H < 22 (diameter > 140 m) for it to be classified as a PHO, although smaller NEOs of 40 m diameter or more can reach the surface too. PHOs are typically also good mining candidates based on their sizes and orbits.

4.4.2 NEA Numbers

Nearly 10,000 NEOs have been identified and cataloged (as of 1 January 2013) primarily from ground-based optical surveys. They range in size up to ~50 km diameter (1036 Ganymed, H = 9.45), but are mostly much smaller, down to a few meters in diameter. 5512 have H < 22, corresponding to a diameter of about 100 m. Current surveys for NEOs are incomplete, with their incompleteness becoming much worse for smaller NEOs. Mainzer et al. (2011b) find that about 80% of 100 m - 300 m diameter NEOs remain undiscovered, with some 2700 now known.

Current NEO surveys have a discovery rate of about 1000 new NEOs a year. For example, in 2012 987 NEOs were found. Of the 2012 discoveries 398 had H < 22. They were roughly evenly divided between Apollos and Amors, with a 6% contribution from Atens [based on Minor Planet Center statistics.]

A careful study of the NEOWISE detections compared with the survey "selection function" - the probability that it will find a given NEO - predicted the total number of NEOs, down to 100 m in diameter to be 20,500 +/-3000 (Mainzer et al. 2011b). This is a factor four or so smaller than previous estimates (Bottke et al.

2002b). This is an important reference number, since asteroids smaller than ~100 m dia. ($H < 22$) are unlikely to be profitable at currently plausible mission costs (Sect. 4.2.2).

Many NEOs smaller than 100 m certainly exist. On a log_{10}(number of NEOs) vs log_{10}(diameter) plot, NEO numbers follow a line of slope no larger than two ($N \propto D^{-\alpha}$, a power law of slope $\alpha = -2$.) Thus there will be no more than 100 times more NEOs 10 times smaller in diameter than any given size. This slope predicts ~80,000 NEOs with diameter greater than 50 meters. Smaller asteroids are hard to find directly, but at the very small end of the size scale (a few meters), the number of NEOs is currently pinned by the rate at which fireballs are detected in the Earth's atmosphere (Brown et al. 2002). In the near future lunar impact monitoring may be able to provide another data point for even smaller, ~ 1 kg, asteroids (Suggs et al. 2008). Having two low mass points would determine the slope of the curve, and so improve our estimates of the numbers at intermediate sizes of a few tens of meters.

4.4.3 NEO Emission of Light

Asteroids are seen in two ways. Their reflected sunlight dominates in the optical and near-infrared (near-IR) bands (~0.5 – ~3 microns). The sunlight that is not reflected is absorbed which heats the asteroid surface, and causes re-radiation at infrared wavelengths. This black body emission dominates from ~3 microns longward for NEOs.

The optically derived size of an asteroid depends on knowing what fraction of the incoming sunlight is reflected, which is called the albedo (or reflectance) of the NEO [Albedo means "whiteness" in Latin.]. Most NEOs are dark, reflecting only a little of the sunlight incident on them. Hence the thermal-IR measurement depends primarily on the size of the NEO, while the optical depends primarily on the albedo.

The geometric albedo, p_V, is the albedo measured at a phase angle of zero. The Bond albedo, A_B, is the fraction of input illumination energy on the object's surface that is reflected. It is what would be measured averaging over all observing phase angles (Harris and Lagerros 2002). NEO Bond albedos also span a range of about five, $A_B = 0.05$ to 0.25. Without knowing more the sizes derived from optical measurement are uncertain by a factor of $\sqrt{5}$ and volumes by a factor 11. This is an unacceptable level of uncertainty for mining purposes. Fortunately albedos within a taxonomic class have narrower distributions, about a factor 1.4 for S-class (Thomas et al. 2011a, Mainzer et al. 2012). The optical is the most technologically developed band, can be done well from Earth, and costs less than the alternative method, and so is easier to apply in bulk.

Most NEOs, being close to 1 AU from the Sun, re-radiate their absorbed solar energy at an equilibrium black body temperature similar to the Earth, about 300 K [K = degrees Kelvin. The Kelvin scale is the same as Celsius, except that it begins at absolute zero, rather than the freezing point of water. Hence 0 K = -273.15 C. K is universally used in astronomy, though C is used in meteoritics]. A 300 K black body peaks at ~10 microns, and is strong throughout the ~5 micron to ~50 micron band. This band is called the "thermal infrared"[The name arises because bodies at common Earthly temperatures emit the peak of their radiation there]. This is a temperature far cooler than most stars, galaxies or other astronomical bodies [For comparison, the surface temperature of the Sun is 6430 K (Cox 1999), which is by no means extremely hot for astrophysics. A 6000 K black body peaks at ~0.6 microns, which is right in the visible band, at a point where the Earth's atmosphere is highly transparent]. So NEOs stand out clearly in this thermal, or mid-, infrared band.

If the surface temperature of an object is known, the observed flux gives the emitting area and so diameter. Obtaining a surface temperature requires at least two wavelengths, and preferably more. Given a size and an H magnitude the albedo can be derived. Thermal inertia of the asteroid surface alters this simple calculation to a small, but significant, extent. Once the asteroid surface is out of direct sunlight it continues to radiate in the thermal-IR, as it is out of equilibrium with its surroundings. The sidewalk is still hot on a summer night, even though there is no more sunlight, for the same reason. This increases the apparent area of the NEO. Careful thermo-physical modeling can take this into account quite well (Mueller et al. 2011). The Near-Earth Asteroid Thermal Model (NEATM, Harris 1998) is often used for this measurement.

Fortunately, the wavelength at which the thermal emission of NEOs peaks lies in the clean, broad (8-13 microns) N-band window of transparency in the atmosphere. However, as this temperature is similar to that of the Earth (or people for that matter), the atmosphere, and the telescope itself, including the mirror, radiate strongly in the thermal infrared. This ubiquitous emission creates a strong background that limits sensitivity in the N-band, when observed from Earth.

Unfortunately, thermal IR surveys are much more challenging technologically. To be sensitive they need to escape the Earth's thermal background and so are best done with space-based telescopes, which are more expensive than optical surveys. But sizes need not be determined for the entire NEO population. If NEOs can be discovered in the optical bands, then a thermal-IR measurement could be postponed to the "remote characterization" phase (Sect. 4.6) when only a promising sub-set of NEOs need be measured.

4.4.4 NEO Sizes and Masses

In most cases NEO sizes are estimated based on their absolute optical magnitudes. An "absolute" magnitude refers to a common scale with all objects placed (artificially) at the same distance and under the same illumination conditions. Observed magnitudes instead depend on how far away the NEA is when it is observed, and whether it is observed fully illuminated, or at a less favorable phase angle. A fully illuminated, full-moon like, phase angle is given the value 0, and a half-moon phase of 0.5. The absolute magnitude used for all asteroids, including NEOs, is the

H magnitude [This H should not be confused with the photometric H band (1.6 microns) used by infrared astronomers (Zombeck 2007), Sect. 4.6.4.2.]

An H magnitude is the magnitude one would theoretically measure from a NEO located 1 AU from the Sun and 1 AU from the telescope, viewed at phase 0 (i.e. fully illuminated [http://www.iau.org/public/nea/]). The H magnitude is a notional reference magnitude only, as Solar System geometry means we never see an NEO under these conditions. Doing so would require a telescope at the solar surface. Fortunately, it is straightforward to convert from magnitudes observed with other geometries to the H magnitude.

H is defined in the Johnson/Cousins V band magnitude (Bessel 2005). The V band is centered on 0.55 microns and spans a 0.09 micron wide band. As the human eye is sensitive over the wavelength range 0.5-0.6 nm (Robinson and Schmidt 1984) the V band is a good approximation to what we would see. As different asteroids, stars and galaxies, have different shapes for their spectra, using bands centered at other wavelengths would produce different magnitudes.

Astronomical magnitudes are logarithmic measures of brightness. This is a natural scale as the response of the human eye is roughly logarithmic. Magnitudes are scaled (Pogson 1857) to match the original Greek/Hellenistic magnitudes for the naked eye stars, as listed in Ptolemy's Almagest (~150 CE) and used continuously thereafter [Astronomy is full of similarly archaic scales and units. Most are retained because they are convenient, and the vast range of scales in astronomy makes no individual unit universally handy. Still, this mongrel usage can be frustrating]. Hence:

$$V = -2.5 \log[f(\text{star})/f(0)].$$

Larger magnitudes correspond to fainter fluxes. A 6th magnitude star (at the limit of naked eye detectability) is 100 times fainter than a 1st magnitude star (one of the brightest few in the sky.) A factor 10 decrease in flux corresponds to 2.5 mag. *increase* in H. f(0) is a zero point flux, defined to be 3.67×10^{-23} W m^{-2} Hz^{-1} for the V band (Zombeck 2007).

At phase angle 0.5 a spherical body is ~10 times fainter than at phase angle 0, not half as bright, as one might naively expect (e.g. Clark et al. 1998). This steep "phase curve" is due to geometry – the same observed area on the NEA is illuminated by less sunlight, due to the acute angle of the surface to the Sun – to shadowing, and to the high degree of coherent backscattering of optical light (Hapke 1990).

The diameter, D (km) of an asteroid of magnitude H and geometric albedo p_V, is given by (Harris and Harris 1997):

$$\log D = 3.1236 - 0.5 \log p_V - 0.2\,H$$

or, equivalently:

$$D = 1329\, p_V^{-0.5}\, 10^{-H/5}$$

Observing a NEO at 1 AU with V = 24.5 at phase = 0.5 implies H = 22. H = 22 corresponds to a diameter of 110 m to 240 m, for albedos of 0.25 and 0.05 respectively [from http://neo.jpl.nasa.gov/glossary/h.html].

NEO sizes can be calculated with much greater accuracy using thermal infrared data, rather than optical data because of their low albedo. Changing A_B from 0.05 to 0.25 only changes the NEO black body emission from 95% to 75%, which changes the size estimate by only about 20%. This gives us a more acceptable volume range of a factor 1.7. If we want to measure accurate NEO sizes remotely, the thermal infrared is the best band to use.

Knowing a size for a NEO only roughly defines its mass. Asteroids have a range of density and often contain large fraction of space that are empty (voids). The fraction of the volume that is a void is called the porosity. The largest (> 300 km diameter) asteroids have lower porosity (≤ 0.1). Smaller asteroid span a wide range of porosity, from 0.1 up to 0.5, or possibly 0.7 (Baer and Matson 2011). This makes masses uncertain by a factor 2 – 3 even when their size is well measured. High porosity asteroids are likely the result of almost destructive collisions. The porosity of 100 m diameter asteroids has been little investigated, but seems likely to include many high values.

The mass of the NEO depends finally on the density of the material it is composed of. Asteroid densities are derived from meteorite samples. Pure Ni-Fe has a density of 7.3 – 7.7 g cm^{-3}, the common silicates in meteorites, pyroxene and olivine have 3.2 – 4.37 g $cm-^{3}$, and clays have 2.2 - 2.6 g cm^{-3} (Britt et al. 2002). This is a range of 3.5 in density. Spectroscopy, and possibly optical photometry, can put the asteroidal material into one of these classes with reasonable confidence (Sect.s 4.6.2, 4.6.1.1). However, in addition, meteorites show micro-porosity, which reduces their densities from these compacted densities (Britt et al. 2002). Without finding a binary (Sect. 4.6.1.2) or measuring the Yarkovsky effect (Sect. 4.5.4.3), both of which are applicable only to a minority of NEOs, local characterization at the NEO is essential to determining a mass to commercial accuracy (Sect. 4.7).

4.4.5 NEO Orbits

All closed orbits are elliptical, and are characterized by six parameters (Table 4.2). Three describe the orbit shape (*a, e, i*) and three describe the phase of the orbit ($\wedge$, ω) and of the object (*M*)[http://www.lns.cornell.edu/ ~seb/celestia/orbital-parameters.html].

Good orbit determinations are essential to claiming an NEO as detected. Unless the orbit determination is at least good enough to predict the location of the NEO accurately enough to find it again the next time it becomes visible (its next "apparition") it will be lost as soon as it is found. In practice, orbits good enough to predict future apparitions accurately for decades ahead (in most cases) are achievable with modest means, one to two meter diameter terrestrial telescopes (Sect. 4.4.1.2), which are fairly common and inexpensive to use.

Table 4.2 The six parameters of an elliptical orbit

Orbit Parameter	Symbol	Meaning
Semi-major axis	a	Longer of the two ellipse semi-axes.
Ellipticity	e	Degree of departure of the orbit from a circle. $e = \sqrt{(1-(b^2/a^2))}$, where b is the semi-minor axis of the ellipse. For an ellipse, $0 < e < 1$
Inclination	i	Inclination (i), in our case relative to the Earth's orbital plane, the ecliptic
Longitude of the ascending node	Ω	Angle at which the NEO crosses the Earth's orbit plane going North, measured from the first point of Ares, where the Ecliptic and the Celestial Equator cross as the Sun moves North.
Argument of pericenter	ω	Angle from the ascending node, Ω, to the pericenter, the point of closest approach to the Sun, measured in the plane of the NEO orbit.
Mean anomaly	M	Angular location of the object in its orbit, which is time dependent.

Orbit parameters are calculated by the Minor Planet Center (MPC) of the International Astronomical Union (IAU) [MPC is located at the Harvard-Smithsonian Center for Astrophysics, 60 Garden St., Cambridge, Massachusetts O2138 USA. (The author's home institution.) http://www.minorplanetcenter.org]. The MPC assigns a logarithmic uncertainty code, U, which ranges from zero to nine, for the uncertainty in position of an asteroid after 10 years [http://www.minorplanetcenter.net/iau/info/UValue.html]. For example, an uncertainty value of six (U=6) corresponds to a 2.08 degree uncertainty after 10 years, which is hard to pick up again. Imaging by follow-up telescopes with few arcminute diameter fields of view needs U≤5, for <2 arcminutes after 1 year. Spectroscopy needs U≤2, to achieve <2 arcseconds after a year. In practice 75% of NEOs have U<2 (Fig. 4.4).

The U values are strong confidence limits, corresponding to 3σ uncertainties in position, so that 99.7% of the time the asteroid will lie within that radius of the predicted position (Gareth Williams 2012, private communication). NEO sky locations vs. time ("ephemerides") are provided by the MPC and by NASA-JPL at their "HORIZONS" site, object by object [http://ssd.jpl.nasa.gov/horizons.cgi#top]. Improved algorithms for deriving orbits, taking into account different uncertainties from different sources and correlated errors within each data source may improve the situation further (Baer and Matson 2011).

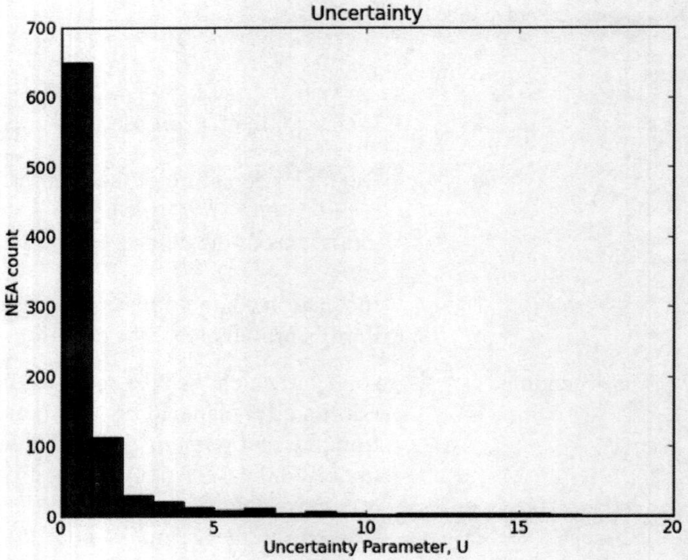

Fig. 4.4 Orbit uncertainty codes, U, for NEOs

4.5 NEO Discovery

Discovering NEOs requires sensitive surveys with telescopes. Ground-based surveys are in the optical band. Space-based surveys use the thermal infrared band. To efficiently use the available observing time, the telescopes must have as wide a field of view as possible. In this way the maximum number of NEOs will be captured in each exposure.

Individual exposures cannot be longer than a few minutes, as NEOs move quite rapidly across the field of background stars, which would create an elongated image in a long exposure. Although this shape identifies a moving object cleanly, the sensitivity of the survey is reduced, as the signal is then spread over many pixels.

No image is perfect and minor blemishes due, for example, to cosmic rays coming through the camera, can produce an apparent detection on one frame that is gone on the next. To guard against such false detections, NEO surveys typically require three to five detections in a row of an object moving against the star field.

This section first describes ground-based current and future surveys, and then space-based current and possible future surveys.

4.5.1 *NEO Optical Ground-Based Surveys*

Ground-based surveys are only practical in the optical band. This is due to the affordability of very large optical detector arrays and the high atmospheric transparency and low background in the optical band.

Ground-based surveys have inherent limitations. Each can view only the hemisphere the telescope is in (North or South). As a result a pair of telescopes is needed to cover the whole night sky at any one time. A survey lasting a decade or more would be able to catch most asteroids in more than one apparition, as 98% of the known NEOs swap hemispheres during their apparitions over a 10-year interval, because of their orbital inclination, and the Earth's tilt out of the ecliptic.

4.5.1.1 The Synodic Period Problem

The most accessible NEOs are those in the most Earth-like orbits in terms of *a, e,* and *i* (Table 4.2, e.g. Elvis et al. 2011). As a result these NEOs tend to have orbital periods close to that of the Earth, and so either catch-up with, or fall-behind, the Earth slowly. This can make the time between their apparitions (their "synodic period") quite long [A synod is simply a meeting.]. In some circumstances it can also lead to several apparitions in rapid succession. Synodic periods of decades are possible, so that a 10-year long survey would miss these slow movers (Vereš et al. 2009). These NEOs are effectively "hiding behind the Sun" much of the time, where we cannot see them from Earth.

4.5.1.2 Telescopes

Telescopes [Professional astronomers never say "scopes". It's a shibboleth] are described mainly by the diameter of their primary mirror [The primary mirror diameter is also called the telescope aperture], as this determines the amount of light each can gather. The cost of a telescope rises quickly with the primary mirror diameter. The ground-based telescopes of interest range from small, about one meter, which are common and cheap enough to be within reach of dedicated amateurs, through moderate aperture, two to four meter, telescopes, which are significant investments (a large university might have one), on up to large, six to ten meter, telescopes, each of which is a major international project. A new generation of 20-40 m telescopes is just entering construction and should see first light in the early 2020s [The three projects are the ELT (Extremely Large Telescope): www.eso.org/public/teles-instr/e-elt.html; GMT (Giant Magellan Telescope): www.gmto.org/, and TMT (Thirty Meter Telescope) www.tmt.org]. A different class of wide field of view (a few to 40 square degrees) telescopes are used for ground-based surveys of large areas of the sky. These have primary mirror diameters in the range 0.5 - 1.2 meters. They are usually Schmidt designs (Schroeder 1999).

Ground-based surveys can only work at night, of course. This ensures that any object discovered is passing by near to or beyond 1 AU [To be precise, beyond the

Earth's current distance from the Sun. As the Earth's orbit is slightly elliptical (e=0.016) this varies from 147x10^6 km to 152x10^6 km (Cox 1999).] Objects near to 1 AU, unless very nearby, will appear close to the horizon at local midnight, or nearly overhead near dawn or dusk. Observing close to the horizon involves looking through a large column of atmosphere ("air mass"), which decreases the observed flux from this object through scattering and absorption ("extinction"), and produces a poorer image quality due to atmospheric turbulence ("seeing"), which further reduces sensitivity. In practice, no surveys look more than ~60 ° from directly overhead, the zenith. At this angle the air mass ~ sec(zenith angle) = 2. Astronomical twilight is the time after sunset or before sunrise when the sky is too bright for sensitive surveys to be performed, and is defined as being when the Sun is 18 deg below the horizon. This constraint, plus the air mass constraint, means that the closest a ground-based survey can come to the Sun's direction is 48 °. The solar elongation angle is the angle of the object from the Sun as viewed from the Earth. In practice, most NEOs are found at Solar elongations >60 ° (fig. 3 of Larson 2007).

Ground-based surveys are also interrupted by weather, and are less sensitive when the Moon is near full. Nearly one week per month is lost to the full Moon, leaving 75% of 365 nights. Bad weather, even at the best sites (Mauna Kea, Hawaii and Northern Chile) takes another 20% of the remaining nights. Other losses, due to instrument and telescope maintenance or failures, typically take another 5%. At best, then, a ground-based telescope operates effectively on about 210 nights a year. Asteroids crossing the Milky Way are also hard to find over a roughly ±10 ° swath of sky, imposing a seasonal variation on asteroid discovery rates.

4.5.1.3 Major NEO Surveys

Table 4.3 lists the main NEO surveys. Note the bias toward the Northern hemisphere. All of them are optical surveys. These surveys have been very effective in virtually completing the census of NEOs larger than one kilometer diameter (H < 17.5) (Harris 2008) [http://neo.jpl.nasa.gov/stats/]. These surveys were instituted by NASA in response to the George E. Brown Congressional mandate. This mandate was motivated by the extreme hazard posed by such a large NEO if one were to impact the Earth. Together they constitute the Spaceguard program.

They all use broad photometric filters to define their band. These filters typically have widths, $\delta\lambda$, about 10% of their wavelength, λ, and so have $\delta\lambda/\lambda \sim 10$. For example, the much-used V-band filter is centered at 0.55 microns and has a width of 0.09 microns. Two main sets of filters are used. The older Johnson/Cousins magnitude system has the U, B, V, R, I filters (centered at 0.36, 0.45, 0.55, 0.66 and 0.81 microns, respectively). A newer system (Fukugita et al. 1996) was defined for the Sloan Digital Sky Survey (SDSS): *u', g', r', I, 'z'* (centered at 0.36, 0.48, 0.62, 0.76, and 0.91 microns respectively). The SDSS filters were carefully chosen to reduce sky background. Larson (2006) provides a good summary of the older surveys. The three largest contributors to NEA discoveries are LINEAR, CSS and SSS. These three surveys each re-purposed an under-utilized telescope, in order to fit within their available budgets.

Table 4.3 Ground-based NEO Surveys

Survey	Location (Latitude)	Telescope	Primary diameter (meters)	Field of view, Area/night (sq.deg)	Limiting magnitude at 5σ, exposure	Operator
Lincoln Near-Earth Asteroid Research LINEAR [http://www.ll.mit.edu/mission/space/linear/]	White Sands, New Mexico, USA (+33°)	GEODSS[a]	1.0	2	V=19.5	Lincoln Labs
Catalina Sky Survey CSS [http://www.lpl.arizona.edu/css/css_facilities.html	Mt. Bigelow Arizona, USA (+32°)	Catalina Schmidt	0.7	8.2 [19.4[b]] ~800	V=20.5 (1 min)	University of Arizona[c]
Siding Spring Survey SSS	Coonabarabran, New South Wales, Australia (-31°)	Uppsala Schmidt	0.5	4.2	V=19-20 (1 min)	Siding Spring Observatory
Pan-STARRS 1[e,] PS-1 [http://pan-starrs.ifa.hawaii.edu/public/]	Haleakala, Maui, Hawai'i, USA (+21°)	PS-1	1.8	7.0	V=24	University of Hawai'i[d]
Palomar Transient Factory[e,] PTF [http://www.astro.caltech.edu/ptf/]	Mt. Palomar, California, USA (+33°)	Oschin Schmidt	1.2	7.8 [40[b]]	r=20.5 (1 min)	Caltech[d]
UNDER CONSTRUCTION						
Asteroid Terrestrial-impact Last Alert System ATLAS [http://www.fallingstar.com/technical.php]	Hawai'i or Arizona	2 sites at 100km separation	0.25 x 4	40 20,000	V~19.1	University of Hawai'i[d]
Large Synoptic Survey Telescope[e,] LSST [http://www.lsst.org/lsst/science/development]	Cerro Pachon, Chile (-30°)	LSST	8.4	9.6	V = 23 (15 sec)	LSST Consortium
Pan-STARRS-2[e]	Haleakala, Maui, Hawai'i, USA (+21°)	PS-2	1.8	7.0	V=24	University of Hawai'i[d]

a. Ground-based Electro-Optical Deep Space Surveillance. b. when upgraded. c. by Steward Observatory with the Lunar and Planetary Lab. (LPL) (Larson et al. 2001). d. for a consortium. e. asteroid detection shares this facility with other projects.

LINEAR dominated NEO discoveries from 1998 until about 2004, after which CSS has dominated. Construction began on upgrades to the CSS in 2012, funded by NASA, and should be completed in late 2013. The field of view of the Catalina Schmidt will be more than doubled, to 19.4 square degrees. The field of view of the CSS 1.5 m is also being increased to 5 square degrees, with a limiting magnitude of V = 21.5 (Christensen et al. 2012).

A new interest among astronomers in transient events, such as supernovae, has led to two new wide-field survey facilities Pan-STARRS (PS-1, Kaiser et al. 2002) and the Palomar Transient Factory (PTF). Both of these are, in principle, excellent for NEO discovery, but neither was designed specifically for that task. The partially funded plan for the "Palomar Transient Factory 2" (PTF-2) upgrade should increase the field of view of the Oschin Schmidt to the full 40 sq. deg (Sect. 4.4.1.4). Construction began in 2012.

Two new, and very different, surveys are planned that will also find NEOs:

- The Asteroid Terrestrial-impact Last Alert System (ATLAS) will consist of a global array of small telescopes that can give a one month warning of an impact from a 300 m diameter NEO, or one week for 50 m. NEOs (Tonry 2009, 2011; Jedicke et al. 2012). ATLAS is designed to scan the entire available sky every night. NASA funding was awarded in 2012, and the detailed design is still evolving.
- The Large Synoptic Survey Telescope (LSST) is also designed to survey the entire available (Southern) sky every night, but much more deeply than ATLAS and in the five SDSS bands, plus the newer Y-band (at 1.04 microns with width 0.15 microns, Hillenbrand et al. 2002). The primary science goal is to find transient objects, with NEOs being one sub-class of such objects. The baseline plan is to make two full images of the accessible sky every night. These images pairs will be 15s long to avoid blurring (Ivezič et al. 2011). LSST should see first light in 2018.

4.5.2 NEO Space-Based Surveys

Space-based survey telescopes overcome many of the limitations of ground-based surveys. A telescope in space can be designed to work at whatever wavelength is optimal for the task at hand. Space-based astronomical telescopes already exploit this freedom by working in the X-ray, [Chandra X-ray Observatory: chandra.harvard.edu/] ultraviolet [Hubble Space Telescope: www.nasa.gov/hubble/], and far-infrared bands (which are 100% absorbed by the atmosphere) [Spitzer Space Telescope: www.spitzer.caltech.edu/; Herschel: sci.esa.int/herschel/], as well as in the near- to mid-infrared bands where the atmosphere, though quite transparent, produces an intense background. Having access to the thermal infrared without the interfering effects of strong emission at the same wavelengths from the telescope itself, provides a huge advantage.

The zodiacal light (e.g. de Pater and Lisauer 2001, chapter 9) is a background that peaks in the thermal IR. Zodi (as it is commonly called) is thermal emission from small interplanetary dust grains at temperatures similar to NEOs, and so is still present in space. These grains are concentrated in the ecliptic plane. This means that low *i* NEOs, which tend to have low delta-v, are the most affected. The zodiacal light is about as bright as the Milky Way. The Milky Way is the other background in space, and is concentrated close to its plane (about ±10 deg., more toward the Galactic Center, totaling ~20% of the sky).

The improved viewing geometry from space is highly beneficial. Without the atmospheric scattering of light, a telescope can observe much closer to the Moon and the Sun. Space-based telescopes can easily look right along the Earth-Sun tangent line (solar elongation = 90°). A solar elongation fixed to near 90° (typically ±20°) allows a spacecraft to have simple fixed solar panels, which lowers cost. If the telescope is equipped with good baffling, solar elongations as small as 40° can be reached.

The two primary limitations on sensitivity in space are usually solar light, scattered by the spacecraft itself, reaching the instruments at the telescope focal plane, and thermal constraints as the spacecraft must operate within a modest range of temperatures. Going out of range on temperature could distort the optics, compromise the electronics, or raise the detector background, which would reduce the signal to noise ratio of the NEO detections.

In most low Earth orbits (LEOs) the Earth blocks most sky locations from view for half the time on each orbit. But there are some almost polar orbits (inclination ~99°) which stay above the Earth's day/night terminator. Using this terminator orbit a telescope that scans around the circle perpendicular to the Sun once per orbit achieves continuous unocculted observations.

4.5.2.1 WISE: The Wide-Field Infrared Survey Explorer

The NASA WISE mission (Wright et al. 2010), operated out of UCLA and Caltech/JPL, used a terminator orbit to perform a one year all sky survey in four thermal infrared bands: 3.4, 4.6, 8 and 22 microns. It was a NASA Explorer class mission, the smallest NASA mission class (e.g. Elvis et al. 2009), which utilized a 40cm telescope. Initially the instrument was cryogenically cooled with solid hydrogen to <12 K. The field of view of WISE is a relatively modest 47 x 47 arcminute (0.6 sq.deg.). WISE has now run out of cryogen and has warmed up to 180K. It is currently in hibernation, lacking the funding to continue operations. Given new funding WISE could continue to survey in the two shortest wavelength bands, 3.4 and 4.6 microns, which remain sensitive.

The NEOWISE project reprocessed the WISE data to find about 500 NEOs (Mainzner et al. 2011b, 2012). Finding moving objects required new, specially adapted software, as the WISE data reduction pipelines were designed to find distant objects that are effectively stationary on the sky. Detector artifacts, in particular those produced by cosmic rays passing through the detector, make moving object detection a non-trivial exercise. The smallest NEO found by NEOWISE was about 100 m diameter. A comparison of NEOWISE sizes with direct radar size measurements shows good agreement (Mainzer et al. 2011a).

4.5.2.2 Potential Space Surveys in the Thermal-IR

While Earth orbiting telescopes are far more effective than those on the ground, if restricted to Earth orbit, they will still have difficulty finding NEOs on the sunward side of the Earth, primarily Atens and Atiras. Two telescopes are in development that address this problem by using solar orbits.

SENTINEL [http://b612foundation.org. The name is not an acronym.] takes a radical approach. SENTINEL mission will place a thermal infrared (5 - 10 micron) telescope in a Venus-like (0.7 AU) orbit. It will point its 0.5 m telescope outward, scanning ~1/3 of the 1 AU orbit at any one time. Just as Venus orbits the Sun in seven months, SENTINEL will too, sweeping over the entire Earth orbit space around 1 AU. This guarantees that any long synodic period NEO will be found quickly. In addition, all the NEOs will be seen near zero phase angle, making them somewhat brighter. As a result, SENTINEL should detect at least 90% of 100 m diameter NEOs, as well as a significant fraction of 50 m diameter NEOs, during its 5.5 year mission. Communications will be a significant challenge. The large distance of SENTINEL from Earth (0.3 to 1.7 AU) limits the telemetry rates. Even though NASA will be providing the Deep Space Network to receive data, full frame images cannot be returned for analysis on the ground. Hence intensive on-board data processing will be required to detect moving objects, including NEOs. Only the small, postage stamp, images around NEO candidates can be beamed back. The SENTINEL mission is being pursued by the B612 Foundation, using philanthropic funding. With a possible launch in 2015, this would be the first privately funded deep space mission.

NEOCam [http://neocam.ipac.caltech.edu/] takes another approach to escaping the Earth-orbiting problem. It intends to put a 0.5m telescope at the Sun-Earth L1 Lagrange point, roughly 10^6 km (~0.01 AU) Sunward of the Earth [http://www.esa.int/esaSC/SEMM17XJD1E_index_0.html]. This Lagrange point requires less delta-v to reach, which should make it less expensive than SENTINEL. In addition, the telescope can be cooled to about 30 K without any cryogen to limit its lifetime (Mainzer 2006). (This is called passive cooling.) The NEOCam mission can reach solar elongations as small as ~40°, opening up most of the 1 AU orbit to surveying, although NEOs at small elongations will be more distant than for SENTINEL, and will be viewed at less favorable phases. Another advantage of the L1 point is that it is close enough to Earth to allow high data rates to be transmitted back, so that ground data processing can be used to select NEOs. This will ensure higher reliability in the reported NEOs. NEOCam is designed to survey in two bands, 5 and 10 microns, allowing a temperature to be determined, rather than assumed, which will lead to more accurate sizes. In a four year mission NEOCam could find two thirds of all NEOs larger than 140 m. NEOCam has been proposed to NASA's Discovery program twice, and has been funded for technology development in 2010 (McMurty et al. 2013).

4.6 Remote Telescopic Characterization

The available astronomy techniques for the characterization of asteroids form a pyramid, or "wedding-cake", ranging from low-cost methods that can be applied broadly, to thousands of asteroids, but return only modest information about the asteroid, to higher and higher cost methods that can thus be applied to smaller and smaller numbers of NEOs.

Any more detailed characterization, beyond simple detection and orbit determination, demands the collection of more photons. This has consequences. More observing time, or a larger primary mirror, or both, are required. This means that far fewer asteroids will be characterized than detected. To date, most characterization has been carried out with terrestrial equipment. Space-based facilities would open up new possibilities.

The resolving power of a spectrum is $R = \lambda/\Delta\lambda$. Here $\Delta\lambda$ is the width of the smallest resolved spectral bin and λ is the central wavelength of that bin. A spectrum that is adequate for asteroid characterization has a resolving power $R \sim 100$. By optical astronomy standards $R \sim 100$ is a very low resolution [Values of a few thousand are commonly achieved, and $R = 100{,}000$ is quite possible. Requiring only low resolving power has the advantage of requiring fewer photons than higher resolving power]. The photometric bands have an even lower $R{\sim}10$. To get the same signal-to-noise in 10 bins each 1% wide needs 10 times as many photons as a simple detection over the whole 10% wide band. That would appear to move characterization from a ~1 meter telescope to a ~3 meter telescope. However, the near-IR has a higher background, which degrades the signal-to-noise there. Fortunately, this can be compensated for as, once an asteroid orbit is known accurately, a telescope can track its path during an observation. This allows integration times to be increased from a few minutes to an hour or longer. With these long integrations, a 3-meter class telescope can indeed characterize many NEOs.

In the following sections I describe each layer of the characterization wedding cake, from various forms of photometry, the easiest technique, up to spectroscopic techniques. The section concludes with a brief look to future space-based opportunities.

4.6.1 Photometry

4.6.1.1 Colors

In astronomy a "color" is the difference, in magnitudes, between two photometric bands. Colors give a coarse view of the shape of the spectrum. The diagnostic value of optical broad band photometry (Sect. 4.5.1.3) for asteroids was demonstrated with the "Eight-Color Asteroid Survey" (ECAS, Zellner et al. 1985). Using this data, Tholen (1989) was able to establish a taxonomy of asteroid types. More recently, Ivezić et al. (2001) showed that using just four of the SDSS bands (*griz*) asteroids can be classified into a few sub-types, if the photometry errors are a few percent, as in SDSS (3%, Carvano et al. 2010). This allowed a first rough classification of NEO types. For example, S-type asteroids can be distinguished from C- or D -type, with reasonable reliability. But SDSS photometry is not good at finer grained typing, such as separating the E-, M- and P-type objects.

4.6.1.2 Optical Variability

NEO rotation periods and aspect ratios can be determined using time series data of their brightness. These are called "light curves" by astronomers. A period is found from the time taken for the shape of the light curve to repeat. For a symmetrical body the period is double the repeat time. The aspect ratio of the asteroid is given roughly from the amplitude of the light curve.

NEO rotation periods range from a few minutes, or even shorter, to about 10 hours or longer. Only smaller NEOs have spin periods shorter than about an hour. Faster rotations rates would throw off loose regolith from their surfaces centrifugally (Pravec and Harris 2000). This division of rotation rates is good evidence that smaller NEOs are more likely to be monolithic, while most larger NEOs are rubble piles. Small NEOs ($H < 20$) rarely have spin periods of less than a few hours. This may be because they are easily spun up by the YORP (Yarkovsky - O'Keefe - Radzievskii -Paddack) effect (Rubincam 2000). The possibility of there being against seeing faint slow rotators should be considered.

If the asteroid is found to be a binary, by seeing eclipses in the light curve, then the masses of the two components of the binary can be determined using Kepler's laws. Only about 15% of NEOs are seen to be binaries at present. However, our sensitivity to small secondaries is not good. The high precision photometry demonstrated by the Kepler mission (Sect. 4.6.4), may be able to find a higher fraction of binaries. This would allow many more NEO masses to be determined cheaply.

If the light curves sample at least 10 phases per period, and span a number of rotation periods with good accuracy (~1-3%) then the NEO gross morphology can be determined using tomographic techniques (Hanuš and Ďurke 2011). Asteroids, especially the smaller ones, do not always spin around one of their principal axes. These non-principal axis (NPA) rotators, or tumblers, present different parts of their surfaces to us on each rotation. Tumblers thus allow better tomographic reconstruction of their three dimensional shapes. Newer data analysis techniques, such as genetic algorithms, are increasingly able to extract structural information. The resulting structures can be then be checked against radar, adaptive optics and interferometric images, or *in situ* spacecraft images, in a few cases, to validate the techniques (Bartczak and Marciniak 2012).

4.6.1.3 Facilities

A dedicated 1m telescope could classify a few hundred NEOs per year into S or C types using photometry in the SDSS bands. At NEO magnitudes, exposure times for photometry are quite short. A minute on a 1 m telescope [http://www.noao.edu/gateway/ccdtime/] should reach V = 21 at 10σ, which is a clear detection, 3% in ~5 minutes, but attaining 1% accuracy requires about an hour.

Several observatories have telescopes at least partially dedicated to asteroid photometry (Table 4.4). Most of these observe in a single broad optical band to measure light curves, and do not obtain multi-band data to classify NEO types.

Table 4.4 Ground-based Characterization Facilities

Observatory	Location	Telescope dia(meters)	Capability
NASA Infrared Telescope Facility (IRTF)	Mauna Kea, Hawai'i	3.0	Optical-near-IR spectroscopy (0.8-2.5 microns)
Palmer Divide Observatory (PDO) [http://www.minorplanetobserver.com/PDO/PDOHome.htm]	Colorado Springs, Colorado	3 x 0.35 1 x 0.5	Photometry (1 band for light curves)
Palomar	Mt. Palomar, California	1 x 1.5 1 x 5.0	Photometry (colors) Spectra (optical-NIR)
Las Cumbres Observatories Global Telescope Network, (LCOGT) [http://www.lcogt.net]	various	2 x 2.0 1 x 1.0 N x 0.4	Photometry (colors or light curves)
Catalina Sky Survey (CSS)	Mt. Lemmon, Arizona	N x 1.0	Astrometry for orbits
Tenagra [http://www.tenagraobservatories.com]	Arizona	1 x 0.81	Photometry (colors or light curves)

The Palmer Divide Observatory (PDO), in California, is dedicated to asteroid characterization. Though formally amateur, PDO attains professional standards and is currently the leading source of asteroid light curves. PDO has discovered more than a dozen binary asteroids, and is funded by NASA, the NSF, and the Planetary Society.

The Palomar Transient Factory (PTF) integrates the larger Mt. Palomar 1.5m P60 (60-inch) and 5m P200 (200-inch) telescopes into the system. This enables rapid follow-up. P60 takes images to obtain multi-color photometry, while the much larger P200 is equipped with "Triplespec" a 1-2.4 micron NIR spectrometer.

The Las Cumbres Observatories Global Telescope Network (LCOGT) is based in Goleta, California. It is building a telescope network around the globe to allow continuous coverage of anything in the night sky. Observing time is open to institutions that have bought into the collaboration.

The Catalina Sky Survey (CSS) is upgrading its 1 m follow-up telescopes (Christensen et al. 2012). Their primary role is to determine more accurate orbits for CSS discoveries. It would be possible to use this facility for measuring light curves.

Tenagra Observatory is a privately owned facility. They sell observing time at ~US$200/hour. The Tenagra II telescope is a 0.81 m with a CCD image fitted with Johnson/Cousins UBVRI filters (Bessel 2005). Specialized filters supplied by the customer can be used. Both light curves and spectra could be measured with this facility.

4.6.2 Optical-NIR Spectra

Asteroid classification into Bus-DeMeo classes (De Meo et al. 2009) is routinely achieved using R~100 spectra spanning the near-IR 0.5 - 2.2 micron bands. The diagnostic absorption features are weak (1% - 5% of the flux at their wavelengths). High signal-to-noise spectra (S/N>10 per pixel) are thus needed to measure these features accurately.

Ground-based optical-near-IR spectroscopy has two limitations. Most NEOs have had just one spectrum taken, so it is unknown whether their surfaces have variegated mineral content.

More worryingly, different spectral shapes have been found on repeated observations of the same asteroid, which put it into somewhat different classes depending on the observation (Moskovitz et al. 2012). The strong, highly variable, NIR background on Earth, and the lack of well-characterized calibration standard stars, are contributing factors to this classification uncertainty. Single spectrum classifications are adequate for many scientific purposes, where large samples average out small errors. But a mining expedition demands more certainty on individual objects. Surface composition uncertainty means that a given expedition will need more potential targets, which may be costly.

4.6.2.1 Facilities

The sensitivity of NIR spectrographs limits the rate at which NEOs are characterized spectroscopically. For example, to obtain 2000 NEO spectra/year would require gathering an average of 10 spectra on every available night. In an average 10 hour night this allows exposure times of 45 minutes, plus 15 minutes per spectrum to acquire the necessary calibration data (primarily spectra of "standard stars" with well-measured spectra). A well-run program on a dedicated telescope can sustain this pace.

The largest ongoing program of optical-near-infrared spectroscopy is the MIT-UH-IRTF Joint Campaign for NEO Spectral Reconnaissance [http://smass.mit.edu/minus.html]. This survey uses the NASA 3.3 m Infrared Telescope Facility and the SpeX spectrograph. In the year May 2011 to May 2012, this program obtained spectra of 84 NEOs in 19 nights. While this rate is much beyond that achieved a decade ago, it is only 10% of the rate at which new NEOs are being discovered. This program uses the SpeX spectrograph on the 3 m NASA Infrared

Telescope Facility (IRTF) on Mauna Kea, Hawaii [http://irtfweb.ifa.hawaii.edu/~spex/]. SpeX can reach targets as faint as V~17.5 (Binzel et al. 2006). Most SpeX spectra are of larger NEOs (H < 15). Exposures of 30 minutes to one hour are normal, tracking the NEO as it moves against the background field of stars.

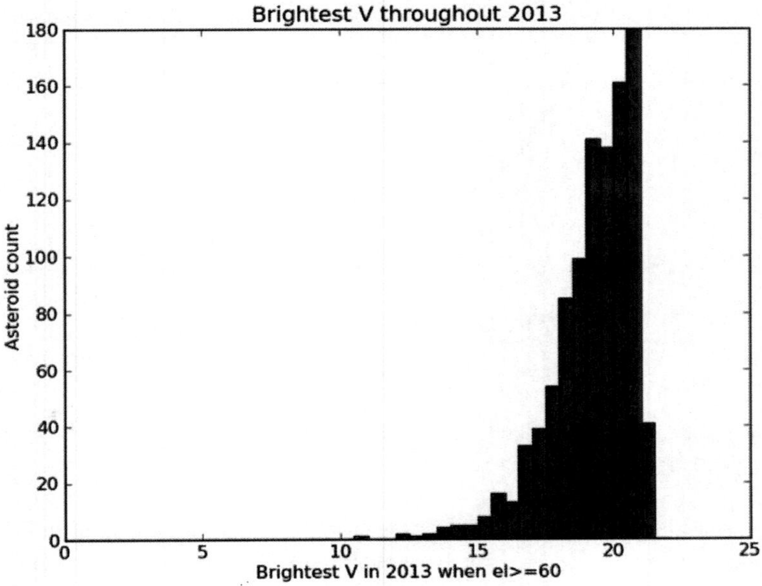

Fig. 4.5 Brightest V magnitude for all NEOs visible in 2013

However, most of the NEO population lies beyond the reach of SpeX. Most NEOs are discovered at V ~ 18 - 21.5 (Fig. 4.5). Of these ~30% get brighter than the V = 17.5 limit of SpeX (Beeson et al. 2013).

Newer NIR spectrographs on larger telescopes can obtain spectra of fainter NEOs. For example, the FIRE instrument [The Folded-port InfraRed Echellette.] on the 6.5 m Magellan telescopes (Simcoe et al. 2010), can reach V = 20 on a NEO (Moskovitz et al. 2012) in ~1 hour exposures. At this limit, 96% of newly discovered NEOs are accessible. However large telescopes are heavily oversubscribed for their available observing time and are operated by consortia, so that little time is given to asteroid observations.

Another advantage of going fainter is that a telescope can then have a defined schedule of observations well in advance, rather than having to react quickly, within a few days. This is because half of NEOs stay brighter than V = 20 for over three months, compared with just about 15% staying brighter than V = 17.5 for the same period (Fig. 4.6)[These numbers are based on MPC statistics from 2011 November 1 to 2012 November 1]. This simplifies the scheduling of spectroscopy after discovery. Observations can then also be scheduled away from full Moon, for increased sensitivity.

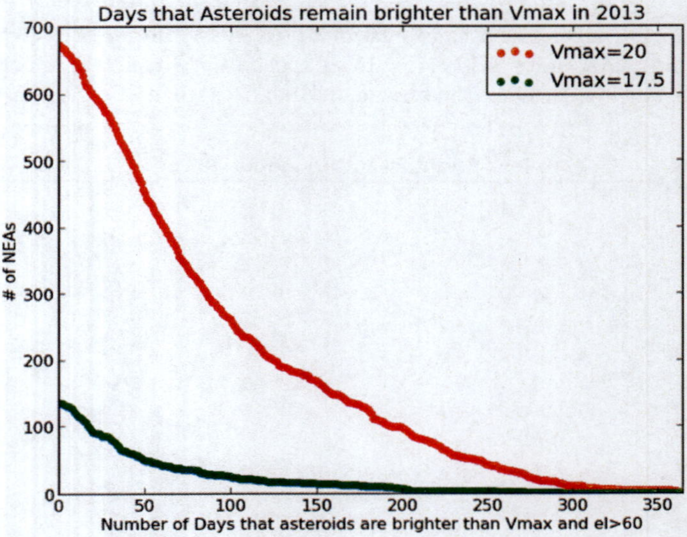

Fig. 4.6 Number of NEOs visible for longer than N days for V<17.5 (SpeX/IRTF) and V<20 (FIRE/Magellan)

Another route to higher sensitivity is new NIR technology. A great increase in sensitivity is, in principle, possible by reducing the NIR sky background (Ellis and Bland-Hawthorn 2008). This background is due to primarily hundreds to thousands of emission lines produced by hydroxyl atoms (OH) high in the atmosphere. These lines are not only extremely bright, they are also variable on both hours and minutes timescales, which complicates the use of even nearby calibration stars. If these lines could be removed, then the background would be reduced by 20-30 dB, equivalent to 5 - 7.5 magnitudes. A one meter telescope could then be as sensitive as a 10 meter telescope is now! Because the OH lines are extremely narrow in wavelength (<0.1nm) it should be possible to selectively remove these wafer thin slices of the spectrum without reducing the signal from the asteroid. Recent work has made progress. FIRE has a resolving power of $R \sim 6000$, so that several hundred OH lines can be excised. Sullivan and Simcoe (2012) report a substantial gain of ~20 dB, but then other background components, not all understood, become important. OH suppression is an active area of development, mainly because of its importance to cosmological studies. Useful improvements can be expected in a few years.

4.6.3 Thermal IR Ground-Based Photometry and Spectroscopy

From the ground there are three thermal IR windows of transparency in the atmosphere: M-band (5 microns), N-band (8-13microns), and Q-band (20 microns).

By far the widest and most transparent is the N-band, which is thus the most sensitive. N-band imaging is sufficient to give a decent asteroid size measurement. To measure a temperature requires just R~10 across the 5 μm wide N-band. Details that reveal the NEO surface composition and dust size distribution emerge at R~30 (Vernazza et al. 2010). Unfortunately very few NEOs come close enough to become sufficiently bright to observe in the themal IR from the ground, so this technique has limited utility for large surveys.

4.6.4 Other Characterization Techniques

4.6.4.1 Radar

Radar can give precise orbits, detailed shapes, and rotation axes of NEOs. Some of the best radar data are spectacular, with an effective resolution of a few meters. However, even with the world's largest radio dishes, radar can only reach the small minority of NEOs that come nearest (0.05-0.1 AU [http://echo.jpl.nasa.gov/~lance/snr/far_asnr18.gif]). As such, for bulk characterization, radar is generally limited to being a calibration tool for other methods (Howell et al. 2012, Mainzer et al. 2011a). The 70 m Goldstone or 300 m Arecibo dishes are the main facilities used for radar astronomy. High power transmitters are needed because the outbound signal is attenuated as $1/r^4$ by the time it returns to Earth. ($1/r^2$ on the outbound journey, and $1/r^2$ on the return.) Arecibo uses a one megawatt transmiter, making fuel costs a significant factor. To gain a factor two in distance requires a factor $2^4 = 16$, increase in power, which rapidly becomes problematic. NASA funded radar work at Arecibo and Goldstone supports the study of 20-30 NEOs per year, though in 2012 fuel cost savings allowed 80 NEOs to be targeted [http://neo.jpl.nasa.gov/neo/2011_AG5_LN_intro_ wksp.pdf, slide 20.]

4.6.4.2 Optical and Infrared Interferometry

Another way to measure the sizes of NEOs, in principle, is interferometry. Interferometry allows us to reconstruct images at the resolution of a single large telescope by combining the signal from several smaller telescopes. The maximum angular resolution of a telescope is limited by diffraction for a fixed wavelength. A larger aperture produces a sharper image at fixed wavelength, and a shorter wavelength produces a sharper image for a fixed aperture size. In interferometry, two small telescopes with a large separation produce interference between their combined light beams as they move in and out of phase scanning across a field. Using multiple telescopes allows the reconstruction of images at the resolution that would be afforded by a single large dish of the same size as the separation of the smaller telescopes. This technique has long been used in radio astronomy. The NRAO Jansky Very Large Array (JVLA) is the premier instrument of this type [https://science.nrao.edu/facilities/vla]. It has baselines up to 27 km long.

Interferometry is inherently harder to implement at the shorter wavelengths of optical or infrared astronomy. CHARA, on Mt. Wilson, California and the European Southern Observatory (ESO) Very Large Telescope Interferometer (VLTI) are the two interferometers operating in the optical and infrared. The VLTI "MID-infrared Interferometric instrument" (MIDI) [http://www.eso.org/sci/facilities/paranal/instruments/midi/], produces thermal infrared (N-band) images at 20 milli-arcsecond resolution, about 10 times better than that of the Hubble Space Telescope in the optical band, but still only about 1.5 km resolution at 0.1 AU, much worse than radar. VLTI-MIDI can image only the occasional NEO, as it is limited to objects brighter than V ~ 14 (1 Jansky at 11.8 microns, Delbo et al. 2009, Matter et al. 2011).

In the near-infrared (1 – 2 microns) the VLTI–AMBER (Astronomical Multi-Beam combinER) instrument has 2 milli-arcsecond resolution, giving a resolution of 144 meters at 0.1 AU, still worse than radar. However, AMBER can only reach a 1.6 micron H magnitude of 8 [This is NOT the absolute H magnitude discussed in Sect. 4.4.3]. CHARA has a similar instrument with a factor ~2 better resolution, and should soon be upgraded to reach H ~ 10. The Magdalena Ridge Observatory Interferometer (MRO-I) promises to be five magnitudes more sensitive than VLTI-AMBER making many NEOs [http://www.mro.nmt.edu/about-mro/ interferometer-mroi/]. First light for MRO-I depends on funding availability.

4.6.4.3 Astrometry

Astrometry is the accurate measurement of the position of objects on the sky. Small changes in NEO orbit parameters are constantly occurring because of the "Yarkovsky effect". In this effect, the rotation of the asteroid causes its black body emission to be strongest away from zero solar elongation, producing a tiny net force (Yarkovsky 1901). The change in NEO semi-major axes (*a*) due to the Yarkovsky effect is only ~10^{-4} AU/Myr (~5 mm s^{-1}). Nevertheless, the effect has been confirmed using precise radar measurements (Chesley et al. 2003). To predict NEO locations decades in advance orbit changes of this magnitude need to be measured. Most asteroids are small, and their larger ratio of area to mass for smaller asteroids makes the Yarkovsky effect more important for them. If they are fast rotators though, the thermal differences will be smoothed out. As most NEOs are small, taking account of the Yarkovsky effect in calculating their orbits can be critical to finding them again. If larger orbit changes are seen, they indicate another force is present, such as mass loss from volatiles or, for very small NEOs, solar radiation pressure.

In principle, if the size, shape and thermal conductivity of the NEO are known, then the size of the Yarkovsky effect can be used to calculate the mass of the NEO. Unfortunately, the prospects for bulk measurement of NEO masses from this technique appear poor. On a decade timescale, normal astrometry can rarely detect this effect. Nugent et al. (2012) were able to detect the Yarkovsky effect in just 54 out of ~1250 NEOs. The ESA Gaia mission will achieve few milli-arcsecond positions with a telescope area equivalent to a 0.5 m diameter mirror [http://sci.esa.int/science-e/www/object/index.cfm?fobjectid=40129]. Even so, Mouret & Mignard (2011) find that Gaia could measure the Yarkovsky drift for only a handful of NEOs. Expensive, one-at-a-time, local characterization must be used instead to measure NEO masses (Sect. 4.7), unless they are among the minority in binaries (Sect. 4.6.1.2). However, Pan-STARRS routinely obtains 1 dex better astrometry than the 1 arcsecond typical of NEO position measurements, and LSST promises to deliver another 1 dex improvement. At this level NEO astrometry just might have lead to good mass measurements when combined with other data, which would be a significant gain for asteroid prospecting.

4.6.5 Space-Based Characterization

Space-based observations have seven major advantages:

(1) Outside the Earth's atmosphere the intense, variable, OH emission lines in the 0.9-2 micron range (Sect. 4.5.2) no longer confound observational data.

(2) Cooling the telescope and instruments to ≤40 K removes the thermal background for all wavelengths of 10 microns or less. The low background makes an enormous difference to sensitivity and survey speed. The NEOWISE survey (Sect. 4.4.2.1) was 300 times faster than the 8 m Gemini telescope in reaching a detection.

(3) A single detector can capture the entire 0.5-10 micron spectral range without atmosphere-induced gaps. Good detectors spanning this range have been developed for NEOCam (McMurty et al. 2013).

(4) As small solar elongations become accessible, and latitude on the Earth is not an issue, many more NEOs are accessible at any one time.

(5) Terrestrial weather is irrelevant, all the time is "night" time, and a full moon is much less important, which all lead to higher observing efficiency.

(6) In some configurations long continuous observations are possible. Continuous observations are helpful in measuring shapes from light curves.

(7) Extremely high accuracy photometry is possible from space. The NASA *Kepler* mission, with a 0.5 m primary, has demonstrated that a space observatory can obtain photometry with100 ppm (0.01%) errors, on a 13th magnitude star (Koch et al. 2010), a factor 100 better than good ground-based photometry. Even the 15 cm MOST satellite can reach 200 ppm, albeit for Procyon, a 1st magnitude star (Walker et al. 2003). For optical photometry, in principle, a space mission could obtain quite detailed NEO morphologies, and find small binary companions (Sect. 4.6.1.2).

4.7 Local Characterization

Asteroid prospecting needs to go beyond astronomy-based techniques in order to perform a detailed assay of resource content. If of the full sequence of remote characterization described in the preceding sections could be fulfilled we would know which NEOs are accessible to current launcher capabilities, and which ones are likely to be rich in volatile materials, or valuable mineral resources. However, the accuracy of these remotely acquired resource predictions are of only astronomical accuracy.

More accurate estimates of the resource content for the target asteroid are clearly needed in order to bring down the risk to a prudent level. To do so, making measurements local to the NEO will be necessary, before mining operations begin.

Planetary science mission techniques have a lot to contribute here. This is a large subject, with many missions from NASA, ESA and JAXA having potentially relevant technology to consider. This section presents only a brief overview.

A few key points are worth emphasizing:

(1) Only a tiny fraction of the ~20,000 NEOs >100m dia. can be visited by spacecraft. In order to choose them well, all the remote characterization techniques must be employed.
(2) Mass is the key number missing from remote characterization. Masses require local measurements by robotic spacecraft (using tracking telemetry), except for binary asteroids (Kistler et al. 2010).
(3) Several other measurements require close approach. X-ray and gamma-ray spectroscopy can detect surface resources. Optical imaging can measure density.
(4) Contact measurements could provide fair samples of the resource content. Boreholes could sample the resource content to depths of several meters, where space weathering and outgassing have not been at play.

4.7.1 Missions to Asteroids

Only three spacecraft missions have visited an asteroid to within 1000 km:

1. NEAR/Shoemaker (NASA) was the first spacecraft to orbit an asteroid, 433 Eros, a 33 x 18 km, S-type Amor, in 1999, and later toucheddown on the surface in 2000 [http://science.nasa.gov/ missions/near/].
2. Hayabusa (JAXA) orbited and touched down on 25143 Itokawa (a 500 x 294 x 209 m, S-type, Apollo) in 2005, and returned a milligram sample of the surface regolith to Earth [http://www.jspec.jaxa.jp/e/activity/ hayabusa.html].
3. DAWN (NASA) visited 4 Vesta (a 525 km diameter main belt asteroid) in 2011 and, as of the date of writing, is on its way to 1 Ceres (the largest, 950 km diameter, main belt asteroid) [http://dawn.jpl.nasa.gov/].

Three new missions are under development:

- NASA is building OSIRIS-REx for launch in 2016 to the C-type asteroid 1999 RQ36 (a 493 m diameter, C-type, Apollo). If successful, it will return 60 grams or more of surface regolith to Earth in 2023 [http://osiris-rex.lpl.arizona.edu].
- JAXA is building Haybusa II for launch in 2014. It is intended to return a sample from 1999 JU3 (dia. ~900 m, C-type, Apollo) in 2020 [http://www.jspec.jaxa.jp/e/activity/hayabusa2.html].
- ESA is studying another sample return mission, Marco Polo-R to the binary asteroid 2008 EV5. (primary dia. 1.6km, secondary dia. 400 m, probably C-type, Apollo, de León et al. 2011). However the mission is not yet approved [https://www.oca.eu/MarcoPolo-R/].

4.7.2 *Instrumentation*

The instrument package on OSIRIS-REx is quite comprehensive and provides good baseline of what tools are currently available [http://osiris-rex.lpl.arizona.edu/sites/osiris-rex.lpl.arizona.edu/files/pdfs/OSIRIS_REx_ infosheet.pdf]. These instruments fall into two classes: those that operate nearby from a few hundred meters above the asteroid surface, and those that require contact with the surface.

4.7.2.1 Nearby Instruments

The OSIRIS-REx payload includes a suite of optical/infrared imaging and spectroscopy instruments (OCAMS, OVIRS, OTES) similar to those used remotely. Their advantage locally, a few hundred meters above the surface, is that they can make high resolution maps of topography, and may reveal compositional inhomogeneities. Three dimensional surface models can be made by combining images from many angles and lighting conditions, using photogrammetry techniques (Gaskell 2012).

Other OSIRIS-REx instruments can only be used locally. Doppler radio tracking of the spacecraft allows the NEO mass, and also rough mass distribution, to be determined (Takashima and Scheeres 2012). A laser altimeter, OLA, can deliver ranging, topography and texture data. REXIS maps elemental abundances using X-ray imaging spectroscopy in the 0.5 – 7 keV [keV = kilo-electron volts. 1 keV corresponds to a wavelength of 1.25 nm.]X-ray detectors sensitive to carbon (0.3 keV) and platinum (~12 keV) lines, which REXIS cannot see, are feasible (Kraft et al. 2012). A gamma-ray spectrometer was used on NEAR/Shoemaker and Dawn (GRS, Trombka et al. 1997), but is not included on OSIRIS-REx. GRS was sensitive in the 0.2 - 10 MeV band and measured isotopic ratios, as well as elemental composition. Gamma-ray instruments have only minimal imaging resolution, which is only important if the target NEO is inhomogeneous. We don't know if this is the case.

4.7.2.2 Contact Instruments

Direct contact with the NEO brings further techniques into play. An impactor sent from a distance, as for the NASA Deep Impact mission to comet Tempel 1 (A'Hearn et al. 2004) can be used to measure surface strength [http://www.nasa.gov/mission_pages/deepimpact/main/ index.html]. Such projectiles may penetrate sufficiently to reach depths where volatiles are present. The plume of vaporized material ejected by the impact can be analyzed spectroscopically from nearby, or remotely.

Landing on the surface allows gentler observations. Hayabusa initially included a small (1kg) NASA supplied rover, MUSES-CN. Despite its small size, it would have carried a multi-band imager, and infrared spectrograph and an alpha/X-ray spectrometer. However, MUSES-CN was cancelled. The wide suite of instruments on the Mars Science Laboratory Rover, Curiosity (Grotzinger et al. 2012) show what is possible for contact instruments. Their cost and a mass precludes launching many copies. Laser created plasmas can probe the composition of mm-sized regions in detail, via spectroscopy, as on the ChemCam instrument (Maurice et al. 2005) [http://www.msl-chemcam.com]. The SAM instrument (Mahaffy et al. 2009) has a mass spectrometer for detailed molecular analysis, but the instrument is mechanically complex [http://msl-scicorner.jpl.nasa.gov/ Instruments/SAM/].

Boring a core sample requires a firm attachment to the NEO surface. On deep regolith or rubble piles this may be problematic (Daniels 2013).

4.7.3 Need for Multiple Small Missions

If we take the asteroids that already have complete remote characterization indicating that they are promising mining candidates, we will know only that they lie within a broad range of possible resource richness.

To find one NEO in the top 10th percentile of the resource we are seeking one would expect that we must make local investigations of about 10. But we could be unlucky. The binomial distribution probability of finding at least one in the top 10th percentile with ten trials is 65% [See, e.g., Bevington P.R. and Robinson K., 1992, *Data Reduction and Error Analysis for the Physical Sciences*, McGraw-Hill, ISBN 0-07-911243-9, Chapter 2, Eq. 2.4]. To reach 90% we must make 22 trials. (There will be a 66% probability there will also be a second good asteroid.) Forty-four trials are needed to reach 99% probability. So two to four dozen good NEO candidates must be investigated to have an acceptable chance of finding one ore-bearing asteroid.

Because the transit time from one of the rare good asteroid candidates to the next is likely to be a year or so, serial investigations of a few dozen by one spacecraft will take too long. We must consider a parallel approach. A fleet, or swarm, of smaller spacecraft could investigate a number of NEOs simultaneously.

The cost of obtaining this information needs to be a modest fraction of the value of the resources ultimately retrieved, perhaps 10%. For the nominal ~US$5 B value of the PGM-rich 100 m diameter NEO considered in Sect. 4.3.1, that would cap the local prospecting cost at ~US$500 M. This means that we need to launch cheap, and surely small, spacecraft.

Each one must then cost far less than, for example, OSIRIS-REx (Lauretta et al. 2012). As a New Frontiers class mission, it will cost about US$800M, not including launch. Robotic characterization missions for mining would not necessarily return samples, leading to greatly reduced costs. If sufficient resource data can be gathered without contact with the NEO, mission complexity and costs can be reduced further. Nonetheless, much smaller spacecraft will still be needed.

This need has already been recognized. It is one goal of the Arkyd satellite series that Planetary Resources is developing, and of the Firefly, cubesat-based, satellites from Deep Space Industries. Their first launches (to LEO) could be as early as 2015. To keep the spacecraft small we will need to define a minimal instrument package. This could be limited to radio tracking and optical imaging. For prospecting, the next highest priority instrument is probably an X-ray imaging spectrometer.

Operating small spacecraft at interplanetary distances poses two primary challenges: propulsion and communication. Propulsion might be provided by rocket-on-a-chip technologies, such as microfluidic electrospray propulsion (MEP) systems (Mueller et al. 1997) using solar power. More radically, radioisotope-powered tuned thermo-photovoltaic systems (RTPV, Howe et al. 2012) might provide higher power electric propulsion, though Solar power currently wins for anticipated technology.

Radio communications across AU distances requires high power transmitters and large (>34m) antennae, such as those of NASA's Deep Space Network, to collect the weak signal [http://deepspace.jpl.nasa.gov/dsn/]. One solution to this challenge might be to set up a "Solar System Internet" with a chain of small satellites acting as transponders around the Sun (Elvis, Landau et al. 2012). Another approach might be to look at the technology that is currently being developed for the Square Kilometer Array. [http ://www.ska.ac.za/download/fact_sheet_ skaint_ eng_2011.pdf]. This technology can track many probes at once. At a goal price of a billion US dollars for a one square kilometer array, and, being a highly modular technology, should scale, so that a 100 m diameter antenna, with ~1% of the area, should cost around $10M. A more careful study (Jones 2005) suggests that an X-band only system could cost US$30 M – US$40 M. More than one system would be needed for continuous communication.

Optical, laser-based, communication provides an alternative, and requires only moderate aperture telescopes as ground stations. But pointing near the Sun would be necessary and this might lead to a high background. Fortunately, the very narrow laser bandwidth greatly mitigates this problem. A careful choice of wavelengths that lie in solar spectral absorption bands [most likely Na H+K (589.29 nm) or Hα (656.28 nm)] can help further. The newly developed optical sieve technology might provide a large optical mirror at low mass for transmitting the laser signal from a small satellite [see FalconSAT-7: http://www.usafa.edu/df/dfe/dfer/centers/ lorc/docs/FalconSAT07.pdf]. The greatest difficulty for optical communications may lie in achieving the sub-arcsecond pointing stability required for the beam to intercept the Earth.

4.8 Implementation

How should this program of NEO prospecting best be carried out? This review has concentrated on techniques, with only minor excursions into costs. Now we must consider both plausible costs and timescales.

A first step would be increasing the ground-based NEO spectroscopic characterization rate to keep up with the current rate of NEO discovery of about 1000 a year. If we could double this rate to 2000/year, then the 20,000 NEOs >100 m diameter could be characterized within a decade, always subject to the synodic period problem.

Currently, however, all large telescopes are designed, and funded, for astronomical research. Their available observing time is already oversubscribed by factors of three or more for this research, so obtaining large amounts of time is unlikely by an established route [E.g. Gemini: see right hand column of graph at http://www.gemini.edu/sciops/ statistics?q=node/11676].

4.8.1 Costs

The most direct way to obtain 2000 NEO spectra/year would be to build a new large telescope. The 10 m Keck telescopes cost ~US$100 M per telescope (Stepp et al. 2002). Using the 10 m Southern African Large Telescope (SALT [http://www.salt.ac.za/]) design could provide a more affordable path for a dedicated ground-based NEO characterization facility. SALT was built for under US$30 M (Year 2000 $), including both an imager and a spectrograph (P. Charles, private communication, 2012). This simplified design uses spherical, modular, optics, and a fixed elevation. The moving secondary means that the aperture is not fully utilized, however. It is derived from the similarly low cost Hobby-Eberly Telescope [http://www.as.utexas.edu/mcdonald/het/het.html].

An alternative would be to re-purpose an old telescope, as was done for several NEO surveys. This may be a quicker and less expensive path than a building a new telescope. Re-purposing has become an option, as funding support for several major telescopes has dried up in recent years. The 4 m UK Infrared Telescope (UKIRT) is actually up for sale, an unprecedented situation [http://www.jach.hawaii.edu/UKIRT/news/UKIRT_AO/ Prospectus.pdf]. UKIRT running costs are given as US$1.238M/year. In addition, a new spectrograph would likely be needed for NEO characterization. Similarly, a 2012 review of the US ground-based astronomy facilities run by the National Science Foundation recommended that the NSF divest itself of several moderate to large aperture optical-near-infrared telescopes, and two major radio telescopes, including the 100 m diameter Green Bank

Telescope (GBT), which might be useful for radar [http://www.nsf.gov/mps/ast/portfolioreview/reports/ast_portfolio_review_report.pdf].

But remote characterization is best done from space, as discussed above. The main limitation to in-space NEO characterization is getting an adequate mirror size at an affordable cost. The thermal infrared *Spitzer Space Telescope* has a 0.85 m primary. *Spitzer* is one of NASA's Great Observatories, and cost ~US$720 M (in 2004 dollars). The optical Kepler mission has a 0.5 m primary and cost ~US$500 M, including 3.5 years of operations. WISE cost ~US$320 M.

The proposed NEOCam mission is a Discovery class mission, and so fits within that cost cap of US$425 M, not including launch [http://discovery.nasa.gov/p_mission.cfml]. A modestly enhanced version of NEOCam could carry a 0.5-10 micron spectrometer and, possibly, also be capable of high enough time resolution to make precise light curves. The remote characterization problem would be almost completely solved with such a mission.

4.8.2 Timescale

How quickly can we complete the characterization of the 20,000 NEOs >100 m diameter? If we start now, the remote characterization phase could be essentially completed within a decade.

The quickest start would be made by re-purposing UKIRT to 100% NEO characterization, notably near-infrared spectroscopy. Increasing the number of nights from ~16 a year to about 200 a year is alone enough to get spectra of ~1000 NEOs a year. UKIRT has the thermal-IR MICHELLE instrument, which could begin work at once. Building a new optical-near-infrared spectrograph based on an existing design would take 2-3 years before it could begin work. A new spectrograph design would take a year or two longer.

Building an exact SALT duplicate on an existing site would take about 4 years. (SALT was built from mid-2001 to mid-2005, Phil Charles, 2012, private communication.) The site needs to allow the construction of a new large telescope, which, unfortunately, probably excludes Mauna Kea. Once operational, such a telescope could gather spectra at ~2.5 times the rate of UKIRT, based on its increased area (and assuming background limited instruments).

A new Explorer class space mission is also likely to take five years from approval to launch. B612 is trying for a shorter build time, closer to three years. The mission itself would then last 3-5 years.

Following spectral characterization, local characterization of optimally chosen NEOs could then begin in bulk. A immediate test program for small interplanetary satellites, targeting the best NEOs available at the time could teach us how to build optimal missions for the bulk phase of the investigation, so that we could scale up rapidly. Planetary Resources and Deep Space Industries are taking this approach, planning on a three-year build time to their first launch. The Arkyd 100 series, though, will not have interplanetary capability. A first local characterization phase, with fully optimized – prospecting-worthy - targets, could well begin when the census is half completed. The first local characterization results from optimal targets could then come in a decade, with the rest following within five years.

4.8.3 Research

Some of the programs outlined in this review have almost reached an "industrial" stage. Astrophysics research is not required to produce NEO orbits and spectra in bulk. On the other hand, the characterization of asteroid composition, whether remotely, locally, or via meteorites, is by no means settled science. For example, the theory of planetisimal formation and differentiation, needs to be developed to a point where it can guide prospecting. Theory always needs guidance from observations and experiment. Observing asteroid belts in other planetary systems, and directly comparing meteorite mineralogy with astronomical spectroscopy of the parent asteroid are just two examples. The implication is that industrial asteroid mining will have a continuing need for academic research, in order to optimize its commercial prospecting and mining returns.

Good basic research is not effectively motivated by profit, however. Practitioners of asteroid science do not see an urgent need to characterize the whole NEO population. In order to attract the interest of research astronomers, potential scientist users of prospecting data need to be convinced that surveys for mining needs will also benefit their scientific efforts (Elvis 2012). Large astronomy surveys initially faced a similar challenge. Most astronomers did not realize how big data would benefit their field. But, today, astronomical research has been transformed by the bulk data acquired by the Sloan Digital Sky Survey (Madrid and Macchetto 2009). Bulk data will similarly transform Solar System research.

Already, the greatly accelerated discovery rate of NEOs initiated by the George E. Brown Congressional mandate to find 90% of all NEOs with diameters greater than 140 m diameter, has revitalized asteroid research, although its root motivation was not scientific. A similar explosion in asteroid compositional data, including plentiful local measurements, will be even more transformative (Elvis 2012). The process of formation of our solar system will surely be greatly clarified, including puzzles about Earth's early history. For example, the origin of the water in the oceans, and the ores in the crust, as well as the potential seeding of Earth with organic molecules which may have sped up, or even enabled, the origin of life, are all topics that will benefit from more asteroid data. Beyond that, there is the possibility of exotic materials, with new properties, that form only in space (Bindi et al. 2012, Ma et al. 2012).

Policies enabling open access will be needed, if we intend to derive maximal science benefit from this new data. Yet some of the data will have commercial value as intellectual property. Significant work will be needed to establish a good set of working rules.

4.9 Summary and Conclusions

"*Everything is science fiction right up to the point that it's science fact*" [Chris Lewicki, Chief Asteroid Miner, Planetary Resources]. This review has attempted to lay out how the prospecting challenges faced by would-be asteroid miners can become science fact. The approach tried to be reasonably complete and quantitative. While the implied programs are substantial, a concerted program over the next decade could complete the program, generating a highly valuable set of NEO prospecting data. This would buy down the risk for all future asteroid mining ventures. Raising substantial private capital for such ventures would then be much easier.

Finally, all the techniques described in this review for mining purposes, apply equally well to the selection of asteroid targets for human exploration, and for discovering *actually hazardous objects* (not merely "potentially hazardous" ones) and diverting them. Throughout, science will accrue great benefit from the vastly enlarged data sets. Whatever our interest in asteroids, these preliminary prospecting steps are unavoidable. This clear alignment of motivations calls for strong and rapid action.

Acknowledgements. Asteroid prospecting is such a broad and new topic that no-one is yet expert on all its aspects. I am in debt to many for their assistance, but any errrors are my own. In particular, I thank C. Beeson, J.-L. Galache, J. McDowell, and M. Trichas for valuable calculations, R. Binzel, J. Brophy, M. Busch, P. Charles, S. Jacobsen, G. Macpherson, A. Mainzer, T. McCoy, T. Spahr, C.A. Thomas, D. Trilling, G. Williams, and many others, for letting me share their deep knowledge of asteroid, meteorites, and telescope and space technology. Finally I thank BC Crandall for careful editing that greatly enhanced my tangled prose. Dr. Martin Elvis is a senior astrophysicist at the Harvard-Smithsonian Center for Astrophysics. All of the opinions expressed here are his own and not those of the Smithsonian Astrophysical Observatory or Harvard University.

References

A'Hearn, M., the Deep Impact Team: The Deep Impact Mission to Comet 9P/Tempel 1. In: 35th COSPAR Scientific Assembly, p. 1667 (2004)

Baer, J., Chesley, S.R., Matson, R.D.: Astrometric Masses of 26 Asteroids and Observations of Asteroid Porosity. Astrophysical Journal 141, 143 (2011)

Bartczak, P., Marciniak, A.: Shaping Asteroids with Genetic Evolution (SAGE). LPI-Co 1667, 6126 (2012)

Beeson, C., Galache, J.-L., Elvis, M.: (in preparation, 2013)

Bessel, M.S.: Standard Photometric Systems. Annual Review of Astronomy and Astrophysics 43, 293–336 (2005)

Bindi, L., Eiler, J.M., Guan, Y., Hollister, L.S., Steinhardt, P.J., Yao, N.: Evidence for the extraterrestrial origin of a natural quasicrystal. Proceedings of the National Academy of Sciences 109, 1396–1401

Binzel, R.P., et al.: The MIT-Hawaii-IRTF Joint Campaign for NEO Spectral Reconnaissance. In: 37th Annual Lunar and Planetary Science Conference, League City, Texas, March 13-17, abstract no.1491 (2006)

Binzel, R.P., et al.: Earth encounters as the origin of fresh surfaces on near-Earth asteroids. Nature 463, 331 (2010)

Binzel, R.P., et al.: Cracking the Space Weathering Code: Ordinary Chondrite Asteroids in the Near-Earth Population. American Astronomical Society, DPS meeting #44, #202.03 (2012)

Bottke, W., Cellino, A., Paolichi, P., Binzel, R.P. (eds.): Asteroids III. University of Arizona Press (2002a) ISBN 0-8165-2281-2

Bottke, W., et al.: Debiased Orbital and Absolute Magnitude Distribution of the Near-Earth Objects. Icarus 156, 399 (2002)

Britt, D.T., Yeomans, D., Housen, K., Consolmagno, G.: Asteroid Density, Porosity, and Structure. In: Bottke Jr., W.F., Cellino, A., Paolicchi, P., Binzel, R.P. (eds.) Asteroids III, pp. 485–500. University of Arizona Press, Tucson (2002)

Brown, P., Spalding, R.E., ReVelle, D.O., Tagliaferri, E., Worden, S.P.: The flux of small near-Earth objects colliding with the Earth. Nature 420, 294–296 (2002)

Burbine, T., McCoy, T.J., Meibom, A., Gladman, B., Keil, K.: Meteoritic Parent Bodies: Their Number and Identification. In: Bottke Jr., W.F., Cellino, A., Paolicchi, P., Binzel, R.P. (eds.) Asteroids III, pp. 653–667. U. Arizona and LPI (2002)

Campbell, A.J., Humayan, M.: Compositions of group IVB meteorites and their parent melt. Geochimica et Cosmochimica Acta 69, 4733–4744 (2005)

Carvano, J.M., et al.: SDSS-based taxonomic classification and orbital distribution of main belt asteroids. Astronomy & Astrophysics 510, A43 (2010)

Chesley, S.R., et al.: Direct Detection of the Yarkovsky Effect by Radar Ranging to Asteroid 6489 Golevka. Science 302, 1739–1742 (2003)

Christensen, E., Larson, S., Boattini, A., Gibbs, A., Grauer, A., Hill, R., Johnson, J., Kowalski, R., McNaught, R.: The Catalina Sky Survey: Current and Future Work. DPA 42, 1013 (2012)

Clark, B.E., et al.: NEAR photometry of asteroid 253 Mathilde. Icarus 140, 53–65 (1998)

Cox, A.N. (ed.): Allen's Astrophysical Quantities, 4th edn. Springer (1999)

Daniels, K.E.: Rubble-pile Near Earth Objects: Insights from Granular Physics. In: Badescu, V. (ed.) Asteroids, vol. 138, pp. 305–323. Springer, Heidelberg (2013)

de León, J., Mothé-Diniz, T., Licandro, J., Pinilla-Alonso, N., Campins, H.: New observations of asteroid (175706) 1996 FG3, primary target of the ESA Marco Polo-R mission. A&A 530, L12 (2011)

Delbo, M., Ligori, S., Matter, A., Cellino, A., Berthier, J.: First VLTI-MIDI Direct Determinations of Asteroid Sizes. Astrophysical Journal 694, 1228 (2009)

DeMeo, F.E., Binzel, R.P., Slivan, S.M., Bus, S.J.: An extension of the Bus asteroid taxonomy into the near-infrared. Icarus 202, 160–180 (2009)

Dunn, T.L., McCoy, T.J., Sunshine, J.M., McSween Jr, H.Y.: A Co-ordinated Mineralogical, Spectral and Compositional Study of Ordinary Chondrites: Implications for Asteroid Spectroscopic Classification. LPI 41, 1750 (2010)

Ellis, S.C., Bland-Hawthorn, J.: The Case for OH suppression at near-infrared wavelengths. Monthly Notices of the Royal Astronomical Society 386, 47–64 (2008)

Elvis, M., et al.: A Vigorous Explorer Program. White paper submitted to the Astro2010 NAS/NRC Decadal Review of Astronomy and Astrophysics, 2009arXiv0911.3383E (2009)

Elvis, M., McDowell, J.C., Hoffman, J., Binzel, R.P.: Ultra-low delta-v objects and the human exploration of asteroids. Planetary and Space Sciences 59, 1408 (2011)

Elvis, M.: Let's Mine asteroids for science and profit. Nature 485, 549 (2012)

Elvis, M., Landau, D., et al.: A Swarm of Micro-satellites for in Situ NEO Characterization. American Astronomical Society, DPS meeting #44, #215.04 (2012)

Elvis, M.: How Many Ore-Bearing Near-Earth Asteroids? Planetary and Space Sciences (submitted, 2013)

Freedman, W.L., et al.: Final Results from the Hubble Space Tele-scope Key Project to Measure the Hubble Constant. Astrophysical Journal 553, 47–72 (2001)

Fukugita, M., Ichikawa, T., Gunn, J.E., Doi, M., Shimasaku, K., Schneider, D.P.: The Sloan Digital Sky Survey Photometric System. Astronomical Journal 111, 1748 (1996)

Gaffey, M.J., Bell, J.F., Brown, R.H., Burbine, T.H., Piatek, J.L., Reed, K., Chaky, D.A.: Mineralogical variations within the S-type asteroid class. Icarus 106, 573–602 (1993)

Gaskell, R.W.: SPC Shape and Topography of Vesta from DAWN Imaging Data. DPS 442, 0903 (2012)

Grady, M.M.: Catalogue of Meteorites, 689 p. Cambridge Univ. Press (2000)

Greenstreet, S., Gladman, B.: High-inclination Atens ARE Rare, DPS 4430505G (2012)

Grotzinger, J.P., et al.: Mars Science Laboratory Mission and Science Investigation. Space Science Reviews 170, 5–56 (2012)

Hapke, B.: Coherent backscatter and the radar characteristics of outer planet satellites. Icarus 88, 407–417 (1990)

Hapke, B.: Space weathering from Mercury to the asteroid belt. Journal of Geophysical Research 106(E5), 10039–10074 (2001)

Hanuš, J., Ďurke, J.: New Asteroid models based on combined dense and sparse photometry. Astronomy & Astrophysics (2011) (submitted)

Harris, A.W.: A Thermal Model for Near-Earth Asteroids. Icarus 131, 291 (1998)

Harris, A.: What Spaceguard did. Nature 453, 1178–1179 (2008)

Harris, A.W., Harris, A.W.: On the Revision of Radiometric Albedos and Diameters of Asteroids. Icarus 126, 450 (1997)

Harris, A.W., Lagerros, J.S.V.: (XXX) Asteroids in the Thermal Infrared. In: Bottke Jr., W.F., Cellino, A., Paolicchi, P., Binzel, R.P. (eds.) Asteroids III, pp. 653–667. U. Arizona and LPI

Hillenbrand, L.A., Foster, J.B., Persson, S.E., Matthews, K.: The Y Band at 1.035 Microns: Photometric Calibration and the Dwarf Stel-lar/Substellar Color Sequence. Publications of the Astronomical Society of the Pacific 114, 708–720 (2002)

Howe, T., O'Brien, R.C., Stoots, C.M.: Development of a Small-Scale Radioisotope Thermo-Photovoltaic Power Source. Nuclear and Emerging Technologies for Space 3029 (2012)

Howell, E.S., Rikin, A.S., Vilas, F., Magri, C., Nolan, M.C., Vervack Jr., R.J., Fernandez, Y.R.: Hydrated silicates on main-belt asteroids: Correlation of the 0.7- and 3-micron absorption bands. EPSC Abstracts 6, 637 (2011)

Howell, E.S., et al.: Combining Thermal and Radar Observations of Near-Earth Asteroids. DPS 441, 1107 (2012)

Ivezić, Ž., et al.: Solar System Objects Observed in the Sloan Digital Sky Survey Commissioning Data. Astronomical Journal 122, 2749 (2001)

Ivezić, Ž., et al.: LSST: from Science Drivers to Reference Design and Anticipated Data Products. arXiv:0805.2366 (2011)

Jedicke, R., et al.: ATLAS: Asteroid Terrestrial-impact Last Alert System. American Astronomical Society. DPS meeting #44, #210.12 (2012)

Jenniskens, P.: Quantitative meteor spectroscopy: Elemental abundances. Advances in Space Research 39, 491–512 (2007)

Jenniskens, P., et al.: The impact and recovery of asteroid 2008 TC3. Nature 458, 485 (2009)

Jenniskens, P., et al.: Radar-Enabled Recovery of the Sutter's Mill Meteorite a Carbonaceous Chondrite Regolith Breccia. Science 338, 1583 (2012)

Jewitt, D.: The Active Asteroids. Astronomical Journal 143, 66 (2012)

Jones, D.L.: Lower-Cost Architectures for Large Arrays of Small Antennas. IEEEAC paper # 1199 (2005), doi:10.1109/AERO.2006.1655810

Kaiser, N., et al.: Pan-STARRS: A Large Synoptic Survey Telescope Array. In: Survey and Other Telescope Technologies and Discoveries. Proceedings of the SPIE, vol. 4836, pp. 154–164 (2002)

Kargel, J.S.: Metalliferous asteroids as potential sources of precious metals. Journal of Geophysical Research 99(E10), 21129–21141 (1994)

Kraft, R., Kenter, A., Murray, S., Elvis, M., Branduardi-Raymont, G., Garcia, M., Forman, W., Geary, J., McCoy, T., Smith, R.: X-ray Imaging Spectroscopy for Planetary Science. American Astronomical Society Division of Planetary Sciences meeting #44, #215.10 (2012)

Kistler, J., et al.: Bulk Densities of Binary Asteroids from the Warm Spitzer NEO Survey. DPS 42, 5709 (2010)

Koch, D.G., et al.: Kepler Mission Design, Realized Photometric Performance, and Early Science. Astrophysical Journal Letters 713, L79–L86 (2010)

Larson, S.M., Hergenrother, C., Whitely, R., Kelly, C., Hill, R.: Upgrading the Catalina Sky Survey and Southern Survey. In: International Workshop on Collaboration and Coordination Among NEO Observers and Orbital Computers, p. 35. Japan Safeguard Association (2001)

Larson, S.: Current NEO Surveys. In: Milani, A., Valsecchi, G.B., Vokrouhlický, D. (eds.) Near Earth Objects, our Celestial Neighbors: Opportunity and Risk Proceeding of IAU Symposium No. 236, pp. 323–328 (2007)

Lauretta, D.S., the OSIRIS-REx Team: An Overview of the OSIRIS-REx Asteroid Sample Return Mission. In: 43rd Lunar and Planetary Science Conference, The Woodlands, Texas, March 19-23. LPI Contribution No. 1659, id.2491 (2012)

Longo, G.: The Tunguska Event. In: Bobrowsky, P.T., Rickman, H. (eds.) Comet/Asteroid Impacts and Human Society, An Interdisciplinary Approach, ch. 18. Springer, Berlin (2007)

Ma, C., Tschauner, O., Beckett, J.R., Rossman, G.R., Liu, W.: Panguite (Ti^{4+},Sc,Al,Mg,Zr, Ca) $1.8^{O}3$, a new ultra-refractory titania mineral from the Allende meteorite: Synchrotron micro-diffusion and EBSD. American Mineralogist 97, 1219–1225 (2012)
Madrid, J.P., Macchetto, D.: High-Impact Astronomical Observatories. Bulletin of the American Astronomical Society 41, 913 (2009)
Mahaffy, P.R., et al.: Sample Analysis at Mars (SAM) Instrument Suite for the 2011 Mars Science Laboratory. In: 40th Lunar and Planetary Science Conference, id.1088 (2009)
Mainzer, A.K.: NEOCam: The Near-Earth Object Camera. DPS meeting #38, Bulletin of the American Astronomical Society 38, 568 (2006)
Mainzer, A.K., et al.: Thermal Model Calibration for Minor Planets Observed with Wide-field Infrared Survey Explorer/NEOWISE. Astrophysical Journal 736, 100 (2011a)
Mainzer, A.K., et al.: NEOWISE Observations of Near-Earth Objects: Preliminary Results. Astrophysical Journal 743, 156 (2011b)
Mainzer, A., et al.: Physical Parameters of Asteroids Estimated from the WISE 3-Band Data and NEOWISE Post-Cryogenic Survey. Astrophysical Journal 760, L12 (2012)
Matter, A., Delbo, M., Ligori, S., Crouzet, N., Tangaa, P.: Determination of physical properties of the Asteroid (41) Daphne from interferometric observations in the thermal infrared. Icarus 215, 47 (2011)
Maurice, S., et al.: ChemCam Instrument for the Mars Science Laboratory (MSL) Rover. In: 36th Annual Lunar and Planetary Science Conference, abstract no. 1735 (2005)
McCoy, T.J., et al.: Group IVA irons: New constraints on the crystallization and cooling history of an asteroidal core with a complex history. Geochimica et Cosmochimica Acta 75, 6821–6843 (2011)
McMurty, C., Lee, D., Chen, C.-Y.A., Demers, R.T., Dorn, M., Forrest, W.J., Liu, F., Mainzer, A., Pipher, J.L., Yulius, A.: Development of Passively Cooled Long-wave Infrared Detector Arrays for NEOCam. Optical Engineering, special topic: Space Telescopes II (submitted, 2013)
Morgan, J.W., Waler, R.J., Grossman, J.N.: Rhenium-osmium isotope systematics in meteorites, I: Magmatic iron meteorite groups IIAB and IIIAB. Earth and Planetary Science Letters 108, 191–202 (1992)
Mouret, S., Mignard, F.: Detecting the Yarkovsky effect with the Gaia missin: list of the most promising candidates. Monthly Notices of the Royal Astronomical Society 413, 741–748 (2011)
Moskovitz, N.: Colors of Dynamically Associated Asteroid Pairs, arXiv:1207.3799 (2012)
Moskovitz, N., Abe, S., Osip, D., Bus, S.J., Abell, P., DeMeo, F., Binzel, R.P.: Characterization of Hayabusa II Target Asteroid (162173) 1999 JU3, DPS 4410204 (2012)
Mueller, J.: Thruster Options for Microspacecraft: A Review and Evaluation of Existing Hardware and Emerging Technologies. In: 33rd Joint Propulsion Conference, Seattle, WA, Paper AIAA 97-3058 (July 1997)
Mueller, T.G., et al.: Thermo-physical properties of 162173 (1999 JU3), a potential flyby and rendezvous target for interplanetary missions. Astronomy & Astrophysics 524, A.145 (2011)
Nelson, M.L., Britt, D.T., Lebofsky, L.A.: Review of Asteroid Compositions. In: Lewis, J., Matthews, M.S., Guerrieri, M.L. (eds.) Resources of Near Earth Space, Tucson. University of Arizona Press (1993)

Nugent, C.R., et al.: Detection of Semimajor Axis Drifts in 54 Near-Earth Asteroids: New Measurements of the Yarkovsky Effect. Astronomical Journal 144, 60–73 (2012)

Ostro, S.J., Campbell, D.B., Chandler, J.F., Hine, A.A., Hudson, R.S., Rosema, K.D., Shapiro, I.I.: Asteroid 1986 DA: Radar evidence for a metallic composition. Science 252, 1399–1404 (1991)

Petaev, M.I., Jacobsen, S.B.: Differentiation of metal-rich meteoritic parent bodies: I. Measurements of PGEs, Re, Mo, W, and Au in meteoritic Fe-Ni metal. Meteoritics and Planetary Science 39, 1685–1697 (2004)

Petit, J.-M., Chambers, J., Franklin, F., Nagasawa, M.: Primordial Excitation and Depletion of the Main Belt. In: Bottke Jr., W.F., Cellino, A., Paolicchi, P., Binzel, R.P. (eds.) Asteroids III, pp. 711–723. U. Arizona and LPI (2002)

Pieters, C.M., et al.: Space weathering on airless bodies: Resolving a mystery with lunar samples. Meteoritics & Planetary Science 35, 1101–1107 (2000)

Pogson, N.: Magnitudes of Thirty-six of the Minor Planets for the first day of each month of the year 1857. Monthly Notices of the Royal Astronomical Society 17, 12 (1857)

Pravec, P., Harris, A.W.: Fast and Slow Rotation of Asteroids. Icarus 148, 12 (2000)

Robinson, S.J., Schmidt, J.T.: Fluorescent Penetrant Sensitivity and Removability - What the Eye Can See, a Fluorometer Can Measure. Materials Evaluation 42, 1029–1034 (1984)

Rubincam, D.P.: Radiative Spin-up and Spin-down of Small Asteroids. Icarus 148, 2 (2000)

Schroeder, D.J.: Astronomical Optics, 2nd edn, ch. 7. Academic Press (1999)

Scott, E.R.D., Wasson, J.T., Buchwald, V.F.: The chemical classification of iron meteorites – VII. A reinvestigation of irons with Ge concentrations between 25 and 80 ppm. Geochimica et Cosmochimica 37, 1957–1983 (1973)

Shepard, M.K., et al.: A radar survey of M- and X-class asteroids II Summary and synthesis. Icarus 208, 221 (2010)

Shoemaker, E.M., Helin, E.F.: NASA CP-2053, pp. 245–256 (1978)

Simcoe, R.A., et al.: The FIRE infrared spectrometer at Magellan: construction and commissioning. In: SPIE, vol. 7735, p. 38 (2010)

Sonter, M.J.: The Technical and Economic Feasibility of Mining the Near-Earth Asteroids. Acta Astronautica 41, 637–647 (1997)

Steel, D.: Two Tunguskas in South America in the 1930's? Journal of the International Meteor Organization 23, 207–209 (1995)

Stepp, L., Daggert, L., Gilletta, P.: Estimating the costs of ex-tremely large telescopes. In: Proc. SPIE. 0277-786X 4840, pp. 309–321 (2002)

Suggs, R.M., Cooke, W.J., Suggs, R.J., Swift, W.R., Hollon, N.: The NASA Lunar Impact Monitoring Program. Earth Moon Planets 102, 293–298 (2008)

Sullivan, P.W., Simcoe, R.A.: A Calibrated Measurement of the Near-IR Continuum Sky Brightness Using Magellan/FIRE, ar-Xiv:1207.0817 (2012)

Takashima, Y., Scheeres, D.: Surface Gravity Fields for Asteroids and Comets. In: 22nd AAS/AIAA Space Flight Mechanics Meeting, Charleston, SC, No. 12-224 (2012); Journal of Guidance, Control and Dynamics (in press)

Tholen, D.J.: Asteroid taxonomic classifications. In: Binzel, R.P., Gehrels, T., Matthews, M.S. (eds.) Asteroids II, pp. 1129–1150. U. Arizona Press, Tucson (1989)

Thomas, C.A., Binzel, R.P.: Identifying meteorite source regions through near-Earth object spectroscopy. Icarus 205, 419–429 (2010)

Thomas, C.A., et al.: ExploreNEOs. V. Average Albedo by Taxonomic Complex in the Near-Earth Asteroid Population. Astronomical Journal 142, 85 (2011a)

Thomas, C.A., et al.: Space weathering of small Koronis family members. Icarus 212, 158 (2011b)

Tonry, J.L.: An Early Warning System for Asteroid Impact. Publications of the Astronomical Society of the Pacific 123, 58–72 (2011)

Tonry, J.: Asteroid Terrestrial-impact Last Alert System (ATLAS). White paper submitted to the NRC: Review of Near-Earth Object Surveys and Hazard Mitigation Strategies (2009)

Trombka, J.I., et al.: Compositional mapping with the NEAR X ray/gamma ray spectrometer. Journal of Geophysical Research 102(E10), 23729–23750 (1997)

Vereš, P., Jedicke, R., Wainscoat, R., Granvik, M., Chesley, S., Abe, S., Denneau, L., Grav, T.: Detection of Earth-impacting asteroids with the next generation all sky surveys. Icarus 203, 472 (2009)

Vernazza, P., Binzel, R.P., Rossi, A., Fulchignoni, M., Birlan, M.: Solar Wind as the origin of rapid reddening of asteroid surfaces. Nature 458, 993 (2009)

Vernazza, P., Carry, B., Emery, J., Hora, J.L., Cruikshank, D., Binzel, R.P., Jackson, J., Helbert, J., Maturilli, A.: Mid-infrared spectral variability for compositionally similar asteroids: Implications for asteroid particle size distributions. Icarus 207, 800–809 (2010)

Vilas, F.: A quick look method of detecting water of hydration in small Solar System bodies. LPSC 25, s. 1439 (1994)

Walker, G.: The MOST Asteroseismology Mission: Ultraprecise Photometry from Space. Publications of the Astronomical Society of the Pacific 115, 1023–1035

Weidenschilling, S.J.: Formation of planetismals and accretion of the terrestrial planets. Space Science Reviews 92, 295–310 (2000)

Willman, M., et al.: Using the youngest asteroid clusters to constrain the space weathering and gardening rate on S-complex asteroids. Icarus 208, 758 (2010)

Wright, E.L., et al.: The Wide-field Infrared Survey Explorer (WISE): Mission Description and Initial On-orbit Performance. Astronomical Journal 140, 1868 (2010)

Yang, J., Goldstein, J.I., Michael, J.R., Kotula, P.G., Scott, E.R.D.: Thermal History and origin of the IVB iron meteorites and their parent body. Geochimica et Cosmochimica Acta 74, 4493–4506 (2010)

Yarkovsky, I.O.: The density of luminiferous ether and the resistance it offers to motions. Bryansk (1901)

Yeomans, D.H.: Near Earth Objects: Finding Them Before They Find Us. Princeton University Press, Princeton (2013)

Zellner, B., Tholen, D.J., Tedesco, E.F.: The eight-color asteroid survey: Results for 589 minor planets. Icarus 61, 355–416 (1985)

Zombeck, M.V.: Handbook of Space Astronomy and Astrophysics, 3rd edn., p. 139. Cambridge University Press (2007)

Chapter 5
What's Out There? Asteroid Models for Target Selection and Mission Planning

Mikko Kaasalainen[1] and Josef Ďurech[2]

[1] Tampere University of Technology, Finland
[2] Charles University in Prague, Czech Republic

5.1 Introduction

Sending probes to asteroids for prospecting and, ultimately, mining, is fundamentally different from purely exploratory space missions. Time, cost-efficiency, and an industrially acceptable rate of success are key issues. We should be able to investigate targets that are easy to reach and are probably suitable for mining purposes; i.e., they contain enough material resources and provide favourable working environments. While such prior selection cannot be foolproof, we increase our chances of success by characterizing the potential targets as well as possible prior to space missions (e.g., Mueller et al. 2011).

The purpose of this chapter is to give an overview of our pre-mission sources of information about asteroids. The data available to us are mostly Earth-based; i.e., from ground or satellite observatories. With sufficient remote-sensing data, we can construct models of individual asteroids. Fortunately, the development of observational techniques and facilities over the past two decades has been nothing short of spectacular, offering us various data modalities on which to base our asteroid models. The development of mathematical methods for analyzing such data has been equally important.

This chapter is designed to be accessible to anyone interested in the detective business of asteroid modelling. One does not have to be an engineer, a scientist, or a mathematician to follow our discourse. We avoid technical details and equations, leaving them to references, simply because otherwise the story would fill an entire book. Our focus is on the big picture: how does one form a portrait of an asteroid without sending anything there – usually without seeing more than a point of light – and how accurate can we expect that portrait to be?

We proceed by taking each observational source in turn: what does it reveal to us and how? The mathematical and methodological thread runs alongside: most of the data are so-called generalized projections of the target (Kaasalainen and Lamberg 2006) that essentially sample radiation from sets of points on its surface in various ways. Reconstructing the asteroid from this information is an inverse problem that requires mathematical analysis in the choice of the solution method and the examination of the uniqueness and stability of the solution. An important concept is that the multiple data modalities are complementary: together they form a picture that is more than the sum of its parts.

In the following, we first discuss the information and parameters we want to retrieve in Sect. 5.2. Then, in Sect. 5.3, we briefly list the data sources and tabulate their information content, after which we discuss the inverse problems related to each data type. Finally, we discuss some special cases of targets and data in Sect. 5.4, and in Sect. 5.5 we sketch a scenario for obtaining a pre-mission portrait of an asteroid.

5.2 Making Asteroid Portraits

5.2.1 Exact Model Parameters

First, we define what constitutes a model of an asteroid in terms of the exact physical and mathematical parameters describing it. For many targets, we can determine only some of these parameters simply because of the lack of data. The main reason for the lack of data is the size of the solar system. Because of their small sizes and long distances from us, most asteroids are only revealed to us as dimensionless points of light. Determining their orbits from these points is a relatively easy task in celestial mechanics and accurate astrometry, but inferring the rest is considerably trickier. Our challenge is to find out an asteroid's

1. Size (volume)
2. State of rotation (period and spin axis direction)
3. Shape
4. Mass and density

In the following, by a model we mean a reconstruction that contains a quantitative estimate of some of the items on the list, which is roughly in the order of increasing difficulty. It is also roughly the order in which these parameters are typically determined. For most asteroids that can be observed in the first place (of order 10^5 to 10^6), we can get an estimate of the rotation state and shape (given enough time), and a rough estimate of the size. For some thousands of these, it is (or will be) possible to obtain very good estimates based on some disk-resolved data. Mass and density estimates require non-negligible gravitational interactions between bodies (binary asteroids or asteroids perturbing each other or space probes). Information on the corresponding orbital configurations is contained in the observational data.

5.2.2 Inferred Properties: Composition and Structure

When only ground-based data are available, composition and structure are most often estimated by inference based on the exact parameters and known or guessed correlations. Here we enter something of a grey area: though based on figures, the description cannot be put in a form of, say, a mathematical function. This is the reason why it was necessary to have a geologist among the Apollo astronauts: no instrument can describe something that an experienced eye can infer. This has as much to do with our collection of concepts and definitions as with the quantity (or lack of it) of numerical information.

When an asteroid's density, size, and shape are known, it is possible to draw inferences from morphology and geophysical considerations. The Lutetia flyby by Rosetta (Sierks et al. 2011) is an example of the interior properties of an asteroid deduced from external ones. The volume of the asteroid could be determined well as it was known very accurately for one half, and the other half was strongly constrained by the combined flyby and groundbased observations. The mass was accurately known from the deflection of the probe trajectory, so the density could be deduced. This turned out to be at the high end of known cases, constraining the structure and composition of the interior. Small asteroids such as 1999 KW4 or Steins spun up by the YORP effect feature an equatorial ridge indicating a rubble-pile structure (Ostro et al. 2006; Keller et al. 2010). A strongly bifurcated or a sharply asymmetric shape indicates two loosely connected bodies or a contact binary (Ostro et al. 2002; Demura et al. 2006). A low density for a target with apparently denser material indicates a porous structure (Britt et al. 2002; Carry 2012).

Next we briefly discuss data sources that contain information on composition in a constraining rather than exact model-defining sense.

Spectroscopy. Spectroscopy is a key data source on the material properties of asteroids. In practice, it is a direct indicator rather than a data source for an inverse problem, although one can measure spectroscopic lightcurves (cf. Nathues et al. 2005).

Asteroid taxonomy is mostly based on spectroscopic data (DeMeo et al. 2009). Different taxonomic classes correspond to different mineralogical composition of the surface and/or different evolution history. However, the same taxonomic type does not in general guarantee the same composition. Ideally, the spectroscopy over the visual and near-infrared range of the spectrum should be available for reliable spectroscopic studies. Spectroscopy may give a "tag" to the asteroid material relating it with meteorites with similar spectra (Carry 2012). Large-scale surveys such as Gaia can obtain low-resolution spectra of hundreds of thousands of asteroids (Delbo et al. 2012).

Polarimetry. Polarimetric data of the light reflected from asteroids are available for many targets, but the main problem from the point of view of modelling is that there is no good and computationally well tractable model of the state of polarization as caused by the entire surface and its structure. This, of course, means that there is no solution for the inverse problem. Correlations between, e.g., polarization and asteroid types have been observed, and these can be used for educated guesses. Also, the ratio between the signal strengths from the two different polarization modes in radar experiments is apparently correlated with the surface roughness at small scales (Ostro et al. 2002).

5.3 Data Sources for Model Construction

Exact model-building information about asteroids can be obtained with various instruments, wavelengths, and measurement types:

1. Photometry (lightcurves)
2. Radar data
3. Images (adaptive optics, space telescopes)
4. Interferometry
5. Thermal infrared radiometry
6. Stellar occultations

Of these, photometry and thermal IR (1 and 5) are disk-integrated (i.e., from pointlike sources) in character; the others contain disk-resolved information. In Table 5.1, we evaluate the model parameter estimates obtainable with each data type alone; however, data fusion improves the estimates considerably (see below and Figs. 5.5, 5.6, 5.7 and 5.9). We have also estimated the number of asteroids for which data for modelling can be obtained with each source within, say, a few tens of years.

Table 5.1 Asteroid properties obtainable from various Earth-based data sources. Large X denotes a good estimate (with a conservative estimate of the obtainable level of data and coverage), small x denotes a rough idea. Small r denotes the ability to resolve a spin ambiguity in some lightcurve and radar solutions. The last column gives the number (order of magnitude) of targets suitable for each data type, and their main target populations (where applicable).

	Shape	Spin	Size, vol.	Mass (binaries)	Targets
Photometry	X	X		Density	10^5
Radar	X (details)	X	X	X	10^2/NEAs
AO images	x	xr	X	X	10^3/large MBAs
Interferometry	x	r	X	x	10^3/large MBAs
Thermal IR			x		10^5
Occultations	x	r	X		10^2

The figures are orders of magnitude; i.e., 10^2 means some hundreds or below one thousand and so on. The overwhelming difference between photometry or thermal IR and others is due to large sky surveys. With good and well-calibrated instruments, sparse photometry (Kaasalainen 2004; Ďurech et al. 2006) makes it possible to obtain a large number of asteroid models within some ten years. More detailed models with data sources at higher resolution require large top-level instruments of which there probably will be only a few in the world at a given time. Thus the constraints on observing time available limit the number of models obtainable in intermediate or high resolution, even if it were technically feasible to resolve thousands of asteroids with the future superlarge telescopes. For the same reason, we do not give, e.g., adaptive optics (AO) images the full capability to model 3D shape and spin. Theoretically this is possible, but in practice very few asteroids will be given enough observing time for the purpose: instead, we are likely to have some complementary snapshots.

Next we discuss the modelling possibilities and analysis of data sources each in turn (see also Sect. 1 of Carry et al. 2012); before this, we emphasize the simultaneous use of the sources in data fusion.

Data Fusion. While lightcurves and radar data are sufficient for asteroid modelling by themselves, they can be augmented with the other, complementary data modes to produce more detailed and reliable models. In some cases, the other data can complete a deficient lightcurve or radar dataset so that modelling is possible in the first place. Data fusion, or the simultaneous use of several data sources in a combined inverse problem, is a powerful method for getting as much out of all available data as possible. The key issue is the optimal weighting of each source (or different subsets of one source); this is called the maximum compatibility estimate (Kaasalainen 2011).

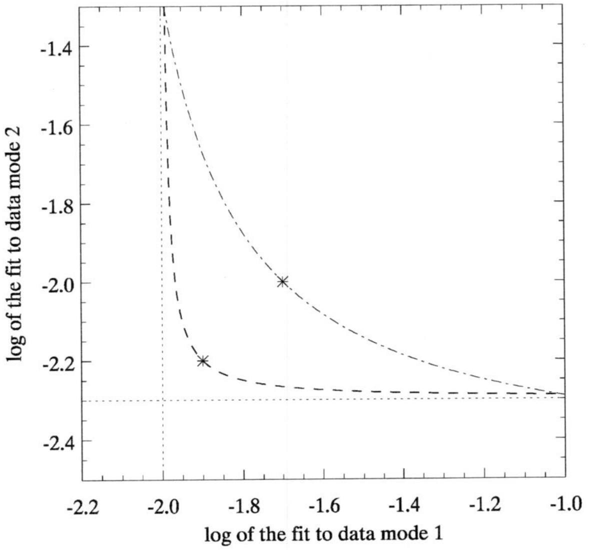

Fig. 5.1 The principle of the maximum compatibility estimate. The ideal solution is given by the point (asterisk) on a curve closest to the intersection of the two straight dotted lines.

The main principle of this approach is shown in Fig. 5.1, where we schematically consider two data modes. The parameters of the mathematical model are adjusted in data fitting. The weight of data mode 2 increases from left to right along a plotted curve; logarithm is used for scale invariance. The dotted lines give the best fits for each data source alone, and the ideal weight and solution corresponds to the point on a curve closest to their intersection. The dashed line describes two compatible models for the data modes: both fit levels are good at the optimum. The dot-dash is an example of two somewhat mutually discrepant models or measurements: separate fits are good, but the optimal solution is a bad compromise. This implies systematic errors in data, or an insufficient mathematical model.

5.3.1 Photometry

The disk-integrated brightness of an asteroid varies in time due to its rotation and motion around the Sun. Light-variations of asteroids have been recorded for about a century, and electric measurements started in the late 1940's; many of the early ones are quite accurate. The mass-production of CCDs in the 1990's allowed routine acquisition of accurate data even with small amateur telescopes. Now photometry is recorded for hundreds of thousands of asteroids in all-sky surveys such as Pan-STARRS and LSST (under construction; Jedicke et al. 2006; Jones et al. 2009). This makes photometry the richest data source on asteroids by far.

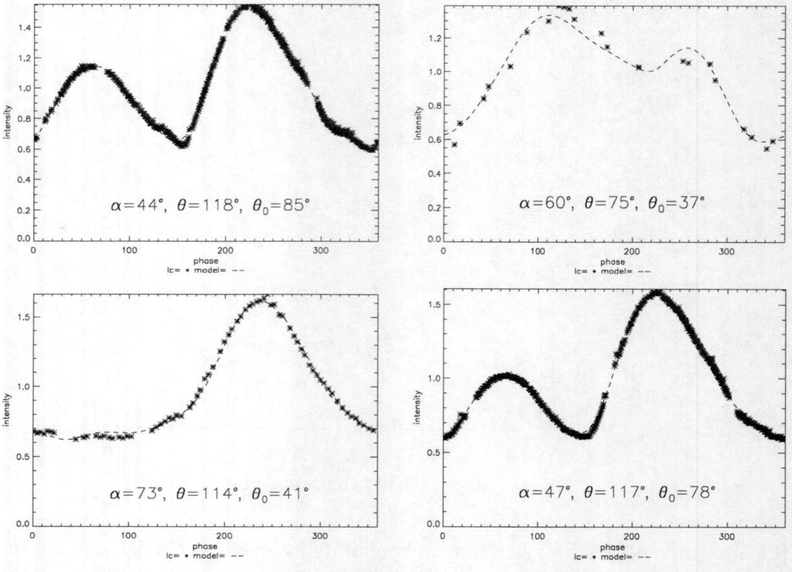

Fig. 5.2 Lightcurves of the asteroid Golevka (asterisks), each for one rotation around its axis, together with the model fit (dashed line). The model is shown in Fig. 5.4. The measurements are plotted in relative intensity.

Depending on the direction of the rotation axis of an asteroid, different parts of its surface are tilted towards the Earth as it moves around the Sun. For example, main-belt asteroids in the plane of the ecliptic show largest variation in the shapes of their lightcurves when their rotation axes lie in that plane. Such a configuration allows the target to be seen from all viewing aspects. In all the inverse problems discussed here, sufficiently wide aspect coverage is necessary for modelling.

Examples of lightcurves obtained at various illumination and viewing geometries are shown in Fig. 5.2. The observations are shown as asterisks, and the dashed line depicts the values obtained from the shape and spin reconstruction. The solar phase angle α denotes the angle between the illumination and viewing directions (the Sun and the Earth) seen from the target; θ gives the viewing direction measured from the asteroid pole, and θ_0 denotes the illumination direction.

For almost a century, lightcurves were not thought to be sufficient for proper general shape modelling: the inverse problem was considered unsolvable. The data were only used for crude rule-of-thumb estimates of the rotation period and axis and the dimension ratios of the body (Magnusson et al. 1989). This misunderstanding stems from the fact that even if one knows the areas of the shadows of an object in all directions, it is impossible to reconstruct the object uniquely (Russell 1906; Kaasalainen et al. 1992). This was thought to apply to asteroid lightcurves, since the brightness of an asteroid at opposition (the Sun behind the Earth seen from the asteroid) is, to a good approximation, proportional to its area projected on the plane of sky. For the same reason, the full moon looks like a disk instead of a ball.

However, this ambiguity vanishes away from opposition; i.e., when shadows are visible on the surface of the asteroid. This enables a rigorous solution of the inverse problem: the unique reconstruction of the target turns out to be possible even at small solar phase angles α. Seen from the Earth, main-belt asteroids reach phase angles up to about 20-30 degrees, and near-Earth targets often more than 90 degrees, so the natural observing geometries allow model solutions for most of the targets for which accurate photometry can be recorded. This mathematical result (Kaasalainen et al. 1992; Kaasalainen and Lamberg 2006) led to practical robust procedures for reconstructing the shapes and spin states of asteroids from their lightcurves (Kaasalainen et al. 2001, 2002, 2004; Ďurech et al. 2009, 2010; Hanuš et al. 2011, 2013). This method is applicable to almost any asteroid, which is far more than with any other observation technique.

While the problem of lightcurve inversion is solvable, some limitations on the solution and the data available are inevitable. In a nutshell, these are:

- Shape models from Earth-based lightcurves are usually limited to convex approximations of the object. Lightcurves do not reveal craters or valleys (Ďurech and Kaasalainen 2003). At this level of resolution, the models are well tested against the ground truth from space probes (Kaasalainen et al. 2001) and laboratory models (Fig. 5.3; Kaasalainen et al. 2005). The shape solution is also stable: it is not particularly vulnerable to data noise or inaccurate assumptions on the light-scattering properties of the surface.

- For targets close to the ecliptic plane, the longitude of the spin axis has two mirror solutions (while the latitudes of these are essentially the same). This applies also to radar (Kaasalainen and Lamberg 2006).
- An asteroid should be seen from a few different observing geometries at least. A main-belt asteroid requires at least three different apparitions (a time interval of some five years), while one apparition may be sufficient for a near-Earth target.
- The size of the target must be inferred from other data types, or by guessing the darkness (albedo) of the surface material.
- The surface albedo is assumed not to vary much over the surface. There is an indicator for the violation of this condition, but only modest indication has been obtained for about one percent of the observed targets. Probe data and physical considerations (accumulation of surface regolith, space weathering) support this as well (Kaasalainen et al. 2001, 2002).

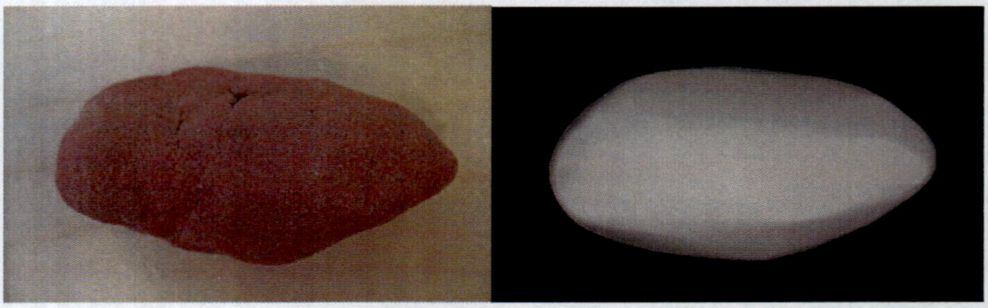

Fig. 5.3 Validation of lightcurve inversion by laboratory measurements (left: target, right: reconstruction from 7 lightcurves). This ground truth also validated the robustness of the inversion against poorly known surface scattering properties.

Using photometry with filters at various wavelengths, it is possible to construct colour maps of the surface with the lightcurve inversion method as well (Nathues et al. 2005). The slight differences in the brightness variations at different wavelengths do not affect the stable shape solution, but they indicate colour variations over the surface. This gives information on the potential differences of the composition of the surface material.

5.3.2 Radar Data

Contrary to other remote-sensing techniques, in radar studies the source of radiation is not the Sun but a powerful radio emitter on the Earth. The pulse is sent towards the target and then the echo is received and analysed; this is the only experiment-like ground-based observation method of asteroids. Because the intensity of the echo decreases with the fourth power of the distance to the target, radar observations are limited to near-Earth asteroids (NEAs) that come close to the Earth or to the largest asteroids of the main belt (MBAs). At present, the only two radar facilities that are capable of these experiments at a sufficient level of signal strength are Arecibo and Goldstone. Future possibilities are offered by combining large new radiotelescopes or telescope systems (e.g., LOFAR and SKA) with transmitters such as EISCAT.

For asteroids that come close to the Earth, radar is the best possibility for detailed modelling. Most main-belt asteroids cannot be measured well with current radars, so the number of well-observed radar targets does not grow fast. In principle, radar can achieve a resolution level of less than ten meters, if the data are very good (the target makes a close approach).

A range-Doppler radar samples the surface of the asteroid in bins that contain the integrated echo power from surface patches that have a common distance from and, due to the rotation, relative speed with respect to the radar. The size of the bins depends on the available resolution level, and each bin is represented as a pixel in a range-Doppler plot (which is not an image of the target). The shape and rotation of the target can be modelled from a set of such plots obtained at various geometries. This inverse problem has, in principle, a unique solution if the geometry coverage determined by orbital and rotational motion is good, just as with lightcurves (Ostro et al. 2002; Kaasalainen and Lamberg 2006). For example, if the rotation axis and the orbit do not allow other than mostly equatorial viewing aspects, a shape ambiguity or instability remains.

As an example of a typical instability, the numerous radar measurements of the asteroid Itokawa had a nominal resolution of some ten meters (the target itself is roughly 500 meters across), but the resulting model (Ostro et al. 2005) failed to portray the shape features of the target that occur at a much larger scale and were later recorded by the space probe Hayabusa (Demura et al. 2006). Thus the actual resolution of radar was not better than that of photometry. Another ground truth was obtained by the Chinese Chang-E probe that flew by the asteroid Toutatis in December 2012. Toutatis has been extensively covered by radar, and the main features of the resulting high-resolution model (Hudson et al. 2003) agree with the probe images. However, the purported resolution of the radar-based model is exaggerated (Xiaoping Lu, personal communication). Radar models should thus not be mistaken for accurate high-resolution representations.

The reliability of a model is always largely dependent on the observing geometries available (constrained by observing schedules or the laws of nature). Any high-resolution models are more vulnerable to instabilities of this kind than the low-resolution lightcurve models that implicitly contain the resolution scale in their representation of the global shape; Figure 5.4 illustrates the difference between the two. They are also more sensitive to assumptions on the scattering properties. Detecting such instabilities and determining the actual resolution scale to avoid over-interpretation of the model obtained is important in all imaging, not just radar, that aims to portray shape details.

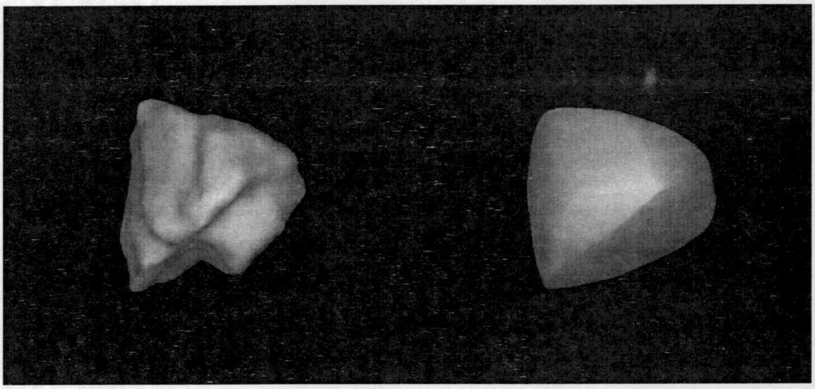

Fig. 5.4 Images of models of the small near-Earth asteroid Golevka (less than 1 km across). Left: radar model (courtesy of Scott Hudson); right: lightcurve model. The radar model contains artificial small-scale details that are retained in the modelling process to enhance the rendering of intermediate-scale nonconvex features. Such features are "gift-wrapped" by the convex lightcurve model that depicts the resolution obtainable from photometry alone.

5.3.3 Adaptive Optics and Other Images

Ground-based telescopes equipped with adaptive optics (AO) are capable of resolving the disks of large main belt asteroids (at least tens of kilometers across). The level of detail is not high as the images do not contain a large number of pixels as shown in Fig. 5.5; however, the main shape features can be resolved. The processing of AO images usually causes the boundary contours of the target to be much better defined than the image pixel brightnesses inside the contour; the latter may contain contrast errors and artificial features. Apart from resolving disks of asteroids, AO techniques have also led to the discovery of small moons of tens of main belt asteroids.

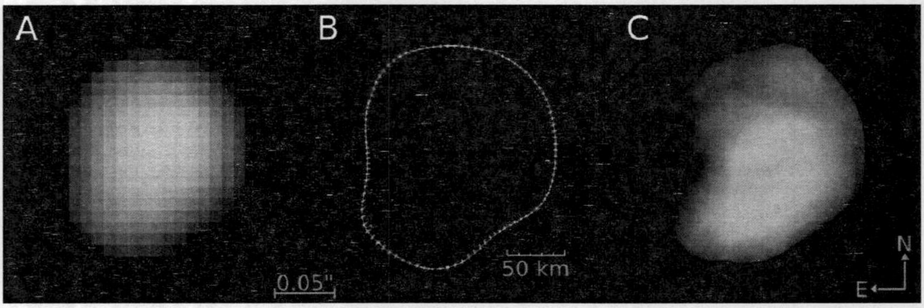

Fig. 5.5 An adaptive optics image of the asteroid Daphne (A), together with the extracted boundary contour (B), and the plane-of-sky image of the model reconstructed with AO and lightcurves (C). (Figure courtesy of Benoit Carry).

Some disk-resolved asteroid images of resolution similar to AO have been taken with the Hubble Space Telescope (HST), but the main development in Earth-based asteroid images is in AO at large new telescopes.

Actual disk-resolved images can currently be obtained of some 200 largest asteroids in principle (Marchis et al. 2006). Most of these are with adaptive optics on large ground-based telescopes. The resolution level is currently not very high; the superlarge future telescopes will enhance it, increasing the number of resolvable targets to some thousands (Merline et al. 2013), but the actual number is likely to remain below this, as discussed earlier. From the point of view of asteroid mining, most of these are large main-belt asteroids not among the potential near-Earth prospecting candidates.

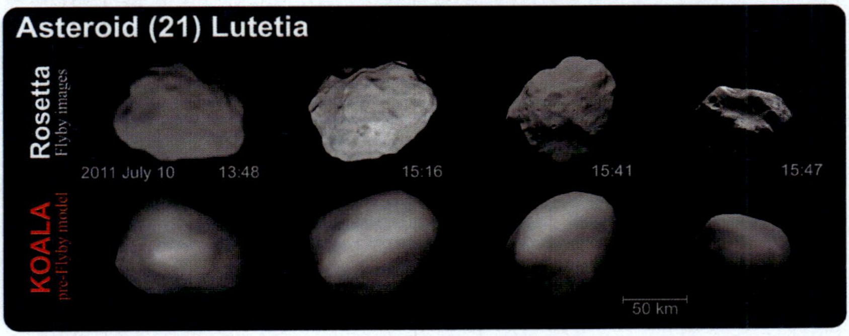

Fig. 5.6 Validation of Earth-based imaging methods. Top row: Rosetta flyby images of the asteroid Lutetia (Sierks et al. 2011). Bottom row: images of the lightcurve- and AO-based surface model at the corresponding viewing geometries (Carry et al. 2012). (Figure courtesy of Benoit Carry).

By combining a number of images obtained at different geometries with other data (especially lightcurves), it is possible to reconstruct a fairly detailed shape model of the target as in Fig. 5.5 (the KOALA algorithm: Carry et al. 2010, 2012; Kaasalainen 2011; Kaasalainen and Viikinkoski 2012). The resulting data fusion model is better than that from any of the complementary sources alone. The power of this approach was validated with the Rosetta flyby of the asteroid Lutetia as shown in Fig. 5.6. The model was made and published before the flyby using lightcurves and adaptive optics images only. The excellent agreement shows that intermediate-sized shape details and features can be reconstructed from ground-based data.

5.3.4 Interferometry

Further increase in angular resolution is possible when using interferometry - the wavefronts from two or more telescopes are combined together. The angular resolution is then equivalent to the resolution of a telescope with the same aperture as the distance of the telescopes. However, this resolution is available only along the line connecting the telescopes, and the resulting observable is a sampling of a transformation of the asteroid's image on the plane of sky. As all data providing angular resolution, interferometry can be used for size estimates (Delbo et al 2009).

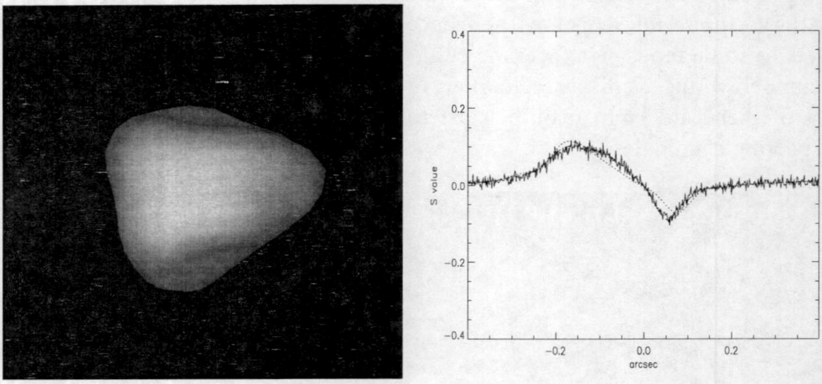

Fig. 5.7 A portrait of the asteroid Eunomia (of order 250 km across), modelled with lightcurves and interferometry from HST/FGS. A sample interferometric curve is shown, together with the model fit (dashed line; the dotted one is for the lightcurves-only model).

So far, there are few interferometric asteroid data of quality sufficient for modelling. These are mainly from HST Fine Guidance Sensor (FGS). Much more data will be available with the new multi-telescope ALMA system (Busch 2009).

In principle, interferometry offers disk-resolved information about the target. The information is scrambled by the response function of the instrument, but can be extracted when solving the inverse problem of simultaneous data sources (Kaasalainen and Viikinkoski 2012; Kaasalainen and Lamberg 2006). An example of HST/FGS interferometry in data fusion is shown in Fig. 5.7. The model from lightcurves alone would have given the fit with the dotted line: interferometry helped to resolve some nonconvex features on the surface of the asteroid Eunomia.

When only a limited number of viewing directions are available, interferometric data are not as rich as AO images. With a large number of interferometric base lines (e.g., thousands in ALMA), the scrambled plane-of-sky image can be sampled densely, and the situation is essentially the same as having an image (AO images are also scrambled by a geometrically simpler point-spread function that has to be removed in the processing). Thus the ALMA data are going to be as important as AO images for modelling several hundreds of large main-belt asteroids. ALMA operates in the thermal infrared, so thermal modelling must be included in the solution of the inverse problem.

5.3.5 Thermal Infrared Data

Thermal infrared observations are much harder to obtain than visual ones because of the absorption bands in the atmosphere and the thermal background contaminating the signal from the asteroid. Therefore, the best observing station for obtaining thermal infrared data is outside the Earth's atmosphere (ISO, IRAS, Akari, WISE, Spitzer; Tedesco et al. 2002; Usui et al. 2011; Mainzer et al. 2011). Observations of asteroid brightnesses in the thermal infrared are the main source of information about asteroid sizes (Harris and Lagerros 2002).

In optical wavelengths, we detect the light from the Sun scattered from an asteroid's surface. However, for dark asteroids, most of the incoming energy is absorbed, heating the surface and subsurface layers, and the surface emits temperature-dependent radiation according to Planck's law. At wavelengths of five micrometers and longer, the emitted heat becomes measurable.

Observations in the thermal infrared can remove the albedo/size ambiguity that is inevitable in visual photometry where one can only surmise the albedo. The temperature of the surface is not very sensitive to the albedo because the albedo is low; i.e., around 95% of the solar radiation energy is absorbed and subsequently emitted, and even large relative changes in the albedo change the temperature and the irradiation flux only a few percent. Thus the amount of emitted radiation is mostly proportional to the surface area.

In order to compute the emitted radiation, the temperature of the surface must be estimated (Harris and Lagerros 2002; Delbo and Harris 2002). In the first approximation, it depends on the thermal inertia (the capability of the material to resist temperature changes). By estimating the thermal inertia, we can infer something about the surface – a low thermal inertia indicates a surface covered by thick and fine regolith, whereas a high inertia indicates bare rock. This has direct influence on the engineering choices of a landing mission.

Reliable thermal infrared data are available for tens of thousands of asteroids (WISE, IRAS, AKARI; Masiero et al. 2011). However, these measurements are usually individual data points covering an interval of months or years. In order to estimate an asteroid's size projected on the plane of sky, one has to know the orientation of the asteroid at the time of observation and its shape, which can be derived from photometry. Thus, by combining optical and thermal data, we can in principle derive scaled shape models of a significant part of the asteroid population.

5.3.6 Stellar Occultations

From time to time asteroids pass in front of distant stars. An observer placed in the path of the shadow sees this as a sudden temporary decrease in the star's brightness. With multiple observers in various parts on the globe, timing the occultations corresponds to measuring a set of chord lengths across the shadow silhouette. The accuracy of the profile points depends only on the accuracy of the timings.

Although these events cannot be planned, only predicted, and the accuracy of such predictions is not perfect, the situation will improve with the precise astrometric catalogue from the Gaia mission (Tanga and Delbo 2007). Occultation measurements are especially important for the faraway trans-Neptunian objects that are faint and cold.

As with AO images and thermal data, if there is only one 'snapshot' of the asteroid, its orientation is not known from that data alone, and we can estimate only the lower limit of the size. When combined with a lightcurve inversion model, even a few occultation chords can determine the right size and pole for the model as shown in Fig. 5.8 (Ďurech et al. 2011). The chords between disappearance and reappearance, obtained at various locations on the globe and plotted on the shadow plane of the occultation, confirm the shape solution from lightcurves (solid contour) and scale it. They also reject the mirror pole solution (dotted contour). Even though the profile points are sparsely sampled, they can also be used in the same way as AO data in the inversion with combined data types.

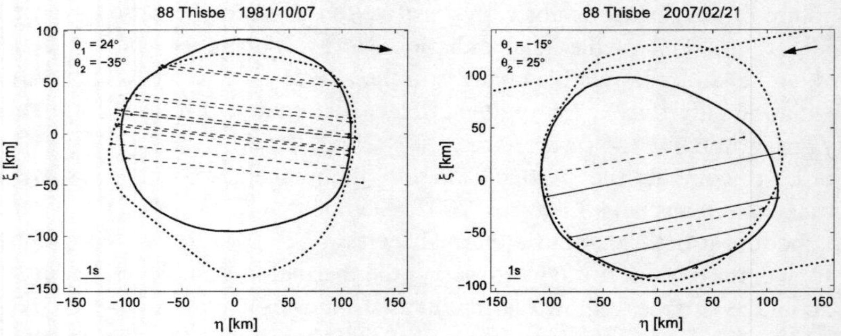

Fig. 5.8 Size and pole determination for the asteroid Thisbe from occultation timings (straight lines). The distance corresponding to the motion of the asteroid in one second in the shadow plane is shown in the plots. The solid contour is the plane-of-sky profile of the lightcurve model with the correct pole; the dotted line is that of the mirror-pole solution.

The practical problem is that we have to wait for a suitable occultation and then place many observers in the path of the shadow. For a successfully observed occultation, an additional benefit is that the resolution of the profile is of the order of kilometers. For large asteroids, this is better than with ground-based imaging techniques.

5.4 Special Cases of Targets and Data

5.4.1 Binaries

The fraction of binary asteroids in the asteroid population is not negligible - according to Pravec et al. (2006) it is about 16% for the near-Earth population, and probably of the same order for small MBAs. A particularly useful aspect about binaries is that we can estimate their densities even if we can only measure their lightcurves. This is because the lightcurves give the orbital period of the system and its approximate relative size scale: how large the distance between the components is relative to their sizes (Scheirich and Pravec 2009). On the other hand, Kepler's third law says that the square of a system's orbital period is proportional to the cube of that distance and inversely proportional to the system's mass. Hence, the density is inversely proportional to the square of the period, and the absolute mass and size scale are not needed for its determination.

For disk-resolved data, we can obtain both the mass and the size scale (and hence the density). Binaries are thus especially important in studying the structure and composition of asteroids.

5.4.2 Complex Rotation

Although most asteroids rotate along the shortest axis in the principal state with the lowest energy, some asteroids exhibit complex 'tumbling' and their rotation state can be described as free precession. These cases can be modelled by the same inversion techniques as 'standard' asteroids with a more complicated description of their rotation (Kaasalainen 2001; Scheirich et al. 2010).

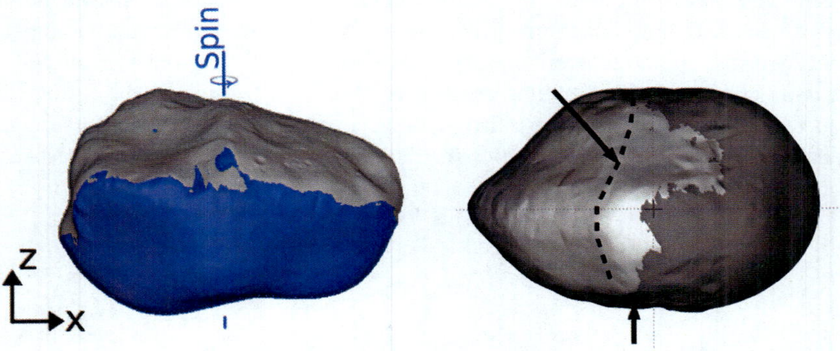

Fig. 5.9 Seeing the dark side. Left: Model of the asteroid Lutetia (120 km across) for the part seen by Rosetta completed by lightcurve and AO data (darker lower half). Right: Steins flyby map (5 km across) completed by lightcurve data (darker right end). The dashed line on the surface emphasizes the ridge-like equatorial feature. Both views are equatorial. For a detailed account of the flyby- and ground-based parts, see Keller et al. 2010 and Carry et al. 2012.

Another deviation from a simple rotation state (fixed rotation axis and rotation period) is caused by the Yarkovsky–O'Keefe–Radzievskii–Paddack (YORP) effect - the net torque caused by thermal radiation from an irregularly shaped asteroid. In the long term, this effect tilts the rotation axis with respect to the orbital plane and changes the rotation period. The change in the direction of the rotation axis is slow compared to the interval of observation. However, the secular change of the rotation period is relatively fast and has already been detected on some targets (Kaasalainen et al. 2007; Ďurech et al. 2008).

5.4.3 Flybys

Space probe flybys allow detailed views of about a half of the target, while the other half remains unseen. This, however, is sufficient for reconstructing the dark side to a better degree than obtainable by ground-based data only, because the directly seen side (that can be mapped with cartographic techiques) constrains the solution strongly (Kaasalainen and Viikinkoski 2012). This technique was used in the Rosetta flybys of Steins (Keller et al. 2010) and Lutetia (Sierks et al. 2011), and is applicable to any flyby. The full models of the Rosetta targets are shown in Fig. 5.9.

5.4.4 Asteroid Tomography

Direct analysis of the interior of an asteroid can only be carried out in situ. Before actually drilling below the surface, tomographic methods and techniques offer a possibility to have a glimpse of the inside of an asteroid. Radio tomography is a technique based on, e.g., the path lengths of the radiation inside the target (Kofman 2007; Pursiainen and Kaasalainen 2013). The experiment can be carried out by transmitters/receivers on the surface and one orbiting the asteroid, allowing a wide coverage of inspection geometries (Fig. 5.10). This can be further developed to include full radar pulse coding, giving a possibility to estimate the changes of the material properties inside and to locate at least the largest cavities or inclusions.

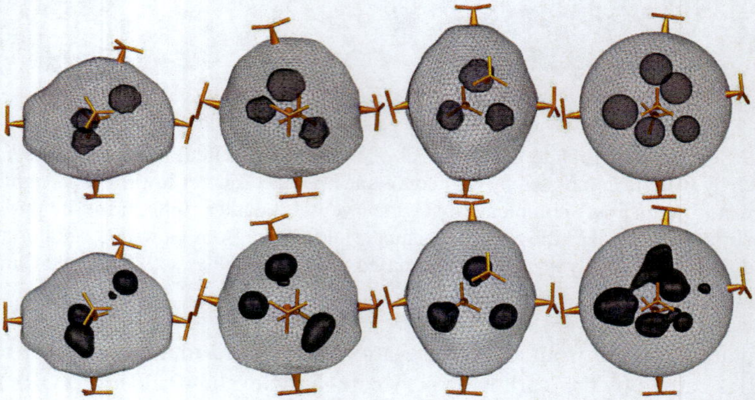

Fig. 5.10 Asteroid prospecting with radio tomography. Top row: inclusions inside 1-km sized asteroids suitable for mining. Bottom: reconstructions from collimated radio beams between an orbiter and the transmitters shown on the surface (not to scale).

Seismic tomography is routinely used for prospecting and mining purposes on earth, and can in principle be applied to asteroids as well (Asphaug et al. 2002).

5.5 Conclusions and Discussion

For mass production of asteroid models, photometry is more or less the only option. Photometric measurements can be made of any target that can be observed in the first place, so millions of asteroids of various sizes and orbits can be reconstructed with lightcurve inversion. The data are from a large number of telescopes of various types (Ďurech et al. 2006, 2007, 2009, 2010; Hanuš et al. 2011, 2013). In practice, a very limited number of data points (hundreds or so per asteroid) from all-sky surveys will be available for the majority of such targets, but this suffices for basic modelling. Of these, the most interesting targets can be observed further to obtain dense additional lightcurves.

More detailed models can be made by combining lightcurves with the other data sources. For main-belt asteroids, adaptive optics images, interferometry from ALMA, and radar data should produce hundreds or even thousands of intermediate-resolution models in this way. Radar can be used to reconstruct the shapes of hundreds of near-Earth asteroids in higher resolution. Structural properties of the asteroid can be inferred from the geometrical model, and spectroscopy gives an idea of the surface material.

We can envisage a top-to-bottom scenario for selecting interesting targets and obtaining increasingly detailed ground-based information about them, from lower to higher resolution. Commercially operated telescopes are probably the fastest and most efficient way to collect data sufficient for modelling the thousands of asteroid candidates for prospecting prior to space missions. Several probes might then fly by a number of most promising candidates, and the data from different sources could be combined to complete individual models. Finally, when a probe flies to an asteroid to stay there, tomographic methods can be used to estimate the interior structure before starting mechanical prospecting and mining operations.

Acknowledgements. This work was supported by the Academy of Finland (Centre of Ecxcellence in Inverse Problems and the project "Inverse problems of stochastic and regular surfaces"), Tampere University of Technology, and the grant P209/10/0537 of the Czech Science Foundation as well as the Research Program MSM0021620860 of the Czech Ministry of Education. We are greatly indebted to Benoit Carry, Marco Delbo, Josef Hanuš, Sampsa Pursiainen, and Matti Viikinkoski for valuable comments and discussions.

References

Asphaug, E., Ryan, E., Zuber, M.: Asteroid interiors. In: Bottke, W., et al. (eds.) Asteroids III, p. 463. U Arizona Press, Tucson (2002)

Britt, D., Yeomans, D., Housen, K., Consolmagno, G.: Asteroid density, porosity, and structure. In: Bottke, W., et al. (eds.) Asteroids III, p. 485. U. Arizona Press, Tucson (2002)

Busch, M.: ALMA and asteroid science. Icarus 200, 347 (2009)

Carry, B., ten colleagues: Physical properties of 2 Pallas. Icarus 205, 460 (2010)

Carry, B., twelve colleagues: Shape modelling technique KOALA validated by ESA Rosetta at (21) Lutetia. Planetary and Space Science 66, 200 (2012)

Carry, B.: Density of asteroids. Planetary and Space Science 73, 98 (2012)

Delbo, M., Harris, A.: Physical properties of near-Earth asteroids from thermal infrared observations and thermal modeling. Meteoritics and Planetary Science 37, 1929 (2002)

Delbo, M., Ligori, S., Matter, A., Cellino, A., Berthier, J.: First VLTI-MIDI direct determinations of asteroid sizes. Astrophys. J. 694, 1228 (2009)

Delbo, M., seven colleagues: Asteroid spectroscopy with Gaia. Planetary and Space Science 73, 86 (2012)

DeMeo, F., Binzel, R., Slivan, S., Bus, S.: An extension of the Bus taxonomy into the near-infrared. Icarus 202, 160 (2009)

Demura, H., nineteen colleagues: Pole and global shape of 25143 Itokawa. Science 312, 1347 (2006)

Ďurech, J., Kaasalainen, M.: Photometric signatures of highly nonconvex and binary asteroids. Astron. Astrophys. 404, 709 (2003)

Ďurech, J., Grav, T., Jedicke, R., Kaasalainen, M., Denneau, L.: Asteroid models from Pan-STARRS photometry. Earth, Moon, and Planets 97, 179 (2006)

Ďurech, J., fourty-one colleagues: Physical models of ten asteroids from an observers' collaboration network. Astron. Astrophys. 465, 331 (2007)

Ďurech, J., ten colleagues: Detection of the YORP effect in asteroid (1620) Geographos. Astron. Astrophys. 489, L25 (2008)

Ďurech, J., ten colleagues: Asteroid models from combined sparse and dense photometric data. Astron. Astrophys. 493, 291 (2009)

Ďurech, J., Sidorin, V., Kaasalainen, M.: DAMIT: a database of asteroid models. Astron. Astrophys. 513, A46 (2010)

Ďurech, J., eleven colleagues: Combining asteroid models derived by lightcurve inversion with asteroidal occultation silhouettes. Icarus 214, 652 (2011)

Hanuš, J., fourteen colleagues: A study of asteroid pole-latitude distribution based on an extended set of shape models derived by the lightcurve inversion method. Astron. Astrophys. 530, A134 (2011)

Hanuš, J., one hundred nineteen colleagues: Asteroids' physical models from combined dense and sparse photometry and scaling of the YORP effect by the observed obliquity distribution. Astron. Astrophys. 551, A67 (2013)

Harris, A., Lagerros, J.: Asteroids in the thermal infrared. In: Bottke, W., et al. (eds.) Asteroids III, U. Arizona Press, Tucson (2002)

Hudson, R., Ostro, J., Scheeres, D.: High-resolution model of asteroid 4179 Toutatis. Icarus 161, 346 (2003)

Jedicke, R., Magnier, E., Kaiser, N., Chambers, K.: The next decade of solar system discovery with Pan-STARRS. In: Proceedings of IAU Symposium, vol. 236, p. 341 (2006)

Jones, R.L., eight colleagues: Solar system science with LSST. Earth, Moon, and Planets 105, 101 (2009)

Kaasalainen, M., Lamberg, L., Lumme, K., Bowell, E.: Interpretation of lightcurves of atmosphereless bodies. I. General theory. Astron. Astrophys. 259, 318 (1992)

Kaasalainen, M., Torppa, J., Muinonen, K.: Optimization methods for asteroid lightcurve inversion. II. The complete inverse problem. Icarus 153, 37 (2001)

Kaasalainen, M.: Interpretation of lightcurves of precessing asteroids. Astron. Astrophys. 376, 302 (2001)

Kaasalainen, M., Torppa, J., Piironen, J.: Models of twenty asteroids from photometric data. Icarus 159, 369 (2002)

Kaasalainen, M.: Physical models of large number of asteroids from calibrated photometry sparse in time. Astron. Astrophys. 422, L39 (2004)

Kaasalainen, M., Lamberg, L.: Inverse problems of generalized projection operators. Inverse Problems 22, 749 (2006)

Kaasalainen, M., Ďurech, J., Warner, B., Yu, K., Gaftonyuk, N.: Acceleration of the rotation of asteroid 1862 Apollo by radiation torques. Nature 446, 420 (2007)

Kaasalainen, M.: Multimodal inverse problems; maximum compatibility estimate and shape reconstruction. Inverse Problems and Imaging 5, 37 (2011)

Kaasalainen, M., Viikinkoski, M.: Shape reconstruction of irregular bodies with multiple complementary data sources. Astron. Astrophys. 543, A97 (2012)

Kaasalainen, S., Kaasalainen, M., Piironen, J.: Ground reference for space remote sensing: Laboratory photometry of an asteroid model. Astron. Astrophys. 440, 1177 (2005)

Keller, H.U., fourty-six colleagues: E-type asteroid Steins as imaged by OSIRIS on board Rosetta. Science 327, 190 (2010)

Kofman, W., fourteen colleagues: The comet nucleus sounding experiment by radiowave transmission (CONSERT): A short description of the instrument and the commissioning stages. Space Science Reviews 128, 1 (2007)

Magnusson, P., seven colleagues: Determination of pole orientations and shapes of asteroids. In: Binzel, R.P., et al. (eds.) Asteroids II, p. 67. U. Arizona Press, Tucson (1989)

Mainzer, A., thirty-six colleagues: NEOWISE observations of near-Earth objects: Preliminary results. Astrophys. J. 743, A156 (2011)

Marchis, F., seven colleagues: Shape, size and multiplicity of main-belt asteroids. I. Keck adaptive optics survey. Icarus 185, 39 (2006)

Merline, W., sixteen colleagues: The resolved asteroid program - size, shape, and pole of (52) Europa. Icarus (in press, 2013)

Masiero, J., seventeen colleagues: Main-belt asteroids with WISE/NEOWISE. I. Preliminary albedos and diameters. Astrophys. J. 741, 68 (2011)

Mueller, M., sixteen colleagues: ExploreNEOs. III. Physical characterization of 65 potential spacecraft target asteroids. Astron. J. 141, 109 (2011)

Nathues, A., Mottola, S., Kaasalainen, M., Neukum, G.: Spectral study of the Eunomia family. I. Eunomia. Icarus 175, 452 (2005)

Ostro, S.J., Hudson, R.S., Benner, L., Giorgini, J., Magri, C., Margot, J.-L., Nolan, M.: Asteroid radar astronomy. In: Bottke, W., et al. (eds.) Asteroids III, p. 151. U. Arizona Press, Tucson (2002)

Ostro, S.J., twelve colleagues: Radar observations of Itokawa in 2004 and improved shape estimation. Meteoritics and Planetary Science 40, 1563 (2005)

Ostro, S.J., fifteen colleagues: Radar imaging of binary near-Earth asteroid (66391) 1999KW4. Science 314, 1276 (2006)

Pravec, P., fifty-six colleagues: Photometric survey of binary near-Earth asteroids. Icarus 181, 63 (2006)

Pursiainen, S., Kaasalainen, M.: Iterative Alternating Sequential method for radio tomography of asteroids in 3D. Planetary and Space Science 82, 84 (2013)

Russell, H.N.: On the light-variations of asteroids and satellites. Astrophys. J. 24, 1 (1906)

Scheirich, P., Pravec, P.: Modeling of lightcurves of binary asteroids. Icarus 200, 531 (2009)

Scheirich, P., ten colleagues: The shape and rotation of asteroid 2008 TC3. Meteor. Planet. Sci. 45, 1804 (2010)

Sierks, H., fifty-seven colleagues: Images of asteroid 21 Lutetia: a remnant planetesimal from the early solar system. Science 334, 487 (2011)

Tanga, P., Delbo, M.: Asteroid occultations today and tomorrow: toward the Gaia era. Astron. Astrophys. 474, 1015 (2007)

Tedesco, E., Noah, P., Noah, M., Price, S.: The supplemental IRAS minor planet survey. Astron. J. 123, 1056 (2002)

Usui, F., twelve colleagues: Asteroid catalog using AKARI: AKARI/IRC mid-infrared asteroid survey. Publ. Astron. Soc. Japan 63, 1117 (2011)

Chapter 6
Earth's Temporarily-Captured Natural Satellites – The First Step towards Utilization of Asteroid Resources

Mikael Granvik[1], Robert Jedicke[2], Bryce Bolin[2], Monique Chyba[2], Geoff Patterson[2], and Gautier Picot[2]

[1] University of Helsinki, Finland
[2] University of Hawaii, USA

6.1 Introduction

Granvik et al. (2012) predict that the Earth is surrounded by a cloud of small temporarily-captured asteroids. These temporarily-captured orbiters (TCOs) originate in the near-Earth-object (NEO) population and are temporarily captured in the potential well of the Earth-Moon system (EMS). Granvik et al. (2012) predict that the largest object in orbit around Earth at any given moment (other than the Moon) has a diameter $D \sim 1$ m (Sect. 6.2). The number of TCOs is inversely proportional to their size such that there are on the order of 10^3 0.1-meter-diameter TCOs in orbit around Earth at any given time.

For the purpose of utilizing resources available in asteroids it is essential to first carry out accurate remote prospecting. The detailed mineralogy of an asteroid can in some cases be derived from spectrometric observations (Gaffey et al. 2002) but in general it is not straightforward to link meteorite types based on mineralogy with asteroid classes based on spectrometric observations. It was, for example, assumed for a long time that M-class asteroids are primarily made of Fe-Ni metal. Later it was realized that the composition of M-type asteroids is more complex and they contain substantial amounts of silicates (see, e.g., Ockert-Bell et al. 2010).

Observational techniques for understanding asteroid mineralogy can be improved by validating the analysis of spectrometric observations of an object with the laboratory-defined mineralogy of samples from the same object. This type of validation is a major motivation for asteroid sample-return missions such as JAXA's successful Hayabusa mission and the planned Hayabusa II, OSIRIS-REx (NASA), and MarcoPolo-R (ESA) missions. But it will take decades and substantial amounts of funding before a large-enough set of samples have been obtained to understand the mineralogy of a meaningful fraction of all asteroid taxonomical classes. Fortunately, there are faster and less expensive alternatives.

In 2008, a small asteroid, 2008 TC_3, was discovered on a trajectory that would lead to a collision with the Earth some 22 hours later. Astrometric, photometric and spectrometric follow-up observations using ground-based telescopes were scheduled and executed in the time leading to the impact. The asteroid entered Earth's atmosphere over the northern Sudanese desert and so-called Almahata Sitta meteorites of the object were later found in the region. According to the spectra of 2008 TC_3 the asteroid belongs to dark asteroids in the C-complex, possibly the F taxonomic class. Analysis of the Almahata Sitta meteorites revealed that the asteroid was made of dark carbon-rich anomalous ureilites, earlier thought to originate in S-class asteroids (Jenniskens et al. 2009). The material did not exist in meteorite collections prior to the 2008 event due to its rarity in interplanetary space or, perhaps, due to its fragility which vaporizes most of the material during the passage through Earth's atmosphere.

The rate of discovery of Earth-impacting small asteroids will increase as the capabilities of telescopic surveys improve (cf. Tonry 2011), but events similar to 2008 TC_3 will nevertheless remain relatively rare. First of all the object needs to be large enough not to completely vaporize in the atmosphere upon impact. Let us assume that 2008 TC_3, with an effective diameter of about a few meters, defines the lower size limit for objects that can produce macroscopic meteorites that survive the passage through the Earth's atmosphere. Objects of this size impact the Earth a few times every year (Brown et al. 2002). About 70% of all meteorites will end up in oceans and a substantial fraction of the remaining 30% will occur in areas that are difficult to reach or otherwise not well suited for meteorite collection such as rain forests, taiga and mountain regions. Thus, an event like 2008 TC_3 is not likely to occur more than once every few years even if future surveys could discover *all* such objects before they enter the Earth's atmosphere. Finally, a major drawback of waiting for a small asteroid impact is that we cannot choose the taxonomic class of the impactor—if we simply wait for impactors it will take many decades to understand the mineralogy for a reasonable fraction of all taxonomic classes. We suggest that the calibration rate can be increased because a suitable source of calibration targets can be found in Earth orbit, the TCOs.

Only one TCO, 2006 RH_{120}, has ever been discovered but recent results suggest that we will start discovering more as survey technology improves. Bolin et al. (2013) look at different possibilities for TCO detection such as the Subaru telescope's Hyper Suprime-Cam, the Large Synoptic Survey Telescope (LSST; anticipated start of science operations in 2021), and meteor cameras (Sect. 6.3). They find, for example, that LSST will discover several TCOs every month given the TCO population model by Granvik et al. (2012). When this happens, TCOs will serve as easily-accessible validation targets (Sect. 6.4) by combining spectrometric observations with sample-return missions. Analyzing the mineralogy of objects of different taxonomic classes in Earth-based laboratories will allow economically sound choices to be made when selecting targets for mining missions. TCOs can also be utilized as test beds for the technology to be developed for mining operations on asteroids such as high-precision automatic navigation and resource mapping using remote-sensing technology.

The option of bringing back entire sub-meter-scale TCOs to Earth-based laboratories would open up a treasure trove for planetary science and thus also provide more accurate information useful for mining operations on asteroids. For example, studying a TCO's interior structure would increase our knowledge of the interior structure of asteroids in general. Based on theoretical considerations it has, for instance, recently been suggested that even small asteroids—previously assumed to be monolithic—may be rubble piles kept together by cohesive forces (D. Scheeres, private communication). Understanding the structure of target asteroids is clearly of key importance for future mining missions.

6.2 Predicted Population Characteristics for Earth's Temporarily-Captured Natural Satellites

Granvik et al. (2012) compute the TCO size-frequency distribution (SFD) using a two-step process. First, the capture probability is estimated as a function of the heliocentric orbital elements. Second, the capture probability is multiplied by a debiased NEO orbital element distribution and scaled using the best available debiased NEO SFD.

The capture probability as a function of heliocentric semimajor axis $a_{\rm h}$, eccentricity $e_{\rm h}$ and inclination $i_{\rm h}$ is computed by integrating 10 million test particles through the EMS that have a reasonable possibility of being captured. A particle is classified as a TCO if it makes one or more revolutions around the Earth in a co-ordinate system co-rotating with the Sun while being energetically bound to the Earth and within three Hill radii of the Earth's center.

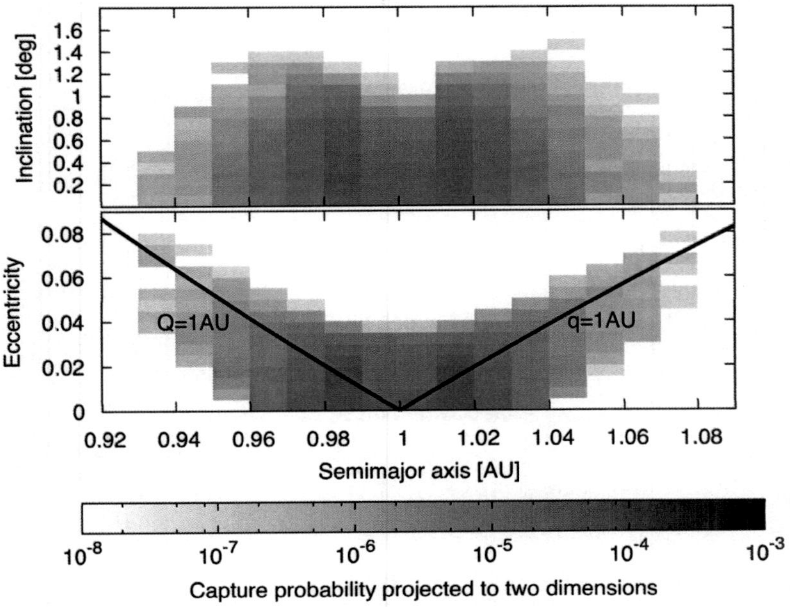

Fig. 6.1 TCO capture probability as a function of pre-capture heliocentric orbital elements. Modified from Granvik et al. (2012).

Particles that will become TCOs have speeds of less than 2.2 km s^{-1} prior to capture at a distance of 4–5 Hill radii from the Earth. The TCOs' heliocentric orbits prior to capture are extremely Earth-like with $a \sim 1$ au, $e \sim 0$ and $i \sim 0°$ (Fig. 6.1).

Once captured—through the Sun-Earth L_1 or L_2 point—TCOs spend most of their time between 1 and 10 lunar distances (LD) from the Earth. The TCOs' speeds relative to the Earth are typically <1.5 km s^{-1} but reach Earth escape speed during close Earth encounters (Fig. 6.2).

TCOs are, on average, captured for 286±18(rms) days and make 2.88±0.82(rms) revolutions around the Earth during this time. Both distributions have very long tails: the longest simulated capture event lasts about 900 years during which the particle makes almost 15,000 revolutions around the Earth.

Figure 6.3 shows the resulting TCO SFD assuming Bottke et al.'s (2002) NEO orbit distribution and Brown et al.'s (2002) Earth-impactor SFD. The latter is generally thought to accurately reflect the SFD of non-impacting small NEOs.

Although the NEO orbit distribution by Bottke et al. (2002) is valid only for relatively large NEOs ($D > 140$ m), most of the uncertainty in the TCO SFD estimate is likely due to the underlying NEO SFD. Granvik et al. (2012) prefer the NEO SFD by Brown et al. (2002) because it was measured for objects in the relevant size range and is consistent with the occurrence of a 2006 RH_{120}-like event once every decade. A steeper NEO SFD such as the one by Rabinowitz et al. (2000) can nevertheless not be ruled out. Brown et al.'s (2002) SFD explains the

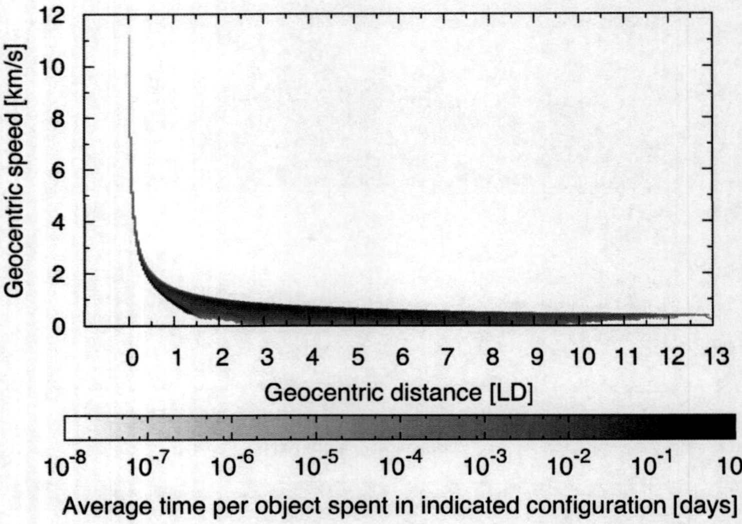

Fig. 6.2 TCO residence time as a function of geocentric distance and speed (r_g and v_g, respectively). The implication is that TCOs will be difficult to detect from the ground because they are moving fast when they are close to the Earth and thus bright. Modified from Granvik et al. (2012).

observed TCO SFD assuming that the surveys' detection efficiency for $D > 2$ m TCOs is 100%. Reducing the assumed detection efficiency increases the steady-state number of TCOs at every size and makes the NEO SFD by Rabinowitz et al. (2000) a viable alternative.

Bottke et al.'s (2002) NEO orbit distribution also has some recently highlighted problems which affect the TCO SFD: NASA's WISE spacecraft detected more low-inclination Atens than Bottke et al. (2002) predict (Mainzer et al. 2012). Recall that low-inclination Atens are among the objects most likely to be captured by the EMS. The discrepancy is most likely caused by a too large integration step in Bottke et al.'s (2002) residence-time integrations (Greenstreet et al. 2012). There is still some inconsistency in the inclination distribution even after accounting for the integration accuracy (S. Greenstreet, private communication). The likely source for the residual discrepancy is the uniform inclination distribution used for the initial conditions by both Bottke et al. (2002) and Greenstreet et al. (2012). In reality the inclination distribution is skewed towards low inclinations (Granvik et al. 2013b). A new NEO model currently in development will soon shed light on this issue.

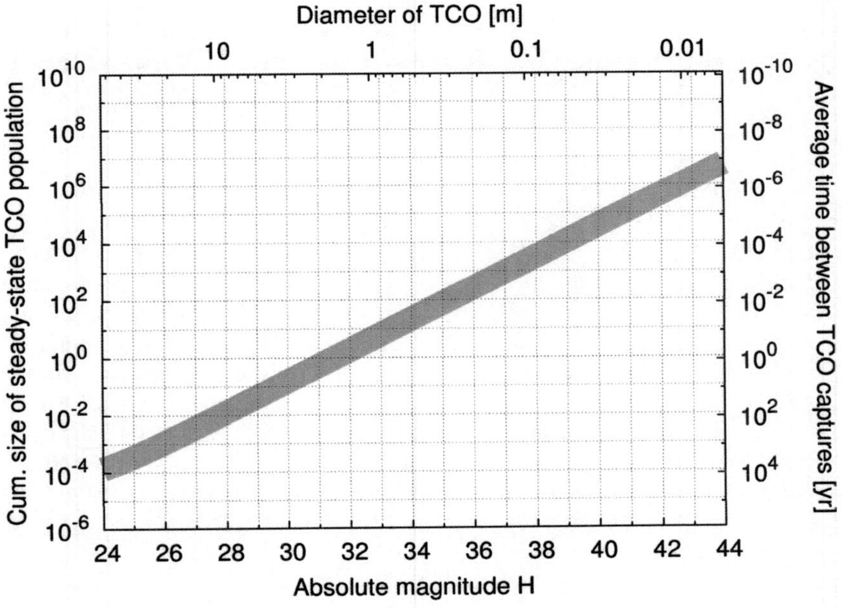

Fig. 6.3 The cumulative steady-state TCO SFD based on Bottke et al.'s (2002) NEO orbit distribution and Brown et al.'s (2002) NEO SFD. The maximum size at which at least one object is captured at any time is $H \sim 32$ (or $D \sim 1$ m). The frequency of capturing a TCO with $H \sim 30$ is about once every decade. The conversion from H magnitude to diameter assumes a geometric albedo of 0.15. The line width reflects the uncertainty in the TCO SFD. Modified from Granvik et al. (2012).

Furthermore, the orbital integrations by Granvik et al. (2012) only take into account gravity. Future work will need to account for non-gravitational forces such as atmospheric drag and solar radiation pressure that will affect the TCO SFD. For example, atmospheric drag aids in the capture of objects that would otherwise not be captured and thus directly increases the capture probability for the fraction of objects that have extremely close encounters with the Earth. We note that for every TCO there are 10^3 asteroids on similar pre-capture orbits that do not get captured (Granvik et al. 2012). Thus accounting for the atmospheric drag may substantially increase the steady-state number of TCOs at every size. Non-gravitational forces affecting TCO trajectories also have an indirect effect on the TCO SFD, because the latter is directly proportional to the average duration of capture. The indirect effect may therefore either slightly reduce or slightly increase the steady-state number of TCOs. The TCO SFD is also affected by break-up of larger asteroids, e.g., during a passage through the atmosphere or through tidal disruption during a close and slow Earth encounter (E. Schunova, private communication).

The uncertainties in the NEO orbit distribution and SFD combined with the gravity-only orbital integrations suggest that Granvik et al.'s (2012) TCO SFD may be interpreted as a lower limit.

6.3 Tracking Earth's Temporarily-Captured Natural Satellites

6.3.1 *Detectability*

Detecting a TCO with $D < 2$ m is not easy. There are about two million 1–2-meter-diameter NEOs that pass within one lunar distance of the Earth each year. Yet only 13 NEOs with $30 < H < 33$ are recorded in the Minor Planet Center's NEO catalogue as of 29 Oct 2012 (i.e., those corresponding roughly to a diameter of about 1–2 m). Given that modern CCD asteroid surveys have been operating for about two decades it seems that the annual probability of detecting a 1–2 m scale NEO within one lunar distance is about 10^{-7}. Thus, it may be somewhat surprising that the few-meter-diameter TCO 2006 RH_{120} was detected at all. However, even though larger TCOs are considerably more rare they are also much easier to detect. Even so, all else being equal, the discovery of a 2-m-diameter object is exceedingly unlikely. Thus, the final contributing factor to the discovery of 2006 RH_{120} is that TCOs approach Earth slowly and remain in orbit for relatively long time intervals compared to close flybys of objects of the same size. Thus, TCOs are more likely to be identified because there is more time to discover them.

Bolin et al. (2013) studied several modes of detecting TCOs while in Earth orbit including 1) optical ground-based a) wide area and b) targeted surveys 2) infrared space-based surveys and 3) radar. Considering that there are currently no infrared space-based surveys we ignore detailed discussion of that opportunity here. Furthermore, considering that we are interested here in detecting TCOs with a maximum amount of time available for a spacecraft mission we concentrate on options 1b and 3 that both attempt to detect the TCOs close to their time of capture when they are passing near L_2.

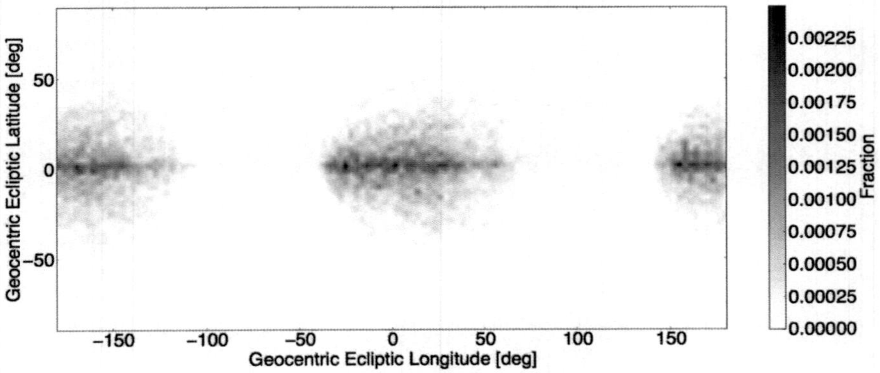

Fig. 6.4 Normalized TCO sky-plane residence distribution at time of capture. There are no constraints on the TCO's apparent magnitude, distance or rate of motion. The ecliptic longitude is centered on the opposition point, that is, the Sun is at ±180°.

Figure 12 of Granvik et al. (2012) suggests that at the moment when TCOs become energetically bound to the EMS they are all located at roughly the same distance, moving at roughly the same speed, and in roughly the same direction. Indeed, Bolin et al. (2013) show that they are at a geocentric distance of 6.3±1.4(rms) LD moving at 0.6±0.1(rms) km s^{-1} at a position angle of 94°±25° (rms). At 6.3 LD and 10° from opposition 1m/0.25m diameter objects have apparent V magnitudes of about 24.7/27.7—the largest objects being barely detectable with the largest aperture telescopes *if* the objects are not trailed too much. Figure 6.4 shows the sky-plane distribution of TCOs at the moment of capture. The distribution on the left and right side of the figure is in the direction of the Sun and not detectable in optical surveys but the distribution towards opposition (center of the figure) is detectable in a targeted survey with the Hyper Suprime-Cam (HSC) on the Subaru telescope (Takada 2010). Since there are about 145 TCOs larger than 25 cm in diameter in orbit at any time and since their typical lifetime is about 9 months we expect that there are about 100 of these objects captured each year near L_2. Thus, a targeted survey near opposition with enough time on a large telescope could detect the largest TCOs at a rate that might be suitable for spacecraft missions.

Figure 6.5 shows the unconstrained (by apparent magnitude, rate of motion, and distance) TCO sky plane distribution. There are strong enhancements at quadrature where the objects tend to be further from Earth (Bolin et al. 2013) and therefore spend more time. These enhancements might be exploited in detecting asteroids with radar. A 100cm/25cm asteroid can be detected with the Arecibo radar facility if it approaches within about 4/0.25 LD and is rotating very slowly. Thus, it is in theory possible for the Arecibo radar system to detect TCOs. The greatest uncertainty in the radar capability is the TCO rotation rate distribution. The rotation rate distribution of meteoroids in this size range is entirely

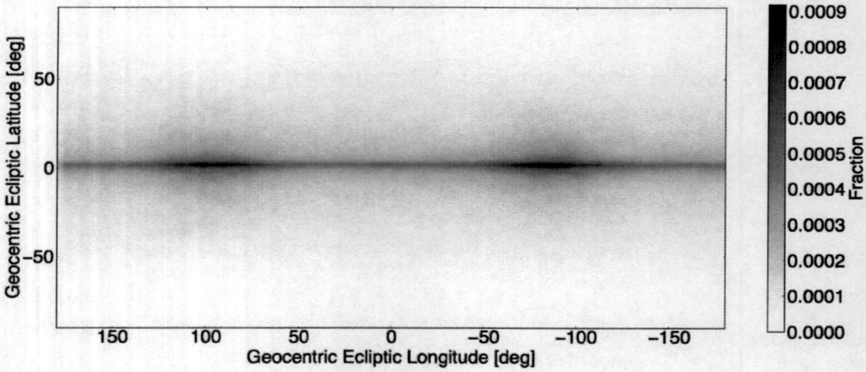

Fig. 6.5 Normalized TCO sky-plane residence distribution with no restrictions on apparent magnitude or rate of motion. The ecliptic longitude is centered on the opposition point, that is, the Sun is at ±180°.

unconstrained with estimates ranging over orders of magnitude. If the TCOs rotate too quickly the reflected radar signal is spread over too wide a frequency range, dropping the signal-to-noise ratio below the detection threshold.

The most straightforward way to detect TCOs is with existing and future ground-based optical surveys as illustrated in Fig. 6.6. There is a strong enhancement towards opposition and along the ecliptic as occurs for most small-body populations in the solar system as viewed from Earth. The enhancements toward quadrature from Fig. 6.5 have disappeared because both the phase-angle effect and the objects' greater distance in that direction reduces the objects' apparent brightness. A survey system with the constraints imposed in Fig. 6.6 would not be effective at regularly discovering TCOs but an LSST-like system with a limiting magnitude $V_{\mathrm{lim}} \sim 24$ might discover one TCO per month (Bolin et al. 2013).

Particularly for human missions to TCOs it would be essential to know the target asteroid many years prior to launching the actual mission. This requirement could be fulfilled by utilizing the fact that pre-capture orbits of TCOs are very similar to the Earth's orbit and they are therefore constrained to a relatively small volume in space. Having an optical or IR telescope on an Earth-like orbit leading or trailing the Earth (cf. Spitzer space telescope) would thus allow the detection of proto-TCOs before they are captured.

In summary, the most likely prospects for detecting TCOs suitable for spacecraft mission targets is probably a concerted targeted survey with a large-aperture telescope and wide-field camera. Alternatively, it is possible that a clever survey strategy from a space-based IR or ground-based radar facility could be optimized for TCO discovery. An extremely high-power version of the so-called Space Fence radar system would also allow the characterization of the TCO population.

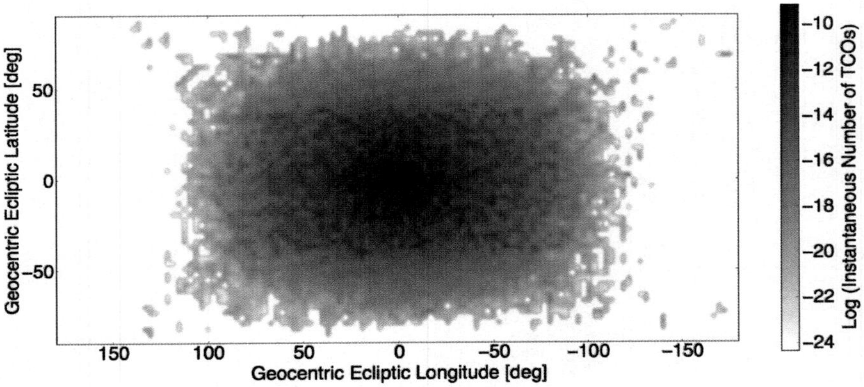

Fig. 6.6 Sky-plane number distribution for TCOs with $H < 38$, apparent magnitude $V < 20$ and rate of motion <15 ° d^{-1}. The ecliptic longitude is centered on the opposition point, that is, the Sun is at ±180°.

6.3.2 *Predictability of Future Locations*

Once a TCO is detected it is important to understand how much follow-up astrometry is necessary to ensure i) that the object is orbiting Earth ii) to allow continued recovery and physical characterization and iii) that the TCO's orbit is known accurately enough to allow the planning of a space mission (Sect. 6.4). Note that we assume that minor corrections to the spacecraft trajectory are allowed after it has been launched or, alternatively, departed the geosynchronous transfer orbit. Therefore the spacecraft trajectory can be planned when a semi-accurate TCO orbit becomes available. A recent case study by Granvik et al. (2013a) suggest that a typical TCO orbit is reasonably well-constrained after only a few consecutive nights of astrometric follow-up observations (see example in Fig. 6.7).

In particular, Granvik et al. (2013a) show that, for 2006 RH_{120} and ten randomly chosen synthetic TCOs, heliocentric Keplerian orbits are incompatible with more than two days of astrometric data. That is, two days after the discovery of a TCO it is clear that the object's orbit is strongly perturbed by a close Earth encounter, either because of a close flyby or because it is orbiting the Earth. In both scenarios it merits additional follow-up observations. Although less than a week of optical astrometry will typically provide an orbital solution that hardly differs from the true orbit, we suggest that radar observations be carried out as soon as possible. Thus constraining the orbit of a discovered TCO to the point required by space missions does probably not take longer than about a week. A preliminary characterization of the object's physical properties such as its type and size need not take much longer using Target-of-Opportunity proposals to large telescopes.

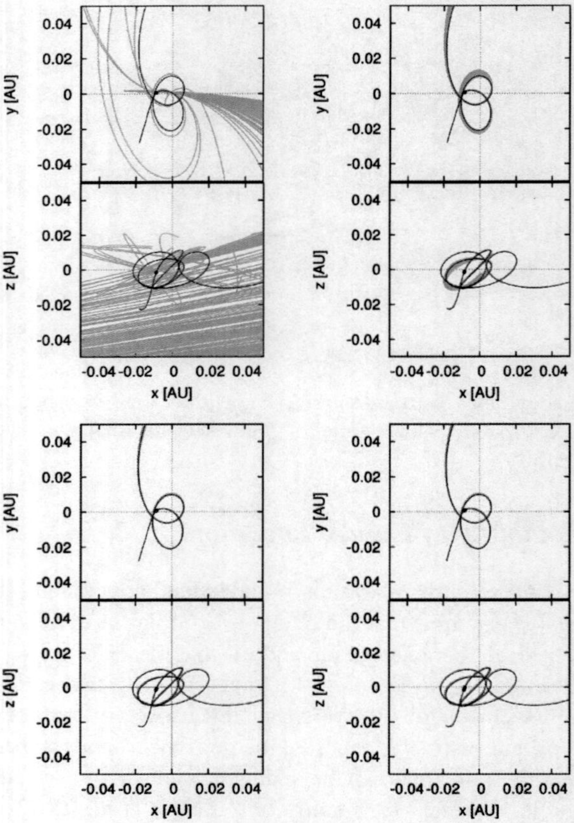

Fig. 6.7 Evolution of a synthetic TCO's orbital uncertainty as a function of increasing observational timespan and number of observations; (top left) 3 detections spanning one hour, (top right) 6 detections spanning 25 hours, (bottom left) 9 detections spanning 49 hours, and (bottom right) 12 detections spanning 73 hours. The black curve shows the true orbit in the xy and xz planes in an ecliptic coordinate system co-rotating with the Sun so that Earth is always in the center (0,0,0) and the Sun is always at about (1,0,0). The gray shaded area shows the extent of all acceptable trajectories and the black dots mark the locations of the synthetic TCO at the times of observation. All trajectories are shown for 500 days beyond the observation epoch.

6.4 Spacecraft Missions to Earth's Temporarily-Captured Natural Satellites

NASA's Near Earth Asteroid Rendezvous (NEAR) and JAXA's Hayabusa are completed missions that involved a rendezvous with an asteroid. NEAR's target was the large NEO (433) Eros, an irregularly shaped asteroid about 30 km in its longest dimension, while the Hayabusa mission targeted the 600 m long NEO (25143) Itokawa. Both missions managed safe landings on their S-class targets.

Spacecraft rendezvous missions with TCOs will provide unique challenges beyond those of the NEAR and Hayabusa missions. The 9-month average capture duration is a major challenge when designing a TCO mission—to make a TCO mission feasible it is essential to minimize the transfer time to the TCO rendezvous point. Contrary to so-called direct trajectory-design methods currently used by space-mission designers, the so-called indirect methods provide second-order conditions for the time-minimality of candidate trajectories, are much more precise and provide a clear geometric understanding of the structure of the trajectory. A detailed understanding of the geometry of the trajectory allows it to be fairly easily perturbed would need arise. The challenge with indirect methods is their sensitivity to initialization and a new algorithm has been developed for the TCO case to tackle this issue (Chyba et al. 2013a,b; see Sect. 6.4.1).

Even if the TCO orbit is known well enough to launch the spacecraft, the meter-scale target may not be easy to identify in the spacecraft's imaging sensors. Navigating close to the TCO may require real-time control from ground-based operators and, once close to the TCO, it may be in a rapid or tumbling rotation state making contact with the object more problematic than with large asteroids.

6.4.1 Low-Thrust Propulsion Transfers

The first step in designing a TCO rendezvous mission is to determine if it is possible given the time constraints and the use of a low-thrust-propulsion spacecraft such as electro-ionic propulsion engines (Ferrier and Epenoy 2001, Geoffroy et al. 1996). Chyba et al. (2013a,b) focus on the initial stage of the mission, i.e., the transfer of the spacecraft to the rendezvous location with the satellite, and assume that the spacecraft is *parked* in a geostationary Earth orbit awaiting discovery of a TCO to initiate the mission.

For simplicity, Chyba et al. (2013a,b) first impose the transfer departing position of the spacecraft on the geostationary orbit to be the point located on the line between the center of the Earth and the Moon. Clearly there is limited control on the location of the spacecraft at the time of detection and a complete study including any departing position is part of the mission design. The selection of the rendezvous location on the TCO orbit is also a large component of the mission design. The initial work by Chyba et al. (2013a,b) is based on known results for low-thrust 2-dimensional transfers to the Earth-Moon L_1 point (Picot 2012). They select the 100 TCOs with the smallest absolute perpendicular distance to the Moon's orbital plane at the time they are nearest the Earth-Moon L_1 point from the 18,096 TCO trajectories of Granvik et al. (2012). This choice is motivated by the fact that near the selected rendezvous location the two-dimensional projection of the TCO's orbit on the Moon's orbital plane provides a good approximation to the three-dimensional orbit for a significant interval of time. However, the final computed transfers are fully three-dimensional trajectories.

Chyba et al.'s (2013a,b) numerical computations of low-thrust time-minimal transfers from the geostationary orbit to TCOs rely on fundamental mathematical results from modern optimal control theory. The motion of a spacecraft in the EMS is approximated as a solution to the restricted three-body problem due to the small eccentricity of the Moon's orbit (Szebehely 1967). This model describes the motion of an object with negligible mass under the influence of the gravitational fields of two planets revolving circularly around their center of mass. Control terms are added to the equation of motion to represent the thrust of the spacecraft that are bounded by the constraints on the engine propulsion power. Taking time as the criterion to minimize, the design of a transfer mission from geostationary orbit to the rendezvous location on the TCO's orbit can be phrased as an optimal control problem.

The transfer time is directly correlated to the engine thrust—the more powerful the thrust the faster we can reach the rendezvous location. Using the planar time-minimal transfer from the geostationary orbit to the Earth-Moon L_1 point computed by Picot (2012) as a reference initial guess, Chyba et al. (2013a,b) compute a three-dimensional reference extremal corresponding to a maximum thrust of 1 N departing from the geostationary orbit to the rendezvous location for each of the selected TCOs. A discrete continuation method on the maximum thrust is then used to determine low-thrust time-minimal transfers.

6.4.2 Orbital Transfer Simulations

The methodology described above is successful in defining orbital transfers for about 40% of the synthetic TCOs. Very recent and still unpublished improvements in the methodology has increased the success rate to virtually 100%. Here we present a representative simulation of a rendezvous mission from the geostationary orbit to a synthetic TCO which is captured for 214 days and makes 1.4 revolutions around the Earth. The distance between the Earth-Moon L_1 point and the rendezvous point is about 0.054 LD. The TCO rendezvous time, that is, the time from the TCO's capture until it passes the rendezvous point, is some 133.3 days.

Table 6.1 provides information about the simulated TCO rendezvous mission given two different values for the maximum thrust, 0.2 N and 1 N. As would be expected, the transfer time with the low-thrust propulsion is about five times longer than with a five times more powerful engine. The search for possible transfers can be refined once the spacecraft's capabilities are determined.

Table 6.1 Characteristics of the simulated TCO rendezvous missions

Thrust [N]	Transfer time [d]	Difference between rendezvous and transfer times [d]
1.0	13.5	119.8
0.2	62.0	71.3

More generally, Chyba et al. (2013a,b) found that the transfer times for 1-N transfers were in the range of 10 to 20 days (Fig. 6.8) while the transfer times for 0.2-N transfers were between 55 and 81 days (Fig. 6.9). The transfer time is less than the time it takes the TCO to evolve from its point of capture to the rendezvous point for a majority of the three-dimensional 0.2-N transfers. This is crucial from a practical stand point since it suggests that it may be feasible to launch a low-thrust time-optimal rendezvous mission if the target TCO is discovered at the time of capture or soon thereafter.

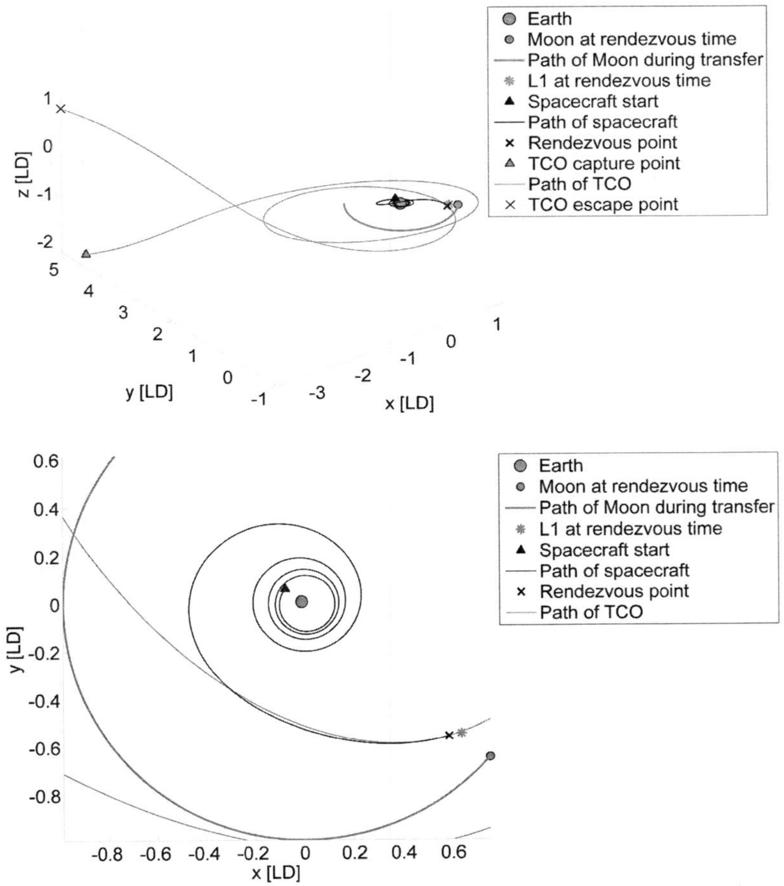

Fig. 6.8 (Top) A wide and (bottom) close up view of a locally time-minimal three-dimensional transfer with 1 N thrust to a TCO in the inertial frame

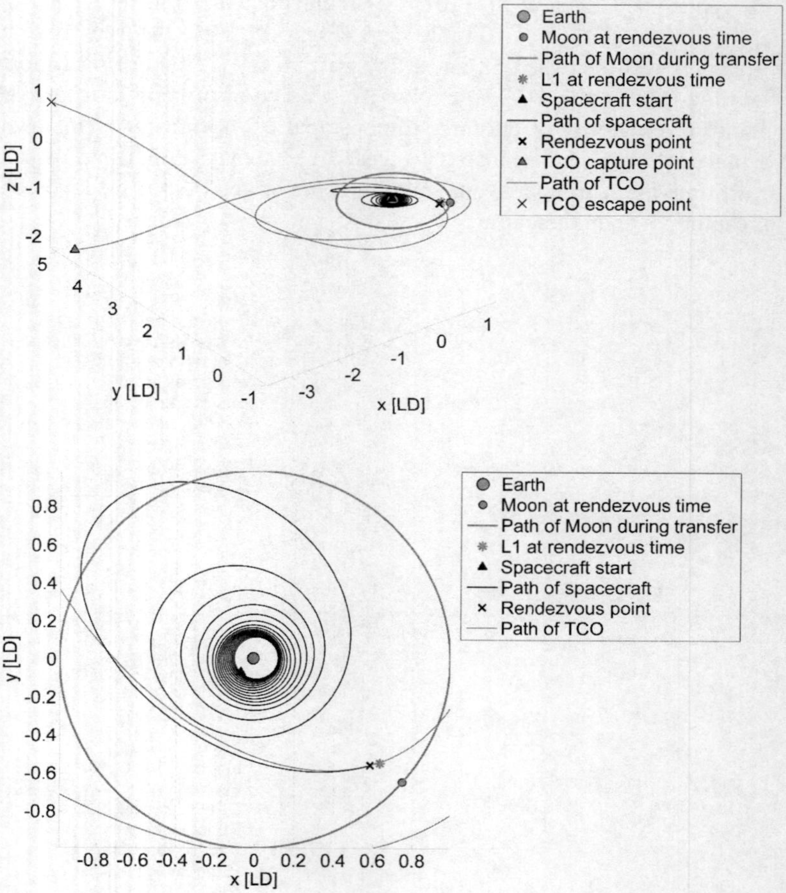

Fig. 6.9 As in Fig. 6.8 but with a maximum thrust of 0.2 N

6.4.3 Other Technological Challenges

The technological challenges of an asteroid-return mission to a TCO have not yet been assessed in detail but there are some challenges that are common for all asteroid sample-return missions. For robotic exploration the main issue is that the current positional accuracy for autonomous navigation is on the order of ten meters. For a meter-sized object the accuracy would need to be improved by 1–2 orders of magnitude. Of course, in Earth-orbit it might be possible to implement real-time control for a TCO rendezvous.

Another challenge is that, although observational selection effects have not been ruled out, there appears to be a lack of small asteroids that rotate slowly according to data obtained with the Arecibo and Goldstone planetary radars

(P. Taylor, private communication). Thus, an asteroid sample return mission would probably need to be equipped with a tool for grabbing and de-spinning fast-spinning objects. E.g., 2008 TC_3 was in a state of non-principal-axis rotation with main rotation periods of 49 and 97 seconds (Jenniskens et al. 2009). Analytical calculations have also shown that even meter-class asteroids may be rubble piles rather than monolithic objects (D. Scheeres, private communication). Grabbing and de-spinning a rubble pile is clearly an even greater challenge.

6.5 Conclusions

Although the small size of a typical temporarily-captured natural Earth satellite would make commercial mining operations unprofitable, we have made the case that studying this population and sending spacecraft to these objects is a natural first step for any project that aims to take advantage of the energy and material resources available in asteroids. The goal of a space mission to a TCO is thus not to extract valuable resources, but to increase our understanding of where valuable resources can be found and test the technology required for extracting those resources.

The TCO population can be utilized in validating orbit and SFD models of NEOs on Earth-like orbits. A mineralogical analysis of TCO samples in Earth-based laboratories allows the calibration of remote-prospecting methods that rely on the compositional classification using spectra. TCOs also provide small-scale platforms for testing the technologies that need to be developed such as accurate automated navigation, and asteroid de-spinning or, alternatively, anchoring methods. Bringing an entire TCO to Earth-based laboratories allows, for example, a detailed analysis of its interior structure which would allow us to test our current theories of the interior structure of asteroids in general. The scientific reward of bringing an entire small asteroid to ground-based laboratories is immense and would most likely lead to advances that also benefit the commercial entities aiming to take advantage of the resources in asteroids.

Some of the unknowns—such as the detectability of TCOs and the existence of viable orbital transfer paths to the chaotic TCO trajectories—have already been assessed in a quantitative manner though numerous aspects still remain to be studied in greater detail. Examples of things that can be studied fairly easily are the lead time from discovery to the escape of the object from the Earth-Moon system and a detailed assessment of the frequency of TCO discoveries produced by the LSST.

There are also several technological challenges that need to be solved before launching a space mission to retrieve a TCO. One of the major questions is whether it is realistic to expect that a mission either be launched within a few months of a TCO discovery or be parked in, e.g., a geosynchronous orbit until a suitable object is discovered. The obvious technological challenges also include the need for a major improvement in the accuracy of automated navigation and the tools to grab and de-spin rotating bodies.

We remain optimistic that these aspects will be studied and solved in the not-too-distant future owing to the current interest in space-based resources in the scientific community, space agencies, commercial entities and, last but not least, the general public.

Acknowledgments. We highly appreciate the thoughtful and constructive criticism offered by the two reviewers, W. F. Bottke and P. Chodas. MG was funded by grant #137853 from the Academy of Finland. BB and RJ were supported by NASA NEOO grant NNXO8AR22G.

References

Bolin, B., Jedicke, R., Granvik, M., Wainscoat, R.: Detecting Earth's natural satellites (in preparation, 2013)

Bottke, W.F., Morbidelli, A., Jedicke, R., Petit, J.M., Levison, H.F., Michel, P., Metcalfe, T.S.: Debiased Orbital and Absolute Magnitude Distribution of the Near-Earth Objects. Icarus 156(2), 399–433 (2002)

Brown, P., Spalding, R.E., ReVelle, D.O., Tagliaferri, E., Worden, S.P.: The flux of small near-Earth objects colliding with the Earth. Nature 420, 294–296 (2002)

Chyba, M., Granvik, M., Jedicke, R., Patterson, G., Picot, G., Vaubaillon, J.: Time-minimal orbital transfers to temporarily-captured natural Earth satellites. In: OCA5 – Advances in Optimization and Control with Applications. Springer Proceedings in Mathematics (accepted, 2013a)

Chyba, M., Granvik, M., Jedicke, R., Patterson, G., Picot, G., Vaubaillon, J.: Designing Rendezvous Missions with Mini-Moons using Geometric Optimal Control. Computational Methods for Optimization and Control, Special Edition of Journal of Industrial and Management Optimization, JIMO (accepted, 2013b)

Ferrier, C., Epenoy, R.: Optimal control for engines with electro-ionic propulsion under constraint of eclipse. Acta Astronautica 48(4), 181–192 (2001)

Gaffey, M.J., Cloutis, E.A., Kelley, M.S., Reed, K.L.: Mineralogy of Asteroids. In: Bottke, W., Cellino, A., Paolicchi, P., Binzel, R.P. (eds.) Asteroids III, pp. 183–204. University of Arizona Press (2002)

Geoffroy, S., Epenoy, R., Noailles, J.: Averaging techniques in optimal control for orbital low-thrust transfers and rendezvous computation. In: Proceedings of the 11th International Astrodynamics Symposium, Gifu, Japan, pp. 166–171 (1996)

Granvik, M., Vaubaillon, J., Jedicke, R.: The population of natural Earth satellites. Icarus 218(1), 262–277 (2012)

Granvik, M., Morbidelli, A., Bottke, W., Jedicke, R., Michel, P., Nesvorny, D., Tsiganis, K., Vokrouhlicky, D.: Source populations for near-Earth objects (in preparation, 2013b)

Granvik, M., Virtanen, J., Jedicke, R., Vaubaillon, J., Chyba, M.: Bayesian orbit-computation methods applied to temporarily-captured natural Earth satellites. In: Wnuk, E., Deleflie, F. (eds.) Proceedings of the International Symposium on Orbit Determination and Correlation (submitted, 2013a)

Greenstreet, S., Ngo, H., Gladman, B.: The orbital distribution of Near-Earth Objects inside Earth's orbit. Icarus 217, 355–366 (2012)

Jenniskens, P., Shaddad, M.H., Numan, D., Elsir, S., Kudoda, A.M., Zolensky, M.E., Le, L., Robinson, G.A., Friedrich, J.M., Rumble, D., Steele, A., Chesley, S.R., Fitzsimmons, A., Duddy, S., Hsieh, H.H., Ramsay, G., Brown, P.G., Edwards, W.N., Tagliaferri, E., Boslough, M.B., Spalding, R.E., Dantowitz, R., Kozubal, M., Pravec, P., Borovicka, J., Charvat, Z., Vaubaillon, J., Kuiper, J., Albers, J., Bishop, J.L., Mancinelli, R.L., Sandford, S.A., Milam, S.N., Nuevo, M., Worden, S.P.: The impact and recovery of asteroid 2008 TC3. Nature 458, 485–488 (2009)

Mainzer, A., Grav, T., Masiero, J., Bauer, J., McMillan, R.S., Giorgini, J., Spahr, T., Cutri, R.M., Tholen, D.J., Jedicke, R., Walker, R., Wright, E., Nugent, C.R.: Characterizing Subpopulations within the near-Earth Objects with NEOWISE: Preliminary Results. Astrophysical Journal 752, 110 (2012)

Ockert-Bell, M.E., Clark, B.E., Shepard, M.K., Isaacs, R.A., Cloutis, E.A., Fornasier, S., Bus, S.J.: The composition of M-type asteroids: Synthesis of spectroscopic and radar observations. Icarus 210, 674–692 (2010)

Picot, G.: Shooting and numerical continuation method for computing time-minimal and energy-minimal trajectories in the Earth-Moon system using low-propulsion. Discrete and Continuous Dynamical Systems – Series B 17(1), 245–269 (2012)

Rabinowitz, D., Helin, E., Lawrence, K., Pravdo, S.: A reduced estimate of the number of kilometre-sized near-Earth asteroids. Nature 403, 165–166 (2000)

Szebehely, V.: Theory of orbits. Academic Press (1967)

Takada, M.: Subaru Hyper Suprime-Cam Project. In: Kawai, N., Nagataki, S. (eds.) American Institute of Physics Conference Series. American Institute of Physics Conference Series, vol. 1279, pp. 120–127 (2010)

Tonry, J.: An early warning system for asteroid impact. Publications of the Astronomical Society of the Pacific 123(899), 58–73 (2011)

Chapter 7
Human Missions to NEO's – A System Perspective

Marco Cenzon[1] and Dragoş Alexandru Păun[2]

[1] Aviospace S.r.l., Torino, Italy
[2] Sofiter System Engineering S.p.A., Torino, Italy

7.1 Introduction

Ever since the discovery of Ceres on January 1st 1801, Near Earth Objects (NEOs) have attracted our attention and fascination. Subsequent proof of their potentially destructive capabilities and the recent improvements in NEO detection rate have given us the impulse to expand our knowledge in order to devise a way of detecting and neutralize possible dangers to our existence. Recent studies however suggest that the pool of materials available in the form of NEOs may also be used both as a source of minerals on Earth as well as a material springboard for future human colonies in the Solar System.

If we consider economic resources typically available in space as being classified into one of four categories, these would be: location, environment conditions, energy and matter. Location is being exploited for example by all communication or Earth observation satellites, which take advantage from their high vantage point increasing their coverage. Low-Earth and geosynchronous orbits host hundreds of revenue generating satellites. Particular environmental conditions such as radiation or gravity might be used to enable biological, physical or chemical processes that would be impossible to recreate on Earth. These phenomena have been studied by space agencies and other organizations from the early beginnings of space exploration. At current, such experiments are being performed on the International Space Station (ISS). In Earth orbit, beyond Earth's atmosphere, solar radiation is abundant and powers most spacecraft and probes. Solar Power Satellites (SPS) have been proposed by different organizations with the purpose of delivering huge quantities of clean, sustainable energy to the Earth's surface. Matter represents the only space resource which is not yet being exploited. The limiting factors for such an endeavor are the capability of finding and extracting the suitable matter as well as its transportation trough space in addition to the economical aspects. The movement of mass trough space is expensive at current technological levels.

The purpose of the present chapter is to give an overview of the state of the art regarding human missions to NEOs, the fundamental challenges that such missions would face and what would the benefits be. Furthermore, the authors propose a flexible architecture aimed at serving both as a human NEO study platform and as an industrial NEO utilization/exploitation platform.

7.1.1 Asteroids. Threats and Opportunities

Asteroids and comets are present in very large numbers in our Solar System. Most asteroids find themselves in the asteroid belt between the orbits of Mars and Jupiter. The belt contains more than 200 asteroids larger than 100 km in diameter and it is estimated that there are more than 750,000 asteroids in the belt with diameters larger than 1 km (NASA JPL 2013). Among all the asteroids and comets, Near Earth Asteroids are those asteroids having a perihelion lower than 1.3 Astronomical Units (AUs) or a Minimum Orbit Intersection Distance (MOID) lower than 0.3 AUs.

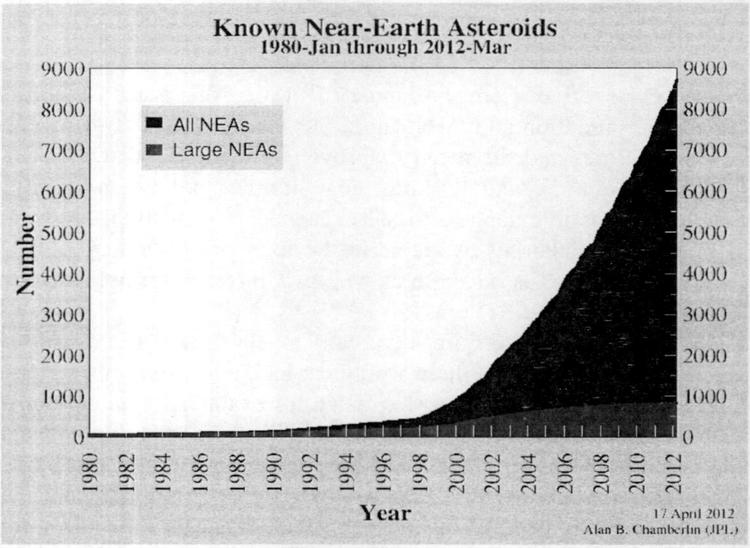

Fig. 7.1 Known NEA's 1980-2012 (NASA JPL 2013)

At current, approximately 9,000 NEAs have been discovered and due to recent improvements in detection capabilities and larger number of studies our rate of discovery is constantly increasing with up to 500 new objects per year (NASA JPL 2013), see Fig. 7.1.

7.1.1.1 The Threat: Potentially Hazardous Asteroids

Potentially Hazardous Asteroids (PHAs) are those asteroids having a potentially threatening close approach to the Earth. Specifically, all asteroids with an Earth MOID lower than or equal to 0.05 AU and an absolute magnitude of 22.0 or less are considered to be PHAs. The asteroid-hunting portion of the NASA WISE mission, called NEOWISE studied PHAs with particular scrutiny. Its findings indicate that there would be roughly between 3,000 and 5,000 PHAs with diameters larger than 100 meters (Barbee 2011). At this point, only an estimated 20 to 30 percent of these objects have been detected. We must as well consider that an asteroid's orbit can be altered by various gravitational and non-gravitational mechanisms and that these mechanisms have the potential of transforming a non-threatening asteroid into a threatening one and vice-versa. This can be seen as a reflection of the dynamic character of the Solar System. For example, Asteroid 1999 RQ36 has a one-in-1,000 chance of hitting the Earth before the year 2200. An analysis of its orbit has predicted that it is most likely to hit us on September 24, 2182 but scientists want to collect a sample of the rock to help forecast its trajectory more accurately. In the eventuality that a mission would be launched, a spacecraft would launch in the year 2106 to map out and collect rock samples from the asteroid (Barbee 2011).

7.1.1.2 The Opportunity: Asteroid Mining

Asteroids seem to be the most affordable extraterrestrial source of industrial elements, due to their relative vicinity and weak gravitational fields. Practically asteroids could represent enormous orbiting repositories of industrial metals. It is sufficient to say that some asteroid impact craters such as the Sudbury basin in Canada, which was created by a 10 km wide asteroid impact approximately 2 billion years ago, represent commercially viable nickel mines on Earth. Metals such as those belonging to the Palladium Group Metals (PGMs) represent critical industrial metals such as rare catalysts with increasing application in the near future (e.g. automotive catalytic converters, which are required by law worldwide). PGM supplies are present in limited quantities and worldwide depletion is possible within the next few decades. Rare Earth Metals form another group of critical resources subjected to increasing demand due to their use in electronic equipment.

NEO resources could incrementally expand our planet's economy if space agencies and companies alike make asteroid mining a possibility, and demonstrate if and how it can be accomplished.

7.1.2 Exploration Scenario Timeline

Human Exploration of Near Earth Objects requires a high degree or preparation and many steps must be achieved before this goal can be reached. As of yet, few low ΔV reachable targets have been identified and more importantly, few asteroids having similar orbits with that of Earths have been discovered (Barbee 2011). However, the early steps into the exploration of NEOs have already been taken. Missions such as NASA's DAWN and ESA's ROSETTA have already performed asteroid flybys collecting very important data and JAXA's Hayabusa mission as well as NASA's Stardust mission have managed to safely return asteroid and comet material back to Earth for study, proving that returning material from NEO's is possible by automated means. Capitalizing on the success of private companies becoming increasingly dominant in the space launch industry, several companies have stated their interest in constructing and executing missions with the specific aim of prospecting NEOs for their material values and future mining missions to relevant asteroids (i.e. Planetary Resources and Deep Space Industries).

A tentative NEO exploitation timeline is being proposed in the Fig. 7.2.

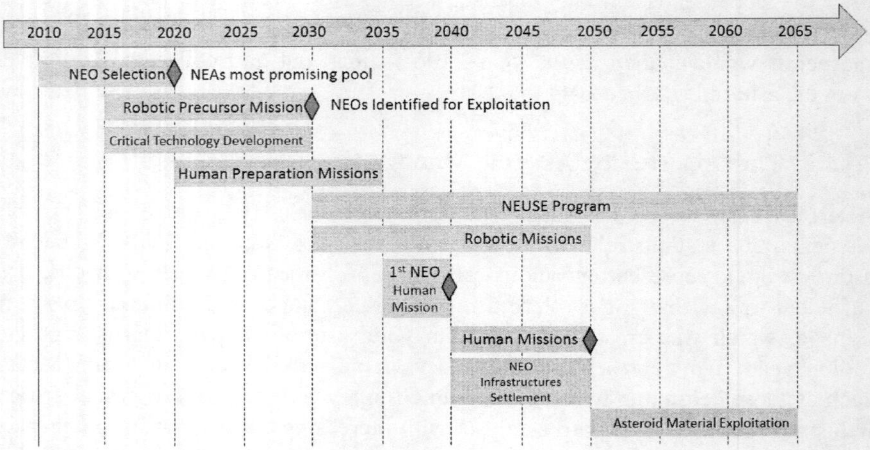

Fig. 7.2 Proposed Asteroid Exploitation Timeline

7.2 Mission Definition

The presented scenario proposes an architecture of human craft and highly autonomous robotic elements able to constantly return valuable resources from various NEOs to Earth and composed of a set of missions with the goal of both sustaining Earth's technological development and future missions beyond Low Earth Orbit by exploiting the NEOs mineral resources. This complex space program will be generically called NEOUSE.

A proposed scenario is that of sending out tens of relatively inexpensive, "hard-landing", mineralogical probes to examine the most economically attractive NEOs. Subsequently, having gained a solid grasp in terms of asteroid geology and composition, it would be possible to confidently identify the most promising sites for profitable extensive exploitation. Human craft will subsequently be sent to the most attractive asteroids of the initial lot to construct the outpost and the necessary infrastructure (e.g. telecommunication and navigation assets or power units). This stage would represent the shift from purely scientific goals to commercial exploitation. Finally, "low-cost" sample-return spacecraft will ferry materials collected by other robotic elements and/or astronauts.

Mark Sonter proposes logic relations that link together aspects that once assessed and solved would define one or more mission concepts (Sonter 2011) (Fig. 7.3).

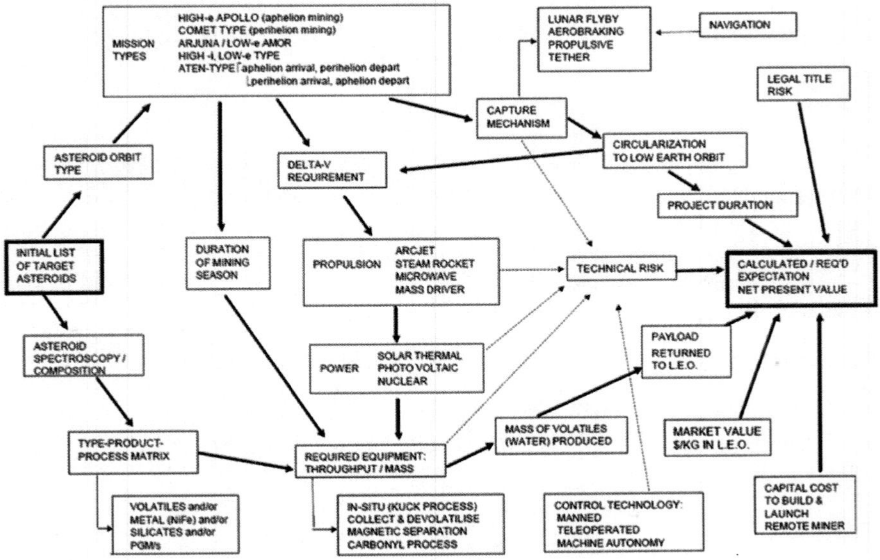

Fig. 7.3 Mission definition, spider scheme (Sonter 2011)

In the frame of this study, it has been preferred to adopt a simpler and more system engineering oriented approach. As such, a list describing the defining elements of the mission has been compiled and is presented below:

1. Definition of the NEOUSE mission objectives.
2. Functional analysis and definition of high level requirements (e.g. contamination issues) and external interfaces.
3. Identification of suitable NEOs and their characteristics (i.e. the most promising sites for profitable material "mining").
4. Preliminary mission analysis and trade-offs at system of systems level and definition of alternatives.
5. Definition of one or more mission architectures and the correlation among the different missions and elements.
6. Identification of the main elements and assets.

7. Determination of key-features of the human rated missions conceived.
8. Risk assessment and identification of critical points, and definition of development plans for associated critical, transversal and enabling technologies not yet available.
9. An economic analysis comparing the estimated missions costs and value of the extraterrestrial materials recovered for the different scenarios.
10. Estimation of a possible development plan for the NEOUSE program.

These points can be logically schematized and arranged such as in Fig. 7.4.

7.2.1 Mission Statement

The mission statement of the NEOUSE program was defined considering its two constituent segments, which are the Robotic and Human respectively. The mission statement of the robotic segment of the program is:

- To continuously transport to Earth high quantities of valuable asteroid materials in the most affordable way possible.

While considering the human segment of the program, the statement is:

- To safely transport and return a human crew to a Near Earth Object with the purpose of enabling the scientific study, sample return and set up of autonomous mining machinery.

7.2.2 Mission Objectives

The NEOUSE objectives are summarized in:

- Short term:

  1. To develop technologies critical to the discovery and study of NEOs
  2. To develop critical technologies for NEAs exploitation.

- Medium term:

  1. To bring back to Earth a significant amount of industrially-valuable asteroid materials
  2. To prove and test capabilities to prevent NEO impacts on Earth
  3. To initialize the market and commercialization of goods from space
  4. To develop elements, technologies and heritage for human Mars and deep space exploration

- Long term:

  1. To sustain the Earth technological development regularly supplying the terrestrial industries with critical extraterrestrial resources.

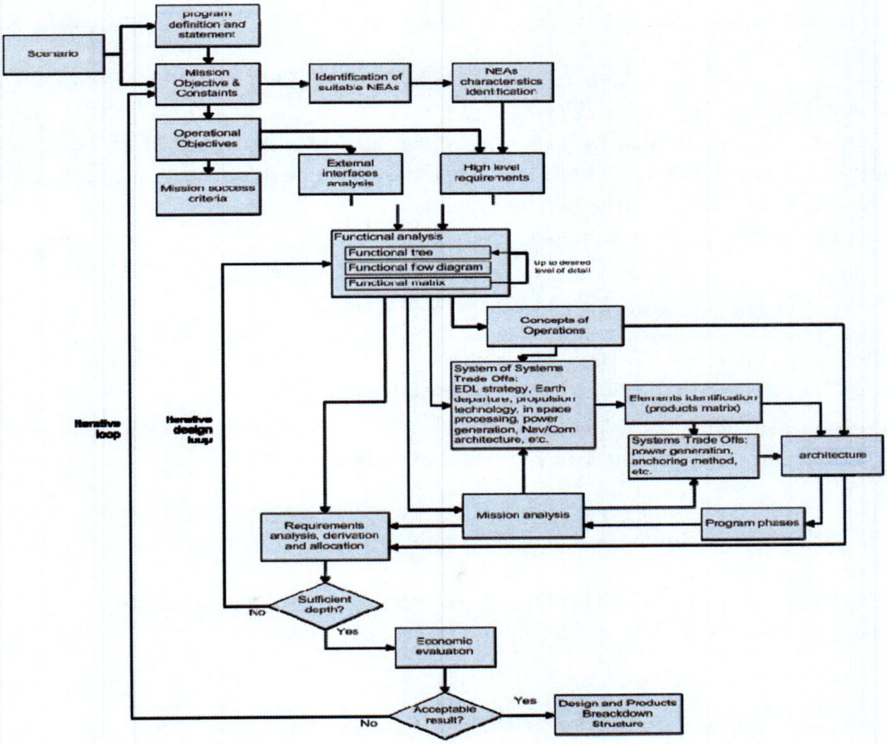

Fig. 7.4 Proposed logic scheme to define an NEOUSE program

7.2.3 *Top Level Requirements*

A complex system such as a human operated spacecraft able to perform flexible long duration missions to Near Earth Objects must be able to satisfy a very complex set of requirements. The purpose of this section is to present both common as well as specific mission segment top level requirements.

7.2.3.1 General Top Level Requirements

R-1 *The system shall withstand the Earth – NEO and return transfers*

A necessity of any space system is that of having a high probability of success, i.e. surviving possible environments from manufacturing, transportation and launch to orbital injection, transit, time on target, return trip and in the case of sample/cargo return and human missions the re-entry into Earth's atmosphere and landing.

R-2 *The system shall be able to safely approach the NEO and perform stationkeeping*
R-3 *The system shall be able to perform scientific and exploration operations*
R-4 *The system shall be flexible*

A complex system such as the one represented by the collection of elements tasked with prospecting and/or mining asteroids must be designed to interface and communicate with a wide variety of systems.

R-5 *The system shall be modular and reconfigurable*

7.2.3.2 Robotic Segment Top Level Requirements

RR-1 *The system shall be fully autonomous*
RR-2 *The system shall allow for remote human control*

7.2.3.3 Human Segment Top Level Requirements

HR-1 *The system shall provide a habitable environment for 4 crew members for the entire period of the mission (nutrition, breathable atmosphere, water and thermal control)*
HR-2 *The system shall provide for safe re-entry into Earth's atmosphere*
HR-3 *The system shall allow for IVA and EVA operations*

7.3 Challenges and Criticalities

A human mission to an asteroid poses a series of technological, biological, psychological and financial barriers. These challenges have been shortly described in the following section.

7.3.1 Deep Space Mission Operations

7.3.1.1 Ground Segment

By "ground segment" we intend a complex Earth based system used for space mission control, data management, processing and storage (if needed). The complexity of such a system is prominently increased when coping with deep space operations because of the large delay in signal transmission-reception (with an order of magnitude of hundreds of minutes). Furthermore, due to mission design issues, the available power is lower than the one used for LEO or NEO missions. An additional difficulty is coming from the need to reduce costs while maintaining on ground personnel 24/7.

In order to overcome transmission problem, mainly due to low power and long distances, the employment of big diameter antennas (15-35 m and in some cases even larger i.e. 72m) is mandatory. For example one of the ESTRACK ground stations or DSN (NASA Deep Space Network) could be a suitable candidate.

7.3.1.2 Guidance System and Maneuver

Due to the large distances and subsequent communication delays, maneuvering automation and high crew "ground-independence" is mandatory.

Relative guidance technologies encompass algorithms that determine the desired trajectories to be followed between vehicles performing rendezvous, proximity operations, and/or docking and capture or debris/asteroids avoidance maneuver. These algorithms must anticipate applicable environmental effects, the nature of the trajectory change/attitude control effectors in use, and the inertial and relative navigation state data available to the guidance algorithms. This kind of technology is currently available (2013) at TRL 3. The existing technologies of interest provide real-time, onboard algorithmic functionality that can compute and manage spacecraft maneuvers to achieve specific trajectory change objectives. Relative guidance aligns well with aforementioned needs because it impacts crewed deep-space exploration, sample return, servicing, and orbital debris mitigation.

7.3.1.3 Safety and Reliability

A topic covering an overwhelming importance is represented by related and overlapping topics of integrated systems health management (ISHM), fault detection and isolation and recovery (FDIR), and vehicle systems management (VSM), which together provide the crucial capability for an autonomous spacecraft to operate safely and reliably. ISHM/FDIR/VSM will improve the reliability of future missions by providing a diagnostic capability that helps ground or crew failure assessment and an automated capability to fix/overcome faults; increase robotic mission flexibility in response to failures; and increase crew safety in the event of a detected need for crew escape and abort. This technology is highly aligned to exploration needs because it will impact many missions, such as deep-space exploration, robotic science missions, planetary landers, and rovers.

7.3.2 Deep Space (Ionizing) Radiation

Acute and long-term crew physiological effects from radiation (Solar Particle Events -SPE- and Galactic Cosmic Rays -GCR-) and crew behavioral health are critical considerations for long-duration missions.

Deep space travels bring astronauts out from the protection of the Van Allen belts. During the Apollo era, the exposure was fortunately limed due the short overall duration of the missions, however a NEO mission as well as a Mars mission would require a much longer travel period (e.g. 6 months), neglecting the in-situ operational phase. This is a constraint for any deep space exploration beyond the Moon and radiation protection is a critical factor in safely transporting humans to a NEO. Prolonged exposure to space radiation is lethal to humans and in the case of Solar Particle Events death can be immediate.

Traditional protections based on hard shields might jeopardize the feasibility of deep space human missions due to mass constraints; new innovative solutions, also at system level, must be identified and verified to sustain human exploration.

The state of the art concepts are:

- Traditional shields with innovative materials (i.e. high hydrogen content). This case considers the physical architecture where water or propellant storages are distributed around the crews "safe haven".
- Genetic selection of astronauts naturally less affected by radiations.
- Active shielding by using induced electromagnetic fields, even if complexity and energy budgets would be significantly affected.
- Accurate and timely SPE alerts.
- Anti-radiation drugs since they are already used with good results in medicine for several scopes (e.g. treat life-threatening injuries, reduce symptoms, and manage pain, cancer treatment and so on),
- Active radiation shielding available at TRL 3-4.

Shorter mission times are to be recommended in all NEO missions as they are at the moment the best available method in limiting the total radiation dose.

7.3.3 Asteroid Environment

Depending on the orbit, asteroids experience impacts of many sizes, from dust grains whose cumulative effect is significant surface erosion to catastrophic collisions that disrupt objects and dissipate debris throughout the asteroid belt, and the solar system.

After billions of years of impacts, the surface of a NEO is expected to be covered by craters and ejecta of numerous impacts, having a powdery or gravelly texture. This layer of broken-up rock is called regolith. Below this layer of incoherent material, there might be a solid rock bed.

The amount of ejecta retained by a body depends mainly on its gravity pull (i.e. different escape velocity), therefore the regolith granularity and the layer thickness are expected to differ on objects of different mass.

Not all the Asteroids are a unique block of rocks, but sometimes it can happen that they are composed of boulders. Best example is Itokawa, which is an unconsolidated rubble pile composed of smaller re-accumulated fragments with no overall structure.

High-energy ultraviolet light and x-rays, beside from the direct effects on humans and spacecrafts, they can cause electric charge in the regolith particles producing electrostatic levitation, which has been already observed on the Moon.

Expected temperature and heat fluxes should be similar to those encountered during the Apollo missions, where the rock under sunlit behaves like heaters.

7.3.4 Approach and Station Keeping

Gravitational field, celestial body shape, surface topography, tumbling and/or spinning velocity are all very important for proximity operations. Any spacecraft maneuvers should be carefully designed considering these aspects; especially the object rotation affects the feasibility of surface approaching and touching concepts and strategies. For example a standard solution based on propulsion would require a complex control and propellant consumptions. Assuming to choose for any reason a high-velocity spinning target, the preferable condition might be a polar landing, where the relative surface velocity is lower as possible.

Rotation also affects the gravitational filed, which is neither spatially regular. The effect is a low gravitational field locally variable in space and in time and a very challenging orbital control (e.g. station keeping or orbital docking)

7.3.5 Mining

Mining on an asteroid with very low gravity attraction is principally like on Earth: to tear off or to cut pieces/particles and then collect them together. On an asteroid what is trivial on earth becomes tricky. Any single low force exercised on a body will push you back unless you are solidly clung to something stable. But many asteroids are just a clump of regolith and disconnected rocks.

Material can be gathered form the surface or the mining system might work as a worm digging tunnels but any mining system shall be design accordingly to the specific gravity and terrain morphology expected.

7.3.6 Contamination Issues from NEAs

When a physical interaction with a celestial body and man craft are foreseen or even just probable under a risk probability, the potential contamination between the two objects has to be considered. The contamination issue in exploring celestial bodies concerns two aspects:

- The problems of spreading Earth-origin biological materials to other planets and space objects in general. Although the existence of life elsewhere in the solar system may be unlikely, the conduct of scientific investigations of possible extraterrestrial life forms, precursors, and remnants must not be jeopardized.
- The threats of bringing to Earth dangerous and unknown extraterrestrial materials or substances; the Earth must be protected from the potential hazard posed by extraterrestrial matter carried by a spacecraft returning from another planet.

On these international agreement is being sought through The Committee on Space Research (COSPAR 2013). Five categories for target body/mission type combinations and their respective suggested ranges of requirements are identified (Table 7.1).

Table 7.1 Planetary protection target/mission classification and requirements (COSPAR 2013)

Category	Target/Mission	Protection Level
I	Includes any mission to a target body which is not of direct interest for understanding the process of chemical evolution or the origin of life.	No protection of such bodies is warranted and no planetary protection requirements are imposed by this policy.
II	Missions to those target bodies where there is significant interest relative to the process of chemical evolution and the origin of life, but where there is only a remote chance that contamination carried by a spacecraft could jeopardize future exploration.	The requirements are for simple documentation only: preparation of a short planetary protection plan is required primarily to outline intended or potential impact targets, brief Pre- and Post-launch analyses detailing impact strategies, and a Post-encounter and End-of-Mission Report which will provide the location of impact if such an event occurs.
III	Missions (mostly flyby and orbiter) to a target body of chemical evolution and/or origin of life interest or for which scientific opinion provides a significant chance of contamination which could jeopardize a future biological experiment.	Requirements will consist of documentation (more involved than Category II) and some implementing procedures, including trajectory biasing, the use of cleanrooms during spacecraft assembly and testing, and possibly bioburden reduction. Although no impact is intended for Category III missions, an inventory of bulk constituent organics is required if the probability of impact is significant.
IV	Missions (mostly probe and lander) to a target body of chemical evolution and/or origin of life interest or for which scientific opinion provides a significant chance of contamination which could jeopardize future biological experiments.	Requirements imposed include rather detailed documentation (more involved than Category III), including a bioassay to enumerate the bioburden, a probability of contamination analysis, an inventory of the bulk constituent organics and an increased number of implementing procedures. The implementing procedures required may include trajectory biasing, cleanrooms, bioload reduction, possible partial sterilization of the direct contact hardware and a bioshield for that hardware. Generally, the requirements and compliance are similar to Viking, with the exception of complete lander/probe sterilization.

Table 7.1 (*continued*)

Un restricted V	Unrestricted Earth-return Missions from solar system bodies deemed by scientific opinion to have no indigenous life forms.	Planetary protection requirements on the outbound phase only, corresponding to the category of that phase (typically Category I or II).
Restricted V	Missions comprise all Earth-return missions.	The highest degree of concern is expressed by the absolute prohibition of destructive impact upon return, the need for containment throughout the return phase of all returned hardware which directly contacted the target body or unsterilized material from the body, and the need for containment of any unsterilized sample collected and returned to Earth. Post-mission, there is a need to conduct timely analyses of any unsterilized sample collected and returned to Earth, under strict containment, and using the most sensitive techniques. If any sign of the existence of a non-terrestrial replicating entity is found, the returned sample must remain contained unless treated by an effective sterilizing procedure. Category V concerns are reflected in requirements that encompass those of Category IV plus a continuing monitoring of project activities, studies and research (i.e., in sterilization procedures and containment techniques).

COSPAR agrees that all samples returned or human exploration (missions of Category V) from planetary satellites and small solar system bodies that must be contained should be treated as potentially hazardous until proven otherwise. Thus the first approach toward the NEAs should prove the absence of any dangerous contamination.

Sample return provisions for contained samples are the same as for Mars. Generally no sample containment and handling is warranted beyond what is needed for scientific purposes. Recently there has been considerable progress in designing sample containment devices to prevent contamination of the asteroid by terrestrial organisms, the transfer of organisms to Earth or the contamination of one asteroid sample with another. Although well-developed, these procedures require cost and complexity for the mission design, therefore they will be implemented only in the first phases of the NEOUSE project. Then when/if the targets are proven to be safe with respect to contamination, the following missions could be considered of Category Unrestricted V. It should be noted that COSPAR Planetary Protection Standards stem from the need to control scientific missions and that new planetary protection rules should be considered in the frame of continuous material return missions.

7.4 Asteroids Selection

7.4.1 Suitable Asteroids

M-type asteroids are composed of crystalline iron-nickel alloy, not in oxidized form, therefore they do not need to be smelted, but just mechanically separated to purify the desired metal. They also contain some cobalt, gold, silver, platinum, the rest of the platinum group metals (PGMs), and other industrial metals. PGMs should be particularly abundant in certain types of NEAs, which are mineralogically similar to "observed fall" meteorites. Highest metal concentrations (i.e. PGM-5 of around 90 ppm) may be found, if we can identify and target an asteroid that is compatible to a group IVB iron meteorite, thought to be the "fractional crystallization of a molten, magmatic, body of metal, presumably an ancient asteroidal core" (Hutchison 2004), even if the evolution of IVB parent-bodies remains cloudy (Walker et al. 2008). In these sources, it may be possible to extract up to 187 parts per million (ppm) of precious metals, which includes Au, the PGM, Re, and Ge. More than 1000 ppm of other metals (Mg, Al, Ti, V, Cr, Mn, Cu, Mo, Pb), semiconductors, and non-metals such as Ag, In, Co, Ga, and As. IVBs have the most extreme compositions in a number of aspects, including the lowest abundances of the volatile elements Ga and Ge and, on average, the highest abundances of refractory siderophile elements such as PGM (Ross 2001). Comparing to Earth sources, profitable terrestrial mines present instead a PGM-5 concentration of only 3-6 ppm, thus more than 10 times lower.

These meteorite types seem the plausible best-case return values for asteroid mining ventures. But this, from a business perspective, is only the beginning.

7.4.2 Identification of the NEA Targets

Moon and Mars missions require more on-orbit mass for landing and liftoff. Mars could offer an atmosphere, valuable for radiation protection, but for the moment, we do not possess the technology necessary to accomplish human exploration to such distances. Asteroids could represent a "profitable" the first step ahead in preparation of a human Mars exploration and settlement.

The most promising sites for "profitable" exploitation have to be identified in terms of:

a. Energy required for repetitive round-trips;
b. Asteroid rotation and orientation of spin axis;
c. Travels duration;
d. Asteroid size;
e. Expected mineralogical and geological composition;
f. Surface morphologies.

7.4.2.1 Accessibility and Travels Duration

The mission Delta-V (ΔV) [Velocity Variation, cumulative index of energy necessary for the orbital maneuvers conceived] required to reach selected low-ΔV target NEAs is not much greater than that needed one to acquire a geosynchronous orbit. Instead the ΔV required to come back to Earth from these targets is much less, and can be imparted gradually, over several weeks, thus very substantially reducing the demands on the propulsion system. Some NEAs are relatively easy to reach, and some 200 of them require a ΔV lower than 6.5 km/s (to Moon landing is 6.3 km/s). It is however relatively simple to inject a spacecraft into an Earth Transfer Orbit with only 1 km/s of ΔV, with the possibility of using just electric propulsion (low thrust but very high specific impulse) due to the very low escape velocity. Scientific missions like Hayabusa have already proved this.

Figure 7.5 shows the NEAs accessibility in terms of ΔV with respect to their aphelion.

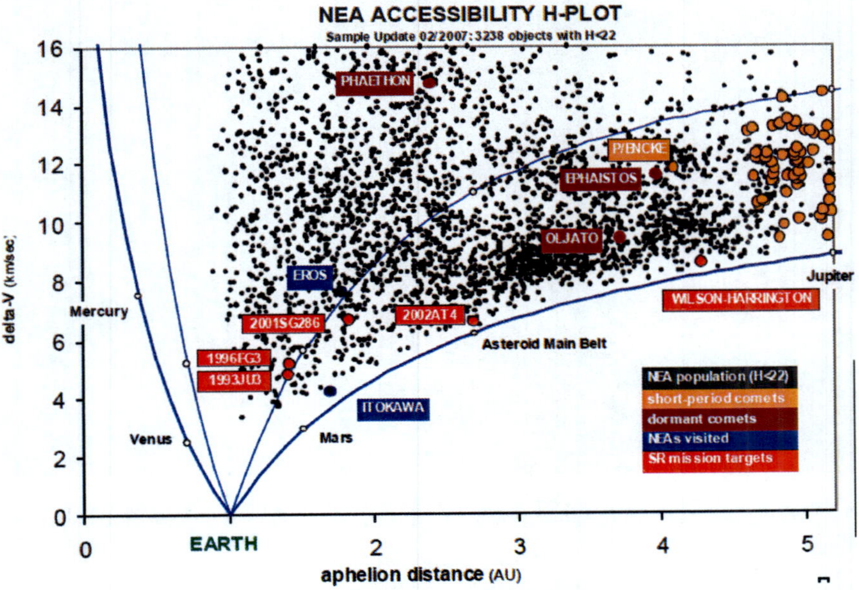

Fig. 7.5 NEA accessibility with respect to aphelion distance (Barucci and Yoshikawa 2007)

The low-eccentricity and low-inclination Apollos and Atens as the regions of interest (Fig. 7.6), for reason of the low ΔV's required for outbound and return trajectories. Among those, low-eccentricity targets may be preferred allowing use continuous-thrusting propulsion (not Hohmann transfer), and extended mining seasons. When designing the orbit there is always a compromise between energy required and travel duration. For example slow spiral return implies longer mining season, and hence less demanding specifications on mining, processing, and propulsion equipment, and on solar collector. Note that spiral return trajectories can be designed to deliver the payload at very small return hyperbolic velocity (V_{hyp}), because the spacecraft trajectory can be made tangent to the Earth's orbit. Such low V_{hyp} implies easy capture into Highly Elliptical Earth Orbit (HEEO) by lunar flyby.

A pareto comparison has been performed to assess the major parameters of a typical NEO mission. NASA considers the total ΔV including the maneuver required to depart a notional 400 km altitude circular Earth parking orbit, the ΔV required to match the NEA's velocity at arrival, the ΔV required to depart the NEA, and the ΔV (if any) required to control atmospheric entry speed at Earth return. Figure 7.7 collects and compares the data of the most accessible NEAs taken from the NASA JPL database (NASA JPL 2013).

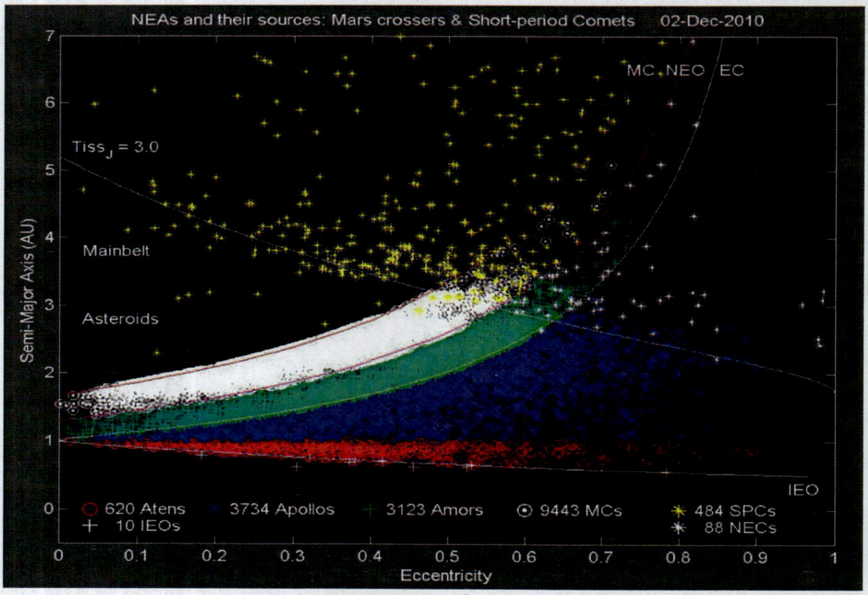

Fig. 7.6 NEO distribution with respect to orbit eccentricity and semi-major axis. (DLR 2011).

It can be seen that the scenario with mission duration of 354 days and a Δ-V of 6 km/s should cope with the all the most easily accessible NEAs known and still unknown.

Other important aspects to be considered are (Friedensen 2011):

- To provide multiple consecutive launch windows to shorten on-orbit dwell time
- Ensuring equivalent mission durations (to ease logistics and margins)
- Enabling similar mission 'portfolios' (scientific and pragmatic goals would be similar for each primary/secondary destination identified);

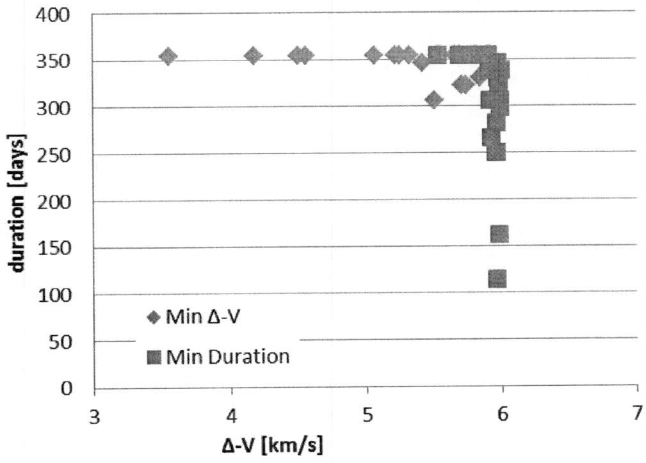

Fig. 7.7 NEA accessibly assessment, mission duration versus Δ-V. Data from http://neo.jpl.nasa.gov/nhats/ (NASA 2010).

7.4.2.2 Asteroid Rotation

Rotation state is a key driver for an asteroid mission design. If the spin occurs on a time scale comparable with that one of the mission, the orientation of the rotation axis and the modalities at which it changes are critical since, as for the period and rotation velocity, they strongly affect also the power and the thermal systems. The derived centrifugal force of 60% of the asteroids is on the order of 10^{-3} m/s^2. This acceleration has to be compared with the gravitational pull.

Usually the pole is the preferred site for the absence of rotation, but a phenomenon to be taken into consideration in determining the axial orientation of an asteroid is the precession that is a gravity-induced, slow and continuous change in the orientation of an astronomical body's rotational axis.

7.4.2.3 Asteroid Size

Size implies different gravity pulls and therefore different close-operation propellant consumption (different orbital maneuvering strategies) and anchoring/mining concepts.

Of course in addition the size is directly proportional to the total amount of gatherable material, bigger is the asteroid and higher will be the amount of resources obtainable.

7.4.2.4 Mineralogical and Geological Composition

Suitable asteroids should have a high concentration of valuable material. A part from this, these materials should be easily harvestable in terms of:

- Extraction/gathering
- Material process

7.4.2.5 Surface Morphologies

Surface operations shall be compatible with the asteroid superficial morphologies in terms of craters beds and rims, regolith average size, regolith layer thickness.

7.5 System Work

7.5.1 Mission Scenario Assumptions and Constraints

Before the definition of a space program such complex and far ahead in time like the NEO-use program, some boundary conditions has to be settled. Careful considerations and/or assumptions should involve: transportation capabilities (availability of heavy lift launchers); political and economic conditions, the technologies maturity and scientist expectations.

7.5.2 Identification of Assets

The architecture elements and how they will work together has to be identified. Hereafter an exemplificative list of elements that might be included in a system of systems (Table 7.2).

Table 7.2 Possible NEOUSE assets

Possible ground segment elements:	
Launchers and space ports Control stations Deep-space navigation antennas Landing sites and logistics Material processing and distribution centres (ISRU) Habitation modules	The necessary number, distribution and capability have to be determined and compared with the actual availability.
Possible space segment elements:	
Eventual scout probes	These are a fleet of probes that go to potentially interesting NEAs and ping them with all kinds of sensors for understanding their chemical, physical, orbital, rotational and gravitational properties. They are miniaturized as far as possible. Scouts just observe with remote sensing Prospectors send back samples, or do on-site analysis Both are built with exchangeable modules put together in various ways to fulfill the mission. They are launched from LEO, and may have the capacity to navigate to more than one NEA.
Asteroids Orbiters for Nav/Com services	Satellites orbiting/flying around or next to the asteroid to provide Nav/Com services among Earth ground segment, the elements on the asteroid surface and the other orbital elements.
Asteroids miners/catchers	Unit down/on the surface of the asteroid, that digs in and breaks the material up enough for transport to the eventual processor unit.
Row material processors	Unit that takes in raw asteroidal material delivered by the diggers and separates it out into: Iron, nickel, cobalt, PGMs, volatiles and rocky slag. All of them leave the processor.

Table 7.2 (*continued*)

Space Tugs	Multipurpose transportation elements able to carry and interface with other elements and material among different orbits. Each Tug will remain in orbit several years performing multiple missions.
Propellant and material containers	Volatiles are tanked, and these tanks also stacked on the outside of the processor for easy pickup by docking tugs.
Earth Entry Capsules	Capsules to be provided by Earth or produced in space by means of asteroid resources, used to re-enter on Earth the collected materials.
Reusable LEO carriers (Space Shuttle-like vehicles)	There can be reusable elements used for transportation purposes between Earth surface and LEO.
LEO infrastructure for refuelling and maintenance	A sort of infrastructure similar to the actual International Space Station. It could be located in LEO or in a lagrangian point.

These or other elements could be picked up to figure out the mission architecture. The selection process should follow a systematic approach from objectives, to functions and requirements, tradeoffs and analysis.

7.5.3 High Level Trade-Offs

Depending on the assumed scenario rationale choices to be performed might be:

- To bring back asteroidal material or whole asteroids
- To bring back raw asteroidal material, or process it on-site to bring back only processed materials, and eventually produce propellant for the return trip.
- Possibility about stopping or mitigate the rotation of the asteroid.
- Asteroid anchoring concept
- Material harvesting/collection concept
- Earth entry strategy: direct, aerocapture, lunar assisted capture, propulsive braking.
- Earth entry, descent and landing strategy: Inflatable Re-entry Descent Technologies (IRDT), lifting body, controlled or passive, parachutes, skip re-entry.
- Propulsion: electrical, chemical, hybrid.

All these choices shall be rationally faced comparing their implications at system level and the ratio between cost and benefits.

7.5.3.1 Bring Back Asteroidal Material or Whole Asteroids

Instead of mining an asteroid far from Earth, it could be thought to bring back to Earth the whole asteroid, and then to mine it in Earth orbit or if it is small to safely re-enter it to the Earth surface, the materials could be extracted much more easily after landing. Safety is the concern of this scenario due to the risk of falling coarse material. Operations have to be carefully programmed.

7.5.3.2 Bring Back Raw Asteroidal Material, or Process It On-Site

Assuming available a sufficient amount of power (e.g. nuclear reactor), processing on site the material will increase the percentage of valuable resources within the returning mass. In first approximation a longer mining/processing period could raise the payback mass ratio. A simple economic analysis could suggest which choice should be the better.

Thus, material processing can enhance the mission performance by:

- mineral processing to increase the net percentage of valuable mass.
- propellant production or material exploitation for propulsion purpose; using part of the retrieved non-terrestrial mass as reaction mass, such as asteroid-derived volatiles, or in-situ-produced propellant. It becomes possible to return to Earth orbit much more mass.
- A simple Earth re-entry capsules manufactured at the M-type asteroids, moulded in one or two pieces of Inconel metal.

7.5.3.3 Possibility about Stopping or Mitigate the Rotation of the Asteroid

Stopping the rotation of the asteroid, if small, might obtain some benefits such as an almost constant sun vector direction, to easily point the solar-powered processing equipment or other thermal equipment, and easier landing/departing operations.

A solution would probably be to use a yoyo-like gadget commonly used to design satellites without using any fuel propellant. Others more exotic concepts are possible, but for bigger asteroids, or for a low-budget operation on a small asteroid, it is probably better to let the asteroid keep spinning and just attach the industry to the pole of rotation using a rotary joint.

7.5.3.4 Asteroid Anchoring Concept

Since gravity cannot assist to keep the rig in place, the mining machinery must always be anchored to the asteroid surface or subsurface in order for forces to be exerted onto the asteroid. The released material must then be efficiently contained and recovered. Containment is important, because the escape velocity for small asteroids may be of the order of 20 cm/s.

Anchoring may be difficult due to the loose rubble structure of many asteroids. Securing the elements is easy on solid consolidate objects but it is likely to be very difficult with low strength or unconsolidated material, such as loose regolith or the hypothesized loose dusty covering of a dormant or extinct comet. The reaction forces created by such mining operations as drilling or scraping may in that case require the operation to be spread over a very wide footprint, if the regolith strength is low, and because of the very low gravity. This may need very wide anchoring area, including the approach of totally surrounding the target body, by wrapping it with a net or membrane.

Possibilities for securing spacecraft to an asteroid would be:

- Assuming a loose substrate
  - to tie the spacecraft down with a rope or a net passing around the entire NEA;
  - to screw in large area augers or screw-plates (assumes that there is a regolith and it is loose enough and compressible enough for a screw to penetrate);
  - to weld tie-downs into massive clasts of metal, ice, or solid silicate rock;
  - to use large area fluked anchors;
  - to burrow completely into the regolith (e.g., using contra-rotating screws).
- Assuming a mechanically competent substrate
  - to glue or adhere to the surface;
  - to clamp against opposing surfaces.
  - to drive in pitons;
  - to fire in harpoons or penetrators which resist extraction.

The most promising concepts should be investigated and tested during the technology development phase.

7.5.3.5 Material Harvesting/Collection Concept

The mining method will depend on the material being sought. If regolith, the method will clearly be very different from that chosen if recovering solid metal; different again, if the ore is high in volatiles and ices. Mining approaches will depend on the material:

- *Loose* material can be scooped, scraped, or shoveled.
- *Friable* but bound material will have to be broken or cut, or somehow disaggregated, before collection.
- *Hard rock* will require drilling, cutting, or blasting. The force required can be exerted against the rock surface, either by impact or by pressurization or by static loading. (e.g. drills, hammers, tunnel borers).
- *Silicates and ices* or hydrocarbons (frozen volatiles) can be melted or vaporized in addition to the other solutions (cut or mechanically mined).

- *Extensive metal* can be cut/machined, melted at high temperature or reacted at a lower one, e.g., using the carbonyl vapor-metallurgical process.

In very-low-gravity it is necessary to:

- ensure the scraper or shovel is held against the surface; and to
- ensure that collected material is effectively retained within the collecting mechanism, and does not float away. Thus, mining on low gravity bodies will require an approach that encloses the regolith being collected.

For these motivations there may be good reasons to use underground mining techniques when mining on asteroids:

- it is easier to generate reaction forces for cutting, drilling, or digging (i.e., it uses standard terrestrial mining technologies);
- the surface layer may be depleted in the desired material (e.g., volatiles may only lay at depth under a lag deposit in a dormant comet);
- it may be easier to contain the material;
- the resulting volume may itself be useful, e.g., for storage, habitat, or plant.

An underground mining technology should not require a large normal reaction force, and should have minimal impact on ground that is suspected to be weak and friable (Gertsch and Gertsch 1997). During tunneling the miner holds itself steady by the walls of the tunnel pushing against the walls or cutting into them. Tunneling prevents consumption of the entire asteroid, but desirable ore veins or cracks can be followed.

Another concept is to have a stationary canopy. In this case a dust kicker goes down to the asteroid and just kicks up the ore. When there is enough material in the canopy, it could be sealed off and moved to the eventual processing site.

7.5.3.6 Earth Entry, Descent and Landing Strategy

There are various possibilities for reducing velocity from hyperbolic to a bound orbit upon Earth return (Gertsch and Gertsch 1997):

- *Propulsive braking*. This method is the simplest in terms of orbital operations, but may be undesirable, as it reduces the quantity of retrievable material that is available for sale.
- *Aerobraking*. It takes advantage of the highest layers of Earth atmosphere to reduce the spacecraft velocity by means of aerodynamic drag enhanced by large surfaces (e.g. solar arrays or IRDT) orthogonal to the velocity direction.
- *Lunar assisted capture*. It uses a lunar flyby to remove the Earth hyperbolic velocity. A velocity reduction in the order of 1.5 km/s is achievable for a single lunar flyby but severe navigation and timing constraints must be met, to ensure the requisite low altitude pass over the Moon at the proper time in its orbit to provide maximum velocity loss.

- *Direct reentry.* No reduction of velocity is foreseen. This is absolutely the simplest approach even if the reentry mechanical and thermal loads are the highest. Past missions demonstrated it is feasible a direct entry trajectory with no manoeuvring system on the entry vehicle. Genesis demonstrated that a soft landing system is not required at all [http://genesismission.jpl.nasa.gov/]. A traditional aeroshell entered Earth's atmosphere and crashed into the desert floor. Although the capsule broke on impact, scientists recovered the sample. Asteroid material bullion delivered by capsule would be bent, dented, and twisted by impact, but would be melted down anyway at the refinery so it does not matter. Genesis has been a resounding successful demonstration of recovery of asteroid material.

The Star Dust capsule re-entry velocity was 12.9 km/s, which has been the fastest ever re-entry speed into Earth's atmosphere ever achieved by a man-made object [http://stardust.jpl.nasa.gov/home/ index.html].

Many solutions are also possible to control the aerodynamic coefficient of drag and lift during reentry and therefore acquire the capabilities of trajectory control which implies to perform a certain level of precision landing and to regulate the mechanical and thermal loads during reentry (Table 7.3). Even Skip reentry is often an interesting choice depending on the mission requirements.

Table 7.3 Solutions to control the aerodynamic coefficient of drag and lift

	Passive	Active
Capsule	Stable geometrical configuration (aeroshell shape and center of gravity)	Active aerodynamic surfaces Controllable center of gravity Reaction control system
Lifting body	-	Active aerodynamic surfaces

Even the last phase before surface touching conceives many different solutions for descent and landing. The common possibilities are hereafter listed:

- Descent technologies:
  - Parachutes
  - Paragliding
  - Retrorockets
  - IRDT (Inflatable Reentry and Descent Technologies)
- Landing technologies:
  - Ground Landing / Splashdown
  - IRDT
  - Airbags
  - Crushable structures
  - Retrorockets

7.5.3.7 Propulsion

The decision about what propulsion technology is the most suitable to be adopted is strictly related to the mission analysis. Table 7.4 reassumes the propulsive technologies available.

Table 7.4 Propulsion options for transportation elements

Propulsion philosophy	Technology	Mission analysis constraints	Characteristics
Electrical	Power source	They depend upon the selected power generation technology (e.g. attitude wrt the sun)	Higher is the power available and higher are the propulsive performance
	Thrust generator device	Low thrust, as with ion-engines, and contour navigation along gravitational "contours". Very promising is the VASIMIR engine technology.	Very high specific impulse
Chemical	Solid	No versatility	Simple, high TRL, low specific impulse.
	Liquid (cryogenic/storable, bi/mono propellant) or Hybrid	-	High TRL
Mixed propulsion	Combination of electrical and chemical propulsive systems.	-	Chemical for escaping LEO, then ion-engines for transfer and insertion into NEA orbit, then again chemical for local micronavigation at the NEA, ion-engines on the way back to Earth, and chemical again for rounding out the aero-braked orbit.
Solar sails	Wide reflecting films Mini Magneto Sphere Plasma Propulsion (M2P2[a])	It implies very complex low-thrust trajectory design.	It uses the solar wind to push the spacecraft. Low thrust, low TRL, no propellant required.

(a) http://www.ess.washington.edu/Space/M2P2/

For now it seems that the mixed propulsion option for a transportation element could offer the greatest benefits.

For the propulsion system of the NEOUSE program, the VASIMR is also advisable (Variable Specific Impulse Magnetoplasma Rocket) [http://www.adastrarocket.com/aarc/VASIMR]. It consists of three linked magnetic cells. Plasma ions are injected from the forward cell, and then accelerated through an ion cyclotron. After the ions have gained enough energy, they are shot out the aft cell to produce thrust. A major reason for choosing the VASIMR system is its ability to use constant power throttling. This enables the fastest possible round trip time for a given amount of propellant. Using the VASIMR, NEOUSE transport elements would be able to make the round trip from the asteroid to low Earth orbit and back to the asteroid, again, in about one year. This makes VASIMR the ideal choice of a propulsion system for this mission.

7.5.4 Identification of Critical and Enabling Technologies

A program such as NEO-use foresees a high number of very complex missions also different among them. The individual and overall risk must be carefully analyzed since the very beginning of the design.

Table 7.5 Critical and enabling technologies list

Technology	Description
High efficiency propulsion	Electrical propulsion or new other propulsive means, even if travelling time is a constraint
Material processing	In situ resources processing to: increase the payback mass manufacture re-entry capsules recover propellant in loco
Electrical power generation	Solar arrays, nuclear reactor, fuel cells, Solar Power Satellite (SPS), etc.
Asteroid anchoring technology	Devices to fix/retain the asteroid surface elements next to the surface.
Radiation protection	Active or passive system providing long term human protection against high energy radiation, particles and solar events.
NEA guidance and navigation	Guidance and control systems for maneuvering and navigating around a NEA proximity.

An important early step is to identify of critical, transversal and enabling technologies in order to concentrate the efforts in research and the investments of resources on Earth. A clear development roadmap shall be established.

In particular is has to be highlighted that NEA mining technologies are largely also the same enabling technologies for a future PHA deflection mission (Table 7.5).

7.5.5 Cargo Return Vehicle

During the NEO-use campaign, the huge quantity of material which will be gathered should be processed to be enriched on the asteroid by in-situ robotic elements. Then, a resulting much smaller and most valuable amount shall be brought to Earth by means of a fleet of Cargo Return Vehicles (CRV)(Fig. 7.8). Neglecting all the other assets (services satellites, mining systems, etc.), this section aims to provide a closer look on what on what would be a solution to the gathered material transportation to Earth. The CRV details presented are taken from the study work of the SEEDS Master course (SEEDS 2009, 2010).

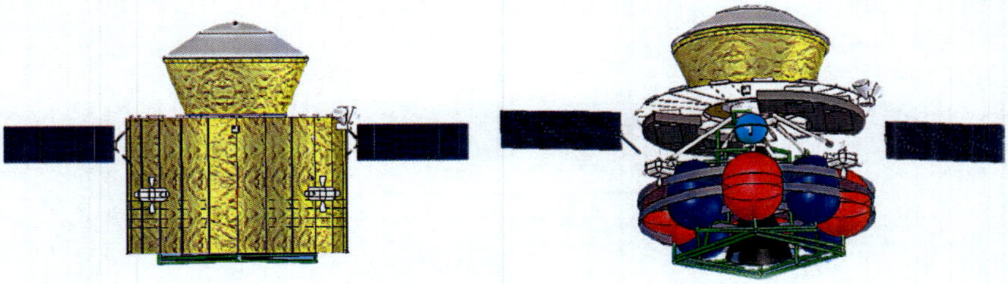

Fig. 7.8 CRV concept

The CRV conceived is a complex system, which is itself composed by three major elements described in Fig. 7.9.

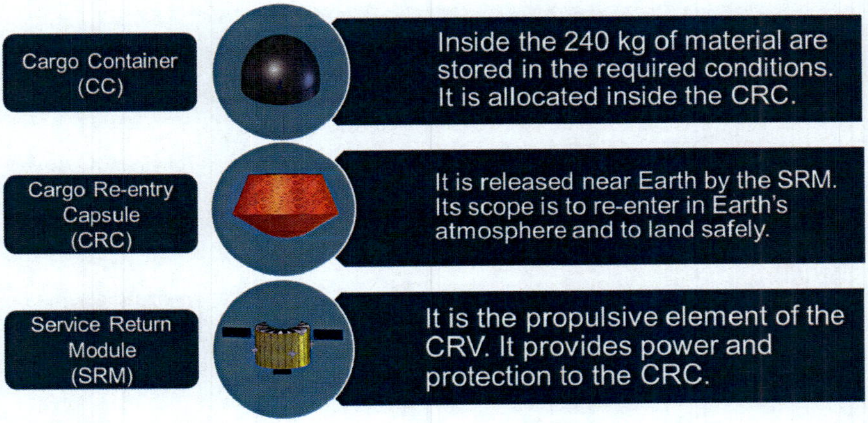

Fig. 7.9 Cargo Return Vehicle element decomposition

CRVs shall be able to reach and touch ground autonomously on the asteroid nearby the local infrastructures, and then the extracted material will be loaded by robotic elements in the CRV, which will bring them safely to Earth. Once in Earth proximity, the CRV will release and inject a Cargo Re-entry Capsule (CRC) with the material into a precise reentry trajectory. Such CRC will land close to Earth asteroid material and retrieval processing facilities, guaranteeing material protection and containment until its recovery.

The CRC is suitable to withstand the thermal and mechanical loads present in a hypersonic atmospheric entry; it would have a loading capacity of 240 kg, with an external diameter of about 2 m.

- A landing error of 50 km would be achievable also with a very simple passive ballistic reentry.
- The descent system will be only made of main parachutes and pilot chutes achieving a maximum ground impact velocity of less than 10 m/s.
- To minimize the mass of the descent system, it has been conceived to release the heat shield in flight.

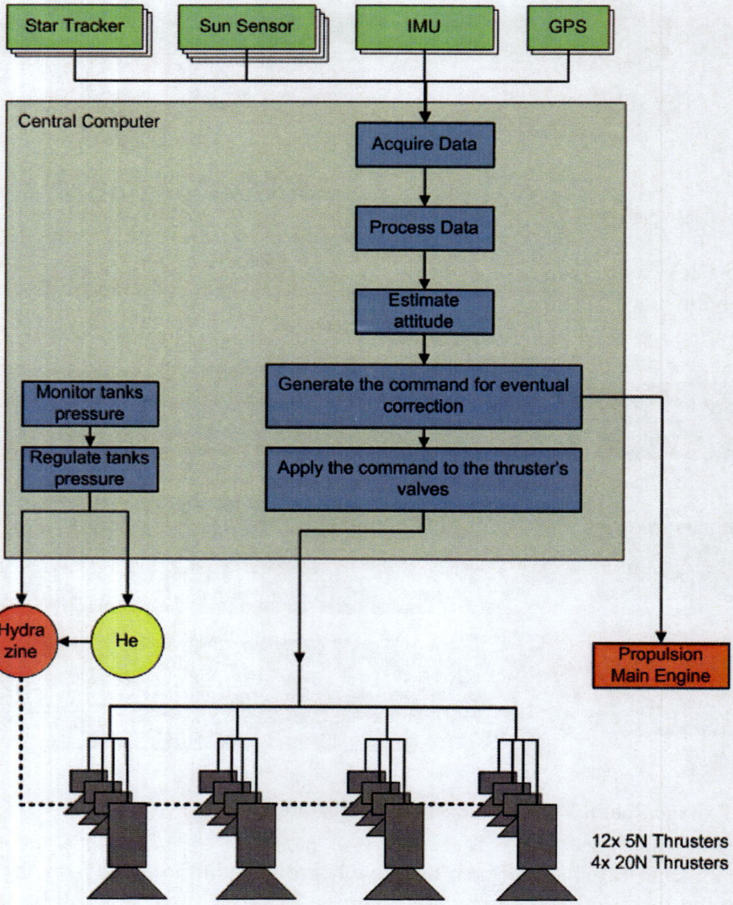

Fig. 7.10 CRV attitude and orbital control system architecture

The transferring module (SRM) has instead to perform different functions, particularly the propulsive one during orbital flights. The control method proposed is 3-axes stabilization, zero momentum with only thrusters; Figure 7.10 summarizes the attitude and orbital control system (AOCS) architecture.

The navigation sensors selected are a star tracker with three optical heads, sun Sensors are a redundant system solar array sun-pointing, two inertial measurement units (IMUs) in cold redundancy and a GPS-like receiver. The GPS system will not be used until the navigation satellites are in line sight and range. It might provide high accuracy position for important maneuvers in Earth proximity like the insertion, the capsule release and the CRM disposal.

The CRV engine shall be able to restart, with thrust vector control capability and throttability. Mono-Methyl-Hydrazine (MMH) as fuel and Nitrogen Tetroxide (NTO) as oxidizer is the selected propellant/oxidizer combination of a Gas Generator (GG) cycle. High pressure Helium is utilized to pressurize MMH and NTO tanks. This combination probably ensures the best performance in terms reliability, storability (years in specific conditions) and specific impulse. MMH/NTO can also be quite easily reignited and throttled.

The driving factor in the size of the vehicle is represented by the fuel tanks and the engine respectively (Fig. 7.11). Because of the main driving elements being linked to the propulsion system, the evolution of the design would heavily relied on the evolution of the propulsion system. Since more than the 60 % of the CRV mass is propellant, some improving design solutions could be:

- staging strategy and/or orbital nodes
- to increase the I_{sp} using different propulsion technologies (e.g. VASIMR).

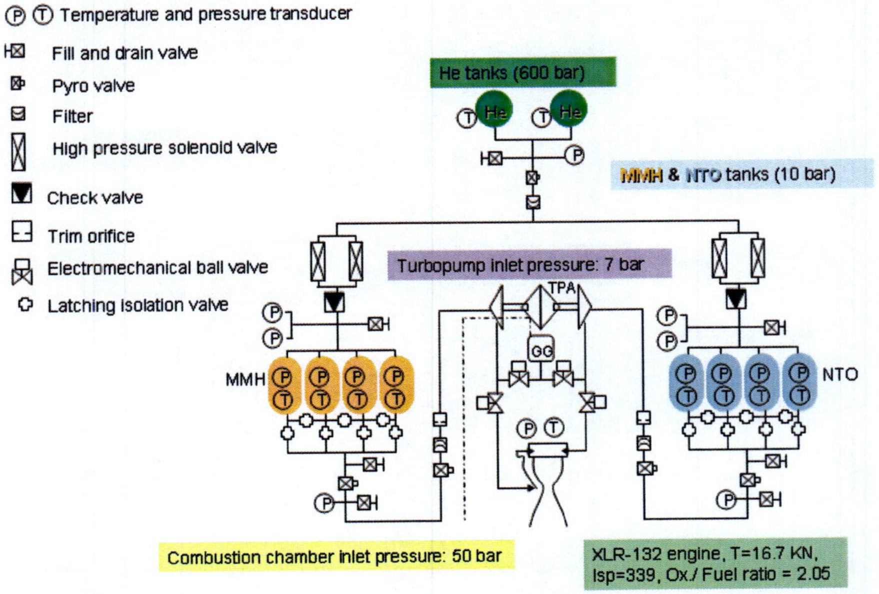

Fig. 7.11 CRV main propulsion system

The system presented has been designed based on a transport capability of 240 kg of payload material. It would be possible for it to be scaled accordingly to different quantities of material. The barrier is given by the performance of propulsive technology adopted. Over that, simply, multiple missions may however guarantee to transport all the mass required and overcome the technological threshold of the single craft.

7.5.6 Human Missions

Although humans are more versatile than robots are, it is difficult to justify the cost and the risk of performing manned missions to NEO if not necessary. Very realistically first precursor missions will be robotic, after which human crews will settle the first outposts and infrastructures. After these phases it is imaginable that only autonomous robotic elements can perform with the NEO mining and material transportation to Earth.

These intermediate human tended missions have to be carefully conceived taking into consideration the peculiar constraints of a crew. As already discussed, with current radiation protection technology, a typical human can only endure ~180 days of deep space before reaching what NASA defines a maximum acceptable risk from radiation: 3% fatal cancer risk at a 95% confidence level. As example of how it would look like, in Fig. 7.12 is reported a manned vehicle for a NEO mission from a Caltech Space Challenge study (Caltech Space Challenge 2011).

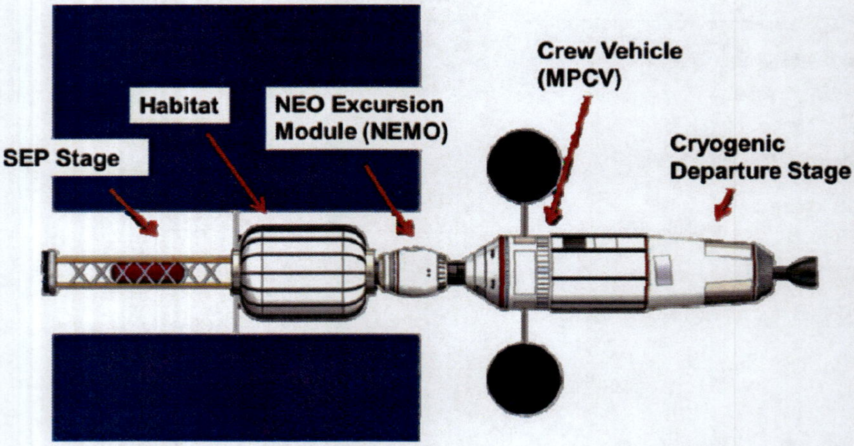

Fig. 7.12 Example of a manned vehicle, for a NEO mission. (Caltech Space Challenge 2011).

Careful selection of the crew size and composition, as well as of the habitable volume and of the activities and sources of entertainment for the astronauts.

It is generally agreed that the optimal number of crew members should be between four and six as a compromise between the resourced needed to sustain the crew life and wellness and the activity that can be performed. On the NEO surface a minimum team of two astronauts will perform each Extra Vehicular Activity (EVA), and will be monitored by the Intra-Vehicle Activity (IVA) crew. A concept proposed was to have a grid of cables anchored to the NEO to which the astronauts could attach themselves and translate to the various locations around the NEO (NASA Exploration Systems Mission Directorate).

A possible configuration of the manned craft may include:

- a crew module - including environmental controls and life support systems (ECLSS), power, logistics storage, medical facilities. Possibly an additional module with additional living space might be required.
- An Excursion Vehicles (defined as a two-people, free-flying vehicle for close operations with the NEO and EVAs.
- Propulsion stages for in-space transportation to and from the NEO.
- Earth return vehicles (atmosphere re-entry capability).

Acknowledgments. The authors kindly thank Francesco Marziani (ALTEC, Torino, Italy) for his contribution in Deep Space Mission Operations and Deep Space Radiation.

References

Barucci, M.A., Yoshikawa, M.: NEO Sample Return Mission Marco Polo; proposal to ESA COSMIC VISION prepared by a joint European Japanese team and supported by 440 confirmed scientists, Moscow (October 3, 2007)

Caltech Space Challenge, Finding NEO: A manned mission to a Near-Earth Object, Caltech, Pasadena, September 11-16 (2011)

COSPAR (2013), `http://cosparhq.cnes.fr/Scistr/Pppolicy.htm`

DLR, The Near-Earth Asteroids Data Base (2011), `http://earn.dlr.de/nea/aaamcspc.jpg`

Barbee, B.W. (ed.): Target NEO: Open Global Community NEO Workshop Report. George Washington University (July 2011)

Hutchison, R.: Meteorites. A Petrologic, Chemical and Isotopic Synthesis. Cambridge University Press (2004)

Sonter, M.: Near Earth Objects as Resources for Space Industrialization. Solar System Development Journal 1, 1–31 (2001)

NASA JPL (2013), `http://neo.jpl.nasa.gov/`

NASA, NASA Exploration Systems Mission Directorate · Explore NOW Report (September 20, 2010)

SEEDS, Executive Summary of the fourth edition of the International Master in Space Exploration and Development Systems (December 2009)

SEEDS, Executive Summary of the fifth edition of the International Master in Space Exploration and Development Systems (2010)

Ross, S.D.: Near-Earth Asteroid Mining Control and Dynamical Systems. Space Industry Report (December 2001)

Gertsch, R., Gertsch, L.: Economic Analysis Tools for Mineral Projects in Space. Space Resources Roundtable (1997)

Friedensen, V.: Exploration's NEA User Team (NUT) Precursor Requirements, presentation to LPSC, NEA User Team belongs to NASA Exploration Systems Mission Directorate, Advanced Capabilities Division (March 8, 2011)

Walker, R.J., McDonough, W.F., Honesto, J., Chabot, N.L., McCoy, T.J., Ash, R.D., Bellucci, J.J.: Origin and chemical evolution of group IVB iron meteorites. Geochim. Cosmochim. Acta 72, 2198–2216 (2008)

Chapter 8
Designing Robots for Gravity-Independent Locomotion

Marco Chacin

Singularity University, California, USA

8.1 Introduction

In recent years, the scientific community has had an increased interest in exploring the asteroids of the solar system (APL 1996, JAXA/ISAS 2003, NASA/JPL 2007). Technological advances have enabled mankind for the first time to take a closer look at these small solar system objects through sensors and instruments of robotic deep space probes. The in-situ study of asteroids can lead to important scientific findings in the effort to map the main asteroid belt. Mapping the belt by spectral classes and knowledge about which region on Earth the meteorites have landed can provide key clues about the origin and evolution of our solar system, even including the geological history of our planet Earth (Fujiwara et al. 2006). However, little attention has been given to locomotion on their surfaces with a mobile robotic system, due to the challenging gravity conditions found in these small solar system bodies.

In small bodies like asteroids, the gravitational fields are substantially weaker than those of Earth or Mars, therefore the likelihood of a robot's unintentional collision with the surface while attempting a movement is substantially higher. In spite of these difficulties in order to maximize the scientific return from any given mission on an asteroid's surface, future missions must have the ability to conduct stable mobility and accurate positioning on the rough terrain.

In this chapter, the focus is on gravity-independent locomotion approaches, technologies and challenges of robotic mobility on asteroids that have the possibility to provoke minimum reactions on the surface that could push the probe with sufficient force to reach escape velocity and drift into space. Recommendations and methods to perform compliant motions during operations by a surface robot under microgravity conditions are presented taking into account the contact dynamics, the natural features of the environment, the friction of the surface and its reaction force.

8.2 Mobility in Microgravity

Asteroids' physical characteristics provide a very hostile environment distinguished by the absence of (almost any) gravity. The effects of the microgravity environment can be approximated for convenience as those on order of 10^{-6} g (Scheeres 2004) (where g is the acceleration due to gravity on Earth). In such an environment, objects basically do not fall, but remain orbiting unless they reach the low escape velocity of the asteroid on order of 20 cm/s (Scheeres 2004), as in the case of the asteroid 25143. To attain stable mobility on these bodies, it is critical to consider the interaction forces between a robot and the asteroid's surface in such microgravity environments.

Relatively little attention from planetary scientists and planetary robotics engineers has been focused on surface mobility on asteroids. As a result, there exists some risk that premature conclusions about the feasibility of stable mobility on asteroid surfaces may be drawn without thorough consideration of all possible alternatives. However, it is clear that in order to increase any scientific return from a mission operating on an asteroid, movement on the surface would require a closer look at stability control against the forces interacting between bodies in a microgravity environment.

Recent developments in robotics which may pose the most feasible solutions to mobility in weak gravity domains of asteroids are discussed below.

8.2.1 Rolling and Hopping Locomotion

Although wheels are not an obvious solution for asteroid surface mobility, at least one study has suggested wheels to be viable in certain cases for rovers with mass less than one kilogram. Baumgartner et al. (1998) reported that the analysis of whether adequate tractive forces can be achieved for rolling mobility depends on the wheel-terrain interaction model employed. Traction sufficient for a small rover to traverse at a rate of 1 cm/sec was shown to be feasible via dynamic simulations revealing traction losses at the beginning of traverses.

Computer-based dynamic simulation studies of hopping and wheeled vehicles have concluded that both types have limited use for asteroid mobility (Behar 1997). Hopping robots were deemed to be of limited utility due to complexity of both thruster control for accurate maneuvers and robot pose estimation. Wheeled robot utility was viewed as limited due to difficulty maintaining wheels on the surface when undesired surface reactions led to long periods of ballistic floating before touching down. Another variation on rolling locomotion for asteroid mobility was proposed by Hokamoto and Ochi (2001) based on a vehicle with a dodecahedron shape and 12 individually actuated prismatic leg joints oriented radially around the body to provide intermittent walking and rolling.

When complexities associated with accurate maneuvers are avoided and replaced by simpler propulsion mechanisms, ballistic hopping is perhaps the simplest means of mobility for reaching discrete patches of asteroid terrain. The aforementioned MINERVA vehicle was the only asteroid hopping rover fully developed for a space flight mission thus far. It was a ~600 g vehicle designed for several asteroid days of autonomous operation involving ballistic hopping with variable hop speed and some control of hop direction depending on its attitude on the surface (Yoshimitsu et al. 2001). Other designs are in development or have been proposed as viable concepts. A 1.3 kg wheeled rover, at one time considered a payload for the Hayabusa mission on which MINERVA flew (Wilcox et al. 2000, Kawaguchi et al. 2003), was proposed with a novel mobility mechanism that also enables ballistic hopping. It further offered a capability to self-right in response to inevitable tumbling during landing from a hop or while traversing the surface at 1.5 mm/sec in low gravity (Tunstel 1999).

Another hopping vehicle intended for asteroids is the Asteroid Surface Probe (Cottingham et al. 2009), an 8 kg battery-powered (100 hours) spherical body of 30 cm diameter that uses thrusters to hop. When stationary, the sphere opens up using 3 petals to expose a science instrument payload and provide the means to self-right the probe (Ebbets et al. 2007). Similar, in principle, is a 12 kg thruster-propelled ballistic free-flyer concept designed by the German Aerospace Center (DLR) as part of a European Space Agency study (Ritcher 1998). Other hopping robots proposed for asteroid exploration include a pyramid-shaped, 533 g prototype with four single degree-of freedom (DOF) flippers at its base to enable hopping plus a lever arm for self-righting (Yoshida 1999), and a spherical 1 kg robot with internal iron ball actuated by electro-magnets to induce hopping (Nakamura et al. 2000).

A recent study concluded that wheeled and hopping locomotion modes in low gravity are comparable in locomotion speed (Kubota et al. 2009). The study considered ideal conditions (e.g., flat terrain and no loss of contact between wheels and terrain). A similar conclusion regarding energy consumption was reached in a comparative study of wheeled and hopping rovers for Mars gravity (Schell et al. 2001).

8.2.2 Locomotion by Crawling and Climbing

Nature offers many existence proofs, in the form of animals and insects, for solutions capable of traversing rough terrain against forces of gravity. Limbed locomotion has received limited consideration in the past for reasons of lower efficiency as compared to wheeled systems, but it is a common alternative that could enable crawling or climbing to achieve viable mobility solutions for planetary rovers.

Related arguments are less persuasive when dealing with the microgravity environment encountered on asteroids. On asteroids, a means to cling to the surface (Yoshida et al. 2002) would offer a critical capability for controlled motion and fine positioning. Limbs can also be beneficial as an active suspension that damps and prevents "bouncing" during traverse or upon landing after a hop. Crawling/climbing locomotion approaches without the use of limbs may also have merit for the asteroid domain. Such "limbless" approaches are briefly discussed below followed by promising technologies and concepts for clinging and gripping to surfaces.

8.2.3 Limbless Crawling

A significant volume of engineering literature exists on research advances for snake-like or serpentine robots. A recent survey is provided in (Transeth et al. 2009). Robots have been developed to execute a variety of locomotion gaits inspired by snakes that could be best suited for effective locomotion under low surface friction conditions (Dalilsafaei 2007) and especially effective for traversal of loose or slippery terrain (Hatto and Choset 2010). Other technologies such as dry adhesion are advancing for robotics applications in combination with mechanisms for climbing. Discussed next are adhesive means of achieving secure surface contact while crawling or climbing based on examples of technology proposed for space and planetary rovers.

8.2.4 Enabling Adhesion Technologies

Dry adhesive and electrostatic adhesion approaches that permit walking or climbing robotic systems to stick to natural surfaces hold promise for gravity-independent locomotion. An example is the Automated Walking Inspection and Maintenance Robot (AWIMR), a concept intended for operation on the exterior of crewed space vehicles or structures in space rather than on planet or asteroid surfaces (Wagner and Lane 2007). The AWIMR engineers established the feasibility of walking on such surfaces with the aid of prototype sticky feet, inspired by gecko feet, using dry adhesive polydimethylsiloxane for adhesion. The robot's sticky feet could walk on any clean, non-fragile surface (of the types found on space vehicle exteriors) and required a certain pull-off force. The AWIMR project also tested electrostatic means of sticking to surfaces, finding that greater shear forces were possible and that 2-3 kV was suitable for locomotion in this case (Wagner and Lane 2007).

Bombardelli et al. (2007) proposed artificial dry adhesives inspired by geckos and spiders for securing ballistically delivered microprobes to asteroid surfaces upon landing. The concept is motivated by the successful fabrication of several engineering prototypes of artificial reusable gecko adhesives. It is reported that the strongest such dry adhesive was recently fabricated using bundles of carbon nanotubes exhibiting four times the stickiness of natural gecko foot hairs (Bombardelli et al. 2007, Ge et al. 2007).

Among the desirable characteristics of synthetic, gecko-like dry adhesion for enabling asteroid traversal by rovers is its effectiveness on many surface types (as its functional basis is Van der Waals forces), its effectiveness in the vacuum of space, the fact that no additional energy is required to maintain an established grip on a surface, and their potential for mimicking the self-cleaning or dust resistant property of natural gecko footpads (Menon et al. 2007, Silva et al. 2008). The applicability of this technology for space and planetary robotic vehicles that would walk or climb on in-space structures and terrestrial surfaces is highlighted in (Menon et al. 2007). What could be considered early phases of suitable asteroid robot designs are briefly described in that work. We next discuss examples of technologies that offer mechanical means for gripping with robot limbs or momentarily anchoring robot limbs to natural surfaces.

8.2.5 *Limbs with Gripping End-Effectors*

Limbed approaches employing gripping end-effectors (Fig. 8.1) as feet/hands can enable walking/climbing locomotion while maintaining contact with asteroid surfaces. The ability to grapple onto surfaces is key to gravity-independent locomotion allowing mobility in any orientation including steeply sloped natural terrain and upside down. Such "grapple-motion" capability enables natural surface traversal by clawing into regolith or forming grasping configurations against rough, hard surfaces of high friction.

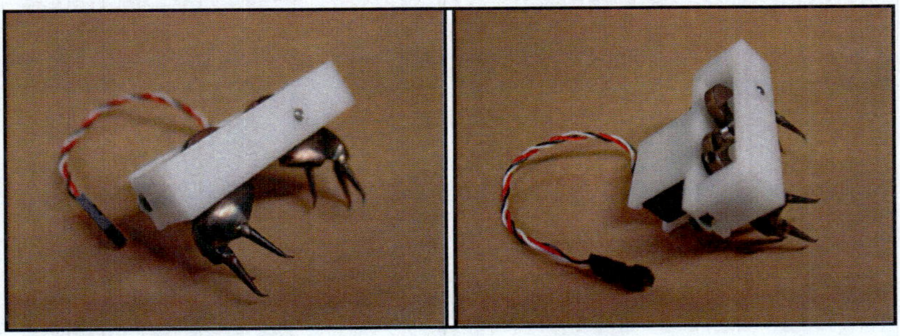

Fig. 8.1 Prototype gripper

During the past decade, prototypes of such limbed systems have been under development for planetary mobility and more recently focused on the problem of climbing steep terrain on Mars. A representative example of the state of the art for such applications is an 8 kg four-limbed planetary rover, LEMUR IIb, for which several types of climbing end-effectors have been investigated (Kennedy et al. 2005). The locomotion functionality for the LEMUR-class of robots evolved (kinematically) from 6-limbed robots for walking on space structures in orbit to 4-limbed free-climbing on steep terrain. Technologies addressed during the development of free-climbing capabilities for LEMUR IIb (e.g., gripping end-effectors, force control, and stability-based motion planning) should be useful for gravity-independent locomotion on asteroids as well.

Recent work at Tohoku University spearheaded limbed locomotion solutions and prototypes that are more specific to asteroid surface mobility and explored the feasibility of statically stable grapple-motion in microgravity (Chacin and Yoshida 2005). The work is motivated by a desire to achieve finer and more deterministic control of robot motion and position. Motion control complexities are handled using a behavior-based control approach in addition to bio-inspired central pattern generators for rhythmic motion and sensor-driven reflexes. Dynamic simulation results showed that static locomotion is feasible when grasping forces on the surface can be achieved (Chacin and Yoshida 2005). A 2.5 kg prototype of the Tohoku asteroid rover (Fig. 8.2) was built using a piercing spike at the tip of each limb to serve as momentary anchors in soft regolith or as contact points of a static grip on hard surfaces when used in combination (Chacin et al. 2006). Crawling gaits feasible for locomotion in microgravity environments using this system are analyzed in (Chacin and Yoshida 2006) for stability (in the sense that they hold the rover to the asteroid surface).

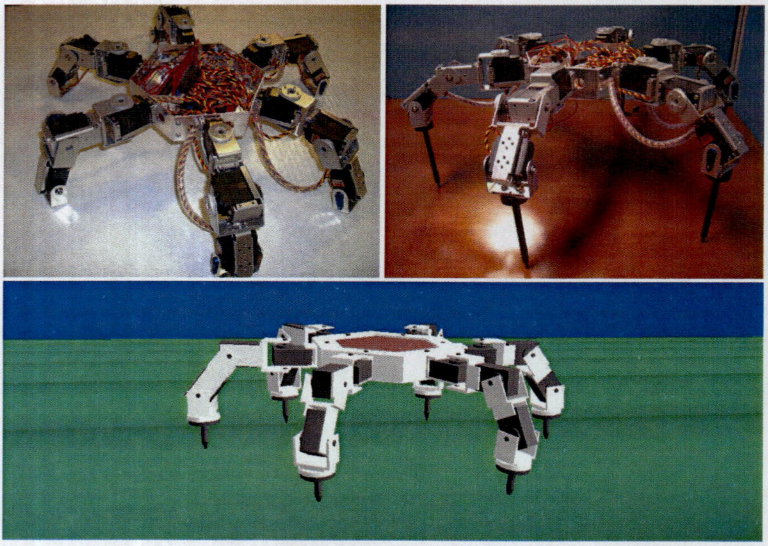

Fig. 8.2 Tohoku University's robot during development (top) and simulated model (bottom)

The next section focuses on this robotic system as an example of an asteroid mobility solution and control approach. It considers issues related to the microgravity environment and its effect on dynamics of robotic systems on asteroids.

8.3 General Assumptions

In a future asteroid exploration mission (Chacin and Yoshida 2005, Chacin et al. 2006, Yoshida et al. 2002), it is expected to have a smart design of a robotic system that would allow scientists to benefit from more accurate instrument positioning capability on the asteroid's surface in microgravity. However, the engineering complexity of this task makes the design of an effective robot with stable locomotion difficult. A feasible robotic design for such a mission would be a small limbed robot deployed over the asteroid to crawl on its rough surface (Chacin and Tunstel 2012).

During regular operations the robot is intended to use the natural features of the environment and the friction of the surface for omni-directional walking having multiple contacts between the limbs and the environment. In a microgravity environment, a planning algorithm can exploit this property by generating motions in contact, or more formally compliant motions (Borenstein 1995). During the execution of a compliant motion, the trajectory of the end tip is reflected according to the sensed forces derived from the contacts.

In this context, the problem of specifying a compliant motion command is similar to the problem of planning using pure position control to orient the end tip. The compliant motion control (Klein and Briggs 1980) allows specification of the forces and velocities to be maintained in the motion frame until the meeting of a set of termination conditions is detected.

Throughout the following analysis, to simplify the discussion, it is assumed that:

1. The object is a rigid body in contact with a rigid link of the robot;
2. Accurate models of the limbs and object are given;
3. Interference between limbs is ignored;
4. Each limb has only one frictional contact point at a fixed location;
5. The z direction of the contact point is always inward of the surface normal;
6. Contact points are known and the mass of each link of the robot is negligible;
7. Dynamic and static frictional coefficients are not distinguished between each other;
8. The motion is quasi-static to suppress any dynamic effect.

Assuming just one frictional point allows us to consider only forces at the contact points. In this way, while executing a compliant command, the robot controller can interpret the sensed forces to automatically generate the corrective actions needed to comply with the task while preserving contact during motion.

Static friction occurs when the relative tangential velocity at a contact point is zero; otherwise, the friction is called dynamic friction. Assumptions 7 and 8 allow considering a "first-order" (or quasi-static) world where forces and velocities are related and also have static friction but no dynamic friction.

8.3.1 Dynamic Model

The model of the robot (Chacin and Yoshida 2005) consists of a central body with a hexagonal shape and six identical limbs (Inoue et al. 2002, Inoue et al. 2001) symmetrically distributed around it. This type of surface mobile robots have the same structure in terms of dynamics equations as free flying or floating robots which do not have a fixed point but have interaction with the ground.

The dynamic motion of the free-flying multi-body system with the presence of the external forces $\boldsymbol{F}_{ex}$ is described as (Yoshida 1997):

$$H\begin{bmatrix}\ddot{x}_b\\ \ddot{\varphi}\end{bmatrix}+\begin{bmatrix}c_b\\ c_m\end{bmatrix}=\begin{bmatrix}F_b\\ \tau\end{bmatrix}+J^T F_{ex} \quad (8.1)$$

where H is the inertia matrix of the robot, x_b - position/orientation of the base; ϕ - articulated joint angles; c_b, c_m - velocity/gravity dependent non-linear terms; F_b - forces/moments directly applied on the base; τ - joint articulated torque; J^T -Jacobian matrix.

In this model, any configuration of the robot can be defined by a set of parameters, the coordinates and orientation of the body, and the joint angles of each limb. The limbs have three links and three actuated revolute joints. Two of these joints are located in the junction of the leg/limb with the central body. The third joint is located at the knee connecting the upper and lower link, which results in each limb having three DOF.

The kinematic relationship around the end points is expressed as follows:

$$\begin{aligned}\dot{x}_{ex} &= J_m\dot{\varphi} + J_b\dot{x}_b\\ \ddot{x}_{ex} &= J_m\ddot{\varphi} + \dot{J}_m\dot{\varphi} + J_b\ddot{x}_b + \dot{J}_b\dot{x}_b\end{aligned} \quad (8.2,3)$$

where J_b and J_m denote the Jacobian of the base (main) body and the Jacobian of a given manipulator (limb) respectively.

8.3.2 Contact Dynamics

In commonly used contact models (Brach 1991, Keller 1986), the relationship of momentum exchange and force-time product assumes infinitesimal impact. However, the infinitesimal impact between two single rigid bodies is a much idealized case. When modeling the ground (natural terrain), it is usually observed that as the stiffness coefficient is lowered, greater penetration in the ground occurs. The lower the damping constants, the longer the vibrations occur. However, a very high increment in stiffness or in damping constants in the limbs' model produces instabilities in simulations due to numerical issues that can be avoided by using a rigid limb, thus reducing the model order. The following discussion looks at how to determine the contact force F_{ex} (Der Stappen 1999).

Where there is friction at a contact point, the friction force at a point acts tangential to the contact surface. We will denote the magnitude of the friction force at the i-th contact as f_i, a magnitude of the normal force as f_n. To specify the tangential acceleration and friction force completely in a three-dimensional system, we would also need to specify the direction of the acceleration and friction force in the tangent plane (Gilardi and Shraf 2002, Yoshida 1999).

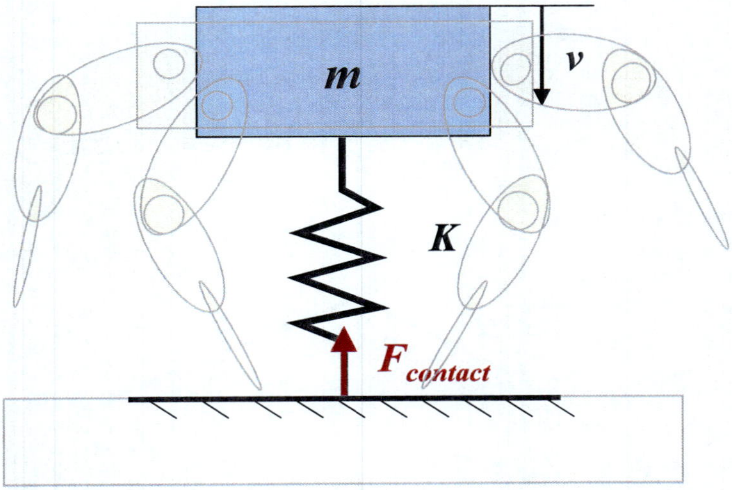

Fig. 8.3 Contact model

In order to assume a purely elastic contact at the end tips (Chacin and Yoshida 2008) (normal (z) direction), the impact phase can be divided into two stages: compression and restitution. During the compression stage, the elastic energy is absorbed by the deformation of the contact surfaces of the impacting bodies. In the restitution stage the elastic energy stored in the compression stage is released back to the bodies making the relative velocity greater than zero. Then, the robot can be modeled as a mass-spring system (Fig. 8.3) and as a result, the following relation between the mass, the velocity and the force should hold:

$$2mv = F\Delta\Delta \tag{8.4}$$

Now, considering the time of contact as a function of the mass of the system m and the stiffness coefficient of the limbs K, we have:

$$\Delta t = \pi\sqrt{\frac{m}{K}} \tag{8.5}$$

From Eqs. (8.4) and (8.5),

$$F_{contact} = \frac{2}{\pi}\left(\sqrt{mK}\right)v$$
$$F_{contact} = \sum_{i=1}^{N} f_i \qquad (8.6,7)$$

where N is the number of limbs in contact at landing.

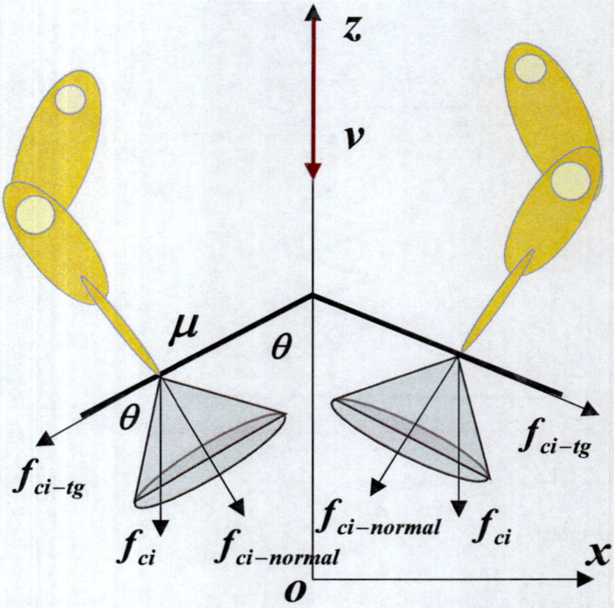

Fig. 8.4 Decomposition of the contact forces

Considering the Coulomb friction in the tangential directions, we have the following general expressions from Fig. 8.4:

$$f_{ci-tg} = f_{ci}\cos\theta$$
$$f_{ci-normal} = f_{ci}\sin\theta \tag{8.8,9}$$

where θ is the angle of the surface normal. Next, the coefficient of friction can be denoted as:

$$\mu = \frac{f_{ci}}{f_{ci-normal}} > \frac{f_{ci}\cos\theta}{f_{ci}\sin\theta}$$
$$\mu > f_{ci}\tan\theta \tag{8.10,11}$$

Considering $\boldsymbol{F}_{contact} = f_{ci}$ in Eq. (8.11) and using Eq. (8.4) we have,

$$\mu > \frac{2mv}{\Delta t}\tan\theta \tag{8.12}$$

And substituting Eq. (8.5),

$$\frac{\mu \tan\theta}{2m}\left(\pi\sqrt{\frac{m}{K}}\right) > v \tag{8.13}$$

Equation (8.13) shows that the considered contact stability will strictly depend on the approach velocity of the robot.

Quasi-static stability is a more general stability criterion than that used in the previous discussion. Under this condition, inertia forces are included but limb dynamics are not separately considered; the masses of the limbs are considered with that of the body. In the previous argument, quasi-static stability is partly assumed, provided all the normal components of the contact points' f forces are positive. Since the contact point cannot support a negative normal force (as shown in Fig. 8.4), the appearance of a negative force indicates that the given limb will lift and, since it cannot provide the required movement about the center of mass, the robot will jump in a microgravity environment like MINERVA (Yoshimitsu et al. 2001).

8.3.3 *Motion Control*

If a robot is to successfully achieve walking with static stability, the control system must ensure that the behavior of the robot does not deviate from the following stable condition:

$$\sum_{i=1}^{m} f_i + m_0 g = 0$$

$$\sum_{i=1}^{m} p_i \times f_i = 0 \tag{8.14,15}$$

To remain balanced, the robot must be able to apply forces with its end tips on the terrain that compensate for gravity without slipping. A necessary condition is that the robot's center of mass lies above the support polygon. But on an irregular surface, the support polygon does not always correspond to the base of the contact points. To compute the support polygon the contact interface (all contact points) is modeled as shown in (Chacin 2007), with

$$\|f_n\| \geq 0$$

$$\|f_t\| \leq \mu \|f_n\| \tag{8.16,17}$$

The body attitude also needs to be controlled in order for the robot to maintain static balance. Assuming that the robot is well balanced in the lateral plane, the body attitude at any given moment is an important determinant of the acceleration experienced by the center of mass (COM) in the sagittal plane. If the desired value of the acceleration output is known at all times throughout the gait cycle, then a corrective action can be generated which will maintain the robot in a stable state. If the body attitude needs to be increased or decreased depending upon the error signal, and assuming that we can determine the desired acceleration in the sagittal plane for all times throughout the gait cycle, we can implement a continuous control system.

8.3.4 Zero Moment Point and Momentum of the System

In the walking robotics community, *Zero Moment Point* (ZMP), which was first introduced by (Vukobratovic et al. 1970), is a key concept to discuss the tip-over stability and gait control of a robot.

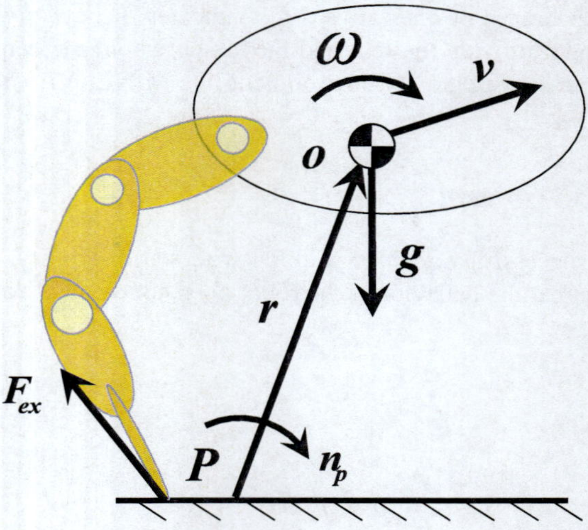

Fig. 8.5 A schematic force/moment model for an articulated surface robot

Figure 8.5 is a schematic drawing to explain ZMP. Point O is the center of gravity (COG) of the entire robot. Let vector $\boldsymbol{r}$ be a position vector from an arbitrary point on the ground P to the point O, and vector $\boldsymbol{l}_i$ be a position vector from the point P to each ground contact point of the limbs. For this model, the following dynamic equilibria hold true

$$\dot{P}_p = \sum f_{exi} + Mg$$
$$\dot{L}_p = n_p + r \times \dot{P}_p + \sum (l_i \times f_{exi}) \tag{8.18,19}$$

where P_p and L_p are linear and angular momentum around point P, and M is the total mass of the robot. The ZMP is a position P at which the component of moment n_p around the horizontal axes, n_{px} and n_{py}, becomes zero. The robot is stable, otherwise if the ZMP stays inside a polygon formed by ground contact points of the limbs. Otherwise the robot starts tipping over.

Gait generation and motion control algorithms for walking robots have been developed based on this concept. In addition, an advanced planning and control algorithm with a special attention to the kinetic momentum has been proposed recently (Kajita et al. 2003).

For a surface limbed robot, the gravity forces exerted on it can be neglected; the non-linear term in Eq. (8.1) then becomes

Integrating its upper set with respect to time, we obtain the total momentum of the system as:

$$L = \int J_b^T F_{ex} dt = H_b \dot{x}_b + H_{bm} \dot{\phi} \tag{8.20}$$

Going back to Eq. (8.2), and eliminating $\dot{\phi}$, we can obtain the following equation:

$$L = \left(H_b - H_m J_s^{-1}\right)\dot{x}_b + H_m J_s^{-1} \dot{x}_{ex} \tag{8.21}$$

In this way, if the system does depart from static stability, then the control system can identify this condition and bring the robot back to the statically stable condition.

8.4 A Generalized Control Algorithm

Since walking is a continuous and cyclic process, we can consider two main types of control systems, a closed-loop control system. A general closed-loop control system for controlling the continuous process of the walking gait can be

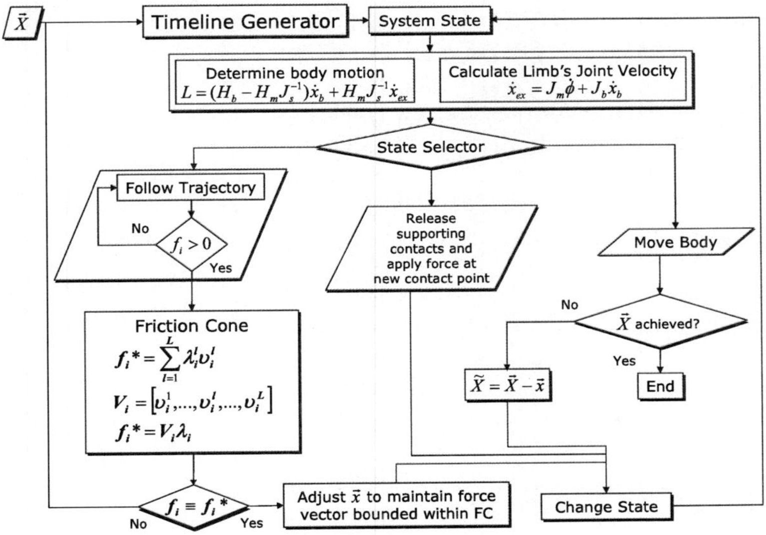

Fig. 8.6 Gait planner detail

considered. However, since the positioning of the end tip itself can be modeled as a discrete process, we use an event driven control system to identify the existence of such states and modify the closed-loop control depending upon the current state of the system.

Given a motion command (a vector X), the motion planning and control algorithm to move in the commanded direction and magnitude is carried out in the following way (Fig. 8.6):

1. Use the gait planner presented in (Chacin and Yoshida 2006) to plan the complete set of limb motions to move in the desired direction.
2. At time t, compute link positions and velocities, recursively from link 0 to n.
3. Set accelerations $\ddot{x}_b$ and $\ddot{\phi}$, and external forces F_b and F_{ex} to zero, then compute the inertial forces recursively from link n to 0. The resultant forces on the coordinates x_b and ϕ are equal to the non-linear forces c_b and c_m, respectively.
4. Plan the end point trajectory of each limb x_{exi} using the kinematics of the ZMP so that it satisfies the stability condition. Then obtain the end point velocity $\dot{x}_{ex}$ along the trajectory.
5. Considering the friction cone estimation (Chacin 2007) and the contact stability conditions shown by Eq. (8.13), determine the robot base motion $\dot{x}_b$ by

$$\begin{bmatrix} \ddot{x}_b \\ \ddot{\phi} \end{bmatrix} = H^{-1}\left\{\begin{bmatrix} F_b \\ \tau \end{bmatrix} + J^T F_{ex} - \begin{bmatrix} c_b \\ c_m \end{bmatrix}\right\} \tag{8.22}$$

6. Calculate the joint velocity of the limbs $\dot{\phi}$ by Eq. (8.2), using $\dot{x}_b$ and $\dot{x}_{ex}$ while considering Eq. (8.17); change the state of the control system. If necessary, adjust x to maintain the force vectors bounded within the friction cones.
7. Adopt the new contact configuration to release the support contacts and apply a permissible contact force at the new contact points. Dynamic exploration can be applied to reduce the surface position/orientation uncertainties. Change the state of the control system.
8. Control the joints along with the solution from step 6 to move the body. Verify if the goal position X has been reached; if it has not, then repeat.

One difference of this algorithm with respect to conventional ones is the consideration of momentum of the robot in step 3. Without this step, the obtained joint motion may have errors from the originally planned end point trajectory, thus may not satisfy the stability condition. Conventionally, a feedback control may be employed to correct these errors. But using Eq. (8.21) in step 3, the error can be compensated in advance.

8.5 Challenges

A rich set of challenges are encountered during development and evaluation of prospective solutions for gravity-independent locomotion on asteroids. The experiments reported here are indicative of a few, but several additional key challenges deserve early attention by researchers. One of them is the mechanics of controlled ballistic hopping on rotating asteroids and in non-uniform gravity fields due to their irregularly shaped bodies. Bellerose et al. (2008) and Bellerose and Scheeres (2008) modeled the dynamics of hopping vehicles to enable hops covering designated distances by computing and controlling initial hop velocity. The model accounts for distance covered by residual bounces as the vehicle comes to rest (considering surface friction coefficient and restitution). A particularly challenging aspect to consider is that some asteroid shapes may have surface locations where a vehicle could stay in equilibrium, thus affecting vehicle dynamics on the surface. Conceivably, a hopping rover could be perturbed away from predicted ballistic trajectories by such equilibria. This can affect exploration objectives by constraining the total area that a rover can safely or reliably traverse to on an asteroid surface when stable and unstable equilibrium locations happen to coincide with surface regions of scientific interest. Purely hopping vehicles that operate primarily at the mercy of small body physics can have limited accessibility of such surface regions. Bellerose's model also provides insight into the effects of non-uniform gravity fields and how centripetal and Coriolis forces due to asteroid rotation may assist or hinder hop performance (Bellerose and Scheeres 2008).

Another key challenge is achieving the ability to land after hopping in such a way as to avoid rebound. Control and robotics techniques can be used to address this challenge. One robot concept employs a spring and linear actuators with horizontal velocity control to achieve this (Shimoda et al. 2003), while other research is experimenting with active grappling of the surface upon landing (Chacin 2007, 2008, 2009). The related challenge, central to gravity-independent locomotion, is maintaining grip or temporary anchoring while controlling force for closure and compliance. The work presented herein and in (Chacin and Yoshida 2009) examines the motion/force control and dynamic modeling germane to the problem of stable crawling and force closure needed to maintain contact/grip with an asteroid surface under microgravity conditions. Experiments reveal the utility of force feedback for maintaining contact during execution of compliant motion. Kennedy et al. (2005) address active force control to achieve anchoring associated with stable free-climbing motion control. Tactile sensing and related motion planning algorithms (Bretl et al. 2003) have been implemented on the LEMUR IIb robot.

The low gravity environment and its effect on surface vehicles present a key challenge for rover localization whether hopping, crawling, or climbing. Determining, updating and maintaining knowledge of rover position and orientation on an asteroid surface can be important for recording spatial context for surface science measurements and for certain mission concepts of operation. Localization approaches for hopping robots have been proposed with some reliance on range measurements to an orbiting or station-keeping mother spacecraft (Yoshimitsu et al. 2001) and via use of more general approaches such as particle filters (Martinez-Cantin 2004), landmark geo-referencing (Fiorini 2005), and visual odometry while tumbling (So et al. 2008, 2009).

Finally, a key challenge is the testing and verification of gravity-independent locomotion systems to ensure confidence in their technology readiness for asteroid missions. This is always a challenge for space systems and particularly those intended for operation in microgravity domains. The testbed described in (Chacin and Yoshida 2008) and its means of emulating reduced gravity are representative solutions for addressing the challenge using relatively affordable technology. Other testbed approaches to emulating reduced gravity in the laboratory include the use of overhead gantry systems with frictionless air-bearing pulleys from which to suspend prototype rovers, and the use of prototype rovers on flat air-tables or mounted on a mobile base with integrated air bearings. Beyond the fundamental feasibility of controlled surface mobility in low gravity fields of asteroids, additional challenges of high relevance and importance remain to be addressed by advanced research and technology development.

8.6 Summary

In this chapter, various approaches to gravity-independent locomotion on weak gravity surfaces of asteroids are discussed along with related technologies. Challenges are also described that affect planning and control of surface exploration robots that use hopping and rolling mechanisms and/or articulated limbs for the ground contact.

Given the focus on gravity-independent locomotion approaches, technologies, and challenges of robotic mobility on asteroids, an in-depth representative example of an asteroid mobility solution and control approach is provided. The control approach considers reaction and friction forces with the asteroid surface and is demonstrated using a prototype robot and laboratory testbed that emulates microgravity. This example considered issues that most solutions must address related to the microgravity environment and its effect on dynamics of robotic systems on asteroids. It works by reacting to the current locations of contact points and estimating the force-closure condition for stable motion. Such a mechanism is central in the control structure.

The control methods proposed are useful to improve the operational performance and efficiency for robots capable of position-based controlled motion on an asteroid. They demonstrated that proper knowledge of the force cone interaction with the surface plays a significant role in the development of proper control procedures that can allow next-generation surface robots to gain proper mobility in a microgravity environment.

References

Brach, R.M.: Mechanical Impact Dynamics: Rigid Body Collisions. John Wiley & Sons, New York (1991)

Baumgartner, E.T., Wilcox, B.H., Welch, R.V., Jones, R.M.: Mobility performance of a small-body rover. In: Proceedings of the 7th International Symposium on Robotics and Applications. World Automation Congress, Anchorage, Alaska (1998)

Behar, A., Bekey, G., Friedman, G., Desai, R.: Sub-kilogram intelligent tele-robots (SKIT) for asteroid exploration and exploitation. In: Proceedings of the SSI/Princeton Conference on Space Manufacturing, Space Studies Institute, Princeton, NJ, pp. 65–83 (May 1997)

Bellerose, J., Girard, A., Scheeres, D.J.: Dynamics and control of surface exploration robots on asteroids. In: Proceedings of the 8th International Conference on Cooperative Control and Optimization, Gainesville, FL, pp. 135–150 (January 2008)

Bellerose, J., Scheeres, D.J.: Dynamics and control for surface exploration of small bodies. In: Proceedings of the AIAA/AAS Astrodynamics Specialist Conference and Exhibit, Honolulu, Hawaii, Paper 6251 (2008)

Bombardelli, C., Broschart, M., Menon, C.: Bio-inspired landing and attachment system for miniaturised surface modules. In: Proceedings of the 58th International Astronautical Congress, Hyderabad, India (2007)

Borenstein, J.: Control and Kinematic Design of Multi-degree-of-freedom Mobile Robots with Compliant Linkage. IEEE Transactions on Robotics and Automation 11(1), 21–35 (1995)

Bretl, T., Rock, S., Latombe, J.C.: Motion planning for a three-limbed climbing robot in vertical natural terrain. In: IEEE International Conference on Robotics and Automation, Taipei, Taiwan, pp. 2947–2953 (2003)

Chacin, M., Tunstel, E.: Gravity-Independent Locomotio: Dynamics and Position-based Control of Robots on Asteroid Surfaces. In: Robotics Systems – Applications, Control and Programming. InTech (2012)

Chacin, M., Yoshida, K.: Multi-limbed rover for asteroid surface exploration using static locomotion. In: Proceedings of the 8th International Symposium on Artificial Intelligence, Robotics and Automation in Space, Munich, Germany (2005)

Chacin, M., Yoshida, K.: Stability and Adaptability Analysis for Legged Robots Intended for Asteroid Exploration. In: Proceedings of the 2006 IEEE International Conference on Intelligent Robots and Systems (IROS 2006), Beijing, China (2006)

Chacin, M., Yoshida, K.: Evolving Legged Rovers for Minor Body Exploration Missions. In: Proceedings of the 1st IEEE/RAS-EMBS International Conference on Biomedical Robotics and Biomechatronics, BioRob 2006, Pisa, Italy (2006)

Chacin, M., Nagatani, K., Yoshida, K.: Next-Generation Rover Development for Asteroid Surface Exploration: System Description. In: Proceedings of the 25th International Symposium on Space Technology and Science and 19th International Symposium on Space Flight Dynamics, Kanazawa, Japan (2006)

Chacin, M.: Landing Stability and Motion Control of Multi-Limbed Robots for Asteroid Exploration Missions. Ph.D. dissertation, Dept. of Aerospace Engineering, Tohoku University, Tohoku, Japan (2007)

Chacin, M., Yoshida, K.: A Microgravity Emulation Testbed for Asteroid Exploration Robots. In: Proceedings of International Symposium on Artificial Intelligence, Robotics, Automation in Space (i-SAIRAS 2008), Los Angeles, CA (February 2008)

Chacin, M., Yoshida, K.: Motion control of multi-limbed robots for asteroid exploration missions. In: Proceedings of IEEE International Conference on Robotics and Automation, Kobe, Japan (May 2009)

Cottingham, C.M., Deininger, W.D., Dissly, R.W., Epstein, K.W., Waller, D.M., Scheeres, D.J.: Asteroid Surface Probes: A low-cost approach for the in situ exploration of small solar system objects. In: Proceedings of the IEEE Aerospace Conference, Big Sky, MT (2009)

Dalilsafaei, S.: Dynamic analyze of snake robot. Proceedings of World Academy of Science, Engineering and Technology (29), 305–310 (2007)

Der Stappen, A.V., Wentink, C., Overmars, M.: Computing form-closure configurations. In: Proceedings of the IEEE International Conference on Robotics and Automation, ICRA, USA, pp. 1837–1842 (1999)

Ebbets, D., Reinert, R., Dissly, R.: Small landing probes for in-situ characterization of asteroids and comets. In: Proceedings of the 38th Lunar and Planetary Science Conference, League City, TX (2007)

Fiorini, P., Cosma, C., Confente, M.: Localization and sensing for hopping robots. Autonomous Robots 18, 185–200 (2005)

Fujiwara, A., Kawaguchi, J., Yeomans, D.K., et al.: The Rubble Pile Asteroid Itokawa as observed by Hayabusa. Report: Hayabusa at asteroid Itokawa. Science 312, 1330–1334 (2006)

Ge, L., Sethi, S., Ci, L., Ajayan, P.M., Dhinojwala, A.: Carbon nanotube-based synthetic gecko tapes. Proceedings of National Academy of Sciences 104, 10792–10795 (2007)

Gilardi, G., Shraf, I.: Literature Survey of Contact Dynamics Modeling. Mechanism and Machine Theory 37, 1213–1239 (2002)

Hatton, R.L., Choset, H.: Generating gaits for snake robots: annealed chain fitting and keyframe wave extraction. Autonomous Robots 28, 271–281 (2010)

Hirabayashi, H., Sugimoto, K., Enomoto, A., Ishimaru, I.: Robot Manipulation Using Virtual Compliance Control. Journal of Robotics and Mechatronics 12, 567–575 (2000)

Hokamoto, S., Ochi, M.: Dynamic behavior of a multi-legged planetary rover of isotropic shape. In: Proceedings of the 6th International Symposium on Artificial Intelligence, Robotics, and Automation in Space, St-Hubert, Quebec, Canada (June 2001)

Inoue, K., Arai, T., Mae, Y., Takahashi, Y., Yoshida, H., Koyachi, N.: Mobile Manipulation of Limbed Robots-Proposal on Mechanism and Control. Preprints of IFAC Workshop on Mobile Robot Technology, pp. 104–109 (2001)

Inoue, K., Mae, Y., Arai, T., Koyachi, N.: Sensor-based Walking of Limb Mechanism on Rough Terrain. In: Proceedings of the 5th International Conference on Climbing and Walking Robots, CLAWAR (2002)

JAXA/Institute of Space and Astronautical Science, Asteroid Explorer HAYABUSA (2003), `http://hayabusa.jaxa.jp/`

Johns Hopkins University Applied Physics Laboratory, Near Earth Asteroid Rendezvous - Shoemaker Mission (1996), `http://near.jhuapl.edu`

Jones, R., et al.: NASA/ISAS collaboration on the ISAS MUSES C asteroid sample return mission. In: Proceedings of 3rd IAA International Conference on Low-Cost Planetary Missions, Pasadena, CA (1998)

Kajita, S., Kanehiro, F., Kaneko, K., Fujiwara, K., Harada, K., Yokoi, K., Hirukawa, H.: Resolved Momentum Control: Humanoid Motion Planning based on the Linear and Angular Momentum. In: Proceedings of the IEEE/RSJ International Conference on Intelligent Robots and Systems (IROS 2003), pp. 1644–1650 (2003)

Kawaguchi, J., Uesugi, K., Fujiwara, A.: The MUSES-C mission for the sample and returnits technology development status and readiness. Acta Astronautica 52, 117–123 (2003)

Keller, J.B.: Impact With Friction. ASME Journal of Applied Mechanics 53, 1–4 (1986)

Kennedy, B., Okon, A., Aghazarian, H., Badescu, M., Bao, X., Bar-Cohen, Y., et al.: LEMUR IIb: A robotic system for steep terrain access. In: Proceedings of the 8th International Conference on Climbing and Walking Robots, London, UK, pp. 595–695 (2005)

Klein, D.P.C., Briggs, R.: Use of compliance in the control of legged vehicles. IEEE Transactions on Systems, Man and Cybernetics 10, 393–400 (1980)

Kubota, T., Takahashi, K., Shimoda, S., Yoshimitsu, T., Nakatani, I.: Locomotion mechanism of intelligent unmanned explorer for deep space exploration. Intelligent Unmanned Systems: Theory and Applications 192, 11–26 (2009)

Martinez-Cantin, R.: Bio-inspired multi-robot behavior for exploration in low gravity environments. In: Proceedings of the 55th International Astronautical Congress, Vancouver, Canada (2004)

Menon, C., Murphy, M., Sitti, M., Lan, N.: Space exploration Towards bio-inspired climbing robots. In: Habib, M.K. (ed.) Bioinspiration and Robotics: Walking and Climbing Robots, pp. 261–278. I-Tech Education and Publishing, Vienna (2007)

Nakamura, Y., Shimoda, S., Shoji, S.: Mobility of a microgravity rover using internal electro-magnetic levitation. In: Proceedings of the IEEE/RSJ International Conference on Intelligent Robots and Systems, Takamatsu, Japan, pp. 1639–1645 (2000)

NASA/Jet Propulsion Laboratory, Dawn Mission (September 2007), `http://dawn.jpl.nasa.gov/`

Richter, L.: Principles for robotic mobility on minor solar system bodies. Robotics and Autonomous Systems 23, 117–124 (1998)

Schell, S., Tretten, A., Burdick, J., Fuller, S.B., Fiorini, P.: Hopper on wheels: Evolving the hopping robot concept. In: Proceedings of the International Conference on Field and Service Robotics, Helsinki, Finland, pp. 379–384 (2001)

Scheeres, D.: Dynamical Environment About Asteroid 25143 Itokawa. University of Michigan, Department of Aerospace Engineering, USA (2004)

Scheeres, D., Broschart, S., Ostro, S.J., Benner, L.A.: The Dynamical Environment About Asteroid 25143 Itokawa. In: Proceedings of the 24th International Symposium on Space Technology and Science, Miyazaki, Japan, pp. 456–461 (2004)

Shimoda, S., Wingart, A., Takahashi, K., Kubota, T., Nakatani, I.: Microgravity hopping robot with controlled hopping and landing capability. In: Proceedings of the IEEE/RSJ Intl. Conference on Intelligent Robots and Systems, Las Vegas, NV, pp. 2571–2576 (October 2003)

Silva, M.F., Machado T, J.A.: New technologies for climbing robots adhesion to surfaces. In: Proceedings of the International Workshop on New Trends in Science and Technology, Ankara, Turkey (November 2008)

So, E.W.Y., Yoshimitsu, T., Kubota, T.: Relative localization of a hopping rover on an asteroid surface using optical flow. In: Proceedings of the Intl. Conf. on Instrumentation, Control, and Information Technology, SICE Annual Conference, Tokyo, Japan, pp. 1727–1732 (August 2008)

So, E.W.Y., Yoshimitsu, T., Kubota, T.: Hopping odometry: Motion estimation with selective vision. In: Proceedings of the IEEE/RSJ International Conference on Intelligent Robots and Systems, St. Louis, MO, pp. 3808–3813 (October 2009)

Tunstel, E.: Evolution of autonomous self-righting behaviors for articulated nanorovers. In: Proceedings of the 5th International Symposium on Artificial Intelligence, Robotics & Automation in Space, Noordwijk, The Netherlands, pp. 341–346 (June 1999)

Transeth, A.A., Pettersen, K.Y., Liljebck, P.: A survey on snake robot modeling and locomotion. Robotica 27, 999–1015 (2009)

Vukobratovic, M., Frank, A., Juricic, D.: On the Stability of Biped Locomotion. IEEE Transactions on Biomedical Engineering 17, 25–36 (1970)

Wagner, R., Lane, H.: Lessons learned on the AWIMR project. In: Proceedings of the IEEE International Conference Robotics and Automation, Space Robotics Workshop, Rome, Italy (2007)

Wilcox, B.H., Jones, R.M.: The MUSES-CN nanorover mission and related technology. In: IEEE Aerospace Conference, Big Sky, MT, USA, pp. 287–295 (2000)

Wilcox, B.H., Jones, R.M.: A ~1 kilogram asteroid exploration rover. In: Proceedings of the IEEE International Conference on Robotics and Automation (ICRA), San Francisco, CA (2001)

Wilcox, B.H., Litwin, T., Biesiadecki, J., Matthews, J., Heverly, M., Morrison, J., Townsend, J., Ahmad, N., Sirota, A., Cooper, B.: ATHLETE: A cargo handling and manipulation robot for the moon. Journal of Field Robotics 24, 421–434 (2007)

Yano, H., Kubota, T., Miyamoto, H., Yoshida, K., et al.: Touchdown of the Hayabusa Spacecraft at the Muses Sea on Itokawa. Report: Hayabusa at asteroid Itokawa. Science 312, 1350–1353 (2006)

Yoshida, K.: A General Formulation for Under-Actuated Manipulators. In: Proceedings of the IEEE/RSJ Int. Conf. on Intelligent Robots and Systems, Grenoble, France, pp. 1651–1957 (1997)

Yoshida, K.: Touch-Down Dynamics Simulation of MUSES-C with a Contact Friction Model. In: Proceedings of 9th Workshop on Astrodynamics and Flight Mechanics, JAXA, Kanagawa, Japan (1999)

Yoshida, K.: The jumping tortoise: A robot design for locomotion on micro gravity surface. In: Proceedings of the 5th International Symposium on Artificial Intelligence, Robotics & Automation in Space, Noordwijk, The Netherlands, pp. 705–707 (June 1999)

Yoshida, K., Kubota, T., Sawai, S., Fujiwara, A., Uo, M.: MUSES-C Touch-down Simulation on the Ground. In: AAS/AIAA Space Flight Mechanics Meeting, AAS/AIAA, Santa Barbara, California, pp. 481–490 (February 2001)

Yoshida, K., Maruki, T., Yano, H.: A Novel Strategy for Asteroid Exploration with a Surface Robot. In: Proceedings of the 3rd International Conference on Field and Service Robotics, Finland, pp. 281–286 (2002)

Yoshimitsu, T., Kubota, T., Akabane, S., et al.: Autonomous navigation and observation on asteroid surface by hopping rover MINERVA. In: Proceedings of the 6th International Symposium on Artificial Intelligence, Robotics & Automation in Space, Quebec, Canada (2001)

Chapter 9
Bio-inspired Landing Approaches and Their Potential Use on Extraterrestrial Bodies

Thibaut Raharijaona[1], Guillaume Sabiron[1,2], Stephane Viollet[1], Nicolas Franceschini[1], and Franck Ruffier[1]

[1] Aix-Marseille University, Marseille, France
[2] ONERA, French Aerospace Laboratory, Toulouse, France

9.1 Introduction

Landing on asteroids and extraterrestrial bodies is a critical stage for future exploration missions. Safe and soft landing on asteroids will be required even though the task is way harder than on the Earth due to the small size, irregular shape and variable surface properties of asteroids, as well as the low gravity and negligible drag experienced by the spacecraft. Optical guidance and navigation for autonomous landing on small celestial bodies have been studied in the past years with a focus on the closed-loop guidance, navigation, and control (GNC) systems (De Lafontaine1992, Kawaguchi et al. 1999).

The complexity of the task is dramatically increased by the lack of atmosphere since absence of drag precludes the deployment of parachutes, in contrast with the situation on Mars (Braun and Manning 2006) and other planets.

Added to this are the lack of reliable terrain and obstacle data bases and the lack of conventional sensing systems such as Global Positioning Systems (GPS) as well. On top of that, prohibitive latency in communication between the Earth and the extraterrestrial body obviously forces the landing to be made autonomously while requiring robust, reliable and powerless sensors. Precise GNC systems are indeed needed for safe and accurate landing in unprepared landing zones. Tough requirements in terms of embedded mass led us to design a very lightweight biologically inspired non-emissive optical sensor that is able to measure the optic flow (OF), that is, the angular velocity (in °/s) of the visual image sweeping backwards across the visual field.

Visual cues seem to be a promising way towards autonomous landing. Recently, several studies have shown various optical techniques such as LIDAR (Light Detection And Ranging) techniques (Parkes and Silva 2002, Parkes 2003) or vision based navigation systems to estimate spacecraft position and velocity parameters (Roumeliotis et al. 2002, Frapard et al. 2002, Cheng and Ansar 2005, Janschek et al. 2006, Trawny et al. 2007, Flandin et al. 2009, Mourikis et al. 2009, Shang and Palmer 2009), to avoid obstacle (Strandmoe et al. 1999) or to control unmanned spacecraft (Valette et al. 2010, Izzo et al. 2011, Izzo and de Croon

2011). In (Valette et al. 2010), the OF regulation principle (Ruffier and Franceschini 2005) was applied to autonomous lunar landing problems using a feedback loop aimed at maintaining the ventral optic flow constant. The approach was tested in simulation experiments using the PANGU software (Planet and Asteroid Natural scene Generation Utility) developed for ESA (European Space Agency) by the University of Dundee (see Parkes et al. 2004; Dubois-Matra et al. 2009; for more information), which is a tool used to simulate visual environment of planetary surfaces. In Izzo et al. (2011), based on numerical simulations, optimal trajectories were calculated in terms of the duration of the landing phase or the fuel consumption while keeping the OF constant. In Mahony et al. (2008), a fully OF-based visual servo control system was developed, in which a large visual field was combined with a centroid in order to estimate the direction of the speed vector in the case of simulated small aerial robotic vehicles. In the extended Kalman filter (EKF) method described in Mourikis et al. (2009), both of the above visual approaches were combined with an inertial measurement unit, and accurate estimates of the lander's terrain-relative position, attitude, and velocity were obtained. In the approach presented here, we focus on OF based means to derive useful information such as the orientation of the spacecraft velocity vector.

Finding means of sensing the OF onboard unmanned aerial and terrestrial vehicles has been a key research topic during the last few decades. Several flight control systems based on OF cues have been constructed so far for performing hazardous tasks such as hovering and landing on a moving platform (Herisse et al. 2012), avoiding obstacles (Barrows and Neely 2000, Griffiths et al. 2006, Beyeler et al. 2009), following terrain (Netter and Franceschini 2002, Ruffier and Franceschini 2003, 2004, 2005; Garratt and Chahl 2008) and tracking a moving target (Viollet and Franceschini 1999, Kendoul et al. 2009, Kerhuel et al. 2012). Some of the studies quoted above were inspired by insects, the impressive flight behavior of which relies on the built-in abilities they have developed and improved over several hundred millions of years, despite their small size and limited neural resources.

Flies, in particular, are agile seeing creatures that navigate swiftly through "unprepared" environments, avoiding obstacles with little conventional avionics. Equipped with "only" about one million neurons and "only" 3000 pixels in each eye, the common housefly, for example, achieves 3D navigation, terrain and obstacle avoidance at an impressive 700 body-lengths per second. All this is achieved, surprisingly, without any connections of the animal to a super-computer and an external power supply. The impressive lightness of the processing system at work onboard a fly or a bee makes any roboticist turn pale once he/she realizes that these creatures actually display many of the behaviors that have been sought for in the field of autonomous robotics for the last 50 years: dynamic stabilization, 3D collision avoidance, tracking, docking, autonomous landing in uncharted landing zones, etc.

Fig. 9.1 Left: Head of the blowfly Calliphora erythrocephala (male) showing the two panoramic compound eyes with their facetted cornea. Right: Each ommatidium contains seven micrometer-sized photoreceptors in the focal plane of each facet-lens, as observed *in vivo* here (natural autofluorescence colors). Six outer receptors (R1-6) mediate motion vision and drive optic flow sensor neurons, while the two central cells (R7, prolonged by an R8 receptor not seen here) are in charge of color vision. Figure from Franceschini et al. (2010).

The front end of the fly visual system consists of a mosaic of facet lenslets (see left hand side in Fig. 9.1) each of which focuses light on a small group of photorececeptor cells (see right hand part in Fig. 9.1). Flies possess one of the most complex and best organized retinae in the animal kingdom. Each of their photoreceptor cells is more sensitive and reliable than any photomultiplier ever built, and the two central cells (R7 and R8) display up to four different spectral sensitivities, from near UV to red (Franceschini 1985).

In the mid 1980's, we started designing a fly-inspired robot to demonstrate how an agent could possibly navigate in a complex environment on the basis of OF (Blanes 1986, Pichon et al. 1989). The Robot-Fly (Fig. 9.2A) was equipped with a widefield curved compound eye embedding an array of 114 fly-inspired Local Motion Sensors (LMS) (Franceschini et al. 1991, 1992, 1997).

The robot was equipped with a planar compound eye and a fly-inspired Elementary Motion Detector neuron's array (Pichon et al. 1989).

The sensor array was used to sense the OF generated during the robot's own locomotion among stationary objects. The 50-cm high "robot-mouche" (Robot-Fly, in English) that we constructed in 1991 (see Fig. 9.2A) was the first OF-based, completely autonomous robot able to avoid contrasting obstacles encountered on its way, while traveling to its target at a relatively high speed (50 cm/s) (Blanes

1991, Franceschini et al. 1991, 1992, 1997). The Robot-Fly was primarily based on ethological findings on real flies, whose most common flight trajectories were shown to consist of straight flight sequences interspersed with rapid turns termed saccades (Collett and Land 1975, Heisenberg and Wolf 1984, Wagner 1986), see also (Schilstra and Hateren 1999, Tammero and Dickinson 2002).

Fig. 9.2 Three of the visually-guided robots designed and constructed at the Laboratory on the basis of biological findings on visuomotor control in flies. (A) The 10-kg Robot-Fly ("robot-mouche" in French) incorporates the compound eye (visible at half-height) for obstacle avoidance, and a dorsal eye ("target seeker") for detecting the light source serving as a goal. This robot (height: 50 cm; weight: 12 kg) reacts to the OF generated by its own locomotion amongst obstacles (Franceschini et al. 1991, 1992, 1997). It is fully autonomous as regards its processing and power resources.(B) The robot OCTAVE (OF-based Control sysTem for Aerial VEhicles) is a 100-gram rotorcraft equipped with a 2-pixel ventral eye sensing the OF on the terrain below. This self-sustained aerial creature is tethered to a light rotating arm that allows only three degrees of freedom: forward and upward motion and pitch. The robot lifts itself and circles around a central pole at speeds up to 3 m/s. It ascends or descends depending on the ventral OF it measures (from Ruffier and Franceschini 2003, 2004, 2005)). (C) The robot OSCAR (OF based Scanning sensor for the Control of Aerial Robots) is a 100-gram, twin-engined aircraft equipped with a two-pixel frontal visual system that relies on visual motion detection and on a microscanning process inspired by the fly (Franceschini and Chagneux 1997). It is tethered to a 2-meter-long nylon wire secured to the ceiling of the laboratory. Vision and rate gyro signals combine onboard enabling OSCAR to fixate a target (a dark edge or a bar) with hyperacuity and to track it at angular speeds of up to 30°/s (Viollet and Franceschini 1999b, 2001). Figure from Franceschini et al. (2009).

Based on the results of electrophysiological findings on fly motion detecting neurons obtained at our Laboratory (Franceschini 1985, Franceschini et al. 1989), we developed a 2-pixel Local Motion Sensor (LMS) and proposed several versions of it over the years (Franceschini et al. 1992, Franceschini 1999, Ruffier et al. 2003, Ruffier 2004, Franceschini et al. 2007, Expert et al. 2011). The processing scheme at work in this sensor was introduced in (Blanes 1986, Pichon et al. 1989), and later called the "facilitate-and-sample scheme" (Indiveri et al. 1996) or the "time-of-travel scheme" (Moeckel and Liu 2007). Other vision-based systems have been used to measure the OF onboard UAVs (Unmanned Aerial Vehicles) (Green et al. 2004, Hrabar et al. 2005, Beyeler et al. 2009, Conroy et al. 2009) and in particular in the range experienced during lunar landing (Griffiths et al. 2006, Kendoul et al. 2009, Watanabe et al. 2009).

Most of these visual systems were quite demanding in terms of their computational requirements and/or their weight or were not very well characterized, except for the optical mouse sensors (Beyeler et al. 2009), with which a standard error of approximately ±5°/s around 25° was obtained in a 280° /s overall range.

More recently we developed at the laboratory the concept of Visual Motion Sensor (VMS) by fusing the local measurement from several 2-pixel LMS to measure the 1-D OF more accurately and more frequently (Expert et al. 2012, Roubieu et al., 2011,2012, Ruffier and Expert 2012).

Few studies have been published so far, to our knowledge, in which OF systems were implemented and tested outdoors onboard an unmanned aircraft subject to vibrations, where the illuminance cannot be easily controlled (see Barrows and Neely (2000)) in the case of linear 1-D motion sensors and see (Griffiths et al. 2006, Tchernykh et al. 2006, Garratt and Chahl 2008, Kendoul et al. 2009) in that of 2-D OF sensors).

It therefore seemed to be worth testing the reliability of the present 1-D OF-based visual sensor on a platform that would experience conditions similar to those of a spacecraft in landing approach in terms of vibration dynamics and OF measurement range. To that aim, the sensor was embedded onboard a free-flying, unmanned helicopter and tested in terms of its resolution, accuracy and sensitivity. Particular efforts were made to adapt the sensor's measurement range [1.5°/s to 25°/s] to the one a spacecraft would experience during a lunar landing approach (in the order of [2°/s - 6°/s]).

The intended control strategy, the reference descent trajectory and the basic equations of the spacecraft dynamics are described in Sect. 9.2 with particular reference to lunar landing approaches. Neuromorphic principles are applied to monitoring and processing the OF during an autonomous vision-based extraterrestrial landing scenario. OF measurements allowing the estimation of the orientation of the velocity vector are underlined. Section 9.3 gives a brief description of the new 1-D visual motion device, outlines the processing algorithm and the implemented electro-optical assembly. Results of the tests performed on the airborne visual sensor during the helicopter outdoor free flight experiments are presented.

9.2 Biomimetic Optic Flow Sensing Applied to Spacecraft Landing

9.2.1 *Autonomous Lunar Landing Strategy*

Lunar landing trajectory has been divided into four different phases in Frapard et al. 1996) (see Fig. 9.3):

1. De-orbit Phase,
2. Approach Phase,
3. Final Descent,
4. Free Fall.

In this research work, a solution to the autonomy problem of the approach phase defined from the high gate (500 m above ground level (AGL)) to the low gate (10 m AGL) is studied. High gate corresponds to the height from which the landing site becomes visible from the spacecraft vision system. Low gate corresponds to the height from which visual contact with the landing site is no longer available due to the dust raised by the thrusters. Initial parameters are a horizontal speed V_{x0} = 150 m/s , a vertical speed V_{z0} = -50 m/s , a pitch angle θ_0 = -60 ° , a ground height h_0 = 500 m and a mass $m_{ldr0} = 10^3$ kg (see Fig. 9.3).

This reference trajectory (Valette et al. 2010) is very similar to the one in the Apollo test case scenario used in (Izzo and de Croon 2011, Izzo et al. 2011).

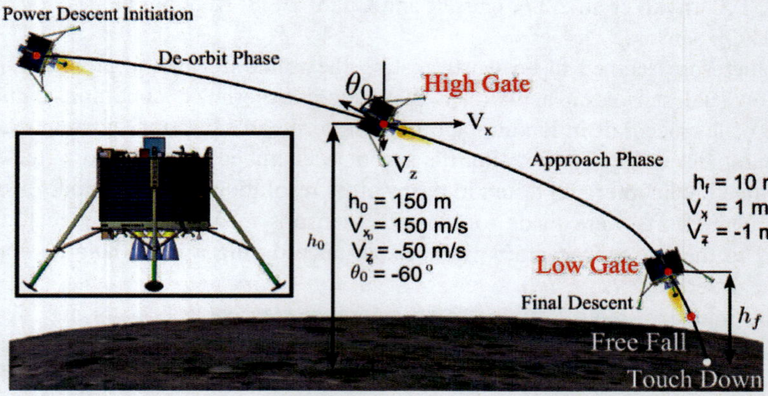

Fig. 9.3 Reference trajectory for lunar landing and 3D representation of the lunar module (courtesy: Astrium). The landing part addressed in this work is the approach phase defined between the high gate (500 m AGL) and the low gate (10 m AGL). The objectives of the lander is to reach the low gate (10 m high) with both vertical and horizontal velocities lower than 1 m/s. Figure modified from (Jean-Marius and Strandmoe 1998, Valette et al. 2010).

Solution to the problem must meet particularly demanding terminal constraints at the low gate (h_f = 10 m) as follows:

$$0 \leq V_{xf} \leq 1 \text{ m/s}$$
$$-1 \leq V_{zf} \leq 0 \text{ m/s}$$

Designing an optimal control strategy for lunar landing also requires considering the propellant consumption.

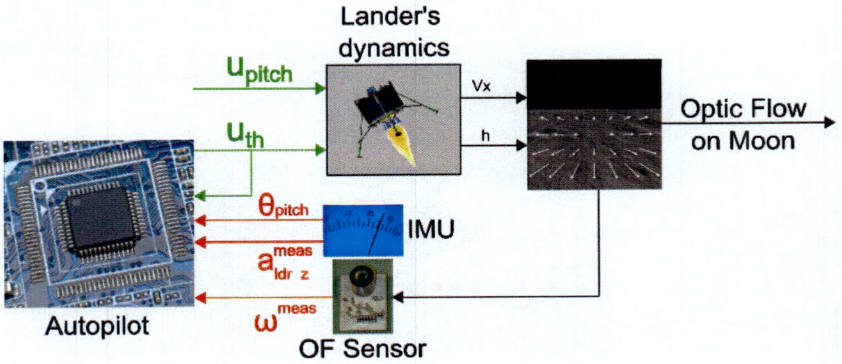

Fig. 9.4 Scheme of the closed loop system based on OF and pitch control. ω^{meas} the measured OF, u_{pitch} and u_{th} are the pitch angle and the main thruster control signal respectively. Concerning the measurements, the IMU assess the pitch angle θ_{pitch} and lander's accelerations both horizontal (a_{ldrx}) and vertical (a_{lrdz}). The mass m_{ldr} is estimated from the sufficiently well known initial mass. Figure modified from Valette et al. (2010)

The main challenge is that the entire state vector is not available from the measurement as can be seen on Fig. 9.4. For instance, velocities and position are neither measured nor estimated, only accelerations, angular position, mass and OF are measured and thus available to feed the controllers. To land safely on the moon the autopilot should be able to reduce the velocity vector magnitude and this is achieved by acting jointly on the lander's pitch and the lander's main thrust, the two available control signals.

9.2.2 *Lander's Dynamic Modeling and Optic Flow Equations*

The autopilot under consideration consists mainly of an OF-based control system operating in the vertical plane (x, z), and controlling the spacecraft's mean thruster force and pitch angle. To stabilize the lander, it is necessary to cope with nonlinearities and the inherent instability.

Since there is no atmosphere on the Moon, the lander experiences neither wind nor drag. In the present model, heave and surge dynamics are coupled via the lander's pitch (see Fig. 9.5). It is worth noting that it is inappropriate to measure $\omega_{45°}$ to determine the direction of the velocity vector, as the value 45° remains close to the focus of expansion where the motion is always null (see Fig. 9.5).

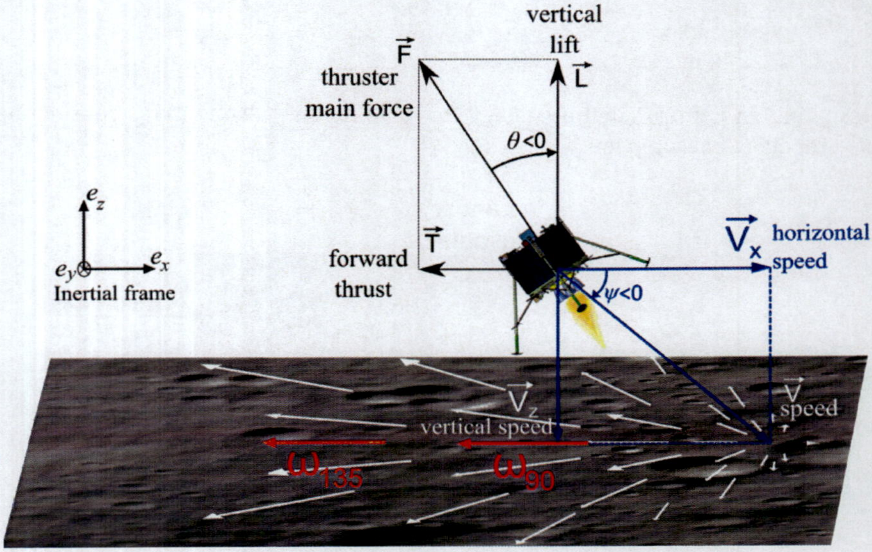

Fig. 9.5 Diagram of the lander, showing its speed vector **V**, the mean thruster force **F** and its projections along the surge axis (X-axis) giving the forward thrust **T**, and along the heave axis (Z-axis) giving the vertical lift **L.**

The dynamic motion of the lander can be described in the time domain by the following dynamic system in the inertial frame associated with the vector basis ($\mathbf{e}_x$, $\mathbf{e}_y$, $\mathbf{e}_z$):

$$\begin{cases} a_{ldrx}(t) = \dfrac{\cos(\theta(t))}{m_{ldr}(t)} \cdot u_{th}(t) - g_{Moon} \\ a_{ldrz}(t) = \dfrac{\sin(\theta(t))}{m_{ldr}(t)} \cdot u_{th}(t) \end{cases} \tag{9.1}$$

where $u_{th} = \|\mathbf{F}\|$ corresponds to the control force applied to the lander, $a_{ldrx,z}$ are the lander's accelerations in the lunar inertial reference frame, m_{ldr} stands for the lander's mass, θ is the pitch angle, *t* denotes the time and g_{Moon} the lunar gravity constant ($g_{Moon} = 1.63$ m/s^2) _{Moon}=1.63m/s^2.

The lander's mass depends directly on the fuel consumption, as given by the following relation:

$$\dot{m}_{ldr} = \frac{-1}{I_{sp}\, g_{Earth}} u_{th}(t) \tag{9.2}$$

where $I_{sp} = 311$ s corresponds to the specific impulse and $g_{Earth} = 9.81$m/s^2 to the Earth's gravity. This means that:

$$m_{ldr}(t) = m_{ldr}(t_0) - \frac{1}{I_{sp}\, g_{Earth}} \int_{t_0}^{t} u_{th}(\varepsilon) d\varepsilon \tag{9.3}$$

where $m_{ldr}(t_0)=10^3$ kg is the lander's mass at high gate level.

Since the initial mass is known and the lander's mass depends linearly on the integral of the lander's thruster control signal, the mass can be computed and assessed at any time during the simulated descent.

The inner pitch control system is modeled as follows:

$$\frac{I}{R}\frac{d^2\theta}{dt^2} = u_{pitch}(t) \tag{9.4}$$

u_{pitch} is the control input signal giving the spacecraft's pitch and θ is measured via an Inertial Measurement Unit, I the moment of inertia of the lander and R its radius.

Once the dynamic model of the spacecraft is defined, one needs to state the OF equations to find what information can be deduced from this visual cue.

The ground-truth OF $\omega_{grd\text{-}trh}$ can be described as the sum of the two distinct components defined by Koenderink and van Doorn (1987), i.e. the translational and rotational OF:

$$\omega_{grd-trh} = \omega_T + \omega_R \tag{9.5}$$

The translational OF ω_T depends on the linear velocity V expressed in the inertial frame, the distance from the ground D in the gaze direction and the elevation angle Φ (i.e., the angle between the gaze direction and the heading direction):

$$\omega_T = \frac{V}{D}\sin(\Phi) \tag{9.6}$$

The rotational OF ω_R depends only on the vehicle's angular Ω_j expressed in the body fixed frame, where j denotes the axis of rotation, and on the elevation angle λ between the gaze direction and the axis of rotation:

$$\omega_R = \Omega_j \sin(\lambda) \tag{9.7}$$

Finally the general equation of the OF in the vertical plane is as follows:

$$\omega_{grd-trh} = \frac{V}{D}\sin(\Phi) + \Omega_j \sin(\lambda) \tag{9.8}$$

Under the assumption that the sensors are embedded on a gimbaled system one can derive the expression of the OF measured in the vertical direction considering a pure translational motion ($\Omega_j=0$). From Eq. (9.8), under the assumption of a practically flat ground (i.e. $D = h / \cos(\pi/2 - \Phi + \Psi)$, $\Phi-\Psi$ denotes the angle between the gaze direction and the local horizontal), and gimballed mounted sensors the ventral optic flow is defined as follows:

$$\omega_{90} = \frac{Vx}{h} \tag{9.9}$$

where $V = V_x/\cos(\Psi)$.

9.2.3 Optic Flow-Based Regulator for Spacecraft Landing

From Eq.(9.1) and modeling the thruster's dynamics by a first order transfer function with $\tau_{thruster} = 100$ ms such that

$$\tau_{thruster} \dot{u}_{th} + u_{th} = m_{ldr} u$$

the Laplace transform of the heave dynamics $Z(s)$ can be written as follows:

$$Z(s) = \frac{1}{s^2}\left[\left(\frac{1/\tau_{thruster}}{1/\tau_{thruster} + s}\cos(\theta)U(s)\right) - g_{Moon}\right] \tag{9.10}$$

where $U(s)$ is the Laplace transform of the control input signal $u(t)$.

The transfer function for the surge dynamics $G_x(s)$ can be written as follows:

$$G_x(s) = \frac{X(s)}{U(s)} = \frac{1}{s^2}\left(\frac{1/\tau_{thruster}}{1/\tau_{thruster} + s}\sin(\theta)\right) \tag{9.11}$$

For the lander model, if a_{th_z} is the vertical thruster's acceleration, we choose the following state vector

$$X = \begin{bmatrix} h \\ V_z \\ a_{th_z} \end{bmatrix}$$

and the defined control input signal u. According to Eqs. (9.10) and (9.11), one can write:

$$\begin{gathered} \dot{a}_{th_z} = \frac{1}{\tau_{thruster}}\left(u - a_{th_z}\right) \\ \dot{V}_z = a_{th_z} - g_{Moon} \end{gathered} \tag{9.12}$$

One can deduce from the Eq. (9.12), the state space representation:

$$\dot{X} = A_p X + B_p u - g_{Moon} \tag{9.13}$$

$$\begin{bmatrix} \dot{h} \\ \dot{V}_z \\ \dot{a}_{th_z} \end{bmatrix} = \begin{bmatrix} 0 & 1 & 0 \\ 0 & 0 & 1 \\ 0 & 0 & \dfrac{-1}{\tau_{thruster}} \end{bmatrix} \cdot \begin{bmatrix} h \\ V_z \\ a_{th_z} \end{bmatrix} + \begin{bmatrix} 0 \\ 0 \\ \dfrac{1}{\tau_{thruster}} \end{bmatrix} . u - \begin{bmatrix} 0 \\ g_{Moon} \\ 0 \end{bmatrix} \tag{9.14}$$

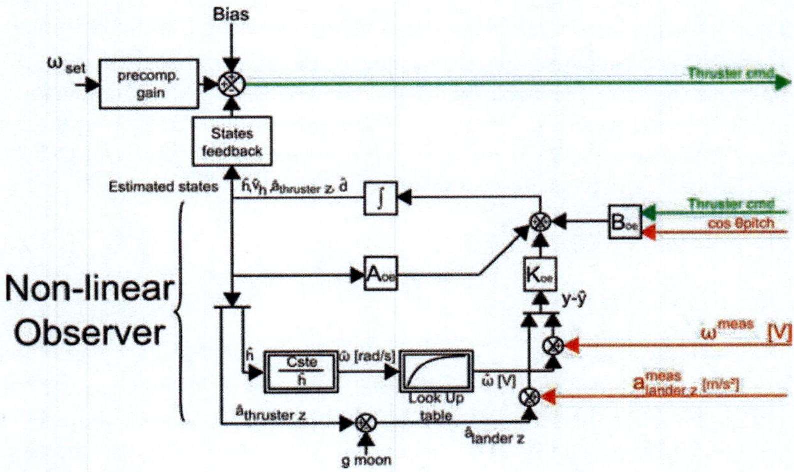

Fig. 9.6 The present autopilot makes use of a nonlinear observer to provide the feedback control scheme with the state estimates. The observer was nonlinear since the ventral OF ω is by definition an inverse function of the controlled variable h. The state estimate of the OF $\hat{\omega}$ in rad/s was computed in terms of the ratio between a constant (the ground speed at a given working point) and the estimated height of the lander $\hat{h}$. The OF was then obtained via a look-up table. The measured acceleration of the lander $a_{lander_z}^{meas}$ served only to improve the state estimation. The estimated acceleration $\hat{a}_{lander_z}$ was obtained by subtracting the approximate gravity on the Moon g_{Moon} from the estimated acceleration induced by the thruster $\hat{a}_{thruster_z}$ Figure from Valette et al. (2010).

The present spacecraft was modeled taking into account the thruster dynamics and a pure double integration between the acceleration and the altitude, using the state-space approach. The autopilot, which operated on the basis of a single OF measurement (that of the ventral OF), consisted of a visuomotor feedback loop driving the main thruster force. Since the vertical lift and the forward thrust are coupled, the loop controls both the heave and surge axes. The pitch angle θ was controlled by an external system which made the lander gradually pitch backwards from -60° to -30° throughout the landing approach. The autopilot presented in Fig. 9.6 was composed of (i) a precompensation gain, (ii) a nonlinear state observer, and (iii) a state feedback gain. The nonlinear state observer estimated the state vector X on the basis of the ventral measured OF ω^{meas}, the lander acceleration, a_{lander_z}, and the lander pitch, θ. The complete regulator combined the estimated states with the full state feedback control loop.

9.2.3.1 State Feedback Control Law Design

The autopilot kept the ventral OF of the simulated spacecraft at the set point ω_{set}. This set point was compared with the product of the estimated state vector $\hat{X}$ (see Fig. 9.6) and the state feedback gain L_{sf} to generate the thruster command. The state feedback gain was calculated using the minimization criterion in the Linear Quadratic Regulator (LQR) method, using the following matrix: A_{sf}=A_p, B_{sf}=B_p and C_{sf}= [K_{lin} 0 0] and the state-cost matrix

$$Q_c = \begin{bmatrix} 7.8\cdot 10^{-4} & 0 & 0 \\ 0 & 0 & 0 \\ 0 & 0 & 0 \end{bmatrix}$$

and R_c= [1]. To compute the C_{sf} matrix, we linearized the expression for the OF near a set point. Here the set point was h_{lin}=200 m, V_{xlin} = 50 m/s, and ω=14.3°/s. The OF was defined as an inverse function of h. We therefore used the slope of the tangent to linearize the expression as follows:

$$K_{lin} = V_{xlin}\cdot \frac{d}{dh}\left(\frac{1}{h}\right)_{h=h_{lin}} = \frac{-V_{xlin}}{h_{lin}^2} \tag{9.15}$$

9.2.3.2 Nonlinear State Observer Design

Since the system is observable, a state observer for $\hat{\hat{X}}$ can be formulated as follows:

$$\begin{cases} \dfrac{d\hat{X}}{dt} = A_0\hat{X} + B_0u + K_0(y - \hat{y}) \\ \hat{y} = C_0\hat{X} + D_0u \end{cases} \tag{9.16}$$

where A_0=A_{sf}, B_0=B_{sf} and

$$C_0 = \begin{bmatrix} & C_{sf} & \\ 0 & 0 & 1 \end{bmatrix}$$

D_0=[0] and K_0 (gain of the observer) was also computed with LQR method, using A_0 and C_0 matrices. As shown in Fig. 9.6, the observer requires the value of the lander's acceleration a_{ldr_z}.

To achieve an integral control, the augmented state vector X_e was thus defined:

$$X_e = \begin{bmatrix} X \\ d \end{bmatrix}$$

where *d* stands for the disturbance. The new state matrices could therefore be written as follows:

$$A_{0e} = \begin{bmatrix} A_0 & B_0 \\ 0\ 0\ 0 & 0 \end{bmatrix}, \quad B_{0e} = \begin{bmatrix} B_0 \\ 0 \end{bmatrix}, \quad C_{0e} = \begin{bmatrix} C_0 & \begin{matrix} 0 \\ 0 \end{matrix} \end{bmatrix} \tag{9.17}$$

The state feedback gain L_{sfe} was equal to $L_{sfe} = [L_{sf}\ 1]$. The observer gain was computed in the case of the extended state using the same method, with the new state matrix (A_{0e} and C_{0e}). The acceleration of the lander $\hat{a}_{ldr_z}$ was estimated by subtracting the lunar gravity g_{Moon} from the estimated engine thrust $\hat{a}_{ldr_z}$ as shown on Fig. 9.6.

The observer is initialized using a rough estimation of the initial height and vertical speed at high gate. The observer tolerates an uncertainty of about 20% in the estimation of the height and vertical speed.

9.2.3.3 Autonomous Landing Simulation Using the PANGU Software Program

To ensure a soft landing, the lander had to reach a distance of approximately 10 meters from the ground (i.e., the low gate) at a residual velocity of one meter per second in both the horizontal and vertical directions. In Valette et al. (2010), thanks to the biomimetic autopilot, the lander reached the low gate with greatly reduced horizontal and vertical speeds approximately equal to the required values as shown in Fig. 9.7. The lunar surface perceived by the lander consisted of gray-scale images generated by PANGU. In the presented simulation, an initial altitude $h_0 = 500$ m, an initial ground speed $V_{x0} = 150$ m/s and an initial vertical speed $V_{z0} = -50$ m/s were adopted. The pitch angle θ was made to decrease exponentially from -60° to -30°. As a consequence, the forward speed decreased quasi-exponentially as well (see Fig. 9.7C), and so did the vertical speed (see Fig. 9.7C) since its integral *h* was reduced quasi-exponentially to hold the measured OF $\omega_{meas} = V_x/h$ around the OF set point value ω_{set} (see Fig. 9.7D).

The spacecraft's simulated approach took 58.4 s, which corresponds to the time required to reach the low gate. The lander reached the low gate at a final ground speed $V_{xf} = 5$ m/s and a final vertical speed $V_{zf} = -4$ m/s; the distance traveled by the lander during the landing was 2660 meters. The final horizontal and vertical speeds are slightly higher than required to strictly satisfy the speeds' criterion at low gate (1m/s).

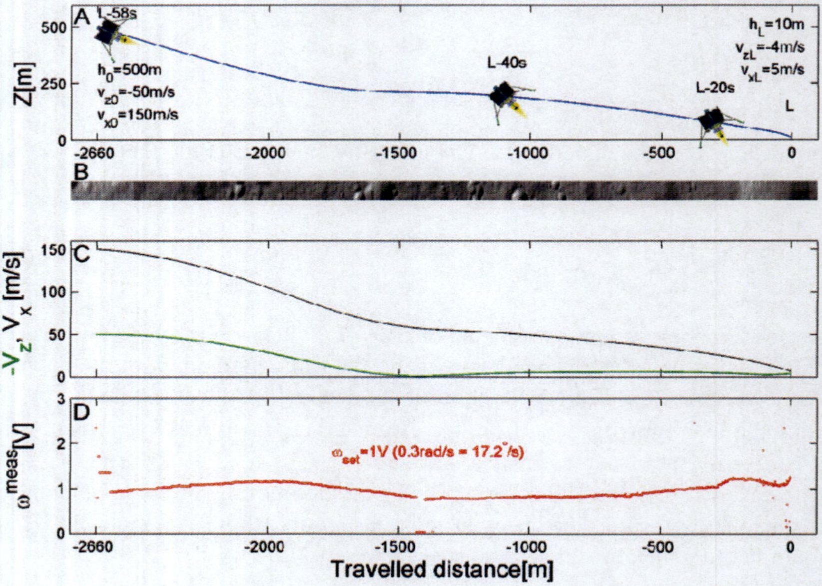

Fig. 9.7 Automatic landing based on a biomimetic OF sensor combined with a bio-inspired strategy. The OF sensor rests on a single pair of photoreceptors (i.e. two pixels), the mean orientation of which was maintained downwards. Figure from Valette et al. (2010).

In Valette et al. (2010), the OF regulation principle was applied to autonomous lunar landing problems using an OF based feedback loop and tested by performing PANGU software simulations. The first simulations presented involved neuromorphic principles applied to monitoring and processing the 1-D OF on an autonomous visual-based extraterrestrial landing scenario. The autopilot generated the landing approach while monitoring only the ventral OF without the need to measure or estimate speed and height, and hence without the need for any bulky and power-consuming sensors. The main drawback with using the ventral OF exclusively is that the vertical dynamics of the lander is not taken into account with the consequence that the low gate cannot be reached with sufficiently low speeds as seen on Fig. 9.7. In the following, further OF measurements are introduced with an aim to recover the dynamics of the lander.

9.2.4 OF Measurements ω_{90} and ω_{135} for Velocity Orientation Estimation and Control

To achieve a fully autonomous lunar landing, a possible way to improve the autopilot presented here would be to design a visual-based pitch control law to replace the fixed reference from -60° to -30° used in the current simulations.

In Izzo et al. (2011) and Valette et al. (2010), the authors have shown the substance of the pitch control law in the design to achieve optimal performances since the system is underactuated.

In the ongoing work, the main idea is to design an autopilot that keeps the main thrust antiparallel to the velocity vector orientation, in order to reduce the lander's fuel consumption as much as possible.

This principle defines pitch angle reference θ_{ref} which is fed into the pitch controller:

$$\theta_{ref} = -\Psi - \frac{\pi}{2} \tag{9.18}$$

where Ψ denotes the angle between the orientation of the speed vector and the local horizontal.

That strategy implies that in order to be able to strongly reduce the lander's speed during the approach phase, one needs to measure or estimate the velocity vector orientation Ψ which is not an easy tasks considering the lack of appropriate sensors onboard the lander. The main question is how to fuse different visual angular speed measurements, to obtain useful information about unavailable measurement of the state vector. It is straightforward to note that the OF cue is related to the orientation of the velocity vector. One can derive the orientation angle Ψ from OF sensors positioned in different directions by fusing $\omega_{90}=V_x/h$ (Eq. (9.9)) and ω_{135}:

$$\omega_{135} = \frac{V}{h/\cos(\pi/4)} \cdot \frac{\sqrt{2}}{2}\left(\cos(\Psi) - \sin(\Psi)\right) = \frac{\omega_{90}}{2}\left(1 - \tan(\Psi)\right) \tag{9.19}$$

Finally one can obtain:

$$\tan(\Psi) = 1 - 2\frac{\omega_{135}}{\omega_{90}} \tag{9.20}$$

It is worth noting that both the horizontal and vertical dynamics are expressed within $\tan(\Psi)$ with $\tan(\Psi) = V_z/V_x$.

Thanks to Eq. (9.20) featuring only visual information, a pitch controller based on OF and pitch measurement θ provided by the IMU could be designed through Eq. (9.4) in order to ensure the colinearity between the lander's main thruster force and its velocity vector orientation.

The low speed visual motion sensors are thus the cornerstones of this autonomous lunar landing strategy. Since both the OF controller and the pitch controller are based on the output signals of OF sensors, it seemed to be worth testing the reliability of such sensors in real-life conditions. This is the aim of the following section.

9.3 VMS-Based OF Measurements Obtained Onboard ReSSAC

The low speed visual motion sensors used to measure the OF are strongly linked to the control signals of the dynamic system. This is why we developed and tested a new VMS dedicated to low angular speed measurements.

9.3.1 Bio-inspired Optic Flow Processing

A mandatory step in the maturation of a technology is to design and embed the previously simulated device on a real-life complex system. In order to validate the feasibility of the theoretical work using the low speed VMS an experimental approach is presented.

A low resolution visual motion sensor based on a 6-pixels array and dedicated to a low speed range has been developed to demonstrate on Earth the feasibility of measuring the 1-D local angular speed on a lunar landing like scenario. We tested the sensor onboard an unmanned helicopter to validate the bio-inspired algorithm at relatively low ground speeds and relatively high ground heights and in the presence of strong natural disturbances (i.e. craft vibrations, uncontrolled illuminance, rough terrain, etc.).

This sensor is an updated version of the 2-pixel Local Motion Sensors designed by the biorobotic team on the basis of neurophysiological findings in flies (Franceschini et al. 1989).

9.3.2 Presentation of the Low-Speed Visual Motion Sensor

The new low-speed visual motion sensor (VMS) consists mainly of a low-cost plastic lens (CAX183 from Thorlabs, focal length 18.33mm, f-number 4.07) placed in front of an off-the-shelf photodiode array LSC from iC-Haus. The latter features six photodiodes, each of which has a large sensitive area (300 x 1600 μm) and an integrated preamplifier. The LSC conveys the six photodiode signals to a hybrid analog/digital processing algorithm which computes the OF value ω_{meas}. A custom-made protective case was added to protect the low-weight sensor and the optical assembly from unfavorable weather conditions.

The new visual motion sensor and its custom-made protective case weighed 29.4 *g*.

Many of the parameters of the original visual motion detecting scheme presented in Blanes (1986), Pichon et al. (1989) have been updated, especially in terms of interreceptor angles and cut-off frequencies of the temporal filters.

The six optical axes formed by the photodiodes are separated by an angle called the interreceptor angle $\Delta\varphi$. By defocusing the lens (i.e., by adjusting the distance between the lens and the photosensors), we obtained a Gaussian angular sensitivity functions for each photoreceptor with a correlation coefficient greater than 99 % ($R^2_{LSC} > 0.990$). These features were assessed by slowly rotating the lens in front of a point light source placed at a distance of 85 cm. The local 1-D angular speed ω_{meas} measured by the sensor was defined as the ratio between the interreceptor angle $\Delta\varphi$ and the time elapsing Δt between the moments when two adjacent photodiode signals reached the threshold. Δt represents the "time of travel" of a any given contrast feature passing from the optical axis of one photodiode to the optical axis of the neighboring one:

$$\omega_{meas} = \frac{\Delta\varphi}{\Delta t} \tag{9.21}$$

In (Expert et al. 2011), the measurement range of the sensor covered a large range of high speeds from 50°/s to 300°/s, whereas the present study focused on low velocities giving a range of 1.5°/s to 25°/s, which is more than tenfold slower. In order to stay in the same range of Δt, whose accuracy of measurement depends on the microcontroller's sampling frequency, we therefore had to narrow $\Delta\varphi$.

The large 18.33mm focal length increases the defocalizing effects of the lens, giving a suitably small mean interreceptor angle of $\overline{\Delta\varphi} = 1.4°$. The second advantage of the defocusing process is that it adds a blurring effect giving each pixel a Gaussian-shaped angular sensitivity function with a half width of similar size $\Delta\rho$=1.4°. The resolution attained here is very similar to that measured in the common housefly compound eye (Kirschfeld and Franceschini 1968):

$$\Delta\varphi = \Delta\rho \tag{9.22}$$

The acceptance angle, defined by $\Delta\rho$, acts as an optical low pass spatial filter. Achieving $\Delta\rho/\Delta\varphi$ ratio of 1 made it possible for the OF sensor to respond to contrasting features of high spatial frequency. With $\Delta\rho = \Delta\varphi = 1.4°$, the overall FOV of the VMS was 10.28°.

The general processing algorithm underlaying the VMS consists of two parts: an analog processing part that converts the six photodiode signals into electrical signals with a high signal-to-noise ratio, and a digital processing part that simultaneously computes five OF values plus the OF median value (see Fig. 9.8). The analog processing part begins with a programmable gain connected to the microcontroller via a SPI communication bus (Ruffier and Expert 2012). A bandpass filter then differentiates the visual signal and acts as an anti-aliasing filter.

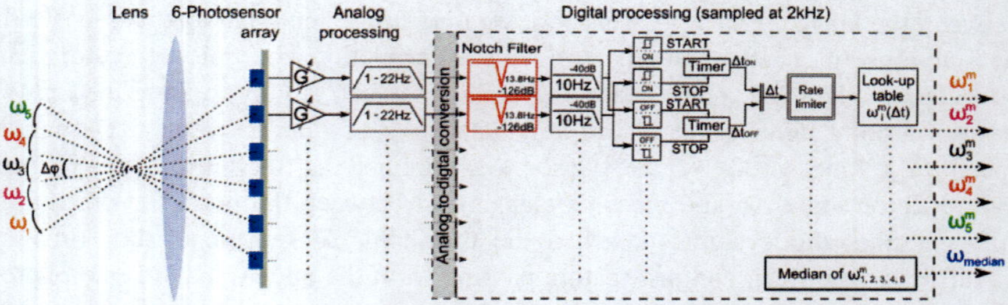

Fig. 9.8 General processing architecture of the low-speed visual motion sensor. First of all, the spatial sampling and low-pass filtering steps are carried out by the defocused lens. The six photodiode signals are amplified with programmable gain to increase the signal to noise ratio, before being filtered by an analog bandpass filter (1-22Hz). The digital stage begins with a second order fixed-point notch filter centered on the helicopter's main rotor frequency (13.8Hz). It is followed by a second order fixed-point low pass (cut-off frequency = 10Hz). A hysteresis thresholding process is associated with the computation of the time Δt elapsing between two adjacent signals (of similar ON or OFF contrast polarity). Lastly, after an outlier filtering step, the output signal of the 1-D visual motion sensor is obtained from a precomputed look-up table and the median value is calculated. Figure modified from (Roubieu et al. 2011).

The digital processing algorithm starts with a second order fixed-point notch filter (rejection factor Q=6.9) whose center frequency was tuned to the helicopter's main rotor frequency (13.5 Hz). Its transfer function has been defined as follows (Orfanidis 1995):

$$H_{notch}(z) = b\frac{1 - 2\cos(\omega_0)\ ...}{...} \tag{9.23}$$

with

$$b = \frac{1}{1 + \frac{\sqrt{1 - G_B^2}}{G_B}\tan(\frac{\Delta\omega}{2})}$$

where $\Delta\omega$ is the full width at a level G_B^2 and ω_0 is the center frequency. We chose $\omega_0 = 2\pi f_s/f_0$, $\Delta\omega = 2\pi\Delta f/f_s$ with Δf= 2 Hz and $G_B^2 = -3dB$.

As the visual angular speed $\omega_{grd\text{-}trh}$ is quite low, the temporal frequency f_t of the visual signals derived from the features of the environments is also quite low, as expressed by the following equation (Landolt and Mitros 2001):

$$f_t = \omega_{grd-trh} \cdot f_{spatial}$$

where $f_{spatial}$ is the spatial frequency associated with the contrasting pattern.

Therefore, a second order fixed-point low pass filter was used to enhance the signal-to-noise ratio by removing the noise remaining at frequencies higher than 10 Hz.

The OF algorithm ("Time of travel scheme") implemented here consists mainly of a hysteresis thresholding process with separate ON and OFF pathways (Blanes 1986, Pichon et al. 1989, Viollet and Franceschini 1999b, Ruffier et al. 2003, Roubieu et al. 2011) followed by the Δt computation, the result of which is fed into a correspondance lookup table. Lastly, the five simultaneously computed OFs ω_i^m are fused by a median operator to increase the refresh rate robustness of the output (Roubieu et al. 2011).

The microcontroller used for this purpose (dsPIC33FJ128GP802) operates at a sampling frequency of 2 kHz, except for the digital filters, which are sampled at a rate of 500 Hz. Special efforts were made to optimize the algorithm, and a computational load of only 17 % was eventually obtained.

9.3.3 Characterization of the Visual Motion Sensor (VMS)

The characteristics of the present visual motion sensor (VMS) were assessed by performing OF measurements under controlled motion conditions (orientation and velocity) outdoors. Pure rotational motion was applied to the sensor at angular speeds ranging from 1°/s to 20°/s using a previously described outdoor set-up (Expert et al. 2011). The triangular response pattern obtained corresponds closely to the reference angular speed (see Fig. 9.9).

It can therefore be said that this new tiny sensor is able to accurately compute the 1-D visual angular speed during a rotational motion within its operating range. The refresh rate is defined as the ratio between the total number of measurements occurring within the acceptable range [1.5°/s-25°/s] and the time elapsing. The mean refresh rate achieved during the dynamic performances evaluation was f_{refresh}= 6.6 Hz: this value depends on the richness of the visual environment, as well as on the actual angular speed.

9.3.4 Free-Flying Results with the Airborne Visual Motion Sensor

The VMS dynamic performance was then studied on a six-DOF UAV during free flight over fields. A Yamaha Rmax helicopter was used in the framework of the ONERA's ReSSAC project. The helicopter characteristics in terms of mass balance have been described in (Watanabe et al. 2010). Its mass (80kg), its flight envelope and the vibration dynamics due to the main rotor's rotational speed presented us with quite a challenging ground-truth OF profile. The flight was performed in South-western France, mid-July at about 5pm on a bright sunny day: the mean illuminance was approximately 10000 lx.

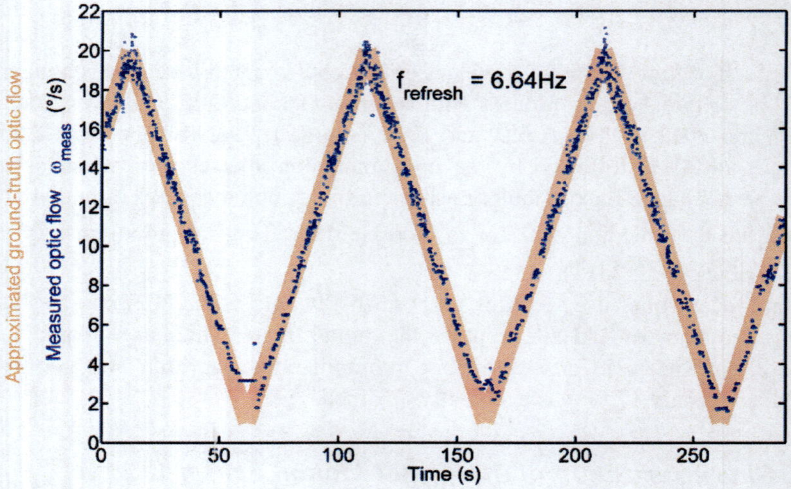

Fig. 9.9 Dynamic outdoor response of the low-speed VMS (blue), as compared with the ground-truth OF (red). The visual motion sensor was rotated by means of a conveyor belt driven by a stepping motor (103H5208-0440 from Sanyo-Denki) (Expert et al. 2011). Rotations from 1°/s to 20°/s were applied to the sensor, which is designed to operate in the 1.5°/s to 25°/s range. The OF measured closely matched the reference signal, with a refresh rate of 6.64 Hz. Since no synchronization signal was available, the ground-truth OF has been roughly synchronized here.

Onboard the ReSSAC helicopter, the 1-D local OF measured is subject to several variations as follows. Since the roll and pitch angles remain small throughout flight, the distance to the ground in the gaze direction D can be approximated as $D \approx h/[\cos(\varphi)\cos(\theta)]$ where φ denotes the roll angle, θ the pitch angle and h the local ground height.

In our case, $\Phi=-\theta+\Psi+\pi/2$ (with the sensor oriented downward, $\Psi < 0$, $\theta<0$), $\lambda = \pi/2$ and $\Omega_j = \Omega_2$, where Ω_2 is the pitch angular velocity defined in the body fixed reference frame, the ground-truth OF (see Eq. (9.8)) is therefore computed as follows:

$$\omega_{grd-trh} = \left(\frac{V}{h} cos(\theta) cos(\varphi) sin\left(\theta + \Psi + \frac{\pi}{2}\right)\right) + \Omega_2 \tag{9.24}$$

During the experiment described below, the ground-truth OF $\omega_{grd\text{-}trh}$ was computed using data from the IMU, the GPS (OEM4 G2 from NovAtel) and the data grabbed by a LIDAR (Sick LDMRS 400001) during previous GPS assisted flights over the same fields.

The low speed visual motion sensor was embedded in the front part of ReSSAC helicopter and pointed vertically downwards with a clear FOV.

Figure 9.10 shows the response of the low-speed visual motion sensor mounted onboard the unmanned ReSSAC helicopter. Despite the complex ground-truth OF, the visual motion sensor responded appropriately to the visual stimuli. The standard deviation of the error between the ground-truth OF $\omega_{grd\text{-}trh}$ and the measured OF ω_{meas} was less than 2.25°/s, which is quite low. The refresh rate $f_{refresh}$ was greater than 7.8 Hz, which is even slightly higher than in the dynamic measurements performed during a rotating motion on ground.

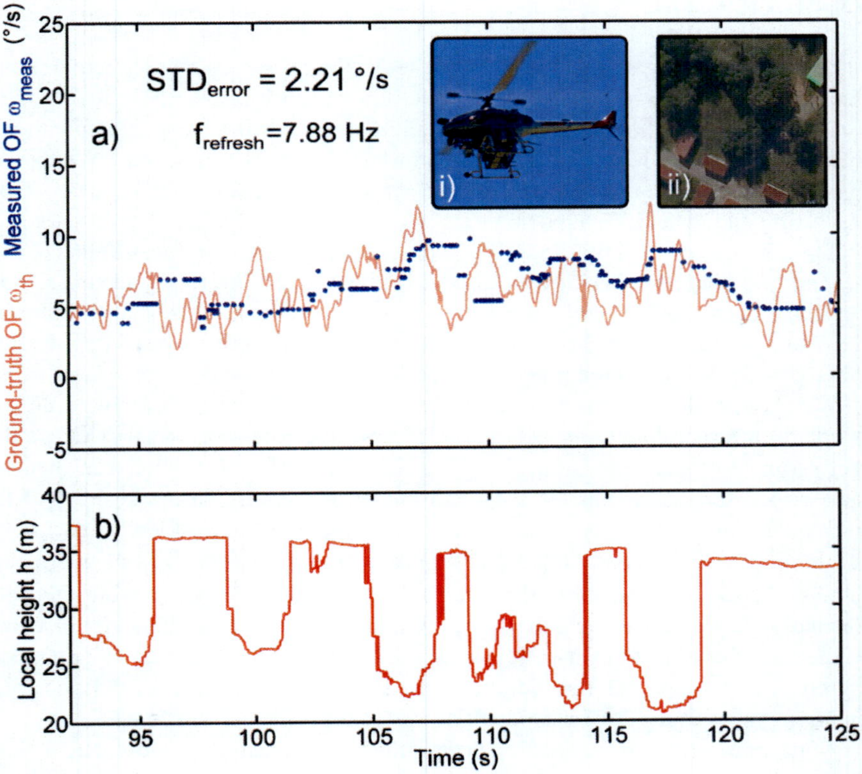

Fig. 9.10 Low-speed visual motion sensor and flight data sensed on-board the ReSSAC UAV. (a) Ground-truth OF (see Eq. (9.24)) (red) and measured OF $\omega_{meas}=\Delta\varphi/\Delta t$ (blue dots). Despite the strong variations mainly due to vibrations, the low-speed visual motion sensor's output closely matched the ground-truth OF, giving a standard deviation of 2.21°/s and a refresh rate of 7.88 Hz. The effects of strong variations in the local height due to the successive trees and houses are directly reflected in the low-speed VMS measurement signal. (i) ReSSAC unmanned helicopter in-flight (ii) Aerial view of the flight environment obtained on http://geoportail.fr. (b) Local ground height measured by combining GPS data and previously mapped LIDAR data. The nominal height was around 40 m. But due to the variable relief (houses, etc.), the local ground height often changed suddenly by 15 meters.

Figure 9.10b, giving the local ground height shows how well the sensor responded to its visual environment. Once again, the low-speed VMS accurately sensed these height variations and yielded similar values to the ground-truth value.

The robust and accurate performances observed during this experiment show that the low-speed visual motion sensor is highly suitable for use in many high-scaled robotic applications.

9.4 Conclusion

In this chapter, we gathered several core technologies to achieve an autonomous landing approach based on low-speed optic flow (OF) sensors.

We presented preliminary results concerning the control scheme, along with the intended control strategy primarily based on the OF regulation scheme.We distinguish two parallel control loops, the one acting on the main thruster, the second one acting on the lander's pitch angle.

The aim is to control the lander's dynamics by exploiting OF measurements and keeping the main thruster force collinear to the velocity vector at all times. The challenge is to provide the lander with a near optimal descent trajectory in terms of its fuel consumption. The main benefit of this biologically inspired approach is that it avoids the estimation process of both height and velocity.

We showed that a sensor assembly based on two OF sensors oriented downwards in different directions (90° and 135° from the local horizontal) yields a direct estimate of the orientation Ψ of the velocity vector. The latter can then be used as a reference signal for the pitch angle controller. The next step will be to develop full simulations featuring the control loops to attain at low gate (h_f = 10 m) the final velocity conditions stated in the reference scenario (V_{xf} =1 m/s, V_{zf} = -1 m/s). The loop controlling the main thruster will be designed using OF cue. Secondly the gimballed setup used in simulations needs to be addressed. Since one of the main benefits of using OF cues is mass saving and simplicity, a gimballed system is not suited for this purpose. Increasing the number of VMS and thus enlarging the sensory FOV is a potential way to achieve such challenge.

After introducing the specific lunar landing approach, we presented a new lightweight visual motion sensor able to compute accurately the OF in the range experienced during the approach phase of a lunar landing. This new VMS has been developed, and then tested both on the ground and in flight onboard an unmanned helicopter flying over an unknown complex outdoor environment and under real-life dynamic and vibratory conditions. Encouraging results of this experiment showed that this sensor is perfectly suited for aeronautics or aerospace applications since it sensed accurately the local 1-D angular speed ranging from 1.5°/s to 25°/s with a quite frequently refreshed measurement.

Acknowledgment. We are most grateful to G. Graton and F. Expert for their fruitful suggestions and comments during this study. We thank P. Mouyon, H.de Plinval, A. Amiez and V. Fuertes for the successful flight of ReSSAC helicopter, T. Rakotomamonjy and A.

Piquereau for their expertise in helicopter dynamics, M. Boyron for his assistance with the electrical design, J. Diperi for his involvement in the mechanical design and J. Blanc for improving the English manuscript.

References

Barrows, G., Neely, C.: Mixed-mode VLSI optic flow sensors for in flight control of a Micro Air Vehicle. In: SPIE: Critical Technologies for the Future of Computing, San Diego, CA, USA, vol. 4109, pp. 52–63 (2000)

Beyeler, A., Zufferey, J., Floreano, D.: Optipilot: control of take-off and landing using optic flow. In: European Micro Aerial Vehicle Conference, vol. 27, pp. 201–2019 (2009)

Blanes, C.: Appareil visuel élémentaire pour la navigation à vue d'un robot mobile autonome. Master's thesis, Master thesis in Neurosciences (DEA in French), Neurosciences, Advisor: N. Franceschini, Univ. Aix-Marseille II, France (1986)

Blanes, C.: Guidage visuel d'un robot mobile autonome d'inspiration bionique. Ph.D. thesis, INP Grenoble, France (1991)

Braun, R., Manning, R.: Mars exploration entry, descent and landing challenges. In: The Proceedings of the IEEE Aerospace Conference, Big Sky, Montana, Pasadena, CA, Jet Propulsion Laboratory, National Aeronautics and Space Administration (2006)

Cheng, Y., Ansar, A.: Landmark based position estimation for pinpoint landing on mars. In: Proceedings of the IEEE International Conference on Robotics and Automation (ICRA), pp. 1573–1578 (2005)

Collett, T.S., Land, M.F.: Visual control of flight behaviour in the hoverfly Syritta pipiens. Journal of Comparative Physiology A: Neuroethology, Sensory, Neural, and Behavioral Physiology 99(1), 1–66 (1975)

Conroy, J., Gremillion, G., Ranganathan, B., Humbert, J.: Implementation of wide-flield integration of optic flow for autonomous quadrotor navigation. Autonomous Robots 27, 189–198 (2009)

De Lafontaine, J.: Autonomous spacecraft navigation and control for comet landing. Journal of Guidance, Control, and Dynamics 15(3), 567–576 (1992)

Dubois-Matra, O., Parkes, S., Dunstam, M.: Testing and validation of planetary vision-based navigation systems with pangu. In: Proceedings of the 21st International Symposium on Space Flight Dynamics (ISSFD), Toulouse, France (2009)

Expert, F., Viollet, S., Ruffier, F.: Outdoor field performances of insect-based visual motion sensors. Journal of Field Robotics 28(4), 529–541 (2011)

Expert, F., Roubieu, F.L., Ruffier, F.: Interpolation based "time of travel" scheme in a visual motion sensor using a small 2d retina. In: The Proceedings of the IEEE Sensors Conference, Taipei, Taiwan, pp. 2231–2234 (2012)

Flandin, G., Polle, B., Frapard, B., Vidal, P., Philippe, C., Voirin, T.: Vision based navigation for planetary exploration. In: Proceedings of the 32nd Annual AAS Rocky Mountain Guidance and Control Conference (2009)

Franceschini, N.: Early processing of colour and motion in a mosaic visual system. Neurosc. Res. Suppl. 2, 17–49 (1985)

Franceschini, N.: De la mouche au robot: reconstruire pour mieux comprendre. In: Bloch, V. (ed.) Cerveaux et Machines, pp. 247–270 (1999)

Franceschini, N., Chagneux, R.: Repetitive scanning in the fly compound eye. In: Göttingen Neurobiology Report, Thieme, vol. 2, p. 279 (1997)

Franceschini, N., Riehle, A., Nestour, A.L.: Directionally Selective Motion Detection by Insect Neurons. In: Facets of Vision, pp. 360–390. Springer (1989)

Franceschini, N., Pichon, J.M., Blanes, C.: Real time visuomotor control: from flies to robots. In: Proceedings of the IEEE Conference on Advanced Robotics (ICAR 1991), Pisa, Italy, pp. 931–935 (1991)

Franceschini, N., Pichon, J.M., Blanes, C.: From insect vision to robot vision. Philosophical Transactions of the Royal Society B: Biological Sciences 337(1281), 283–294 (1992)

Franceschini, N., Pichon, J.M., Blanes, C.: Bionics of visuo-motor control. In: Gomi, T. (ed.) Evolutionary Robotics: From Intelligent Robots to Artificial Life, pp. 49–67. AAI Books, Ottawa (1997)

Franceschini, N., Ruffier, F., Serres, J.: A bio-inspired flying robot sheds light on insect piloting abilities. Current Biology 17(4), 329–335 (2007)

Franceschini, N., Ruffier, F., Serres, J.: Obstacle avoidance and speed control in insects and micro-aerial vehicles. Acta Futura 3(4), 15–34 (2009)

Franceschini, N., Ruffier, F., Serres, J.: Biomimetic Optic Flow Sensors and Autopilots for MAV Guidance. In: Encyclopedia of Aerospace Engineering, p. E309 (2010)

Frapard, B., Champetier, C., Kemble, S., Parkinson, B., Strandmoe, S., Lang, M.: Vision-based gnc design for the leda mission. In: Proceedings of the 3rd International ESA Conference on Spacecraft GNC, Noordwijk, The Netherlands, pp. 411–421 (1996)

Frapard, B., Polle, B., Flandin, G., Bernard, P., Vétel, C., Sembely, X., Mancuso, S.: Navigation for planetary approach and landing. In: Proceedings of the 5th International ESA Conference on Spacecraft GNC, Rome, Italy (2002)

Garratt, M., Chahl, J.: Vision-based terrain following for an unmanned rotorcraft. Journal of Field Robotics 25, 284–301 (2008)

Green, W., Oh, P., Barrows, G.: Flying insect inspired vision for autonomous aerial robot maneuvers in near-earth environments. In: IEEE International Conference on Robotics and Automation (ICRA), vol. 1, pp. 2347–2352 (2004)

Griffiths, S., Saunders, J., Curtis, A., Barber, B., McLain, T., Beard, R.: Maximizing miniature aerial vehicles. Robotics & Automation Magazine (IEEE) 13, 34–43 (2006)

Heisenberg, M., Wolf, R.: Vision in Drosophila. Springer, New York (1984)

Herisse, B., Hamel, T., Mahony, R., Russotto, F.X.: Landing a vtol unmanned aerial vehicle on a moving platform using optical flow. IEEE Transaction on Robotics 28(1), 77–89 (2012)

Hrabar, S., Sukhatme, G., Corke, P., Usher, K., Roberts, J.: Combined optic-flow and stereo-based navigation of urban canyons for a uav. Bio-inspired landing approaches and their potential use on extraterrestrial bodies. In: The Proceedings of the International Conference on Intelligent Robots and Systems (IROS), pp. 3309–3316 (2005)

Indiveri, G., Kramer, J., Kocj, C.: System implementations of analog vlsi velocity sensors. IEEE Micro 16(5), 40–49 (1996)

Izzo, D., de Croon, G.: Landing with time-to-contact and ventral optic flow estimates. Journal of Guidance, Control, and Dynamics 35(4), 1362–1367 (2011)

Izzo, D., Weiss, N., Seidl, T.: Constant-optic-flow lunar landing: Optimality and guidance. Journal of Guidance, Control, and Dynamics 34, 1383–1395 (2011)

Janschek, K., Tchernykh, V., Beck, M.: Performance analysis for visual planetary landing navigation using optical flow and dem matching. In: Proceedings of the AIAA Guidance, Navigation and Control Conference and Exhibit (2006)

Jean-Marius, T., Strandmoe, S.E.: Integrated vision and navigation for a planetary lander. Technical report, AEROSPATIAL, Espace et Defense, Les Mureaux-France. ESA, Estec (1998)

Kawaguchi, J., Hashimoto, T., Misu, T., Sawai, S.: An autonomous optical guidance and navigation around asteroids. Acta Astronautica 44(5), 267–280 (1999)

Kendoul, F., Nonami, K., Fantoni, I., Lozano, R.: An adaptive vision-based autopilot for mini flying machines guidance, navigation and control. Autonomous Robots 27, 165–188 (2009)

Kerhuel, L., Viollet, S., Franceschini, N.: The vodka sensor: A bioinspired hyperacute optical position sensing device. IEEE Sensors Journal 12(2), 315–324 (2012)

Kirschfeld, K., Franceschini, N.: Optische eigenschaften der ommatidien im komplexauge von Musca. Kybernetik 5, 47–52 (1968)

Koenderink, J., van Doorn, A.: Facts on optic flow. Biological Cybernetics 56, 247–254 (1987)

Landolt, O., Mitros, A.: Visual sensor with resolution enhancement by mechanical vibrations. Autonomous Robots 11(3), 233–239 (2001)

Mahony, R., Corke, P., Hamel, T.: A dynamic image-based visual servo control using centroid and optic flow features. Journal of Dynamic Systems, Measurement, and Control 130(1), 1–12 (2008)

Moeckel, R., Liu, S.C.: Motion detection circuits for a time-to-travel algorithm. In: IEEE International Symposium on Circuits and Systems (ISCAS), New orleans, LA, USA, pp. 3079–3082 (2007)

Mourikis, A.I., Trawny, N., Roumeliotis, S.I., Johnson, A.E., Ansar, A., Matthies, L.: Vision-aided inertial navigation for spacecraft entry, descent, and landing. IEEE Transactions on Robotics 25(2), 264–280 (2009)

Netter, T., Franceschini, N.: A robotic aircraft that follows terrain using a neuromorphic eye. In: IEEE/RSJ International Conference on Intelligent Robots and Systems, IROS 2002, vol. 1, pp. 129–134 (2002)

Orfanidis, S.J.: Introduction to signal processing. Prentice-Hall, Inc., Upper Saddle River (1995)

Parkes, S., Dunstan, M., Matthews, D., Martin, I., Silva, V.: Lidar-based gnc for planetary landing: Simulation with PANGU. In: Proceedings of the DASIA (Data Systems in Aerospace), pp. 18.1–18.12 (2003)

Parkes, S., Martin, I., Dunstan, M., Matthews, D.: Planet surface simulation with pangu. In: Proceedings of the 8th International Conference on Space Operations, SpaceOps (2004)

Parkes, S.M., Silva, V.: Gnc sensors for planetary landers: a review. In: The Proceedings of the DASIA (Data Systems in Aerospace), pp. 1–9 (2002)

Pichon, J., Blanes, C., Franceschini, N.: Visual guidance of a mobile robot equipped with a network of self-motion sensors. In: Mobile Robots IV, SPI, vol. 1195, pp. 44–53 (1989)

Roubieu, F., Expert, F., Boyron, M., Fuschlock, B., Viollet, S., Ruffier, F.: A novel 1-gram insect based device measuring visual motion along optical directions. In: Proceedings of the IEEE Sensors Conference, Limerick, Ireland, pp. 687–690 (2011)

Roubieu, F.L., Serres, J., Franceschini, N., Ruffier, F., Viollet, S.: A fully-autonomous hovercraft inspired by bees; wall-following and speed control in straight and tapered corridors. In: IEEE International Conference on Robotics and Biomimetics (ROBIO), Guangzhou, China (2012)

Roumeliotis, S., Johnson, A., Montgomery, J.: Augmenting inertial navigation with image-based motion estimation. In: Proceedings of the IEEE International Conference on Robotics and Automation (ICRA), vol. 4, pp. 4326–4333 (2002)

Ruffier, F.: Pilote Automatique Biomimetique Systeme générique inspiré du contrôle visuomoteur des insectes pour: le décollage, le suivi de terrain, la réaction au vent et l'atterrissage automatiques d'un micro-aeronef. Ph.D. thesis, INP Grenoble, France (2004)

Ruffier, F., Expert, F.: Visual motion sensing onboard a 50-g helicopter flying freely under complex VICON-lighting conditions. In: Proceedings of the International Conference on Complex Medical Engineering, Kobe, Japan, pp. 634–639 (2012)

Ruffier, F., Franceschini, N.: Octave, a bioinspired visuo-motor control system for the guidance of micro-air vehicles. In: Rodriguez-Vazquez, A., Abbott, D., Carmona, R. (eds.) Proceedings of the Conference on Bioengineered and Bioinspired Systems, SPIE, Maspalomas, Spain, Bellingham, USA, vol. 5119, pp. 1–12 (2003)

Ruffier, F., Franceschini, N.: Visually guided micro-aerial vehicle: automatic take off, terrain following, landing and wind reaction. In: Proceedings of the IEEE International Conference on Robotics and Automation (ICRA 2004), Coimbra, Portugal (2004)
Ruffier, F., Franceschini, N.: Optic flow regulation: the key to aircraft automatic guidance. Robotics and Autonomous Systems 50, 177–194 (2005)
Ruffier, F., Viollet, S., Amic, S., Franceschini, N.: Bio-inspired optical flow circuits for the visual guidance of micro-air vehicles. In: Proceedings of the IEEE International Symposium on Circuits and Systems Bio-inspired Landing Approaches and their Potential use on Extraterrestrial Bodies (ISCAS), Bangkok, Thailand, vol. 3, pp. 846–849 (2003)
Schilstra, C., Hateren, J.H.: Blowfly flight and optic flow. 1. Thorax kinematics and flight dynamics. J. Exp. Biol. 202(Pt. 11), 1481–1490 (1999)
Shang, Y., Palmer, P.: The dynamic motion estimation of a lunar lander. In: The Proceedings of the 21st ISSFD, Toulouse, France (2009)
Strandmoe, S., Jean-Marius, T., Trinh, S.: Toward a vision based autonomous planetary lander. In: AIAA, AIAA–99–4154 (1999)
Tammero, L.F., Dickinson, M.H.: The influence of visual landscape on the free flight behavior of the fruit fly drosophila melanogaster. Journal of Experimental Biology 205, 327–343 (2002)
Tchernykh, V., Beck, M., Janschek, K.: An embedded optical flow processor for visual navigation using optical correlator technology. In: Proceedings of the IEEE/RSJ International Conference on Intelligent Robots and Systems, Beijing, pp. 67–72 (2006)
Trawny, N., Mourikis, A.I., Roumeliotis, S.I., Johnson, A.E., Montgomery, J.: Vision-aided inertial navigation for pin-point landing using observations of mapped landmarks. Journal of Field Robotics 24, 357–378 (2007)
Valette, F., Ruffier, F., Viollet, S., Seidl, T.: Biomimetic optic flow sensing applied to a lunar landing scenario. In: Proceedings of the IEEE International Conference on Robotics and Automation (ICRA 2010), Anchorage, Alaska, pp. 2253–2260 (2010)
Viollet, S., Franceschini, N.: Biologically-inspired visual scanning sensor for stabilization and tracking. In: Proceedings of the IEEE/RSJ International Conference on Intelligent Robots and Systems, vol. 1, pp. 204–209 (1999a)
Viollet, S., Franceschini, N.: Visual servo system based on a biologically inspired scanning sensor. In: Sensor Fusion and Decentralized Control in Robotics II. SPIE, vol. 3839, pp. 144–155 (1999b)
Viollet, S., Franceschini, N.: Super-accurate visual control of an aerial minirobot. In: Autonomous Minirobots for Research and Edutainment, AMIRE, Padderborn, Germany, pp. 215–224. Heinz Nixdorf Institute (2001)
Wagner, H.: Flight performance and visual control of flight of the free-flying housefly (musca domestica l.) i. Organization of the flight motor. Philosophical Transactions of the Royal Society of London. Series B, Biological Sciences 312, 527–551 (1986)
Watanabe, Y., Fabiani, P., Le Besnerais, G.: Simultaneous visual target tracking and navigation in a gps-denied environment. In: Proceedings of the International Conference on Advanced Robotics (ICAR), Munich, Germany, pp. 1–6 (2009)
Watanabe, Y., Lesire, C., Piquereau, A., Fabiani, P., Sanfourche, M., Le Besnerais, G.: The ONERA ReSSAC unmanned autonomous helicopter: Visual air-to-ground target tracking in an urban environment. In: Proceedings of the American Helicopter Society 66th Annual Forum, Phoenix, AZ, USA (2010)

Chapter 10
Electric Power System Options for Robotic Miners

Simon D. Fraser

Graz University of Technology, Austria

10.1 Introduction

Exploration and exploitation of asteroids is a technological and economical challenge envisioned and pursued by public and private organizations across the world. Space exploration is driven by science, space exploitation is in the end driven by commercial - maybe strategic - considerations. While the goals of exploration and exploitation may thus be different, the technological challenges and approaches are often very similar.

One of the primary technological challenges in designing mission elements for surface operations on asteroids is power system design. A safe and in terms of available energy and output power sufficiently dimensioned power system is essential to fulfill the goals of each system and sub-system applied in such an endeavor. This is, of course, true for all different types of mission elements and includes stationary elements, such as a processing centre for extracting certain metals out of minerals, mobile systems, such as robotic miners and transportation systems, or portable systems, such as power tools.

The ambient conditions present on the surface of asteroids are a very challenging and sometimes even prohibitive environment for many of the proven and innovative power system elements currently being considered and developed for future power generation and/or energy storage applications in space exploration and exploitation missions.

Spacecraft power system design for asteroid surface exploitation applications, where terms such as *revenue* and *profitability* – hardly ever considered with the same relevance in previous and current space exploration programs of public space agencies – are suddenly in the primary focus of mission design, therefore requires a very careful evaluation of the possibilities and limitations of each individual power system element. The goal is to find a single power system technology, respectively a hybrid system consisting of more than one individual power system technology, providing the energy storage capacity and the electric output power profile required in a specific application in the best possible way.

In comparison to space exploration missions, where gravimetric and volumetric figures as well as operational safety (no maintenance is possible), performance degradation, and compatibility with operating conditions are of central importance, does power system design for asteroid exploitation also have to consider investment and profitability of the system over the designated operational lifetime.

The sum of all this provides a very challenging environment for power system engineering. As no underlying mission profile or detailed application specifications are available as of today, it is the purpose of this chapter to provide a high-level discussion about the most relevant power system options for mobile and robotic miners designed to collect rubble on the surface of asteroids.

10.2 Environmental Considerations

The term *asteroid* traditionally refers to small solar system bodies orbiting around the Sun. Starting from this very basic definition, the term 'asteroids' has evolved to only refer to minor planets, or planetoids, present in the inner solar system. In terms of resource exploitation of particular relevance is not the majority of asteroids present in the asteroid belt between Mars und Jupiter, but the so-called Near-Earth Objects (NEOs). As of June 2012, a cumulative figure of 8,971 NEOs have been discovered (NASA NEO Discovery Statistics 2012). This figure is expected to increase significantly with time as more and more NEOs are discovered and documented.

NEOs can be classified by their orbital parameters, most importantly by their peri- and aphelion distance. The *perihelion distance (q)* is the point in the orbit of an object where it is nearest to the Sun. The *aphelion distance (Q)* is the point in the orbit of an object where it is furthest to the Sun. The *semi-major axis (a)* is one half of the major axis of the elliptic orbit. The *orbital period (P)* is the time an object requires to complete one full orbit around the Sun.

In terms of orbital elements, NEOs are asteroids and comets with a perihelion distance of less than 1.3 astronomical units (1 AU equals 149.6×10^9 m); Near-Earth Comets (NECs) are further restricted to include only short-period comets with orbital periods of less than 200 years (NASA NEO Groups 2012). The vast majority of NEOs are asteroids, and are referred to as Near-Earth Asteroids (NEAs). NEAs are further divided into three different groups (Atens, Apollo, Amor) according to their perihelion and aphelion distances as well as their semi-major axes.

A brief categorization of NEOs in terms of groups and definitions based on orbital parameters extracted from the NASA NEO Discovery Statistics (NASA NEO Groups 2012) is given in Table 10.1.

What makes NEOs interesting is not only their vicinity to the Earth orbit, but also the fact that they contain many resources required in terrestrial industry as well as in future space exploration. Due to these resources, NEOs are not only in the focus of scientific investigations, but are also considered very relevant in terms of exploitation (Nelson et al. 1993; Lewis and Hutson 1993; Ross 2001; Gerlach 2005; Landis et al. 2009; Matloff and Wilga 2011; Sanchez and McInnes 2012).

Table 10.1 Categorization of Near-Earth Objects (NASA NEO Groups 2012)

Group	Description	Definition
NECs	Near-Earth Comets	q<1.3 AU P<200 years
NEAs	Near-Earth Asteroids	q<1.3 AU
Atiras	NEAs whose orbits are contained entirely within the orbit of the Earth	a<1.0 AU Q<0.983 AU
Atens	Earth-crossing NEAs with semi-major axes smaller than the Earth's	a<1.0 AU Q>0.983 AU
Apollos	Earth-crossing NEAs with semi-major axes larger than the Earth's	a>1.0 AU q<1.017 AU
Amors	Earth-approaching NEAs with orbits exterior to Earth's but interior to Mars'	a>1.0 AU 1.017<q<1.3 AU

The composition of NEOs can be inferred from spectral reflectivity studies. Based on this information, three categories of NEOs can been distinguished (Ross 2001):

- **C-Type:** water-bearing with very high contents of opaque, carbonaceous material ('C' stands for *carbonaceous*)
- **S-Type:** anhydrous rocky material, consisting of silicates, sulphides and metals ('S' stands for *stony*)
- **M-Type:** high radar reflectivity characteristics of metals ('M' stands for *metallic*)

Carbonaceous C-type NEOs are very interesting for life support systems and in-situ propellant production due to their water content and potential for hydrocarbon production. S- and M-type NEOs, on the other hand, are very interesting with respect to metallic iron-nickel alloys, ferrous sulphide minerals and olivine, for instance. Trace amounts of rare metals, such as Platinum group metals (PGM - ruthenium, rhodium, palladium, osmium, iridium, and platinum) as well as non-metals such as arsenic, selenium and germanium, for instance, can also be found (Ross 2001).

With these resources in mind, NEOs could play a very important role in the future of space exploration by providing metals for building space structures and by providing water, hydrogen and oxygen for life support systems as well as propellants for interplanetary spacecraft.

Sanchez and McInnez developed a resource map providing estimates about the mass of material resources in near Earth space as a function of energy investment. According to their results, a considerable mass of resources is available and can be exploited at relatively low levels of energy (Sanchez and McInnes 2012). This makes resource exploitation a both feasible and interesting scenario.

An overview of key facts to be considered in the evaluation of power system options for surface operation on NEOs is compiled in Table 10.2.

Table 10.2 NEO fact sheet

Property	**Comment**
Surface temperature	The surface temperature of the Moon is >100K and <400K; the surface temperature of NEOs is in the order of the Moon's, and can even exceed the lunar temperature range
Atmosphere	No atmosphere present, vacuum conditions (volatile materials available within comets can vaporize and stream out of the nucleus when comets approach the inner solar system; this stream of material leaving the nucleus may also carry dust away from the surface)
Surface gravity	Less than 1‰ of the Earth's gravity field of 9.81 m/s^2 (lunar gravity: 1.62 m/s^2)
Rotation rate	Depends on the specific NEO, but can be much faster or slower than the Earth's rotation rate

Many of these conditions are also faced in lunar surface exploration. Synergies in technological approaches are thus present.

NEOs vary greatly in size. Some are 10s of kilometers in diameter, most measure less than 1 kilometer in diameter. Most NEO's are irregularly-shaped and the surface of larger objects shows craters documenting the impact of smaller objects.

An image of asteroid 243 Ida and its moon Dactyl is shown in Fig. 10.1 (NASA Photojournal 1996). This image was taken on August 28, 1993, during a fly-by of the Galileo spacecraft. Ida is an irregularly-shaped asteroid having a length of approximately 52 km and is classified as an S-class 'stony' asteroid. Ida is a main-belt asteroid and its orbit therefore lies between Mars and Jupiter. Due to the orbital parameters, Ida is therefore not considered as primary candidate for resource exploitation.

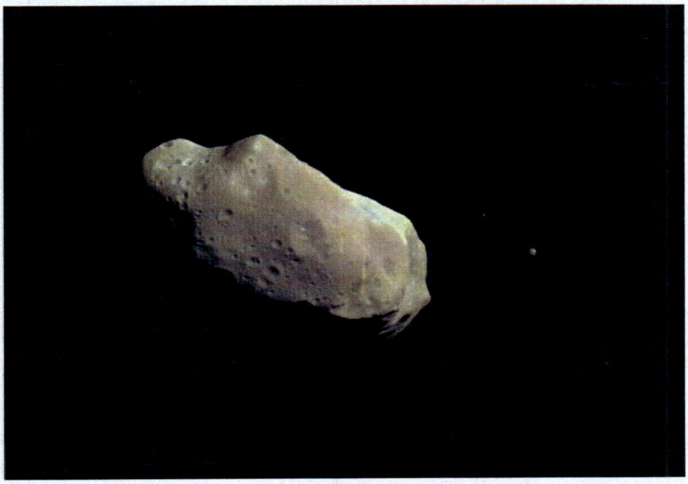

Fig. 10.1 View of the asteroid 243 Ida and moon Dactyl acquired by the Galileo spacecraft on August 28, 1993 (Image Credit: NASA/JPL)

10.3 An Overview of Relevant Power System Elements

A power system is a combination of different elements designed to provide an output power profile matching the required load profile in the best possible way. Leaving aspects such as power conditioning, management and distribution aside, the key question with a spacecraft – or in this case a robotic miner – is how to assure that the required output power levels can be provided when they are required; only then, the robotic miner will be able to perform as desired.

Within this section, relevant power system elements are briefly presented and discussed with respect to their possibilities and limitations in the investigated application in robotic miners operated on the surface of asteroids, more specifically NEOs.

An overview of major categories of power system technologies is presented in Table 10.3. Each power system technology is briefly presented and discussed in the following.

Table 10.3 An overview of power system technologies

Power System Technology	Examples
Solar energy utilization	o Photovoltaic power systems o Solar thermal power systems
Nuclear power system technologies	o Radioisotope generators o Fission reactors

Table 10.3 (*continued*)

(Electro)chemical power system technologies	o Supercapacitors o Primary batteries o Secondary batteries o Fuel cells o Flow batteries o Combustion power systems
Physical power system technologies	o Flywheel energy storage systems o Superconducting magnetic energy storage systems
In-situ propellant production	o Regenerative H_2/O_2 fuel cells

10.3.1 Solar Energy Utilization

The solar irradiance received by an object orbiting the Sun is determined by the distance between the object and the Sun, and whether or not an atmosphere and/or any solid particles or objects absorbing and/or reflecting sunlight are present before the sunlight reaches the surface of the object.

The irradiance available at a certain distance from the Sun (more precisely: from the center of the Sun) is proportional to the square of the distance between the object and the Sun. The average air mass zero (AM0) irradiance, I_{AM0}, is the mean irradiance received outside the Earth's atmosphere and has a value of 1,367.6 W/m² (NASA Earth Fact Sheet 2010). This value is, for instance, applied in predicting the output power of solar cells installed aboard satellites orbiting the Earth.

Considering that the solar spectrum received on the surface of an asteroid is not changed by absorption or reflection within an atmosphere, the irradiance received on the surface of an asteroid ($I_{asteroid}$) is therefore either higher or lower than I_{AM0}, depending on the distance from the Sun.

The mean irradiance received on an asteroid is shown in Fig. 10.2 as a function of the mean asteroid/Sun distance.

NEOs are therefore exposed to a level of irradiance similar to that available in Earth orbit; this would not be the case if main belt asteroids were considered, where the irradiance is significantly lower at mean asteroid/Sun distances in excess of two or even three AUs. The utilization of solar energy is therefore an interesting and relevant option considering that the vast majority of satellites operating in Earth orbit is powered by photovoltaics and a backup system of secondary batteries.

The lack of an atmosphere is also advantageous in terms of solar energy utilization on NEOs, particularly with stony and metallic asteroids. Volatile materials vaporizing and streaming out of the nucleus and possibly carrying dust particles have to be considered with surface operations on comets; this phenomenon would also have to be considered with solar energy utilization.

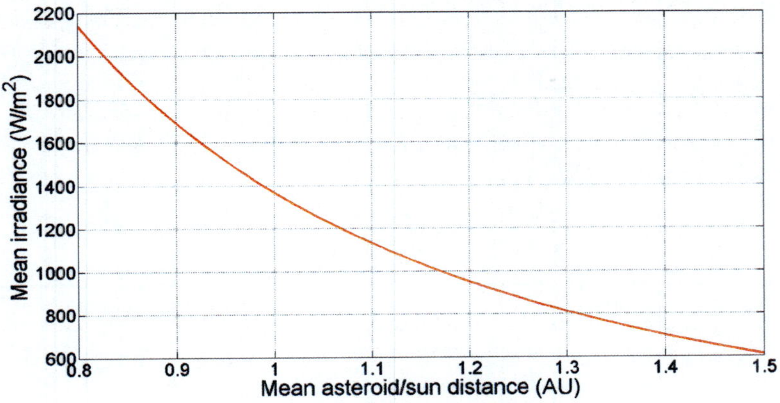

Fig. 10.2 Mean irradiance received on an asteroid as a function of the mean asteroid/Sun distance

Solar energy utilization is far less attractive with asteroids located in the main asteroid belt, with the majority of these objects being located between 2 and slightly over 3 AU.

Solar power generation is briefly discussed with photovoltaic and solar thermal power systems in the following.

Photovoltaic Power Systems

Photovoltaic panels utilize the photovoltaic effect to directly create a voltage and an electric current in semiconductor material exposed to light. Photovoltaic panels are static systems without any moving parts, which makes them perfectly suited for applications in space. Photovoltaic cells are either mounted onto rigid or flexible arrays to allow for optimal sun-pointing.

Rigid planar arrays consist of photovoltaic cells mounted onto a rigid substrate; individual panels can be hinged together. Panels providing an area-specific mass density of 1.3 kg/m^2 have been considered for GEO and LEO applications (Reddy 2003).

Flexible planar arrays, on the other hand, are lightweight thin film composite structures. The solar panel substrate consists of a graphite fibre reinforced plastic composite enveloped with sheets of Kapton. Such flexible solar panels provide an area-specific mass in the order of 0.65 kg/m^2 (Reddy 2003).

Concentrator and fold-out arrays have also been developed. The former applies lenses or mirrors to concentrate the incoming irradiance onto small strips or pieces of solar cells with the aim of providing an economical utilization of costly high-performance cells. The latter reduces the space during launch, transfer and landing and also protects the photovoltaic cells.

Different types of solar cells are available. An overview of relevant single crystalline and thin film cells is compiled in Table 10.4. The major disadvantage of solar cells applied in terrestrial applications, which is the low efficiency in utilizing solar energy in comparison to solar thermal systems, is significantly reduced with the availability of both efficient and expensive multi-junction solar cells offering solar-to-electric conversion efficiencies in excess of 25%, up to 30% at Beginning of Lifetime (BoL)

Table 10.4 Overview of solar cell types (Reddy 2003)

Type of solar cell	Efficiency at BoL
Single crystalline solar cells	
Si cells (silicon wafer with back surface reflector)	13.2 %
GaAs/Ge single junction cells	18.5%
$GaInP_2$/GaAs/Ge triple junction cells	27%
Thin film cells	
Amorphous silicon cells	10%
Copper Indium Gallium Diselenide (CIGS) cells	12.6%

Solar Thermal Power Systems

Solar thermal power systems do not directly convert light into electric energy, but rather convert light into thermal energy, and subsequently convert the thermal energy into electrical energy via a thermal-to-electric conversion system (Mason 1999). In solar thermal power system applications, solar energy is normally bundled onto a receiver element utilising a system of mirrors and lenses. Solar thermal power systems can also be considered with non-concentrating solar absorbers, as for instance studied by Badescu and co-workers, where low-temperature heat is utilised to drive a heat engine in a small Mars surface application (Badescu et al. 1999, 2000).

Depending on the level of concentration, very high temperatures can be achieved in solar thermal systems. The thermal energy thus available can be converted into mechanical energy by applying Stirling, Rankine or Brayton cycle heat engines. This basically resembles what happens in large-scale terrestrial power generation. In addition, a number of static heat-to-electric conversion technologies are either already applied or currently considered for future applications in space power generation. These static technologies include thermoelectric converters, space-proven in radioisotope thermoelectric generators for decades, as well as innovative alkali-metal thermal to electric converters (AMTECs), thermophotovoltaic converters and thermionic converters.

Potential advantages of solar thermal power systems are the high solar-to-electric conversion efficiency, small degradation effects and lower capital investment compared to photovoltaic systems. Disadvantages are that most solar thermal power systems have to be aligned with the Sun as they apply concentrating elements, and – particularly the dynamic thermal-to-electric power systems – also apply moving parts in their conversion process, with all the problems potentially associated with a non-static system operated in a remote space application.

In summary, photovoltaic systems have a long history in space applications, which solar thermal power systems do not yet have. Solar thermal power systems do have potential advantages in performance and/or costs; these potential advantages will have to outweigh the increased complexity of machinery and system in order to make solar thermal power systems an option for future power generation applications in space exploration and exploitation missions. Up to date, however, these potential advantages have not yet led to the application of solar thermal power systems, neither in space exploration applications, nor in the commercial satellite market.

10.3.2 Nuclear Power System Technologies

Historically, the use of nuclear fission and radioactive decay for spacecraft power generation has been extensively investigated. A considerable number of systems has actually been built and launched; radioisotope systems are also currently applied, e.g. in the Curiosity rover landed on Mars on August 6, 2012 (NASA Mars Science Laboratory Curiosity Rover 2012).

Power sources based on nuclear fission and radioactive decay are currently also actively developed for future applications in space exploration missions at a wide range of different output power levels.

Synergies are present with the application of RHUs, radioactive heating units (e.g. applied in the Mars rovers Sojourner and the twin Mars Exploration Rovers), where radioisotope heating units are applied without a thermal-to-electric conversion system as well as with next-generation nuclear space propulsion systems.

The application of nuclear fusion might be an option in the distant future, but is of no practical relevance for space power generation applications at this moment. Radioisotope generators and nuclear fission reactors are therefore briefly discussed as relevant options for near-future asteroid mining operations in the following.

Radioisotope Generators

Radioisotope Thermoelectric Generators (RTGs) use the heat released by the decay of a suitable radioactive material to generate electricity. The radioactive material is normally applied in the form of one or more pellets of plutonium-238. Other isotopes could also be used, but for space applications only those with low levels of penetrating radiation are relevant to avoid heavy shielding for protection of sensitive equipment on board. Noteably, the European Space Agency is pursuing americium-241 as a potential isotope for European space power systems (Summerer and Stephenson 2011).

The radioactive material releases high energy radiation, which in a first step has to be converted into thermal energy by enclosing the radioisotopes in an appropriate container. Shielding has to be applied to keep radiation in. This aspect is not only of relevance in manned missions, but also to prevent the interference with scientific instrumentation.

The radioactive decay of the isotope thus provides a certain level of heat, which in turn can either be directly converted into electrical energy utilizing a thermoelectric conversion system, or via an intermediate conversion into mechanical energy utilizing Stirling engines, for instance. The former provides a very robust system with no moving parts or engine wear, the latter provides higher heat-to-electric conversion efficiencies and thus the possibility to either increase the electric output power for a given quantity of radioisotope material, or to reduce the amount of radioisotope material required for a given output power, but again at the cost of increasing the overall system complexity.

RTGs have been successfully used over decades in terrestrial and space applications. Terrestrial systems have been primarily applied in remote power generation applications where a regular supply of fuel for a generator was impractical. RTGs have also been successfully applied in a wide range of space applications, particularly in missions to the outer solar system, where the irradiance is weak and photovoltaic systems are therefore limited. RTGs have also been applied in the supply of relatively low-power mobile systems such as surface rovers, where photovoltaic panels are impractical due to their size.

Radioactive decay is not constant with time, but has a certain half life which has to be considered in power system design. Output power levels can, however, be very accurately predicted as a function of isotope mission time, as presented in Table 10.5.

Table 10.5 Data of relevant isotopes (extracted from Blanke et al. 1960)

Isotope	Half-life (years)	Thermal output
Strontium-90	29.1 years (EPA 2012)	0.935 W/g
Polonium-210	138 days	143 W/g
Plutonium-238	89.6 years	0.545 W/g
Americium-241	433 years	0.115 W/g
Curium-242	162.5 days	120 W/g
Curium-244	18.4 years	2.73 W/g

Current RTGs provide system efficiencies of less than 10% at specific power rates of about 5 W/kg (with respect to the whole RTG and not with respect to the isotope mass). Applying advanced thermal-to-electric conversion technologies, such as Stirling engines, the efficiency of future radioisotope power systems is to be increased to about 30% and the specific power is to be increased to about 10

W/kg (Surampudi 2011). This will equally reduce the isotope mass and the heat rejection area of the power systems considerably, and also make radioisotope power systems more attractive for higher electrical output power levels.

Radioisotope power systems can thus be considered as primary candidates for robotic or unmaintained systems requiring a few hundred watts (electric) or less over long operating periods when batteries and fuel cells are not competitive due to mass and volume constraints anymore, and when photovoltaic panels are not practical due to size limitations.

Advanced radioisotope power systems applying more efficient heat-to-electric conversion technologies are therefore certainly an option to be considered for robotic systems operated on the surface of asteroids.

Nuclear Fission Reactors

RTGs are proven and reliable sources of electric power with output levels of a few hundred watts and below. At the moment it is not considered to develop radioisotope power systems providing 10s or evens 100s of kilowatts of electric output power. Nuclear fission reactors, on the other hand, have been considered for providing such output power levels in the past, and are currently also considered and actively developed for future applications in space exploration.

Small fission reactors have been developed since the late 1950s within the US nuclear rocket program, pursuing the development of a Nuclear Engine for Rocket Vehicle Applications (NERVA). At the time, the US space reactor program developed the SP-100 reactor. This system was a 2 MWt (thermal output power) reactor unit with a thermoelectric heat-to-electric conversion system delivering up to 100 kWe (electric output power) (Nuclear Association 2012).

The Soviet space program also developed small fission reactors with thermoelectric conversion systems, and later the Topaz nuclear fission power source with thermionic conversion system (World Nuclear Association 2012).

The "Vision for Space Exploration" presented by President Bush in 2004 has renewed the interest in human exploration of Moon and Mars. The vision proposed lunar surface missions in the 2020s, and Mars surface missions in the 2030s. Power requirements for human bases are expected to be in the range from 25 to 100 kWe during the early build-up phases, and up to 1 MW when the base becomes fully operational. Nuclear fission systems are considered as the most mass-efficient means of providing high power for surface missions (NASA fission surface power 2010).

In general, nuclear fuel provides an energy density far superior to any chemical fuel or electrochemical energy storage technology. The comparison of the energy content available in fusion and fission fuels compared to a chemical fuel (hydrogen/oxygen to be used for power generation in a heat engine or fuel cell) and an electrochemical energy storage technology presented in Table 10.6 clearly shows that the different options are separated by many magnitudes of order. Efficiency in terms of the heat-to-electric conversion rate is not considered in this table. The mass of the auxiliary systems (heat-to-electric conversion system, shielding, etc.), which will be drastically different for a fission system compared to a secondary battery system, for instance, is also not considered.

Table 10.6 Theoretical fuel energy density of fusion and fission fuels versus chemical fuel (hydrogen/oxygen) and electrochemical energy storage technology (Houts et al. 2001)

Fuel	Energy density
Fission	8.2×10^{13} J/kg
D-D fusion	8.8×10^{13} J/kg
D-T fusion	3.4×10^{14} J/kg
D-He3 fusion	3.5×10^{14} J/kg
H_2/O_2 (HHV)[1]	1.6×10^{7} J/kg
Secondary battery[2]	7.2×10^{5} J/kg

[1]Higher heating value (HHV) of hydrogen: 141.80 MJ/kg; 7.94 kg of oxygen required for stoichiometric combustion of 1 kg of hydrogen

[2]Energy density of a secondary battery assumed to be 200 Wh/kg; based on Li-Ion technology, the theoretical energy density of the Lithium anode alone would be 11 kWh/kg

NASA has recently implemented a project to develop a small nuclear fission system within the so-called Fission Surface Power (FSP) technology program. Current FSP plans apply uranium dioxide (UO_2) fuel pins, a sodium or a sodium-potassium heat transport system and Stirling or Brayton cycle heat engines for thermal-to-electric conversion (NASA FSP Handout 2010).

Fig. 10.3 Artist's concept of a Fission Surface Power System (Image Credit: NASA)

A single FSP unit could provide a net electrical output power of 40 kW with a design life of 8 years. An artist's concept of a Fission Surface Power System embedded in lunar regolith and with folded as well as unfolded radiators is shown in Fig. 10.3 (NASA Science News 2009).

10.3.3 (Electro)Chemical Power System Technologies

Supercapacitors

Supercapacitors, also referred to as electric or electrochemical double-layer capacitors, store electrical energy in the electrical double layer at the interface between carbon electrodes and an electrolyte. Single supercapacitors, although being operated at low voltages only, reach energy contents of a few Wh, and even a few kWh can be achieved by combining a number of individual supercapacitors in a module.

Supercapacitors can accept very high input and output power levels, with supercapacitor modules reaching some 100 kW currently under development. Lifetime and charge/discharge efficiency are also very favourable. Superacapacitors are thus an option for intermediate energy storage and peak power applications, but not for large energy storage applications where a high specific energy content in terms of Wh/kg is required.

Primary Batteries

Primary batteries are non-rechargeable electrochemical cells designed for single-use only. They provide superior specific energy and power as well as energy and power densities compared to rechargeable batteries, but usually at lower power rates. Due to the fact that they cannot be recharged, their practical use is limited to applications where a simple and robust power source with medium- or long-term energy storage capacity is required at output power levels far less than provided by a supercapacitor. Practical applications are thus normally also limited to small applications; primary batteries are hardly ever used in multi-kWh applications.

Secondary Batteries

Secondary batteries can be recharged; the specific energy of the electrochemical cell is thus not only available for a single discharge cycle, as in case of a primary battery, but can be accessed multiple times.

Secondary batteries are well-developed and can be considered as the standard electrical energy storage option in terrestrial and space applications. A number of different cell chemistries and battery module sizes are space-qualified. Large battery module development benefits from terrestrial efforts aiming at electric and hybrid electric vehicle applications. This increases the understanding of cell performance and degradation, and provides engineering solutions for thermal management of large battery modules.

The reader is referred t*o Moon: Prospective Energy and Material Resources* (Fraser 2012) for a discussion about performance figures of advanced secondary battery modules relative to the performance of fuel cell systems. Based on development goals for terrestrial electric vehicle development, the useable specific energy content of advanced secondary battery modules with large energy storage capacity (>10 kWh) was derived as 144 Wh/kg in the cited article, and the energy density was derived as 216 Wh/liter. These figures are derived from the nameplate capacity of terrestrial electric vehicle battery development goals after considering that the battery can only be discharged to a certain minimum depth of discharge to assure a sufficiently long lifetime, and furthermore considering a limited discharge efficiency due to resistive losses and internal overvoltages. Satellites in low earth orbits have, due to the number of cycles (5,000/year), a significantly lower depth of discharge (DoD) in the order of 20% compared to geo-stationary satellites with 1,350 cycles covering a life span of 15 years in orbit at a DoD of up to 80%.

Considering current and predicted performance figures, secondary batteries will remain one of the most important elements in the power system element portfolio for space applications, and will certainly also be applied in future resource extraction applications on asteroids. The only question is if they will be the primary energy storage system and applied in multi-kWh modules, or if they will only be applied for backup and in smaller applications (e.g. in portable systems such as power tools).

Fuel Cells

Fuel cells are electrochemical energy conversion systems that are capable of directly and continuously converting fuels and oxidants into electrical output power.

The major difference between fuel cells and batteries is that a fuel cell (i.e.: the electrode and electrolyte assembly) is not a closed system with educts and products permanently contained within a closed package like in batteries, but the system is split into an energy conversion subsystem where the electrochemical reactions take place, referred to as the *fuel cell stack*, and separate tanks for fuel and oxidant supply as well as reaction product(s). A significant number of peripherals is required as well for reactant, water and thermal management. The size of the fuel cell stack (= the active electrode area) essentially determines the maximum electrical output power and the corresponding conversion efficiency. The size of the fuel and oxidant tanks determines how much energy can be drawn from the system before it has to be refuelled or recharged; the latter can only be done if a regenerative fuel cell system is considered.

A small fuel cell stack in combination with large fuel and oxidant tanks provides a power system with a specific energy and an energy density far superior to secondary batteries, considering that the lower heating value (LHV) of hydrogen is 33.33 kWh/kg. When considering mass and volume of the oxidant required in the electrochemical reaction, as well as the tanks required for educt storage and the fuel cell stack itself, this difference is significantly reduced but still remains substantial as long as the ratio of stored energy versus output power is large enough (Fraser 2009, 2012).

In reality, energy densities of >1 kWh/kg have been demonstrated, e.g on the US space shuttle orbiter, versus the nameplate capacity of a secondary lithium battery being in the order of some 300 Wh/kg at the very beginning of life.

The major difference between fuel cells and heat engines (e.g. internal combustion engines or Stirling engines) is that the energy conversion process from the chemical energy of the fuel into electrical energy does not proceed via the intermediate step into thermal and mechanical energy but directly, and thus in a very efficient way and not limited by the Carnot cycle.

Basically, three different types of fuel cell systems can be distinguished:

- **Primary fuel cell systems:** These fuel cell systems have to be refuelled from an external source, as suggested by the word primary borrowed from primary (i.e. non-rechargable) batteries as opposed to secondary (i.e. rechargable) systems
- **Regenerative (secondary) fuel cell systems:** Secondary regenerative fuel cell systems can be recharged by operating the integrated electrolyser module, thus providing the possibility to store and at a later time release electrical energy
- **Unitised regenerative or reversible (secondary) fuel cell systems:** these systems are a special case of a secondary fuel cell system where fuel cell and electrolyser are not operated in two separate electrochemical cells, but in the same cell. However, the roundtrip efficiency is lower with significant impact on reactant storage and thermal subsystem

There are different types of fuel cells available, some of them operating at low (terrestrial near-ambient) temperatures, some of them operating at elevated temperatures of 700°C and above.

Fuel cell systems can also be operated with different types of fuels. Hydrogen is the primary fuel of choice; hydrocarbon fuels can be applied with external reforming systems and partial removal of carbon mono- or dioxide (in the case of low-temperature fuel cells) or directly with a synthesis gas (in the case of high-temperature fuel cells). Special low-temperature fuel cells are even designed to be directly operated with liquid hydrocarbon fuels, primarily methanol, but in some cases also ethanol.

Depending on the specific application considered, the best combination of fuel, oxidant, type of fuel cell and fuel cell system configuration can thus be chosen. The strong advantage of being able to, on the one hand, decouple energy storage and power generation, and on the other hand to utilize the advantageous possibilities of storing energy in a gaseous and/or liquid/cryogenic fuel make fuel cells a very important option for many different power generation applications in future space exploration and exploitation. The renewed interest in terrestrial fuel cell systems since the 1990s furthermore provides synergies in terms of material science, diagnostics and engineering know how that can now also be successfully translated into next-generation and space-qualified fuel cell systems.

REDOX Flow Batteries

REDOX (reduction/oxidation) flow batteries are reversible fuel cells in which all electro-active components are dissolved in the electrolyte. As with hydrogen/oxygen fuel cells, output power and energy capacity are completely decoupled and are determined by the size of the electrochemical conversion system (= the active electrode area) for power and the volume of the tanks for energy. Examples of REDOX flow batteries are, for instance, vanadium and polysulfide bromide REDOX flow batteries. The polysulfide bromide REDOX flow battery has been extensively discussed for terrestrial applications and is utilised with the Regenesys system (Grant 2002).

Major advantages of REDOX flow batteries are the aforementioned decoupling of power and energy, long cycle life, quick response times, and no need for equalising individual cells. Some types of REDOX flow batteries also offer easy state-of-charge determination, low maintenance and a high tolerance to overcharge/overdischarge. Disadvantages of REDOX flow batteries are the increased complexity compared to conventional secondary battery technology, and the rather low energy density.

Flow batteries are considered for large terrestrial energy storage application in the order of 1 kWh up to 10 MWh required e.g. for load leveling, intermediate energy storage from renewable sources, for discharging during periods of peak demand and as an uninterrupted supply.

Up to date, REDOX flow battery technology has not yet been applied in space applications, and the author does not know any effort aiming at the development of a space-qualified system. As of today, high-performance secondary lithium batteries and primary and/or regenerative fuel cell systems are more promising in terms of future space applications than REDOX flow batteries.

Hybrid Flow Batteries

Hybrid flow batteries are similar in design to REDOX flow batteries, but one or more electro-active component is deposited as a solid within the system, whereas all electro-active components are dissolved in the electrolyte in a REDOX flow battery.

Hybrid flow batteries therefore have one battery electrode and one fuel cell electrode. The term *hybrid* refers to this mix of fuel cell and battery characteristics. The quantity of the electro-active components dissolved in the electrolyte will thus be determined as a function of the electro-active components deposited on the battery electrode, thus the battery electrode size. Power and energy are therefore not fully decoupled in hybrid flow batteries. Examples of hybrid flow batteries include the zinc-bromine, zinc-cerium and lead-acid flow batteries.

Up to date, hybrid flow battery technology has not yet been applied in space applications, and the author does not know any effort aiming at the development of a space-qualified system. As of today, high-performance secondary lithium batteries and primary and/or regenerative fuel cell systems seem to be more promising in terms of future space applications than hybrid flow batteries.

Combustion Power Systems

Combustion is a process in which a fuel and an oxidant undergo exothermic chemical reactions which produce heat and convert the educts into one or more reaction products. The combustion reaction can occur with a flame or flameless; the former, for instance, occurs in a conventional combustion engine; the latter can be achieved by applying catalytic combustion technology.

The heat released by the combustion reaction can be converted into electrical energy by applying one of the many different heat-to-electric conversion technologies available. The range of heat-to-electric conversion technologies includes electric generators coupled to internal and external combustion engines. Gasoline engines, diesel engines, wankel engines, and gas turbines are typical examples of internal combustion engines. Steam turbines and Stirling engines are well-known external combustion engines.

Heat-to-electric conversion can, however, also be done without the intermediate step of first converting heat into mechanical energy of a shaft driving an electric generator. Thermoelectric generators, alkali-metal thermal to electric converters (AMTECs) (Tournier and El-Genk 1999; Schock et al. 2002), thermophotovoltaics (Coutts 1999) and thermionic conversion systems can directly convert heat supplied to the system into electrical energy (Fraser 2001).

Combustion power systems could utilize robust and proven heat-to-electric technologies in space applications. The availability of fuel cell systems for direct fuel-to-electric conversion, however, provides a space-proven and highly efficient energy conversion alternative avoiding a combustion process and rather going via an electrochemical conversion route. Advanced heat-to-electric energy conversion technologies for space applications are thus primarily developed for radioisotope systems and not for converting the heat of a combustion process into an electric output power.

Synergies with solar thermal power generation as well as radioisotope power systems will soon provide mature and space-qualified heat-to-electric conversion technologies. As of today, however, the application of these technologies in converting the heat of combustion of a fuel/oxidant combination into mechanical energy does not seem to be a preferred option in future space exploration and exploitation missions. This option may become more relevant in a scenario where in-situ resource utilization becomes an issue, for instance with water being extracted from a C-class asteroid. In this case, however, a combustion-based conversion process would again have to offer advantages over fuel cell technology. This will most likely not be possible in terms of conversion efficiency, but rather in terms of robustness and safety only, considering a static heat-to-electric conversion system.

10.3.4 Physical Power System Technologies

Chemical and electrochemical energy storage technologies are currently considered as the method of choice in many terrestrial applications, with electrochemical technologies being the preferred option in space applications. There are, however, also physical energy storage technologies considered and investigated for space applications. Two of these technologies are briefly discussed in the following.

Flywheel Energy Storage Systems
Flywheels can store energy as kinetic energy in the rotating mass of a rotor. The amount of energy stored increases with the square of the rotational speed; the rotation speed of the rotor is limited by the tensile strength of the material used. Rotors of advanced flywheel energy storage systems are designed to rotate at up to 100,000 revolutions per minute, thus offering the possibility of storing considerable amounts of energy in a mass- and volume-efficient way.

Flywheel energy storage systems have thus also been investigated as energy storage option for the International Space Station. Prototype systems have been developed which offered some 25-30 Wh/kg of the theoretical limit of 2700 Wh/kg energy storage capacity computed for systems with carbon nano fiber rotors (Lyons et al. 2010).

As of today, however, the development of flywheel energy storage systems for space applications is not funded anymore (Lyons et al. 2010).

Superconducting Magnetic Energy Storage Systems (SMES)
SMES store energy in the magnetic field of a coil made from special alloys. By cooling the conducting wires down to cryogenic conditions, the electrical resistance of the material is practically reduced to zero, making it possible to conduct very high currents without electrical losses, but at the cost of maintaining the system at cryogenic temperature.

SMES are very favorably in terms of energetic efficiency and provide very high input and output power levels. The energy storage capacity of current SMES is, however, very limited. This makes SMES only an option in niche applications at this time, despite having a comparably small temperature difference between cryogenic and ambient conditions present during nighttime conditions on NEOs, particularly when a relatively slow rotation rate is present.

10.3.5 In-Situ Propellant Production

The Earth launch mass is one of the decisive figures to be considered in the planning of every space mission. This is true for space explorations missions, and this is obviously to an even larger extent true for space exploitation missions, where the return on investment is not primarily considered in terms of scientific findings, but in a dominant economical dimension. Each option reducing the Earth launch mass and nevertheless fulfilling the mission profile without adding a large risk penalty is therefore welcome.

Two concepts have been established in mission planning for the exploration of Moon and Mars: *in-situ resource utilization* (ISRU) and *in-situ propellant production* (ISPP). ISRU and ISPP aim at reducing the total system mass that has to be transferred from the Earth onto the surface of Moon or Mars by utilizing resources that are readily available and easily accessible in-situ, thus directly at the landing site or in its vicinity. This could be surface regolith with Moon and Mars, and – in case of Mars – primarily the atmosphere containing more than 95% carbon dioxide (Zubrin and Wagner 1996; Hoffman and Kaplan 1997, Pipoli et al. 2002, Fraser 2009).

In case of C-type asteroids, an often very significant amount of water is available. By decomposing this water into hydrogen and oxygen, propellants could be produced directly on the surface of the asteroid. This would reduce the amount of propellants that have to be launched from the Earth, as the propellants for the return flight could be produced directly in situ on the surface of the asteroid; this, of course, at the cost of launching a water extraction, decomposition and propellant storage system.

Water could also be used as consumable in life support systems and, when considering power system options, could also be used as fuel and oxidant in power generation applications. Hydrogen and oxygen could thus be burned in internal combustion engines and gas turbines. Hydrogen and oxygen could also be produced and consumed in electrochemical cells, with the electrochemical cells producing hydrogen and oxygen being referred to as *electrolyzers*, and the electrochemical cells consuming hydrogen and oxygen being referred to as *fuel cells*. The former requires an input of electrical energy to decompose water, the latter generates electrical energy and produces water.

The utilization of water available in situ is thus an option for intermediate storage of electrical energy, producing hydrogen and oxygen with an electrolyzer when excess power is available and consuming hydrogen and oxygen in a fuel cell for power generation when, for instance, the photovoltaic panels do not provide a sufficient output power level.

10.4 Power System Options

The ambient conditions present on the surface of asteroids are, as briefly discussed above, very challenging or even prohibitive for many of the proven and innovative power system elements considered for energy storage and/or power generation applications in future space exploration and exploitation missions.

Spacecraft power system design for asteroid surface applications therefore requires a very careful evaluation of the possibilities and limitations of each individual power system element in order to find a single element - or a hybrid system of multiple power system elements - satisfying the energy storage and power generation requirements of a specific asteroid surface exploration and/or exploitation system in the best possible way; and this not only with respect to gravimetric and volumetric figures, but also with respect to operational safety, performance degradation, and compatibility with operating conditions.

The best power system configuration can obviously only be determined on the basis of a specific mission profile, and the power system specification inferred from this mission profile.

10.4.1 Mission Profile and Operation of Robotic Miners

Within this chapter, the term *miner* refers to a robotic system used to extract resources, particularly minerals, from the ground of an asteroid. This could be done by scraping or digging rubble off the surface with or without the assistance of electromagnets to collect magnetic materials. Miners are considered to be mobile systems capable of relocating themselves if the mineral resources in vicinity of the extraction mechanism have been extracted. Due to the low gravity present, maneuvering and excavating could not be done as easily (since gravity-assisted) as done on Earth, Moon and Mars, but would have to include some sort of a hold-down mechanism either installed with the mobile system or the asteroid surface itself (Garrick-Bethell and Carr 2007).

Miners are further considered to be operated in combination with a static installation referred to as the *processing facility* where the minerals are processed. This scenario could be considered as a kind of strip mining, where material is scraped off the surface and collected. Shaft mining, where material is extracted from subsurface resources utilising a system of shafts and cavities, would rather require a system of conveyor bands and not vehicle-like units.

The stationary processing facility could also include a *launch site* where extracted resources could be launched to the Earth (e.g. in a scenario where precious PGM metals are extracted), to the Moon (e.g. in a scenario where water is collected for life support systems aboard a lunar habitat) or to a space station orbiting the Earth.

The processing facility could furthermore also be connected to a stationary *power station*. This power station could provide electric energy to the processing facility; the power station could also be used to recharge or refuel mobile and portable systems such as robotic miners. Mobile and portable systems would thus not have to be self-sustainable in either carrying a sufficient supply of energy right from the start or generating sufficient energy during their operational phase. Mobile and portable systems could thus be built lighter and more robust at the cost of requiring frequent visits to the stationary structure for recharging/refuelling.

If these frequent visits were necessary to unload material to the stationary processing facility anyway, this would not necessarily be a burden or limiting factor in the operational performance of mobile and stationary systems.

10.4.2 Commented Summary of Power System Options

It is beyond the scope of this article to propose designs and specifications for robotic miners operating on the surface or in subsurface shafts of asteroids. The goal of this chapter is to provide a general overview of relevant power system technologies, and to provide a brief and comparative summary about relevant power system technologies in the light of the application in a mobile and robotic miner.

This comparative summary of power system technologies is provided in Table 10.7. Each power system element discussed in Sect. 10.3 of this chapter is briefly summarized with a short comment and a personal assessment of the author whether or not the respective power system technology will be relevant in terms of an onboard power supply of a mobile and robotic miner.

Table 10.7 Commented summary on power system elements

Power System Element	Summary
	Solar energy utilization
Photovoltaic power systems	Very relevant option for stationary power generation. Large surface area (>4 m^2/kWp even for 25% TJ cells operated on NEOs) is a limiting factor for mobile applications. Intermediate energy storage during nighttime required, depending on rotation rate of asteroid
Solar thermal power systems	Currently not likely to be considered
	Nuclear power system technologies
Radioisotope generators	Very relevant option for small mobile systems. State-of-the-art systems weigh 200 kg/kW, advanced systems having 100 kg/kW are planned. Major advantage: provides continuous output power over years
Fission reactors	Currently being developed for stationary power generation applications in the 10s and 100s of kW output power range; impractical for mobile applications due to shielding requirements and large radiator area
	(Electro)chemical power system technologies
Supercapacitors	Relevant for high power and low energy storage
Primary batteries	Emergency power and single-use equipment
Secondary batteries	Primary option for energy storage, particularly in mobile systems. Useable specific energy and energy density considerably smaller than with H_2/O_2 systems; approx. 150 Wh/kg for battery system versus >1000 Wh/kg for H_2/O_2 fuel cells (Fraser 2012)
Fuel cells	Relevant option providing energy storage capacity (e.g. regenerative fuel cell for intermediate storage of photovoltaic power system) and as primary (not rechargeable) fuel cell in mobile applications
Flow batteries	Currently not likely to be considered
Combustion power systems	Internal combustion engines not likely to be chosen due to availability of fuel cell systems; synergies with external combustion engines (e.g. Stirling engines); mainly for propulsion
	Physical power system technologies
Flywheel energy storage systems	Currently not likely to be considered
SMES	Currently not likely to be considered
	In-situ propellant production
Regenerative H_2/O_2 fuel cells	Perfect match for ISPP, as same fuel/oxidant can be used for power generator and rocket motor

Based on the power system requirements of a specific mobile mining unit, one or more of these individual power system technologies have to be appropriately dimensioned and implemented in order to provide the required output power profile with the level of safety and redundancy demanded by the operator.

10.5 Conclusions

Primary and secondary batteries, photovoltaics and radioisotope thermoelectric generators have been the power system elements of choice with mission elements designed for lunar and Mars surface operation in the past. These power system technologies are also relevant with respect to future applications in asteroid exploitation.

Fuel cells, already considered but not applied with the Apollo Lander Module, are an interesting option if power consumption and operational period are challenging, and if the comparably low specific energy of batteries makes the mass of the battery module a limiting factor. By including a fuel cell system operating on hydrogen and oxygen, the same propellants used with rocket motors and possibly also produced on the surface of a (C-class) asteroid using In-Situ Propellant Production technology could also be used for power generation purposes. This would provide interesting synergies and also reduce the mass penalty of including a dedicated fuel/oxidant processing and refueling facility required if ISPP is not considered in the return flight.

Nuclear fission reactors, having been considered and developed since the pioneering days of spaceflight in the 1960s, have already seen a strong increase of research funding in the recent years and are considered as central elements in current mission plans for human exploration of Moon and Mars. The presence of a stationary nuclear fission reactor continuously providing a certain level of output power adds a degree of freedom in the power system design for mobile and portable systems.

To summarize, power system engineers will be able to consider a greater variety of power system technologies for their specific applications in the future, than there was available in the past. Depending on average and peak electric output power requirements, mission profile and mission duration, a prudent combination of different power generation and energy storage technologies will provide power systems that are lightweight and compact, and at the same time safe and robust.

Power system technology will therefore certainly be able to comply to the requirements and challenges of the upcoming asteroid exploration and exploitation applications. As always in engineering, the key will be to investigate and optimize until the best possible solution is identified.

References

Badescu, V., Popescu, G., Feidt, M.: Model of optimized solar heat engine operating on Mars. Energy Convers. and Manage. 40, 1713–1721 (1999)

Badescu, V., Popescu, G., Feidt, M.: Design and optimisation of a combination solar collector–thermal engine operating on Mars. Renew. Energ. 21, 1–22 (2000)

Blanke, B.C., Birden, J.H., Jordan, K.C., Murphy, E.L.: Nuclear battery-thermocouple type summary report. United States Atomic Energy Commission Research and Development Report, US Government contract No. AT-33-1-GEN-53 (1960), http://www.osti.gov/bridge/servlets/purl/4807049-6bvOmJ/4807049.pdf

Coutts, T.J.: A review of progress in thermophotovoltaic generation of electricity. Renew. Sustain. Energy Rev. 3, 77–184 (1999)

EPA (2012), http://www.epa.gov/rpdweb00/radionuclides/strontium.html

Fraser, S.D.: Non-nuclear power system options for a mission to mars and derived terrestrial applications. Diploma thesis, Graz University of Technology (2001)

Fraser, S.D.: Fuel cell power system options for Mars surface mission elements. In: Badescu, V. (ed.) Mars: Prospective Energy and Material Resources, pp. 139–174. Springer, Heidelberg (2009)

Fraser, S.D.: Fuel cell power system options for lunar surface exploration applications. In: Badescu, V. (ed.) Moon: Prospective Energy and Material Resources, pp. 377–404. Springer, Heidelberg (2012)

Garrick-Bethell, I., Carr, C.E.: Working and walking on small asteroids with circumferential ropes. Acta Astronautica 61, 1130–1135 (2007)

Gerlach, C.L.: Profitably exploiting near-earth object resources. In: 2005 International Space Development Conference. National Space Society, Washington, DC (2005)

Grant, I.: TVA's Regenesys energy storage project. In: 2002 IEEE Power Engineering Society Summer Meeting. Tennessee Valley Authority, Chattanooga (2002)

Hoffman, S.J., Kaplan, D.I. (eds.): The reference mission of the NASA Mars exploration study team. NASA Special Publication 6107 (1997)

Houts, M., Van Dyke, M., Godfroy, T., Pedersen, K., Martin, J., Dickens, R., Salvail, P., Hrbud, I., Rodgers, S.L.: Options for development of space fission propulsion system. In: Space Technologies Applications International Forum Conference, Albuquerque, NM, United States (2001)

Landis, R.R., Abell, P.A., Korsmeyer, D.J., Jones, T.D., Adamo, D.R.: Piloted operations at a near-Earth object (NEO). Acta Astronaut. 65, 1689–1697 (2009)

Lewis, J.S., Hutson, M.L.: Asteroidal resource opportunities suggested by meteorite data. In: Lewis, J.S., Matthews, M.S., Guerrieri, M.L. (eds.) Resources of Near-Earth Space, pp. 523–542. University of Arizona Press (1993)

Lyons, V.J., Gonzalez, G.A., Houts, M.G., Iannello, C.J., Scott, J.H., Surampudi, S.: DRAFT Space Power and Energy Storage Road map (2010), http://www.nasa.gov/pdf/501328main_TA03-SpacePowerStorage-DRAFT-Nov2010-A.pdf

Matloff, G.L., Wilga, M.: NEOs as stepping stones to Mars and main-belt asteroids. Acta Astronautica 68, 599–602 (2011)

Mason, L.S.: A solar dynamic power option for space solar power. In: 34th Intersociety Energy Conversion Engineering Conference, Vancouver, British Columbia, Canada (1999)

Nelson, M.L., Britt, D.T., Lebofsky, L.A.: Review of asteroid compositions. In: Lewis, J.S., Matthews, M.S., Guerrieri, M.L. (eds.) Resources of Near-Earth Space, pp. 493–522. University of Arizona Press (1993)
NASA Earth Fact Sheet (2010), http://nssdc.gsfc.nasa.gov/planetary/factsheet/earthfact.html
NASA NEO Discovery Statistics (2012), http://neo.jpl.nasa.gov/stats/
NASA Fission surface power (2010), http://www.grc.nasa.gov/WWW/TECB/fsp.htm
NASA FSP Handout, Fission Surface Power System Technology for NASA Exploration Missions (2010), http://www.grc.nasa.gov/WWW/TECB/FSP_Handout.pdf
NASA Mars Science Laboratory Curiosity Rover (2012), http://mars.jpl.nasa.gov/msl/mission/rover/
NASA NEO Groups (2012), http://neo.jpl.nasa.gov/neo/groups.html
NASA Photojournal, PIA00069: Ida and Dactyl in Enhanced Color (1996), http://photojournal.jpl.nasa.gov/catalog/?IDNumber=PIA00069
NASA Science News (2009), http://science.nasa.gov/science-news/science-at-nasa/2009/15may_stirling/
Popoli, T., Besenhard, J.O., Schautz, M.: In situ production of fuel and oxidant for a small solid oxide fuel cell on Mars. In: Wilson, A. (ed.) Space Power, Proceedings of the Sixth European Conference (2002)
Reddy, M.R.: Space solar cells - tradeoff analysis. Sol. Energ. Mat. Sol. C 77, 175–208 (2003)
Ross, S.D.: Near-earth asteroid mining. Caltech Internal Report (2001), http://www.nss.org/settlement/asteroids/NearEarthAsteroidMining%28Ross2001%29.pdf
Sanchez, J.P., McInnes, C.R.: Assessment on the feasibility of future shepherding of asteroid resources. Acta Astronaut. 73, 49–66 (2012)
Sanchez, J.P., McInnes, C.R.: Synergistic approach of asteroid exploitation and planetary protection. Adv. Space Res. 49, 667–685 (2012)
Schock, A., Noravian, H., Or, C., Kumar, V.: Design, analyses, and fabrication procedure of Amtec cell, test assembly, and radioisotope power system for outer-planet missions. Acta Astronaut. 50, 471–510 (2002)
Summerer, L., Stephenson, K.: Nuclear power sources: a key enabling technology for planetary exploration. Proceedings of the Institution of Mechanical Engineers, Part G: Journal of Aerospace Engineering 225, 129–143 (2011)
Surampudi, S.: Overview of the space power conversion and energy storage technologies (2011), http://www.lpi.usra.edu/sbag/meetings/jan2011/presentations/day1/d1_1200_Surampudi.pdf
Tournier, J.-M., El-Genk, N.: Performance analysis of Pluto/Express, multitube AMTEC cells. Energ. Convers. Manage. 40, 139–173 (1999)
World Nuclear Association, Nuclear reactors for space (2012), http://www.world-nuclear.org/info/inf82.html
Zubrin, R., Wagner, R.: The case for Mars: the plan to settle the red planet and why we must. Touchstone, New York (1996)

Chapter 11
Rubble-Pile Near Earth Objects: Insights from Granular Physics

Karen E. Daniels

North Carolina State University, Raleigh, USA

11.1 Introduction

Most Near Earth Objects (NEOs) are composed of fractured rock, sometimes highly fractured and porous, and they have come to be known as rubble piles (Britt 2001; Fujiwara et al. 2006). The constituent particles, ranging from millimeters up to tens of meters, are weakly held together as an aggregate by a combination of both gravitational and van der Waals forces, which can be of comparable strength (Scheers et al. 2010). Future missions to these rubble NEOs, whether human or robotic, will need to operate in such a way that they can safely and successfully probe a fragile object. Of key importance is the ability to predict and control the circumstances under which the NEO material will remain intact or become unstable during activities such as digging, sample-collection, anchoring, or lift-off.

Rubble-like materials are members of a broader class of what have been named granular materials (Jaeger et al. 1996). These are commonly defined as a collection of macroscopic particles that interact through classical mechanics: force balance, collisions, friction, inelasticity, etc. Natural and industrial examples include sand/gravel, agricultural grains, and pharmaceutical powders; idealized particles such as glass spheres are also frequently considered as laboratory models. There are extensive laboratory experiments and computer simulations on these materials (both with and without gravity), all of which are leading towards an improved theoretical understanding of their dynamics. However, a comprehensive theory of granular materials remains elusive, and most techniques have a limited range of validity within a large parameter space. This chapter reviews the current state of knowledge of granular materials, and provides guidance about how to apply this knowledge to rubble-pile asteroids.

On any mission to collect samples from a NEO, it will be necessary to design both reliable techniques for anchoring to the surface, and methods for collecting and retrieving samples. A sense of the challenges is given by the following representative examples, which begin with the approach of the mission hardware. As a human or robotic explorer attempts to slow its approach to the surface of the NEO,

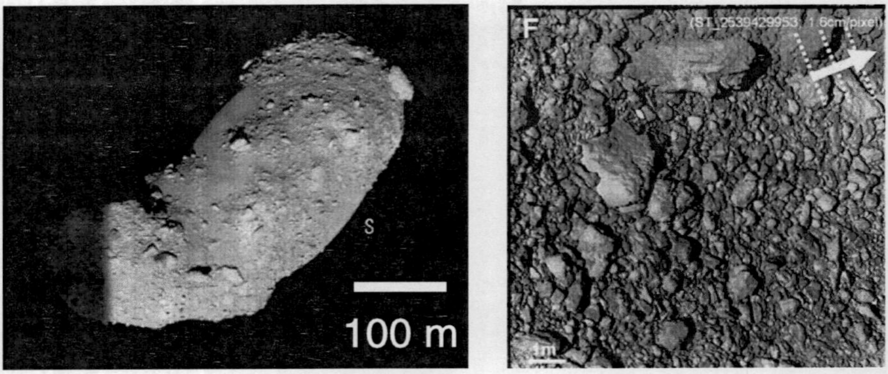

Fig. 11.1 Images of rubble pile asteroid 25143 Itokawa, observed by spacecraft Hayabusa, the first mission to return asteroid samples to Earth, illustrating size and granular texture at two different scales. Sources: Fujiwara et al. (2006) (left) and Miyamoto et al. (2007) (right).

any firing of thrusters will disrupt the surface material. Without strong gravitational forces to return the material to the NEO, the dislodged particles will remain in the vicinity of the explorer as a hazard. Even once the explorer is located at the surface, further movements can similarly dislodge particles. Furthermore, if the explorer needs to anchor to the fragile surface, any thrust into the surface will cause it to move in the opposite direction: away from the very surface that it is trying to probe. If the explorer requires digging (whether for the purposes of anchoring or sample-collection), this will necessarily disrupt the surface material (Miyamoto et al. 2007; Tsuchiyama et al. 2011), again creating a population of dislodged, hazardous particles in the vicinity.

The granular materials that comprise rubble NEOs have much in common with the geological granular materials studied in Earth's gravity, but with the key distinction that gravitational forces do not as easily return disturbed particles to the ground. For example, asteroid 25143 Itokawa (shown in Fig. 11.1), has a mass of 3.5×10^{10} kg (Fujiwara et al. 2006). At a distance of 150 m from the center of mass, the gravitational acceleration is only $10^{-5}g$; this means that any dislodged particles that travel faster than 17 cm/s will escape the asteroid; this speed is known as the escape speed.

No motion on or near the NEO surface will be simple, and although one mission has already been successful (Fujiwara et al. 2006), no established protocols yet exist for this new class of explorations. Ongoing studies of lunar and martian regolith (Rickman et al. 2012) and blast protocols (Metzger et al. 2009) will provide some guidance, but the much-lower gravitational forces are likely to limit their transfer of knowledge. In this chapter, we describe the current state of knowledge as it relates to the problem of digging and anchoring on rubble NEOs, drawing on studies performed both within Earth's gravity and in zero gravity. Finally, we close with some concrete recommendations for safe, successful exploration and mining operations on NEOs.

11.2 Resisting Applied Forces: Force Chains

Dry granular materials differ from conventional solids in that cohesion due to interatomic and intermolecular forces plays little role in determining the bulk and shear response of the aggregate. Instead, the material resists deformation via the friction and elasticity at each interparticle contact, together with the impossibility of two particles passing through each other. The interparticle contacts have long been known to create a highly heterogeneous distribution of forces, first visualized in the 1950s using the photoelastic (birefringent) properties of glass (Dantu 1957). Figure 11.2 shows a modern implementation of the same effect, using 2,500 disks cut from Vishay PhotoStress supported by a plastic floor and subjected to biaxial stresses: either isotropic compression or pure shear. In both cases, chain-like structures carry the majority of the force in the system, but this force network is far more isotropic in the compressed sample than in the sheared sample. The sheared sample is strongly anisotropic and shows a clear signature of the principal stress axis.

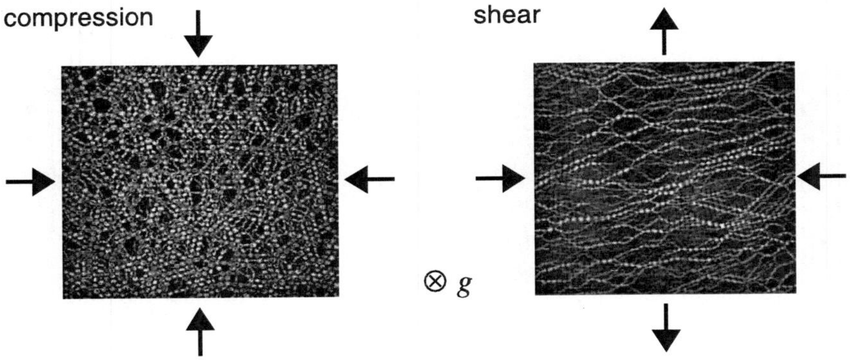

Fig. 11.2 Images of a quasi-two-dimensional granular material composed of disk-shaped, birefringent particles arranged in a singular, horizontal layer and subject to (left) isotropic compression and (right) pure shear in a uniform gravitational potential. By viewing the particles with polarized light, particles subject to larger stresses are brighter and show more optical fringes. Figure adapted from Majmudar and Behringer (2005)

The properties of this force chain network have been the subject of much recent research. Notably, the probability distribution function of individual forces within a static system can be highly non-Gaussian (Liu et al. 1955) and depends strongly on the type of loading (see Fig. 11.2 and Majmudar and Behringer (2005)). While circular/spherical particles have been the focus of much of the quantitative research on force distributions, the existence of such heterogeneous networks is a robust feature that holds even for non-circular particles (Geng et al. 2001). Furthermore, the weaker gravitational potential within asteroids will not destroy the effect: the images in Fig. 11.2 were obtained within a constant gravitational potential, resting upon a horizontal surface. Through the use of an air-table that gently floats flat particles (Lechenault and Daniels 2010; Puckett and Daniels 2013), it would be possible to investigate dynamics within a two-dimensional version of a zero-gravity environment. When the load on a granular material is increased, for instance by a digging or anchoring process, the force and contact network will necessarily rearrange. As long as all particles maintain torque and force balance, the granular material can remain intact. However, even for small changes this typically cannot be accomplished through linear (elastic) processes alone, and particle and force rearrangements are necessarily present (Cates et al. 1998). If the disturbance is sufficiently large, as in the case of a high-energy impact, then the material can become "fluidized" and transiently exist in a collisional, flowing state (Asphaug et al. 1998; Daniels et al. 2004; Pica Ciamarra et al. 2004; Clark et al. 2012). In each of these studies, whether laboratory or simulated, the propagation of stress or failure within the material is observed to have a branching structure reminiscent of the underlying force chain network.

The degree to which a granular material is in a gas-like (collision-dominated) vs. liquid-like (sustained contacts) state can be described by the non-dimensional inertial number $I = \dot{\gamma} d (P/\rho)^{-0.5}$. Here, $\dot{\gamma}$ is the imposed shear rate, d is the particle diameter, P is the confining pressure, and ρ is the mass-density of the particle material. This can be thought of as a competition between two time-scales: $\dot{\gamma}^{-1}$ is the macroscopic timescale over which shear-deformation takes place (creating gaps), and $d/(P/\rho)^{-0.5}$ is the microscopic timescale describing the time it takes for a particle to fall into a gap of size d (Campbell 2005; Forterre and Pouliquen 2008). In the context of rubble-pile asteroids, the downward pressure on particles arises from gravitational acceleration ($10^{-5}g$) and this microscopic timescale would instead be $(d/g)^{-0.5}$. For particles ranging in size from 10 cm to 10 m, this corresponds to timescales of 30 sec to 5 min. Therefore, under all but the slowest of shear rates, I will tend to take large values, corresponding to the gas-like, rapid-flow limit. Each of these situations (solid-like, liquid-like, gas-like), and the transitions between them, will be discussed in more detail below.

11.3 Solid-to-Fluid Transition

In order to explore a granular asteroid, it will be important to control its response: whether the surface material remains solid-like (good for anchoring, bad for digging), or transitions to a fluid-like state (good for digging, bad for anchoring). Several frameworks have been developed in recent years that address both the two regimes (solid-like, fluid-like) and the transition between them. Because flowing granular materials are always compressible, the term fluid-like will be used when both gas-like (collisional) and liquid-like (sustained contacts) materials are under discussion. Below, we review several key approaches and indicate which are applicable to NEOs in microgravity and which will require further research in order to be applicable.

One key descriptive variable underlying many of these studies is the extent to which the granular material fills the available space. In the physics literature, this is commonly referred to as the packing fraction or packing density ϕ, defined as the fraction of space occupied by particles. Within the geotechnical community, it is more common to report the inverse (1- ϕ), called the void fraction or porosity. Below, we will use the convention from the physics literature and refer to ϕ. Conceptually, ϕ is important because a loose packing (low ϕ) has fewer inter-particle contacts to stabilize particles (Wyart 2005), and more possibilities for particles to rearrange than does a dense packing (high ϕ) (Edwards and Oakeshott 1989, Song et al. 2008). In a survey of 26 asteroids, Baer et al. (2011) found that for asteroids with effective diameters below 300 km, bulk porosities in the range of 50% to 70% are quite common; this corresponds to $\phi = 0.3$ to 0.5. As will be discussed below, these values fall below what is needed to constrain the motion of particles relative to each other.

The connection between applied stress and changes in packing density dates back to a classic paper by Reynolds (1885). When a dense granular material is subjected to shear, the individual particles cannot pass through each other; instead, the material must be allowed to dilate in order to make space for the shear to occur. Conversely, a very loose granular material will collapse under shear, due to gravity. Within the geotechnical community, this relationship has been formalized as critical state soil mechanics. In particular, the point at which the internal friction between the particles can no longer support a load is referred to as the Mohr-Coulomb criterion (Schofield and Wroth 1968, Nedderman 1992). There exist both continuum models that account for either dilation or compaction as averaged properties of a granular flow (Campbell 2005), and experiments that quantify the transition at a particle scale by using x-ray radiography and tomography (Kabla and Senden 2009, Métayer et al. 2011) in Earth's gravity. In a microgravity environment, there will be no restoring force to drive compaction, so these behaviors may be significantly modified.

A sheared NEO will likely not self-compact, and instead the constituent particles will only be slowed by dissipative effects (inelastic collisions, friction). Further investigation using simulations (Baran and Kondic 2006) or experiments (Murdoch et al. 2013) are necessary to know what the dominant effects will be as gravity is reduced. Understanding the transition between unstable and stable configurations has been a particular focus in recent years, a process that has come to be known as jamming (Liu and Nagel 1998). Two recent reviews (Liu and Nagel 2010, Hecke 2010) provide details about this approach, which takes as its starting point studies of frictionless spheres in zero gravity, interacting only when in contact. The central idea is that the average number of constraints (contacts) per particle, the natural vibrational modes of the aggregate, and the rigidity of the system are all fundamental manifestations the same physics. This has allowed for the understanding of such features as how the shear modulus of a jammed material changes as a function of its packing density. As a spherical material is packed denser than its critical (lowest stable) packing fraction ϕ_c, its shear modulus G

increases according to the function $G \propto (\phi - \phi_c)^\alpha$. The scaling exponent α is set by the exponent in the functional form of the contact force law between two particles, and $\phi_c \approx 0.64$ for disordered, frictionless, spherical particles (Hern et al. 2003). Similar scaling relations are known for other quantities such as the bulk modulus, pressure, and the mean number of particle contacts (Liu and Nagel 2010, Hecke 2010).

More recently, the jamming approach has been extended to frictional (Somfai et al. 2007, Henkes et al. 2010) or elliptical (Mailman et al. 2009, Zeravcic et al. 2009) particles, for which ϕ_c takes a different value. Two important complications are the fact that friction always introduces history-dependence (the frictional force opposes the direction of motion), and non-spherical particles can have more complicated relationships between ϕ and the number of interparticle contacts. Therefore, the scaling relations between the bulk/shear modulus and the packing density are no longer the simple power laws known for the frictionless, spherical particle case. Promising new directions are studies that investigate when shear rigidity is either enhanced or degraded as a function of the preparation protocol or the choice of shear-direction (Bi et al. 2011; Dagois-Bohy et al. 2012). Ultimately, it may be possible to use the vibrational modes of the system as an empirical measure (Owens and Daniels 2013) of how near a granular system is to the point at which the material yields to external stresses and begins to flow.

11.4 Granular Gases

When sufficient energy is injected into a granular system that the individual particles have enough kinetic energy to overcome dissipative and attractive forces, they can exist in a gas-like state in which the behaviors are collision-dominated instead of contact-dominated. Such a situation can quite readily occur in and near NEOs during probing or capture, particularly as material becomes dislodged from the surface. For packing densities low enough that the material is collision-dominated, it is possible to understand the dynamics of the granular material using extensions of the kinetic theory of ordinary gases. A review article (Goldhirsch 2003) and book (Brilliantov and Pöschel 2004) detail the methods used to provide a quantitative theory, although this has been most commonly considered without gravitational attractions and between spherical particles.

One predicted feature is that granular gases undergo a clustering instability as they 'cool' through collisions in free space (Goldhirsch and Zanetti 1993), each pair of particles losing kinetic energy during each collision. The mechanism for this instability is that particles located within any small density fluctuation will (due to the increased density) suffer more collisions than neighboring regions. This causes that region to lose energy at a greater rate than the neighboring regions. Therefore, any initial density fluctuation will grow rather than decay; this effect has been observed in numerical simulations (Goldhirsch and Zanetti 1993).

Were it to also be present in real granular materials, it would predict a highly heterogeneous environment once any granular material was dislodged on a NEO.

To date, there have been relatively few experimental studies of granular gases in zero- or micro-gravity, due to the difficulty and expense of obtaining high-quality environments for long enough duration. While expensive, Earth-based experiments are possible through the use of parabolic flights, drop-towers, and sounding rockets. In studies of a dilute gas of steel spheres, Leconte et al. (2006) found significant differences with theoretical predictions from kinetic theory that remain unresolved. Murdoch et al. (2013) found that an important effect on Earth, the presence of convective-like flows in sheared granular materials, is strongly suppressed in microgravity, and Harth et al. (2013) made quantitative measurements of the translational and rotational motions of particles in a dilute rod-like granular gas. Intriguingly, unlike what was predicted for spherical particles, no clustering effect was observed. Each of these experiments illustrates the limitations of the current theoretical predictions. Further experiments probing the collisional dynamics of frictional, non-spherical particles in a microgravity environment would provide new information about what to expect once an asteroid is fluidized by impacts or digging.

11.5 Simulation Techniques

While continuum models of granular materials (Goldhirsch 2003, Jop et al. 2006, Kamrin and Koval 2012) only span narrow regions of parameter space, discrete numerical simulations are able to span the full range of solid-like to gas-like behavior. The classic Cundall and Strack (1979) method established the discrete (or distinct) element method (DEM), which models individual particles according to Newton's laws; these techniques are analogous to molecular dynamics simulations, but with macroscopic interactions instead of microscopic. Several reviews of up-to-date DEM methods have been published (Pöschel and Schwager 2005, Vermeer et al. 2001), and most typically allow for the modeling of spherical particle with interparticle friction, Hertzian (elastic) contact forces, and inelastic losses during collisions. Inherent in the design of such simulations is a choice between event-driven protocols, which are fast but only permit instantaneous binary collisions, and fully-resolved soft-sphere models which permit multiple sustained contacts. While event-driven simulations work well for dilute (low-ϕ) systems, densely-packed aggregates require the inclusion of sustained contacts. More recently, DEM simulations have been extended to permit a super-set of sphere-like shapes that includes ellipses (Zeravcic et al. 2009, Mailman et al. 2009) or conjoined circles/spheres (Gravish et al. 2012, Phillips et al. 2012), but angular shapes remain a challenge due to difficulties in modeling contact force laws for arbitrary curvatures. It is important to distinguish DEM techniques from smoothed-particle hydrodynamics (SPH) (Asphaug et al. 1998), in which a continuum fluid is divided into a set of discrete elements whose properties are smoothed over a predetermined lengthscale that need not correspond to a granular lengthscale.

In order to capture the dominant interactions, simulations of granular materials on asteroids require not only sustained, elastic contacts, but the inclusion of both long-range gravitational forces and cohesion. Numerical simulations on the stability of and collisions between asteroids have begun to include all of these effects (Scheeres et al. 2010, Sánchez and Scheeres 2011, Richardson et al. 2011, Ringl et al. 2012, Schwartz et al. 2012), but such advances are quite recent. To date, such studies have primarily focused on asteroid-scale dynamics, rather than localized disturbances appropriate to digging and anchoring applications.

To judge the need to include various of these terms in simulations, it is helpful to consider the magnitude of various key energies, particularly as compared with any kinetic energy imparted to either individual particles or the collective (e.g. centripetal effects). The magnitude of the gravitational energy for two identical particles of mass m is Gm^2/r, where r is the distance between them and G is the gravitational constant. If both particles have a mass density ρ and a radius R, then this energy is proportional to R^5; this predicts a significant particle-size dependence. In contrast, the kinetic energy grows proportional to R^3. The elastic energy stored between two spherical particles compressed by an amount δ is given by the Hertzian contact law (Johnson 1985), and is of magnitude $ER\delta$, where E is the Young's modulus of the particle material. While dry, convex granular materials can sustain shear or compressive loading, they cannot support a tensile load without the presence of cohesive forces between the particles. Because asteroids are observed to have a tensile strength that resists breakup under rotation (Trigo-Rodriguez and Blum 2009), some type of cohesive force can be inferred. While the energy associated with van der Waals adhesion may seem small – the magnitude is set by the Hamaker constant (10^{-19} to 10^{-20} J for typical materials) – Scheeres et al. (2010) showed that it can dominate over electrostatic or radiative-pressure effects for small objects. In addition, this paper provides scaling arguments that illustrate how to use experiments on Earth-based powders to simulate and infer behaviors in rubble-pile asteroids.

11.6 Particle Properties

The granular material studied in many of the above examples has most commonly been composed of hard (little deformation), frictional, dry (non-cohesive), and smooth (often circular/spherical) particles. In numerical simulations, the choice of smooth, frictionless, cohesionless particles facilitates computations, since the Hertzian contact forces are known analytic functions of the interparticle force in the limit of small deformation (Johnson 1985). Within this framework, governing properties such as particle size and elastic modulus are easily non-dimensionalized for comparison. Experiments have often chosen similarly-idealized particles, both to facilitate direct comparisons and to make experiments more repeatable. Nonetheless, recent numerical and laboratory studies have begun to investigate the roles of deformation (Saadatfar et al. 2012), polydispersity (distribution of particle sizes) (Tsoungui et al. 1998, Wackenhut et al. 2005, Muthuswamy and Tordesillas 2006, Voivret et al.

2009), particle shape (Azéma and Radjaï 2010, 2012; Torquato and Jiao 2012), and cohesion (Nowak et al. 2005, Richefeu et al. 2006, Herminghaus 2005, Strauch and Herminghaus 2012). In all cases, the results differ in nontrivial ways from experiments/simulations on monodisperse spherical particles.

Rubble pile NEOs are composed of hard, dry, polydisperse (see Fig. 11.1), non-spherical particles (Fujiwara et al. 2006, Michikami and Nakamura 2008, Yano et al. 2006), so comparison to the granular physics literature must proceed carefully. Regolith samples returned from the surface of 25143 Itokawa were also found to be highly polydisperse, with grain sizes ranging from micro- to millimeter scales (Tsuchiyama et al. 2011). It remains an open question how particle size and shape distributions (Herrmann et al. 2004, Pena et al. 2007, Saint-Cyr et al. 2012) influence the point at which a granular material transitions from solid-like to fluid-like behavior.

11.7 Interacting with a Rubble NEO

Safely interacting with rubble NEOs will require engineering within the properties and constraints described above. While self-gravitational and van der Waals effects have so far seen little investigation in Earth-based studies due to the inaccessibility of the regime, enough is known about general properties to point the way towards promising, and away from dangerous, modes of interaction.

As noted above, the escape speeds for rubble NEOs can be quite low: cm/s is typical. For comparison, the 17 cm/s speed calculated for asteroid 25143 Itokawa is about half a km/hr, two orders of magnitude lower than a child playing softball can pitch. Importantly, all motions at the surface of the asteroid must move at speeds below this limit if both the constituent particles and the probe hardware are to remain gravitationally bounded.

There is a second speed scale, set by the speed of sound within the asteroid body, the consequences of which are less clear. Since the rigidity of the fragile aggregate vanishes on approach to the jamming point (Liu and Nagel 2010, Hecke 2010), the speed of sound vanishes as well. Thus, for an asteroid only slightly above the jamming point (compressed slightly by gravitation), even very small perturbations will likely create supersonic shocks which propagate through the asteroid body (Gomez et al. 2012). As the material is compressed behind the propagating shock wave, the speed of sound in that region will increase. Meanwhile, collisions between the particles are dissipating energy; as discussed in the context of granular gases, it remains unclear to what degree clustering instabilities are to be expected for frictional, non-spherical particles. Whether this combination of effects leads to a stable (local damage) or unstable (propagating damage) situation is unknown. SPH simulations (Asphaug et al. 1998) of an 8 m sphere impacting an intact, non rubble-pile asteroid, suggest that damage propagates within (but not between) intact bodies. Laboratory or DEM simulations would be able to model the dynamics of an impinging probe and thereby estimate how large of an impact would be sustainable in the presence of real particle-contacts. There is likely a speed or energy threshold associated with the transition to local or global failure.

It has been commonly advocated that quickly firing a harpoon into a NEO would be an effective way to either tether it or produce ejecta to be collected. However, it remains to be determined whether fast or slow insertion is safer, more efficient, or more effective. For example, one goal might be to insert a probe or anchor into the asteroid while imparting as little momentum and breakup as possible. Therefore, it is illustrative to consider what distinguishes fast (firing) insertion from slow (digging). Fast firing would be defined as any speed at which the gravitational relaxation time $(d/g)^{-0.5}$ exceeds the reciprocal shear rate ($\dot{\gamma}^{-1}$). For a 10 cm particle at the surface of 25143 Itokawa, this timescale would be approximately 30 seconds. As such, all but the slowest of insertion speeds will be too fast for any relaxation of particle positions to take place during insertion, and will meet with significant resistance. From experiments, it is known that the resistance the projectile encounters is strongly dependent on how tightly-packed the material is (how difficult it is to rearrange) (Albert et al. 1999, Geng and Behringer 2005, Constantino et al. 2011). Therefore, one should additionally consider slow/digging techniques. Recent experiments (Wendell 2011) on flexible diggers have shown that that in Earth's gravity, thin/flexible objects can be more-efficient diggers than thick/stiff ones (see Fig. 11.3). For a given maximum digging force, probe flexibility was observed to allow for deeper digging. Similar optimization-style studies could examine what shape/flexibility/speed of a probe displaces the fewest particles in a zero gravity or self-gravitating environment.

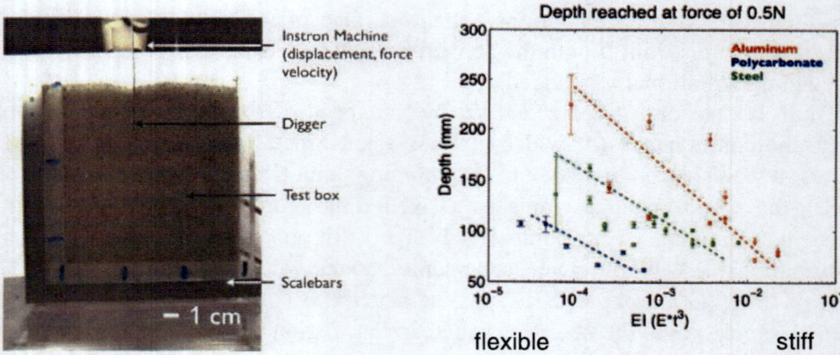

Fig. 11.3 In tabletop experiments in Earth's gravity (Wendell 2011), flexible diggers made of thin (0.1–1 mm) strips of polycarbonate, aluminum, steel were driven into a bed of dry 1–2 mm glass beads until a force threshold was reached. The maximum obtainable depth for 0.5 N of force is plotted at right; this depth was observed to decrease with the strip stiffness *EI*, where *E* is the material-dependent Young's modulus and *I* is the thickness-dependent bending moment.

Once a probe is inserted, it may be desirable to use it as an anchor, either to tether a spacecraft or to drag the asteroid through space. In this case, the engineering goal would be to design a system that can resist the largest force before failure of its hold. Because a dry, convex granular aggregate has no tensile rigidity of its own, it would be necessary to induce such behavior by a change in material properties. Two possibilities, already explored on Earth, are liquid films and magnetism. It is known that even very thin fluid layers can dramatically change granular collisions (Donahue et al. 2010), and this dissipative effect may be beneficial to anchoring. Furthermore, the presence of even microscopic fluid layers on the surface of particles would lead to liquid bridges which would further stabilize the granular material and provide tensile strength (Herminghaus 2005). A key difficulty in utilizing liquids would be to find a material that was not volatile on the timescales required for its use. A magnetically-induced rigidity would rely on using an electromagnet as the probe, so that the magnetic cohesion could be turned on and off as desired, by analogy to ferromagnetic shock-absorbers and valves. However, such a solution would require that the asteroid be composed of magnetically susceptible material (for instance, in M-type asteroids). Reliable identification of a particular asteroid's composition would of course be definitively settled by a sampling mission: a classic chicken and egg problem! A final concern is that any jamming of the surface particles on a NEO which aids the anchoring, would at the same time hinder digging efforts.

11.8 Conclusion

While this chapter has focused on rubble-pile asteroids that are thought to be completely granular in character, other asteroids such as Eros (Veverka 2000, Robinson et al. 2002, Li et al. 2004) are thought to have solid cores with a surface regolith composed of a granular material. Exploration of either type of asteroid will require working with granular materials in a microgravity environment. We have already sent one successful mission (Hayabusa) to asteroid 25143 Itokawa, and future missions for asteroid sample return are in various stages of planning. NASA is preparing for the 2016 launch of the Origins Spectral Interpretation Resource Identification Security Regolith Explorer (OSIRIS-REx) mission to study a carbonaceous (C-type) asteroid. As part of this mission, they plan to obtain and return a regolith sample. Similarly, JAXA is considering a second Hayabusa mission, this time to visit an C-type rocky asteroid and return samples from the surface. Finally, ESA is currently assessing the feasibility of the MarcoPolo-R mission, which also aims to visit a NEO. Eventually, it is likely that NASA (KISS 2012) or another organization will plan a mission to entirely capture a NEO; this would be altogether more challenging to successfully complete.

Because the population of NEOs is heterogeneous and largely uncharacterized, advance planning for particular asteroid or regolith materials will only partially predict the observed dynamics. The design (and ultimate success) of robust interaction protocols for digging, sample-collection, anchoring, and lift-off will depend on laboratory experiments and simulations which take place ahead of time.

Controlled experiments on granular materials in microgravity remain quite limited in both number and scope (Leconte et al. 2006, Chen et al. 2012, Murdoch et al. 2013, Harth et al. 2013). It is notable that two of these studies (Murdoch et al. 2013, Harth et al. 2013) were funded by European programs in which students apply to perform experiments using existing microgravity facilities. This means that while there is a population of young scientists ready to address many of the problems laid out in this chapter, there has been not yet been a sustained research program to address these issues. While drop-towers, catapults, sounding-rockets, and parabolic flights can all provide platforms for short experiments, developing tools for slow, gentle manipulation of space-based granular materials (both idealized and realistic) will required longer access to microgravity. The International Space Station is well-suited to host long-term experiments, and the burgeoning commercial spaceflight industry can provide intermediate timescales in a suborbital environment. Facilities and funds with which to perform future investigations will determine whether we are able to improve our ability to understand and predict the dynamics of rubble-pile asteroids, but the ground is laid for opportunities to make significant advances.

References

Albert, R., Pfeifer, M.A., Barabási, A., Schiffer, P.: Slow Drag in a Granular Medium. Physical Review Letters 82(1), 205–208 (1999)

Asphaug, F., Ostro, S.J., Hudson, R.S.: Disruption of kilometre-sized asteroids by energetic collisions. Nature 393, 437–440 (1998)

Azéma, E., Radjaï, F.: Stress-strain behavior and geometrical properties of packings of elongated particles. Physical Review E 81(5), 1–17 (2010)

Azéma, E., Radjaï, F.: Force chains and contact network topology in sheared packings of elongated particles. Physical Review E 85(3), 1–12 (2012)

Baer, J., Chesley, S.R., Matson, R.D.: Astrometric Masses of 26 Asteroids and Observations on Asteroid Porosity. The Astronomical Journal 141(5), 143 (2011)

Baran, O., Kondic, L.: On velocity profiles and stresses in sheared and vibrated granular systems under variable gravity. Physics of Fluids 18(12), 121509 (2006)

Bi, D., Zhang, J., Chakraborty, B., Behringer, R.P.: Jamming by shear. Nature 480(7377), 355–358 (2011)

Brilliantov, N.V., Pöschel, T.: Kinetic Theory of Granular Gases. Oxford University Press, Oxford (2004)

Britt, D.: Modeling the Structure of High Porosity Asteroids. Icarus 152(1), 134–139 (2001)

Campbell, C.S.: Stress-controlled Elastic Granular Shear Flows. Journal of Fluid Mechanics 539, 273–297 (2005)

Cates, M.E., Wittmer, J.P., Bouchaud, J.P., Claudin, P.: Jamming, Force Chains, and Fragile Matter. Physical Review Letters 81, 1841–1844 (1998)

Chen, Y.-P., Evesque, P., Hou, M.-Y.: Breakdown of Energy Equipartition in Vibro-Fluidized Granular Media in Micro-Gravity. Chinese Physics Letters 29(7), 074501 (2012)

Clark, A., Kondic, L., Behringer, R.: Particle Scale Dynamics in Granular Impact. Physical Review Letters 109(23), 238302 (2012)

Costantino, D., Bartell, J., Scheidler, K., Schiffer, P.: Low-velocity granular drag in reduced gravity. Physical Review E 83(1), 2009–2012 (2011)

Cundall, P.A., Strack, O.D.L.: Discrete Numerical-model for Granular Assemblies. Geotechnique 29(1), 47–65 (1979)

Dagois-Bohy, S., Tighe, B., Simon, J., Henkes, S., van Hecke, M.: Soft-Sphere Packings at Finite Pressure but Unstable to Shear. Physical Review Letters 109(9), 09570 (2012)

Daniels, K.E., Coppock, J.E., Behringer, R.P.: Dynamics of meteor impacts. Chaos 14, S4 (2004)

Dantu, P.: Contribution l'étude méchanique et géométrique des milieux pulvérulents. In: Proceedings of the Fourth International Conference on Soil Mechanics and Foundation Engineering, London, pp. 144–148 (1957)

Donahue, C., Hrenya, C., Davis, R.: Stokes's Cradle: Newton's Cradle with Liquid Coating. Physical Review Letters 105(3), 034501 (2010)

Edwards, S.F., Oakeshott, R.B.S.: Theory of Powders. Physica A 157, 1080–1090 (1989)

Forterre, Y., Pouliquen, O.: Flows of Dense Granular Media. Annual Review of Fluid Mechanics 40, 1–24 (2008)

Fujiwara, A., Kawaguchi, J., Yeomans, D.K., Abe, M., Mukai, T., Okada, T., Saito, J., Yano, H., Yoshikawa, M., Scheeres, D.J., Barnouin-Jha, O., Cheng, A.F., Demura, H., Gaskell, R.W., Hirata, N., Ikeda, H., Kominato, T., Miyamoto, H., Nakamura, A.M., Nakamura, R., Sasaki, S., Uesugi, K.: The Rubble-Pile Asteroid Itokawa as Observed by Hayabusa. Science 312(5778), 1330–1334 (2006)

Geng, J.F., Howell, D., Longhi, E., Behringer, R.P., Reydellet, G., Vanel, L., Clement, E., Luding, S.: Footprints in Sand: The Response of a Granular Material to Local Perturbations. Physical Review Letters 8703, 35506 (2001)

Geng, J., Behringer, R.: Slow drag in two-dimensional granular media. Physical Review E 71(1), 011302 (2005)

Goldhirsch, I., Zanetti, G.: Clustering instability in dissipative gases. Physical Review Letters 70(11), 1619–1622 (1993)

Goldhirsch, I.: Rapid granular flows. Annual Review of Fluid Mechanics 35(1), 267–293 (2003)

Gómez, L., Turner, A., van Hecke, M., Vitelli, V.: Shocks near Jamming. Physical Review Letters 108(5), 058001 (2012)

Gravish, N., Franklin, S., Hu, D., Goldman, D.: Entangled Granular Media. Physical Review Letters 108(20), 208001 (2012)

Harth, K., Kornek, U., Trittel, T., Strachauer, U., Höme, S., Will, K., Stannarius, R.: Granular gases of rod-shaped grains in microgravity. Physical Review Letters 110(14), 144102 (2013)

Henkes, S., Shundyak, K., van Saarloos, W., van Hecke, M.: Local contact numbers in two-dimensional packings of frictional disks. Soft Matter 6(13), 2935–2938 (2010)

Herminghaus, S.: Dynamics of wet granular matter. Advances in Physics 54(3), 221–261 (2005)

Herrmann, H.J., Astrøm, J.A., Mahmoodi Baram, R.: Rotations in shear bands and polydisperse packings. Physica A: Statistical Mechanics and its Applications 344(3-4), 516–522 (2004)

Jaeger, H.M., Nagel, S.R., Behringer, R.P.: Granular Solids, Liquids, and Gases. Reviews of Modern Physics 68, 1259–1273 (1996)

Johnson, K.L.: Contact Mechanics. Cambridge University Press, Cambridge (1985)

Jop, P., Forterre, Y., Pouliquen, O.: A constitutive law for dense granular flows. Nature 441(7094), 727–730 (2006)

Kabla, A.J., Senden, T.J.: Dilatancy in Slow Granular Flows. Physical Review Letters 102(22), 228301 (2009)
Kamrin, K., Koval, G.: Nonlocal Constitutive Relation for Steady Granular Flow. Physical Review Letters 108(17), 178301 (2012)
KISS, Asteroid Retrieval Feasibility Study. Technical Report, Keck Institute for Space Studies, California Institute of Technology, Jet Propulsion Laboratory (April 2012)
Lechenault, F., Daniels, K.E.: Equilibration of granular subsystems. Soft Matter 6(13), 3074 (2010)
Leconte, M., Garrabos, Y., Falcon, E., Lecoutre-Chabot, C., Palencia, F., Evesque, P., Beysens, D.: Microgravity experiments on vibrated granular gases in a dilute regime: non-classical statistics. Journal of Statistical Mechanics: Theory and Experiment 2006(07), P07012 (2006)
Li, J., A'Hearn, M.F., McFadden, L.A.: Photometric analysis of Eros from NEAR data. Icarus 172(2), 415–431 (2004)
Liu, A.J., Nagel, S.R.: Nonlinear dynamics: Jamming is not just cool any more. Nature 396, 21–22 (1998)
Liu, A.J., Nagel, S.R.: The Jamming Transition and the Marginally Jammed Solid. Annual Review of Condensed Matter Physics 1(1), 347–369 (2010)
Liu, C.H., Nagel, S.R., Schecter, D.A., Coppersmith, S.N., Majumdar, S., Narayan, O., Witten, T.A.: Force Fluctuations in Bead Packs. Science 269(5223), 513–515 (1995)
Mailman, M., Schreck, C.F., O'Hern, C.S., Chakraborty, B.: Jamming in Systems Composed of Frictionless Ellipse-Shaped Particles. Physical Review Letters 102(25), 255501 (2009)
Majmudar, T.S., Behringer, R.P.: Contact force measurements and stress-induced anisotropy in granular materials. Nature 435(7045), 1079–1082 (2005)
Métayer, J.-F., Suntrup III, D.J., Radin, C., Swinney, H.L., Schröter, M.: Shearing of frictional sphere packings. Europhysics Letters 93(6), 64003 (2011)
Metzger, P.T., Immer, C.D., Donahue, C.M., Vu, B.T., Latta, R.C., Deyo-Svendsen, M.: Jet-Induced Cratering of a Granular Surface with Application to Lunar Spaceports. Journal of Aerospace Engineering 22(1), 24–32 (2009)
Michikami, T., Nakamura, A.M.: Size-frequency statistics of boulders on global surface of asteroid 25143 Itokawa. Earth Planets Space 60, 13–20 (2008)
Miyamoto, H., Yano, H., Scheeres, D.J., Abe, S., Barnouin-Jha, O., Cheng, A.F., Demura, H., Gaskell, R.W., Hirata, N., Ishiguro, M., Michikami, T., Nakamura, A.M., Nakamura, R., Saito, J., Sasaki, S.: Regolith migration and sorting on asteroid Itokawa. Science 316(5827), 1011–1014 (2007)
Murdoch, N., Rozitis, B., Nordstrom, K., Green, S.F., Michel, P., De Lophem, T., Losert, W.: Granular Convection in Microgravity. Physical Review Letters 110(1), 018307 (2013)
Muthuswamy, M., Tordesillas, A.: How do interparticle contact friction, packing density and degree of polydispersity affect force propagation in particulate assemblies? Journal of Statistical Mechanics: Theory and Experiment 9 (2006)
Nedderman, R.M.: Statics and Kinematics of Granular Materials. Cambridge University Press, Cambridge (1992)
Nowak, S., Samadani, A., Kudrolli, A.: Maximum angle of stability of a wet granular pile. Nature Physics 1(1), 50–52 (2005)
Owens, E.T., Daniels, K.E.: Acoustic measurement of a granular density of modes. Soft Matter 9(4), 1214–1219 (2013)
O'Hern, C., Silbert, L., Liu, A., Nagel, S.: Jamming at zero temperature and zero applied stress: The epitome of disorder. Physical Review E 68(1), 011306 (2003)
Peña, A.A., García-Rojo, R., Herrmann, H.J.: Influence of particle shape on sheared dense granular media. Granular Matter 9(3-4), 279–291 (2007)

Phillips, C., Anderson, J., Huber, G., Glotzer, S.: Optimal Filling of Shapes. Physical Review Letters 108(19), 198304 (2012)
Pica Ciamarra, M.P., Lara, A.H., Lee, A.T., Goldman, D.I., Vishik, I., Swinney, H.L.: Dynamics of Drag and Force Distributions for Projectile Impact in a Granular Medium. Physical Review Letters 92(19), 194301 (2004)
Pöschel, T., Schwager, T.: Computational Granular Dynamics: Models and Algorithms. Springer, New York (2005)
Puckett, J.G., Daniels, K.E.: Equilibrating Temperature-like Variables in Jammed Granular Subsystems. Physical Review Letters 110(5), 058001 (2013)
Reynolds, O.: On the dilatancy of media composed of rigid particles in contact. Philosophical Magazine 20, 469 (1885)
Richardson, D.C., Walsh, K.J., Murdoch, N., Michel, P.: Numerical simulations of granular dynamics: I. Hard-sphere discrete element method and tests. Icarus 212(1), 427–437 (2011)
Richefeu, V., El Youssoufi, M., Radjaï, F.: Shear strength properties of wet granular materials. Physical Review E 73(5), 051304 (2006)
Rickman, D., Immer, C., Metzger, P., Dixon, E., Pendleton, M., Edmunson, J.: Particle Shape in Simulants of the Lunar Regolith. Journal of Sedimentary Research 82(11), 823–832 (2012)
Ringl, C., Bringa, E.M., Bertoldi, D.S., Urbassek, H.M.: Collisions of Porous Clusters: a Granular-Mechanics Study of Compaction and Fragmentation. The Astrophysical Journal 752(2), 151 (2012)
Robinson, M.S., Thomas, P.C., Veverka, J., Murchie, S.L., Wilcox, B.B.: The geology of 433 Eros. Meteoritics & Planetary Science 37(12), 1651–1684 (2002)
Saadatfar, M., Sheppard, A.P., Senden, T.J., Kabla, A.J.: Mapping forces in a 3D elastic assembly of grains. Journal of the Mechanics and Physics of Solids 60(1), 55–66 (2012)
Saint-Cyr, B., Szarf, K., Voivret, C., Azéma, E., Richefeu, V., Delenne, J.-Y., Combe, G., Nouguier-Lehon, C., Villard, P., Sornay, P., Chaze, M., Radjaï, F.: Particle shape dependence in 2D granular media. Europhysics Letters 98(4), 44008 (2012)
Sánchez, P., Scheeres, D.J.: Simulating Asteroid Rubble Piles With a Self-Gravitating Soft-Sphere Distinct Element Method Model. The Astrophysical Journal 727(2), 120 (2011)
Scheeres, D.J., Hartzell, C.M., Sánchez, P., Swift, M.: Scaling forces to asteroid surfaces: The role of cohesion. Icarus 210(2), 968–984 (2010)
Schofield, A., Wroth, P.: Critical State Soil Mechanics. McGraw-Hill, New York (1968)
Schwartz, S.R., Richardson, D.C., Michel, P.: An implementation of the soft-sphere discrete element method in a high-performance parallel gravity tree-code. Granular Matter 14(3), 363–380 (2012)
Somfai, E., Van Hecke, M., Ellenbroek, W.G., Shundyak, K., Van Saarloos, W.: Critical and Noncritical Jamming of Frictional Grains. Physical Review E 75(2), 20301 (2007)
Song, C., Wang, P., Makse, H.A.: A phase diagram for jammed matter. Nature 453(7195), 629–632 (2008)
Strauch, S., Herminghaus, S.: Wet granular matter: a truly complex fluid. Soft Matter 8, 8271–8280 (2012)
Torquato, S., Jiao, Y.: Organizing principles for dense packings of non-spherical hard particles: Not all shapes are created equal. Physical Review E 86(1), 011102 (2012)
Trigo-Rodriguez, J.M., Blum, J.: Tensile strength as an indicator of the degree of primitiveness of undifferentiated bodies. Planetary and Space Science 57(2), 243–249 (2009)
Tsoungui, O., Vallet, D., Charmet, J.-C., Roux, S.: Partial pressures supported by granulometric classes in polydisperse granular media. Physical Review E 57(4), 4458–4465 (1998)

Tsuchiyama, A., Uesugi, M., Matsushima, T., Michikami, T., Kadono, T., Nakamura, T., Uesugi, K., Nakano, T., Sandford, S.A., Noguchi, R., Matsumoto, T., Matsuno, J., Nagano, T., Imai, Y., Takeuchi, A., Suzuki, Y., Ogami, T., Katagiri, J., Ebihara, M., Ireland, T.R., Kitajima, F., Nagao, K., Naraoka, H., Noguchi, T., Okazaki, R., Yurimoto, H., Zolensky, M.E., Mukai, T., Abe, M., Yada, T., Fujimura, A., Yoshikawa, M., Kawaguchi, J.: Three-dimensional structure of Hayabusa samples: origin and evolution of Itokawa regolith. Science 333(6046), 1125–1128 (2011)

van Hecke, M.: Jamming of soft particles: geometry, mechanics, scaling and isostaticity. Journal of Physics. Condensed Matter: An Institute of Physics Journal 22(3), 33101 (2010)

Vermeer, P.A., Diebels, S., Ehlers, W., Herrmann, H.J., Luding, S., Ramm, E.: Continuous and Discontinuous Modelling of Cohesive-Frictional Materials. Springer, Berlin (2001)

Veverka, J.: NEAR at Eros: Imaging and Spectral Results. Science 289(5487), 2088–2097 (2000)

Voivret, C., Radjaï, F., Delenne, J.-Y., El Youssoufi, M.: Multiscale Force Networks in Highly Polydisperse Granular Media. Physical Review Letters 102(17), 2–5 (2009)

Wackenhut, M., McNamara, S., Herrmann, H.: Shearing Behavior of Polydisperse Media. European Physical Journal E 17(2), 237–246 (2005)

Wendell, D.: Transport in Granular Systems. Phd Thesis, Massachusetts Institute of Technology (2011)

Wyart, M.: On the Rigidity of Amorphous Solids. Annales De Physique 30(3), 1–96 (2005)

Yano, H., Kubota, T., Miyamoto, H., Okada, T., Scheeres, D., Takagi, Y., Yoshida, K., Abe, M., Abe, S., Barnouin-Jha, O., Fujiwara, A., Hasegawa, S., Hashimoto, T., Ishiguro, M., Kato, M., Kawaguchi, J., Mukai, T., Saito, J., Sasaki, S., Yoshikawa, M.: Touchdown of the Hayabusa spacecraft at the Muses Sea on Itokawa. Science 312(5778), 1350–1353 (2006)

Zeravcic, Z., Xu, N., Liu, A.J., Nagel, S.R., van Saarloos, W.: Excitations of ellipsoid packings near jamming. Europhysics Letters 87(2), 26001 (2009)

Chapter 12
Asteroids: Anchoring and Sample Acquisition Approaches in Support of Science, Exploration, and *In situ* Resource Utilization

Kris Zacny[1], Philip Chu[1], Gale Paulsen[1], Magnus Hedlund[1], Bolek Mellerowicz[1], Stephen Indyk[1], Justin Spring[1], Aaron Parness[2], Don Wegel[3], Robert Mueller[4], and David Levitt[5]

[1] Honeybee Robotics, Pasadena, CA, USA
[2] NASA Jet Propulsion Laboratory, Pasadena, CA, USA
[3] NASA Goddard Space Flight Center, Greenbelt, MD, USA
[4] NASA Kennedy Space Center, Kennedy Space Center, FL, USA
[5] Cadtrak Engineering, San Anslemo, CA, USA

12.1 Introduction

The goal of this chapter is to describe technologies related to asteroid sampling and mining. In particular, the chapter discusses various methods of anchoring to a small body (a prerequisite for sampling and mining missions) as well as sample acquisition technologies and large scale mining options. These technologies are critical to enabling exploration, and utilization of asteroids by NASA and private companies.

12.1.1 Types of Near Earth Objects

The term "asteroid" refers to any of a small class of solar system bodies that are in various orbits around the Sun. These include asteroids within the asteroid belt between Mars and Jupiter, asteroids co-orbital with a moon or planet (such as Jupiter's Trojans), and Near Earth Asteroids. In order to be more inclusive in this chapter, we also consider comets, and icy bodies with a wide range of orbital periods, ranging from a few years to hundreds of thousands of years. Comets are made up mostly of water, ice, and some dust particles, while asteroids are classified by their characteristic spectra, with the majority falling into three main groups: C-type (carbon rich), S-type (stony), and M-type (metallic). Both asteroids and comets fall within the larger category of "small bodies".

Near-Earth Objects (NEOs) are comets and asteroids that have been pulled by the gravity of nearby planets into orbits close to Earth. NEO asteroids are also referred to as Near Earth Asteroids, or NEAs in order to distinguish them from asteroids within the Asteroid Belt or Trojan asteroids. What makes NEOs very enticing is that they come close to Earth and could be within relatively easy reach for access by spacecraft. That is quite an important consideration given that traveling from Earth to the Asteroid Belt could take several years.

12.1.2 Motivations for Taking Samples from NEOs

NEOs are of interest to us for two reasons: for scientific study, and as a source of resources (Tsiolkovskii 1903; Lewis 1996). So far, all missions to NEOs were motivated by scientific exploration (Veverka et al. 2001; Yano et al. 2006; Glassmeier et al. 2007). However, given recent advancement in various space technologies, their value for resource mining is becoming of more interest. A significant portion of that value is derived from their location; the resources contained in NEOs do not need to be lifted from the surface of the Earth in order to be utilized in space. To help represent this, a new term was coined: *In situ* Resource Utilization (ISRU). ISRU facilitates planetary exploration by drawing needed resources, such as water, from the local environment. Comets and asteroids are therefore of great interest as a source of raw materials. Currently, the economics of extracting resources from these bodies, processing them *in situ*, and bringing the valuable material back to Earth is speculative. Whether this would be profitable or not depends on a number of assumptions which themselves are only based on expert opinions and not concrete data. Alternatively, there might also be some economic value in processing the resources *in situ* and using these processed resources in space rather than bring them to earth, but this is also quite speculative.

Assuming it will one day be profitable to mine, process, and use materials in space, asteroids can provide a great deal of resources. Raw materials from M-type asteroids could be used in developing various space structures. Water and carbon-based molecules from comets and C-type asteroids could be used to support life and in generating liquid hydrogen and oxygen for chemical propulsion to enable further exploration and colonization of our solar system. In addition, water offers shielding against galactic cosmic rays. Although the exploitation of asteroids has been discussed for a long time, only recently have private companies such as Planetary Resources and Deep Space Industries announced they aim to do this (Wall 2013).

Transporting water from the Near Earth Objects (NEOs) could be very profitable given that launch costs to Lower Earth Orbit (LEO) are on the order of $3,000-40,000/kg (Wilhite et al. 2012). Some major markets for water could include human consumption (e.g. International Space Station, Space Hotels) or refueling of spacecraft and satellites, but the only real sustainable market for water in space would be for chemical propulsion (LOX/Hydrogen). Water for human consumption can be mostly recycled, while water for fuel is a consumable.

Extracting water on Asteroids is much easier than processing metals. To extract precious metals in space new technology has to be developed that works in microgravity, and even potentially in vacuum. Terrestrial methods of mineral extraction require water, various chemicals, and gravity, and hence cannot be easily adapted to the space environment. Extracting water, however, only requires heating ice-bound regolith and capturing the produced water vapor – hence it is quite feasible to achieve with modest technology investment (Zacny et al. 2012a). From the resource extraction standpoint, the carbonaceous C-type asteroids are most desirable; as they contain a mixture of volatiles, organic molecules, rock, and metals (Gaffey et al. 2002; Lodders 2010).

There are at least two exploitation options for mining asteroids: either an entire asteroid could be captured and brought back to the Earth's or Moon's vicinity, or the desirable resource could be extracted and processed *in situ*. Whether to bring an entire asteroid or to process resources *in situ*, will depend on the size of the target asteroid. Smaller asteroids would be easier to capture, de-spin and bring to earth's vicinity while large asteroids would either have to be entirely processed *in situ* or only a fraction of the asteroid could be returned. In addition, very large asteroids or smaller M-type asteroids are more likely to survive atmospheric entry and impact earth because they are less likely to break up in the atmosphere. A recent study found that it is feasible to capture, de-spin, and bring to high lunar orbit, a 7 meter diameter asteroid weighing in excess of 500 metric tons. Such a mission would cost approximately $2.6B and would not require any new technology development (Brophy et al. 2012). Another study concluded that it is also possible to retrieve a 2 meter diameter asteroid to the International Space Station (Brophy et al. 2011).

A capture and return mission is attractive as an early asteroid resource exploitation mission. Having an asteroid in Earth's proximity would allow testing and verification of various material processing technologies. In addition, visits by astronauts could be conducted on a weekly, rather than annual basis and various autonomous or telerobotic technologies could be further demonstrated and improved upon. Once all the required technologies for an industrial scale asteroid mining operation have been developed and validated it might be more cost effective to send asteroid miners and refiners to various targets and only bring back the processed material.

In addition to private investment, national priorities also influence the pace of asteroid exploration. In 2010, President Obama directed NASA to get astronauts to a NEA by 2025, and then on to the vicinity of Mars by the mid-2030s. To reach these destinations, NASA has been developing the largest rocket since Saturn V called the Space Launch System, as well as a crew capsule called Orion. The SLS-Orion system is scheduled to begin launching astronauts in 2021 to a yet to be determined asteroid. This heavy launch capability will also enable launching larger asteroid mining missions.

12.2 Past Missions

Table 12.1 summarizes space missions to small bodies to date, including their cost (where available), and science returned. Missions highlighted in the table are missions which included surface operations. It can be seen that science returned per dollar is relatively low and after spending billions of dollars we still do not know much about the majority of asteroids. In addition, out of over a dozen missions to asteroids and comets, only two managed to touch the surface: the Near Earth Asteroid Rendezvous (NEAR) and Hayabusa missions. These two missions are described in more detail in the sections following. Three other missions: Rosetta (en route), Hayabusa 2, and OSIRIS-Rex (currently under development), also plan to perform *in situ* operations.

Table 12.1 Overview of small bodies missions to date, along with mission cost and results

Mission and Body visited	Agency, Launch Date	Mission Description (relevant to small bodies)	Cost, if readily available.
International Cometary Explorer (ICE)	NASA, 1978	Carried an X-Ray spectrometer and a Gamma burst spectrometer. Flew through the tail of the Giacobini-Zinner comet, and observed Halley's Comet from afar.	$3 Million ops-only add-on to an existing mission.
Vega 1 and Vega 2	SAS, 1984	Gathered images of Halley's Comet after investigating Venus.	
Sakigake	ISAS, 1985	Carried instruments to measure plasma wave spectra, solar wind ions, and interplanetary magnetic fields. Made a flyby of Halley's Comet.	
Suisei	ISAS, 1985	Carried CCD UV imaging system and a solar wind instrument for a flyby of Halley's Comet.	
Giotto	ESA, 1985	Carried 10 instruments to explore Halley's Comet, and provided data despite taking damage. Went on to explore comet Grigg-Skjellerup as well.	
Galileo	NASA, 1989	Carried 10 instruments. Flew by 951 Gaspra and 243 Ida, discovered Ida's moon Dactyl, and witnessed fragments of the comet Shoemaker-Levy 9 crash into Jupiter.	$1.6 Billion

Table 12.1 (*continued*)

Mission and Body visited	Agency, Launch Date	Mission Description (relevant to small bodies)	Cost, if readily available.
Near Earth Asteroid Rendezvous (NEAR) Shoemaker	NASA, 1996	Characterized asteroid Eros using imagers, spectrometers, a magnetometer, and a rangefinder. Although not originally planned to do so, NEAR-Shoemaker landed on Eros.	$220.5 Million
Deep Space 1	NASA, 1998	Carried technology experiments. Flew by asteroid 9969 Braille and comet 19P/Borrelly.	$152.3 Million
Stardust	NASA, 1999	Carried instruments for imaging and dust analysis. Flew by asteroid 5535 AnneFrank, comet Wild 2, and comet Tempel 1. Returned sample material from comet Wild 2.	$199.6 Million
Comet Nucleus Tour (CONTOUR)	NASA, 2002	Carried instruments for imaging, spectrometry, and dust analysis. Spacecraft was lost.	$135 Million
Hayabusa	ISAS, 2003	Landed on the asteroid Itokawa and returned sample material to Earth.	$170 Million
Rosetta	ESA, 2004	Flew by asteroid 2867 Steins and 21 Lutetia. Observed Deep Impact. Mission plans to put a lander on comet 67P/Churyumov-Gerasimenko.	~$1.2 Billion
Deep Impact	NASA, 2005	Carried instruments for imaging and spectrometry. Hit the comet Tempel 1 with an impactor and observed the collision. Will continue to study comets and asteroids as the EPOXI mission.	$330 Million
Dawn	NASA, 2007	Carries an imager, spectrometer, and gamma ray and neutron detector. Currently observing the asteroid Vesta, plans to move on to the asteroid Ceres.	$446 Million

Table 12.1 (*continued*)

Mission and Body visited	Agency, Launch Date	Mission Description (relevant to small bodies)	Cost, if readily available.
Hayabusa 2	JAXA, 2014 (planned)	Plans to create an artificial crater on asteroid 1999 JU3and return samples that have not been exposed to sunlight and solar winds.	\$367 Million
OSIRIS-Rex	NASA, 2016 (planned)	Plans to study C-type asteroid 1999 RQ36 and bring >60 grams of surface sample back to Earth.	\$750 Million

Non-contact instruments provide great scientific value, but they are no substitute for contact instruments, or a returned sample. So much more can be learned if the mission can land a spacecraft and analyze samples *in situ,* or even more importantly, bring the samples back to Earth. Terrestrial laboratories allow far more analysis capability than can be packed into a spacecraft's payload. In addition, these returned samples can be studied by future generations with technology that has not been invented yet. Anchoring and sample acquisition are critical technologies enabling *in situ* exploration of small bodies.

12.2.1 Near Earth Asteroid Rendezvous (NEAR) Shoemaker

The Near Earth Asteroid Rendezvous (NEAR) Shoemaker spacecraft (Veverka et al. 2000) was not originally planned to make physical contact with an asteroid. It was at the end of its life when the mission team decided to take the risk of making a soft landing on the surface of *433 Eros.* If something went wrong (i.e. the spacecraft crashed, or damaged its fragile solar panels and protruding antennae) there would have been relatively few negative repercussions since the mission had already completed its intended goals.

The spacecraft was successfully brought to a 1.9 m/s touchdown on the rocky surface and demonstrated that soft landing on an asteroid is possible (Veverka et al. 2001). Fig. 12.1 shows the final resting place of the NEAR Shoemaker spacecraft on asteroid *433 Eros*. Note that from a distance of 200 km, it is very difficult to resolve any surface details. The second image taken from a distance of only 250 m above the surface revealed the surfaces covered in fine regolith with protruding rocks.

This lack of *a priori* knowledge of the surface conditions presents a significant challenge to missions intending to touch down or sample asteroids. For known surface conditions, the sample acquisition equipment can be tailored to deliver the highest probability of success within the mass, volume, and power constraints of the spacecraft. For unknown surface conditions, the sample acquisition equipment cannot be as effectively tailored. If the NEAR Shoemaker mission had required landing and surface operations (e.g. sample acquisition) as part of its baseline, it would have been quite difficult to decide which anchoring system to use. Selecting and developing an anchoring system that would work in any formation is extremely difficult. If a harpoon were to be used, there is some chance that it might have hit one of the rocks, bounced back and impacted the spacecraft, damaging its solar panels, antennae or structures. If the anchoring system used some type of grippers, they would have been unsuitable for loose soil.

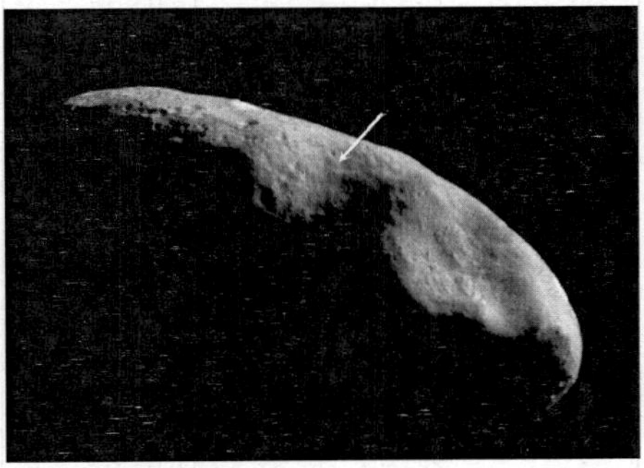

The location of NEAR Shoemaker's landing site from an orbital altitude of 200 kilometers (see the tip of the arrow). Mosaic of images 015246034-015246840. Courtesy NASA/JHU/APL.

NEAR Shoemaker's image of the surface taken from a range of 250 meters. The area imaged is 12 meters (39 feet) across. The cluster of rocks at the upper right measures 1.4 meters (5 feet) across. Image 0157417133. Courtesy NASA/JHU/APL

Fig. 12.1 Images of the asteroid 433 Eros taken by the NEAR Shoemaker spacecraft

12.2.2 Hayabusa

The second asteroid landing was performed by the Japanese Aerospace Exploration Agency's (JAXA) Hayabusa spacecraft. Its target was the S-type asteroid 25143 Itokawa. Hayabusa was launched towards its destination in 2003, rendezvoused, landed, and collected samples in 2005, then returned samples to Earth in 2010 (Kawaguchi et al. 2008). The spacecraft did not perform a sustained landing, but rather performed touch-and-go surface operation. During the brief surface encounters, the sampling system acquired approximately 1500 grains, mostly smaller than 10 microns.

Hayabusa also carried a tiny mini-lander named "MINERVA" (MIcro/Nano Experimental Robot Vehicle for Asteroid). However, MINERVA was released at a higher altitude than intended while the Hayabusa spacecraft was ascending. As a result, the MINERVA lander escaped Itokawa's gravitational pull and tumbled into space (Normile 2005).

Currently JAXA is developing a follow-on mission named Hayabusa 2 to investigate and sample asteroid 1999 JU3.

12.3 Current and Future Missions

Table 12.1 includes two missions that are currently under development: Hayabusa 2 and OSIRIS-Rex. These two missions share a common goal: Acquire sample material from an asteroid, and return that material to Earth. Another sampling mission is the Rosetta mission. The spacecraft already been launched and will acquire subsurface samples from a comet for *in situ* analysis. These missions are further described in the sections that follow.

12.3.1 Rosetta

The objective of the Rosetta mission is to rendezvous with comet 67/P Churyumov-Gerasimenko and acquire scientific data via an orbiting spacecraft and Philae lander (Biele et al. 2002). The 96 kg Philae lander will be the first spacecraft ever to make a soft landing on the surface of a comet nucleus. When it touches down on the comet, the Rosetta lander will use three different techniques to absorb the impact of the landing and secure itself to the surface. As shown in Fig. 12.2 and 12.3, these three techniques are self-adjusting landing gear, harpoons, and ice screws in the landing pads (Ulamec et al. 2006). These three techniques are employed in rapid succession: First the self-adjusting landing gear absorbs the energy of impact, then the ice screws engage the (possibly soft) surface, and finally the harpoons engage the (possibly hard) surface.

Initial fixation of the lander to ground will be achieved by deploying three passive “ice screws”, or one in each foot of the spacecraft. The initial impact energy from the <1.5 m/s descent should be sufficient to push the screws into ground. Each of the ice screws is coupled via a cable mechanism to the two feet of the self-adjusting landing gear in such a way that half of the impact force is led to the ice screw and the other half is equally distributed to the two feet, that can move independently (up or down) of each other.

Upon touchdown, the default harpoon will fire automatically and penetrate up to 2.5 m into cometary surface. Immediately after the firing, the cable will be tightened up to 30 N in 8 seconds. The 30 N is well below the rewind cable system strength of 100 N and the anchor cable breaking strength of 330 N. An identical second harpoon is used as a back-up in case the first harpoon fails to secure the spacecraft to the surface.

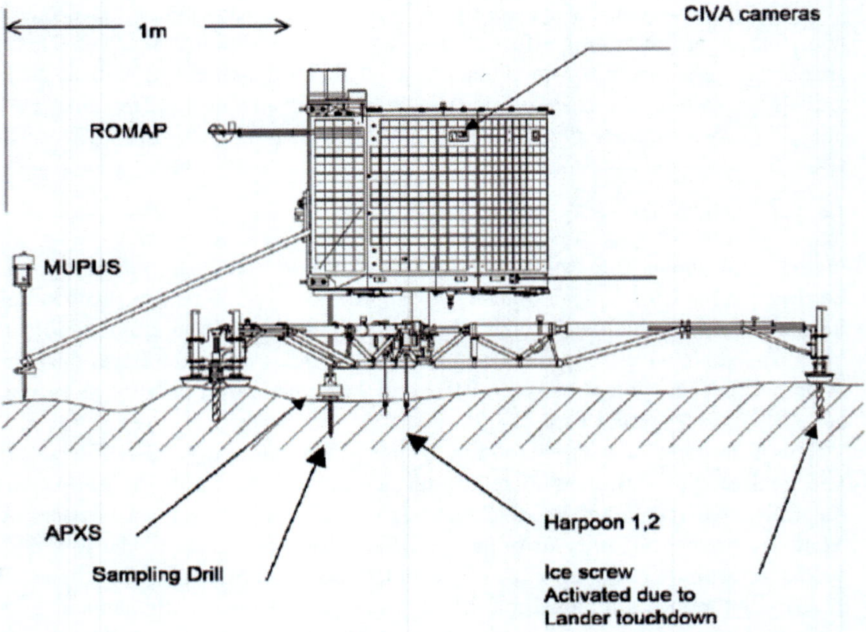

Fig. 12.2 Rosetta Philae lander (Biele et al. 2009)

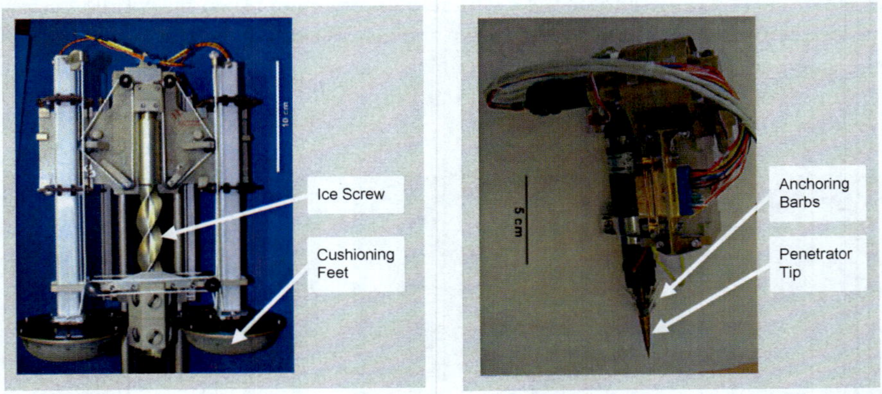

Fig. 12.3 Philae ice screw (left) and harpoon (right). (Biele et al. 2009).

The harpoon design uses both a sharp point and shovel flukes in order to ensure acceptable anchoring in surfaces that have both high tensile strength and low density. The point can penetrate the icy surface of the comet, which is expected to have high tensile strength, and the shovel flukes will develop holding power in the low-density material beneath this crust.

It should be noted that ice screws will not be able to penetrate hard material, while harpoons' anchoring effect will be negligible in very soft material. This dual anchoring approach aims to address the challenges associated with a lack of knowledge of comet surface properties by including anchoring mechanisms suitable to a range of surface types.

12.3.2 OSIRIS-REx

Figure 12.4 shows the Apollo asteroid 1999 RQ36. The asteroid is a potential Earth impactor with a mean diameter of approximately 500 meters. Data acquired from observations by the Arecibo Observatory Planetary Radar and the Goldstone Deep Space Network suggest that the asteroid might impact the Earth during one of its 8 close encounters between 2169 and 2199, with a probability of an impact of 0.07% or less (Andre et al. 2009).

This asteroid is the target of the OSIRIS-REx mission (OSIRIS-REx 2012). The goal of the mission is to return surface samples to Earth for further study. Asteroid 1999 RQ36 has been selected as the mission's target not because of its relatively high probability to impact the Earth, but rather due to the low ΔV required to reach it. The spacecraft is scheduled to launch in 2016, reach the asteroid in 2019, and return samples to Earth only in 2023, 7 years after the launch.

The mission will not land on the asteroid. Instead, the sampler will be deployed from a long and slender robotic arm, approach the surface at 0.1 m/s, fluidize regolith using gas and collect sample in approximately 5 seconds. The pneumatic sampler looks like an automobile air filter; a minimum of 60 grams will be trapped inside the filter part while some powder will also get stuck to the sticky surface of the sampler. After acquisition, the sampler will be inserted into the earth return capsule – the same design used on the Stardust mission.

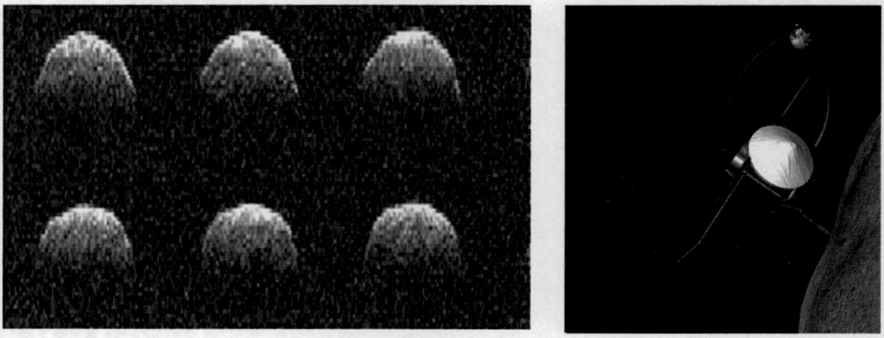

Fig. 12.4 OSIRIS-REx asteroid sample return mission. Left: Doppler imagery of the asteroid target: 1999 RQ36 (NASA's Goldstone Radar). Right: OSIRIS-REx spacecraft. The spacecraft will employ a "touch and go" sampling approach.

It is interesting to note that $750M is being spent to go to an asteroid about which relatively little is known. That is quite risky of course, but this is a reality of Asteroid missions: we do not know much until we get there. The sampling mission must to be able to perform successfully on a range of potential surface materials.

12.3.3 Hayabusa 2

Hayabusa 2 is a successor spacecraft to Hayabusa, with a similar goal of bringing samples back to Earth from asteroid 1999 JU3, a carbonaceous or C-type asteroid (Kawaguchi 2008; Campins et al. 2009). Hayabusa 2 would launch in 2014 or 2015, arrive at the asteroid in 2018, conduct a series of investigations and operations as shown in Fig. 12.5, and then return to Earth towards the end of 2020. Hayabusa 2 will create an artificial crater by striking the asteroid with a copper plate accelerated by explosives, and then sample the freshly exposed material from the floor of the crater. This more pristine material is scientifically valuable in that it has not been exposed to sunlight and solar winds.

Hayabusa 2 is more ambitious than the original Hayabusa mission. It will carry three landers to the asteroid instead of one carried by Hayabusa. The first two are based on the detachable MINERVA lander that was developed and built for the first Hayabusa mission, but missed the asteroid surface and drifted off into space. The third lander is called MASCOT (Mobile Asteroid Surface Scout). It is a standalone lander developed by the German Aerospace Center (DLR).

Fig. 12.5 Concept of Japan's Hayabusa 2 spacecraft. Hayabusa 2 would hurl an impactor into the asteroid 1999 JU3, sample the exposed crater material and bring it back to earth. Credit: JAXA/A. Ikeshita. http://www.space.com/14759-asteroid-sample-mission-hayabusa-2.html

Hayabusa 2's copper impactor will be deployed from the spacecraft and slam into the asteroid to create a large crater. To prevent potential damage of the spacecraft by the crater ejecta, Hayabusa 2 will hide on the other side of the asteroid during the impact, while a deployed standalone camera will record the impact. Two samples will be acquired prior to the impact event, then Hayabusa 2 will attempt to land at the fresh crater and acquire a third sample for transport back to Earth.

12.4 Small Bodies Surface Environment

Due to the close proximity required to detect surface features, most of the information we have about surface features and surface properties of small bodies comes from the past missions that visited these small bodies. In particular, from the Deep Impact (DI) mission we learned that the surface of the comet Mathilde is highly porous, with porosity estimated to be around 60% (Figs. 12.6 and 12.7). The shear strength of the surface soil was also found to be very weak, in the range of 1-10 kPa (Richardson et al. 2007). That presents a challenge to any anchoring system, since the spacecraft has to anchor itself to something that has consistency of fluffy snow or flour.

The Rosetta mission's Philae lander is tasked with anchoring itself and then sampling a comet 67P in 2014. It will use its harpoons and ice screws to anchor itself in an environment where a local gravity on the surface is of the order of 3×10^{-4} m/s^2. It has been estimated that the nominal bulk density of 67P is 100-370 g/cc with an upper limit of 500-600 g/cc. (Hilchenbach et al. 2004). For comparison purposes, the density of freshly fallen snow is 160 g/cc and that of compacted snow is 480 g/cc.

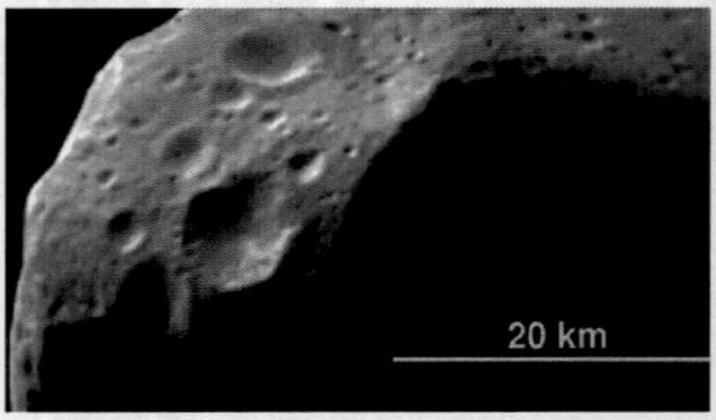

Fig. 12.6 hock dissipation is evident in craters formed in porous materials. The large craters on the porous asteroid Mathilde (a C-Type Asteroid) are packed closely together with little evidence of shock-induced disturbance of adjacent craters. (Images courtesy NASA).

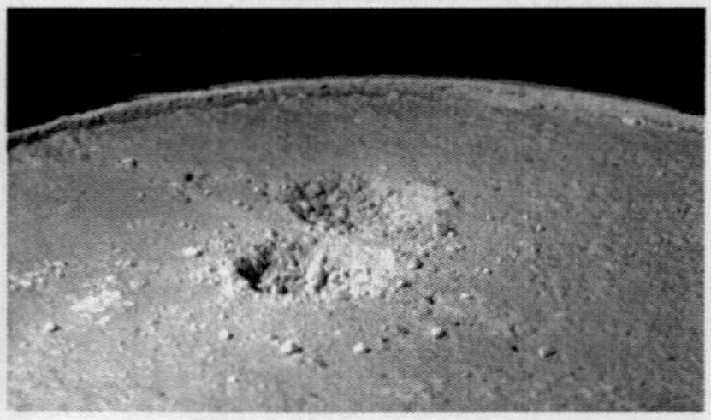

Fig. 12.7 material with porosity of 60%. Images of the first crater (furthest from camera) before and after the second impact showed no noticeable damage caused by the second impact, even though the crater rims were nearly touching. This means that the porous material efficiently damps the shock pressure (Britt et al. 2002).

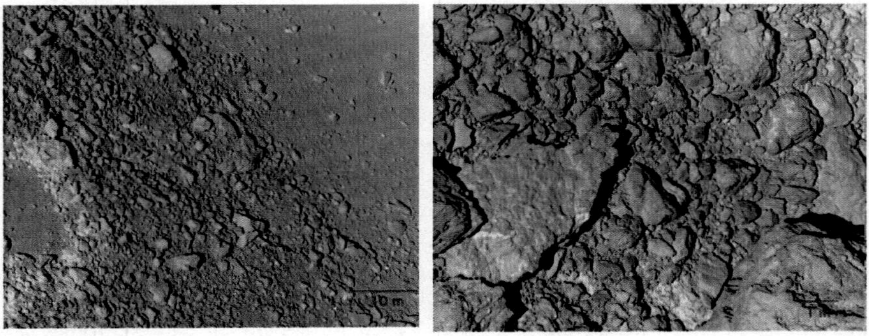

Fig. 12.8 Itokawa (S-type asteroid). Note the scale bars in bottom right. (Images courtesy of JAXA).

The Hayabusa spacecraft took pictures of the Itokawa S-type asteroid as shown in Fig. 12.8. This example demonstrates that the surface properties of asteroids are quite variable: ranging from fluffy powder, and loose gravel, all the way to solid rock. Itokawa is now considered to be a "rubble pile" body because of its low bulk density, high porosity, boulder-rich appearance, and shape. The presence of very large boulders points to the early collisional breakup of a preexisting parent asteroid (Fujiwara et al. 2006). It is quite clear that due to the high variability of the surface terrain, a spacecraft attempting to anchor itself to the asteroid Itokawa would face different challenges depending upon its actual landing site.

12.5 Anchoring Concepts for Small Bodies

For an anchoring system to be effective and low-risk, there are a number of criteria it must meet. Firstly, the main purpose of the anchor is to react all forces and torques caused by the movement of the robotic arm or other deployment or sampling systems on the spacecraft. Since there will be little to no prior knowledge of surface properties, the anchoring system must be able to function in a range of surface types, including powder, gravel piles, and rocks. If the mission requires sample return, the anchor must provide the capability to free the spacecraft, either by detaching from the surface or detaching from the spacecraft. If the mission requires multiple landings, the anchor must be reusable or lightweight and simple enough that multiple sets of anchors could be integrated with a spacecraft.

Landing on a small body can pose significant risk to a spacecraft, especially if that body has a high spin rate or is tumbling. In these cases, the spacecraft must possess a highly capable Guidance, Navigation and Control (GNC) system to time the spacecraft's descent with the small body's rotation, while keeping the spacecraft in proper orientation. This is especially important if large solar panels or other protruding structures are present. If the spacecrafts' altitude is not maintained, the solar panels might impact the surface and be permanently damaged.

To mitigate the risk associated with landing, missions may opt for a *touch and go* concept of operations instead. For example Hayabusa, Hayabusa 2, and OSIRIS-REx all include *touch and go* operations. Hayabusa, Hayabusa 2, and Rosetta all delegate landing attempts to a daughter craft – MINERVA, the MINERVA derivatives, MASCOT, and Philae. Hayabusa's loss of MINERVA during the probe's deployment on 12 November 2005 illustrates the risk of landing. The Hayabusa spacecraft was ascending and at a higher altitude than intended when releasing MINERVA. As mentioned earlier, MINERVA escaped Itokawa's gravitational pull and missed the surface.

The only successful demonstration to date of a spacecraft landing on a small body is NEAR Shoemaker's landing on *433 Eros*. However, it is worth noting that the mission team had gained tremendous risk tolerance by virtue of already completing their baseline mission goals. For all practical purposes, the risk associated with losing the spacecraft was minimal and the team had nothing to lose! Their gamble paid off; the spacecraft successfully landed on *433 Eros,* demonstrating a soft landing on an asteroid and providing valuable information about the surface. While none have been demonstrated through successful landing on a small body, a number of different anchoring concepts have been proposed over the years, and/or developed to various technology readiness levels or TRLs (Mankins 2005).

Table 12.2 summarizes some of these anchoring concepts. One of the primary requirements for any spacecraft anchoring system is the capability to engage a wide range of surfaces, unless two or three different anchoring concepts are employed, each designed to address a specific set of surface conditions. This multi-surface requirement is quite challenging. The vast majority of universal anchor systems rely heavily on spacecraft resources such as fuel for thrusters or momentum from control moment gyroscopes or CMGs. However, these approaches could be ideal as a temporary solution within a sequential stepwise approach, allowing time for the deployment of more permanent anchors.

Table 12.2 Comparative assessment of anchoring methods

Method	Description	Advantages	Disadvantages	Applicable Surfaces
Thrusters	Fire thrusters to push the spacecraft to the surface	Uses existing spacecraft technology. Good as a backup option or to enable deployment of permanent anchor	Requires extra fuel and hence might be good for short stays or during sample acquisition only.	Any

Table 12.2 (*continued*)

Method	Description	Advantages	Disadvantages	Applicable Surfaces
Reaction Wheels	Spin up reaction wheels to counteract reaction forces from sampling systems (e.g. drill) or deploying robotic arm.	Uses existing spacecraft technology. Reusable approach Good as a backup option or to enable deployment of permanent anchor	May require large reaction wheels to achieve proper stability.	Any
Spacecraft momentum	When a spacecraft moves towards the asteroid at certain velocity, any sampler deployed in the direction of the asteroid surface will be reacted against spacecraft forward momentum.	Uses existing spacecraft technology The approach could be used multiple times Good as a backup option or to enable deployment of permanent anchor	The operation time on the surface highly limited. The approach might be good for touch and go mission.	Any
Grippers	Anchor sharp "fingers" or microspines positioned opposite each other. System under development by NASA JPL	Offers strong anchoring forces in rocks Could be reused	Works only on solid surfaces. Will not work in pebbles or soil. Requires additional hardware (grippers) and power during deployment	Rocks

Table 12.2 (*continued*)

Method	Description	Advantages	Disadvantages	Applicable Surfaces
Harpoon/Nail Gun	Fire a harpoon into a surface and use a winch to pull a spacecraft towards the surface. Rebound taken up by spacecraft momentum energy or a free mass ejected in the opposite direction to the harpoon. System developed for Rosetta Philae lander	Could generate high anchoring forces if regolith properties allow. Rosetta Philae lander future heritage	Requires additional hardware (i.e. harpoon). May not work in harder rocks, and very loose gravel or soil. If harpoon hits a rock it may rebound towards the spacecraft. Non reusable (i.e. tether has to be cut for the spacecraft to move to another location or return to earth)	Gravel/soil
Drill/Auger	Deep fluted augers driven into the subsurface	Offers strong anchoring forces Could be reused Can use two counter-rotating drills to off-set reaction torque Can use tapered augers to assist with removal Can use deep flutes to engage large surface area	Need initial reaction compensator during deployment Requires additional hardware (drill) and power during deployment May not work in rocks if they are too hard and reaction compensator during initial deployment is not strong enough.	Rocks and more consolidated gravel/soil

Table 12.2 (*continued*)

Method	Description	Advantages	Disadvantages	Applicable Surfaces
Self-opposing Systems	Spikes or drill bits penetrate the subsurface at oblique angles providing a bracing force.	Takes advantage of rock roughness, porosities (holes) in a rock or cavities between rocks Could be used with a harpoon or auger	Needs more than one anchor to provide the bracing effect. Subject to same penetration limitations as component subsystems.	Any
Fluid System	Fluid is injected underneath the footpad onto and into the subsurface and hardness. Similar to Velcro except it engages subsurface.	Could work on any surface. One anchor is sufficient.	Non-reusable.	Any
Envelopment	The target is encircled using cables or even a complete bag.	No need to penetrate the surface.	Large relative to other anchoring concepts.	Relatively small and/or highly consolidated bodies.
Magnetic Anchoring	Magnetic pad is used to attach to ferromagnetic surface.	No need to penetrate the surface.	Not applicable to non-ferromagnetic bodies	Ferromagnetic bodies

The following sections describe some of the anchoring concepts in more details.

12.5.1 Hard Rock Drilling

Drilling is often the method of choice for penetrating hard rocks. In a vast majority of applications, drilling employs the durning of a hardened bit forced into the rock, abrading small particles. Once the drill bit becomes dull, however, the rate at which the drill bit penetrates the rock drops dramatically unless an ever-increasing downforce (in the case of vertical downward drilling) is applied, creating higher and higher frictional heat. The amount of applied downforce is referred to as "weight on bit" (WOB). In a low-gravity environment, WOB can be severly limited and must be provided either by spacecraft thrusters or an anchored "base" for the drill to push against. Rotary-percussive drilling action can be employed to circumvent the need for prohibitively high WOB. Percussive or hammer systems consume more power, but reduce the required WOB by an order of magnitude, especially in hard rocks (Zacny et al. 2013).

The feasibility of drilling into small bodies with low WOB has been tested using relevant analog rocks as stand-ins for likely asteroid surface materials as shown in Fig. 12.9 (Bar Cohen and Zacny 2009). Such testing demonstrates that the feasibility of drilling into a small body with low WOB is dependent upon the strength of the materials comprising the small body.

For example: Drilling in low-strength materials such as plaster or limestone is feasible using a commercially available drill with a 1.6 mm diameter bit and as little as 5 Newtons WOB. Plaster and limestone have an unconfined compressive strength (UCS) of 8 MPa and 40 MPa, respectively, which would be representative of a materials in a C-type asteroid. However, higher strength materials such as those that would be representative of an S-type asteroid cannot be drilled under those same conditions. For instance, 120 MPa basalt is representative of the upper range of S-type asteroid materials, and cannot be penetrated under the conditions listed above. This is not to say that it is infeasible to drill an S-type asteroid, but it does indicate that the drilling system must be designed with the target material in mind.

Fig. 12.9 Low WOB drilling tests in basalt

12.5.2 Hard Rock Hammer Nailing

Another approach to setting an anchor in consolidated or unconsolidated formations is to hammer an anchor into the surface. Launching an anchor into unconsolidated formations is relatively easy, however setting an anchor in rocks might be difficult.

As with drilling, preliminary feasibility tests yield interesting results: Testing was performed using a 3.8 mm nail and traditional hammer as well as with an off the shelf nail gun (Fig. 12.10), and the same three rock types as for drilling (8 MPa plaster, 40 MPa limestone and 120 MPa basalt). Hammering a nail worked as long as the nail was perpendicular to the surface, but any deviation from vertical resulted in side forces and moments that had a tendency to bend the nail. Stiffer nails would resist buckling, but underline the need to design for off-axis loading. The nail did not penetrate basalt or limestone, but managed to penetrate plaster. The nail gun, on the other hand, was powerful enough to drive a short nail into all three rock types. However, the nail gun also required significant preload, of the order of 10s of Newtons prior to impact. The rebound energy within the nail gun was absorbed by an internal spring.

Fig. 12.10 Nailing experiments. From top (left to right) and bottom: Hammering into 120 MPa basalt, Hammering into 8 MPa plaster, Using a nail gun.

These tests have shown that as long as a nail is constrained so as to prevent buckling, it could be successfully impact driven into rocks as hard as basalt. However, during the anchor setting, the spacecraft has to use some other means of providing a reaction force, e.g. by firing of thrusters in the opposite direction.

12.5.3 Fluid Anchor

In the fluid anchor approach, a wetting fluid (e.g. foam, cement, epoxy etc.) is injected onto a surface or into the soil via a hollow spike beneath the footpad. If applied to the surface, the goal of a fluid anchor is inject an adhesive cushion between the rock surface and the spacecraft footpad, and in turn provide an anchor. If injected into the ground, the fluid would go deeper into the loose gravel or soil, allowing the anchor to engage a larger volume of asteroid material forming a composite footing (glue mixed with soil and gravel).

To free the spacecraft from such an anchor would require that either an entire footpad is detached (in this case, the spacecraft could have a set of 2 or 3 footpads per leg) or the footpad could be warmed up to 'melt' the adhesive underneath and disengage the anchor. One of the benefits of this approach currently under investigation by Honeybee Robotics is that the anchor deployment does not exert any force that requires reaction by the spacecraft. That is, upon soft touch down, the fluid can be discharged and almost instantly glue the spacecraft to the surface.

Applying epoxy-like substances in the harsh environment of space has been demonstrated. In July of 2005, astronauts applied a pre-ceramic polymer sealant impregnated with carbon-silicon carbide powder known as NOAX (Non-Oxide Adhesive eXperiment) to a number of test coupons during an Extra Vehicular Activity on board the Space Shuttle mission STS-114 (see Fig. 12.11). This material has the initial consistency of peanut butter before it is worked into potential cracks and crevices of the Shuttle's Reinforced Carbon-Carbon panels in areas such as the wing leading edge, which sees the highest temperatures during atmospheric re-entry.

Fig. 12.11 Astronaut Soichi Noguchi, STS-114 applies a sealant to a number of test coupons during an Extra Vehicular Activity. Photo courtesy NASA.

12.5.4 *Self-opposing Systems*

Several concepts developed to date employ a common strategy of using multiple instances of an anchoring mechanism to simplify the reaction loads required of the spacecraft. Somewhat like the guide lines on a tent, these multiple anchors pull against each other, providing a balanced net anchoring force. The independently developed examples presented here all employ this general strategy, but use somewhat different means to engage the target surface.

12.5.5 *Self-opposing Drills (Cadtrak Engineering's Low Gravity Anchoring System)*

Cadtrak Engineering developed a novel low-gravity anchoring system which could anchor a sampling tool as shown in Fig. 12.12, or an entire spacecraft as shown in Fig. 12.13. This device would decrease the preload requirement, peak reaction forces and vibration levels on a deployment device, and would significantly reduce the mass and complexity of the spacecraft propulsion system. It employs multiple inclined anchors to generate a net anchoring force perpendicular to the surface while at the same time balancing out any forces transverse to the surface.

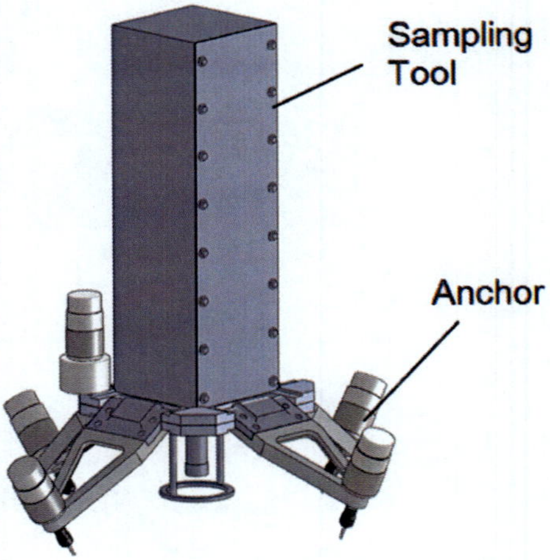

Fig. 12.12 Cadtrak Engineering anchor integrated with a sampling drill

In the absence of an anchor on an asteroid or comet mission the preload on a sampling tool must come from the propulsion system. For example, a sampling mission that requires a 100 N preload for 15 minutes from a spacecraft whose propulsion system has a specific impulse (I_{sp}) of 350 s, the mass of propellant alone would be 26.2 kg. As an alternative, this particular anchoring system can be set in less than 30 seconds and require 20 Newtons of preload. The propellant mass used to set the anchor would equate to 0.2 kg. If the anchoring system weights 5 kg, the net mass savings would be 21 kg.

The anchor uses multiple anchor arms acting in a coordinated manner as to keep the forces in equilibrium horizontally, thus keeping the system in place. The base of each arm is hinged on the tool body or spacecraft, and the far end of each arm contains a small pilot drill. The anchor is deployed by rotating the arms toward the subject rock and drilling into the rock a short distance. The arms are connected through a novel gearing arrangement which allows them to be driven by a single actuator and still conform to any surface profile. Each anchor drill can be engaged with as little as 10 Newton of Weight-on-Bit (WOB). When any two opposing anchor drills have penetrated the rock, the anchor is set, creating a stable platform for sampling or other *in situ* science operations (e.g. deploying an instrument using a robotic arm). The anchor system is compliant to large surface variations and placement can be accomplished with less precision and less preload than that of a sampling tool. The anchor hold down strength could be verified prior to commencement of *in situ* operations. The anchor also allows multiple uses

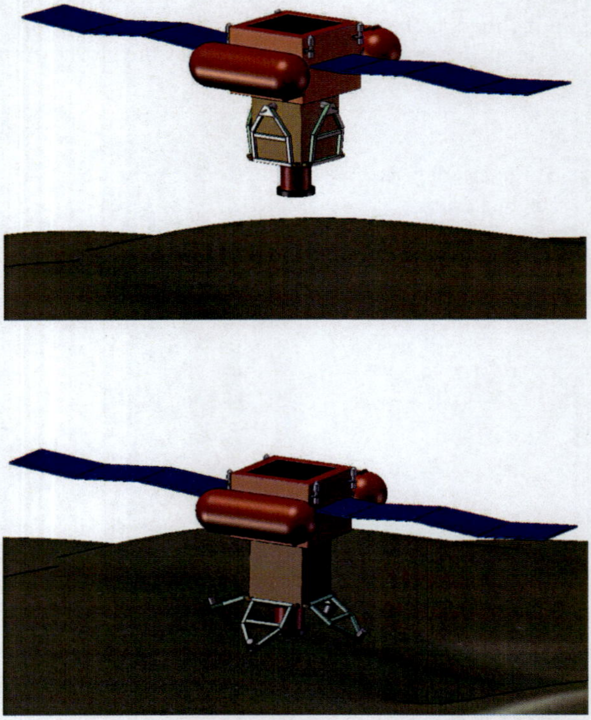

Fig. 12.13 Cadtrak anchor stowed and deployed during a spacecraft landing on an asteroid or comet

and could therefore be used as temporary system for a spacecraft making multiple landings on an asteroid.

Cadtrak developed a bench top anchor testbed which incorporated an anchor platform and simulated sampling system attached to a vertical slide. The setup included several anchor arms with a drill motor and ball-end diamond burr. The 2 mm diameter drill bits ended with a spherical ball with integrated diamonds. The anchoring was accomplished by simultaneously driving the anchor drill motors and the arm actuator until the anchor drills penetrated the rock to a set depth. The individual drill bits would generally encounter the rock surface one at time and hover at the rock surface until all of the bits have engaged the rock. The differential gearing system of the arms ensures that power is always transferred to the free arm or all the arms as shown in Fig. 12.14. This enables the mechanism to accommodate varying surface topography. For a flight system, the anchors might include proximity sensors or contact switches to indicate when the target depth has been reached.

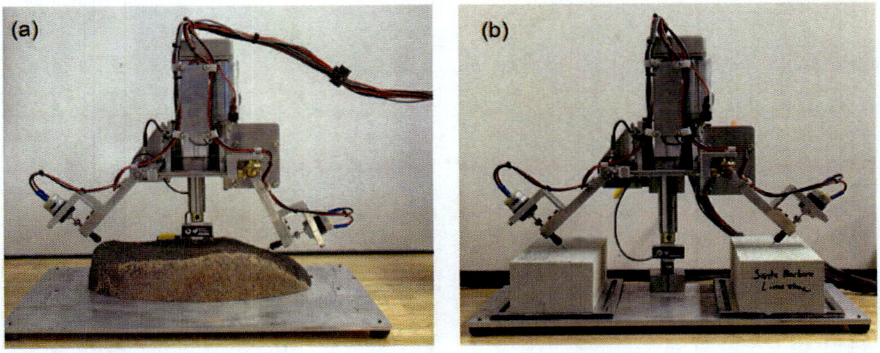

Fig. 12.14 (a) Anchor testbed shown conforming to arbitrary rock surface. (b) Anchor testbed shown with standardized rock samples for weight-on-bit and pull-out tests.

Data from a number of tests in limestone, basalt, sandstone, and kaolinite is shown in Fig. 12.15. It was determined that at a depth of approximately 3 mm, the pullout force for a single anchor arm reached 200 N in basalt and around 100 N in kaolinite rock. When pullout occurred the rock fractured along shear planes forming small craters (Fig. 12.16). In general the pull-out strengths were at least 10 times higher than the WOB requirement to drill the anchor holes.

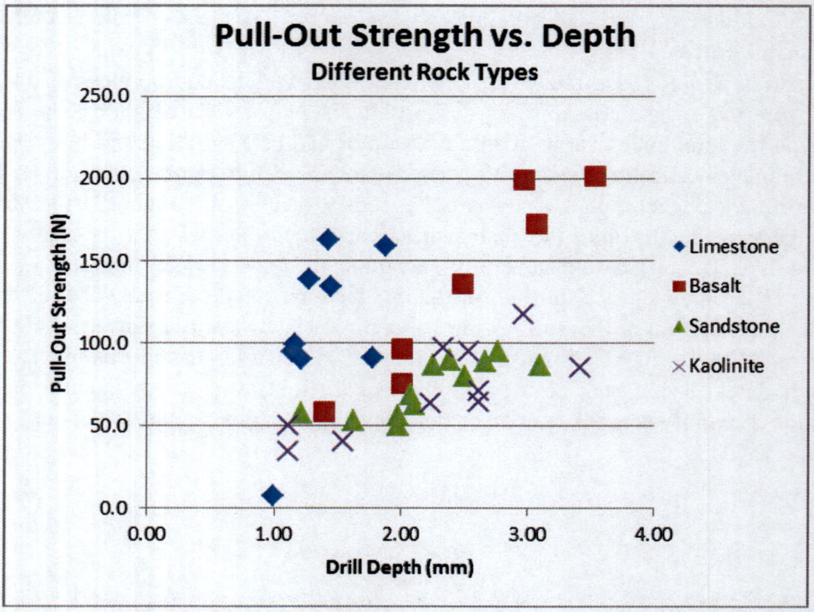

Fig. 12.15 Pull-out strength vs. drill hole depth for different rock types

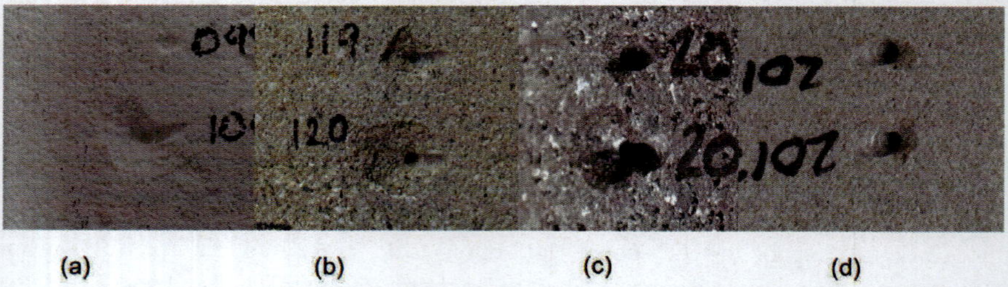

Fig. 12.16 Pull-out craters for various rock samples. (a) Kaolinite (b) Sandstone (c) Saddleback Basalt and (d) Santa Barbara Limestone rocks.

12.5.6 Self-opposing Multi-mode Anchor (Honeybee Robotics' Bracing Anchor)

The bracing system uses two or more multi-mode rock and soil anchors positioned at an oblique angle to the surface as shown in Fig. 12.17, resulting in a net force component along the asteroid surface. This resultant force braces the spacecraft to the surface. The advantage of this approach is that during the anchors' deployment only the force componenet in a vertical direction has to be overcome by, for example, firing rocket thrusters in the opposite direction.

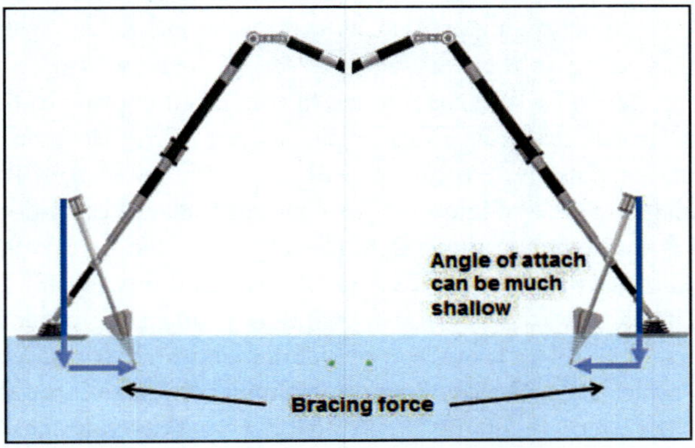

Fig. 12.17 The bracing anchor engages the surface at an oblique angle

Honeybee Robotics' bracing approach uses a long "drill" with a three tier system – each designed for a different surface condition (Fig. 12.18).

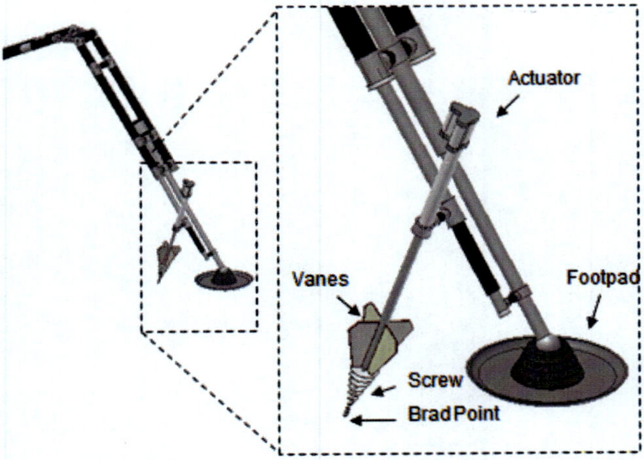

Fig. 12.18 A concept of a bracing anchor with a 3-tier system for 3 different surface conditions

First, the tip of the drill has a sharp Brad point whose purpose is to exert maximum pressure on the surface and ultimately find purchase in small-scale surface features like cracks, crevices, or valleys (if large rocks are present). This is similar to the approached used by microspine anchors. Further up is a self-tapping auger thread form (tapered screw auger) that will draw the anchor into rubble or gravel piles, if present. Lastly, in the weakest of materials such as powder, the bit's non-rotating vanes at the end will distribute lateral forces attempting to take advantage of the material's shear strength. Because the anchor's bit is designed to engage in all possible surface materials: rocks, gravel piles, and loose soil, it considerably reduces risk related to lack of knowledge of asteroid surface conditions.

Upon landing and when commanded, the anchor would be driven into the surface by simultaneously rotating the anchor's "bit" and moving in a linear fashion toward the surface. Bit rotation and translation would be provided by a single reversible actuator. Following surface operations and prior to spacecraft take-off, the anchoring system would disengage as each actuator would retract the bit to a safe position. The anchors should be deployed very slowly because the strength increases with strain-rate resulting in values about an order of magnitude higher (or even more) than the quasi-static strength for the same material (Biele et al. 2009).

This particular anchoring concept has been applied to a NASA Discovery-class mission concept called Amor shown in Fig. 12.19. The goal of the mission was to rendezvous, land, and explore the C-type triple near-Earth asteroid (NEA) system 2001 SN263 (Jones et al. 2011).

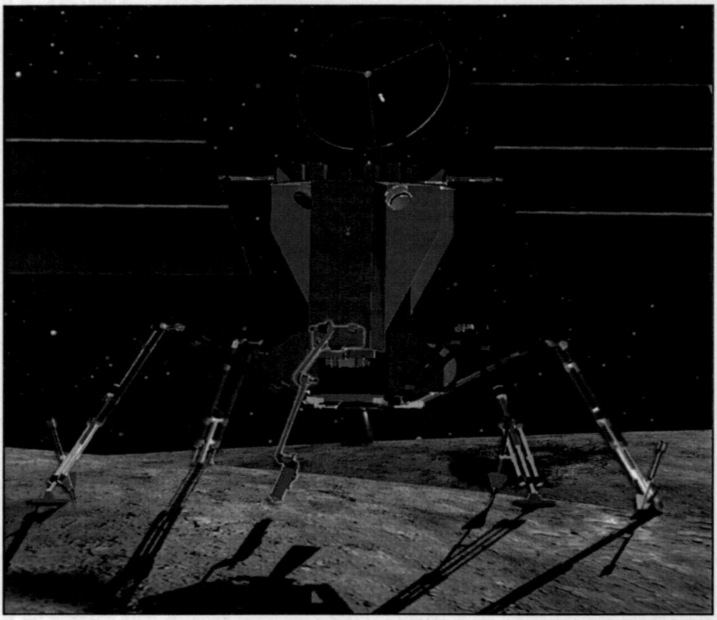

Fig. 12.19 Universal Hybrid Anchor for a lander mission to explore the C-type triple Near-Earth Asteroid System 2001 SN263 (Jones et al. 2011)

12.5.7 *Self-opposing Tines (JPL's Microspine Anchors)*

The NASA Jet Propulsion Laboratory (JPL) has developed a self-opposing anchoring system based on small tines called microspine anchors (Parness et al. 2012a). Microspine toes were initially invented at Stanford University for the RiSE climbing robot, which could scale the exteriors of buildings that used rough materials like brick, stucco, cinder block, and adobe (Asbeck et al. 2006; Spenko et al. 2008). NASA JPL has extended this technology for use on natural rock surfaces by first, using new configurations of opposing microspines that can resist forces in any direction, second, using a hierarchical design that complies to the rock at multiple length scales, and third, substituting materials and mechanisms that are appropriate for the extreme environment of space.

A microspine toe consists of one or more steel hooks embedded in a rigid frame with a compliant suspension system made of elastic flexures or spring elements (Figs. 12.20 and 12.21). By arraying tens or hundreds of microspine toes, large loads can be supported and shared between many attachment points. Since each toe has its own suspension structure, it can stretch and drag relative to its neighbors to find a suitable asperity to grip. The suspension system also works to passively distribute the overall load across an array of toes.

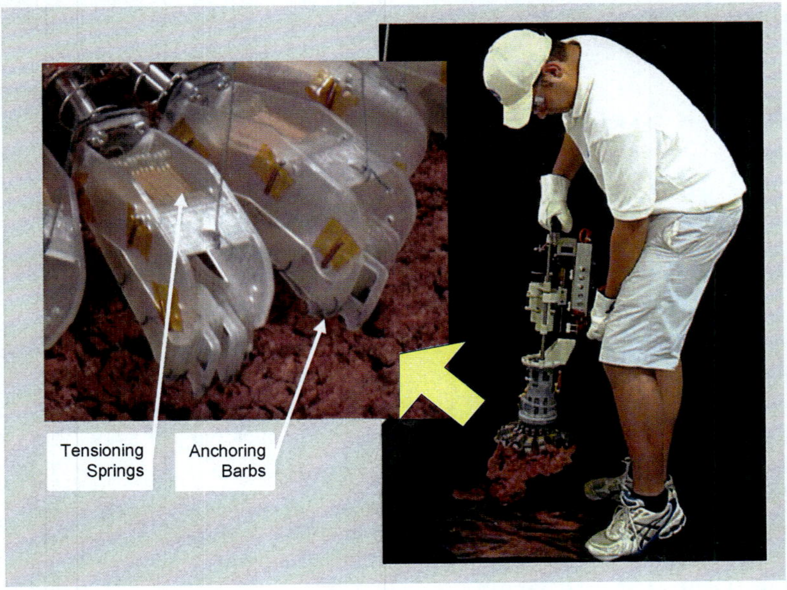

Fig. 12.20 A microspine anchor integrated with a rotary percussive coring drill to produce a sample acquisition instrument that can obtain a subsurface core from consolidated rock without requiring any externally applied forces

For gravity independent rock-climbing and drilling, the omnidirectional anchors use a radial arrangement of microspines with a centrally tensioning degree of freedom. The hierarchical compliance system contains 16 carriages that conform to cm-scale roughness. Each carriage contains 16 microspines, which conform to mm-scale roughness and below. A torsion spring biases each of these carriages into the rock face regardless of gravitational orientation so that the toes will drag across the rock surface and establish a grip, even in an inverted configuration. The radial symmetry creates a secure anchor that can resist forces in any direction away from the surface. Figure 12.21 shows many of the important components of the gripper with an explanation of the function of each.

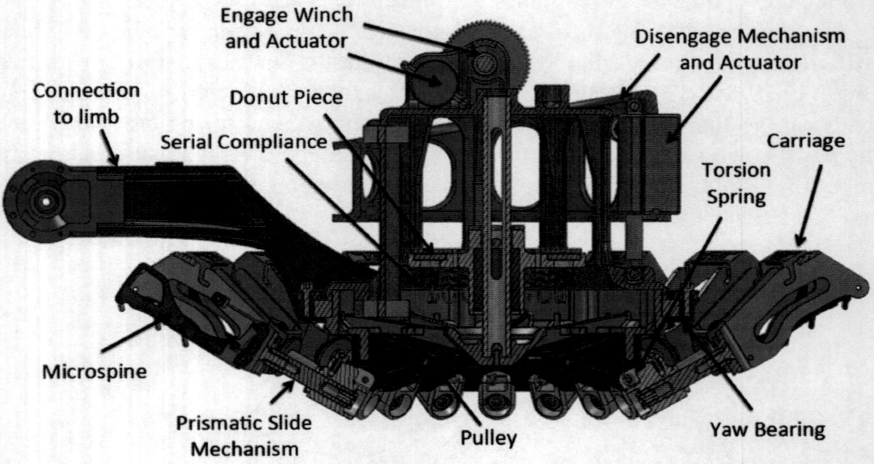

Fig. 12.21 Details of microspine-based anchor system

These anchors support loads in excess of 180 Newtons both tangent and perpendicular to the surface when used on rough, consolidated rocks like vesicular basalt and a'a lava rock. Anchor strength falls off as the roughness of the rock decreases due to the decreased number of potential asperities to grip. Anchor strength values in excess of 100 N were common on Bishop Tuff, and values of 50 N were consistently achieved on a smoother saddleback basalt sample. For unconsolidated materials like pebbles and sand, negligible (<10 N) anchoring forces were measured (Parness et al. 2012c).

A microspine anchor was integrated with a rotary percussive coring drill to produce a sample acquisition instrument that can obtain a subsurface core from consolidated rock without requiring any externally applied forces (see Fig. 12.20). The instrument is self-contained; redirecting the load path back into the rock, with forces reacted by the microspine gripper. The drill uses an additional two actuators, one to activate the rotary-percussive motion, and a second to feed the drill into the rock. Compression springs are used in series with the feed actuator to preload the drill bit into the rock with approximately 50-100 N of WOB (Parness et al. 2012b).

The microspine-based drills successfully cored in multiple configurations including drilling into the ceiling, into a vertical wall, and, using the Astronaut microspine drill, lifting a rock using the anchor and then drilling into it while the rock is lifted. These tests were performed on multiple rock types including vesicular basalt and a'a. Bore speed was dependent on the WOB, drill speed, and material properties of the rock, but nominally ranged from 15-45 mm/min. A carbide-tipped coring bit created a 20 mm diameter borehole to depths ranging from 25-82 mm for the inverted drill test, and 15 mm for the vertical and horizontal drill tests. The retained core samples measured 12 mm in diameter and usually were extracted in several broken pieces, but with stratigraphy maintained. While a broken core may not always be desirable, it does eliminate the need to perform a core breakoff.

During the drilling process, failure most often occurred during hole-start. The bit would sometimes wander before achieving a good hole-start. Occasionally, this caused the microspine anchor to lose grip. This was accentuated by the built-in compliance in the microspine anchor, which must resist the wander. However, this compliance also acts to dampen the vibrational forces, and is essential to the load sharing within the gripper.

12.5.8 Magnetic Anchoring

If an asteroid is metal-rich, highly consolidated, and magnetized, a magnetic anchor could be effective. The asteroid Gaspra, discovered by Galileo mission, is an example of such an asteroid (Kerr 1993).

12.5.9 Envelopment

Envelopment blurs the distinction between anchoring to an asteroid and grabbing a sample from it. In an envelopment -based mission's architecture, the spacecraft grabs the whole asteroid! In general, the suitability of the envelopment concept depends upon the size and makeup of the asteroid, and the ability of the spacecraft to control it once enveloped. An envelopment system might use cables that span around the small body (see Fig. 12.22) or long and skinny legs to embrace the small body in the same way a spider captures its prey. If an asteroid is relatively small, an entire body could then be captured inside a bag as shown in Fig. 12.23.

Fig. 12.22 Deep Space Industries (DSI) concept of a Harvestor™-class asteroid collection mission. (Image Credit: DSI).

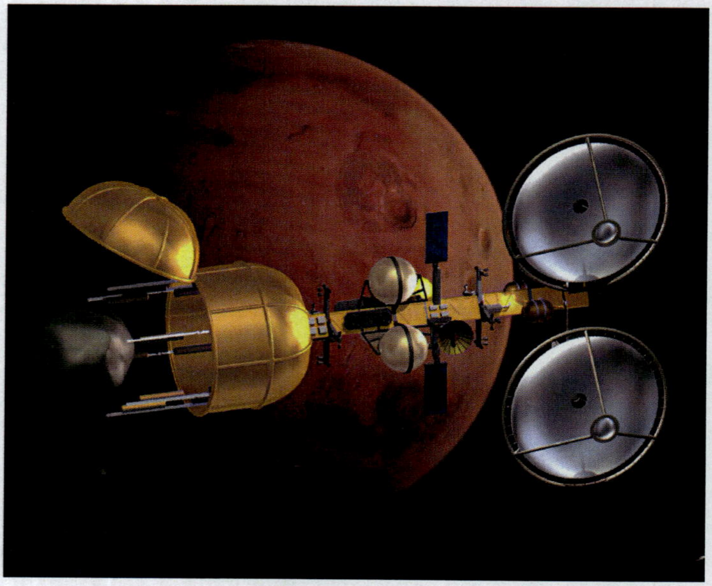

Fig. 12.23 Illustration of an asteroid retrieval spacecraft in the process of capturing an asteroid (Courtesy Honeybee Robotics and V Infinity Research)

12.6 Small Bodies Sampling and Excavation Approaches

There are two main reasons to acquire material from an asteroid or a comet: for science investigations, and for extracting resources. The two different motivations impose different performance requirements on the system acquiring the material.

For the purposes of science investigations, comets and asteroids are remnants from the solar system formation and can offer clues to the chemical mixture from which the planets formed. Science investigations normally require relatively small (on the order of grams) but pristine samples with no forward contamination. Hence sampling systems have to be designed to withstand various sterilization techniques, such as Dry Heat Microbial Reduction. For a sample return mission, the sample (in most cases) must be placed in a hermetically sealed container and kept within a specified temperature range at all times. The thermal requirement is of particular importance to prevent the loss of volatiles and/or possible chemical reactions.

For the purposes of resource mining and extraction, the amount of material to be retrieved and processed is much larger. In general, there are two options for resource extraction: transport the raw material to a processor, or transport the processor to the raw material. In the first case either a fraction or an entire asteroid could be brought back to cislunar space such as the Earth Moon Lagrange Point 1 (EML1). In the second option, the asteroid material could be processed *in situ* and only the useful material brought back. There are advantages and disadvantages to both approaches. It should be noted that for the purpose of commercial ISRU, only NEA's are considered, because of their proximity to Earth.

Processing material *in situ* and returning the final product means substantial savings in rocket fuel. This approach would also be desirable if the resource would be required for a mission to continue exploration of other planets rather than return to Earth. However, in this case, the mining and material processing systems must be very robust and fully autonomous.

Bringing an entire asteroid, some fraction of it, or even an ore concentrate to cislunar space would require more fuel and would take more time. However, it would also allow for teleoperation or human operation and the testing and verification of a number of extractions and processing technologies. The latter would be particularly advantageous since systems could be easily fixed if broken. In 2011 the Keck Institute for Space Studies (KISS) sponsored a study to investigate the feasibility of returning an entire 7 m asteroid weighing approximately 500 tons to the vicinity of the Earth. A 500 ton, C-type asteroid may contain up to 200 tons of volatiles such as water and carbon-rich compounds (100 tons of each), 90 tons of metals (83 tons iron, 6 tons nickel, and 1 ton cobalt), and 200 tons of silicate residue which is similar to the lunar surface material. The study found that it is feasible to capture and retrieve such an asteroid at a cost of approximately $2.6B (Brophy et al. 2012).

Capturing a small asteroid and bringing it back to Earth's vicinity might be the best first step in mining asteroids. Mining and processing technologies, as well as concepts of operation, could be tested and further developed within reasonable reach of Earth. Once robust technologies for mining asteroids are validated, it might be more cost effective to process all the materials *in situ*. The decision on whether to bring material (e.g. metal or water) back to Earth's vicinity or use it *in situ* for developing new spacecraft components ultimately depends on the cost of each approach. Furthermore, the current lack of pertinent data and information makes modeling these approaches particularly challenging. It is, however, safe to assume that bringing material to Earth will never be as cost effective as mining it even from great depths on Earth. Currently several South African gold mines are 4 km deep and heading to 5 km to tap into new gold reserves. Mining gold from these depths, even at grades as low as a few grams per ton, is still highly profitable. Commodities mined in space will have to compete economically against commodities mined on Earth. The intended location of end use becomes important to the relative economic appeal of resources mined in space. If the resources are to be used in space, then space-based sources become more attractive. Technologies to extract and process materials that will be widely useful in space (e.g. aluminum or titanium to create space structures) would be more valuable than technologies to extract materials that may command a high price on Earth, but are less useful in space. By way of analogy, 1 kg of water in a desert could be more valuable than 1 kg of gold.

Over the past few decades a number mining and sampling technologies have been developed (Bar-Cohen and Zacny 2009; Zacny et al. 2008; Ball et al. 2007). A vast majority of the exploration technologies focused on asteroids are in conceptual stages and only a limited number of them have been breadboarded, tested, and validated even under terrestrial conditions. Some sampling approaches have been tested on past missions or will be tested on future missions (Marchesi et al. 2001; Yano et al. 2002; Fujiwara and Yano 2005). In general, the progress has been slow because of the difficulty and costs associated with testing in reduced gravity and vacuum. The following sections describe a range of promising sampling, mining, and processing technologies. It should be noted that the list of presented methods is not all inclusive, but rather aims to give the reader an idea of the range of approaches.

12.6.1 Hayabusa

Hayabusa was the first mission to return sample material from another celestial body surface other than the Moon. Due to multiple malfunctions of the attitude control devices, the sampler did not work as designed. However, the mission did ultimately succeed in retrieving sample material from the Near Earth Asteroid (25143) Itokawa (Fig. 12.24).

The Hayabusa spacecraft used a small 5 gram tantalum pellet fired at 300 m/s into the surface to acquire a few grains of samples as shown in Fig. 12.25 (Barnouin-Jha et al. 2004). The material ejected by the pellet's impact was collected using a bi-impact sampler designed as a single collection system suitable for a range of target materials: metal-silicate hard bedrock, regolith layers with gravel, and micro-particles. The sampler consisted of a 1 meter long horn made of aluminum. The horn diameter at the tip was 15 cm. The goal of the horn was to direct sample into a sample chamber.

Fig. 12.24 A graphic of Hayabusa spacecraft acquring a sample from the Itokawa S-type asteroid. Courtesy JAXA

This approach was chosen because the mission planners could not know *a priori* what the surface properties would be – hard and consolidated or soft and powdery? This is unlike the Moon, Mars, or Venus, for example, which have been visited many times, within a relatively short period. In addition, information about the asteroid target is very limited even with a substantial ground observation campaign effort. This means that only once the spacecraft arrives at the target, will it be possible to perform detailed examination.

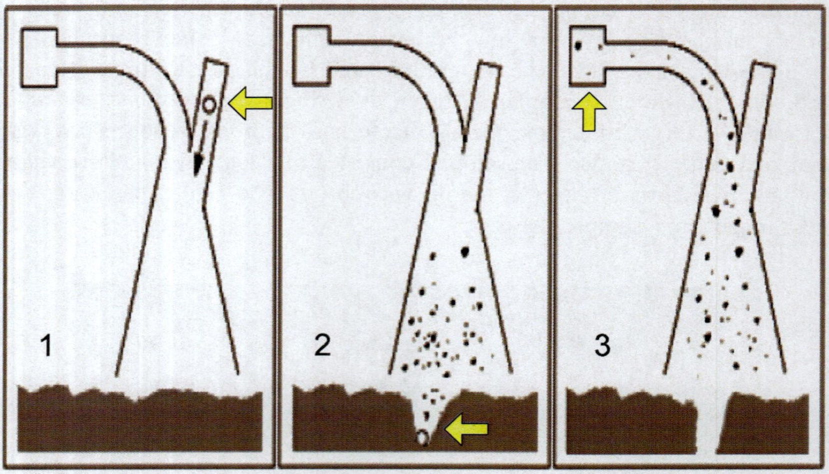

Fig. 12.25 Hayabusa sample acquisition sequence. (1) Pellet is launched at the asteroid surface. (2) Pellet strikes surface, scattering material. (3) Some of that material is captured for return. (Barnouin-Jha et al. 2004).

To determine the characteristics of a sample for planning the sampling operation, a number of impact experiments were performed during the development stage in various analog materials such as heat-resistant bricks, 200-mm glass beads, and lunar regolith simulant. Tests were performed at a normal and oblique impact angles as well as at 1g and in micro gravity by using a 140 m tall vacuum drop tower. Results indicated that sampler could acquire several hundred milligrams to several grams per shot. For oblique impacts at 45° or greater, however, the collection mass was less than 100 mg per shot.

For the nominal sampling procedure at the asteroid, once the tip of the sampler horn touched the surface, a pair of 5 gram tantalum pellets were fired at a surface at 300 m/s. These pellets would impact the surface, throwing ejecta into conical horn. On the actual mission, anomalies that occurred during the sampling operation prevented the spacecraft from firing the projectiles and capturing regolith material (Yano et al. 2006). However, during the first of the two attempts, some surface particles made their way up the horn and into a sample chamber when the horn touched the surface.

12.6.2 *Rosetta*

The goal of the European Space Agency's Rosetta mission is to study the comet 67P/Churyumov–Gerasimenko. The Rosetta mission consists of two spacecraft: the Rosetta orbiter and the Philae lander. The mission was launched on 2 March 2004 and will reach the comet by mid-2014. In November of 2014, the Philae lander is scheduled to land on the comet and perform investigations of its surface. To enable investigations of the comet, the lander is equipped with a sampling drill, named the sampling, drilling, and distribution (SD2) subsystem. The SD2 weighs 5 kg and can penetrate up to 250 mm below the surface and capture samples at predetermined depths. The samples, up to tens of mm^3, then can be transported to a carousel which distributes them into various onboard instruments (Finzi et al. 2007).

SD2 was designed and built by Galileo Avionica and consist of three sub-systems: a tool box, a carousel, and a local control unit (Fig. 12.26).

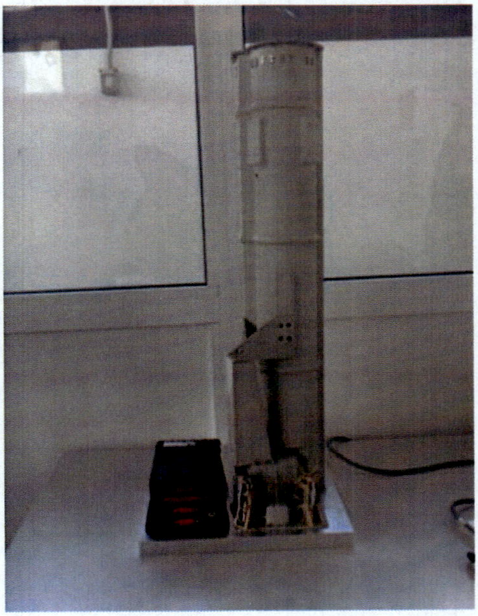

Fig. 12.26 Rosetta Drill, Sample & Distribution, called SD2 (SD2, 2012). Courtesy ESA

The tool box contains the actual drill and the sampler and can rotate about its vertical axis. The drill has two degrees of freedom: the Z-axis and rotation. To enable autonomous operation, the drill head is fitted with a compact force and torque sensor. The drill has been designed to penetrate material with strength ranging from fluffy snow to materials with a strength approaching a few MPa. The average drilling power is in the range of 10 Watts. The drill can also withstand storage temperatures to -160°C and can operate at temperatures down to -140°C.

Upon reaching the target depth, a sample is captured inside a drill bit, the drill is removed from the borehole, and the sample is delivered into a carousel. The carousel consists of 26 ovens and mates with scientific instruments (Finzi et al. 2007).

12.6.3 The Sample Acquisition and Transfer Mechanism (SATM) Drill

The Sample Acquisition and Transfer Mechanism (SATM) is a four-axis, highly instrumented drilling system that features sample preparation and handling system and also sample return containers. A prototype was developed and tested by Honeybee Robotics to validate the performance requirements for the NASA ST/4 Champollion mission (Fig. 12.27).

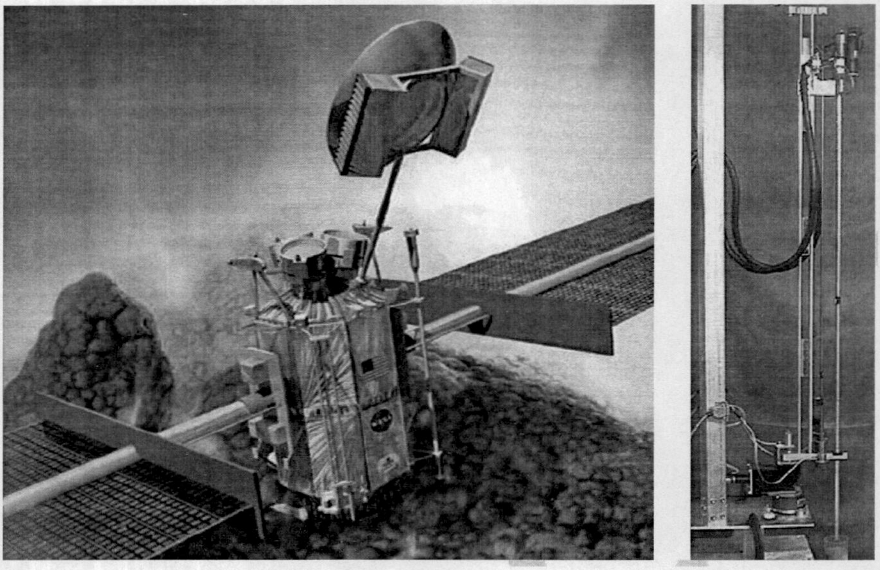

Fig. 12.27 SATM system. (a) Artistic impression of SATM on Champollion spacecraft. Image: NASA. (b) Prototype system developed and tested by Honeybee Robotics.

The drill was designed to acquire subsurface samples from a comet at selectable depths up to 1.2 m with little cross-contamination. The sample size is continuously adjustable between 0.1 and 1.0 cm^3 to cater for a variety of analytical instruments' requirements. The SATM creates a 13 mm diameter borehole. The mass of the prototype shown in Fig. 12.28 is 9 kg and its volume envelope is 60 x 60 x 138 cm.

The SATM sample is always delivered as fine powder regardless of the material type sampled (i.e., consolidated or unconsolidated). Powder samples can be transported and transferred to instruments or vessels such as chemical analysis ovens, a microscope/IR spectrometer, and a sample return container located at the base. To maintain the sample temperature to within 5 °C of its natural environment, the SATM drills at low speeds.

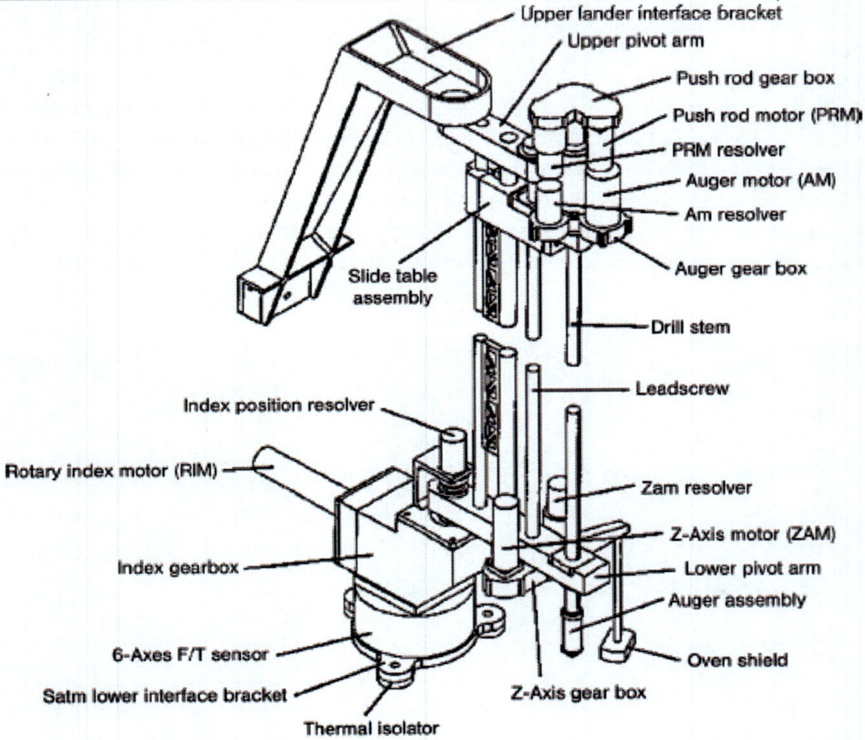

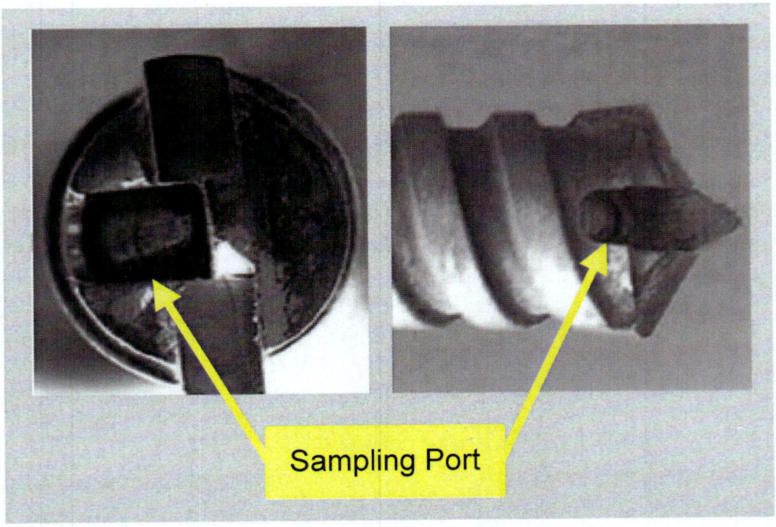

Fig. 12.28 SATM system. Top: System components. Bottom: Sample inlet feature.

Major components of the SATM design are shown in Fig. 12.28. The images on the right show the sample inlet feature of the drill tip. This door can be open at a desired depth to allow powder cuttings to flow into the sample chamber. The system also features a positive sample-eject mechanism within its sample chamber to ensure that samples are delivered to the *in situ* instruments. Samples can also be presented for analysis via a sapphire window located on the side of the drill stem. SATM can accommodate the bonding of a small cesium-137 source at the drill tip to permit density measurements. The drill tip can also be used as a tool to open and close the sample return container; eliminating the need for a separate actuator.

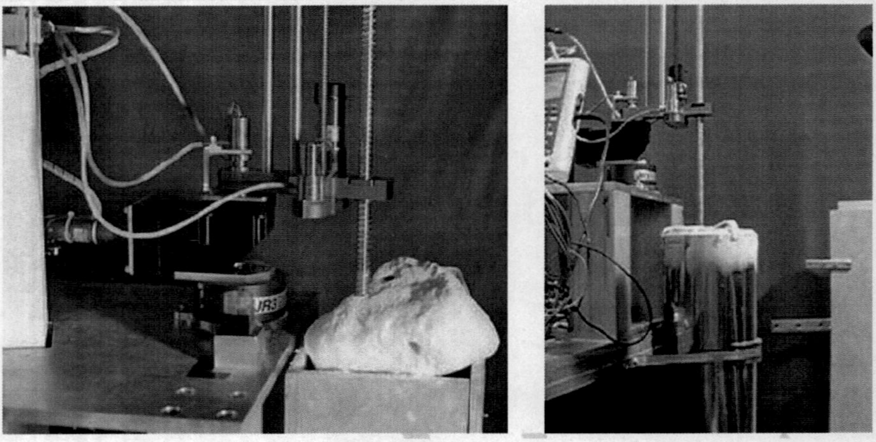

Fig. 12.29 Prototype testing in laboratory with (a) limestone and (b) cryogenic regolith stimulant

Specialized control algorithms were developed to allow autonomous adaptive operation in a low-gravity environment. The algorithms could also be adjusted for off-normal drill approach angles to minimize bit wandering. Laboratory tests conducted in limestone, basalt, and a cryogenic regolith simulant (see Fig. 12.29 have shown that a total energy of 25 W-hr is required to sample limestone (40 MPa UCS) at a rate of 0.88 cm/min with an auger speed of 194 rpm, a WOB of 55.6 N, and a drilling torque of 325 mNm. Limestone is an adequate ice simulant since the strength of ice at cryogenic temperature is similar to that of limestone.

12.6.4 Touch and Go Surface Sampler

The "Touch and Go Surface Sampler" (TGSS) developed by Honeybee Robotics, can drill and acquire a sample of regolith (up to 50 cc) or weak consolidated materials (UCS < 10 MPa) while the cutters penetrate to a depth of 1 to 4 cm. The system is reusable, and can store samples inside individual containers for *in situ* analysis or sample return. The TGSS consists of a high-speed sampling head attached to the end of a flexible shaft (Figs. 12.30 and 12.31). The sampling head rotates its counter rotating cutters at speeds of 5000 to 8000 rpm and consumes 20 W to 30 W of power. The mass of the current prototype is 450 grams, with a volumetric envelope of 50 mm x 75 mm x 150 mm (excluding the center drill bit).

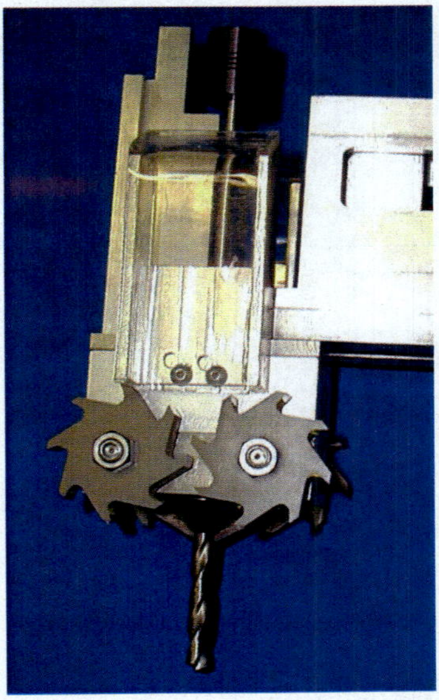

Fig. 12.30 Touch and Go Surface Sampler (TGSS) developed by Honeybee Robotics

The TGSS consists of three subsystems: a deployment mechanism, a sampling head, and a sample containment subsystem. The deployment mechanism deploys the sampling head by extending a boom to the surface. The sampling head contains 5 high-speed cutters (center drill side mounted toothed wheels) driven by a single motor. These high-speed cutters throw up surface material on contact and two guides mounted above the cutters direct sample debris/chips into a removable sample chamber. The sample containment subsystem transfers and seals multiple samples. The sampling head has a removable sample chamber on top of the cutters.

A prototype was developed and tested in a laboratory ambient environment on various target materials. The TGSS was demonstrated to sample regolith at a rate of 30 cc/sec and consolidated chalk with strength of 10 MPa at a rate of 0.5 cc/sec. A number of microgravity tests have shown that the TGSS can sample both consolidated and unconsolidated material, and includes a sample canister changeout system that allows sampling of multiple sites with minimal cross contamination.

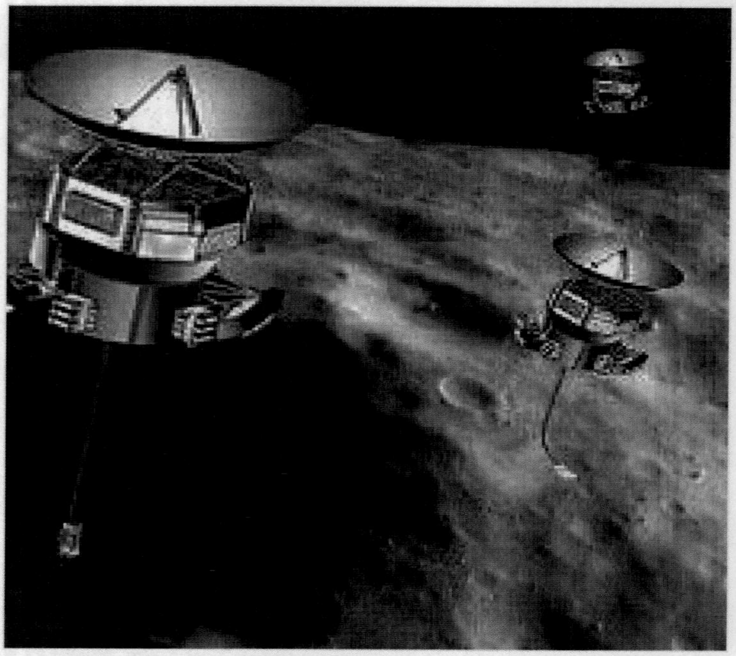

Fig. 12.31 A concept of a Touch and Go Surface Sampler (TGSS)

12.6.5 Brush Wheel Sampler

Another Touch and Go concept with brush-wheel mechanisms as shown in Fig. 12.32 rather than cutters has been developed at NASA Jet Propulsion Laboratory (Bonitz 2012). The main advantage of using brush wheels (as opposed to cutting wheels or other, more complex mechanisms) is that upon encountering soil harder than expected, the brushes could simply deflect and the motor(s) could continue to turn. That is, sufficiently flexible brushes would afford resistance to jamming and to overloading of the motors used to rotate the brushes, and so the motors could be made correspondingly lighter and less power hungry. Of course, one could select the brush stiffness and motor torques and speeds for greatest effectiveness in sampling soil of a specific anticipated degree of hardness. In simplest terms, such a mechanism would contain brush wheels that would be counter-rotated at relatively high speed. The mechanism would be lowered to the ground from a spacecraft or other exploratory vehicle. Upon contact with the ground, the counter rotating brush wheels would kick soil up into a collection chamber. Thus, in form and function, the mechanism would partly resemble traditional street and carpet sweepers.

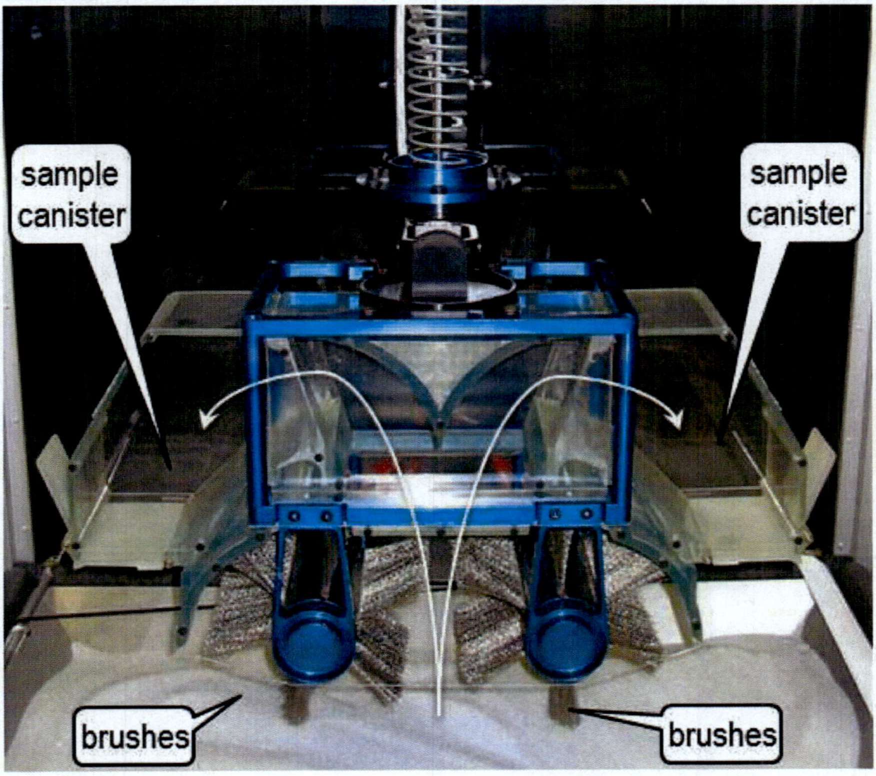

Fig. 12.32 2nd generation prototype of Brush Wheel Sampler (BWS) used for testing in micro-gravity environment aboard NASA's micro-gravity aircraft in 2004 and in vacuum on earth in 2009 (Bonitz 2012)

12.6.6 Sample Return Probe

A slightly different approach for small body sampling includes a standalone small spacecraft with a sample acquisition system. The concept includes several sampling probes that travel to a small body onboard a parent spacecraft. After arriving at the small body of interest, establishing an orbit, and selecting a site of interest, one of a probes will detach from the parent spacecraft, spin stabilize using an attitude control system, and propel itself towards the surface. Upon impact, the probe will collect a sample and transport the sample into its upper stage where it will later be hermetically sealed. The upper stage of the probe with the collected sample will then detach from the rest of the probe body and take off from the surface using the same attitude control system used to guide it to the surface. The probe with a sample inside it will then dock with the mother spacecraft and hand-off the hermetically sealed sample (Fig. 12.33). Multiple probes could be used in this way to insure mission success or to sample multiple locations.

The main advantage of this approach is that the sampling system is completely independent of the spacecraft. As such, the dangers associated with proximity operations in more conventional approaches, are eliminated. Developing tiny spacecraft is quite feasible. For example the Hayabusa mission carried the 591 gram and approximately 10 cm tall by 12 cm in diameter MINERVA lander. Hayabusa 2, on the other, hand will carry at least one MINERVA type lander as well as another small lander named MASCOT (Mobile Asteroid Surface Scout) from the German Aerospace Center. Given the small size of sample return probes, technologies developed for nano-satellites could be directly applicable.

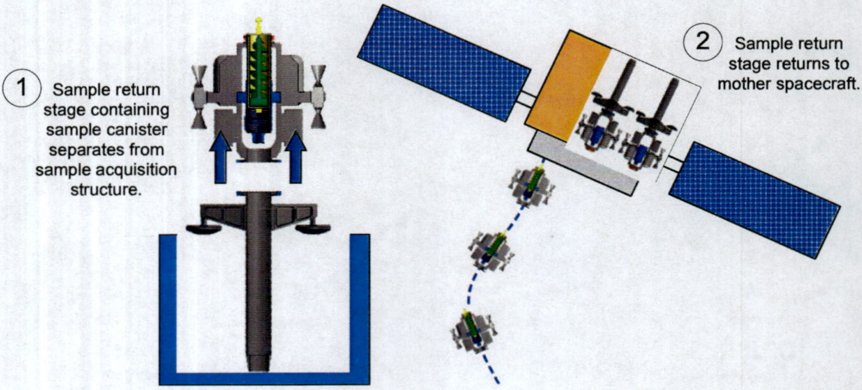

Fig. 12.33 A standalone sample return probe. Upon sample acquisition the upper stage releases and rendezvous with the mother spacecraft.

12.6.7 Harpoon Samplers

A harpoon sampler provides a means to rapidly collect samples from microgravity bodies at distances defined only by the length of the tether system. Such systems do not require landing on the target or a means to hold the spacecraft to the surface. The time to acquire a sample using a projectile could range from seconds to minutes, and is therefore compatible with a slowly moving science platform. This allows samples to be collected from specific interesting regions, such as inside a crevasse or vents of an active comet. Harpoon samplers can be fired into the surface of a small body, capture a sample during the course of penetration into the subsurface, and reeled back into the spacecraft using a tether. All these operations could be accomplished at a relatively safe distance from the asteroid.

A number of potential concepts for capturing and retrieving a sample exist (Bar-Cohen and Zacny 2009; Nuth 2011). These vary from a harpoon dropped from the spacecraft to lowering a mechanism that could fire a sampling tip into the surface using compressed air or stored mechanical or chemical energy. A number of harpoon acceleration concepts could impart enough energy on the sampling tip to accomplish sampling in relatively consolidated formations. Of course a drawback to any high powered harpoon is that if the formation is very soft, the harpoon will penetrate much deeper than expected, making it more challenging to retrieve the sample.

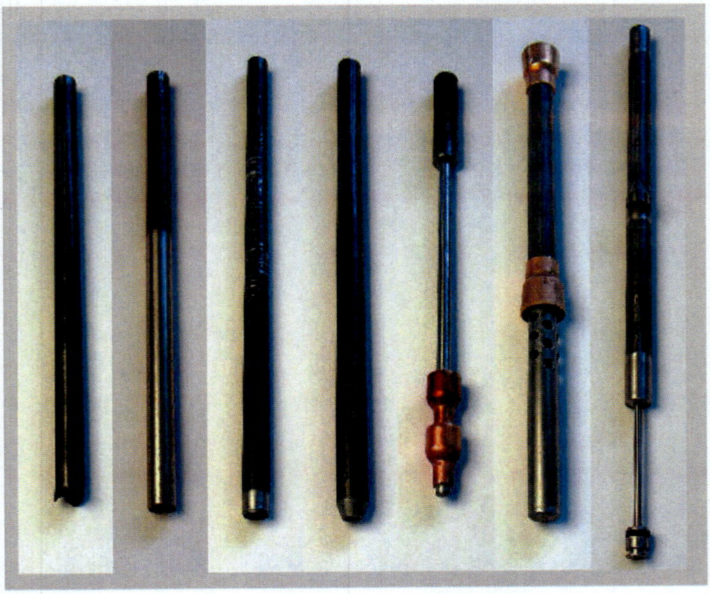

Fig. 12.34 Honeybee Robotics Harpoon Breadboards

Honeybee Robotics developed a number of harpoon concepts that could be deployed in a variety of formations (see Fig. 12.34). The final harpoon breadboard was tested with cryogenic ice at approximately -150° C. The tests were conducted in order to determine the ability of the tip to sample ice when impacted at a 45° and at a 0° angles. During these tests, the harpoon tip successfully captured samples of cryogenic ice when impacted at up to 45° off vertical.

NASA Goddard Space Flight Center has also developed a projectile-based sample acquisition system (SAS) for comet sampling. The system consists of a launcher, a tether payout and retrieval system, and a Sample Retrieving Projectile (SaRP). Goddard's 3rd generation sample retrieving projectile (SaRP) prototype shown in Fig. 12.35 consists of an outer sheath and an inner sample collecting cartridge. The sample cartridge uses a spring loaded rotating knife-edged seal to contain the collected sample. This prototype has been tested and consistently collects and retrieves several hundred grams of sample (Wegel and Nuth 2013).

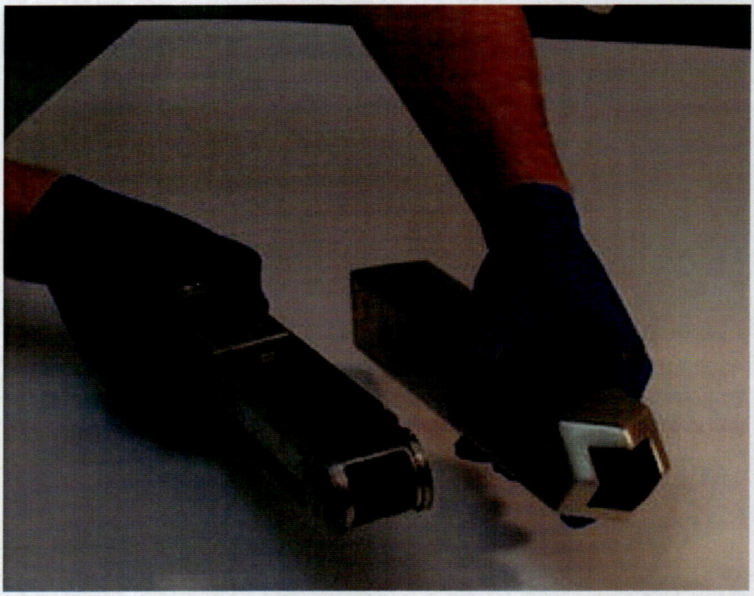

Fig. 12.35 NASA GSFC Sample Retrieving Projectile (SaRP). The photo shows a prototype sampler tip (right) and the sample collection cartridge (left).

12.6.8 Pneumatic Approaches

Many terrestrial applications use vacuum cleaners for picking up dirt. The principle lies in creating a lower pressure at the back end of a pick-up hose than at the front, and thereby forcing the outside air to flow in and loft particles along the way. Such a system will not work in the vacuum of space. However, one can create a differential pressure by injecting gas into the regolith and then guide this gas, as it escapes from the regolith, into appropriate pick up tubes (Zacny et al. 2004, 2008, 2010). Figure 12.36 shows two potential approaches. The first one relies on injecting pressurized gas into the top few centimeters of regolith and then capturing the regolith propelled upwards by the escaping gas into a transfer tube. The second approach entails a self-enclosed tube with injector holes. Once the regolith is acquired into the tube, gas is injected into a tube and lofts the captured regolith above the gas injector holes up the tube. Some of this gas will escape into the surrounding vacuum, reducing the excavation efficiency. The exact volume of gas lost will be a function of the permeability of regolith, the depth of the external tube buried inside the regolith and the depth of an injector nozzle inside the regolith.

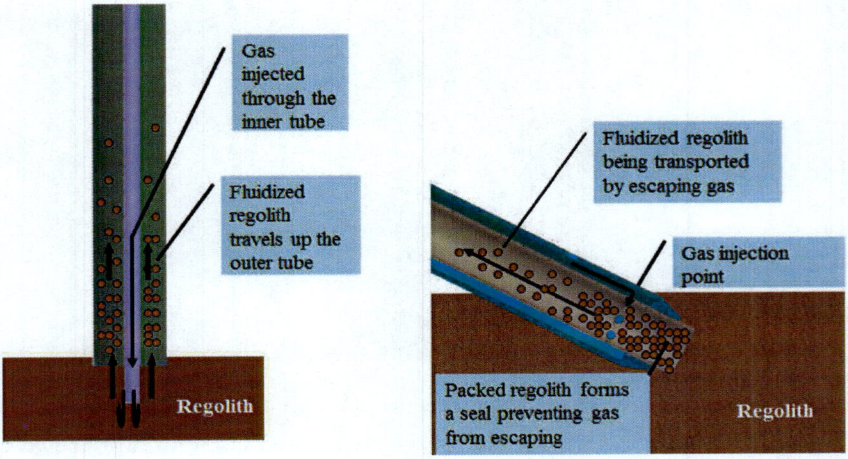

Fig. 12.36 Left: Plunge method of pneumatic lift/excavation. Right: Traverse method of pneumatic lift/excavation.

The pneumatic approach can be ideally suited for obtaining both small samples for scientific analysis, as well as a bulk sample for mining and processing of resources. The working gas could be supplied by electrolyzing water into its constitutive elements: Hydrogen and Oxygen. Since the pneumatic system consist of fixed nozzles and a series of tubes for providing of gas for mining and guiding of excavated regolith to a storage container or storage area potentially hundreds of meters away it has no moving parts such as motors, bearings and so forth, and hence is well suited for the dusty environment. By adjusting the pressure and flow rates, it is possible to differentiate smaller and larger particles, allowing optimization tailored for specific ISRU processing systems (Zacny et al. 2008). If only smaller particles are removed from the surface, it may remove the need for communition of the regolith, and thus reduce the energy consumed in the regolith processing stage. In addition, since smaller particles have larger surface area to volume ratios, the extraction efficiency will naturally increase.

Pneumatic excavation in the context of space applications is not a new concept. David McKay at NASA Johnson Space Center has proposed pneumatic excavation for lunar mining (as shown in Fig. 12.37) and evaluated the feasibility of pneumatic transfer for the movement of lunar regolith at lunar gravity conditions (and atmospheric pressure) on NASA's KC-135 reduced gravity aircraft (Sullivan et al. 1994). They found that the choking velocity (in the vertical transfer) and the saltation velocity (in the horizontal transfer) at lunar gravity were reduced to 1/2–1/3 of the velocity required at 1 g (choking and saltation velocities are the minimum gas velocities required to keep particles aloft).

Fig. 12.37 Lunar pneumatic mining with a production-scale system: LUNOX Pilot Plant. NASA image: S91-25382 created by Pat Rawlings.

Additional tests in vacuum conditions have shown that 1 g of gas (air) at 101 kPa absolute (i.e. at atmospheric pressure) can successfully lift 6000 g of soil particles at high velocity at 1g gravity (Zacny et al. 2008, 2010). Tests conducted at various pressures suggested that gas lofting efficiencies increase as the ambient pressure drops, reaching a maximum value at approximately 1 mTorr.

For bulk regolith mining, a potential approach might use a system that has been initially developed for lunar regolith mining (Zacny et al. 2010). The pneumatic regolith miner is similar to a conventional vacuum cleaner; however instead of creating suction at the nozzle mouth, a compressed gas is injected, moving the captured soil within the nozzle up the tube and through the cyclone separator into a soil bin. Figure 12.38 shows a pneumatic excavator integrated onto the NASA Ames research Center K10-mini platform. The system has been successfully tested in a 3 meter long bed filled with GRC-1 soil simulant within a 3.5 long vacuum chamber (Zacny et al. 2008).

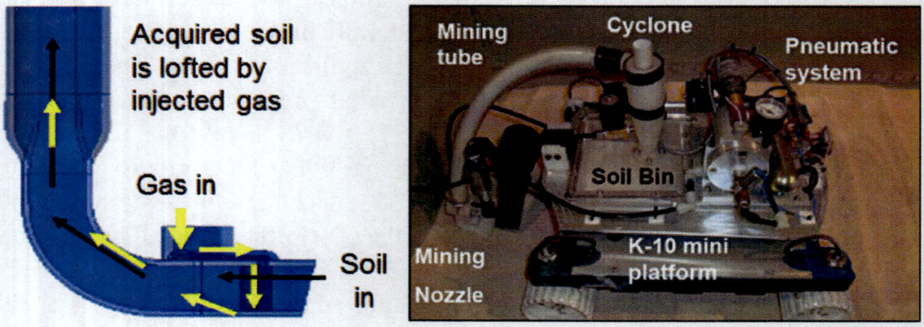

Fig. 12.38 Components of the pneumatic mining rover

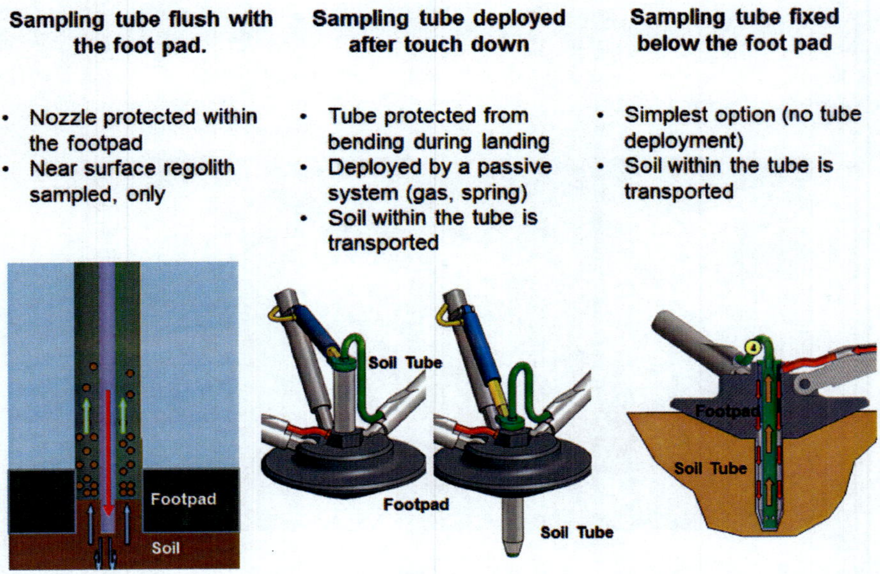

Fig. 12.39 Various approaches to sample acquisition using pneumatic systems embedded inside a lander footpad

For acquisition of small samples for scientific analysis, the pneumatic system could be integrated, for example, within each of the footpads of a lander. Sampling tubes could either be fixed or deployable, flush with the footpad or sticking beneath the footpads (Fig. 12.39). One would use deployable tubes if there is some risk that the lander will not contact the surface perfectly vertically. If only near surface regolith is of interest, a tube that is flush with the footpad would be the method of choice (Zacny et al. 2008, 2012).

With this type of deployment a level of redundancy is built into the system. For example, in case one of the legs lands on a rock the other two or three pneumatic tubes (if the lander has 4 and not 3 legs) will still be functional. Upon landing, the tubes within each of the legs will fill up with regolith. With one puff of gas, the captured soil can be lofted to a sampling chamber onboard of the spacecraft. Hence, this sampling system requires just one valve to open and close the gas cylinder and one actuator to open/close a sampling chamber.

Pneumatic sample acquisition will be demonstrated for the first time on the OSIRIS-RE-x asteroid sample return mission (Fig. 12.40). Just recently, NASA has selected the OSIRIS-REx mission to travel to a near-Earth carbonaceous asteroid (101955) 1999 RQ36, study it in detail, and bring at least 60 grams of sample material back to Earth (OSIRIS-Rex, 2012). The sampling operation will be conducted using a Touch-And-Go Sample Acquisition Mechanism (TAGSAM) system. Upon contacting the surface, an annular jet of nitrogen pointed at a surface fluidizes the regolith. This dusty gas escapes through a filter element within the round sampler. The filter then captures regolith and lets the nitrogen escape into space. During this time, the surface contact pads also collect fine-grained material. After sample acquisition, the sampler is placed inside a Sample Return Capsule for return to Earth.

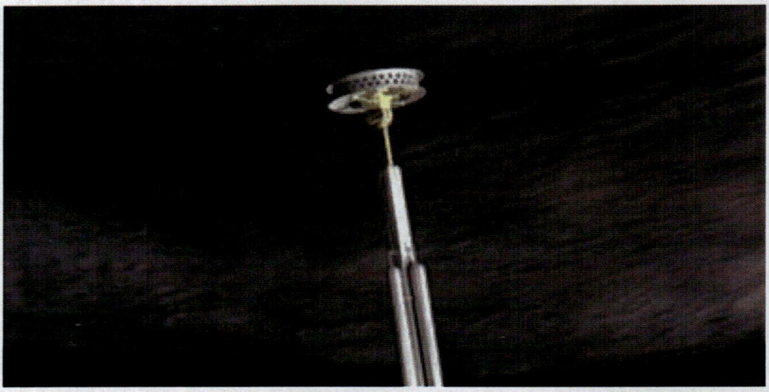

Fig. 12.40 OSIRIS-Rex asteroid sample return mission will use pneumatics to capture at least 60 grams of asteroid material. Credit: NASA/Goddard/University of Arizona.

12.6.9 Mobile In situ Water Extractor System

Many approaches to extracting water from frozen soil follow a 'terrestrial' mining approach; they consist of mining ice or ice-bearing soil, transferring the feedstock to a water recovery plant, and then extracting and storing of the water. The Mobile *In situ* Water Extractor (MISWE) eliminates the transfer or processing of feedstock steps and in turn simplifies the water extraction process. The MISWE approach is to integrate both the mining and the water extraction systems into a single unit, integrated with the drill. The water extraction process follows three steps: 1) mining the soil using deep fluted auger, 2) extracting the water from soil within the flutes, and 3) discarding the soil. Hence only the water is transported back, while dry soil is left behind.

A single MISWE reactor consists of the Icy-Soil Acquisition and Delivery System (ISADS), and the Volatiles Extraction and Capture System (VECS). The ISADS is a deep fluted auger that drills into the ice or icy-soils and retains material on its flutes. Upon material acquisition, the ISADS is retracted into VECS and sealed. The VECS consists of a cylindrical heat exchanger and volatiles transfer system (a reactor). The material on the deep flutes is initially heated and resultant water vapor is bled into a water collection canister by a one way valve where it condenses. The heat from the water collection canister can be circulated back into the reactor via a heat exchanger. After water extraction is complete the ISADS is lowered towards the ground and spun at high speed ejecting the dry soil via centrifugal action. At the same time, the collected water is pumped from the canister into a holding tank. The MISWE then moves to the next location and the operation is repeated. Once the water tanks are full, the spacecraft is launched back towards Earth or other destination.

Since the regolith is not actually transferred, there is no need for a transfer system and associated mechanisms. Also, if a spacecraft is powered using Radioisotope Thermal Generators (RTG) or more efficient the Advanced Stirling Radioisotope Generator (ASRG), the heat generated by the unit can be transferred to the reactor. For example, the RTG on the 2011 MSL Curiosity rover generates ~120 Watts electrical power and > 1 kW heat.

Figure 12.41 shows a concept of the Asteroid MISWE with 8 water reactors attached to each leg of the lander. The reactors are placed at oblique angle to provide anchoring. Hence, reactors have dual use: anchor and water extraction.

To determine feasibility of this water extraction approach, approximately 50 tests were completed in vacuum chamber as shown in Fig. 12.42 (Zacny et al. 2012). The MISWE breadboard, in the optimum configuration, has been demonstrated to extract 19 grams of water from icy soil. The water extraction efficiency was 92% and the remaining 8% of water was lost and never captured. The power usage during the 30 minute process was 34 Watts. This translates to the energy usage of 17 Wh or 0.9 Wh/gram of water, or ~80% energy efficiency.

Fig. 12.41 A concept of the Asteroid water extraction system called Mobile *In situ* Water Extractor (MISWE) with 8 water reactors attached to each leg of the lander. The reactors are placed at oblique angle to provide anchoring. Hence, reactors have dual use: anchor and water extraction.

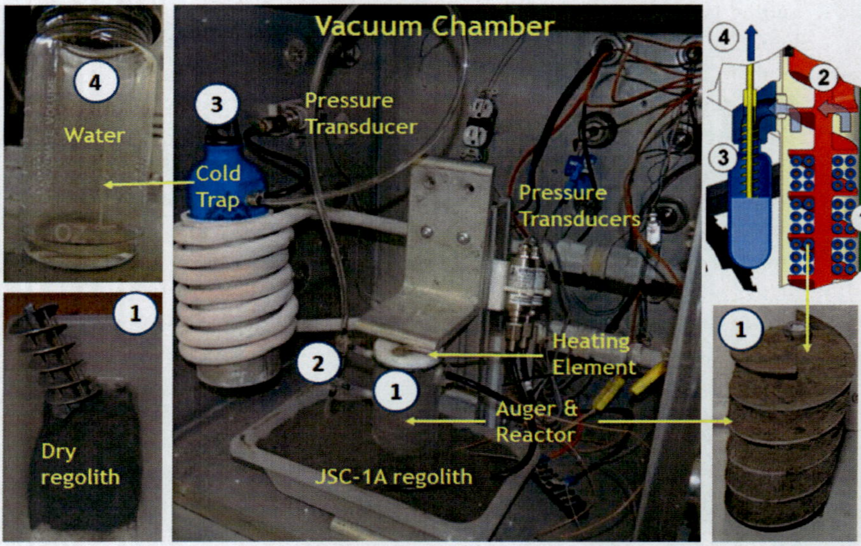

Fig. 12.42 MISWE water extraction system being tested in vacuum chamber: 1. Icy soil is collected between auger flutes. 2. Auger is heated, releasing water vapor from soil. 3. Water vapor condenses on cold finger and collects inside a canister. 4. Liquid water is pumped out of the canister to holding tank.

During the tests it was observed that a soil's temperature can be used to monitor the drying cycle. Once the temperature starts increasing it indicates that the soil is dry and the heat is no longer absorbed by the water sublimation process. Hence the heating process can be terminated making the extraction system more efficient. To make the process even more efficient, the power and duration of the applied heat and dwell time after the heating cycle could also be varied.

A MISWE reactor with a 1 meter long and 20 cm diameter auger will be able to recover ~3 kg of water every 40 minutes from regolith with ~10 wt% water. Assuming that it takes another 20 minutes for the remaining tasks (drilling to 1 meter depth, discarding of dry soil), a total of 3 kg of water can be recovered every hour. Thus the mass of water that the Asteroid MISWE system with 8 reactors can recover per hours is 8 x 3 kg per hour = 24 kg per hour or 16 tons per month. At a very conservative cost of water at EML1 of $10K/kg, this translated to a value of extracted water of $160M.

12.6.10 Percussive and Vibratory Systems

Percussive and vibratory systems are employed in circumstances where reduction in excavation forces is of primary importance (Craft et al. 2009; Zacny et al. 2009, 2012; Green et al. 2013). They are ideally suited to applications where the excavator is very light, e.g. small robots and/or low gravity environments. The main drawback to these systems is that percussive or vibratory mechanism requires additional power and in turn energy, which of course taxes the spacecraft energy supply.

The difference between percussive and vibratory operation is that the former uses hammer blows (i.e. a scoop or other digging end effector is periodically impacted by some kind of a hammer), while the latter uses for example off-centered spinning masses to induce vibrations along various planes. It should be noted that percussive systems also vibrate, but vibrations lies along the direction of the hammer blow. There are numerous ways of producing vibrations and hammering such as cam-spring, voice coil, magnetic and so on. Figure 12.43 shows an example of vibratory scoop developed by Honeybee Robotics, at the end of a robotic arm.

The reduction in digging forces in soils is attributed to reduction in friction angle because of soil dilation (increase in volume). That is vibrating or percussive scoops fluidize soil, which are therefore easier to penetrate. If soils have substantial cohesion, percussive systems are probably better suited since these tend to impart various energy impacts to crusty soils. Note that in icy soils, where ice is firmly bonded to soil particles, neither of the systems will be able to succeed due to the high strength of ice-bound soil.

Fig. 12.43 Vibratory scoop can substantially reduce excavation forces

12.6.11 The Regolith Advanced Surface Systems Operations Robot (RASSOR)

The Regolith Advanced Surface Systems Operations Robot (RASSOR) excavator robot developed at NASA Kennedy Space Center and shown in Fig. 12.44 is a teleoperated mobility platform with a space regolith excavation capability. This more compact, lightweight design (<50 kg) has counter rotating bucket drums, which results in a net-zero reaction horizontal force due to the self-cancelation of the symmetrical, equal but opposing, digging forces.

This robot can operate in extremely low-gravity conditions, such as on the Moon, Mars, an asteroid, or a comet. In addition, the RASSOR system is designed to be easily transported to a space destination on a robotic precursor landing mission. The robot is capable of traversing over steep slopes and difficult regolith terrain, and has a reversible operation mode so that it can tolerate an over-turning incident with a graceful recovery, allowing regolith excavation operations to continue.

Fig. 12.44 The Regolith Advanced Surface Systems Operations Robot (RASSOR)

The RASSOR excavator consists of a mobility platform with tread belts on the port and starboard sides that are each driven by electrical motors, but it could also operate with a wheel system to further reduce mass. Two batteries are mounted in a “saddlebag” configuration on either side. Two counter-rotating bucket drum digging implements are held by a rotating cantilever mechanism at the fore and aft ends of the mobility platform. The cantilever arms are raised and lowered to engage the bucket drum into the soil or regolith. A variable cutting depth is possible by controlling the angles of the cantilever arms.

The unit has three modes of operation: load, haul, and dump. During loading, the bucket drums will excavate soil/regolith by using a rotational motion whereby scoops mounted on the drum's exteriors sequentially take multiple cuts of soil/regolith while rotating at approximately 20 revolutions per minute. During hauling, the bucket drums are raised by rotating the arms to provide a clearance with the surface being excavated. The mobility platform can then proceed to move while the soil/regolith remains in the raised bucket drums. Finally, when the excavator reaches the dump location, the bucket drums are commanded to reverse their direction of rotation to the opposite spin from digging, causing the gathered materials to be expelled out of each successive scoop. It can also stand up in a vertical mode to deliver regolith over the edge of a hopper container. For ore recovery from captured regolith, the buckets could be designed to form a reactor. In particular, once regolith is captured, the bucket drum can be sealed and the regolith inside it can be heated to recover valuable volatiles, such as water-ice.

The RASSOR can operate with either side up in a reversible mode and it can flip itself over. This means the unit can drive directly off of the deck of a lander to deploy in low gravity, eliminating a deployment mechanism, which saves mass and increases reliability due to decreased complexity. The RASSOR system is scalable and may be mounted on mobility platforms of various sizes.

12.7 Conclusions

For decades, asteroid mining has been a very popular subject of study (Ross 2001; Sonter 1998; Lewis 1996). However, only recently a number of studies demonstrated that capturing a small asteroid and bringing it back to cislunar space is feasible with a current state of technology (Brophy at al. 2011, 2012).

In addition, the *In situ* Resource Utilization (ISRU) community has been very active in developing various planetary mining and processing technologies. The community published their research at various meetings and hence there exists a wealth of information on the subject. More information could be gained, for example, by tapping into proceedings from the following meetings: Space Resource Roundtable, Planetary and Terrestrial Mining Sciences Symposium, the AIAA Space Conference and Exposition, and the ASCE Earth and Space Conference.

However, when examining available literature, it becomes clear that there has been an extensive technology development for the Moon and Mars, but very little for asteroids. Some of the Moon-focused technologies could also be applied to a microgravity environment (i.e. asteroids), but each technology has to be treated on a case by case basis.

This chapter attempted to consolidate asteroid focus technologies related to anchoring, mining, and excavation. One notable commonality among these technologies is the low level of maturity. Only a few have even been tested in a relevant environment. If NASA or a commercial sector is planning to explore these bodies, the excavation and processing technologies would have to be developed and tested from scratch. In technology development, all potential investors are of course concerned about the associated cost to increase the maturity of the technology. However, it takes many years to develop and mature a space-rated technology so it can be robust for commercial applications (i.e. multiple rather than a single operation). This timeline cannot be shortened by simply investing more money. Hence, it would be more cost effective in a long run to have a steady research and development effort even at a modest level of funding, rather than nothing for many years followed by a massive cash investment to meet a deadline.

Useful Websites

NASA Asteroid and Comet Watch http://www.nasa.gov/mission_pages/asteroids/main/index.html

JPL Near Earth Object Program; http://neo.jpl.nasa.gov/

NASA National Space Science Data Center Asteroid Photo Gallery; http://nssdc.gsfc.nasa.gov/photo_gallery/photogallery-asteroids.html

National Space Society's Asteroid Page; http://www.nss.org/settlement/asteroids/

Planetary Resources; www.planetaryresources.com

Deep Space Industries; http://deepspaceindustries.com/

References

Andrea, M., Chesley, S., Sansaturio, M., Bernardi, F., Valsecchi, G., Arratia, O.: Long term impact risk for (101955) 1999 RQ36. Icarus 203(2), 460–471 (2009)

Asbeck, A.T., Kim, S., Cutkosky, M., Provancher, W., Lanzetta, M.: Scaling hard vertical surfaces with compliant microspine arrays. Int. J. Robot. Res. 25(12), 1165–1179 (2006)

Ball, A., Garry, J., Lorenz, R., Kerzhanovich, V.: Planetary Landers and Entry Probes. Cambridge University Press (2007)

Bar-Cohen, Y., Zacny, K. (eds.): Drilling in Extreme Environments Penetration and Sampling on Earth and Other Planets. John Wiley & Sons, New York (2009)

Barnouin-Jha, O.S., Barnouin-Jha, K., Cheng, A.F., Willey, C., Sadilek, A.: Sampling a Planetary Surface with a Pyrotechnic Rock Chipper. In: Proc. IEEE Aerospace Conference, March 6-13 (2004)

Bonitz, R.: The Brush Wheel Sampler - a Sampling Device for Small-body Touch-and-Go Missions. In: 2012 IEEE Aerospace Conference, Big Sky, MT, March 3-10 (2012)

Campins, H., et al.: Spitzer Observations of spacecraft target 162173 (1999 JU3). Astronomy and Astrophysics 503, L17–L20 (2009)

Fujiwara, A., et al.: The Rubble-Pile Asteroid Itokawa as Observed by Hayabusa. Science 312, 1330–1334 (2006)

Gaffey, M.J., Cloutis, E.A., Kelley, M.S., Reed, K.L.: Mineralogy of Asteroids. In: Bottke Jr., W.F., Cellino, A., Paolicchi, P., Binzel, R.P. (eds.) Asteroids III, pp. 183–204. University of Arizona Press, Tucson (2002)

Biele, J., et al.: Current status and scientific capabilities of the Rosetta lander payload. Advances in Space Research 29, 1199–1208 (2002)

Biele, J., et al.: The putative mechanical strength of comet surface material applied to landing on a comet. Acta Astronautica 65, 1168–1178 (2009)

Biele, J., Ulamec, S.: Capabilities of Philae, the Rosetta Lander. Space Sci. Rev. 138, 275–289 (2008)

Britt, D.T., Yeomans, D., Housen, K., Consolmagno, G.: Asteroid Density, Porosity, and Structure. In: Bottke Jr., W.F., Cellino, A., Paolicchi, P., Binzel, R.P. (eds.) Asteroids III, pp. 485–500. University of Arizona Press, Tucson

Brophy, J.R., Gershman, R., Landau, D., Yeomans, D., Polk, J., Porter, C., Williams, W., Allen, C., Asphaug, E.: Asteroid Return Mission Feasibility Study. AIAA-2011-5665. In: 47th AIAA/ASME/SAE/ASEE Joint Propulsion Conference and Exhibit, San Diego, California, July 31-August 3 (2011)

Brophy, et al.: Asteroid Retrieval Feasibility Study. Keck Institute for Space Studies, California Institute of Technology, Jet Propulsion Laboratory (2012), `http://kiss.caltech.edu/study/asteroid/asteroid_final_report.pdf`

Craft, J., Wilson, J., Chu, P., Zacny, K., Davis, K.: Percussive digging systems for robotic exploration and excavation of planetary and lunar regolith. In: IEEE Aerospace Conference, Big Sky, Montana, March 7-14 (2009)

Finzi, E., Zazzera, B., Dainese, C., Malnati, F., Magnani, P., Re, E., Bologna, P., Espinasse, S., Olivieri, A.: SD2 - How to Sample a Comet. Space Science Reviews 128, 281–299 (2007)

Fujiwara, A., et al.: The Rubble-Pile Asteroid Itokawa as Observed by Hayabusa. Science 312, 1330–1334 (2006)

Fujiwara, A., Yano, H.: The asteroidal surface sampling system onboard the Hayabusa spacecraft. Aeronautical and Space Sciences Japan, 8–15 (2005)

Glassmeier, K., Boehnhardt, H., Koschny, D., Kührt, E., Richter, I.: The Rosetta Mission: Flying Towards the Origin of the Solar System. Space Science Reviews 128, 1–21 (2007)

Green, A., Zacny, K., Pestana, J., Lieu, D., Mueller, R.: Investigating the Effects of Percussion on Excavation Forces. J. Aerosp. Eng. (2013), doi:10.1061/(ASCE)AS.1943-5525.0000216

Hilchenbach, M., Rosenbauer, H., Chares, B.: First contact with a comet surface: Rosetta lander simulations. In: Luigi, C., et al. (eds.) The New Rosetta Targets. Observations, Simulations and Instrument Performances. Astrophysics and Space Science Library, vol. 311, p. 289. Kluwer Academic Publishers, Dordrecht (2004)

Jones, T., et al.: Amor: A Lander Mission to Explore the C-Type Triple Near-Earth Asteroid system 2001 SN263. 42nd Lunar and Planetary Science Conference, The Woodlands, Texas, March 7-11, p. 2695. LPI Contribution No. 1608 (2011)

Kawaguchi, J., Fujiwara, A., Uesugi, T.: Hayabusa – Its technology and science accomplishment summary and Hayabusa-2. Acta Astronautica 62, 639–647 (2008)

Kerr, R.: Magnetic Ripple Hints Gaspra Is Metallic. Science 259, 176 (1993)

Lewis, J.S.: Mining the Sky, Untold Riches from the Asteroids, Comets, and Planets. Helix Books (1996) ISBN 0-201-47959-1

Lodders, K.: Solar System Abundances of the Elements. In: Goswami, A., Eswar Reddy, B. (eds.) Principles and Perspectives in Cosmochemistry. Lecture Notes of the Kodai School on 'Synthesis of Elements in Stars' held at Kodaikanal Observatory, India, April 29-May 13. Astrophysics and Space Science Proceedings, pp. 379–417. Springer, Berlin (2010)

Mankins, J.: Technology Readiness Level (2005), `http://www.hq.nasa.gov/office/codeq/trl/trl.pdf`

Marchesi, M., Campaci, R., Magnani, P., Mugnuolo, R., Nista, A., Olivier, A., Re, E.: Comet sample acquisition for ROSETTA lander mission. In: Proceedings of the 9th European Space Mechanisms and Tribology Symposium, Liège, Belgium. Compiled by Harris, R.A. ESA SP-480, September 19-21, pp. 91–96. ESA Publications Division, Noordwijk (2001)

Normile, D.: Rover Lost in Space. Science 310, 1105 (2005)

OSIRIS-Rex (2012), `http://www.nasa.gov/topics/solarsystem/features/OSIRIS-rex.html` (accessed January 10, 2012)

Parness, A.: Microgravity Coring: A Self-Contained Anchor and Drill for Consolidated Rock. In: IEEE Aerospace Conference, Big Sky, MT, USA (2012)

Parness, A., Frost, M., Thatte, N., King, J.: Gravity-Independent Mobility and Drilling on Natural Rock Using Microspines. In: IEEE ICRA, St. Paul, MN, USA (2012)

Parness, A., Frost, M., King, J., Thatte, N.: Demonstrations of Gravity-Independent Mobility and Drilling on Natural Rock Using Microspines, video. In: IEEE ICRA (2012), `http://www.youtube.com/watch?v=0KUdyBm6bcY`

Richardson, J., Melosh, H.J., Lisse, C.M., Carcich, B.: A ballistic analysis of the Deep Impact ejecta plume: determining Tempel 1's gravity, mass and density. Icarus 190, 357–390 (2007)

Ross, S.: Near-Earth Asteroid Mining. Space Industry Report (December 14, 2001), `http://www.esm.vt.edu/~sdross/papers/ross-asteroid-mining-2001.pdf`

SD2 (2013), `http://www.aero.polimi.it/SD2/?SD2` (accessed January 10, 2013)

Sonter, M.: The Technical and Economic Feasibility of Mining the Near-Earth Asteroids. In: 49th IAF Congress, Melbourne, Australia, September 28-October 2 (1998)

Spenko, M., Haynes, G., Saunders, J., Cutkosky, M., Rizzi, A.: Biologically inspired climbing with a hexapedal robot. J. Field Robotics 25, 223–242 (2008)

Sullivan, T., Koenig, E., Knudsen, C., Gibson, M.: Pneumatic conveying of materials at partial gravity. Journal of Aerospace Engineering 7, 199–208 (1994)

Tsiolkovskii, K.: The Exploration of Cosmic Space by Means of Rocket Propulsion. Nauchnoe Obozrenie (Scientific Review) Magazine (5) (1903) (in Russian)

Yano, H., et al.: Touchdown of the Hayabusa Spacecraft at the Muses Sea on Itokawa. Science 312, 1350–1353 (2006)

Ulamec, S., et al.: Rosetta Lander—Philae: Implications of an alternative mission. Acta Astronautica 58, 435–441 (2006)

Ulamec, S., Biele, J.: Surface elements and landing strategies for small bodies missions – Philae and beyond. Advances in Space Research 44, 847–858 (2009)

Veverka, J., et al.: NEAR at Eros: Initial imaging and spectral results. Science 289, 2088–2097 (2000)

Veverka, J., et al.: The landing of the NEAR-Shoemaker spacecraft on asteroid 433 Eros. Nature 413, 390–393 (2001)

Wall, M.: Is Space Big Enough for Two Asteroid-Mining Companies? SpaceNews (January 22, 2013)

Wegel, D., Nuth, J.: NASA Developing Comet Harpoon for Sample Return (2013), `http://www.nasa.gov/topics/~solarsystem/features/comet-harpoon.html` (accessed on January 13, 2013)

Wilhite, A., Arney, D., Jones, C., Chai, P.: Evolved Human Space Exploration Architecture Using Commercial Launch/Propellant Depots. In: 63rd International Astronautical Congress, Naples, Italy, October 1-5 (2012)

Yano, H., Hasegawa, S., Abe, M., Fujiwara, A.: Asteroidal surface sampling by the MUSES-C spacecraft. In: Proceedings of Asteroids, Comets, Meteors, pp. 103–106 (2002)

Zacny, K., Chu, P., Avanesyan, A., Osborne, L., Paulsen, G., Craft, J.: Mobile. *In situ* Water Extractor for Mars, Moon, and Asteroid. *In situ* Resource Utilization. AIAA Space 2012, Pasadena, CA, September 11-13 (2012)

Zacny, K., Huang, K., McGehee, M., Neugebauer, A., Park, S., Quayle, M., Sichel, R., Cooper, G.: Lunar soil extraction using flow of gas. In: Proceedings of the Revolutionary Aerospace Systems Concepts-Academic Linkage (RASC-AL) Conference, Cocoa Beach, FL, April 28-May 1 (2004)

Zacny, K., Paulsen, G., Szczesiak, M., Craft, J., Chu, P., McKay, C., Glass, B., Davila, A., Marinova, M., Pollard, W., Jackson, W.: LunarVader: Development and Testing of a Lunar Drill in a Vacuum Chamber and in the Lunar Analog Site of the Antarctica. J. Aerosp. Eng. (2013), doi:10.1061/(ASCE)AS.1943-5525.0000212

Zacny, K., Bar-Cohen, Y., Brennan, M., Briggs, G., Cooper, G., Davis, K., Dolgin, B., Glaser, D., Glass, B., Gorevan, S., Guerrero, J., McKay, C., Paulsen, G., Stanley, S., Stoker, C.: Drilling Systems for Extraterrestrial Subsurface Exploration. Astrobiology Journal 8, 665–706 (2008)

Zacny, K., et al.: Pneumatic Excavator and Regolith Transport System for Lunar ISRU and Construction. Paper 2008-7824, AIAA Space 2008 (2008)

Zacny, K., et al.: Investigating the Efficiency of Pneumatic Transfer of JSC-1a Lunar Regolith Simulant in Vacuum and Lunar Gravity During Parabolic Flights. AIAA Space 2010 (2010)

Zacny, K., Beegle, L., Onstott, T., Mueller, R.: MarsVac: Actuator free Regolith Sample Return Mission from Mars. Abstract 4263, Concepts and Approaches for Mars Exploration, Houston, TX, June 12-14 (2012)

Zacny, K., Mueller, R., Galloway, G., Craft, J., Mungas, G., Hedlund, M., Fink, P.: Novel Approaches to Drilling and Excavation on the Moon. AIAA-2009-6431, AIAA Space, Conference and Exposition, Pasadena, CA, September 14-17 (2009)

Zacny, K., Mueller, R., Paulsen, G., Chu, P., Craft, J.: The Ultimate Lunar Prospecting Rover Utilizing a Drill, Pneumatic and Percussive Excavator, and the Gas Jet Trencher. AIAA Space 2012, Pasadena, CA, September 11-13 (2012)

Chapter 13
Closed-Cycle Pneumatics for Asteroid Regolith Mining

Leonhard E. Bernold

University of New South Wales, Sydney, Australia

13.1 Introduction

While it is believed that over 1.1 million Asteroids are circling the Sun, NASA counts 853 Near-Earth Asteroids of at least1 kilometer in diameter. These celestial bodies are without atmosphere and most of them without ice. Important for economic mining is that the dry Asteroids contain fine regolith similar to the Moon.

Bates and Jackson (1980) defined regolith as

> "a general term for the layer or mantle of fragmental and unconsolidated rock material, whether residual or transported and of highly varied character, that nearly everywhere forms the surface of the land and overlies or covers bedrock."

During 4.6 billion years, space weathering, bombardments by meteorites, solar and galactic particles, radiation and frequent quakes pulverized the "rocky" lunar surface that is now covered with a flower-like dust, leaving wonderful boot prints. Miyamoto et al. (2007) presented Japan's Hayabusa mission to the 300 m diameter asteroid Itokawa "covered with unconsolidated millimeter-size and larger gravels" with patches of fine regolith in between. The researchers concluded that the finer particles had migrated below the rocks. McKay and his colleagues concluded that the same method of grain sorting happened on the Moon

"The process of regolith formation can be divided roughly into two stages. During the early stage, shortly after bedrock is first exposed and the regolith is still relatively thin (less than a few centimetres), both large and small impacts can penetrate the regolith and excavate fresh bedrock …it continues to be progressively modified by micrometeoroid impacts and by high-energy solar and cosmic charged particles…. Subsequent impacts may turn the regolith over (gardening) and bring young, buried regolith to the surface ... and the maturing process continues." (McKay et al. 1991)

The result of this gardening is a fine grained regolith that is between 4 m and 15 m in depth. On the other hand, Kargel (1994) reported that observations of "Asteroids Ida and Gaspra suggest that there is indeed a layer of pulverized regolith, probably at least 100 m deep." McKay et al. (1991) also presented a grain-size distribution where approximately 90% is smaller than 500 μm and 50% smaller than 100 μm. Tsuchiyama et al. (2007) surmised that the comparatively lower percentage of fines below 100 μm was due to "(i) smaller grains having higher ejection velocity and therefore higher loss rates after impacts, (ii) selective electrostatic levitation of smaller grains and and/or (iii) size-dependent segregation by vibration (Brazil-nut)." This granular convection effect, where larger particles or stones slowly rise to the surface during seismic events, is part of the gardening effect discussed by McKay et al. (1991).

Another characteristic important for mining regolith is its density, g/mm^3, given either as bulk or relative density. Both increase with depth where particles are packed together ever tighter. Carrier et al. (1991) confirmed that "the in situ relative density of lunar soil has been found to be about 65% (medium to dense) in the top 15 cm, increasing to more than 90% (very dense) below a depth of 30 cm". The extremely high density at greater depth can not be explained with the increasing weight of the overburden compressing the lower layers. Carrier et al. (1991) experimented with lunar simulants and concluded that this phenomenon has to be the result of "Post-depositional vibrations of the material (e.g., by seismic waves associated with meteorite impacts) that produce a material with an initial relative density of 90% or more." (Critical to predicting the density of regolith on asteroids is the observation that the rapid increase of density with depth can not be the result of self-compaction alone. In other words, lower gravity on Asteroids may not automatically lead to less densified regolith, as the seismic vibrations may be very effective in compacting regolith.

Closely related to relative density of a soil are its porosity, diffusivity and permeability. The latter two are related to quantity of gas that flows through porous medium such as dense regolith. This characteristic is of interest for terrestrial issues and is being studied extensively. For example, a group of researchers in Denmark recently looked at the cause-and-effect of these two coefficients in silt and volcanic ash (dust). Deepagoda et al. (2011) found that "soil compaction more than soil type was the major control on gas diffusivity and to some extent also on air permeability." To calculate the flow rate of a gas in m^3/sec, Darcy's law is being used in its simplest form. Volumetric flux, a function of permeability, is multiplied with the pressure gradient divided by the gas viscosity.

13.2 Mining Drives Human Prosperity

Mining for precious material, be it amber, flint or copper, is as old as our civilisation. Gold rushes around the world opened new areas for agriculture and created wealth. In modern times, we still mine copper, iron or platinum once brought to Earth by debris from asteroids collisions. The drive to ever higher production led to larger and large equipment and more efficient methods. Today, surface mining is preferred to underground mining, even though large efforts have to be expended to strip large amounts of material from a buried ore deposit. Of course, Hollywood loves underground mining with its promise of quick riches, dark tunnels, deep shafts and deadly explosions. However, the timber supported narrow tunnels seen in the films have been replaced with long-wall miners, feeders, remote controlled crushers, conveyors and even robot trucks.

Will we extend our state-of-the-art mining technology to the Moon and Asteroids? Will the machines and tools perfected for terrestrial use be as efficient? Reviewing history of human exploration should provide us with some hints. The near catastrophic beginning of England's penal colony in Sydney Australia winks from the past, alerting of the danger of assuming that what has been designed to fit one environment could be transferred to another. "They had very little building material and the government had provided only a very limited supply of tools, which were of a bad quality. With the local trees being huge, and the wood hard, these tools were soon blunt or broken and building slowed" (Dunn and McCreadie 2005). Eighteen years earlier, the British explorer James Cook had mapped the east coast of Australia. Similar to the present mapping spacecraft, Cook had sketches of the coast made from his ship, the Endeavour. However, what looked like normal elm or pine was, in fact, the extremely hard eucalyptus, or gum tree, unknown in England. This short story, which ended when the ships of the Second saved the First Fleet from starvation, highlights the historical importance of knowing and utilizing In-Situ-Resource Utilization (ISRU) when we venture into space to mine the Asteroids.

13.3 Terrestrial Approach to Asteroid Mining

Carrier et al. (1991) warned that the density of lunar regolith bellow 30 cm had a relative density between 90-100%. Although such highly compacted soil does not challenge the heavy diesel powered dozers or excavators on Earth, the high transportation cost and the lower gravity makes the brute-force approach unfeasible. Between 1989 and 1993, a small research group at the University of Maryland (UMCP) began studying the problem of energy efficient surface mining on the Moon (Bernold and Sundareswaran 1990, Bernold 1992). After compacting 0.5 t of lunar soil simulant to the needed 90% relative density, having to make use of a large hydraulic press and a heavily steel reinforced container, terrestrial excavation methods were used unsuccessfully. For example, a scaled robot backhoe excavator (Bernold 1991), simulating a 6 times larger bucket on the Moon, was unable to penetrate the dense soil-surface. The design of an earth-based experiment has to consider the changed conditions existing on an Asteroid. This process is referred to as ensuring similitude. For space excavation, the most important factors to be considered, in addition to gravity, are the cohesion and friction angle of the regolith and the lunar simulant (Bernold 1991, Zacny et al. 2010). Because Carrier et al. (1973) was able to measure, we know that the peak cohesion of the returned samples is between 0.1-1.0 kN/m^2, which is considered extremely low. On the other hand, the peak friction angles had a spread between 30° to 50°. The basalt based simulant used for the tests had a peak cohesion value of 14 kN/m^2, an amount that is also considered minimal, and a friction angle of 39°, which lies well between the boundaries of the tested regolith. Thus, it is believed that a high level of similitude was achieved with the designed set-up for the excavation focusing on the reduction of size and mass to 1/6.

A second excavation tool was used, a dragline bucket also simulating the 1/6 g condition on the Moon with a bucket that was 1/6 its lunar size made of lightweight bronze material. It only scratched the compacted surface, while a third tool, a clamshell bucket, was not able to "clam" any soil. It became apparent that a different approach to soil loosening was needed.

An extremely efficient, safe and highly compact technology to loosen dense material is related to Nobel, who invented dynamite as a secure, powerful source of chemical energy that could be translated into mechanical force. In fact, over 250 explosives were used for every Space Shuttle flight to separate parts with special detonators performed without any miss-firings in the vacuum of space. The hypothesis of the research group at UMCP was that small amounts of the safe explosives detonated inside a hole drilled vertically into the lunar simulants would drastically decrease the amount of force and energy needed to extract the soil material. Figure 13.1 presents three pictures from that effort.

As explosives, 0.25 or 1.0 g of pentaerythritol tetranitrate (PETN) were used. The result of the chemical reaction is a large volume of gas that, if constrained by solid, will create pressure that can be used to perform work. One such application is to insert and stem the PETN into a small hole that was drilled into a large rock. Properly designed, the pressure created by the gas is large enough to crack and break it into smaller pieces. In the vacuum of an Asteroid, the effect would be very similar to what would happen on Earth with 1 atm.

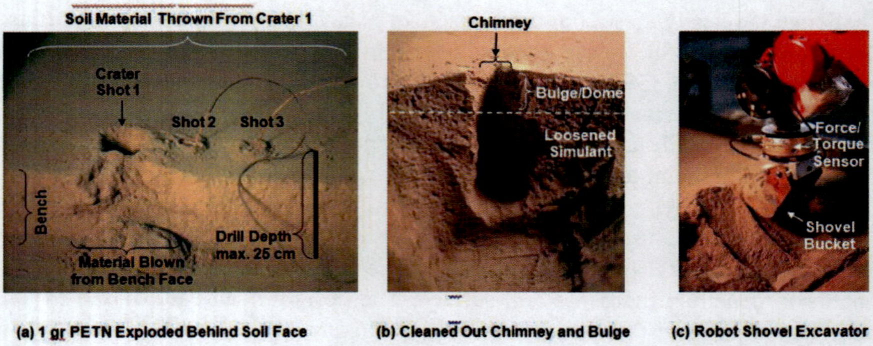

Fig. 13.1 Loosening and Excavation of Lunar Soil Simulant with 100% Relative Density

In order to place the PETN at the proper depth without impacting the density of the soil around it, small vertical shafts had to be created to bury and stem the small amount of explosives together with the tiny blasting cap and wire, as depicted in Fig. 13.1(a). The scene shows a 20 cm high bench after 1 gr of PETN was ignited at 25 cm depth behind the face of the bench. Two more shots are ready to be blasted. Figure 13.1(b) presents the a cross-sectional view of the chimney, dome and small crater created by a deep 1 gr PETN charge far from a bench. As indicated, the loosening effect resulted in significant bulge on the surface while some of the material was ejected through the chimney. The effects of different PETN charges were measured using a robotic arm equipped with a force/torque sensor behind an excavator bucket (Bernold 1992). While following a programmed excavation path, the energy used and the amount of soil excavated were measured after each test. As expected, the explosive loosening resulted in drastic force and energy reductions (Goodings et al. 1992, Lin et al. 1994).

A side problem, which led to a new thinking, was the creation of the small and vertical borehole necessary to bury the small PETN charges. Various mechanical drills were tested, with none resulting in a long borehole with a desired small diameter. Eventually, the team found that a drinking straw attached to a small vacuum cleaner was able to rapidly create perfect holes rapidly reaching the desired depths. Looking around, one finds that construction has adopted this non-mechanical method of excavation to create holes in the ground to visually locate buried utilities. It's referred to as suction, or vacuum, excavation. The main problem that remains is to find a light-weight and low energy method to transport soil from the place of extraction to the processing. The following section introduces a conveying system that has many characteristics extremely desirable to establishing an Asteroid miner.

13.4 Principle of Pneumatic Conveying

The term "pneu" always reminds me of my Swiss German times, as there is no such word in German. It was adopted from French, meaning a car tire, although the term pneumatic has been adopted from the Greek word *pneuma* meaning *wind* or *breath.* Of course, the effectiveness of wind to move things was welcomed by seafaring merchants, but was also feared in the desert where sand storms buried caravans and towns. Sand dunes and snow drifts are both example of pneumatic conveying of solids. The principles of the concept have long been adopted by man to build efficient and reliable transportation capabilities. "Pneumatic conveying system is a century-old technology of pipeline transportation that has been proved to be successful for conveying mail, telegraph, document, cash and other lightweight materials in some of the United State Cities. Besides, the system is also used to transport granular materials and powder over long distances." (Cohen, 1999) In fact, this technology is widely being used to move particles ranging from powders to pellets with bulk densities from 15 to 3000 kg/m^3 (0.9 to 185 lb/ft^3). Modern pneumatic conveying defines a method of moving solids suspended in or forced by a gas moving from a high to a low pressure are forced through horizontal and vertical pipes. Common solid particles that can be up to 40 mm (1.6 in) are cement, flour, sugar, saw dust, plastic pellets, beans, cereals and grain. In dry and dilute conditions, such material is blown at a velocity of 5-30 m/s (16 – 100 ft/s) over long distances.

In a pneumatic conveying system, most of the energy is used for the transport of the air itself. The energy need for fast pneumatic conveying is relatively high, but this is often outweighed by easy handling and, in well designed systems, dust free solutions.

Airflow is the result of a pressure difference in the pipe system. This is achieved by either creating high pressure at the intake side or low pressure at the output side. The fundamental physical principle of this operation is the ideal gas law:

$$pV = nRT \tag{13.1}$$

where: p = pressure of the gas, V= volume of the gas, n = amount of substance of the gas (in moles), R = gas constant (8.314 $J{\cdot}K^{-1}mol^{-1}$), T = absolute temperature (in K).

Boyle's law refined the general law for a non-fluid condition where the temperature stays constant

$$p_A V_A = p_B V_B \tag{13.2}$$

Equation (13.2) indicates that the value of pV for condition A in a closed cycle is equal to that for condition B. In other words, if the pressure between A and B goes down, the volume must go up along a non-linear convex curve.

Both pressurizing and vacuuming tanks manipulate n, the amount of gas moles. With the help of a pump molecules are either added or removed from a tank with a fixed volume at a fixed temperature. Removing molecules results in a lower p, while adding molecules to the tank will increase the pressure. Connecting a 120 kPa tank to a pipe system with the outside pressure of 101.3 kPa (1 atm), the air with the higher pressure or high n will reduce its pressure p from 120 to 101.2 kPa while, according to Boyle's law, the air will expand its volume V when leaving the tank. Creating pressure differences between two ends of a pipe offers the possibility to build a closed system able to transport small particles for a long distance. Pneumatic transportation of many different kinds of materials and objects has gained acceptance for terrestrial applications because of its simplicity, reliability, containment, opportunity for automation, and economics.

13.5 Suction Intake Mechanism

Many different technologies exist to load the material into the pneumatic system at the high pressure side. The goal is to supply the particles fed from 1 atm into the flowing air of the pipe without losing differential pressure. One such example is the rotary airlock feeder. Still, it is sometimes more desirable to use an open pipe or hose to suck up solids that are stored on the ground at1 atm. A good example is the vacuum cleaner where a pump creates a low pressure and the surrounding air pushes the material into and along the pipe to a cyclone or air-filter that separates the solids. A modification of this concept is the Venturi based air injector intake system shown in Fig. 13.2.

An air injector takes advantage of the Venturi effect created when a pipe with a flowing gas inside changes its diameter. This important phenomenon can be explained by considering the volumetric flow rate (Q = velocity times the surface area of pipe) that has to stay constant and Bernoulli's equation for fluid dynamics (Eq. (13.3)).

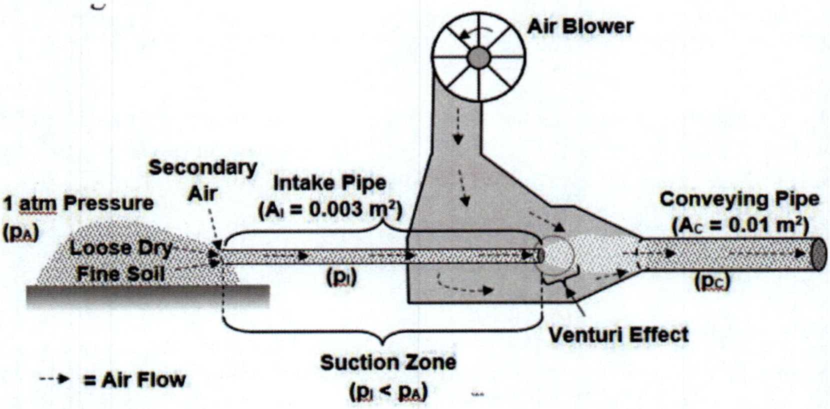

Fig. 13.2 Charging Pneumatic Conveyor with an Air Injector

Figure 13.2 presents a situation where loose dry fines is pushed into the intake nozzle of a pipe by the high pressure of 1 atm (p_A). This intake process is helped by the turbulence and high speed of the secondary air. The air blower injects air into an inline Venturi chamber which feeds into a conveying pipe with an assumed cross-sectional area $A_C = 0.01\ m^2$. If a blower creates a speed of $v_C = 12$ m/s (40 ft/s) inside the conveying pipe with $A_C = 0.01\ m^2$, then the flow rate $Q_C = 0.12\ m^3/s$ (4.24 ft^3/s). Because Q_C has to match the Q_I at the intake the air velocity at the intake with the smaller area is $v_I = Q_C/A_I = (0.12\ /0.003)$ m/s = 40 m/s (131ft/s). To calculate the pressure differences between the two pipes, the Bernoulli's equation for dynamic fluids can be used:

$$\frac{v^2}{2} + gz + \frac{p}{\rho} = \text{const} \tag{13.3}$$

where : v = speed of fluid/gas, g = gravitational acceleration, z = elevation above a reference plane, p = static pressure, ρ = density of fluid/gas (air density at 20 °C = 1.2041 kg/m^3).

Since the elevation difference is zero, this function can be used in the following simplified form:

$$\rho\frac{v^2}{2}+p=\text{const} \tag{13.3'}$$

Thus :

$$\rho\frac{v_I^2}{2}+p_I=\rho\frac{v_C^2}{2}+p_C;$$

$$p_C-p_I=\frac{\rho}{2}\left(v_I^2-v_C^2\right)=\frac{1}{2}\cdot 1.2041\frac{\text{kg}}{\text{m}^3}\left(1600\frac{\text{m}^2}{\text{s}^2}-144\frac{\text{m}^2}{\text{s}^2}\right)=876\text{ Pa}$$

It is apparent that increasing the velocities in the two pipes, by increasing the blower speed or changing the pipe diameter, will change the dynamic and total pressures. This, in turn, will affect the production of the overall system.

13.6 Closed Cycle Pneumatic Conveying

Zacny et al. (2009) wrote that

> "pneumatic regolith mining system is akin to terrestrial vacuum cleaner in that gas momentum is used to move the solid particles into a bin. … On the Moon, there is no air, and in turn conventional blower-type systems will not work. However, compressed gas can be delivered at the nozzle itself, to 'push' the acquired soil up through the tubing and into the storage bin."

Prototype systems were stationary (Zacny et al. 2008) but could be mounted on a rover travelling across the surface (http://goo.gl/UjB0L). The problem with dust being kicked up and the limit of how much can be mined by a lightweight vehicle needing power replenishment makes it desirable to consider a stationary mining system.

Figure 13.3 displays the concept of a closed-cycle mining system, powered by one blower, delivering regolith from a point of extraction to a fixed processing plant. The glass pipes are made in-situ, utilizing the abundant silicates through an extrusion process.

Closing the gas cycle is enabled by the use of a cyclone separator, possibly put in series, removing all the fine particles from the gas stream. Different than in Fig. 13.2, air does not flow from the surrounding atmosphere but, instead, the contained gas moves from the cyclone to the low pressure return pipe leading back to the blower.

Because of a lack of atmospheric pressure in vacuum, the static pressure value in Bernoulli's equation equals 0 Pa. This will not impact the efficiency of the conveying system, as it depends slowly on the pressure differences. Finally, further Venturi injectors may be installed along the pipe in order to accelerate and elevate the particles.

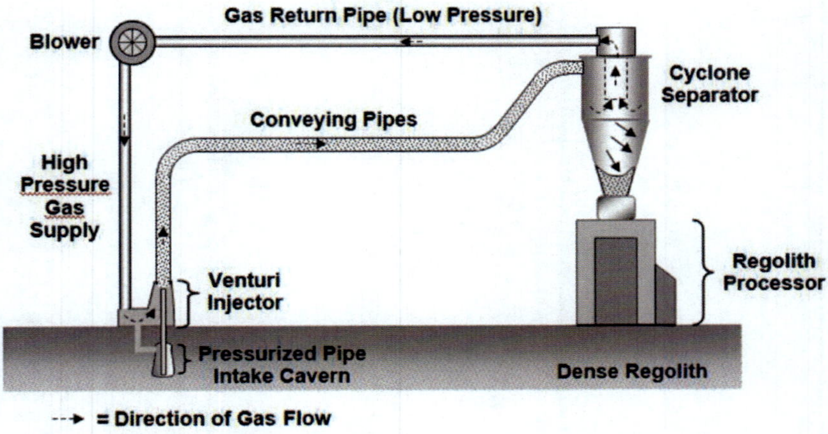

Fig. 13.3 Concept of a Closed Cycle Pneumatic Regolith Conveying

13.7 Creating a Subsurface "Atmosphere"

It was discussed earlier that regolith on the Moon has been, over millions of years, vibrated to 90 to 100 % relative density. The same situation can be expected on asteroids as they were bombarded by meteorites small and large. As a consequence, gas diffusivity and permeability, both impacted by density and the pressure gradient, will be low. Naturally, maintaining a low pressure inside a cavern deep inside the compacted regolith will minimize the loss of gas through the pores of the soil.

Creating a pressurizable seal between soil or rock and a pipe was of key importance in grouting layers of soil or rock around boreholes or permanent anchors. The objective is to fill the spaces in loose or cracked ground with a grout to increase its density and reduce permeability without excavation. For this purpose, one or two inflatable packers mounted around a centre pipe, as shown in Fig. 13.4(a), are lowered into the drilled hole and positioned at the desired depth. Through a separate tube gas or a liquid is pumped into the sleeve causing it to expand and press against the wall of the bore hole. The goal is to create a seal that will stop the grout that will be injected through another tube into the space below.

While terrestrial packer systems are used to create spaces that can be injected with high pressure grout, one can easily see that the physical principle can be transferred into an environment with 0 atm. As Fig. 13.4(b) demonstrates, the tight seal created by the packer sleeve allows the space below to be pressurized in order to force the grout into the cracks that need to be closed. The following section will discuss how the principles were used to assemble a prototype pneumatic mining system that could be used to experiment under terrestrial conditions.

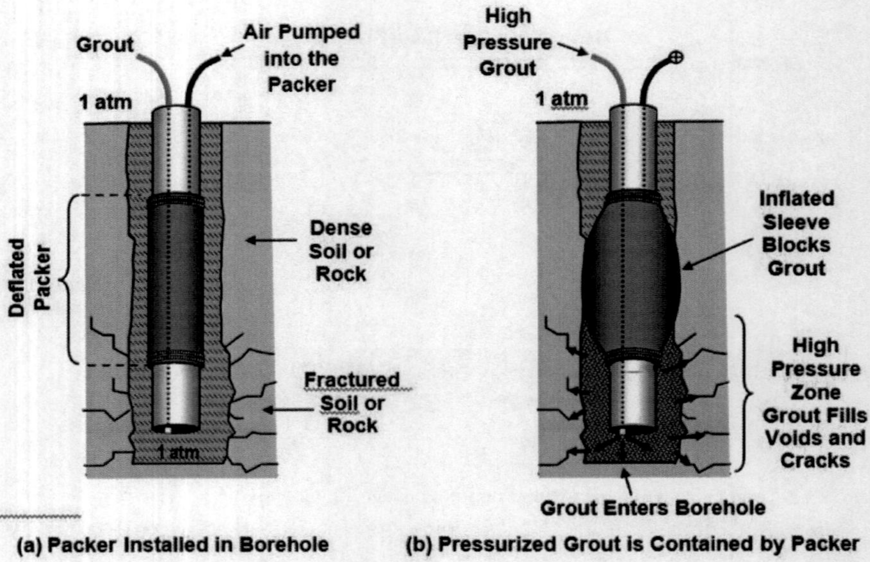

Fig. 13.4 Principle of the Inflatable Borehole Packer

13.8 An Experimental Pneumatic Regolith Mining System

Conducting scaled experiments on space technology under terrestrial conditions consider requires consideration of two types of similitude. The first addresses the sizing of the model in relation to the actual condition, while the second concerns itself with the differences between Earth and the space object, such as an Asteroid. For pneumatic conveying the two main parameters are: a) gas velocity, and b) gravity force. As was discussed earlier, the velocity of the gas inside a closed-cycle system is a function of the pressure gradient and the cross-sectional area of the pipe. The effect of a reduced gravitational acceleration on particles within a pneumatic system was studied by Sullivan et al. (1994) for both horizontal and vertical transport. By installing a test apparatus into NASA's KC-135 aircraft, the team was able to validate the predicted values for chocking velocities of in lunar gravity at 1/2 - 1/3 of the velocity needed at 1 g. The question is still open regarding how to simulate the micro-gravitational acceleration g_A on an Asteroid on a soil particle with a mass m ($F_{gA} = m \cdot g_A$) on Earth. For the purpose of this initial study basaltic simulant particles with sizes smaller than 75 μm were used.

The prototype mining system displayed in Fig. 13.5 consists of seven modules: 1) Vibratory simulant compactor, 2) gas supply to extractor, 3) simulant extractor, 4) pneumatic conveyor, 5) filter separator, 6) low pressure gas return, and 7) blower.

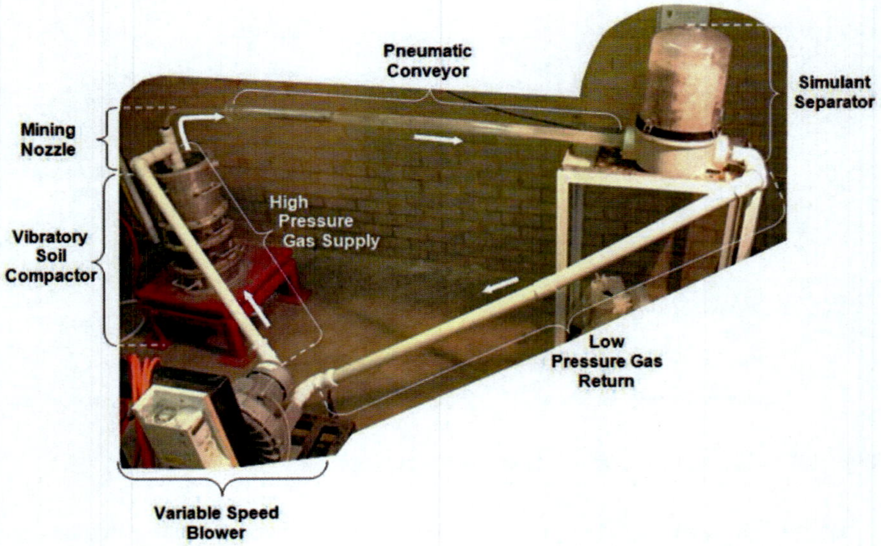

Fig. 13.5 Prototype Regolith Miner

The closed-cycle mode of the air movement is indicated by the arrows starting at the variable speed blower on the bottom. Fed by the return air from the simulant separator, positioned on a table to the right, it "pushed" the air into the pipe leading to the mining nozzle installed into the compacted simulant. Also noticeable is the fact that the conveying pipe is larger than the tube used to lift the soil vertically out of the sub-surface cavern.

The following section will elaborate on the method used to prepare the simulant for the extraction experiments.

13.9 Vibratory Compaction

In a study by Bernold (1994), it was found that vibratory compaction inside a mould with a surcharge is an efficient method to compact the fine non-cohesive regolith simulant. Figure 13.6(a) presents the initial design of a container mounted on a vibrating table with a desired surcharge of 175 kg providing a standard load pressure of 0.14 kg/cm^2 during vibration. Due to money limitations, the design was adapted so that available container parts and weights could be utilized. A new 1hp vibrating table capable of 3,000 vibrations/minute (VPM) created a centrifugal force of 4.92kN. Figure 13.6(b) shows the system that was put together.

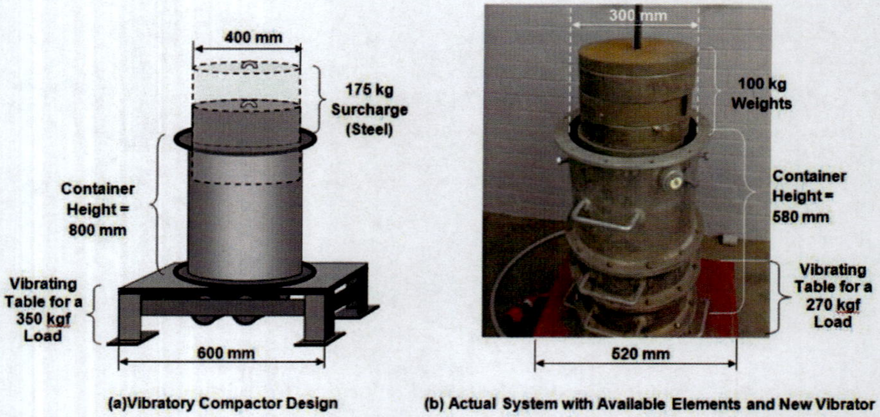

Fig. 13.6 Design and Fabrication of the Simulant Compactor

Basalt fines and dust were mixed to represent an average lunar simulant, and filled into a deep container mounted on a vibrator. As observed by Bernold (1994), vibrating fine simulants will increase its density until the smaller particles have filled up the space between larger particles, thus increasing both the absolute and relative density.

Vibrating beyond this point, in time may eventually lead to a desegregation of the soil, especially when it has too much of the very fine particles. In order to find the densification curve of the prepared simulant, several time-based tests were conducted to compare vibrating time with the settlements of the weights into the container. The results of these measurements are shown in Fig. 13.7.

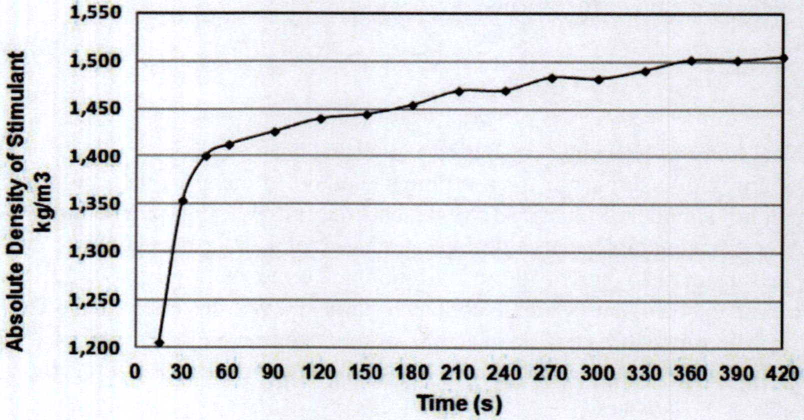

Fig. 13.7 Time Dependent Vibratory Compaction of Regolith Simulant

As expected, the increase in compaction stabilizes after a certain time and eventually reaches a platform. Figure 13.7 graphs the first 7 minutes which, considering earlier tests, represents a soil with 90% relative density. Thus, this simulant mix will achieve a density of 1,750 kg/m^3. However, for the purpose of the mining experiments, it was decided to vibrate for 7 minutes to a density of approximately 1,500 kg/m^3.

13.10 Installation of Pneumatic Soil Extractor

Similar to the packer for soil or rock grouting in terrestrial applications, a starter hole that is slightly larger than the service tube has to be created to a depth where the packer can be inflated safely. The goal is to find a soil layer that is sufficiently dense in order to "endure" the pressure exerted by the pressure of the inflating packer sleeve. Of course, the packer sleeve forces exerted onto the surrounding soil are directly related to the internal pressure selected for the efficient pneumatic conveying.

As we learned earlier, the top layer of regolith on an Asteroid will be loose with rapid densification further down. By using an appropriate drill on an Asteroid, the creation of a starter hole will be left to a small mine installation manipulator that would create an opening, similar to the one shown in Fig. 13.8(a).

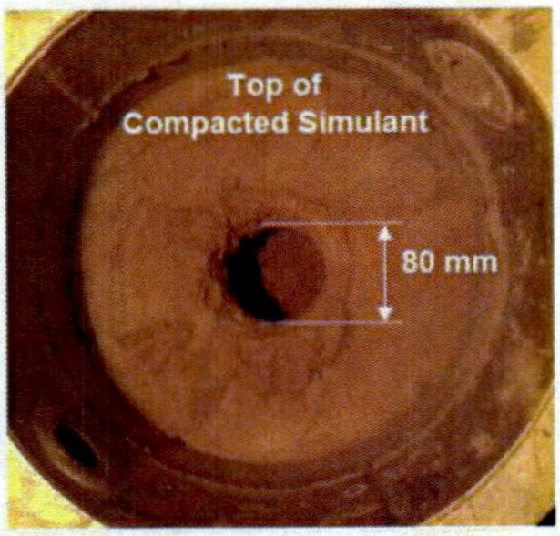

(a) Bird's View of Completed Starter Opening

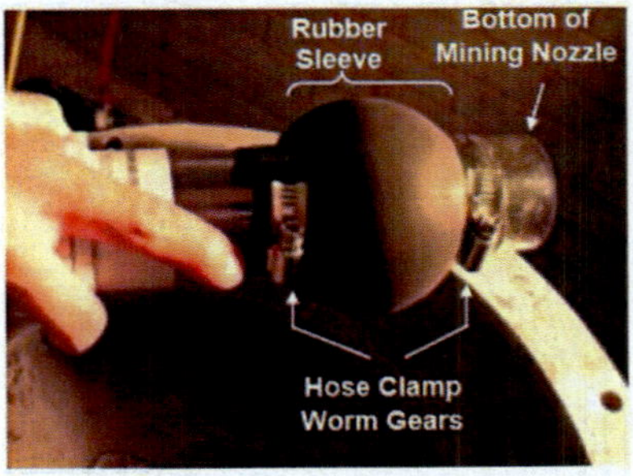

(b) Pressure Test of Packer Sleeve

Fig. 13.8 Preparations to Install Mining Nozzle

Finally, the mining nozzle can be installed into the starter hole with the air supply pipe attached to the mining nozzle and the lift tube feeding into the conveying pipe. Figure 13.9 presents a sketch and a picture of the completely installed extractor.

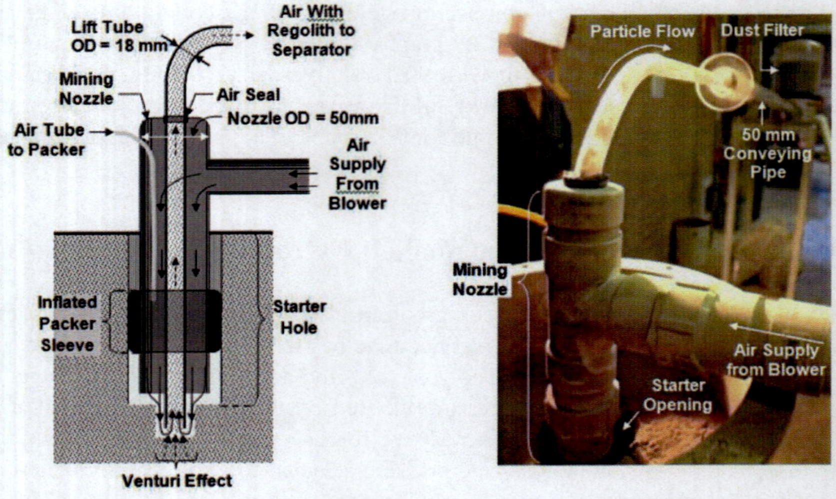

(a) Cross-Section of Installed Mining Nozzle (b) Soil Extractor Readied for Experimentation

Fig. 13.9 Design and Final Assembly of the Mining Nozzle

Figure 13.9(b) shows a picture of the prototype with the vertical tube, made of clear plexiglass that is bent into a 90 degree elbow. The end of the lift tube feeds into the 50 mm diameter horizontal conveying pipe which, in turn, is connected to the dust filter.

Inflating the packer sleeve and joining of the pipes to the blower and the air filter closes the air flow cycle. The next section will present the results of the experiments conducted to assess the performance of the miner.

13.11 Performance of Mining Nozzle Designs

The goal of the first test with the prototype was to compare a basic straight end with that of a conical funnel extension. For these basic tests, the lift tube was lowered in increments of 30 mm with the bottom of the starter opening defined as depth 0 mm. After each advancement, the packer sleeve was inflated and the blower turned on until the air in the lift tube was clear. This was followed by the removal of the entire mining nozzle to measure the size of the additional cavity that was created, photographic documentation, and finally the weighing of the removed simulant using an electronic scale. Because the 450 mm deep drilled shaft was only 75 mm in diameter on the top section and 27 mm at the bottom, precise measurements were impossible. As will be discussed, however, the correlation of the projected values with the accurately measured masses for each 3 cm increment provided an evaluation of its accuracy. Future tests with larger diameter nozzles should allow the use of a scanning device to measure the contour.

Preliminary tests to measure static and dynamic pressure to calculate the air velocity had shown that the atmospheric pressure would supply sufficient air to extract and lift the simulant particles into the larger conveying pipe. As a result, the blower created an under- pressure (below 1 atm) on the filter side, which varied with the speed of the blower. Since the goal of the experiment did not include a testing of mining production, the blower was generally kept at 25 Hz, resulting in an air speed of approximately 40 km/hr inside the conveying pipe with a cross sectional area of 1520 mm^2. The same air velocity is used in the industry for the dilute phase conveying of fluidizable, mildly abrasive material such as silica flour, fly ash or cement. Of course, cement and silica flour have particle sizes that match that of basaltic dust used to simulate asteroid regolith.

13.11.1 Geometries of Cavity Formations

While it was expected that the conical funnel would result in a larger diameter cavity, a marked difference in the resulting shaft was observed. Figure 13.10 offers views of the openings created by the two nozzles.

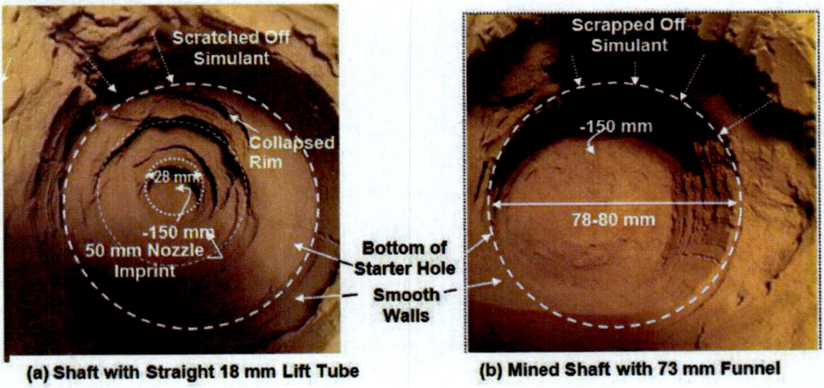

Fig. 13.10 Bird's-Eye-View Comparison of Excavations Using Different Nozzle Shape

Immediately apparent are the chafes in the smooth upper walls ending at the dashed circle, marking the bottom of the starter hole. These grooves were non-intended by-products of the imperfect fabrication of the packer, in particular, the use of hose clamps with worm gear and steel band to build the packer sleeve (see Fig. 13.8). Pressurizing the packer sleeve to reinflating the packer after each increment caused the hose clamp to scratch the smooth wall and the rubber sleeve, making new indentations. Of course, inserting the long nozzle into the narrow opening led to accidental damages, as well. Naturally, each time the loosened simulant fell into the cavity bellow it was picked up by the miner. The discussion of the mining data will show that the effect of these inadvertent damages to the vertical wall can be recognized easily. At each level, an opening with a minimum of 28 mm was created and partially abraded by the air flowing to the tube entrance after it had been lowered to the next level.

Both mining operations reached the desired depth of 150 mm. Striking are the totally different wall surface shapes. The straight 18 mm diameter tube, depicted in Fig. 13.10(a), created a wall with deep undercutting resembling washed out cliffs at ocean shores. This observation can be explained by the incremental lowering of the suction tube by 30 mm and the resulting turbulences. The inverted funnel, with its 73 mm large opening pointing downwards, produced an almost perfectly smooth wall surface presented in Fig. 13.10(b). One recognizes that the funnel only allowed about 3 mm space around the perimeter for the air to reach the low pressure zone at the stem of the conical funnel mouth. This ring of air flowing and abrading the soil from the wall represents an area of 970 mm^2. By disregarding the small pressure differences between the conveying pipe and the mining nozzle, one can approximate the air velocity around the perimeter as 40 km/hr (1519 mm^2 /970mm^2) = 62 km/hr. In a sense, this speed constitutes an equilibrium state between the simulant density and the force of the flowing air. It is understood that more and better observations are needed to explore this fact in more depth.

13.11.2 Comparing Output

The final section of the chapter discusses the output of the mining operation with measurements every 30 mm presented in Fig. 13.11.

Again, the nozzles using the straight-end tube and the funnel are being compared. As indicated, the Y-axis represents the computed output as the mass of simulant excavated after the suction nozzle was lowered by 30 mm. The beginning level was set at the bottom of the starter opening. Thus, the data shown at a drilling depth of 3 cm represents the amount of soil that was mined between 0 and 30 mm.

For each nozzle design, three values are given. The solid line plots the weight increase of the conveying pipe and the filter measured using an electronic scale. The dashed lines represent predicted values computed from the newly created cavity in mm^3 and two densities, 1500 and 1700 kg/m^3.

Predictably, the use of the funnel resulted in a significantly larger output per 30 mm increment, with an average of 0.33 kg compared to0.06 kg.. As mentioned earlier, the effect of the accidental damages to the smooth shaft surfaces are becoming apparent in that the weight measurements are inexplicably higher at certain depths. A point in case is when the straight tube results are compared at a depth of 9 cm. The electronic scale showed a value indicating an amount of simulant double that calculated using the volume as the base. This discrepancy can be explained by comparing the bird's-eye- view shots of the shaft at 6 and 9 cm. It shows that the first rim, created by an undercut, had collapsed just before or during the installation of the nozzle at 9 cm depth. Figure 13.10(a) indicates the fallen off rim with the possibility of additional small collapses.

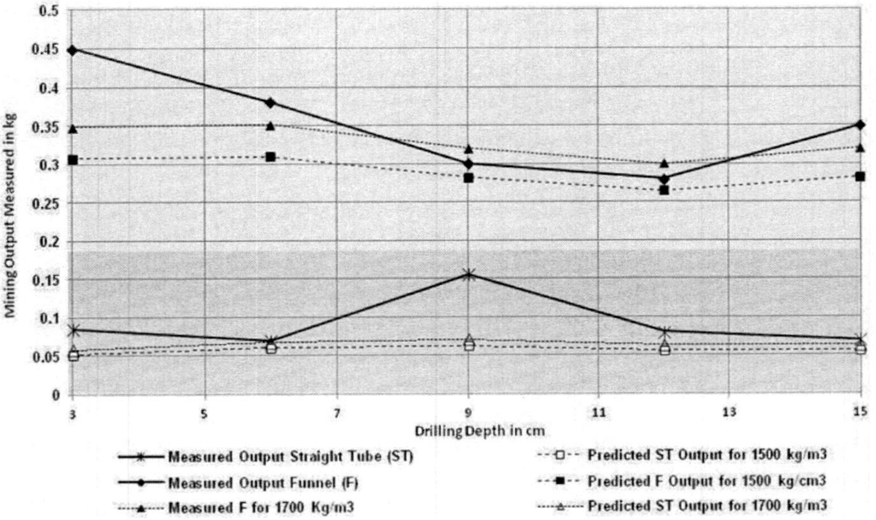

Fig. 13.11 Simulant Extraction During Incremental Advancements

Similar effects of dropped soil material are observable for both methods when measuring the first output at 3 cm. Most pronounced is the additional 0.15 kg which must have fallen from the wall during the first packer installation. A similar drop has possibly occurred for the straight tube case, but only very little fell into the area that the tube reached.

In summary, the conducted mining experiments with two different suction nozzle designs proved the efficacy of the pneumatic mining of soil simulants at 90-100% relative density. They also showed that with changes to the nozzle, large output changes can be expected. It also showed that the small scale prototyping limits the accurate measurements, a drawback that has to be overcome with careful observations.

13.12 Conclusion

The design and development of Asteroid miners is hampered not only by a lack of design standards, on which engineers depend on Earth, but also a vague understanding of relevant characteristics of soil and rock to be mined. On the other hand, data gathered from the Moon and some Asteroids indicate that the lunar conditions might provide useful information about what can be expected. Based on this belief, the research presented in this paper uses the geo-technical knowledge that was gained from the activities during and after the Apollo program to establish similitude between Earth and an Asteroid of unknown size.

Early research work on lunar mining, which was guided by terrestrial "thinking", proved that the brute-force approach to mining the fine, but dense, regolith will be extremely expensive and difficult to make sustainable. Even the use of energy efficient explosives to loosen the compacted simulant had only a minor effect on energy needed for digging and transporting the soil. Furthermore, the amount of levitated dust created using mechanical earth-moving technologies would create high risks to moving equipment, as well as to humans.

This paper presents an alternative approach. Pneumatic closed-cycle regolith mining on Asteroids takes advantage of several ISRU's, such as the highly densified fine regolith, lack of clay and the presence of large amounts of silicates in the soil. Similar to the Moon, we might find that the gardening effect on Asteroid surfaces did create areas with tightly compacted fine soil. Data from the Asteroid Itokawa adds more certainty to this thought. Its extremely low permeability is considered another ISRU resource, as it offers the opportunity to create a cavity below surface that can be pressurized to establish an artificial atmosphere. Although experiencing a different amount of gravity acceleration, the physical laws of gases, the Venturi effect and Bernoulli's formulation of fluid dynamics can be adapted. Applying the principle laws of physics to the dense fine regolith simulant a closed-cycle pneumatic mining prototype system was built and tested. Similitude of the soil was established by using only the fine basalt dust which also has similar low cohesion.

The experimental testing of two different mining nozzle designs in compacted simulant highlighted the problem related to measuring accurately in small scale models. Nevertheless, the pneumatic miner, set to an air-velocity of 40 km/hr, easily removed the soil around the nozzle and lifted it into the conveying pipe. Incremental advancements of the nozzle into the formed cavity, allowing the weighing of the removed simulants and the photographing of the progress, confirmed the consistent performance of both nozzles. As expected, the larger size funnel type nozzle led to a larger output than the straight end tube. However, the shape of the conical funnel forced the air flow to its rim where it created a circular ring opening, causing an increase in velocity (smaller area) until a balance of the abrasive and the resistive force was found.

The results of the presented work indicate that pneumatic mining of the Asteroids offers a viable technology able to take significant advantage of ISRUs. Lower gravity, low permeability of the dense regolith, and solar energy are as useful as the abundantly available silicates that can be heated and extruded into conveying pipes of any shape. The needed gas can be processed from the soil or brought as a liquid from Earth.

Acknowledgement. I would like to acknowledge the work done by my Honours student Hock Keong Tan. He had jumped into the deep end of pneumatic mining and kept treading water until he reached some sturdy ground. On the way, he diligently learned how to execute the experiments and to interpret the data.

References

Bates, R.L., Jackson, J.A. (eds.): The Glossary of Geology, 2nd edn., p. 751. American Geological Inst., WA (1980)

Bernold, L.E., Rolfsness, S.L.: 'Earthmoving' in the Lunar Environment. In: SPACE 1988, Albuquerque, NM, USA, pp. 202–216 (1988)

Bernold, L.E., Sundareswaran, S.: Laboratory Research on Lunar Excavation. In: Proc. SPACE 1990, pp. 305–314. ASCE, Albuquerque (1990)

Bernold, L.: Experimental Studies on the Mechanics of Lunar Excavation. J. Aerospace Eng. ASCE 4(1), 9–22 (1991)

Bernold, L.E.: Principles of Control for Robotic Excavation. In: Proc. SPACE 1992, May 31-June 4, pp. 1401–1412. ASCE, Denver (1992)

Bernold, L.E.: Compaction of Lunar-Type Soil. J. Aerospace Eng. ASCE 7(2), 175–187 (1994)

Carrier III, W.D., Mitchell, J.K., Mahmood, A.: The nature of lunar soil. J. Soil Mech. and Found. Div., ASCE 99(10), 813–832 (1973)

Carrier III, W.D., Olhoeft, G.R., Mendell, W.: Physical Properties of the Lunar Surface. In: Heiken, G., Vaniman, D.T., French, B.M. (eds.) Lunar Sourcebook, A User's Guide to the Moon, pp. 475–594. Cambridge Univ. Press, NY (1991)

Cohen, R.: The Pneumatic Mail Tubes: New York's Hidden Highway and its Development (1999), `http://www.coneysstamps.com/files/PneumaticTubes2-09.pdf` (downloaded October 12)

Deepagoda, T.C., Moldrup, P., Schjønning, P., Wollesen de Jonge, L., Kawamoto, K., Komatsu, T.: Density-Corrected Models for Gas Diffusivity and Air Permeability in Unsaturated Soil. Vadose Zone J. 10(1), 226–238 (2011)

Dunn, C., McCreadie, M.: The founders of a nation, Australia's first fleet-1788 (2005), `http://www.ulladulla.info/historian/ffstory.html` (downloaded October 10)

Goodings, D.J., Lin, C., Dick, R., Fourney, W.L., Bernold, L.E.: Modelling the Effects of Chemical Explosives for Excavation on the Moon. J. Aerospace Eng. ASCE 5(1), 44–58 (1992)

Kargel, J.S.: Asteroid: Sources of Precious Metals. J. Geophysical Res. 99(E10), 129–141 (1994)

Lin, C.P., Goodings, D.J., Bernold, L.E., Dick, R.D., Fourney, W.L.: Model Studies of Effects on Lunar Soil of Chemical Explosions. J. Geotechnical Eng. ASCE 120(10), 1684–1703 (1994)

McKay, D.S., Heiken, G., Basu, A., Blanford, G., Simon, S., Reedy, R., French, B.M., Papike, J.: The Lunar Regolith. In: Heiken, G., Vaniman, D.T., French, B. (eds.) Lunar Sourcebook, A User's Guide to the Moon, pp. 285–356. Cambridge Univ. Press, NY (1991)

Miyamoto, H., Yano, H., Scheeres, D.J., Abe, S., Barnouin-Jha, O., Cheng, A.F., Demura, H., Gaskell, R.W., Hirata, N., Ishiguro, M., Michikami, T., Nakamura, A.M., Nakamura, R., Saito, J., Sasaki, J.: Regolith Migration and Sorting on Asteroid Itokawa. Science 316, 1011–1014 (2007)

Sullivan, T., Koenig, E., Knudsen, C., Gibson, M.: Pneumatic conveying of materials at partial gravity. Aerospace Eng. ASCE 7(2), 199–208 (1994)

Tsuchiyama, A., Uesugi, M., Matsushima, T., Michikami, T., Kadono, T., Nakamura, T., Uesugi, K., Nakano, T., Sandford, S.A., Noguchi, R., Toru Matsumoto, T., Matsuno, J., Nagano, T., Imai, Y., Takeuchi, A., Suzuki, Y., Ogami, T., Katagiri, J., Ebihara, M., Ireland, T.R., Kitajima, F., Nagao, K., Naraoka, H., Noguchi, T.: Three-Dimensional Structure of Hayabusa Samples: Origin and Evolution of Itokawa Regolith. Science 333(26), 1125–1231 (2011)

Zacny, K., Mungas, G., Mungas, C., Fisher, D., Hedlund, M.: Pneumatic Excavator and Regolith Transport System for Lunar ISRU and Construction. In: Proc. AIAA SPACE Conf. & Exp., AIAA-2008-7824, San Diego, California, September 9-11 (2008)

Zacny, K., Craft, J., Hedlund, M., Wilson, J., Chu, P., Fink, P., Mueller, R., Galloway, G., Mungas, G.: Novel Approaches to Drilling and Excavation on the Moon. In: AIAA SPACE Conference & Exposition, pp. 6431–6443 (2009)

Zacny, K., Mueller, R.P., Craft, J., Wilson, J., Hedlund, M., Cohen, J.: Five-Step Parametric Prediction and Optimization Tool for Lunar Surface Systems Excavation Tasks. In: Proc. ASCE Earth and Space 2010, Honolulu, HI, USA, March 15-17 (2010)

Chapter 14
Extracting Asteroidal Mass for Robotic Construction

Narayanan Komerath, Thilini Rangedera, and Scott Bennett

Georgia Institute of Technology, Atlanta, GA, U.S.A

14.1 Introduction

Near Earth Objects (NEOs) offer convenient low-gravity sources of the resources needed to extend a permanent human presence beyond earth, and build a Space-based economy. In previous work (Wanis 2005), we considered the concept of automatically forming closed wall shapes such as cylinders in Space from solid material of random shape and multidisperse size distribution, using "Tailored Force Fields". This addresses the primary obstacle to building radiation-shielded, 1-G habitat modules in orbit. The mass required for long-term radiation shielding comes from low-gravity locations such as NEOs. Vanmali et al (2005) laid out requirements for such a mission to an NEO imagined to be at earth-sun L-4. Rangedera at al (2005) considered the conceptual design of a robotic machine to do surface excavation on a typical NEO. The present chapter is derived from Rangedera et al, presented at a Space Systems conference in Atlanta in 2005, but modified to consider returning the mass for processing to Earth orbit, instead of building habitats.

Previous work on extraterrestrial resource exploitation has considered harpoons to anchor a drilling craft onto the surface of a NEO in order to drill out material for mining. Muff et al (2004) proposed a nuclear-powered bucket-wheel excavator. The vehicle would roll along the lunar surface and scoop up material with a bucket wheel. This machine would use its own weight to apply downward pressure to dig into the ground with the teeth of each bucket. Such a device is suitable for loose sandy material found on planetary surfaces, filling the role of a lifting machine rather than a cutter. The NASA Deep Impact mission demonstrated automatic adjustment of its navigation at high speed to intercept a comet. This demonstrates a solution to part of the problem of robotic NEO rendezvous. Barucci et al (2008) describe the Marco Polo mission to return samples from a NEO. Our interest is in developing a solar-powered solution that can then be used repeatedly away from earth for long-term resource exploitation. The craft thus considered has to be autonomous and versatile. The Rockbreaker shown in Fig. 14.1 is a multipurpose robotic craft designed primarily to cut rocky material to construct habitats. The craft is to independently rendezvous with a NEO and attach itself. The craft uses plasma jets and laser cutters in order to cut out 20 cm cubic blocks and make them float into a helical cloud away from the NEO. The design is suited to extra-terrestrial resource extraction from low-gravity NEOs. In this chapter, we specialize the mission to a 'typical' choice of Near Earth Object.

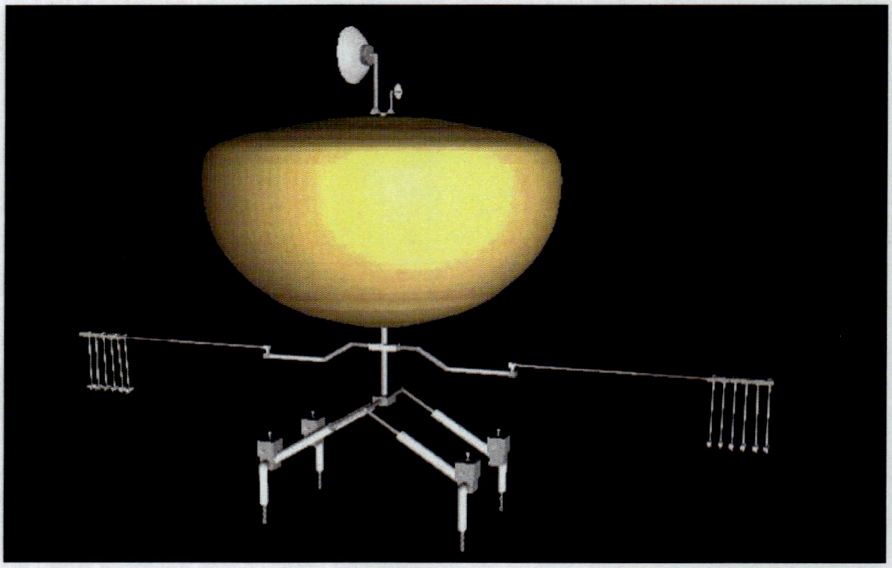

Fig. 14.1 Conceptual Drawing of the Rockbreaker robotic NEO resource extractor craft

14.2 NEO Population

NEO sizes range from dust-sized fragments to objects tens of kilometers in diameter (Bottke 2002). The lower limit of detectability is generally on the order of a few hundred meters diameter. Jones et al (2002) discussed short-duration missions to small NEOs of several meters in diameter, which could be reached with very low energy requirements. Mainzer et al (2012) studied NEO subpopulations that might pose threats to Earth using a Wide-field Infrared Survey Explorer spacecraft. Roughly 4700 NEOs have been found with diameter greater than 100m, which pose a possible threat to Earth, of which around 100 are accessible with mission ΔV comparable to that for lunar missions, though the mission time may be on the order of years (Christou 2003). Alotabi et al (2010) developed a roadmap on asteroid mining technologies and applications. Christou (2003) provides guidance in selecting NEO mission destinations. Objects in orbits similar to that of Earth are the easiest and fastest to reach. An orbit eccentricity range of $0.3<e<1.2$, and an inclination limit of 5°, along with limits on object size, still leaves at least a few dozen possible NEOs. One possibility is 1996 FG3.

Asteroid mining and resource extraction requires numerous steps (Alotabi 2010). Even after detailed prospecting using robotic spacecraft, a number of specialized tools and processing systems must be available to extract the desired substances, prior to returning them to Earth. On the other hand, other needs of a space-based economy require mass of any kind that can be obtained from low-gravity sources such as NEOs. For instance building radiation shields for habitats or counter-masses for tether systems are applications where waste from mineral processing can be utilized. Thus we conceptualized a system that can return large masses to Earth orbit, where detailed processing can be distributed over several facilities. This reasoning leads to the choice of an architecture where a large mass is returned without detailed triage as to immediate usefulness.

Alotabi (2010) considered several architecture options for NEO resource mining, and selected one where the entire NEO was returned to low Earth orbit for mining and extraction. This limits the choice to NEOs that are small enough to be so returned by attaching propulsion units and propellant tanks to the NEO. It also creates a substantial monolithic object that poses risks if it ever encounters Earth's atmospheric drag. However, the difficulties in conducting the multiple exploration and sample return missions and the mechanical and chemical processing involved in resource extraction, might drive such a drastic choice.

We reasoned that mass return can occur over several years, once a continuous system is set up. The consumables of the mining craft must be replenished and mass return packages must be sent in time to replace those that are sent back to Earth. But this system allows also for the resource extractor craft to move from one NEO to another if the orbit energies and transit times are reasonable. The requirements for the Rockbreaker are given in Table 1, modified from Rangedera et al (2005). Given that most early NEO missions will be near the orbit of Earth, and these missions are not time-critical unlike human spaceflight, solar sail propulsion is an attractive choice. Since the solar sail is a low, continuous thrust propulsion system, it is difficult to find the best possible trajectories without detailed numerical optimal control methods. To circumvent this problem, Rangedera et al (2005) simply used the results obtained by solar sail trajectory experts, and scaled the sail area for the required payload, to match the sail loading for which those results were obtained.

14.3 Requirements Definition

Dachwald and Seboldt (2005) used artificial neural networks (ANNs) along with evolutionary neurocontroller algorithms (EAs) to optimize for solar sail craft trajectories. We use their calculations of a solar sail mission to 1996FG3 "InTrance". A total sail craft mass of 148 kg (75 kg payload and 73 kg sail assembly) would have a 2500 m^2 sail area. Characteristic acceleration was on the order of $0.14 mm/s^2$. The transfer time was 4.15 years starting from earth escape. The characteristic acceleration is due to the solar radiation pressure (SRP) acceleration acting on the solar sail that is oriented perpendicular to the sun-line at 1 AU. Technology advances can be used to increase sail area for the same mass, thereby increasing the characteristic acceleration and reducing the transfer time significantly (Barucci 2008). Alternatively, the total craft mass can be reduced. The Dachwald/ Seboldt design envisages total sail craft loading of $0.0592 kg/m^2$.

Scaling the sail area needed for the same trajectory, we see that the sail area needed for our mission would be 833,000 m^2 and the total craft mass for a 25,000 kg Rockbreaker craft would be 49,333 kg. Recent experiments (Wilcox 2000) have demonstrated a nested sail deployment concept which promises area density of 0.01 kg/ m^2. With this technology, the sail mass can be reduced, and the sail area required is less than 510,000 sq.m. The total mass at Earth escape would be just over 30,000 kg, consisting of a 25,000 kg Rockbreaker craft, and a 5080 kg solar sail assembly.

Table 14.1 Mission requirements of the Rock Breaker

Requirement	Assumptions / choices
Single unit launch from Earth.	Avoid on-orbit assembly costs.
Travel to L-4 using solar propulsion	Minimizes launch mass and takes advantage of the 1 A.U. destination.
Rendezvous with a low-gravity NEO	NEO size is at most only a few kilometers.
Attach itself to the NEO in order stay anchored while cutting	Removable attachment means suitable for many sorts of surfaces.
Cut out a large amount of mate rial in the form of discrete blocks, within a short time and loosen the blocks enough to float apart at controlled speeds.	Discrete blocks is a conservative choice; blasting is presumed to be unacceptable because of poor control on ejection velocity
Detach itself and move to another location on the same NEO, or leaving the NEO.	Craft used for repeated quarrying/ assembly/ resource extraction, with possible refill of cutting / maneuvering gas.

14.4 Systems

14.4.1 Power Generation and Transmission

As the craft reaches its destination, its systems are powered up. The 0.5 sq. km solar sail is formed into a solar collector, focusing sunlight directly to the power supplies of the laser system, and on high-intensity solar cells to power other systems. Efficiencies of up to 38% have been achieved in the laboratory (Saiki 2005) in converting broadband sunlight directly to 1064-nm laser radiation using a Cr_2-doped Nd-fiber laser. This breakthrough is well suited to our application, since the primary tool used by the Rockbreaker is a Nd-fiber laser cutting tool.

14.4.2 Cutting System

The system conceived to cut material from the surface of the NEO consists of six lasers each of 25KW, conveying pulsed beams to 60 nozzles arranged at the "fingertips", five to each arm, of 12 cutting arms, each laser beam sheathed in an annular plasma jet. We propose evolved versions of the above laser to perform most of the cutting functions of the Rockbreaker. Laser cutting tools have very low tool attrition, enabling maintenance-free operation for long periods with a variety of materials. These lasers offer the highest power density available in the industry and are two orders of magnitude faster than rotary drills for rock drilling. The Nd fiber laser offers 10 W/kg and long operating lifetime, exceeding 100,000 hours, compared to 10,000 and 25,000 hours respectively for Nd-YAG and CO2 lasers. A single fiber strand of a few microns can already generate and deliver over 1kW; this is expected to rise considerably. A typical 700W fiber laser generates a beam intensity of more than 50MW/cm^2. A beam power density of less than 1kW/cm^2 sufficed to cause thermal spalling of sandstone and shale, both of which have similar densities to silicon dioxide. The major uncertainty about NEOs is their composition. Given this uncertainty, the latent heat of melting of silicon dioxide was used in our calculations, assuming that 25% of the energy required for melting sufficed to crack the material. Fusion cutting, where laser energy melts and cracks the material, and a gas jet blows out the debris, requires only 10% of the power and 3% of the time for vaporization cutting.

14.4.3 Plasma Cutting Mechanism

A plasma jet sheath around the laser beam blows away the material being heated and fractured by the laser, and exposes fresh material to the laser beam. Non-metallic NEO materials are amenable to a non-transferred plasma jet, where the torch nozzle becomes the anode. The pressures, standoff distance and nozzle expansion ratio are parameters used in optimizing the cutting trench width and the stagnation pressure exerted on the cut blocks (see Fig. 14.2) to move them in desired directions (including the original objective of floating them away from the NEO for the construction application).

The electrodes are consumables that require periodic replacement; currently at less than 1000 hours. Generating a thin jet sheet in vacuum presents substantial difficulties, for which the aerospike nozzle provides a partial solution. The storage volume and mass on the craft limit the amount of gas that can be carried for the plasmajet, so that minimizing the mass flow rate is critical. Electrode attrition and gas mass severely restrict use of the plasmajet.

14.4.4 Hybrid Aerospike Cutting System (HACS)

A hybrid laser and plasma cutting system is proposed for the Rock-Breaker design. The main cutting tool that breaks up the surface material is the laser, while the plasma jet finishes the process by removing waste material from the trench and pushing the block out of the trench. The HACS is lowered from the craft on a vertical pillar. HACS has two main booms and can extend from the craft using a telescoping feature. At the end of each boom are 6 arms to lower 10 cutting nozzles each to the ground. Contained in each arm is a laser. The two arms rotate around the pillar in the vertical axis. The outcome is a spiral-cutting pattern. The rotor design, spiral pattern, and telescoping arms maximize the surface area covered by the HACS, and thus the amount of material that can be cut when the craft is anchored in one location. Vanmali (2005) described an optimized cutting sequence to generate the most cut blocks in minimum time. This is matched to the rotating/telescoping arm and multiple-fingers design.

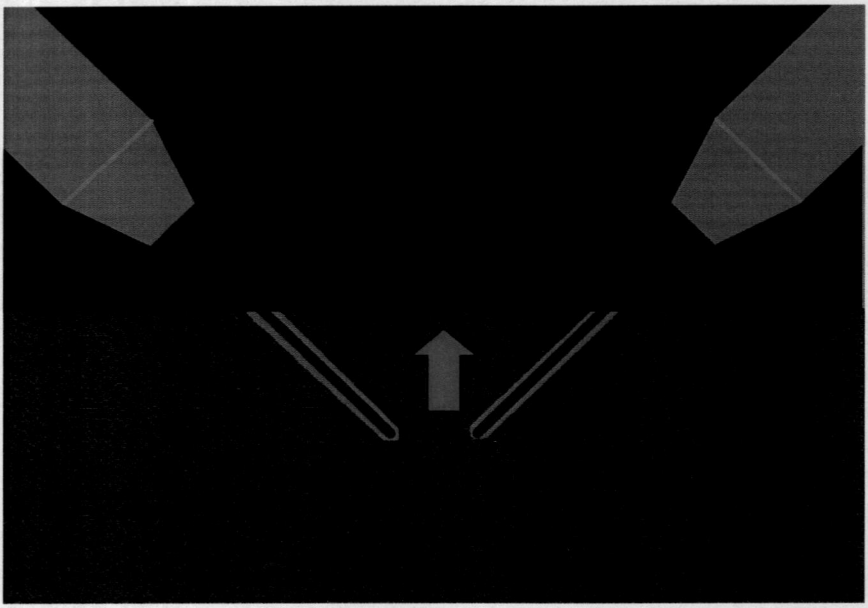

Fig. 14.2 HACS operation, showing force exerted on the blocks by the plasmajet

14.4.5 Truncated Linear Aerospike Cutting Nozzle

A linear truncated aerospike nozzle (Huzel 1967) design is employed, similar in concept to that in the X-33 vehicle. The linear design stacks several modules to produce an effective "knife edge". The laser beam is directed through the truncated base into the recirculation zone downstream, which adjusts itself to the pressure of the jet, forming a virtual spike. The laser beam is thus sheathed in a high-speed plasma jet, while the flow immediately in contact with the laser delivery lens remains at a low speed. The jet sheath and the recirculating base flow region protect the delivery lens from abrasion.

14.4.6 Integrated Rendezvous-Anchoring Maneuver-System (IRAMS)

NEO rendezvous is complicated by NEO spin and tumbling motions, which will demand a considerable magnitude of maneuvering ΔV. Reactions from any cutting operation must be absorbed by anchoring the vehicle on the surface of the NEO. We use a concept derived from the Impact Screwdriver that converts an axial, impulsive load to torque, matching the dynamics of the system to a pulsed solid-grained plasma thruster. Each leg of the vehicle is provided with an IRAMS pulsed plasma thruster which can be oriented in any direction for maneuvering. At the terminal stage of rendezvous, each leg approaches the NEO surface with a large-pitch, deep-thread drill-bit or auger facing the surface. At impact, the inertia of the system translates into the first torque. Following landing, the plasma thruster operates in short bursts, with the reaction driving the solid fuel grain into the torque hammer groove, causing high torque on the spring-loaded threaded tool (Fig. 14.3). Ten to twenty centimeters of travel into the surface, depending on its hardness and brittleness, will suffice to anchor the craft in place.

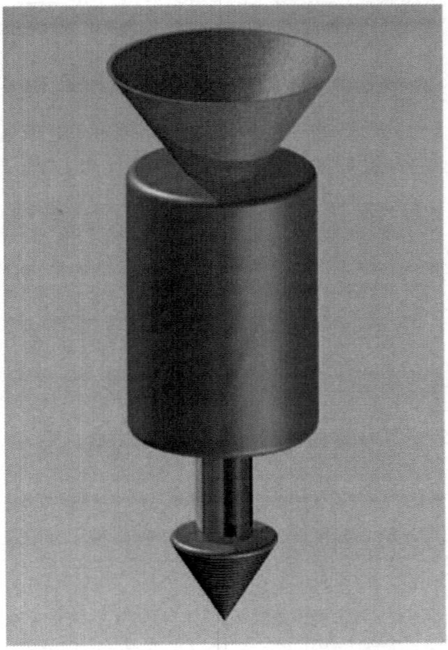

Fig. 14.3 Impact Driver

Pulsed, solid-grained plasma thrusters offer precise pulsed operation, cold start, intermittent duty cycles and reliable, storable solid grain. They are thus the best choice for long-term operation. Again, the solid grain is a consumable, but it is not expected to be a limiter of the craft's lifetime. Being a mass-spring system, the IRAMS can be optimized. We used a dynamical model of the IRAMS to obtain the natural frequency combination that would produce the highest amplitude. As shown in Vanmali (2005) the time between pulses depends on the natural frequency of the system, which should be matched with the thrust pulse rate to minimize power loss from thrust to torque output.

14.5 Auxiliary Systems

14.5.1 Active Beam Dynamics

The pulsed thrust of the plasmajets from the 60 independent nozzles will be phase-controlled to cancel out all but the lowest-order modes of bending of the Rockbreaker's arms during the cutting operation. This will also be used to control the arms during use as manipulators.

14.5.2 Sensor Systems

Sensor systems special to this craft, beyond those required for a standard spacecraft, include ground penetrating radar (GPR), which has been used on Earth to map geophysical properties (Olhoeft 2002) under the surface. Its role on the Rock-Breaker is to scout out favourable areas on the surface of the NEO before anchoring onto a location. The system to be used has an antenna frequency range of 0.5 - 1 GHz for medium to shallow penetration with a high-resolution radar survey. Optical sensors will be needed to control the temperature of the materials being heated by the Rockbreaker, as well as to see obstacles and samples.

14.5.3 Propellant Tank

The plasma jets feed off an argon (Hypertherm 2012) propellant tank. The tank structure mass is estimated at 5% of the total propellant mass. The plasma jet in its current application has an estimated power range of 1-10kW and current not exceeding 200A (Bauchire 2004). For these limitations, the mass flow rate range is 0.03-1.2g/s (Kelly 2004; Gibbs 2002).

14.5.4 Robotic Manipulator Operations

The robotic arms have enough degrees of freedom to be ideal as manipulator arms for use in Space-based construction. The four deployable legs with the IRAMS enable precise maneuvering of the craft.

14.5.5 Command and Control

The Hayabusa mission showed the extreme challenges in command and control encountered while attempting a mission to a Near Earth Object. For the purposes of the concept description here, we must assume that the craft operates under fully autonomous control for approach and landing, once commands based on observations have been programmed into the system during the approach phase. This no doubt requires advances in the confidence level of fully autonomous spacecraft operations, which have been made possible with the success of the DARPA Orbital Express missions demonstrating rendezvous, circumnavigation and component exchange between spacecraft in orbit. We do not project that command and control functions will add significantly to the mass budget at the present fidelity of that estimate.

14.6 Sample Return to Earth Orbit

The collector for materials ejected from the NEO by the Rockbreaker, is conceptualized as a drogue parachute, attached to the solar sail. Initially the chute is anchored to the Rockbreaker, and is partially deployed using gas inflation. The chunks of material drifting up from the NEO then provide the momentum to open the chute fully. The return trip could use one of several propulsion options. Given the low delta-v needed to return to earth orbit, a solar-sailed trajectory is once again an option. In addition, the mass collected provides a good source of mass-driver propellant. For initial thrust, the solar sail would be reconfigured to focus sunlight onto a heater, raising the pressure in a gas tank. The gas would drive a substantial fraction of solids through a nozzle, providing thrust. Following this phase, the solar sail would be configured for the return trajectory.

14.7 Mass Estimation

In Table 2, the mass of the various components is estimated for the Rockbreaker mission, for only the forward mission, i.e., including cutting and sending materials floating into orbit, but not the capture and return to Earth. The operating duration and mass excavation capability were calculated in Vanmali (2005) using the requirements to cut up enough blocks of SiO2-class material to form a closed cylinder 50m diameter and 50 m long, with a 2m thick wall. The mass to be excavated was twice this amount, since 50% wastage was assumed. The mass of material thus sent into Space was over 1.4 million kilograms. The number of nozzles, and the power delivery rate, was calculated to generate this amount of material with 19 earth-hours of continuous operation at Earth-sun L-4. These parameters are adequate as a starting design for a more general-purpose excavator/ resource extractor craft.

The majority of the mass is the laser/plasma cutting system, which is based on lasers currently available, assuming 10W/kg. The HACS is built with a network of hollow aluminum alloy rods strong enough to withstand the force of the plasma jets at the ends with active compensation as indicated above. Its weight is determined by its density (2700kg/m^3) and the length of each section. The present Space Shuttle Manipulator Arm (Canadarm2) has a mass of 410 kg, and is sized to move the entire STS Orbiter (120,000 kg) at acceptable rates for rendezvous with the ISS, so the HACS estimate is reasonable and conservative. Solar sail construction targets an aereal density of 0.01kg per square meter. The mass of argon gas was found using the mass flow rate of 0.03g/s feeding to 60 nozzles over a period of 460 hours. The propellant mass required for drilling, anchoring and re-anchoring is included in the IRAMS component; the amount required for moving the Rock-Breaker from the rendezvous point to NEO surface and other maneuvering tasks is in the thruster propellant mass.

Table 14.2 Component Mass Breakdown

Component	Mass (kg)
IRAMS, Thrusters (propellant included)	1000
HACS	2000
Power conversion system (other than lasers) incl. high-intensity solar cells.	1200
Plasma/Laser Cutting System	15000
Propellant Tank (argon)	200
Argon gas	4000
Communications antennae	50
Sensor suite	300
Manipulator arms	500
Protective Covering	200
Thruster Propellant (for maneuvering)	500
Total payload	24950
Solar Sail	5080
Booster + auxiliary propellant	7500
Total propulsion package	12580
Component	Mass (kg)

The mass estimate in Table 2 is broken into two packages. The first is the basic Rock-Breaker craft, which is the payload to be propelled by the solar sail. The second is the solar sail package itself, along with a 7500 kg boost package to move the craft from LEO to a transfer orbit where the sail can be deployed. The boost package mass is ample for an electric thruster system to send the craft into an earth escape trajectory.

The material return system is not included in the above mass estimate. This is conceptualized as being sent in a separate solar-sailed package along with the Rockbreaker, but will be sized for the particular needs of the material return. In other words, the package may not be sized to bring the entire 1.4 million kilogram capacity of the Rockbreaker. Propulsion options vary depending on the urgency of the return, and may vary from completely solar sailed, to designs that use some of the collected mass in a mass driver, to systems based on ion propulsion. As a rough estimate, we believe that mass ratios (initial mass of the system at takeoff from the NEO, divided by final arriving mass at Earth orbit) can be as low as 1.5.

14.8 Cost Estimate

An approximate cost estimate can be developed from the case of the Mars Science Laboratory. The complexity of this mission is comparable to the proposed solar sail craft. The large mass difference between the Mars Science Laboratory and the Rockbreaker is mostly in the mining package, so the cost of development should be comparable. The launch cost assumes the use of a heavy lift vehicle such as

SpaceX's Falcon Heavy which is sized to take a payload of 50000kg into LEO. The cost of the launch vehicle is between 80-125 US FY2012 \$M, the latter of which was used for a conservative launch estimate. The NASA June 8, 2011 audit report was used for the formulation, development, and operational cost scheme. The values presented in the audit were scaled for inflation and multiplied by an uncertainty factor of 1.5.

Table 14.3 Cost Estimate

Element	Cost (US FY2012 \$M)
Launch Cost	125
Formulation	803
Development Cost	1413
Operational Cost (per year)	101
Propulsion Package	288
Total (10 year operation)	3643

Finally the solar sail propulsion package estimate was conducted from the cost of Cosmos1. Cosmos1 had a total mission cost of approximately 4.0 FY2006 \$M (Leonard 2004). It was assumed that 75% of this cost was development. Then the cost per area of the sail craft was determined with this development cost and scaled up to meet the area of the Rockbreaker's sail. This was then multiplied by an uncertainty factor of 1.3. Based on this low fidelity cost model, the total cost of the mission is not unreasonable for a mission that would be part of developing the infrastructure for manufacturing and construction in orbit. Table 14.3 shows the result of this first order cost estimation.

14.9 ΔV Considerations

The ΔV required for a solar sail mission is greater than a typical mission where the Hohmann Transfer is used. If one considers the 1996FG3 NEO as the point of interest, mission ΔVs can be computed. Using the technique described by Shoemaker (1978), the minimum possible ΔV was calculated for the considered mission. The ΔV required to reach 1996FG3 is 6.609km/s. The calculated ΔV considered the spacecraft to be in LEO and use both an injection orbital maneuver and a rendezvous orbital maneuver. If the spacecraft is not in LEO but has already escaped Earth's gravitational influence, the required ΔV is less. Under ideal conditions, the ΔV required for return is equal to the ΔV required for the rendezvous orbital maneuver. The calculated value for the rendezvous orbital maneuver was 3.033km/s.

The computed ΔVs are for low thrust propulsion systems, such as a solar sail. The required ΔVs can be obtained by means of solar sail given a long enough period of time for radiation pressure acceleration. The ΔV required for a return maneuver can be obtained instantaneously through a mass driver system to allow the spacecraft to easily maneuver from one NEO to another or to return to LEO.

14.10 Relevant Past Mission

The HAYABUSA mission (Kawaguchi 2006) launched in 2003 demonstrated the technologies required for a successful sample return Itokawa (1998SF36), an NEO in the Apollo group, of approximately 500m diameter. HAYABUSA used low thrust ion engines for primary propulsion in cruise phases. HAYABUSA was also equipped with autonomous navigation and guidance software using optical measuring devices. Itokawa was detected by an onboard star camera (STT). Once the optical navigation cameras were within range of Itokawa, the craft was able to autonomously navigate to Itokawa. On September 12th, 2005 HAYABUSA successfully rendezvoused with Itokawa. The "touch-and-go" sampling method employed by HAYABUSA shot a small projectile at the surface of the asteroid and collects the ejected fragments through a sample retrieval horn. HAYABUSA then lifted off from the asteroid to avoid potential collision between the solar array and the asteroid's surface.

The HAYABUSA mission overcame many challenges. The surface of Itokawa was littered with boulders and very steep areas, making landing site selection difficult. Shadows on the surface of the asteroid as well as orbital disturbances from attitude control systems made autonomous optical-guided descent difficult. The craft used manual control to capture a single sample with approximately 100 particles from the asteroid (Kawaguchi 2006).

14.11 Conclusions

This chapter describes the approach to design a new kind of spacecraft for extraterrestrial construction and resource exploitation applications. Solar propulsion, beamed microwave power, and fiber laser and plasma jet cutting tools, and an Integrated Rendezvous, Anchoring and Maneuvering System are explored, and their issues considered.

The main conclusions are:

- Solar sail primary propulsion appears to be well suited to the NEO resource mining and return application.
- With direct solar-pumped Nd-fiber lasers for material extraction, the solar sail of a single craft would suffice to act as a solar collector and provide continuous power for operations.
- With the new generation of launch vehicles announced under the Moon-Mars initiative, a Rockbreaker and the propulsion package for boost to an earth-escape trajectory can be launched assembled in a single launcher to low earth orbit.

- Capturing the mass for return to Earth appears to be relatively simple. Options for propelling the mass back to Earth orbit vary depending on the urgency of the return. They may range from a slow, inexpensive solar-sail system, to systems that expel part of the collected mass in a mass driver engine, to a set of ion engines.
- As much as 1.4 million kilograms may be collected, even with 50 percent wastage, using the consumables aboard the Rockbreaker as conceived.

Acknowledgements. This work was initially funded through a Phase 2 NIAC grant (Dr. Robert Cassanova, Technical Monitor), and through President's Undergraduate Research Assistantships from Georgia Tech (NS). Assistance from several students and alumni of the Experimental Aerodynamics Group at Georgia Tech is gratefully acknowledged, in particular, that of Ravi Vanmali for calculations on the dynamics of the IRAMS.

References

Alotabi, G., Boileau, J., Bradshaw, H., Criger, B., Chalex, R., 40 co-authors: Asteroid mining Technologies Roadmap and Applications. Final Report, Space Studies Program, International Space University. Publications Library, Parc d'Innovation, Strasbourg Central Campus (2010), http://Mendeley.com

Barucci, M.A., et al.: Marco Polo: A near Earth object sample return mission. In: 39th Lunar and Planetary Science Conference, League City, Texas, USA, March 10-14 (2008)

Bauchire, J.M., Gonzalez, J.J., Gleizes, A.: Modeling of a DC Plasma Torch in Laminar and Turbulent Flow. Plasma Chemistry and Plasma Processing 17, 409–432 (1997)

Bottke, W.F., Morbidelli, A., Jedicke, R., Petit, J., Levison, H.F., Michel, P., Metcalfe, T.S.: Debiased Orbital and Absolute Magnitude Distribution of the Near-Earth Objects. Icarus 156, 399–433 (2002)

Christou, A.A.: The Statistics of flight opportunities to accessible Near-Earth Asteroids. Planetary and Space Science 51, 221–231 (2003)

Dachwald, B., Seboldt, W.: Multiple Near-Earth Asteroid Rendezvous and Sample Return Using First Generation Solar Sailcraft. Acta Astronautica 57, 864–875 (2005)

Gibbs, G., Savi, S.: Canada and the international space station program: Overview and status. Acta Astronautica 51, 591–600 (2002)

Huzel, D.K., Huang, D.H.: Design of Liquid Propellant Rocket Engines. NASA SP 125 (1967)

Hypertherm. Hypertherm Inc. Product Line (2012), http://www.hypertherm.com (viewed November 30, 2012)

Jones, T.D., Davis, D.R., Durda, D.D., Farquhar, R., Gefert, L., Hack, K., Hartmann, W.K., Jedicke, R., Lesis, J.S., Love, S., Sykes, M.V., Vilas, F.: The Next Giant Leap: Human Exploration and Utilization of Near-Earth Objects. In: Sykes, M.V. (ed.) The Future of Solar System Exploration. ASP Conference Series, vol. 272, pp. 141–160 (2003-2013)

Kawaguchi, J., Fujiwara, A., Hashimoto, T.: MUSESC(HAYABUSA): The mission and results. In: Proceedings of the IEICE General Conference, BK-2-1 (2006)

Kelly, H., Mancinelli, B., Prevosto, L., Minotti, F.O., Marquez, A.: Experimental Characterization of a Low-Current Cutting Torch. Brazilian Journal of Physics 34(4B), 1518–1522 (2004)

Leonard, D.: Planetary Society's Cosmos 1 Solar Sail Ready for Flight (2004), http://www.space.com/524-planetary-societys-cosmos-1-solar-sail-ready-flight.html

Mainzer, A., Masiero, J., Bauer, J., McMillan, R.S., Giorgini, J., Spahr, T., Cutri, R.M., Tholen, D.J., Jedicke, R., Walker, R., Wright, E., Nugent, C.R.: Characterizing Subpopulations within the Near Earth Objects with NEOWISE: Preliminary Results. The Astrophysical Journal 752(2), 110 (2012)

Muff, T., Johnson, L., King, R., Duke, M.B.: A Prototype Bucket Wheel Excavator for the Moon, Mars and Phobos. In: Proceedings of STAIF 2004, Institute for Space and Nuclear Power Studies, p. 214 (February 2004)

Olhoeft, G.R.: Applications and frustrations in using ground penetrating radar. IEEE Aerospace and Electronics Systems Magazine 17(2), 12–20 (2002)

Rangedera, T., Vanmali, R., Shah, N., Zaidi, W., Komerath, N.M.: A Solar-Powered Near Earth Object Resource Extractor. In: Proceedings of the 1st Space Systems Engineering Conference, Atlanta, GA (December 2005)

Saiki, T., Motokoshi, S.: Development of Solar-Pumped Lasers for Space Solar Power. In: IAC-05-C3.4-D2.8.09, 56th International Astronautical Congress, Fukuoka, Japan (October 2005)

Shoemaker, E.M., Helin, E.F.: Earth-Approaching Asteroids as Targets for Exploration. In: Asteroids: An Exploration Assessment, NASA CP-2053, pp. 245–256 (January 1978)

Vanmali, R., Li, B., Tomlinson, B., Zaidi, W., Wanis, S., Komerath, N.: Conceptual Design of a Multipurpose Robotic Craft for Space Based Construction. In: AIAA Paper 2005-6733, SPACE 2005 Conference, Long Beach, CA (August 2005)

Venkatramani, N.: Industrial plasma torches and applications. Current Science 83(3), 254–262 (2002)

Wanis, S., Komerath, N.: Advances in Force Field Tailoring for Construction in Space. In: IAC05-D1.1.02, 56th International Astronautical Congress, Fukuoka, Japan (October 2005)

Wilcox, B.H.: Nesting-Hoop Solar Sail. NASA Tech Briefs 24(9) (September 2000)

Chapter 15
Curing of Construction Composite Materials on Asteroids

Alexey Kondyurin

University of Sydney, Australia

15.1 Constructions on Asteroid

Human activity on asteroids will need constructions. It could be human habitat, base for mining machines, space station for communication, observation or deep space missions. The first constructions can be delivered from Earth, but an ability of space carriers are limited that limits the size and mass of the delivered constructions. Therefore, the extensive and long exploitation requires a technology to create new constructions on asteroid surface, under asteroid surface or near asteroid.

The most promising way to build the construction in space environment is a polymerization of composite material, which consists of fiber-filled composites and a reactionable matrix applicable on asteroid when the space construction should be working during a long period of time. For construction the fabric impregnated with a long-life matrix (prepreg) is prepared in terrestrial conditions and, after folding, can be shipped in a space ship to orbit and kept folded on board of the ship. On asteroid the prepreg is carried out into free space and unfolded by, for example, inflating an internal pocket. Then a reaction of matrix polymerization initiates. The temperature for the initiation of reaction could be achieved by Sun light irradiation and/or additional heaters. When the polymerization reaction is complete, the durable frame of construction can be used.

In this case, there are no any restrictions of the frame size and form of future space construction, there is no necessity for some launch vehicles for the creation of large space station, solar cell panel or antenna, there is not a dangerous and complicate procedure to join separate launch vehicle on asteroid.

The creation of large space construction based on polymerization process in free space has long history of development of inflatable space structures, the study of the polymer materials in LEO and GEO free space conditions, the study of the polymerization process in microgravity on board of space stations and the study of polymerization process in hermetic shells for future inflatable structures. The history of inflatable structures for space application was started from "Echo", "Explorer", "Big Shot" and "Dash" balloon satellites in 1960s (Wilson 1981). Echo satellite was build from balloon of Mylar with Al-coating for good reflection of light. Using Earth experiments and flight results the technology of deployment of inflatable big structures was developed. Based on success of balloon satellites flights the new projects of inflatable structures for antennas, reflectors, Lunar and Mars houses and bases, airlocks, modules based on light polymer films were presented from 60s years (Cadogan and Scarborough 2001; Cadogan et al. 1998; Grahne and Cadogan 1988). Some of them were used as, for example, airlock of cosmonaut Alexey Leonov for first travel into free space from space ship in 1965. From these time the inflatable structures based on new unique materials were developed and used for space application (Allred 2002; Bar-Cohen 2001; Cadogan et al. 2002; Darooka and Jensen 2001; Darooka 2001; Grossman and Williams 1990). The world leaders of space inflatable structure production are American companies ILC Dover and L'Carde, Inc., which work in close connection with NASA (Veldman and Vermeeren 2002). In 2006 and 2007 Bigelow Airspace company (http://www.bigelowaerospace.com/index.php) has launched two orbiting prototypes Genesis I and II. These inflatable constructions have been successfully tested on Earth orbit almost 7 years for now. NASA has engaged Bigelow Airspace to create the inflatable construction as a module of International Space Station. The inflated cylinder module will have 4 m length and 3 m diameter.

To increasing of durability of inflatable structures the method of rigidization could be used after complete deployment of structure in space. Some methods of rigidization are discussed: rigidization due to chemical reaction of soft polymer matrix by thermal or UV-light initiation of reaction (curing, polymerization), or by inflation of gas reaction; mechanical rigidization due to stressed aluminium layer of deployed shell; foam inflation; passive cooling below T_g of material; evaporation of liquid swells from gels (Grahne and Cadogan 1988; Cadogan et al. 1998a, 1999; Cassapakis and Thomas 1995; Derbes 1999; Guidanean and Williams 1998; Kato et al. 1989; Sandy 2000). In some cases a combination of hard and rigidizable structures was developed (Simburger et al. 2002; Willey et al. 2001). All of these methods were used in Earth laboratory experiments. But despite on huge funding of the investigations during long time, only one real mechanism of rigidization was tested in real space conditions - Aluminium stressed layer (Freeland and Veal 1998; Semenov et al. 2000).

The main attention of space experiments in space stations and satellite missions was applied to solid (cured) polymer materials. The effects of free space environment on polymer materials were analyzed during and after Low Earth Orbit (LEO) space flights when polymers were exposed by atomic oxygen, VUV light, X-rays, electrons and ions flows, thermal cycling and high vacuum (SETAS, LDEF, MEEP, SARE, AORP, DSPSE, ESEM, EuReCa, HST, MDIM, MIS, MPID and MISSE missions). The investigations of polymer materials in deep space flight missions were not found.

In literature we found only one space program connected with the polymerization of epoxy composite in free space: "Polymer curing experiment", Consort-02 (1989), Consort-03 (1990), Consort-04 (1992) missions and "Polymer composite curing experiment", Joust-01 (1991) in NASA. The complete description of experiment results was not founded excluding some notes, that the experiment was failed because the heating units was not operated due to extremely low temperature of electric battery in space.

The skepticism of polymerization processing in free space is due to specific conditions of free space environment for polymer materials. The conditions of free space have great destructive influence on polymer materials, especially, on liquid polymer matrix. In the free space the uncured composite material is treated by high vacuum, sharp temperature changes, plasma of free space formed by cosmic rays, sun irradiation and atomic oxygen (on low Earth orbit), micrometeorite fluency and microgravitation (Briskman et al. 2001; Kondyurin 2002).

The polymerization processes in free space is not considered in space agencies as a real possibility because of absence of the sufficient investigations of the polymerization processes in free space environment. On the other side, the investigation of the polymerization processes in free space environment are not supported and not carried out because the creation of the large space construction is not listed in priority tasks in space agencies due to "impossibility" to create such construction. Except the political and financial issues, the scientific and technological reasons of "impossibility" do not exist. We consider here an applicability of the polymerization technology in deep space on asteroid surface.

15.2 Conditions of Space Environment on Asteroid

We know a little about asteroids. Few space flight missions were directed to investigate asteroids. The experimental data are not sufficient to analyze the space environment near the asteroids. However, because the conditions of the environment near the asteroid surface are important for understanding of the polymer behavior there, we can consider known space environment conditions near Earth and Moon, where the environmental measurements were done experimentally during space flights and extrapolate it to the asteroid environment.

The first and main significant factor of space environment is vacuum. The Low Earth Orbit (LEO) for most man's flights is varied between 300 and 400 km of altitude where the pressure of residual atmosphere can be measured, for example, to 10^{-3}-10^{-5} Pa (Lee and Chen 2000), or 10^{-5} Pa (Walter 1987), or even $2.47 \cdot 10^{-7}$ Pa (ECSS Space Environment Standard 2000). Different results are explained by strong dependence of the residual atmosphere on altitude, Earth season and daytime, sun irradiation, local configuration and materials of space ship, and activity of spacecraft engine. The pressure near space ship or space construction depends on time from Earth start. During flight a desorbing of gas, venting trapped volumes, releasing dust and ice particles increase the pressure near new space construction. The contamination of the virgin atmosphere (without influence of artificial space construction) at 400 km altitude is following: O (86.6%), He (9.6%), N_2 (1.5%), H (1.3%), O_2 (0.01%), Ar (0.00001%) by (ECSS Space Environment Standard 2000) data.

With distance of space ship from Earth, the residual atmosphere decays and at distance of 36000-42000 km corresponding to Geostationary Earth Orbit (GEO) the virgin pressure without disturbance by spacecraft equals to 10^{-9}-10^{-11} Pa. The pressure near satellite or spacecraft can be significant higher up to 10^{-3}-10^{-5} Pa.

The virgin pressure on the Moon surface is about 10^{-9} Pa at night (http://nssdc.gsfc.nasa.gov/planetary/factsheet/moonfact.html). The composition of Moon atmosphere is (particles per cubic cm): Helium 4 – 40 000; Neon 20 – 40 000; Hydrogen – 35 000; Argon 40 – 30 000; Neon 22 – 5 000; Argon 36 – 2 000; Methane - 1000; Ammonia - 1000; Carbon Dioxide – 1000. The day and night pressure is different due to heating and degassing of lunar soil. The measured pressure in the area near the spacecraft or other man activity is much higher due to degassing of the spacecraft materials and exhausted gases of spacecraft engine.

The virgin pressure near asteroid can be expected similar as near the Moon about 10^{-9} Pa on night side of the asteroid. However, the pressure near the spacecraft or any other human made construction is expected higher, similar as it was observed on LEO, GEO and Moon surface of 10^{-3}-10^{-5} Pa.

Second important factor is temperature. Our imagination of temperature behavior based on Earth experience is wrong. The space construction is heated by the Sun from one side and irradiates the heat from all sides. If the construction does not have internal heat source and limited thermo-conductivity, the dark side of the construction is cooled to -150…-200^{0}C.

The total solar irradiation at the Earth distance from Sun equals to 1362-1367 W/m^2 (Walter 1987). The solar irradiation level depends on season (position of Earth on solar orbit) and it varies from 1316 W/m^2 at minimal solar energy flux (summer solstice) to 1428 W/m^2 at maximal solar energy flux (winter solstice) (ECSS Space Environment Standard 2000). The level of de-irradiation of sun light by Earth surface and its atmosphere equals to 240 W/m^2. Temperature of spacecraft surface depends on an orientation to Sun, absorption and emission indexes of the surface and internal heat sources (Favorskii and Kadaner 1972). Experimentally measured temperature of the spacecraft surface on LEO varied in wide range by different data: -56…+77^{0}C by (Teichman et al 1992), -90…+120^{0}C by (Barbashev 1982), -100^{0}C…+200^{0}C by (Fu and Graves 1985) and (de Groh and Morgana 2002), -150…+150^{0}C by (Haruvy 1990). For far space mission as NGST mission (halo orbit, $1.5{\cdot}10^{-6}$ km from Earth) the estimated temperature equals to -223…+122^{0}C by (Dever et al. 2002).

The temperature of the construction on Moon surface depends on sun irradiation, altitude position on the Moon, an orientation to the Sun, shadowing of lunar rocks, absorption and emission indexes of the wall surface and internal heat sources. The temperature of the spacecraft surface landed on the Moon varies in diapason of -150…+150°C (Kondyurin 2012).

The temperature of the spacecraft surface on asteroid will depend on similar factors as on Moon and Earth orbit. However, the sun irradiation flux depends on distance of asteroid from the Sun. The orbits of Earth and Moon have low eccentricity and the temperature in perihelion and aphelion positions varies non-significantly. The asteroid orbits have usually high eccentricity orbit or even more complicate irregular shape of the orbits. The temperature variations during "asteroid year" are higher and depend on asteroid orbit. When spacecraft is approaching asteroid in Perihelion position, the temperature of the irradiated spacecraft will be high due to short distance to the Sun. When the spacecraft approach the asteroid in Aphelion position, the temperature of the Sun irradiated spacecraft will be low. If the temperature of spacecraft (T_e) on the Sun side is known on Earth orbit, the temperature of the spacecraft under the same irradiation conditions near asteroid (T_a) can be calculated as following:

$$T_a = T_e \cdot \sqrt{\frac{D_{Earth-Sun}}{D_{asteroid-Sun}}} \tag{15.a}$$

An example of the sun irradiated spacecraft temperature on some asteroids in Perihelion and Aphelion positions is shown in Table 15.1.

Table 15.1 The calculated temperature of the spacecraft irradiated by Sun light nears the asteroids in Perihelion and Aphelion positions under the same irradiation conditions as the spacecraft on Earth orbit heated up to 100 ^{0}C (marked with *)

Asteroid	Distance from the Sun, million km		Temperature, °C	
	Perihelion	Aphelion	Perihelion	Aphelion
Adonis	66	495	283	-66
Apollo	97	343	187	-25
Apophis	112	164	155	86
Bacchus	105	218	169	39
Cerberus	86	237	215	26
Eros	169	267	75	9
Golevka	148	600	99	-85
Hephaistos	54	595	345	-84
Hermes	93	402	196	-44
Icarus	28	294	581	-5
Itokawa	143	254	106	16
JM8	142	668	106	-95
Midas	93	266	196	9
Nereus	143	303	106	-9
Phaethon	21	360	716	-30
Sisyphus	131	436	123	-53
Toutatis	140	617	109	-88
VE68	64	153	293	99
YU55	98	244	185	22
Earth	147	152	100*	100*

The spacecraft is considered on asteroids under the same Sun irradiation and thermo conductivity conditions as heated up to 100 ^{0}C on Earth orbit. The temperature of spacecraft on different asteroids varies from +716 ^{0}C for Phaethon in Perihelion position, to -88 ^{0}C for Toutatis in Aphelion position. This temperature difference is extremely high for the polymer materials and curing reactions.

The space plasma is created by the galactic and Sun protons, electrons, neutrons and heavy particles with wide diapason of energy from some eV to GeV; infrared, visual, ultraviolet, vacuum ultraviolet and X-ray photons. The most investigations of polymers in space flights were done in LEO, where the fluent of atomic oxygen (AO) is a most significant factor for polymers in compare with other factors of space plasma (Walter 1987). The estimation of average AO flux on LEO (near 300 km altitude) during real experiments with polymer materials equals to $2.88 \cdot 10^{13}$ at/cm^2/sec average value by (Walter 1987), $3.88 \cdot 10^{13}$ at/cm^2/sec in LDEF mission by (Klein and Lesieutre 2000), 10^{14} in MISSE mission by (de Groh et al. 2001), $5 \cdot 10^{14}$ at/cm^2/sec theoretical value and $4.3 \cdot 10^{14}$ at/cm^2/sec on Kapton equivalent by (Pippin 1999) for ESEM mission, $5 \cdot 10^{13}$ at/cm^2/sec by (Connell 1999) and 10^{12}-10^{15} at/cm^2/s by (Kiefer et al. 1999) for ESEM mission too, 10^{13}-10^{15} at/cm^2/s by (Czaubon et al. 1998). In Habble mission (595 km altitude) the AO flux equals to $6.86 \cdot 10^{11}$ at/cm^2/sec by (Dever et al. 1998). The flux of AO varies due to Sun activity, season, position, longitude-latitude and altitude of space ship, variations of Earth atmosphere and outgassing processes of space ship materials. The model approximation developed in (ECSS Space Environment Standard 2000) shows the AO flux at 400 km altitude of $2 \cdot 10^{11}$ atoms/cm^2/s. The asteroid environment does not include AO flux. There is no experimental data on effect of cosmic rays in asteroids on polymers. For using of the flight results on polymers and for prediction of the polymer behavior near the asteroid, we have to consider the environmental effect on LEO.

Vacuum ultraviolet (VUV) irradiation is the part of solar spectra of irradiation. The intensity of VUV light is low, but effect of VUV light on polymers is significant higher than visual and UV light. At Earth orbit the level of VUV light can be estimated about $4 \cdot 10^{11}$ photons per cm^2 in second for 121.6 nm wavelength (Koontz et al. 1991). The Sun irradiation corresponds to 0.75 μW/cm^2 in VUV diapason of 100-150 nm wavelength by (Koontz et al. 1989) and 11 μW/cm^2 in VUV diapason below 200 nm wavelength by (Lura et al. 2003). The level of X-ray on Earth orbit equals to $2.3 \cdot 10^{-9}$ W/cm^2 for 1-8 A wavelength and $1.43 \cdot 10^{-10}$ W/cm^2 for 0.5-4 Å wavelength (de Groh and Morgana 2002). The most flux of X-rays is directed from sun and less from stars.

The energetic spectrum of electron and ion fluxes at LEO and GEO is sufficiently complicate. The energetic spectrum and flux of charged particles depend on kind of particle, seasons, Sun activity. The energy of charged particles varies in diapason from 0.1 eV to GeV.

Most flight experiments with polymers were done on LEO, where significant part of the particles flux is shielded by Earth magnetic field. The density of electrons at LEO altitude of 400 km equals to 10^5 e/cm^3 (night side) and 10^6 e/cm^3

(day side) by (Walter 1987) with energy of 0.1 eV by (Koontz et al. 1991). The total electrons flux at GEO mission equals to 10^9 e/cm^2/sec for electrons with energy of 0-12 keV by (Lai and Della-Rose 2001). The electron density on GEO mission of 1.12 e/cm^3 at average energy of $1.2 \cdot 10^4$ eV and the ion density of 0.236 ion/cm^3 with average energy of $2.95 \cdot 10^4$ eV are used for simple analysis of plasma on GEO missions in (Purvis et al. 1984). The most frequent ions are hydrogen ions (90%), when other 10% correspond to heavier ions.

The asteroids don't have sufficient magnetic field to change trajectories of the charged particles. Therefore, all flux of particles as in deep space will bombard the spacecraft. The galaxy rays flux does not depend on orbit of asteroid, while the solar wind intensity depends on orbit and position of the asteroid on the orbit. In average the intensity of solar wind depends on distance from the Sun with square function. However, the intensity of solar wind can change significantly in a case of solar events. The variation of flux particles from the Sun can be 2-3 orders of magnitude that make estimation of the flux during spacecraft flight very problematic.

The asteroid gravity is very low and can be ignored in comparison with Earth gravity. Therefore all gravity effects in polymerization technology could be ignored.

Meteorite fluency has strong effect on brittle materials. It can cause cracking and mechanical destruction of material and whole construction. In the case of polymer on Earth orbit the influence of meteorite fluency is low due to low meteorite flux. Erosion of polymer at meteorite fluency corresponds to 0.1 nm/year in deep space that is not significant for 20-30 years of exploitation of the composite.

The composition of asteroids is not well known. Assuming a general similarity of all celestial bodies, the composition of Earth and Moon can be considered as examples. Lunar regolith contains of oxygen, silicon, iron, calcium, aluminum, magnesium, titanium, sodium, chromium, manganese, potassium and traces of other elements of sulfur, carbon, nitrogen, hydrogen, helium^4 and helium^3. The presence of these elements and its oxides on asteroid surface and in nearby do not have specific interaction with polymer matrix and it cannot disturb the curing reaction.

This analysis shows, that vacuum, space plasma and temperature changes are most critical factors for polymerization processing on asteroids.

15.3 Curing Reactions in Vacuum

If the pressure is lower than the vapour pressure of individual components of the polymer matrix, the components evaporate. Due to evaporation a concentration of the reactionable components decreases and the curing reaction cannot be completed. This effect cannot be ignored when the curing reaction is considered in space environment.

The space agencies released the standard tests for outgassing of the volatile components. For example, ESA standard ECSS-Q-70-02A requires the polymer material is vacuumed up to 10^{-3} Pa at temperature of 125°C for 24 hours. The total mass loss (TML), recovered mass loss (RML) and collected volatile condensable material (CVCM) have to be tested. The acceptable materials must show RML < 1,0 % and CVCM < 0,10 %. The standard requires a stable solid material for the test. The uncured prepreg is unstable and contains liquid components. Therefore, these outgassing standards cannot be applied. Another standards and requirements for curable materials in space vacuum should be developed.

There are a number of vacuum simulations in laboratory experiments with polymers. The pressure in simulate space environment experiments is varied: 10^{-7}-10^{-8} Pa (Grossman et al. 1999), 10^{-5} Pa (Bilen et al. 2001; Iwata et al. 2001), $7{\cdot}10^{-5}$ Pa (de Groh et al. 2001), $2.33{\cdot}10^{-4}$ Pa (Koontz et al. 1989), $6.65{\cdot}10^{-4}$ Pa (Dever et al. 2002a), 10^{-4} Pa (ECSS Space Environment Standard 2000; Lura et al. 2003; Gonzales et al. 2000), 73 Pa (Golub and Wydeven 1988). The temperature in experiments is usually varied from -150 to +150 ^{0}C and even +800 ^{0}C. The combination of high vacuum and high temperature is used for analysis of outgassing processes in materials for space applications.

The main influence of such vacuum is observed as evaporation of low molecular mass fractions. The rate of evaporation into high vacuum is described by Langmuir formula:

$$W = P \cdot \sqrt{\frac{M}{T}} \tag{15.1}$$

where M - molecular mass of vapour of fraction, T - temperature, P - equilibrium vapour pressure of fraction from Klausius-Klapeyron equation (Kroshkin 1969). At long time vacuum action the stockhiometric of low weight molecular components in matrix can be changed (Kondyurin et al. 1998, 2001). The changes of active components concentration in oligomers composition are important for creation of plastic with high durability.

Most of polymer composites certified for space application are based on epoxy resins compositions with carbon, organics or glass fibers. The curing includes the chemical reaction of epoxy group with amines, amides, anhydrides, metal-organic complexes or carboxyl acids as hardener. A wide number of different epoxy matrix compositions were developed since discovery of epoxy resins for construction materials. In the curing reaction two or more components of low molecular mass take part and form crosslinks between molecules. General schemes of reactions are following:

primary amine hardener

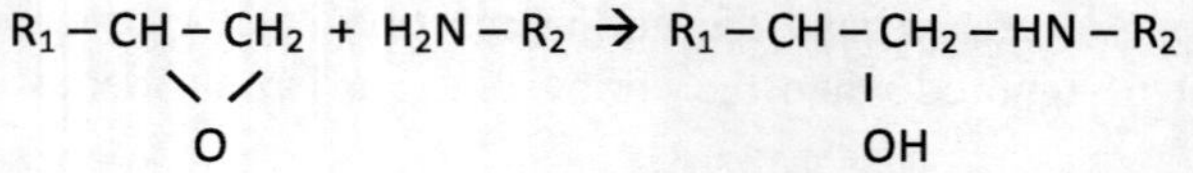

(15.2)

secondary amine, amide and urethane hardeners

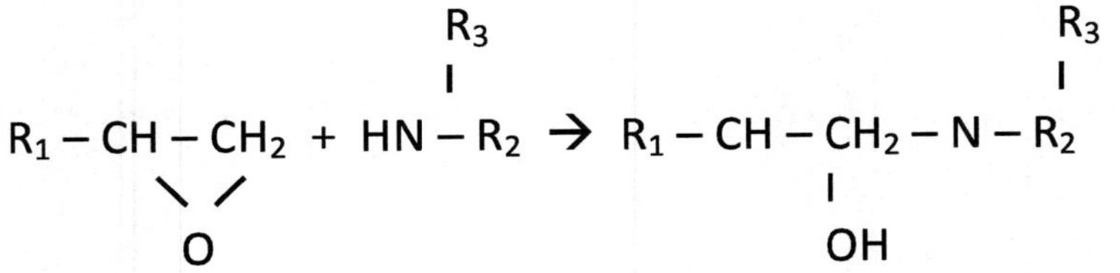

(15.2)

tertiary amine hardener

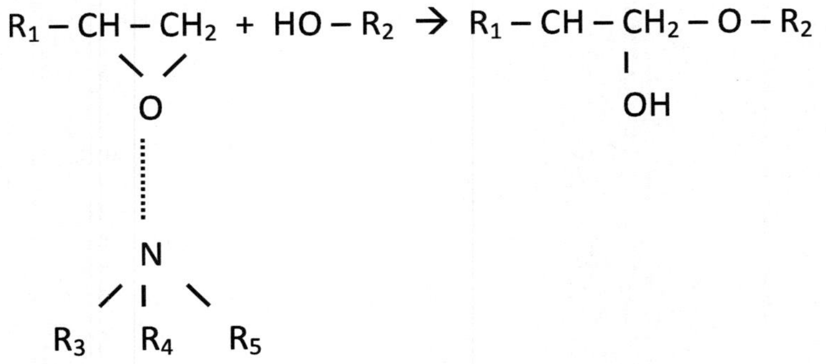

(15.3)

These reactions require two or more components with definite concentrations. If uncured prepreg is placed into vacuum, the components of the prepreg evaporate. The rate of evaporation depends on molecular mass of the component, intermolecular interactions with neighbor molecules and temperature. If the prepreg is thick enough, the evaporation rates depend on diffusion of the components to the surface of the prepreg.

The evaporation rates of the components are different. Therefore, after exposure in vacuum the concentration of the components is drifting with time from initial concentration. In worse case, the curing reaction could not be completed due to deficit of highly volatile components. In such case, the prepreg becomes incurable and the required exploitation characteristics of the composite could be not achieved.

When the evaporation rate of some component is too high, the cavitation effect is observed. The matrix contains bubbles and even foams. The mechanical strength of cured composite drops down.

However, a number of compositions curable in vacuum have been found. In experiments we tested various epoxy matrices consisted of high and low evaporation rate components. The examples of the representative composites cured in vacuum are shown on Figs. 15.1 and 15.2.

These composites were not covered from the top and the evaporation of low molecular components occurred freely into vacuum. When the curing composite is placed between hermetic sealings (for example, protective film), the bubbling is stronger and the matrix is completely transferred to a foam. This effect was observed in real flight experiment on LEO.

When the evaporation rates of all components are too high, the polymer matrix could disappear, the fibers could be naked and the composite material could not be cured.

Our experimental and theoretical investigations of the evaporation kinetics and curing kinetics in vacuum at various temperatures show that the compositions with suitable evaporation rates of the components curable under high vacuum conditions could be found (Kondyurin 1997, 2011, 2012; Kondyurin et al. 2004, 2009, 2009a, 2010; Kondyurina et al. 2006).

Fig. 15.1 Microphoto of composite 1 cured in vacuum. The epoxy matrix consisted of low evaporation rate components. The cured matrix is smooth and binds fibres without defects. The strength of composite cured in air and vacuum is identical.

Fig. 15.2 Microphoto of composite 2 cured in vacuum. The epoxy matrix consisted of high evaporation rate hardener. The cured matrix is foamed. Different size bubbles are observed. The strength of composite cured in vacuum is significantly less than the cured in air.

The general principles of the selection are following:

1. The molecular mass of components should be high enough to provide low evaporation rate. However, the high molecular mass of the components means high viscosity. The viscosity of the matrix should be low enough to keep elasticity of the prepreg at unfolding. Therefore, the selection of the components requires an optimization of the molecular mass and viscosity to provide unfolding ability and low evaporation rate.

2. The evaporation rates of the components should be identical to exclude deficit of active components at evaporation.

3. Any solvent should be excluded from any stage of the prepreg preparation. Even small traces of the solvent or other low molecular components can give foam structure of the matrix at curing in vacuum.

4. The multifunctional components with 3 or more active groups in one macromolecule are preferable. One link of the molecule to hardener during first stage of the curing reaction can decrease the evaporation rate significantly, when other active groups can react later to provide required mechanical strength.

5. The geometry of the prepreg should permit low molecular components to evaporate freely. Any closed volumes and covers must be excluded. If some area of the prepreg is covered, the evaporated fractions can form bubbles and foam structure. The biggest thickness of the polymer matrix in prepreg depends on volatility of the components. It limits the thickness of the polymer matrix in the prepreg.

With these principles, we selected and cured epoxy matrix compositions in high vacuum up to 10^{-3} Pa without foam. A number of suitable compositions are described in (Kondyurin 2012). The example of 5 mm thick composite plate with low evaporation rate epoxy matrix cured in high vacuum is shown on Fig. 15.3.

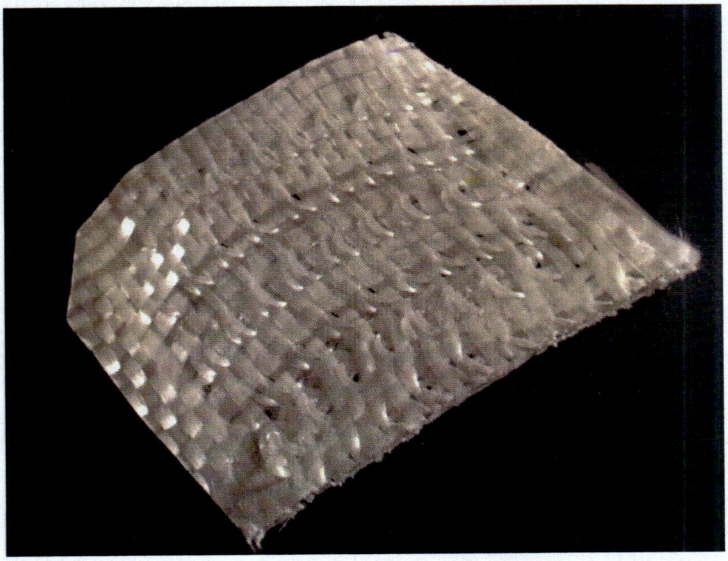

Fig. 15.3 Glass fiber composite with epoxy matrix cured in vacuum

Therefore, the high vacuum factor of the asteroid environment is not a problem, if the composition of polymer matrix is selected and used correctly.

15.4 Effect of Cosmic Rays and Sun Wind

The effect of space radiation on polymer materials was investigated intensively in space flight near Earth on LEO. As observed in these flight missions the effect of atomic oxygen flux (AO) is most significant factor of free space environment at LEO. During exposition of polymer in AO flow the processes of etching and chemical reactions leads to formation of etched oxidised surface layer. The simulation of AO effect on polymers was done in plasma and ion beam reactors. For uncured prepreg, a period of composition in liquid state is critical due to short macromolecules of resin and hardening agent. In this case the coming oxygen atom breaks chemical bond of backbone and the shortened molecular parts of macromolecule can be quickly released.

The first period of polymerization reaction is very important for erosion activity of high energy particles. On other hand, our experiments showed, that plasma action influences on curing reaction by the way of formation of free radicals and movement of treated layer into bulk layers of liquid composition. In this case, the free radical macromolecules can take part in polymerization reaction with formation of additional crosslinks in complete bulk layer of composite. Therefore, the two opposite process compete during plasma in curing epoxy matrix: degradation (scission) and evaporation of short parts of macromolecules, and crosslinking of the macromolecules due to free radical reactions.

The free radicals are chemically very active and take part in a chain of chemical reactions. For example, following reactions have been observed in irradiated hydrocarbon polymers:

The moving of free radical along the macromolecule

$$\text{-CH}\bullet\text{-CH}_2\text{-CH}_2\text{-CH}_2\text{-} \rightarrow \text{-CH}_2\text{-C}\bullet\text{H-CH}_2\text{-CH}_2\text{-} \rightarrow \text{-CH}_2\text{-CH}_2\text{-CH}\bullet\text{-CH}_2\text{-} \quad (15.4)$$

the jumping of free radical between the macromolecules

$$\text{-CH}_2\text{-CH}\bullet\text{-CH}_2\text{-} + \text{-CH}_2\text{-CH}_2\text{-CH}_2\text{-} \rightarrow \text{-CH}_2\text{-CH}_2\text{-CH}_2\text{-} + \text{-CH}_2\text{-CH}\bullet\text{-CH}_2\text{-} \quad (15.5)$$

When two free radicals meet, they form a double bond

$$\text{-CH}\bullet\text{-CH}_2\text{-CH}_2\text{-} + \text{-CH}\bullet\text{-CH}_2\text{-CH}_2\text{-} \rightarrow \text{-CH=CH-CH}_2\text{-} + \text{-CH}_2\text{-CH}_2\text{-CH}_2\text{-} \quad (15.6)$$

and crosslink

$$\begin{array}{l} -CH_2\text{-}CH\bullet\text{-}CH_2\text{-}CH_2\text{-} \\ \quad\quad\quad + \\ -CH_2\text{-}CH\bullet\text{-}CH_2\text{-}CH_2\text{-} \rightarrow \\ \\ -CH_2\text{-}CH\text{-}CH_2\text{-}CH_2\text{-} \\ \quad\quad\quad | \\ -CH_2\text{-}CH\text{-}CH_2\text{-}CH_2\text{-} \end{array} \tag{15.7}$$

The free radical could break a hydrocarbon backbone

$$-CH\bullet\text{-}CH_2\text{-}CH_2\text{-}CH_2\text{-} \rightarrow -CH{=}CH_2 + \bullet CH_2\text{-}CH_2\text{-} \tag{15.8}$$

and oxygen containing backbone

$$-CH\bullet\text{-}O\text{-}CH_2\text{-}CH_2\text{-} \rightarrow -CH{=}O + \bullet CH_2\text{-}CH_2\text{-} \tag{15.9}$$

The reactions (15.6)-(15.9) produce stable groups. The reaction (15.7) gives crosslinking of the polymer. This reaction has the same strengthen effect as the curing reaction. The reactions (15.8) and (15.9) give the break of the macromolecule, as opposition reaction for the curing. The ratio of crosslinking/scission reactions depends on chemical structure of the macromolecules, environmental gases, temperature, kinds and density of free radicals.

Besides that, the epoxy group can directly react with free radicals (Klyachkin 1992; Mesyats 1999):

$$\begin{array}{l} R_1{-}CH{-}CH_2 + \bullet R_2 \rightarrow R_1{-}CH{-}CH_2{-}R_2 \rightarrow \\ \quad\quad \diagdown\ \diagup \quad\quad\quad\quad\quad\quad\quad | \\ \quad\quad\ \ O \quad\quad\quad\quad\quad\quad\quad\ \ O\bullet \\ \\ R_1{-}CH{-}CH_2{-}R_2 + \bullet R_3 \rightarrow R_1{-}CH{-}CH_2{-}R_2 \\ \quad\quad | \quad\quad\quad\quad\quad\quad\quad\quad\quad\quad\quad | \\ \quad\ \ O\bullet \quad\quad\quad\quad\quad\quad\quad\quad\quad\ \ O{-}R_3 \end{array} \tag{15.10}$$

As a result of free radical reactions, the polymer matrix disappears from the fibers (if etching is prevailed) or remains and deforms due to internal stresses (if crosslinking is prevailed). Both cases have been observed in the experiments.

The effect of space plasma was observed in laboratory experiments on curing of the epoxy matrix prepreg in glow discharge plasma and ion beam.

The effects of additional crosslinking up to carbonization of the surface layer were observed and investigated with a number of experimental methods such as gel-fraction analysis, spectroscopy, mechanical and thermomechanical analysis (Kondyurin and Bilek 2008). These effects depend on plasma and ion beam conditions as well as on composition of the epoxy matrix.

On asteroid the high energy electrons and ions with VUV irradiation are the main destructive factors of free space environment for polymer. The high energy particles penetrate into polymer much deeper, than at AO flux, and cause more defects in macromolecules. The structure changes of polymer matrix are available in bulk layers in this case.

The VUV and high energy electrons have similar effect to generate the free radicals in polymers. Such radicals can take part in curing reaction and increase a curing rate. The kinetic rate of curing reaction has non linear dependence on a concentration of active groups in reactionable mixture. Therefore, the actions of AO, VUV, electron beam together with mixing processes of liquid matrix on kinetic of curing reaction can not be predicted by combination of experiments with separate kinds of irradiation. Taken into account the non-homogeneity distribution of intensity of ions, VUV light and electron in polymer, the behavior of polymer matrix during curing depends on thickness of matrix and presence of fibers. The theoretical study of radiation space environment factors is sufficiently difficult. In this case the real space experiments have a great significance for estimation of space environmental effects.

Our experiments showed, that the kinetic effect of the high energy particles is significant for modification of the bulk layers of liquid composition. The macromolecules get mechanical impulse from the upcoming particles and move into deep layer. With this movement, the radical products of the macromolecule destructions move to the bulk layer of the resin, where they are able to react with virgin macromolecules.

Fig. 15.4 Microphoto of carbon fibre composite cured in plasma discharge. The amount of epoxy matrix on fibres is low. The cured matrix is smooth.

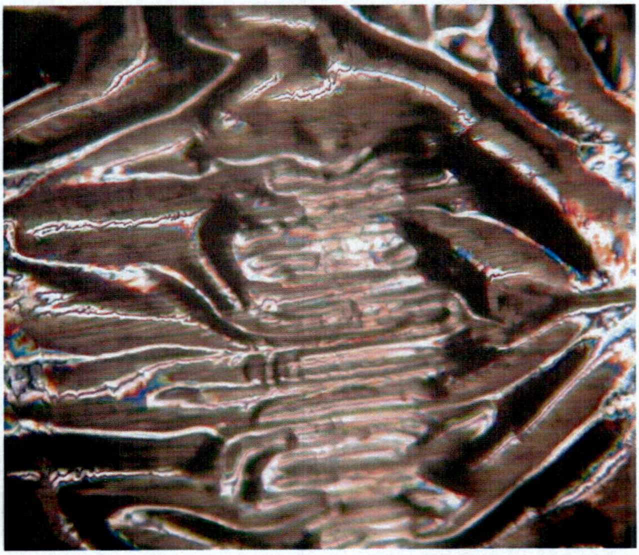

Fig. 15.5 Microphoto of carbon fibre composite cured in plasma discharge. If the amount of epoxy matrix is high, the top layer of matrix becomes deformed.

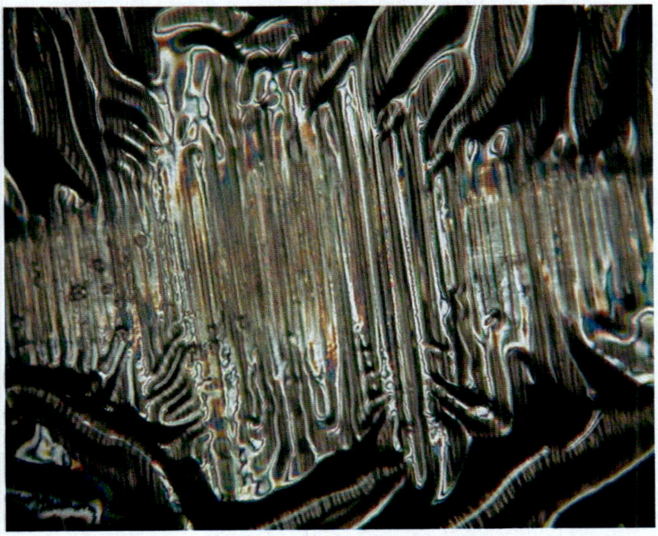

Fig. 15.6 Microphoto of carbon fibre composite cured in plasma discharge. If the amount of epoxy matrix is distributed non-uniformly (for example, in case of satin fabric), the top layer of matrix is deformed non-uniformly.

Fig. 15.7 Microphoto of carbon fibre composite cured in ion beam. The part of epoxy resin is etched and the top of some fibres becomes naked. In the case of thick layer, the matrix is deformed.

Fig. 15.8 Microphoto of carbon fibre composite cured in ion beam with fluence of $5*10^{15}$ ions/cm^2. The part of epoxy resin is etched and some fibres become naked. In the case of thick layer, the matrix is deformed.

The effect of cosmic rays on curable composites was observed in stratospheric flight with altitude higher than ozone layer. In 2010 first experiment with uncured compositions in stratosphere (40 km altitude) have been done during stratospheric balloon flight (Kondyurin et al. 2010a, 2013). The cassette with cured (control) and uncured epoxy matrix of different compositions was exposed in stratosphere during 3 days.

Fig. 15.9 Cassette with cured (control) composite and uncured composites (prepreg) on GPS antenna bar of the payload before the launch. NASA stratospheric flight in Alice Springs, Australia.

During flight the residual atmospheric pressure was 2-4 Torr that is less than the vapour pressure of the epoxy resin components (10-20 Torr). Because of flight altitude was higher than the ozone layer, vacuum UV light and the cosmic rays of high flux bombarded the uncured epoxy matrix of glass and carbon fiber composites during the flight. Due to absence of AO flux, the conditions during stratospheric flight were closer to the conditions in deep space, than the conditions of LEO.

After the flight, the samples were investigated with a number of structure methods. The effects of free radical reactions, additional crosslinking, oxidation and depolymerization due to cosmic rays were clear observed in the flight samples (Kondyurin et al. 2010a, 2013).

Fig. 15.10 Telemetry image of the cassette with uncured epoxy composition in stratosphere (40 km altitude)

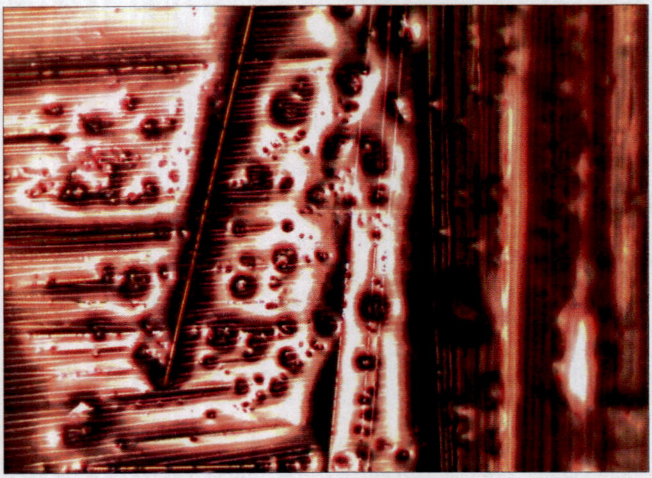

Fig. 15.11 Microphoto of carbon fibre composite cured in stratosphere (27 km altitude). Some part of epoxy matrix is bubbled due to low pressure in stratosphere during curing.

It was shown, the mechanical properties of the flight samples satisfies to the exploitation properties of the composites and are not far from the properties of the control ground samples for some of the selected compositions. Similar radiation effects are expected in curable composite at curing on asteroids.

The second stratospheric flight with uncured compositions was done in 2012 in South Australia. The composition was cured completely during the flight. The effect of cosmic rays on crosslinking of epoxy matrix was observed as well as the effect of low molecular weight component evaporation. The microphotograph of the composite surface shows bubble formation in the epoxy matrix where the layer of the matrix is thick (Fig. 15.11).

Thus, the cosmic rays from Universe and from the Sun play role of additional hardener and it can accelerate the curing effect if the choice of the polymer matrix components is correct.

15.5 Curable Constructions for Asteroids Exploitation

An exploitation of asteroids is viewed a fantastic idea now. However, some projects are discussed and reviewed in scientific journals and in present book. The human presence on asteroid needs long-time base with self-regulated ecological system for food, water and air regenerations. All equipment, life-support systems, storage and crew life space need sufficient volume for the base and spaceship. There is no technology, that can provide a launch of such large volume and mass construction from Earth. Therefore, the large space construction must be assembled in Earth orbit or on asteroid. A real way to assemble the large space construction is to use polymerization technology for construction composite materials in space environment.

When human presence is not planned, the large space constructions can be assembled on asteroids for different purposes as following:

- telescope scientific mission. The asteroid is unique celestial body to carry a high sensitive space telescope. A large size antenna can be unfolded and cured on the asteroid surface to provide high sensitivity of measurements and communication.
- signal transducer for deep space communication. The large size antenna can be unfolded and cured on the asteroid surface to provide transportation of the communication signals for deep space missions.
- mining base. When robotic missions are planned for asteroid mining, the storage, shielding and different structured elements can be assembled and fixed with the polymerization technology.

These and other projects of asteroid exploitation become more realistic if the polymerization technology is considered to create large space constructions as an alternative to present ready-to-use launching technologies.

Acknowledgments. Author gratefully acknowledges NASA, ESA, RFBR and Humboldt Foundation for the project funding.

References

Allred, R., Hoyt, A.E., McElroy, P.M., Scarborozgh, S., Cadogan, D.P.: UV rigidizable carbon-reinforced isogrid inflatable booms. AIAA-2002-1202 (2002)

Bar-Cohen, Y.: Transition of EAP material from novelty to practical applications – are we there yet? In: SPIE's 8th Annual International Symposium on Smart Structures and Materials, Newport, USA, March 5-8, Paper No. 4329-02 (2001)

Barbashev, E.A., Dushin, M.I., Ivonin, Y.N., Kozin, V.I., Nikishin, E.F., Panshin, B.I., Perov, B.V.: Some results of tests of polymer materials after exposition in conditions of free space. In: Space Technology and Material Science, Moscow, Nauka (1982)

Bilen, S., Domonkos, M., Gallimore, A.: Simulating ionospheric plasma with a hollow cathode in a large vacuum chamber. J. of Spacecraft and Rockets 38, 617–621 (2001)

Briskman, V.A., Yudina, T.M., Kostarev, K.G., Kondyurin, A.V., Leontyev, V.B., Levkovich, M.G., Mashinsky, A.L., Nechitailo, G.S.: Polymerization in microgravity as a new process in space technology. Acta Astronautica 48, 169–180 (2001)

Cadogan, D., Stein, J., Grahne, M.: Inflatable composite habitat structures for Lunar and Mars exploration. In: 49th International Astronautical Congress, Melbourne, Australia, September 28-October 2, IAA-98-IAA.13.2.04 (1998)

Cadogan, D., Grahne, M., Mikulas, M.: Inflatable space structures: a new paradigm for space structure design. In: 49th International Astronautical Congress, Melbourne, Australia, September 28-October 2, IAF-98-I.1.02 (1998a)

Cadogan, D.P., Lin, J.K., Grahne, M.S.: Inflatable solar array technology. AIAA-99-1075 (1999)

Cadogan, D.P., Scarborough, S.E.: Rigidizable materials for use in Gossamer Space Inflatable structures. AIAA-2001-1417 (2001)

Cadogan, D.P., Scarborough, S.E., Lin, J.K., Sapna, G.H.: Shape memory composite development for use in Gossamer space inflatable structures. AIAA 2002-1372 (2002)

Cassapakis, C., Thomas, M.: Inflatable structures technology development overview. AIAA-95-3738 (1995)

Connell, J.W.: The effects of low-Earth orbit atomic oxygen exposure on Phenylphosphine oxide-containing polymers. Final report. Evaluation of Space Environment and Effects on Materials (ESEM). Appendix D, NASA Technical Report (1999)

Czaubon, B., Paillos, A., Siffre, J., Thomas, R.: Mass spectrometric analysis of reaction products of fast oxygen atoms-material interactions. J. of Spacecraft and Rockets 35, 797–804 (1998)

Darooka, D.K., Jensen, D.W.: Advanced space structure concepts and their development. AIAA-2001-1257 (2001)

Darooka, D.K., Scarborough, S.E., Cadogan, D.P.: An evaluation of inflatable truss frame for space applications. AIAA 2001-1614 (2001)

de Groh, K.K., Banks, B.A., Hammerstrom, A.M., Youngstrom, E.E., Kaminski, C., Marx, L.M., Fine, E.S., Gummow, J.D., Wright, D.: MISSE PEACE Polymers: An International Space Station Environmental Exposure Experiment. In: Proceedings of the Conference on ISS Utilization - 2001, Cape Canaveral, Fl, AIAA 2001-4923 (2001); also in NASA TM-2001-211311

de Groh, K.K., Morgana, M.: The Effect of Heating on the Degradation of Ground Laboratory and Space Irradiated Teflon FEP. NASA TM-2002-211704 (2002)

Derbes, B.: Case studies in inflatable rigidizable structural concepts for space power. AIAA-99-1089 (1999)

Dever, J., de Groh, K.K., Townsend, J.A., Wang, L.L.: Mechanical Properties Degradation of teflon FEP Returned from the Hubble Space telescope. NASA report 1998-206618, AIAA-98-0895 (1998)

Dever, J., Semmel, C., Edwards, D., Messer, R., Peters, W., Carter, A., Puckett, D.: Radiation durability of candidate polymer films for the next generation space telescope sunshield. NASA TM 2002-211508 and AIAA-2002-1564 (2002)

Dever, J.A., Pietromica, A.J., Stueber, T., Sechkar, E., Messer, R.: Simulated space vacuum ultraviolet (VUV) exposure testing for polymer films. NASA TM-2002-211337 and AIAA-2001-1054 (2002a)

ECSS Space Environment Standard, ECSS E-10-04, Guide for LEO mission, ECSS-Q-70-04 (outgassing), ESA (2000)

Favorskii, O.N., Kadaner, Ya, S.: About heat transfer in space. Visshaya Shkola, Moscow (1972)
Freeland, R.E., Veal, G.R.: Significance of the inflatable antenna experiment technology. AIAA-98-2104 (1998)
Fu, J.H., Graves, G.R.: Thermal Environments for Space Shuttle Payloads. In: AIAA Shuttle Environment and Operation II Conference Proceedings, p. 18 (1985)
Golub, M.A., Wydeven, T.: Reactions of atomic oxygen (O(3P)) with various polymer films. Polymer Degradation and Stability 22, 325–338 (1988)
Gonzales, R.I., Phillips, S.H., Hoflund, G.B.: In situ oxygen atom erosion study of polyhedral oligomeric silsesquioxane-siloxane copolymer. J. of Spacecraft and Rockets 37, 463–467 (2000)
Grahne, M.S., Cadogan, D.P.: Inflatable solar arrays: revolutionary technology? ILC Dover, Inc., 1999-01-2551 and Sasakawa International Center for Space Architecture, SICSA outreach, Special Design Project Issue 1, No.7 (1988)
Grossman, E., Lifshitz, Y., Wolan, J.T., Mount, C.K., Hoflund, G.B.: In situ erosion study of Kapton using novel hyperthermal oxygen atom source. J. of Spacecraft and Rockets 36, 75–78 (1999)
Grossman, G., Williams, G.: Inflatable concentrators for solar propulsion and dynamic space power. Journal of Solar Energy Engineering 112, 299 (1990)
Guidanean, K., Williams, G.T.: An inflatable rigidizable truss structure with complex joints. AIAA-98-2105 (1998)
Haruvy, Y.: Risk Assessment of Atomic-Oxygen-Effected Surface Erosion and Induced Outgassing of Polymeric Materials in LOE Space Systems. ESA Journal 14, 109–119 (1990)
Iwata, M., Ohnishi, A., Hirosawa, H., Tohyama, F.: Measurement and Evaluation of Thermal Control Material with Polyimide for Space Use. J. of Spacecraft and Rockets 38, 504–509 (2001)
Kato, S., Takeshita, Y., Sakai, Y., Muragishi, O., Shibayama, Y., Natori, M.: Concept of inflatable elements supported by truss structure for reflector application. Acta Astronautica 19, 539–553 (1989)
Kiefer, R.L., Orwold, R.A., Harrison, J.E., Ronesi, V.M., Thibeault, S.A.: The effects of the space environment on Polyetherimide films. Final report. Evaluation of Space Environment and Effects on Materials (ESEM), Appendix C, NASA Technical Report (1999)
Klein, T.F., Lesieutre, G.A.: Space environment effects on damping of polymer matrix carbon fiber composites. J. of Spacecraft and Rockets 37, 519–525 (2000)
Klyachkin, Y.S., Trushnikov, V.A., Kondyurin, A.V., Imankulova, S.A.: Study of the nature of interaction of EPDM-40 rubber with an epoxy adhesive. J. Adhesion Science and Technology 6, 1137–1145 (1992)
Kondyurin, A.V.: Building the shells of large space stations by the polymerisation of epoxy composites in open space. Plasticheskie massy 8: 25. Translated in Int. Polymer Sci. and Technol. 25(4), T/78 (1997)
Kondyurin, A.: Large size station on Mars surface by the way of polymerization of composite polymer material. In: Fourth Canadian Space Exploration Workshop. Science Payloads for Mars, Abstracts, Ottawa, Canada, November 15-16 (2002)
Kondyurin, A.: Direct Curing of Polymer Construction Material in Simulated Earth's Moon Surface Environment. Journal of Space Craft and Rockets 48, 378–384 (2011)
Kondyurin, A.: Curing of composite materials for an inflatable construction on the moon. In: Badescu, V. (ed.) Moon. Prospective Energy and Material Resources, vol. 102, pp. 503–518. Springer, Heidelberg (2012)
Kondyurin, A., Mesyats, G., Klyachkin, Y.: Creation of High-Size Space Station by Polymerization of Composite Materials in Free Space. J. of the Japan Soc. of Microgravity Appl. 15, 61–65 (1998)

Kondyurin, A., Kostarev, K., Bagara, M.: Polymerization processes of epoxy plastic in simulated free space conditions. Acta Astronautica 48, 109–113 (2001)

Kondyurin, A., Lauke, B., Richter, E.: Polymerization Process of Epoxy Matrix Composites under Simulated Free Space Conditions. High Performance Polymers 16, 163–175 (2004)

Kondyurin, A., Bilek, M.: Ion Beam Treatment of Polymers. Application aspects from medicine to space. Elsevier, Oxford (2008)

Kondyurin, A.V., Komar, L.A., Svistkov, A.L.: Modelling of curing of composite materials for the inflatable structure of a lunar space base. Mechanics of Composite Materials and Constructions 15, 512–526 (2009)

Kondyurin, A.V., Nechitailo, G.S.: Composite material for Inflatable Structures Photocured under Space Flight Conditions. Cosmonautics and Rockets 3(56), 182–190 (2009a)

Kondyurin, A.V., Komar, L.A., Svistkov, A.L.: Modelling of curing reaction kinetics in composite material based on epoxy matrix. Mechanics of Composite Materials 16, 597–611 (2010)

Kondyurin, A., Kondyurina, I., Bilek, M.: Composite materials with uncured epoxy matrix exposed in stratosphere during NASA stratospheric balloon flight (2010a), http://arxiv.org/pdf/1008.5236

Kondyurin, A., Kondyurina, I., Bilek, M., de Groh, K.: Composite materials with uncured epoxy matrix exposed in stratosphere during NASA stratospheric balloon flight. NASA TM 216512 (2013)

Kondyurina, I., Kondyurin, A., Lauke, B., Figiel, L., Vogel, R., Reuter, U.: Polymerisation of Composite Materials in Space Environment for Development of a Moon Base. Advances in Space Research 37, 109–115 (2006)

Koontz, S., Leger, L., Albyn, K., Cross, J.: Vacuum ultraviolet radiation / atomic oxygen synergism in materials reactivity. J. of Spacecraft 27, 346–348 (1989)

Koontz, S., Albyn, K., Leger, L.: Atomic oxygen testing with thermal atom systems: a critical evaluation. J. of Spacecraft 28, 315–323 (1991)

Kroshkin, M.G.: Physical-chemical bases of space studies. Mashinostroenie, Moscow (1969)

Lai, S.T., Della-Rose, D.J.: Spacecraft charging at Geosynchronous altitudes: new evidence of existence of critical temperature. J. of Spacecraft and Rockets 38, 922–928 (2001)

Lee, C.-H., Chen, L.W.: Reactive probability of atomic oxygen with material surfaces in low Earth orbit. J. of Spacecraft and Rockets 37, 252–256 (2000)

Lura, F., Hagelschuler, D., Abraimov, V.V.: The complex simulation of essential space environment factors for the investigation of materials and surfaces for space applications. KOBE. DLR paper, Berlin, Germany (2003)

Mesyats, G., Klyachkin, Y., Gavrilov, N., Kondyurin, A.: Adhesion of Polytetrafluorethylene modified by an ion beam. Vacuum 52, 285–289 (1999)

Pippin, H.G.: Final report on analysis of Boeing specimens flown on the effects of space environment on materials experiment. Boeing Phantom Works (1999)

Purvis, C.K., Garrett, H.B., Whittlesey, A.C., Stevens, N.J.: Design guidelines for assessing and controlling spacecraft charging effects. NASA TP-2361 (1984)

Sandy, C.R.: Next generation space telescope inflatable sunshield development. ILC Dover, Inc. (2000)

Yu, S., Efremov, I., Blagov, V., Cherniavskiy, A., Yu, K., Tziganko, O., Medzmariahvili, E., Kinteraya, G., Bedukadze, G., Datashvili, L., Djanikashvili, M., Khatiashvili, N.: Space Experiment REFLECTOR on Orbital Station MIR. In: European Conference on Spacecraft Structures, Materials and Mechanical Testing. ESTEC, Noordwijk, The Netherlands (2000)

Simburger, E.J., Matsumoto, J., Lin, J., Knoll, C., Rawal, S., Perry, A., Barnett, D., Peterson, T., Kerslake, T., Curtis, H.: Development of a multifunctional inflatable structure for the powersphere concept. AIAA 2002-1707 (2002)

Teichman, L.A., Slemp, W.S., Witte Jr., W.G.: Evaluation of selected thermal control coatings for long-life space structures. NASA TM-4319 (1992)

Veldman, S.L., Vermeeren, C.A.J.R.: Inflatable structures in aerospace engineering - an overview. ESA paper (2002)

Walter, H.U.: Fluid sciences and materials science in space. A European Perspective. Springer, Berlin (1987)

Willey, C.E., Schulze, R.C., Bokulic, R.S., Skullney, W.E., Lin, J.K.H., Cadogan, D.P., Knoll, C.F.: A Hybrid Inflatable Dish Antenna System for Spacecraft. AIAA 2001-1258 (2001)

Wilson, A.: A history of balloon satellites. J. of the British Interplanetary Society 34, 10–22 (1981)

Chapter 16
Architecture for an Asteroid-Mining Spacecraft

Haym Benaroya

Rutgers University, New Jersey, USA

16.1 Introduction and Background

The purpose of this chapter is to propose a spacecraft concept that may be a good generic framework for operations on a variety of asteroid types, shapes and sizes, that can be performed in a straightforward way, regardless of whether mining operations are performed on-site or the body is moved into orbit about, say, the Moon.

The promise of treasure has energized those proposing the mining of asteroids. The backbone of future solar and interstellar economies will depend on the resources that the asteroids can provide, as all of us who read science fiction know well. But with the beginning of the modern manned space program, sending mining machines to the asteroids has been viewed as critical to being able to live beyond the Earth gravity well. Any viable in-situ resource utilization (ISRU) framework depends on accessing and processing the asteroidal elements.

Pignolet (1980) briefly summarizes the benefits of asteroid resource recovery, in particular, as a foundation for space industrialization. Asteroids are mentioned as source materials for the construction of large space structures, a main source of water in space, and the slag – a byproduct of smelting ore with uses in that process – to be used, for example, as radiation shields. He refers to the literature that suggested asteroid retrieval using solar sails or mass drivers. Also mentioned were rotary pellet launchers, a variant mass driver.

Pearson (1980) proposes a technique for recovering asteroid resources by equipping it with a rotary rocket propulsion system. Proposed is a rapidly spinning, tapered tube of high strength material driven electrically by solar or nuclear power.

> "Pellets of asteroid material would be released from the tube ends with velocities of a few km/s, achieving specific impulses comparable with the best chemical rockets."

The ejected pellets could be placed on trajectories to near Earth space for capture, or used as reaction mass to bring the asteroid into orbit around the Moon or Earth for complete processing.

O'Leary (1982) provides an overview of space industrialization. As many of us were optimistic, and remain so about space exploration and settlement, the following quote from the preface leads us to emphasize that we must view space as a multi-generational activity:

"Near-term programs will begin during the 1980s and will help to form an evolutionary foundation for a lunar or asteroidal materials processing enterprise that could begin as early as 10 years from now. Political rather than technical considerations will probably determine the pace of development of space industrialization. Convincing economic cases can be made for a wide variety of facets of space industrialization, but, for the most part, it is still early for industry to take the necessary risks."

Although industry is getting its space legs, much of that assessment holds true three decades later.

Gertsch (1992) provides an overview of asteroid mining, their material properties, how an asteroid prospect is identified, the trade-offs between manned and automated missions, and a conceptual mining method. One conclusion is

"because it appears to be easier and cheaper to accomplish, the lunar mine is probably a better first project to exploit nonterrestrial materials than is the asteroid mine."

This is quite interesting in today's political and technological environment where the Moon has lost favor, even though it is clearly the best choice against all measures for the settlement of space and for the needed technology development and physiological understanding that would lead us to Mars and the asteroids.

Lewis (1992) summarizes the composition of asteroids and the list of the potential asteroids of the day that may be suitable for mining.

Sonter (1997) suggests near Earth asteroids (NEA) as primary targets for resources to support space industrialization. This chapter discusses the engineering and mission-planning choices and examines the concept of probabilistic net present value to optimize asteroid mining designs.

Gertsch et al. (1997) discusses the effects of the space environment on the mining technologies developed on Earth, and how they translate to NEO mining. Included are the following overviews – NEO site preparation – anchoring and tethering, motion control, restraints to contain the body while it is mined, operating platforms, and bagging – NEO Mining Methods – for various classes of rocks, including techniques for whole-NEO rubblization, - and Marketing NEO Resources.

Ross (2001) discusses the resources available from NEAs as well as the engineering aspects of possible mining project designs, including a survey of mission plans, mining and extraction techniques that may be used. He emphasizes,

"... it is likely to be very difficult with low strength or unconsolidated material, such as loose asteroidal regolith or the hypothesized loose dusty covering of a dormant or extinct comet. ... This may need very wide-area anchoring, over an extended footprint, including the approach of totally surrounding the target body, by wrapping it with a net or membrane."

Huebner and Greenberg (2001) review methods for determining bulk properties and geologic structures of NEOs.

Kecskes (2002) describes four scenarios that may cause the rise of an asteroid-based technical civilization. There are (1) a private development of an asteroid-based colony, (2,3) an asteroid-based colony is created and eventually used as a kinetic energy weapon, and (4) an asteroid is moved to Earth orbit and used as a platform for generating and transmitting solar power.

Garrick-Bethell and Carr (2007) discuss the challenges to operations on the surface of an asteroid, especially astronauts attempting to walk on the surface. Suggested is the use of tie ropes that have been placed around the asteroid. These provide downward forces on an astronaut, permitting mining operations such as drilling, excavation, hammering and documenting materials in safety.

Taylor et al. (2008) and Taylor and Benaroya (2009) point to the Earth business model for oil drilling in extreme environments as a basis for the colonization of the comet Wilson-Harrington for habitation and development. In this approach Wilson-Harrington is drilled with tunnels and shafts, with the goal of creating a volume large enough to provide a shielded, habitable environment for 10,000 people that will be self-sustaining.

Blair and Gertsch (2010) provided an overview of asteroid mining, discussing requirements and constraints, as well as suggesting that asteroid mining and planetary defense are natural allies. First required will be the site preparation: anchoring, NEO motion control, placement of body/fragment restraint system, construction of operations platform and the processing system.

Misra (2010) assesses the economic feasibility of asteroid selection for mining and settlement using net present value (NPV) procedures. That asteroid would be orbited by a modified toroidal structure.

Sanchez and McInnes (2012) discuss the shepherding near Earth objects (NEO) for use as sources of useful materials for the industrialization of space. The energy savings as compared to launching the same materials from Earth is vast. The chapter estimates the probability of finding an object that could be transferred for certain delta V budgets, resulting in a resource map that provides an assessment of the mass of material resources near Earth space as a function of energy investment.

Hasnain et al. (2012) develop and implement an algorithm to determine trajectory characteristics necessary to move NEAs into capture orbits around the Earth. Based on considerations of the time to return on investment, the time of retrieval is constrained to ten years. 23 asteroids are recommended for future capture.

A recent study by the Keck Institute for Space Studies (2012) explored asteroid retrieval in some detail. Included in the study by 34 participants are target discovery, flight system design and mission design. An asteroid retrieval spacecraft is offered for the capture of a 7 m, 500-ton asteroid, whereby it is captured into a bin.

The Autonomous NanoTechnology Swarms (ANTS) program of NASA and its contractors have focused on distributed systems for various applications, such as for surface mobility and the Prospecting Asteroid Mission (PAM). While the distributed architecture philosophy is similar to the proposed architecture of this chapter, we are suggesting a distinct concept.

16.2 Challenges of Operations on Asteroids

A number of architectures for accessing asteroidal resources have been proposed. These resources can be tapped at the source, or the asteroid can be moved to a more useful location, such as lunar orbit. An advantage of being in lunar orbit is that the lunar gravitational field can be used to advantage to control the mining operations on the asteroid. It has been proposed to move asteroids to Earth orbit, but the obvious risk is that instead of orbiting the Earth, the asteroid may enter the atmosphere. While the probability of such an event may be small, the resulting outcome would be so costly that the overall risk cost may be too high with current or near-term technological capabilities.

The key engineering challenge to mining operations on an asteroid is the almost nonexistent gravitational field. This is not to minimize all the other difficulties, but zero-*g* operations impact all aspects of the work involved. We present a summary of the environmental issues in the following (Benaroya and Bernold 2008):

- Micro/nano-gravity environment impacts all operations – as well as how dead and live loads are modified in a design framework
- Safety and reliability – effects on design procedures and factors of safety; design robustness is critical for operations in unique and severe environments
- Regolith and rock mechanics – effects on operations as well as foundation design, load-bearing capacity of a medium depends on the confining stress surrounding it; many classes of asteroids are so inhomogeneous that foundations are not possible
- Internal air pressurization is a bigger issue for habitable volumes; for robotic miners this is not as important, but beneficiation processes may have stringent requirements
- Shielding against radiation and micrometeoroids – galactic and solar radiation are a design consideration for electromechanical systems
- Vacuum effects on construction – drilling, blasting, and outgassing of materials; blasting on small bodies is an unlikely method due to exceedingly high pressures and unpredictable ejecta
- Dust particles – are they charged and do they cling depends on the particular asteroid
- Temperature cycles and gradients between dark and light sides – if the asteroid is far from the sun then the gradients will not be a problem
- Very low temperature effects on materials, especially exotic materials – outgassing may be a concern
- Thermal management in a vacuum – likely spacecraft-type heat rejection systems except with a larger load due to mining operations.

Some of these issues are significant if we assume that humans are part of the mining operation in-situ. As noted below, we view it very likely that asteroid miners will be autonomous with minimal human oversight in delayed-time. However, a sophisticated robotic asteroid miner will need to be designed to withstand all of the environmental factors listed above.

16.3 Spacecraft Structures

There will be little similarity between structures designed for asteroid mining operations and those envisioned for the surface of the Moon or Mars, whether fixed or rover. The difficulties associated with operations on a body with almost zero gravity (and of great inhomogeneities) are similar to those encountered by astronauts working on the space station in Earth orbit, where microgravity exists. But while in orbit around the Earth, advantage can be taken of being in orbit and of the gravity field of the central body. By this we mean that the gravitational pull can be taken advantage of in the maneuvering of the spacecraft around the asteroid, much in the way that gravitational assist is used in long-distance trajectories to save fuel. An asteroid orbiting about the sun does not provide that advantage.

While mining operations can begin while orbiting or secured to an asteroid, and we can envision an apparatus that permits operation of mining tools tethered to an orbiting craft, it does not appear that orbital mining of an asteroid is desirable. Depending on the size of the asteroid, various possibilities have been proposed in the literature for traversing and performing mining operations. The "right" answer has not been found – there may not be a single space miner type. Rather, a robust and adaptable design is needed.

Our general criteria for such a craft are:

- Its design should be robust in the face of a broad spectrum of asteroid sizes and shapes, and spin rates. In particular, the spacecraft should be adaptable to a reasonable range of dimensions and aspect ratios so that custom designs are not needed except in unique instances.
- The craft should generally be comparatively small and not a behemoth.
- The essential design should be suitable regardless of whether the asteroid is processed in-situ or is shepherded to a local orbit. On-site processing would necessarily require a larger craft.
- It is assumed that the craft is autonomous. It is unreasonable, even in the near-term, to expect astronauts in space suits to work in the almost zero-*g*, radiation-rich environment while processing a rock in space. Even robotically assisted mining cannot justify using astronauts in that venue. Additionally, the designs of mining craft that include life support would clearly increase costs and complexity of the mining venture, delaying operations and increasing risks.

16.4 A Deployable Concept

For purposes of our discussion, we consider near- to medium-term mining activities rather than futuristic and science-fictionish solutions lacking connection present-day technologies and projections. Near-term is defined here as 20 years, medium-term as 40 years, and long-term 60 years. It is not possible to predict very far into the future on matters such as these.

Specific mechanical design issues/constraints for the proposed spacecraft are:

- Power systems (solar or nuclear)
- Materials processing equipment
- Mechanisms for deployable containment devices
- Dynamics and control of craft
- Packaging of components into a compact body
- Thermal rejection systems
- Shielding of equipment.

Since the craft will ideally be comparatively small, compact power technologies are imperative. Deployable solar sail technology is a possibility for inner solar system operations, and solar thermal propulsion is a possibility. Nuclear power systems will likely be needed for outer solar system operations and for projects expected to last a significant amount of time, or those requiring enough power to move asteroids to closer orbits.

Various technologies have been proposed to drill, crush, dissolve and eject parts of asteroids as part of a framework to process them and move the end products to market or for further processing. A suite of technologies is needed. The design and deployment of NASA's *Curiosity* demonstrates a capability to package a variety of instruments into a relatively small craft. While it is expected that the asteroid miner will be larger, we can also expect that the modules within such a craft will be *Curiosity*-sized or smaller.

We can be confident that mechanism quality and robustness of the asteroid miner will determine whether the concept is viable. So much of the mining operations and miner survivability will depend on mechanisms that attach the craft to the asteroid, mechanisms that operate on the asteroid, and mechanisms that maintain the systems. Project viability will rest on a family of creatively designed and synergistic mechanisms.

Dynamics and control of spacecraft are always a crucial. When spacecraft must operate around and on a body with the characteristics and environment of an asteroid, the challenges to spacecraft design and control become additionally challenging. Not only will dynamics and control bring the asteroid miner to the body, once the miner is attached and mining operations begin, there will be dynamics and control issues. Mining operations will affect asteroid dynamics. As material is processed, asteroidal mass properties will change, requiring an adaptive control system. If mass is ejected, either for moving the body to a closer orbit, or to propel mined asteroid volumes to another location, the asteroid will react. These behaviors must be understood and adaptively controlled.

The spacecraft and its equipment must be properly shielded against radiation damage. Electronics can be hardened and redundancy is needed. It is unclear how serious an issue micrometeorite impacts are in such operations. It is likely that unintended impacts from rocks loosened from the asteroid are a more serious concern.

Reliability and maintainability of the spacecraft underlies all the above. At this time it is unclear how to design for the extreme asteroid mining environment. When drilling on Earth or in the oceans access to replacement parts and technicians is assured. How that translates a robotic asteroid site is not currently known. Some ideas can be gleaned from mining and drilling activities on Earth in cold regions, and on Mars with our limited but valuable experience.

16.4.1 Tetrahedral Miner

For moderate-size asteroids with essentially no gravitational pull, we require a new structural paradigm. Concepts have been proposed for strapping machine and human to the asteroid. We propose here a multi-component spacecraft/miner that decouples when in the vicinity of the asteroid, creating a volume within which the asteroid is placed. Then, the partitioned spacecraft elements, which are connected via ultra-strong carbon nanotube ribbons, pull together and eventually touch down on as many facets of the asteroid as there are spacecraft elements.

In this position, the component miners can begin operations while secure to the body. Depending on the size of the asteroid, it may be possible to deploy connecting ribbons on which cable cars can traverse and perform mining operations. It may be desirable to provide elevated cables for transport above the asteroid surface, cable car-like.

We describe the spacecraft concept that can satisfy the constraints listed above. Elements of this concept can be found in a few of the listed references, but the overall concept of a deploying spacecraft is, as best as can be determined, new.

The essential concept is encapsulated in Fig. 16.1 below, where a mining spacecraft in the shape of a cube is split into six tetrahedral elements. The concept would be the same if the craft is rectangular or oddly shaped, and is split into an arbitrary number of arbitrary shapes that can be packaged into a reasonable, transportable volume. The cube split into tetrahedral elements is just easy to refer to here. But the tetrahedral element is a shape that provides more options for meshing with an odd shaped asteroid. Since we expect these cubes to be relatively small, fitting inside the payload fairing. However, we do not minimize the challenges of packaging the fuel tanks, antennas, and the sensitive elements.

Deployable structures have an extensive history. They are challenging, but an important solution to complex engineering problems in difficult environments. Also, the tetrahedral configuration has been seen in a variety of applications, unlike

Fig. 16.1 Cube spacecraft concept shown with tetrahedral components exploded. Each of these tetrahedral elements is connected to the others. Once in position around asteroid, the elements move out radially creating a volume within which the asteroid is placed. Then these tetrahedral elements pull together until all elements are in contact with the surface of the asteroid. Inside each element is specialized equipment that can perform anticipated operations on the asteroid.

the one proposed here, where it provides optimality, for example, in Herr and Horner (1980), Clark et al. (2007) and Capo-Lugo and Bainum (2009).

Suppose we could design the aforementioned spacecraft miner so that its components can be stowed in the tetrahedral elements of a cube. These elements are tightly stowed for the flight to the asteroid. When the cube approaches its target it will split open into its component elements, which are connected to each other via ribbons of great strength, such as nanotube ribbons. As a possible maneuver for the asteroid capture, a spinning cube can be used to transform spinning energy into the opening motion mentioned above without the need for additional power. Conservation of angular momentum is the guiding principle. Also, spinning the spacecraft to match the spin speed and axis of the asteroid will ease the capture.

The tetrahedral elements will open up as far as needed to encompass the asteroid and, once the asteroid is encased within the volume enclosed by the elements, they will pull together until each element has lodged itself onto a facet of the asteroid and locked. The nanotube ribbons will then have been wrapped around the surface of the asteroid. If needed, a fine-mesh net could also be deployed around an asteroid composed of smaller rocks. Figures 16.2 and 16.3 illustrate the concept.

In this final locked configuration, drilling, processing, and ejecting machines and elements will begin to operate, as per design, to process the asteroid. If the design purpose were to move the asteroid to a local orbit, then these tetrahedral elements would be equipped with thrusters and/or deployable solar sails to nudge the asteroid to that location.

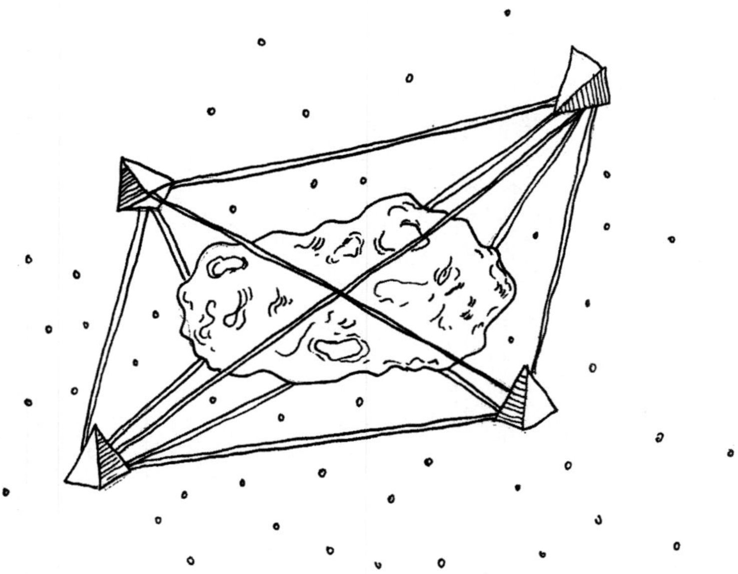

Fig. 16.2 Here we see the tetrahedral elements in their deployed, connected configuration, having encased the asteroid. (Copyright 2012 Ana Benaroya, used with permission.)

Certain structural and dynamics design issues are obvious at this juncture:

- The orbital mechanics of projecting this cube spacecraft to the chosen asteroid needs to be designed, providing it enough spin energy that can be transferred to each tetrahedral element, thus allowing the asteroid to be surrounded, as detailed above.
- The ribbons connecting the tetrahedral elements need to be designed for strength, deployment, and the connections as do the "winches" used to pull them all to the asteroid face. The ribbons can be used for communications as well as transport.
- The placement of the spacecraft and mining components within the tetrahedral elements needs to be optimized. Robustness in design suggests redundancy in tetrahedral capabilities. Thus, if one or two elements fail, others can still proceed with the majority of the mission.

Based on this framework, we are proceeding with a conceptual design that will address some of the issues listed above. There are likely a number of solutions that satisfy the above criteria and constraints.

Fig. 16.3 The tetrahedral elements have been winched to the asteroid surface. Each element is planted face down as per design. Drilling or other processing can proceed through either the face on the asteroid or through an exposed face. Cables connect the elements; in one case a cable is suspended, allowing cable car-like operations. (Copyright 2012 Ana Benaroya, used with permission.)

16.5 Concluding Statement

We have proposed a spacecraft architecture that identifies the key issues of asteroid capture and mining. Clearly, the goal of placing robotic miners at asteroids is beyond our current technological capabilities. The lesser (or is it greater) capability of placing people on asteroids for purposes of mining is beyond our current capabilities. (With people onsite, life support becomes a major burden. Being completely robotic places large burdens on automation.) We are able to send rockets to asteroids, have some minimal interactions with the rock resulting in small sample return and small-scale local processing, for example, the 2003 launch of the *Hayabusa* spacecraft to the asteroid *Itokawa*, arriving in 2005, and returning samples to Earth in 2010. The framework proposed here is meant to apply to the generation of spacecraft whose goal is to integrate with the target asteroid and begin operations, whether mining or relocation. While it is difficult to accurately predict when a new technology can go online, it is not unreasonable to expect that the proposed architecture can be ready within a decade.

Acknowledgements. I am grateful for the comments of the reviewers, in particular, that pointed me in the direction of the ANTS and Keck studies. I also thank the editor of this volume for taking on the challenges of creating such a volume on such a timely subject.

References

Benaroya, H., Bernold, L.: Engineering of lunar bases. Acta Astronautica 62, 277–299 (2008)

Blair, B.R., Gertsch, L.S.: Asteroid mining methods. Presentation at the SSI Space Manufacturing 14 Conference. NASA Ames Research Center (2010)

Capo-Lugo, P.A., Bainum, P.M.: Deployment procedure for the tetrahedron constellation. J. of Mechanics of Materials and Structures 4, 837–854 (2009)

Clark, P.E., Curtis, S.A., Rilee, M.L., Cheung, C.Y., Wesenberg, R., Brown, G., Cooperrider, C.: Extreme mobility: Next generation tetrahedral rovers. Space Technology and Applications International Forum – STAIF, CP880 (2007)

Garrick-Bethell, I., Carr, C.E.: Working and walking on small asteroids with circumferential ropes. Acta Astronautica 61, 1130–1135 (2007)

Gertsch, R.E.: Asteroid mining. In: McKay, M.F., McKay, D.S., Duke, M.B. (eds.) Space Resources – NASA SP-509, vol. 3, pp. 111–120 (1992)

Gertsch, R., Gertsch, L.S., Remo, J.L.: Mining near-Earth resources. In: UN International Conference on Near Earth Objects, Annals of the New York Academy of Sciences, vol. 822, 511–537 (1997)

Hasnain, Z., Lamb, C.A., Ross, S.D.: Capturing near-Earth asteroids around Earth. Acta Astronautica 81, 523–531 (2012)

Herr, R.W., Horner, G.C.: Deployment tests of a 36-element tetrahedral truss module. Second Annual Technical Review, NASA Langley (1980)

Huebner, W.F., Greenberg, J.M.: Methods for determining material strengths and bulk properties of NEOs. Adv. Space Res. 28, 1129–1137 (2001)

Keck Institute for Space Studies, Asteroid Retrieval Feasibility Study, Cal Tech, JPL, Pasadena (2012)

Kecskes, C.: Scenarios which lead to the rise of an asteroid-based technical civilization. Acta Astronautica 50, 569–577 (2002)

Lewis, J.S.: Asteroid resources. In: McKay, M.F., McKay, D.S., Duke, M.B. (eds.) Space Resources – NASA SP-509, vol. 3, pp. 59–78 (1992)

O'Leary, B.O.: Space Industrialization, vol. 1. CRC Press (1982)

Pearson, J.: Asteroid retrieval by rotary rocket. In: AIAA 18th Aerospace Sciences Meeting, AIAA-80-0116 (1980)

Pignolet, G.: Retrieving asteroids. L5 News (1980), `http://www.nss.org/settlement/L5news/1980-retrieving.htm`

Ross, S.D.: Near-Earth asteroid mining. Space Industry Report (2001), `http://www.esm.vt.edu/~sdross/papers/ross-asteroid-mining-2001.pdf`

Sanchez, J.P., McInnes, C.R.: Assessment on the feasibility of future shepherding of asteroid resources. Acta Astronautica 73, 49–66 (2012)

Sonter, M.J.: The technical and economic feasibility of mining the near-Earth asteroids. Acta Astronautica 41, 637–647 (1997)

Taylor, T.C., Benaroya, H.: Developing a space colony from a commercial asteroid mining company town. In: Bell, S., Morris, L. (eds.) Living in Space, Aerospace Technology Working Group Book, pp. 155–126 (2009)

Taylor, T.C., Grandl, W., Pinni, M., Benaroya, H.: Space Colony from a Commercial Asteroid Mining Company Town. In: AIP Conf. Proc., vol. 969, pp. 934–941 (2008)

Chapter 17
Near Earth Asteroids – Prospection, Orbit Modification, Mining and Habitation

Werner Grandl[1] and Akos Bazso[2]

[1] Consulting architect, Tulln, Austria
[2] University of Vienna, Austria

17.1 Introduction

The number of known Near Earth Asteroids (NEAs) has increased continuously during the last twenty years. We now understand the role of asteroid impacts for the evolution of life on Earth (Alvarez et al. 1980). To ensure that mankind will survive as a species in the long run, we have to face the "asteroid threat" seriously.

On one hand we will have to develop methods of detection and deflection for Hazardous Asteroids, on the other hand we can use these methods to modify their orbits and exploit their resources. Rare-earth elements, rare metals like platinum group elements, etc. may be extracted more easily from NEAs than from terrestrial soil, without environmental pollution or political and social problems.

To change the orbit of NEAs it is necessary to develop advanced propulsion systems to move the huge masses of these asteroids, e.g., Deuterium-Helium-3 fusion engines.

For the mining process an Earth orbit beyond the Moon should enable us to keep the rate of mining advance equal to the rate of cargo shipping between the NEA and the Earth-Moon System.

When the mining is finished some asteroids with a diameter of more than 400 m can be used to build *rotating toroidal habitats* inside the remaining hull. Oxygen, Hydrogen and Carbon can be extracted from C-Type and similar asteroids.

To use asteroid resources will be a crucial step of human evolution, it will definitely establish human civilization in space. To be independent from planet Earth is a "life insurance" for the human species in case of global disasters caused by nature itself, like super volcanoes, ice ages, novae and other cataclysmic events that we may expect in the far future.

17.2 Near Earth Asteroids

This section deals with Near-Earth Asteroids (NEAs) and their potential for human activities in space. We start by giving a review on the origin of asteroids in general in the context of solar system formation. Then the physical properties together with minerals and raw materials related to NEAs are presented. Following that we give details on the classification of NEAs into groups based on orbital characteristics, and finally we consider the accessibility and potential target objects for future missions.

17.2.1 Context

Contemporary models of planet formation provide a scenario for forming planets in our solar system and in extra-solar systems, and they are also capable to explain the origin of asteroids and comets (Morbidelli et al. 2012; Weidenschilling 2011; Alibert et al. 2010).

The basic steps consist of the growth of interstellar dust particles, embedded into the gas of the nebula forming the proto-planetary system, from nanometer scales to centimeter sized particles. By continuous growth processes (e.g., sticking and collisions) these particles form meter-sized boulders, which in turn tend to grow further on different time-scales. Once these planetesimals reach sizes of the order of one kilometer, their gravitational attraction boosts their size, as they tend to collect smaller objects and exhibit a "runaway growth" (Aarseth et al. 1993). Later the proto-planets are formed by the biggest of these planetesimals, reaching sizes of some 1000 km. The left-over debris from this process, myriads of small objects ranging in size from some meters to hundreds of kilometers, forms the population of asteroids and comets in our solar system.

But the picture would not be complete without additional processes that have been identified to contribute to the characteristics of the asteroid population today. First of all, not all asteroids are pristine objects from the very early days of the solar system but quite a lot of them are of later origin, formed by dynamical processes during the evolution of the solar system. Such processes include collisions between asteroids, tidal disruption, and rotational instabilities (Wyatt et al. 2010).

As for the first and most important of the aforementioned processes, one has to add that although asteroids in the main-belt, between the orbits of Mars and Jupiter, can achieve dynamical life-times of the order of millions or even billions of years (Morbidelli and Nesvorny 1999), sometimes collisions occur between them. The fate of the colliding asteroids depends on parameters such as the relative velocity in the collision, the mass of the involved objects, and the composition (Leinhardt et al. 2000; Durda et al. 1998). While collisions between a relatively big asteroid and a smaller object might just form an impact crater on the former, a collision between similar sized objects can well lead to a catastrophic disruption of both, forming a swarm of small debris with the potential for further collisions to other objects. One might wonder why there are few of those planetesimals of typically 1000 km size that should be abundant during solar system formation. Today Ceres, the biggest asteroid (or dwarf planet) in the main-belt with roughly 1000 km diameter, Vesta and Pallas, next in size with about 500 km diameter, seem to be the only remnants of that early phase of the solar system. Other asteroids with sizes larger than 100 km are expected to have suffered from numerous impacts and collisions, which could have decreased their sizes considerably (Asphaug et al. 1998).

17.2.2 Physical Properties and Mineralogy

An important point for the understanding of the different types of asteroids is to consider their origin. As the proto-planets gathered more and more mass, they were struck by frequent impacts on the surface. In their interior radioactive isotopes with short half-life (e.g., Al-26) produced enough heat to completely melt them, thus leading to a separation of heavier and lighter elements (Moskovitz and Gaidos 2011).

The geochemical analysis of meteorites found on Earth shows certain typical groupings of elements (Heide and Wlotzka 1995, Mittlefehldt 2003). One such group are the "siderophile elements" that are related to nickel-iron. Another group is related to FeS (iron-sulfide), characterized by the "chalcophile elements"; besides there is also the group of "lithophile elements" that are related to oxygen, and are enriched in the silicate parts of meteorites and asteroids. Typically asteroids are depleted in volatile elements, such as hydrogen and helium, the primary components of the proto-solar nebula.

While the heavy elements, such as nickel or iron together with siderophile elements, sank to the center of the molten proto-planet and formed the metallic core, the lighter lithophile elements remained in the mantle and formed various silicate minerals. The crust finally formed from the uppermost layers of the mantle and contained the lightest minerals.

Table 17.1 Geochemical groups for meteorites with typical elements occurring in mineral associations

Group	Elements (selection)
Siderophile	Fe, Ni, Co, Cu, Au, Pd, Pt, Os, Ir
Chalcophile	Fe, Ag, Cd, In, Th, Pb, Bi, and S, Se, Te
Lithophile	Rb, Cs, Be, Al, Sc, rare-Earth elements, Th, U, Ti, Nb, Ta, Cr, Mn

This shell structure, a metallic core, silicate mantle and crust, accounts for the different types of asteroids observed today as well as the types of collected and studied meteorites. Collisions with low speeds would remove the crust of the objects and produce low density fragments, while high speed impacts excavate additional material from the mantle and can mix different silicate minerals – like those found in achondrite meteorites. Hyper-velocity impacts and catastrophic collisions can lead to the destruction of the whole parent body and the release of fragments from the previous core, which can be found today as iron and stony-iron meteorites.

After a collision with sufficiently low velocity the fragments can re-assemble under the action of the mutual gravity and form a loosely bound “rubble-pile”. This model can explain the unusually low bulk densities of some asteroids, having densities of <1 – 1.5 g/cm3, while values around 3 g/cm3 for solid bodies (composed of silicate minerals) are to be expected (Baer and Chesley 2008, Britt et al. 2002). An asteroid with this rubble-pile structure would raise additional difficulties for a mining mission, as the internal structure may not permit a deep mining.

17.2.3 Asteroid Classification

Classically, there are three major classes of meteorites: (i) stone meteorites, (ii) stony-iron meteorites, and (iii) iron meteorites, with each class consisting of a number of subclasses (Heide and Wlotzka 1995, Krot et al. 2003). This implies that asteroids, from which the meteorites are stemming, have the same characteristics. However, based on photometric and spectroscopic observations of asteroids, a different classification has evolved in astronomy. By observing the reflected sun-light from the asteroid's surface and comparison to laboratory spectra of known minerals on Earth, the surface composition of an asteroid may be determined.

Two examples of widely used classification schemes are the Tholen and the SMASS classes (Tholen 1989; Bus and Binzel 2002). The common feature of both is that they define three broad categories which contain the majority of objects. The C-group includes carbonaceous objects, i.e. carbon rich asteroids with appreciable water content and some organic molecules. This group comprises about 75% of all asteroids; they are most numerous in the outer main belt (Gradie et al. 1989). The S-group contains silicaceous asteroids, and makes up for 17% of asteroids. The third is the X-group, including generally metal-rich asteroids, but also objects with quite different compositions. Thus there is a link between the astronomical and geochemical classes: C-group is equivalent to carbonaceous chondrites, S-group to stony meteorites, and X-group to iron meteorites.

17.2.4 Asteroid Populations

The first asteroid, Ceres, has been discovered by Piazzi only in 1801. He undertook an observational survey of the “gap” between the planets Mars and Jupiter, aiming to find the missing planet believed to be present on the basis of the Titius-Bode law. Since then more objects were discovered using ground-based as well as space-based facilities. It became clear that there are certain “patterns” in the distribution of these small objects, which can be explained in part by dynamical processes (Knezevic et al. 1991; Dvorak et al. 1993), but also as an imprint of the early structure of the solar system.

In the inner solar system, intersecting the orbits of the terrestrial planets Mercury, Venus, Earth, and Mars, there are the NEAs, and we will return back to them later. The overwhelming majority of asteroids are located in the main-belt, a region ranging from 2 to 4 AU (an Astronomical Unit is the Earth's distance from the Sun). Millions of objects of all sizes orbit the Sun and occasionally get into orbits that can lead them into the inner solar system, where they can replenish the NEA population. Further outwards, between the orbits of the giant planets, the population of Centaurs can be found. These objects have some features in common with comets. Then beyond the orbit of the outermost planet Neptune, the Kuiper-belt objects are located, the most prominent being Pluto. These are icy bodies, with high water content and generally low density that are probably unaltered since the very beginning of the solar system. At the outer border of our solar system the Oort cloud contains billions of comets, some of which occasionally penetrate into the inner solar system.

17.2.5 Groups of Near Earth Asteroids

After this short account on the inventory of minor objects let us return to the NEAs. From the observational point of view NEAs are not fundamentally different from main belt asteroids. There are, however, some distinctive points about them. First, NEAs are generally smaller than their counterparts in the main belt. The largest NEA, Ganymed, is about 32 km in diameter, and – like their main belt counterparts – the majority of NEAs have diameters below 2 km (Durda et al. 1998).

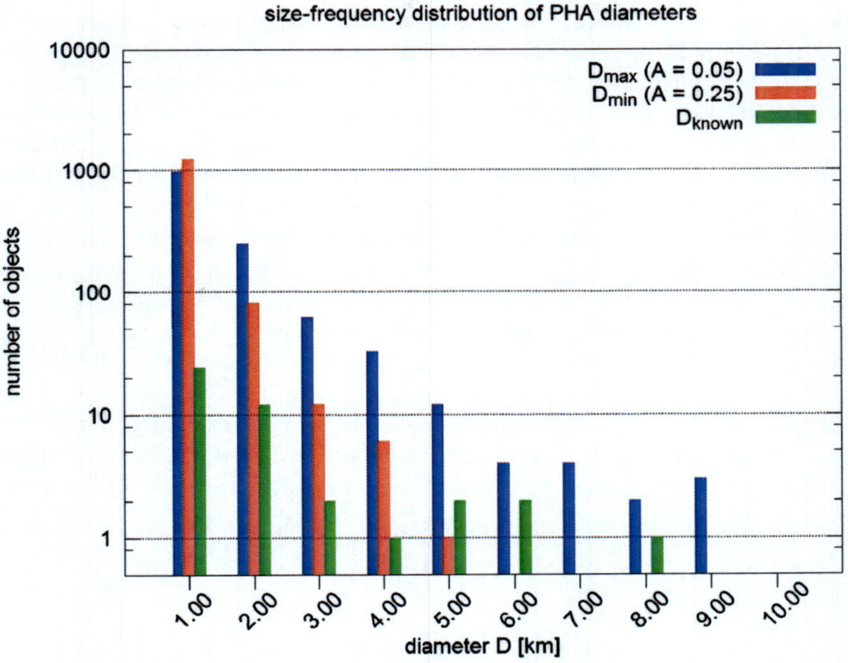

Fig. 17.1 Graph of the size distribution of potentially hazardous objects depending on their diameter. The diameters are estimated based on absolute magnitudes and two different albedos: A=0.05 (blue), A=0.25 (red); the green boxes represent objects with known diameters.

In the past decade automatic telescopes with wide field-of-view detectors have monitored the sky which has led to many new detections. The purpose of these surveys (e.g., LINEAR, Catalina Sky Survey, etc.) is to detect virtually 100% of all NEAs with diameters of 1 km or larger (Stokes et al. 1998).

There are three main classes of NEAs, the Amors, Apollos, and Atens (Shoemaker et al. 1979). The classes are defined on a dynamical basis, depending on the orbital elements of an asteroid, which are subject to changes over timescales of thousands to millions of years (Milani et al. 1989).

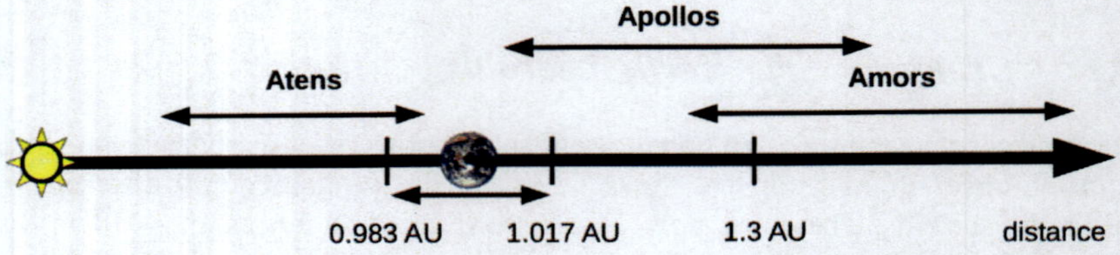

Fig. 17.2 The three major classes of Near Earth Asteroids. The arrows indicate the typical distance range of the group.

The Amor class asteroids are a kind of intermediate population between main-belt and Earth crossing asteroids. While they have typical semi-major axes (average distances from the Sun) that would place them into the main-belt, nevertheless they cross the orbit of Mars and approach the Sun below 1.3 AU [astronomical unit (AU) = average Sun-Earth distance, 150 million kilometers]. These distances still put them in a safe region where they do not present any immediate risk to Earth, unless they do not become Apollo type NEAs. The Apollo group differs from the Amors in the minimum distances from the Sun, which does not exceed 1.017 AU. This value is the Earth's farthest point from the Sun, so that the Apollos may really intersect the Earth's orbit and could have collisions. The third group, the Atens, comprises asteroids that have average distances of less than 1 AU, i.e. they move inside the Earth's orbit. Atens move out to distances above 0.983 AU, crossing the Earth orbit in its Sun-nearest point. Just like the Apollos they can have collisions, but given their higher velocities they are more dangerous. This fact is compensated by the relatively low number of Atens as compared to the Apollos, which represent currently the most numerous NEA group [http://neo.jpl.nasa.gov/stats/].

17.2.6 NEA Accessibility and Possible Targets

Although they are termed as NEAs, only a few of these objects really approach the Earth close enough to present some danger. Think of the Amor group objects that still stay about 0.3 AU (or 45 million km) away from Earth.

For a NEA to be a possible target for robotic or human exploration, it has to fulfill certain criteria. Based on the NASA NHATS [*N*ear-earth objects *H*uman space flight *A*ccessible *T*argets *S*tudy (NHATS), http://neo.jpl.nasa.gov/nhats/] database we have chosen several objects. For our purpose potentially hazardous asteroids (PHAs) seem to be a favorable choice for the following reasons: (i) they have frequent close-encounters to Earth, (ii) the minimum orbit intersection distance (MOID) is less than 0.05 AU (~7.5 million km), (iii) they have diameters exceeding 150 meters. Additionally to the previous points the necessary velocity change (delta V, ΔV) for a space probe is below 12 km/s to reach targets from the NHATS list.

A mission to a PHA would have the additional value of testing various deflection strategies in case that one day the mitigation of an asteroid on collision orbit with Earth will have to be performed.

Table 17.2 List of NEA (PHA) objects as potential candidates for mining with their physical and orbital parameters. For the calculated mass value a spherical shape with homogeneous density of 3 g/cm3 has been assumed.

Designation	Diameter [m]	Mass [kg]	Semi-major axis [AU]	Eccentricity	Spectral type
2004 MN4	270	3.092×10^{10}	0.922	0.191	Sq
1982 DB	330	5.645×10^{10}	1.489	0.360	Xe
1998 SF36	330	5.645×10^{10}	1.324	0.280	S
2005 YU55	400	1.005×10^{11}	1.157	0.430	C
2008 EV5	450	1.431×10^{11}	0.958	0.084	S
1982 XB	500	1.963×10^{11}	1.835	0.446	S
1999 RQ36	493	1.882×10^{11}	1.126	0.204	C

The masses and densities of these objects are poorly determined with exception of Apophis and Itokawa, which received special attention (Chesley 2006; Bancelin et al. 2012; Yoshikawa 2004). We limit the candidate's diameters to values between 150 – 500 meters, as smaller objects may not be rewarding targets for mining, and much larger objects require vast amounts of energy and propellant for orbital maneuvers. By assuming a bulk density of 3 g/cm^3 we calculate an upper bound for the mass of these objects.

17.3 Robotic Prospection of NEAs

Our knowledge about asteroids has been greatly expanded thanks to spacecrafts sent to some objects. These missions were either short duration fly-bys, long-term orbiting or sample return projects. Currently only two of these missions examined a NEA, while main-belt asteroids were more common targets (Shevchenko and Mohamed 2005).

17.3.1 Completed Missions

The first mission to an asteroid was the fly-by of the Galileo space probe to Gaspra and thereafter to Ida. On the way to Jupiter this spacecraft passed the main-belt asteroid 951 Gaspra in the year 1991 (Veverka et al. 1994), and 243 Ida in 1993 (Belton et al. 1996). From the data taken it turned out that both are S-type objects with irregular shapes, and that Ida possesses a moon named Dactyl. Among other interesting physical properties, such as albedo, rotation period and mass, the surface minerals were determined to be olivine and pyroxene, and the mean densities are 2.6 g/cm3 which suggest a link to ordinary chondrites.

A few years later the Near-Earth Asteroid Rendezvous (NEAR-Shoemaker) spacecraft was targeted to investigate for the first time a NEA. In 1997 it passed the main-belt asteroid 253 Mathilde (Clark et al. 1999), and took pictures of this very dark object. The low albedo and low mean density of only 1.3 g/cm3 indicate that Mathilde could be a carbonaceous chondrite type object. After a fly-by to the NEA 433 Eros in 1998, the spacecraft finally began to orbit this asteroid in the year 2000. Eros is an Amor object and the second largest NEA. During the following year the surface and the gravity field were mapped, giving a mean density of 2.67 g/cm3, again compatible with an ordinary chondrite composition (Miller et al. 2002).

At about the same time the Deep Space 1 mission had a successful fly-by to asteroid 9969 Braille (Buratti et al. 2004), again a main-belt object. The dimensions, albedo and spectrum of this asteroid were obtained, where the latter indicates a similarity to 4 Vesta. The mission continued towards the comet 19P/Borelly.

The similar mission Stardust had also the aim to investigate a comet (81P/Wild-2) but additionally had a fly-by to main-belt asteroid 5535 Annefrank. The mission helped to determine the dimensions of this non-spherical body and its albedo.

The Japanese mission Hayabusa (formerly MUSES-C) targeted the NEA 25143 Itokawa, an Apollo object. In 2005 it was the first mission to successfully orbit this NEA, to take samples from its surface and return them to Earth. From the collected small dust particles the laboratory analysis yielded the result that Itokawa is an LL-type ordinary chondrite with a cosmic-ray exposure age of 8 million years, meaning a quite recent origin derived from a former parent body (Nakamura et al. 2011).

Finally, the European mission Rosetta passed by the big main-belt asteroid 21 Lutetia on its way to the comet Churyumov-Gerasimenko.

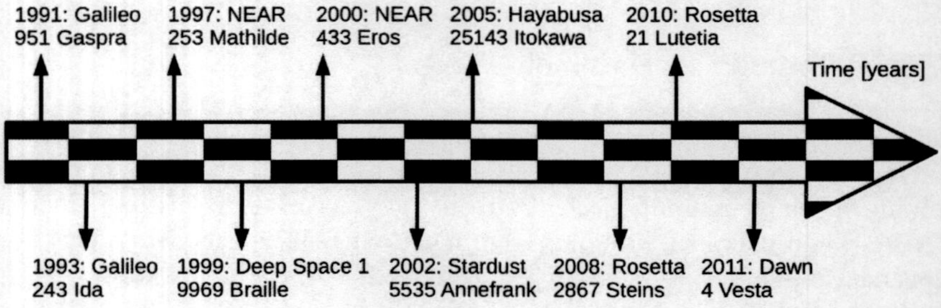

Fig. 17.3 Timeline for completed missions to main-belt and Near-Earth asteroids. The spacecraft's and asteroid's name are indicated together with the year.

17.3.2 Future Missions

In the near future the NASA mission OSIRIS-REx is planned to launch in 2016 and target the asteroid 1999 RQ36 (Lauretta et al. 2011). Once arrived in the 2020s it will orbit the C-type asteroid and attempt a landing maneuver for collecting surface samples. If successful the samples will be returned to Earth giving the great opportunity to study a carbonaceous object and its composition, and to compare it with meteorites found on Earth.

In summer 2012 the Dawn spacecraft finished its reconnaissance of 4 Vesta and is heading for 1 Ceres, where it should arrive in 2015. This mission gives valuable data about the two largest main-belt asteroids, which have strikingly different composition.

17.3.3 Mission Concept for NEA Prospection and Mining

For the future exploitation of NEAs as proposed in this chapter, we must elaborate standard probes and methods to investigate the asteroid we may choose for mining. Currently missions such as Dawn or OSIRIS-REx require a budget of $500M (resp. $800M), for being engineered and built for a unique purpose. These costs could be decreased significantly by a series of probes all standardized and similar in function.

After the decision has been made to venture the mining of a NEA, some preliminary steps will have to be made. In the first phase the target needs to be well characterized and its physical properties need to be determined thoroughly. Whenever this is possible ground-based observations (optical and radio measurements) will constrain the shape and rotational state of the asteroid; a spectrum of the target will enable the mission planners to derive the surface properties (spectral type). If the orbit of the NEA is known with sufficient accuracy, these measurements can be timed around the phase of minimal geocentric distance; otherwise the next window of opportunity depends on the orbital period of the NEA, typically 0.7 – 3 years. As a consequence of these requirements a reasonable time scale seems to be 10 years for the preliminary work.

Concluding the first phase, in the adjacent second phase a small orbiter would be sent to the target NEA (preferably on a direct trajectory), imaging and mapping the surface as well as probing the gravity field, which is essential to determine the mass, density and porosity with good accuracy. Density and porosity are important parameters to exclude asteroids which are very loosely bound rubble-piles that cannot be used for mining. This phase – disregarding the flight-time – can be completed in one year (given the results of the Dawn mission at Vesta); eventually some sort of simple sample return process can follow. The total time allocated for this phase would be another 10 years, including the preparations, flight-time and active mission.

In the third and main phase a group of space tugs (see the respective section) would approach the target and connect to it. Subsequently the orbit of the NEA would be modified using the carried propellant or by converting appropriate surface material into fuel. The ultimate goal is to move the target asteroid into an Earth orbit where it can be mined safely by a manned mining station. The asteroid would be placed either into orbit beyond the lunar orbit, or into a libration orbit about the Earth-Moon Lagrangian points L4 or L5. The duration of this final phase depends strongly on the orbital parameters and the mass of the asteroid. Therefore only a rough assessment can be given here. Provided that the proposed propulsion systems are technically feasible, a mission time of 10 – 20 years in envisioned before the asteroid reaches an orbit around the Earth.

Given the long duration of an individual mission, we propose a "pipeline" concept. While the observations of NEAs can be done in parallel, the precursor missions of the next phase can be launched in short intervals, giving time for technical corrections and upgrades. In this way a continuous data flow is established, and there are no idle times for the mission crew. Also for the last phase a parallel approach seems more adequate than a serial one.

17.4 Orbital Maneuvers

17.4.1 List of Maneuvers

For an unmanned space probe to reach a NEA the flight-time is not the primary objective. Rather one should aim for a low fuel consuming trajectory, requiring as few orbital maneuvers as possible. The reverse problem – moving an asteroid from a solar orbit into an Earth orbit – is somewhat trickier; a number of transfer mechanisms would be available (Chobotov 1991).

- Hohmann transfer

Assuming an initially circular orbit for the NEA – this assumption is generally not met, as many NEAs have moderate to quite high eccentricities – a simple Hohmann transfer from the original NEA orbit to Earth orbit can be applied. If the engine bursts are considered to be instantaneous velocity "kicks", the engines have to deliver a high peak performance (high thrust). Low thrust engines, on the other hand, can gradually change the orbit and operate more efficiently, but the necessary change in velocity will be up to 141% higher, and the mission duration will be longer (Hohmann 1960).

- Bi-elliptic transfer

A bi-elliptic transfer involves two elliptic transfer orbits, but only one half of each is needed. The engines have to fire three times, first for leaving the original orbit, second to change the elliptic transfer orbits, and finally for arriving to the final orbit. Under certain circumstances the bi-elliptic transfer requires less fuel than a Hohmann transfer, although the travel time is generally higher (Chobotov 1991).

- Changing the orbital plane

Another problem with some NEA's orbits is that they have high inclinations relative to the Earth's orbital plane. In such a case before successful insertion into

Earth orbit the inclination has to be decreased. To save fuel, the changes of trajectory (plane changes) should be applied when the magnitude of the velocity is the smallest, i.e. at apocenter in the most distant orbital point.

17.4.2 Swing-By Maneuver

Today planetary flybys not only offer an opportunity to gather data and take pretty pictures, but also allow a spacecraft to gain extra speed. The physical principles governing the swing-by (or gravity assist, gravitational slingshot) maneuver are simple: momentum conservation and energy conservation. The flyby transfers a tiny amount of the planet's momentum to the spacecraft which in turn is accelerated, while the planet is decelerated. Depending on the encounter geometry the spacecraft is either accelerated or decelerated [http://www.dur.ac.uk/bob.johnson/SL] (Johnson 2003).

As demonstrated by the Galileo and Cassini missions, the braking due to the swing-by saves fuel. We propose to modify a NEA's orbit in such a way to arrange for a close-encounter with the Moon and to exploit the swing-by to put the asteroid into Earth orbit. The mass-ratio NEA/Moon is about 10^{-12} such that no negative effects on the orbital stability of the Moon are expected.

17.4.3 Future Propulsion Systems

As is visible from Table 17.2 the masses of the proposed NEA targets are higher than 10^{10} kg. These huge masses pose difficulties to deal with using conventional propulsion systems. There is a clear need for advanced propulsion systems that are able to deliver high thrust and specific impulse. Such a system could be the Bussard Fusion System, also known as the quiet-electric-discharge (QED) engine (Bussard 1997, 2002).

This system uses electrostatic fusion devices to generate electrical power. The fuel consists of Deuterium and Helium-3 that are fusing to Helium-4 plus protons releasing 18.3 MeV of energy per reaction. The charged protons escape from the confinement; their kinetic energy can be converted to electricity or be used directly as a plasma beam for generating thrust. The advantage of the Deuterium-Helium reaction is the low neutron production rate (via Deuterium-Deuterium reaction), as neutrons are unavailable for generating thrust; the disadvantage is that Helium-3 is rather rare on Earth. However, it is more abundant on the Moon (via solar-wind deposition), so that it first has to be gathered from there, which could raise the mission costs. For the reaction a specific energy of 3.5×10^{14} J/kg can be computed (Bussard 1997, 2002), i.e. orders-of-magnitude higher than for any existing propulsion system.

We have calculated for the objects from Table 17.2 the differences in Kepler energy between the NEA's current orbit and the Earth's orbit to estimate the amount of (specific) energy needed for the transfer (Roy 1988, chap. 11.3).

Then the energy needed is this difference multiplied by the object's mass, where we have assumed a bulk density of 3 g/cm3 as mentioned above. The given ΔV values were taken from a table [http://echo.jpl.nasa.gov/~lance/delta_v/] by Benner (JPL).

Table 17.3 gives details on the fuel mass, which is estimated by dividing the energy column by 3.5×10^{14} J/kg for the De-He-3 reaction.

Table 17.3 Energy and fuel mass required for transfer of a NEA into Earth orbit. The deltaV values were taken from a table of JPL.

Designation	Mass [kg]	Energy [J]	Fuel mass [kg]	DeltaV [km/s]
2004 MN4	3.092×10^{10}	1.161×10^{18}	3317	5.687
1982 DB	5.645×10^{10}	8.265×10^{18}	23614	4.979
1998 SF36	5.645×10^{10}	6.164×10^{18}	17611	4.632
2005 YU55	1.005×10^{11}	6.094×10^{18}	17411	6.902
2008 EV5	1.431×10^{11}	2.778×10^{18}	7937	5.629
1982 XB	1.963×10^{11}	3.984×10^{19}	113829	5.490
1999 RQ36	1.882×10^{11}	9.392×10^{18}	26834	5.087

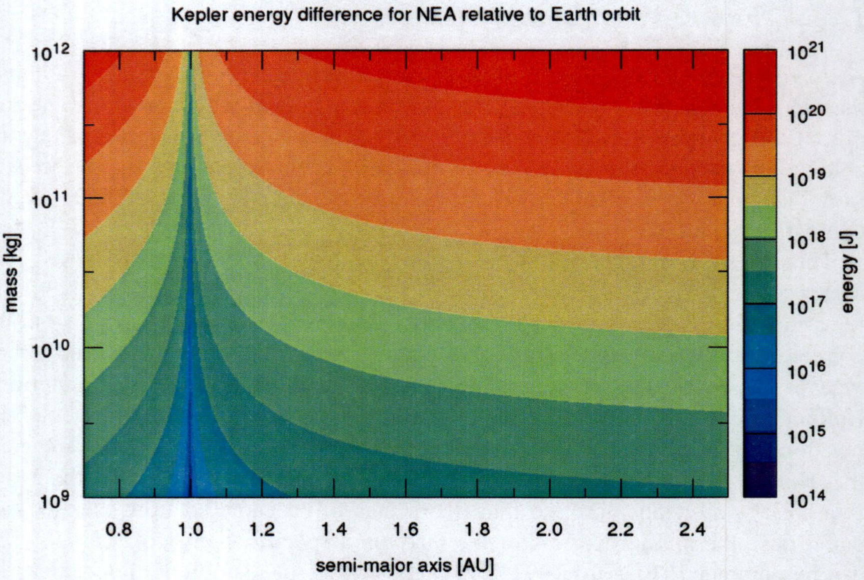

Fig. 17.4 Graphical representation of the relation between a NEA's orbit and mass vs. the energy required to transfer it to a geocentric orbit.

17.4.4 Example Application: 2008 EV5

The Near-Earth object with the designation 2008 EV5 is classified as an Aten group asteroid, with a mean diameter of 450 meters that belongs to spectral type S (stony asteroid). Our mass estimate (using a bulk density of 3g/cm3) is 1.4×10^{11} kg. To bring it to an Earth-like orbit (a = 1 AU, e ~ 0) the energy required is estimated to be of the order of 2.8×10^{18} J; using the presented Bussard fusion system the amount of fuel would be ~8000 kg.

The figure 17.6 shows the spatial orientation of the orbits of 2008 EV5 and the Earth. The dark red line represents the part of the asteroid orbit below the ecliptic plane. The line of nodes is also plotted, as well as the resulting orbits, when the asteroid velocity is increased (decreased) by +/– 1 km/s. As Table 17.3 shows the total ΔV for this NEA is approximately 5.5 km/s. By applying the thrust from the Bussard fusion system the velocity is increased just enough to orient the line-of-nodes parallel to the Earth's. The current aphelion distance is 1.038 AU which is large enough to allow a close encounter with the Earth-Moon system to happen when crossing the nodes. If the orbit is arranged in such a way that the maximum approach to Earth is still in a safe distance beyond the Moon's orbit, but close enough to it to take advantage of a swing-by maneuver, then the remaining ΔV can be supplied through gravity assist from the Moon (Figure 17.5).

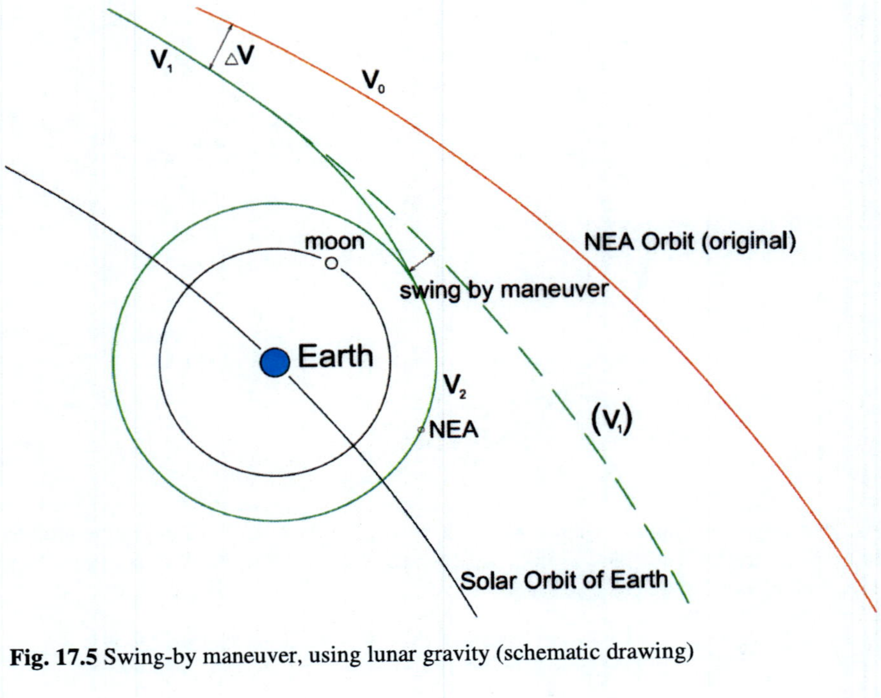

Fig. 17.5 Swing-by maneuver, using lunar gravity (schematic drawing)

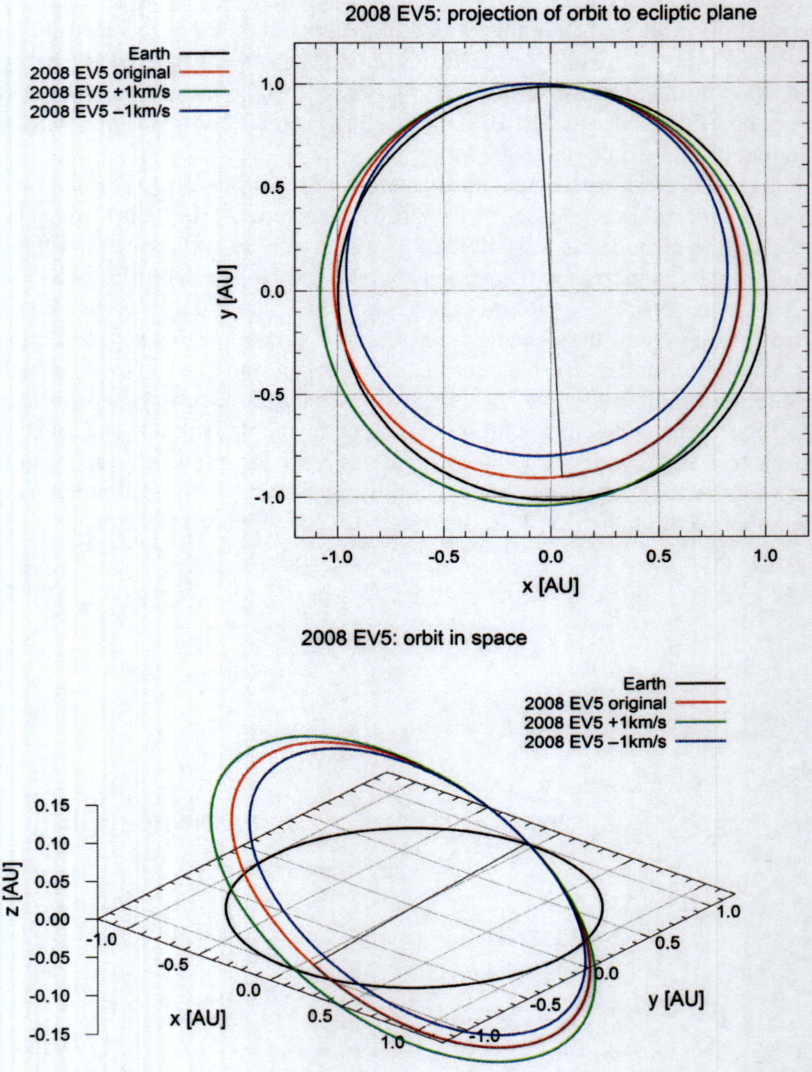

Fig. 17.6 Projection onto the ecliptic plane (above) and spatial view (below) of the orbit of 2008 EV5 together with Earth's orbit. Additionally the modified NEA orbits are shown when the velocity is changed by 1 km/s.

17.5 Asteroid Space Tugs

To realize a "pipeline" mission concept, as discussed in section 17.3.3, it is not only necessary to develop and test the Bussard Fusion System but also to create an unmanned *space tug* which is propulsed by Bussard engines and can carry the required fuel. All components of this spaceship should be designed for series production to reduce costs in the long run. Therefore all parts should be modular and easy to assemble in Low Earth Orbit (LEO). As a main supporting structure we propose cylindrical modules built of aluminum trapezoid sheeting, stringers, frames and bulkheads. Considering possible heavy lift rockets, which could be derived from existing launchers like Ariane 5, we may limit the size to 24 m length and 8 m diameter. The maximum payload for one module, containing a fusion engine, a drilling anchor and technical equipment, should be approx. 20 to 25 tons.

In LEO four modules will be assembled to a space tug as shown in Fig. 17.7. The fuel is filled into the cylinders in space to reduce the launch mass on Earth and because Helium-3 is extracted from lunar soil. This implies that a lunar base (Grandl 2012) has to be established before sending space tugs to the asteroids.

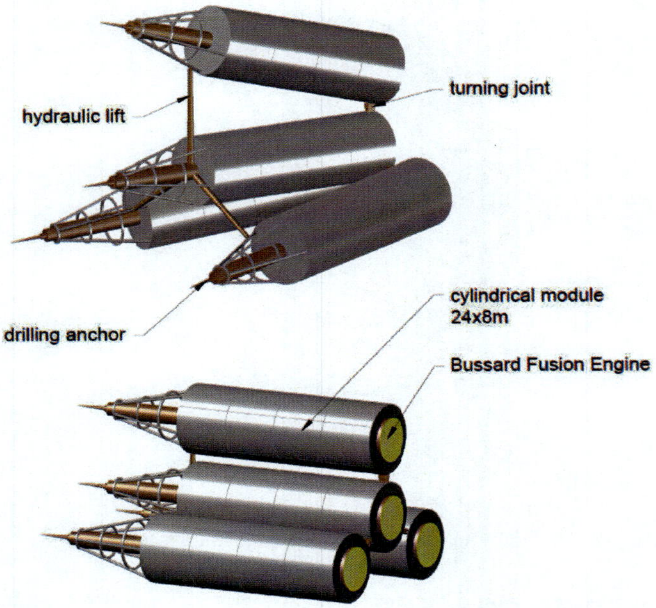

Fig. 17.7 Asteroid Space Tugs (preliminary design)

For the target 2008EV5 as an example the primary fuel mass is approx. 8000 kg (table 17.3). But we have to add the fuel needed by the tugs to reach the NEA. We have also to consider some fuel reserve for adjusting maneuvers after the final swing-by. For these additional thrusts to stabilize the NEA we need a second space tug, which can apply counterforces to the first tug. So we assume a total fuel mass of approx. 9000 kg for the 2008EV5 mission. The missions to some of the NEAs in Table 17.3 will require much more fuel. In those cases we can add external tanks between the modules.

Thus we send a couple of Asteroid Space Tugs to the selected asteroid. When the tugs will reach the NEA they will deploy their drilling anchors and penetrate the target as shown in Figure 17.8. Once fixed to the asteroid rock the tugs can apply forces in any direction by firing their engines. The first tug, which carries the main fuel quantity, applies the primary force for the orbital maneuvers. The second one adjusts the flight track by short engine thrusts.

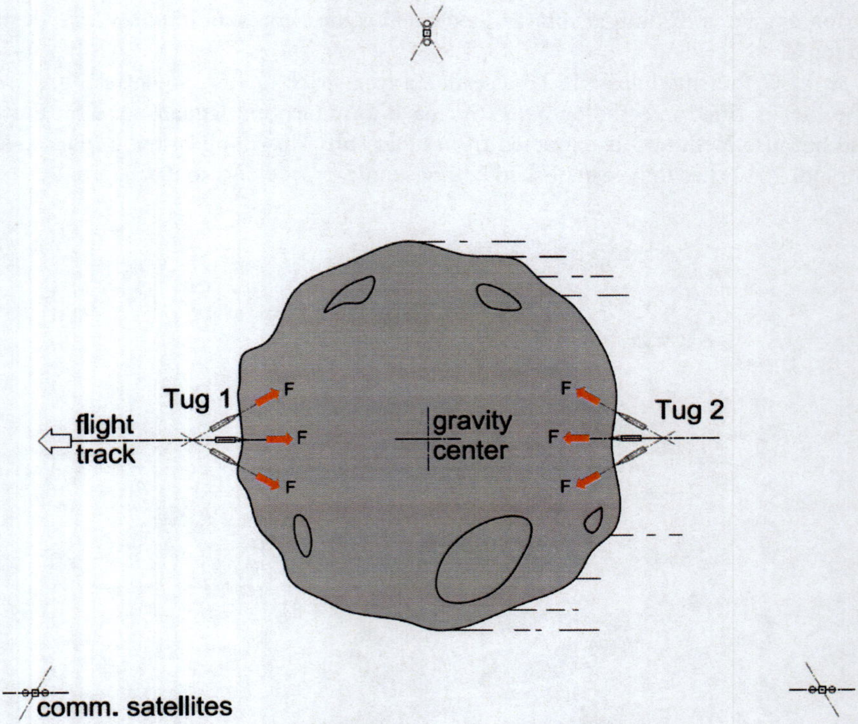

Fig. 17.8 Asteroid, guided by space tugs and communication satellites

All the maneuvers would be supervised by a mission control center on Earth. Therefore the tugs are accompanied by three small communication satellites, which orbit the NEA and enable remote control from Earth or from the Lunar Base.

17.6 Mining and Processing

When the NEA has been stabilized in Earth orbit beyond the Moon, a *manned mining station* is docked to the asteroid. The station is built of cylindrical modules and nodes. It contains a drilling machine, called the *Active Mining Head* (Taylor, Zuppero, Germano, Grandl, 1995), conveying, processing machinery, storage and docking modules. Rotating habitat modules provide a small artificial gravity for the crew. Electric current is provided by solar panels and a small nuclear battery.

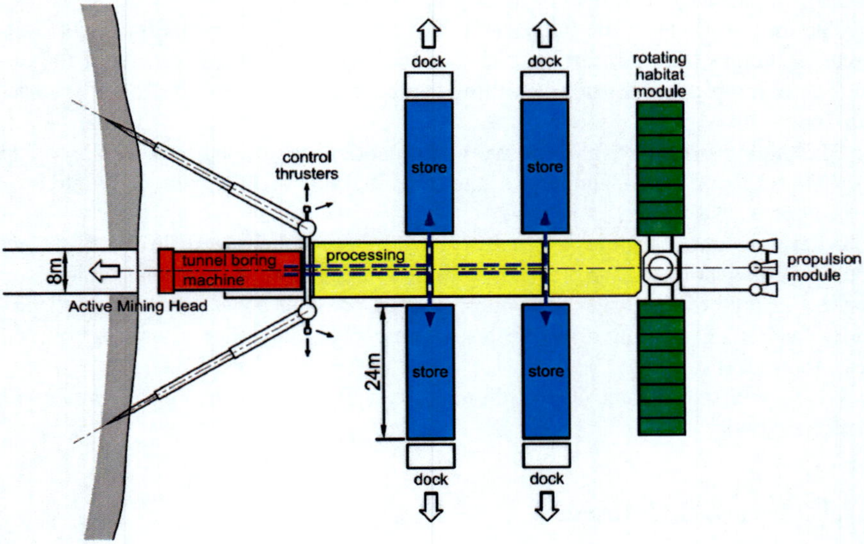

Fig. 17.9 Manned Mining Station (schematic drawing)

Mining an asteroid will be much different from mining on Earth because of low or zero gravity. On one hand it is easier to dig tunnels and caves because one needs less and smaller ring beams, rock bolts, etc., to stabilize the rock. On the other hand in a low gravity environment it is much more difficult to convey the excavated rock particles, known as "muck", which are chipped away by the Active Mining Head (AMH). On the asteroid's surface these rock chips can float away and be lost easily.

For this reason we prefer underground mining instead of open cast mining. The asteroids which we choose for mining, e. g. 2008 EV5, should have a minimum bulk density of approx. 2 g/cm3. At lower density the drilling could destabilize the structural constitution of the soil and cause dangerous fractures in the NEA.

The Active Mining Head is a flexible drilling, digging and tunnel boring machine. The drill bits are furnished with cutting elements of sintered artificial diamonds. The AMH must work slower, smoother and more precisely than in a terrestrial mine, not to disturb the structural stability of the asteroid rock. First it drills a *main central tunnel* of 8 m diameter to the center of the asteroid, and then it excavates step by step a spherical cave up to 50 % of the NEA's volume.

The excavated cavern is permanently filled with a pressurized gas. Thus the muck can be removed easily in a flexible vacuum conveyor tube.

In the *manned mining station* the mined material is processed, stored and prepared for transport. Unmanned *cargo ships* transport the material to LEO or the Lagrange Points of the Earth-Moon system for further industrial use, e.g., in metallurgical plants. The rate of mining advance, muck removal and storage must be kept equal to the rate of cargo shipping, which can be done more easily, when the NEA is in an Earth orbit than in its original solar orbit.

The inner surface of the tunnel and the cave cannot be lined by the use of concrete (shotcrete) like on Earth because of the lack of air and water. After the end of the mining process a smooth walling can be made by laser sintering of the rock surface.

The thickness of the NEA's remaining crust depends on the bulk density of the asteroid rock and on the diameter of the celestial body and the cave. It should be at least approx. 30 to 50 m.

After finishing the mining process and the laser sintering of the inner surface the remaining hollow asteroid can be used as a shelter for industrial facilities or for the storage of products like water, oxygen or other gases. The rock hull provides shelter against micrometeorites, cosmic rays, solar flares and last but not least thermal insulation.

NEAs with more than 400 m diameter can be used for *human colonies* with artificial gravity.

17.7 Asteroid Colonies

During the 20th century a number of proposals to build space colonies have been made, e.g., by Ziolkovsky, Noordung, Ehricke, O´Neill (1989) or Germano and Grandl (1993). It was obvious that the construction of a "big" space colony would depend on the use of the in situ resources of outer space, especially the Moon and the asteroids. It will be also necessary to establish heavy industry in space before building the first colony.

As a first step of space colonization we can use NEAs to carry colonies. In 1995 a paper on commercial resource development and utilization of the comet 4015 Wilson-Harrington (1979 VA) was presented (Taylor, Zuppero, Germano, Grandl, 1995).

After finishing the mining process, the caverns of the comet would be used to build a rotating toroidal habitat. In this concept some essential technologies like the AMH (mentioned above) have been described.

Fig. 17.10 Habitat structure inside the comet 4015 Wilson-Harrington (design: A. Germano and W. Grandl, 1995)

17.7.1 A Prototype Asteroid Colony : 2008EV5

After having exploited up to 50 % of a NEA's volume by mining, we can build a *rotating toroidal colony* inside the remaining cave. The shell of the hollow asteroid provides a good shelter against meteorites, cosmic rays and solar flares. The radius of the rotating torus should be at least 100 m, to minimize the Coriolis acceleration (Puttkamer 1987).

Thus we select an asteroid with a minimum diameter of 400 m, e.g., 2008 EV5, for a prototype asteroid colony.

The torus can be built of prefabricated pneumatic elements. After assembling the elements the whole torus is inflated. Once sheltered by the shell of the NEA we can use big glass panels for illumination. Natural sunlight can be beamed into the center of the cave by an array of parabolic mirrors. The focused sunlight is distributed by a central mirror cone into the torus. The parabolic mirror array is designed as a free floating structure outside the asteroid, with independent rotation to collect the sunlight. The central mirror cone is furnished with small parabolic facets to distribute the light.

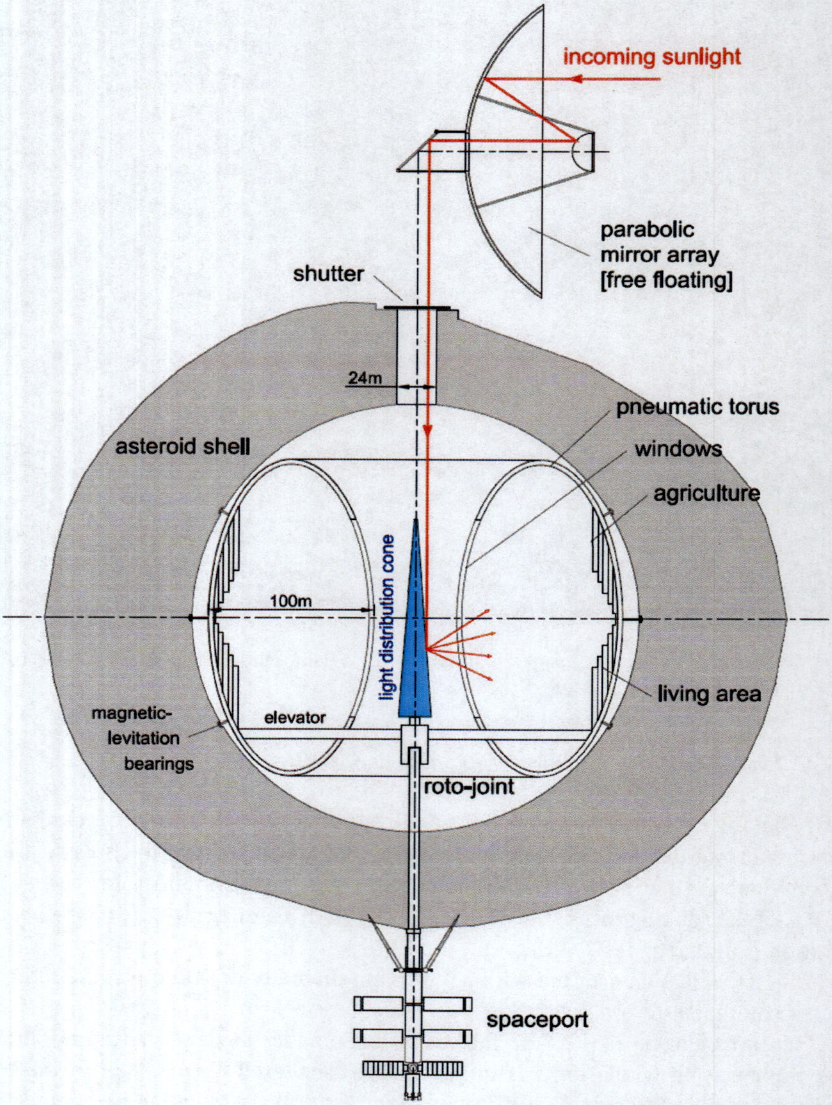

Fig. 17.11 Prototype Asteroid Colony, cross section

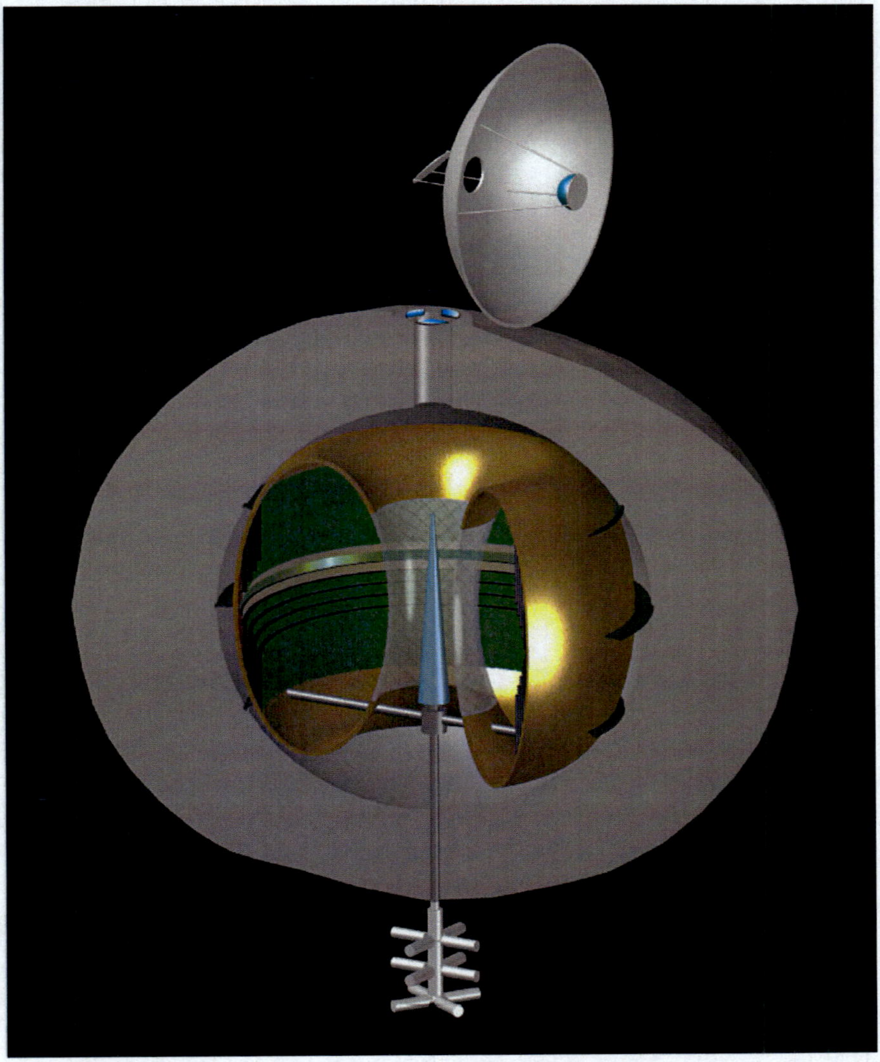

Fig. 17.12 Prototype Asteroid Colony, cross section / perspective view; The rotating torus is driven by electromagnetic bearings in the vacuum cavern of the asteroid. The free floating mirror array in space beams the sunlight into the cave, where the light is finally distributed by a central mirror cone.

The entire toroidal structure will be rotating approximately 3 to 4 rounds per minute by the use of circular electromagnetic bearings (magnetic levitation) and provide up to 90 % of terrestrial gravity. In case of a failure the habitat will be supported by additional mechanical bearings. The Coriolis acceleration will be approx. 0.05 G, still comfortable for humans.

The non rotating *spaceport* at the asteroid's surface provides docking and storage modules and is derived from the former *manned mining station*.

The inner surface of the torus is used for housing, gardening, agriculture (permaculture) and a public center. Vertical farms and aquaculture provide nutrition for approx. 2000 inhabitants. All interior buildings and furniture are light weight constructions, partially made by 3D plotting and similar methods. Lightweight materials like carbon fibers and foamed metals will be used.

To avoid thermal overloading of the cave, a system of heat absorbers and heat exchangers will be installed inside and outside of the asteroid shell. Thus the surplus heat can be absorbed, transformed into electrical power or radiated from the surface as microwaves.

Oxygen and water are extracted from asteroid material during the mining process or in industrial plants in space. The goal is to generate a "closed water cycle" in a nearly self sufficient space colony.

17.8 Future Scope

In the year 2200 some dozens of asteroid colonies may orbit Earth beyond the Moon. Those mined NEAs, which are too small to carry rotating colonies, would be used for industrial facilities, farming and to store products like water, oxygen, etc.

Then some hundred thousands of people could settle in the Earth-Moon system and constitute a new society beyond planet Earth. Solar power satellites and industrial facilities would orbit Earth and Moon, producing rare products for the increasing mankind. Space elevators would transport materials and products between Earth and Low Earth Orbit. Spaceships would start from Lunar Orbit and the Lagrange Points to Mars, the Asteroid Belt and the moons of Jupiter to search for scientific knowledge and to use the resources of the entire Solar System for the benefit of mankind.

Earth then could become a "green planet", because most of the mining and heavy industry processes would be deployed in space.

Arrays of Space Telescopes could survey the Solar System to detect and help to deflect hazardous celestial bodies before they approach the Earth-Moon system.

Acknowledgments. The authors wish to thank Prof. Rudolf Dvorak, Vienna Institute of Astronomy, for his generous support. Special thanks to Clemens Böck, our young design engineer, for his technical advice and the excellent drawings.

References

Aarseth, S.J., Lin, D.N.C., Palmer, P.L.: Evolution of Planetesimals. II. Numerical Simulations. ApJ 403, 351 (1993)

Alibert, Y., Broeg, C., Benz, W., Wuchterl, G., 22 coauthors: Origin and Formation of Planetary Systems. Astrobiology 10, 19–32 (2010)

Alvarez, L.W., Alvarez, W., Asaro, F., Michel, H.V.: Extraterrestrial cause for the Cretaceous-Tertiary extinction. Science 208, 1095–1108 (1980)

Asphaug, E., Ostro, S.J., Hudson, R.S., Scheeres, D.J., Benz, W.: Disruption of kilometre-sized asteroids by energetic collisions. Nature 393, 437–440 (1998)

Baer, J., Chesley, S.R.: Astrometric masses of 21 asteroids, and an integrated asteroid ephemeris. Celestial Mechanics and Dynamical Astronomy 100, 27–42 (2008)

Bancelin, D., Colas, F., Thuillot, W., Hestroffer, D., Assafin, M.: Asteroid (99942) Apophis: new predictions of Earth encounters for this potentially hazardous asteroid. A&A 544, A15 (2012)

Belton, M.J.S., Chapman, C.R., Klaasen, K.P., Harch, A.P., Thomas, P.C., Veverka, J., McEwen, A.S., Pappalardo, R.T.: Galileo's Encounter with 243 Ida: an Overview of the Imaging Experiment. Icarus 120, 1–19 (1996)

Britt, D.T., Yeomans, D., Housen, K., Consolmagno, G.: Asteroid Density, Porosity, and Structure. Asteroids III, 485–500 (2002)

Buratti, B.J., Britt, D.T., Soderblom, L.A., Hicks, M.D., Boice, D.C., Brown, R.H., Meier, R., Nelson, R.M., Oberst, J., Owen, T.C., Rivkin, A.S., Sandel, B.R., Stern, S.A., Thomas, N., Yelle, R.V.: 9969 Braille: Deep Space 1 infrared spectroscopy, geometric albedo, and classification. Icarus 167, 129–135 (2004)

Bus, S.J., Binzel, R.P.: Phase II of the Small Main-Belt Asteroid Spectroscopic Survey: A Feature-Based Taxonomy. Icarus 158, 146–177 (2002)

Bussard, R.W.: System Technical and Economic Features of QED-Engine Drive Space Transportation. In: 33rd AIAA/ASME/SAE/ASEE Joint Propulsion Conference, AIAA 97-3071 (1997)

Bussard, R.W.: An Advanced Fusion Energy System for Outer-Planet Space Propulsion. Space Technology and Applications International Forum 608 (2002)

Chesley, S.R.: Potential impact detection for Near-Earth asteroids: the case of 99942 Apophis (2004 MN4). In: Daniela, L., Ferraz-Mello, S., Angel, F.J. (eds.) Asteroids, Comets, Meteors. IAU Symposium, vol. 229, pp. 215–228 (2006)

Chobotov, V.A.: Orbital mechanics. AIAA Education Series. AIAA, Inc., Washington, DC (1991)

Clark, B.E., Veverka, J., Helfenstein, P., Thomas, P.C., Bell, J.F., Harch, A., Robinson, M.S., Murchie, S.L., McFadden, L.A., Chapman, C.R.: NEAR Photometry of Asteroid 253 Mathilde. Icarus 140, 53–65 (1999)

Durda, D.D., Greenberg, R., Jedicke, R.: Collisional Models and Scaling Laws: A New Interpretation of the Shape of the Main-Belt Asteroid Size Distribution. Icarus 135, 431–440 (1998)

Dvorak, R., Müller, P., Kallrath, J.: A Survey of the Dynamics of Main-Belt Asteroids. I. A&A 274, 627–641 (1993)

Germano, A., Grandl, W.: Astropolis-Space Colonization in the 21st Century. In: Faughnan, B. (ed.) Proceedings of the Eleventh SSI-Princeton Conference on Space Manufacturing 9: The High Frontier, Accession, Development and Utilization, May 12-15, pp. 252–268. American Institute of Aeronautics and Astronautics (1993)

Gradie, J.C., Chapman, C.R., Tedesco, E.F.: Distribution of taxonomic classes and the compositional structure of the asteroid belt. In: Binzel, R.P., Gehrels, T., Matthews, M.S. (eds.) Asteroids II, pp. 316–335 (1989)

Grandl, W.: Building the First Lunar Base– Construction, Transport, Assembly. In: Badescu, V. (ed.) Moon-Prospective Energy and Material Resources, pp. 633–640. Springer (2012)

Knezevic, Z., Milani, A., Farinella, P., Froeschle, C., Froeschle, C.: Secular resonances from 2 to 50 AU. Icarus 93, 316–330 (1991)

Krot, A.N., Keil, K., Goodrich, C.A., Scott, E.R.D., Weisberg, M.K.: Classification of Meteorites. Treatise on Geochemistry 1, 83–128 (2003)

Lauretta, D.S., Drake, M.J., OSIRIS-REx Team: OSIRIS-REx - Exploration of Asteroid (101955) 1999 RQ36. AGU Fall Meeting Abstracts (2011)

Leinhardt, Z.M., Richardson, D.C., Quinn, T.: Direct N-body Simulations of Rubble Pile Collisions. Icarus 146, 133–151 (2000)

Milani, A., Carpino, M., Hahn, G., Nobili, A.M.: Dynamics of planet-crossing asteroids - Classes of orbital behavior. Icarus 78, 212–269 (1989)

Miller, J.K., Konopliv, A.S., Antreasian, P.G., Bordi, J.J., Chesley, S., Helfrich, C.E., Owen, W.M., Wang, T.C., Williams, B.G., Yeomans, D.K., Scheeres, D.J.: Determination of Shape, Gravity, and Rotational State of Asteroid 433 Eros. Icarus 155, 3–17 (2002)

Mittlefehldt, D.W.: Achondrites. Treatise on Geochemistry 1, 291–324 (2003)

Morbidelli, A., Lunine, J.I., O'Brien, D.P., Raymond, S.N., Walsh, K.J.: Building Terrestrial Planets. Annual Review of Earth and Planetary Sciences 40, 251–275 (2012)

Morbidelli, A., Nesvorny, D.: Numerous Weak Resonances Drive Asteroids toward Terrestrial Planets Orbits. Icarus 139, 295–308 (1999)

Moskovitz, N., Gaidos, E.: Differentiation of planetesimals and the thermal consequences of melt migration. Meteoritics and Planetary Science 46, 903–918 (2011)

Nakamura, T., Noguchi, T., Tanaka, M., Zolensky, M.E., 18 coauthors: Itokawa Dust Particles: A Direct Link Between S-Type Asteroids and Ordinary Chondrites. Science 333, 1113 (2011)

O'Neill, G.K.: The High Frontier- Human Colonies in Space. Space Studies Institute Press, Princeton (1989)

von Puttkamer, J.: Der Mensch im Weltraum- eine Notwendigkeit. Umschau-Verlag, Frankfurt am Main (1987)

Roy, A.E.: Orbital motion. Institute of Physics Publishing, Bristol (1988)

Shevchenko, V.G., Mohamed, R.A.: Spacecraft exploration of asteroids. Solar System Research 39, 73–81 (2005)

Shoemaker, E.M., Williams, J.G., Helin, E.F., Wolfe, R.F.: Earth-crossing asteroids - Orbital classes, collision rates with earth, and origin. In: Gehrels, T. (ed.) Asteroids, pp. 253–282 (1979)

Stokes, G.H., Viggh, H.E.M., Shelly, F.L., Blythe, M.S., Stuart, J.S.: Results from the Lincoln Near Earth Asteroid Research (LINEAR) Project. In: AAS/Division for Planetary Sciences Meeting Abstracts #30. Bulletin of the American Astronomical Society, vol. 30, p. 1042 (1998)

Taylor, T., Zuppero, A.C., Germano, A., Grandl, W.: IAA- 95 – IAA.1.3.03. In: Commercial Asteroid Resource Development and Utilization. 46th International Astronautical Congress, Oslo, Norway (1995)

Tholen, D.J.: Asteroid taxonomic classifications. In: Binzel, R.P., Gehrels, T., Matthews, M.S. (eds.) Asteroids II, pp. 1139–1150 (1989)

Veverka, J., Belton, M., Klaasen, K., Chapman, C.: Galileo's Encounter with 951 Gaspra: Overview. Icarus 107, 2–17 (1994)

Weidenschilling, S.J.: Initial sizes of planetesimals and accretion of the asteroids. Icarus 214, 671–684 (2011)

Wyatt, M.C., Booth, M., Payne, M.J., Churcher, L.J.: Collisional evolution of eccentric planetesimal swarms. MNRAS 402, 657–672 (2010)

Yoshikawa, M.: Orbital Evolution of (25143) ITOKAWA, the target asteroid of HAYABUSA (MUSES-C) mission. In: Paillé, J.P. (ed.) 35th COSPAR Scientific Assembly. COSPAR Meeting, vol. 35, p. 3689 (2004)

Chapter 18
Available Asteroid Resources in the Earth's Neighbourhood

Joan-Pau Sanchez and Colin R. McInnes

University of Strathclyde, Glasgow, UK

18.1 Introduction

Any envisioned future for space exploration involves both a growth in large space structures and a human presence in space. Some possible examples of future space endeavours are large space solar power satellites, space tourism or more visionary human space settlements. This, of course, implies a much larger mass of material in-use in space, for both structural mass and life support. The traditional approach to deliver material into orbit has always been to overcome the Earth's gravity well, which is, arguably, not the most effective means if we bear in mind that the energy cost to reach low Earth orbit (LEO) is already "half-way to anywhere".["Once you get to Earth orbit, you're halfway to anywhere in the solar system."-Robert A. Heinlein].

The utilisation of resources already in space is however not a new solution to decrease the large costs of access to space, but was instead conceived already by the first rocketry pioneers as a logical step towards space colonisation (Tsiolkovsky 1903). Asteroids, and particularly near Earth asteroids, are an especially appealing source of resources because of both their accessibility from Earth and their potential wealth (Lewis 1996). The shallow gravity well of these objects also becomes an asset, which combined with the accessibility of their orbits, ensures that their resources can potentially be placed in a weakly-bound Earth orbit at a lower energy cost than material delivered from the surface of the Earth, or even the Moon. A myriad of different materials could then be transported and utilised in space. Water and other volatiles, for example, could potentially be extracted from hydrated carbonaceous asteroids and be utilised for life support and propellant (Nichols 1993). Humans need approximately 3 litres of water daily. Even if a significant fraction of this water is recycled, current water recycling systems require periodic resupply. Water and other volatiles could also be used as a rocket propellant. The possibility of stationing orbital refuelling depots, for example at the Earth-Moon L1 point, to fuel interplanetary missions towards Mars and the outer planets may reduce mission costs enormously. Another interesting use of water is for radiation shielding purposes, for which it is known to have a high efficiency. Metals for space structures or semiconductors for electronics applications should also be found in specific asteroid classes. Even for highly processed materials such as silicon wafers for solar cells, the launch costs are still the largest cost fraction of their use in space. Rare metals, such as the platinum group metals, could potentially even find interest in terrestrial markets due to their relative scarcity. Finally, even raw unprocessed material may be of value for radiation shielding.

Asteroids and comets, and in particularly, near-Earth Objects (NEOs), are also of strategic importance for science. These celestial objects could potentially uncover the mysteries of the formation, evolution and composition of the solar system. Fundamental debates such as the origin of water on Earth or panspermian theories could be settled by asteroid in-situ science. Lastly, Earth impacting asteroids and comets have permanently altered the characteristics of our planet to varying degrees, and will continue doing so in the long-term (Chapman 2004).

The growing interest in these objects has translated into an increasing number of missions to NEOs, such as sample return missions through NASA' StarDust or JAXA's Hayabusa, future sample returns such as OSIRIS-REx, impactor missions such as Deep Impact or possible deflector demonstrator missions as Don Quixote [http://www.esa.int/ SPECIALS/ NEO/SEM9ST59CLE_0.html]. Asteroids are also of primary importance for near-term human exploration beyond LEO. Under NASA's flexible path plan (Review of U.S. Human Spaceflight Plans Committee, 2009), asteroids are the only "surface" that can be visited without requiring the capabilities necessary to land and take-off from a deep gravity well such as that the Moon and Mars.

The debate on the potential for future exploitation of NEOs, and possible synergies with science (Elvis 2012) and planetary protection, is clearly intensifying in recent years. Evidence can be found in the growing body of scientific literature on the topic of asteroid exploitation, as well as in recent mission proposals, such as the Keck asteroid retrieval mission (Brophy and et.al 2012), and in the interest from the commercial space sector in their potential resources [http://www.planetaryresources.com].

This chapter will attempt to provide an insight into the feasibility of future asteroid exploitation. We will discuss how much near-Earth asteroid material is known, how much remains to be discovered, and, more importantly, how much can be easily accessed for future asteroid exploitation missions. The analysis presented here attempts to assess these questions by analysing the volume of Keplerian orbital element space from which the Earth can be reached under a certain specific energy threshold and then mapping this onto an existing statistical near Earth object model (i.e., (Bottke et al. 2002, Mainzer et al. 2011)). The specific energy required to transport resources back to the Earth can then be defined by a multi-impulsive transfer, requiring an impulse to both phase the asteroid with the Earth and reduce the Minimum Orbital Intersection Distance (MOID) below a minimum critical distance. This allows a final second insertion burn at the periapsis of the hyperbolic encounter with Earth. A resource map can then be developed, estimating the amount of material available as a function of energy investment to access different types of resources.

The specific energy, or energy per unit mass, to transport a given amount of resources back to Earth is proportional to the square of the change in orbital speed,

Δv, necessary to insert these resources into a trajectory leading to the Earth's vicinity. Hence, by using Δv as a figure of merit, we investigate here two different scenarios for future exploitation of material; the transport of extracted material to the neighbourhood of the Earth and the transport of the entire asteroid for processing. The first scenario requires less energy to transport resources, since less mass is transported, while it requires that the mining operations occur in-situ. This implies either long duration crewed missions, with the complexity that this entails, or, if the mining is performed robotically, the need for advanced autonomous systems due to both the communication delay between the asteroid and Earth and the complexity of mining operations. Transport of the entire asteroid to the Earth's neighbourhood requires moving a large mass, with the difficulty that this involves, but allows more flexible mining and science operations in the Earth's vicinity. Recent advances in space propulsion technologies (Brophy 2011), combined with efficient orbit transfer techniques (Sanchez et al. 2012), are rendering this apparently more ambitious scenario a more likely option for near-term missions.

Ideally, one would like to compute the cost of transporting resources from an initial asteroid orbit to Earth by optimising an impulsive or low-thrust trajectory. This is a highly complex numerical process, to which more complexity can be added by considering options such as multiple gravity assists and manifold dynamics (Ross 2006). Here, however, we seek a statistical order of magnitude estimate of the total amount of asteroid resources likely to exist in the Earth's neighbourhood. Thus, we compute the transportation cost of material likely to exist, and not necessarily just currently surveyed objects. An analytical bi-impulse transfer model is thus used in order to define the region in orbital elements space from which resources are accessible given a limiting Δv cost, and to map this onto a near-Earth asteroid model to understand the availability of material. This, of course, provides a conservative estimate of the transportation cost.

18.2 Asteroid Transport Energy Costs

The bi-impulse transfer used here to estimate the transfer costs is composed of interception and insertion manoeuvres. The initial interception manoeuvre is necessary in order to transfer the asteroid into a trajectory that will arrive in close proximity to the Earth, as well as provides the necessary phasing for an encounter with the Earth to occur. At the Earth encounter, a second burn, the insertion manoeuvre, provides the final insertion that ensures the orbital energy with respect to the Earth equal to zero, or a temporarily capture orbit (Sanchez and McInnes 2011). This final insertion can thus be approximated as the difference between the hyperbolic and parabolic velocities at the pericentre of the Earth fly-by:

$$\Delta v_{insertion} = \sqrt{\frac{2\mu_{\oplus}}{r_p} + v_{\infty}^2} - \sqrt{\frac{2\mu_{\oplus}}{r_p}} \tag{18.1}$$

where v_∞ is the hyperbolic excess velocity of the asteroid at the encounter with Earth, r_p is the pericentre altitude of the fly-by, which is assumed to be at 200 km altitude, and $\mu_\oplus$ the Earth's gravitational constant. The asteroid's hyperbolic excess velocity at the Earth encounter can be conveniently expressed as a function of the semi-major axis a, eccentricity e and inclination i of the asteroid by means of Opik's encounter theory (Opik 1951) as:

$$v_\infty = \sqrt{\mu_\odot \left(3 - \frac{1}{a} - 2\sqrt{a(1-e^2)}\cdot\cos i\right)} \tag{18.2}$$

where $\mu_\odot$ is the Sun's gravitational constant and, together with the asteroid's semi-major axis, must be expressed in Astronomical Units (AU) units of length.

Regarding the initial interception maneuvre, we note that even Earth-crossing asteroids, which by definition have periapsis and apoapsis radius smaller and larger than 1 Astronomical Unit (AU) respectively, do not generally cross the orbit of the Earth, unless coplanar with its orbit plane. Thus, unless the elliptical Earth-crossing orbit has a very particular orientation (or is coplanar with the Earth's orbit), an interception maneuvre is always necessary to transfer asteroid resources to Earth. The interception model implemented here assumes a maneuvre such that the asteroid's orbit orientation is modified to allow the asteroid to come close enough to the Earth to undergo a hyperbolic fly-by. The change of orientation is modeled as an instantaneous change of argument of periapsis ω of the orbit, a rotation of the orbit within its orbital plane, or a plane change maneuvre, thus a rotation of the inclination angle i of the orbital plane (see Fig. 18.1).

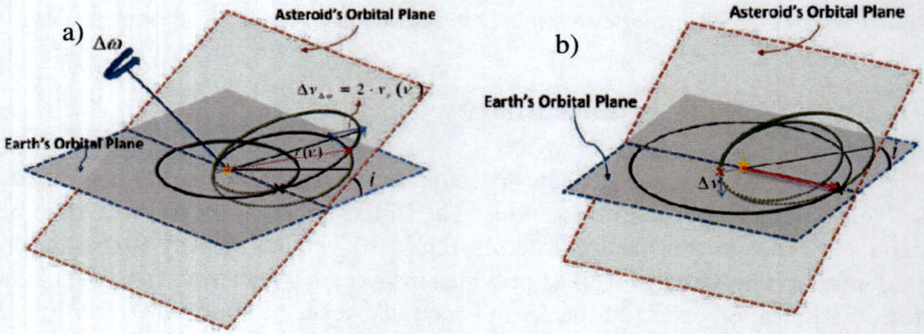

Fig. 18.1 Schematic representations of the interception manoeuvres. a) Manoeuvre providing a change of orientation $\Delta\omega$ of an orbit. Alternatively, a rotation of the orbital plane by Δi is also considered for convenient cases. b) Manoeuvre representation for a non-Earth crossing asteroid.

In order to compute the required change of the argument of periapsis $\Delta\omega$, we first need to identify the four different periapsis orientations ω_{enc} that allow an asteroid with Keplerian elements $\{a,e,i\}$ to intersect the Earth's orbit:

$$\omega_{enc} = \{\pi - \theta_{enc} \quad \theta_{enc} - \pi \quad \theta_{enc} \quad -\theta_{enc}\} \tag{18.3}$$

where θ_{enc} is the true anomaly of the asteroid's orbit such that the Sun is at 1 AU distance, thus:

$$\theta_{enc} = \pm\cos^{-1}\left(\frac{p-1}{e}\right) \tag{18.4}$$

where p is the asteroid's semilatus rectum.

Hence, for an initial argument of periapsis ω_0, the orbit orientation is required to be changed by $\Delta\omega = |\omega_0 - \omega_t|$, where ω_t is either the closest ω_{enc} value to ω_0, or a close value that allows a fly-by with a given specific minimum orbit intersection distance (MOID) that ensures a fly-by with a given perigee altitude (Sanchez and McInnes 2012a). Note that the four values of ω_{enc} enforce a MOID equal to zero, which is not necessary to ensure a fly-by.

The Δv cost of rotating the orbit argument of the periapsis by a given angle $\Delta\omega$, as described by Fig. 18.1a, can be computed as:

$$\Delta v_{\Delta\omega} = 2\sqrt{\frac{\mu_S}{p}}e\sin\left(\frac{\Delta\omega}{2}\right) \tag{18.5}$$

In general, Eq. (18.5) provides a conservative estimate of the transfer cost. In some occasions, for low inclinations and high eccentricities, the manoeuvre described by Eq. (18.5) becomes a very poor option, and a simple change of plane such that the asteroid is inserted into the Earth's orbit plane becomes a better option to change the orbit orientation. For these cases the manoeuvre cost is:

$$\Delta v_{inc} = 2v_{LoN} \cdot \sin(\frac{i}{2}) \tag{18.6}$$

where v_{LoN} corresponds to the asteroid's velocity at the line of nodes.

Hitherto we have only considered Earth-crossing asteroids. Non-Earth crossing asteroids, those with either $a<1$ and $r_a<1$ or $a>1$ and $r_p>1$, can also be transferred to the Earth's neighbourhood by applying a manoeuvre at one of the apsidal points, such that the opposite apsis changes in such a way that an intersection with the Earth's orbit occurs (see Fig. 18.1b).

Given a non-Earth crossing asteroid $\{a,e,i,\omega\}$, the apsidal manoeuvre necessary to obtain an Earth intersecting trajectory can be computed as:

$$\Delta v = \sqrt{\mu_s}\left|\sqrt{\frac{2}{r_m} - \frac{1}{a_f}} - \sqrt{\frac{2}{r_m} - \frac{1}{a}}\right| \tag{18.7}$$

where r_m is the apsidal point at which the impulsive manoeuvre is performed, thus the periapsis for asteroids with $a<1$ and the apopsis when $a>1$, a_f is the semi-major axis of the transfer trajectory, given by:

$$a_f = \frac{a(1 \pm e)}{(1 \pm e_f)} \tag{18.8}$$

where e_f is the eccentricity of the transfer trajectory, which is defined as:

$$e_f = \frac{r_m - 1}{\cos\theta \pm r_m} \tag{18.9}$$

All these formulae allow an estimate of the transport cost of moving asteroid material from its initial orbit to an Earth-bound orbit, but also define a 4-D envelope in the $\{a,e,i,\omega\}$-space for which transfers to Earth are ensured to be below a given Δv limit. Thus, for example, Fig. 18.2 shows the $\{a,e,i\}$-Keplerian element space defined by a Δv threshold of 2.37 km/s, which corresponds to the Moon's escape velocity, and so defines an accessible $\{a,e,i\}$-space that can be exploited at equivalent energy to that necessary to exploit the resources of the Moon.

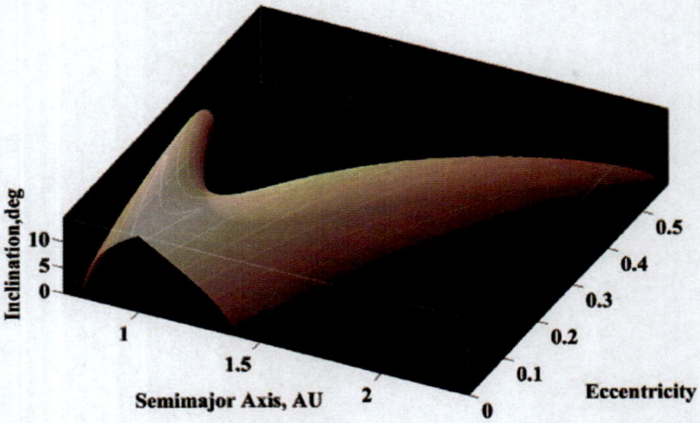

Fig. 18.2 Accessible Keplerian space at a Δv threshold of 2.37 km/s as defined by the analytical transport model

By 1st September 2012, there were slightly more than 9000 known near-Earth objects [http://neo.jpl.nasa.gov/stats/], from which 660 are found within the $\{a,e,i,\omega\}$-space that ensures a Δv transport cost smaller than 2.37 km/s, again equivalent to the transport cost from the lunar surface, by means of the formulae described above. We are however interested in the portion of the population that remains undiscovered, and so a NEO orbital distribution model described in (Bottke et al. 2002) is used instead. The Bottke et al. model is constructed by numerically propagating in time thousands of test bodies and computing steady state densities in different Keplerian regions. Estimating the fraction of this orbital distribution within a 4-D volume, as described above, will allow us to estimate the fraction of the NEO population, and thus the amount of resources, accessible for given a Δv threshold.

18.3 Available Resources

Some 9,144 objects catalogued as Near Earth Objects (NEOs) are known today (as 1st September 2012). By convention, an object is considered a NEO if its perihelion is smaller than 1.3 AU and its aphelion is larger than 0.983 AU. This is a very broad definition which includes predominantly asteroids, but also a small fraction of comets (i.e., 92 comets). NEOs are then the closest objects to the Earth and therefore the obvious first targets for any resource exploitation mission (excluding the Moon).

An analysis of the transportation costs for all the surveyed asteroids, by means of the model described above, already shows a series of interesting target objects for future exploitation missions that require a Δv in the order of a few hundred meters per second. Fig. 18.3 summarises the number of objects that could be exploited for a given level of Δv. The figure also shows different colour bands for each different type of transfer considered as part of the bi-impulse transfer model (i.e., change of ω versus change of i, Earth-crosser type of asteroid or non-Earth crosser).

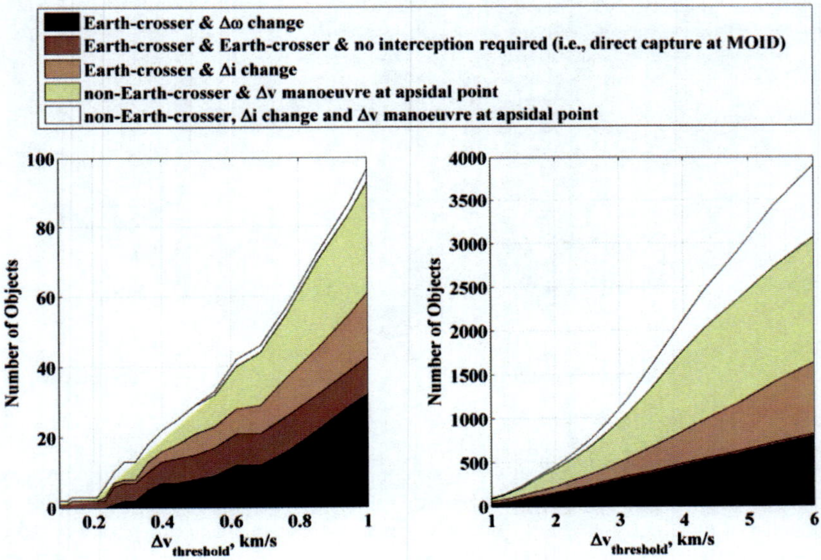

Fig. 18.3 Number of known objects whose resources are accessible for a given Δv threshold. The set of manoeuvres are briefly described in the chapter, further details can be found at (Sanchez and McInnes 2012a).

The purpose of the transfer models however is not to compute the transport cost of existing objects, for which more complex and less conservative transfers should be considered (Sanchez et al. 2012), but to delimit the Keplerian element space regions from which asteroids could be transported to Earth, as shown by Fig. 18.2. Once the accessible Keplerian element space is known, an estimate of the total amount of resources can be made by means of a NEO model capable of providing the number and size of objects expected to be found within these regions.

18.3.1 NEO Orbital and Size Distribution

In order to estimate the amount of available resources, we thus require a sound statistical NEO population model. The NEO model used here is composed of two separate parts: an orbital and a size distribution. The NEO size model is based on NEOWISE observations (Mainzer et al. 2011), whereas the NEO orbital distribution is based on the theoretical distribution model published in (Bottke et al. 2002).

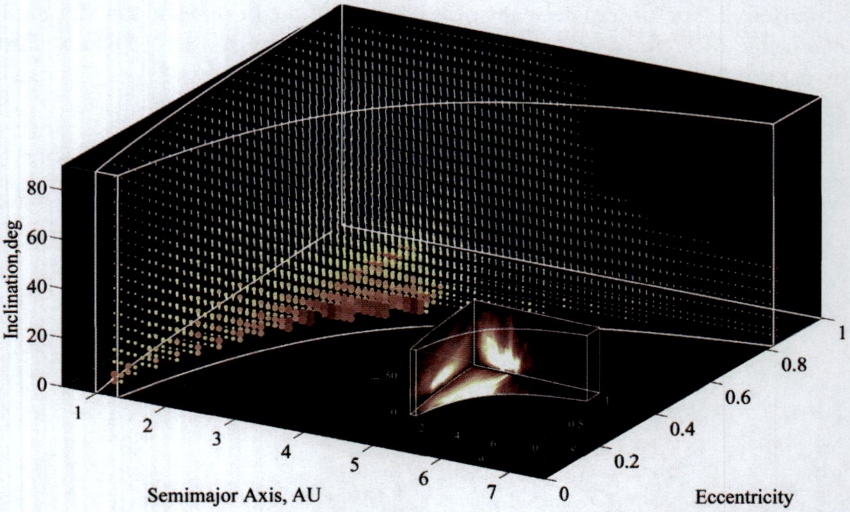

Fig. 18.4 (Bottke et al. 2002) NEO distribution. The figure shows a representation of the NEO density function $\rho(a,e,i)$. The 4th dimension, i.e., the density ρ at a given point (a,e,i), is represented by a set of grid points coloured and sized as a function of the value ρ. A smaller set of axis represent the projection of the total value of ρ onto the planes a=0.5 AU, e=1 and i=0 deg. Note that the colour code has been inverted for the smaller projection figure

The size distribution model allows us to estimate the probability of an object to have a given size. As noted before, it is based on the latest estimates published as a result of NEOWISE Space infrared Survey observational campaing (Mainzer et al. 2011), which indicate a slight deviation from the previously assumed single slope power law population distribution (Stokes et al. 2003). On the other hand, the NEO orbital distribution used is based on an interpolation from the theoretical distribution model by (Bottke et al. 2002). The data used was very kindly provided by W.F. Bottke (personal communication, 2009).

Bottke et al. (2002) built an orbital distribution of NEOs by propagating in time thousands of test bodies initially located at all the main source regions of asteroids (i.e., the ν6 resonance, intermediate source Mars-crossers, the 3:1 resonance, the outer main belt, and the trans-Neptunian disk). By using the set of asteroids discovered by Spacewatch at that time, the relative importance of the different asteroid (or comets) sources could be best-fitted. This procedure yielded a steady state population of near Earth objects from which an orbital distribution as a function of semi-major axis a, eccentricity e and inclination i can be interpolated numerically. Figure 18.4 shows a representation of Bottke's NEO density function $\rho(a,e,i)$. The remaining three Keplerian elements, not shown in Fig. 18.4, the right ascension of the ascending node Ω, the argument of periapsis ω and the mean anomaly M, can be assumed to be uniformly distributed random variables (Stuart 2003).

18.3.2 Accessible Resource Mass

The total mass of the population of near Earth objects, larger than 1 meter diameter, is estimated to be approximately 5×10^{16} kg (Sanchez and McInnes 2012a). The question that arises now is how much of this mass can be easily accessed and exploited, which can be estimated by means of the bi-impulsive transfer model and (Bottke et al. 2002). Figure 18.5 shows the portion of the aforementioned total asteroid mass that can be accessed with a given Δv threshold. The same figure also shows, with a thinner dashed line, the fraction of the accessible mass that has been currently surveyed. The latter is only an approximate estimate of the total mass of surveyed objects (as on 1st September 2012) computed from visual magnitude data (Bowell et al. 1989) and average density values (Chesley et al. 2002).

Figure 18.5 now provides a tool to measure the availability of resources, defined here as the total asteroid mass (without distinguishing between different materials (Lewis and Hutson 1993)), and the difficulty of accessing them. Thus for example, the figure shows that at the same cost of accessing lunar resources (i.e., 2.37 km/s) the total pool of asteroid mass available is of the order of 10^{14} kg. This of course is orders of magnitude lower than the total pool of resources at the Moon (i.e., the mass of the Moon, ~7.36×10^{22}kg), although it is still a very significant amount.

The main advantage of asteroid resources however is that asteroid material can be exploited across a wide spectrum of Δv, rather than at a minimum threshold, as is the case for lunar material (i.e., 2.37 km/s). Thus, for example, with 100 m/s of Δv budget, approximately 8.5×10^9 kg of asteroid resources could potentially be exploited. Note however, from Fig. 18.5, that while for higher Δv thresholds the fraction of surveyed asteroids is only partially completed, many very low Δv exploitable targets still await discovery.

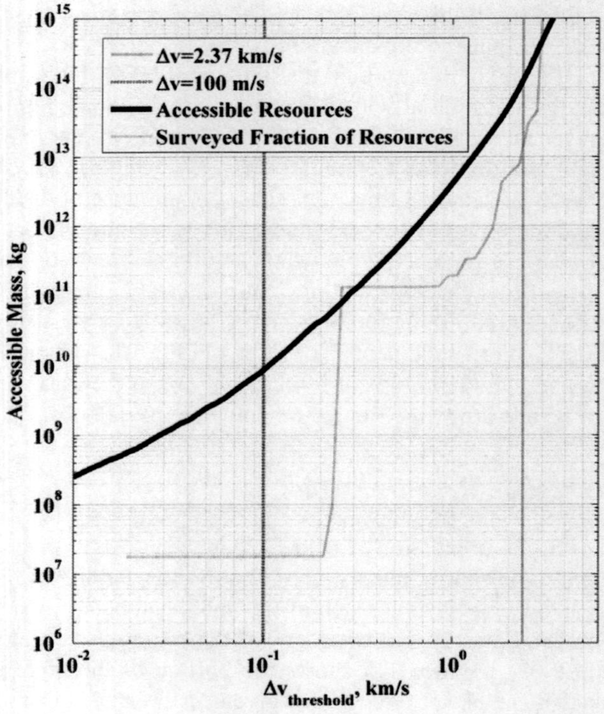

Fig. 18.5 Estimate of the total and surveyed amount of accessible asteroid resources as a function of Δv threshold

18.4 Asteroid Resource Map

One of the important issues not yet resolved by the results shown in Fig. 18.5 is the number of missions that are required to exploit all or part of the accessible pool of asteroid resources. This issue is of key importance, since if a given resource is spread over a large number of very small objects, gathering all of them may become a cumbersome task, and therefore not economically worthwhile. This section will attempt to resolve this issue by discussing the results shown in Fig. 18.6, also referred as asteroid resource map. The map also provides an interesting insight into current and near term capabilities to retrieve asteroid material from the Near Earth Object population.

Figure 18.6 is composed of three different elements. The principal element is the statistical estimates of the diameter of all the population of objects that should be accessible with a given Δv transport costs. These represented by the median asteroid diameter of the largest, the tenth largest, the hundredth largest and the thousandth largest object of the population of asteroids accessible with a given Δv threshold. These can be computed by, first, integrating Bottke's NEO density model (Bottke et al. 2002) within the volume of accessible space defined by the analytical transport formulae (Sanchez and McInnes 2012a) and the parameter Δv threshold, which yields the probability P of an asteroid to be accessible with a given Δv mission budget. We then assume that each single asteroid has the same probability P to be found in the accessible region, which allows us to define the probability to find k asteroids within a population of n asteroids as described by the binomial distribution. In this particular case, P is a very low probability and n is a very large number of asteroids, thus the Poisson distribution (a limiting case of the binomial distribution when n tends to infinity) represents a very good approximation of the statistical behaviour of the problem. Therefore, the probability $g(k,\lambda)$ to find k asteroids when the expected number is λ can be described by:

$$g(k,\lambda) = \frac{\lambda^k e^{-\lambda}}{k!} \tag{18.10}$$

The expected number λ, or average number of accessible asteroids, can be calculated as:

$$\lambda(D_{\min}) = \Delta N(D_{\min} < D \leq D_{\max}) \cdot P \tag{18.11}$$

where ΔN is the total number of asteroids with diameters larger than $D_{\min}$ and smaller than $D_{\max}$, which is fixed at 32 km diameter as the largest existing NEO 1036 Ganymed, and P the probability to find objects within a given accessible orbital region. An estimate of the population of objects with diameters within $D_{\min}$ and $D_{\max}$ can now be obtained by means of the object size distribution as suggested by the NEOWISE observations (Mainzer et al. 2011). As mentioned earlier, the object size distribution modelled here is based on a multi-power law distribution (Stokes et al. 2003, Mainzer et al., 2011) that matches the latest population estimates from NEOWISE for objects larger than 100 m (i.e., 981 objects larger than 1 km, 20,500 objects larger than 100 m) (Mainzer et al. 2011), as well as Brown et al. (Brown et al. 2002) estimates for objects smaller than 10 meters.

An integration such as:

$$\int_{n_{NEA}}^{\infty} g(k,\lambda) \cdot dk \tag{18.11'}$$

yields the probability of finding at least n_{NEA} asteroids when the expected value, or average, was λ. By finding then the value of λ that yields an accumulative probability of 50%,

$$\int_{n_{NEA}}^{\infty} g(k,\lambda(D_{\min})) \cdot dk = 50\% \tag{18.12}$$

we can then estimate the median diameter of the smallest object in the n_{NEA} set. This procedure can also be repeated with accumulative probabilities of 95% and 5% to obtain the 90% confidence region, which is also shown in Fig. 18.6.

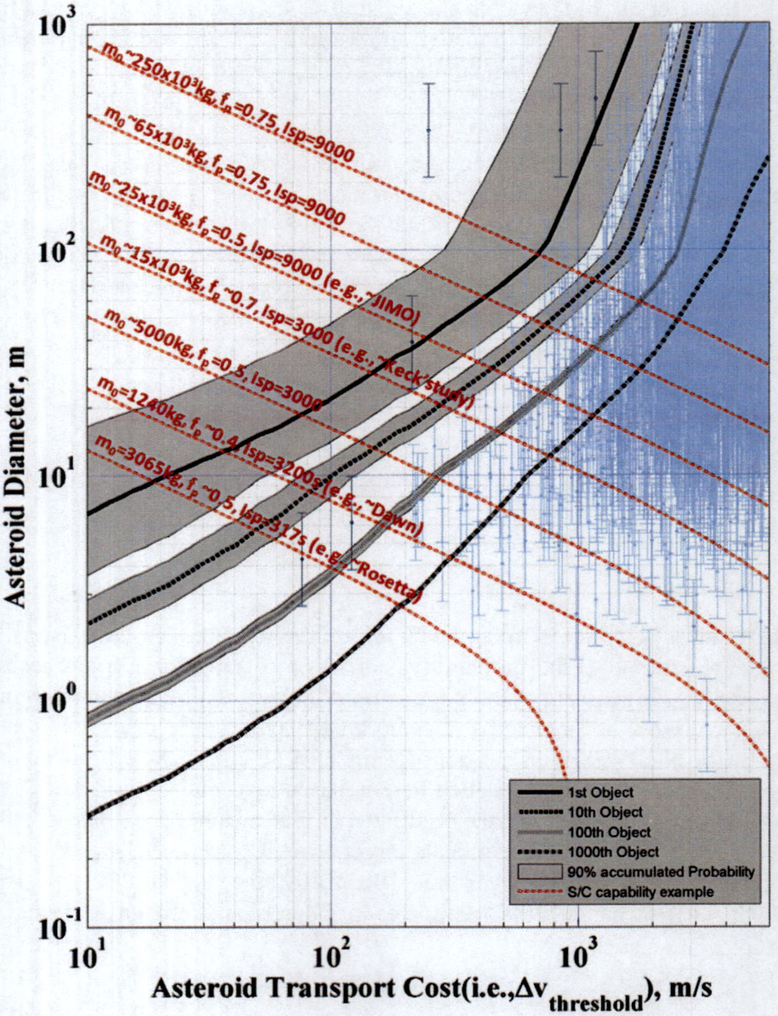

Fig. 18.6 Expected size of the accessible population of asteroids as a function of Δv budget. The figure shows the median asteroid size of four representative objects of the entire population: the largest, the tenth largest, the hundredth largest and the thousandth largest. The overlapped grey regions represent the 90% confidence region of each one of these objects. Note that the 90% confidence region accounts only for the statistical variance of a population of objects perfectly matching the population estimates (Stokes et al. 2003, Mainzer et al. 2011, Brown et al., 2002), not by epistemic uncertanties on the assumed distributions. For further details on the calculations of the figure, the reader should refer to more detailed work (Sanchez and McInnes 2012a).

The information in the asteroid resource map can be read as follows: let us set, for example, the Δv transport budget, or threshold, at 100 m/s. The largest accessible object at this Δv has a 50% probability of being equal to or larger than 22 meters diameter, while we can say with 90% confidence that its size should be between 47 meters and 14 meters (i.e., gray area around median size line). The following set of data in the decreasing ordinate axis refers to the 10th largest object found within the region of feasible capture given by a Δv threshold of 100 m/s, whose median value of the diameter is 10 meters. Thus, a population of 10 objects larger or equal to 10 meters should be expected to be accessible within the 100 m/s budget. The 100th largest object is foreseen to have a diameter of 4 meters and 1000th largest of 1 meter. This provides an overview of all the objects that could be exploited with a Δv of 100 m/s or lower.

It can then be said that the asteroid resource map provides a snapshot of the median asteroid size population expected to be found within the Keplerian element space that ensures transport at a given Δv cost. It is interesting to compare the statistical estimates shown in Fig. 18.6 with the growing catalogue of Near Earth Objects. For this purpose, we have also superimposed in Fig. 18.6 the accessibility (i.e., Δv) of all known asteroids computed by means of the analytical transport formulae. A size estimate of each object in the list is provided by (Bowell et al. 1989):

$$D = 1329\,km \times 10^{-H/5}\, p_v^{-1/2} \tag{18.13}$$

where the absolute magnitude H is obtained from the Minor Planet Centre database, and the albedo p_v is assumed to range from 0.06 to 0.40 and an average of 0.154 (Chesley et al. 2002). This comparison shows how the statistical estimates adjust rather well to the existing population of objects within the sizes for which a high survey completion has been achieved.

Finally, the map can be used to gain some more insight in the current and near term capability to retrieve and exploit the accessible asteroid population of asteroids. For this, a series of isolines have also been superimposed in the figure indicating the transport capability of a given mass in space m_0, propellant mass fraction f_p and specific impulse Isp of the propulsion system. Given a space system such as (m_0, f_p, Isp), and assuming outbound (Earth-to-asteroid) and inbound (asteroid-to-Earth) trajectories with cost as given by the analytical transport formulae, we can use the rocket equation to compute the largest asteroid that a (m_0, f_p, Isp)-spacecraft could bring back to Earth (assuming all asteroids have average density 2,600 kg/m^3 and spherical shape (Chesley et al. 2002)). Some of the considered (m_0, f_p, Isp) sets are based on prior spacecraft systems such as the ESA Rosetta [http://www.esa.int/esaMI/Rosetta/index.html]and NASA Dawn spacecraft (Rayman et al. 2006) or in mission proposals such as the recent Keck study proposal (Brophy et.al. 2012) or the cancelled NASA Jupiter Icy Moon Orbiter (JIMO) (Langmaier et al. 2008).

18.5 Discussion

According to the results shown in Fig. 18.6, a full spectrum of asteroid sizes awaits to be harvested at different energy levels (i.e., Δv costs). Several missions have already attempted to return samples from this population (e.g., Hayabusa spacecraft at Itokawa (Kawaguchi et al. 2008)) and others are under development (e.g., OSIRIS-Rex at 199RQ36). Figure 18.6 provides the energy-cost framework, or resource map, of future asteroid mining ventures, and thus can be used as a tool for initial feasibility studies and mission planning efforts. Thus, for example, the asteroid retrieval capability isolines show that current spacecrafts, such as Rosetta or Dawn (Rayman et al. 2006), could actually provide a sufficient impulse to bring back asteroids with diameters of order 10 meters. Some already proposed missions, such as the Keck asteroid retrieval mission (Brophy et.al. 2012) and JIMO (Langmaier et al. 2008), would enhance that capability to the 30 meters diameter range.

Hence, given the low transport cost expected for the most accessible objects of the Near Earth Asteroid population, the transport of the entire object to Earth's vicinity can be seen as a feasible option for current or near-term technology. The ultimate value of asteroid retrieval missions would greatly depend on both asteroid type and key technologies, such as autonomous mining operations. In the near-term however this type of mission serves as a stepping stone towards more ambitious mining scenarios, providing a unique opportunity for technology demonstrator missions with high scientific return. One however would expect that, as technologies are developed, a minimal level of processing occurs in-situ (distilling, magnetic or SRP separation, etc) and thus a higher fraction of useful material would be transported back to Earth.

It is important to remark that the transport costs for asteroid retrieval missions are here estimated by means of an analytical bi-impulsive transfer which clearly provides only an approximate estimate of the Δv associated with the transport of material. More complex trajectories to model the material transport, such as multiple Earth fly-bys, lunar gravity assists or manifold dynamics, should be expected to decrease the costs of transferring asteroid material to Earth's vicinity. The results presented here are therefore a conservative representation of the resources available. For example, as shown in (Sanchez and McInnes 2011), ballistic escape and capture trajectories with excess velocities of up to 1 km/s are possible from Earth. A trajectory approaching the Earth with an excess velocity v_∞ of 1 km/s would require a Δv of 45 m/s with a patched-conic approximation to become Earth-bound (i.e., inserted into a parabolic orbit). This implies that all objects considered here as accessible with less than 45 m/s may be suitable targets for ballistic capture. Indeed, as discussed in the other chapters, many objects can already be identified that can potentially be captured at a much lower cost than that expected by means of a patched-conic approximation such as the one used here.

For more ambitious scenarios of asteroid exploitation, again, as new techniques and technologies become available, one can envisage multiple concepts that may reduce even further the costs of accessing the available pool of asteroid resources. Such as, for example, benefiting from asteroid-extracted propellants, such as water for solar thermal propulsion systems or by grazing the Earth's atmosphere to reduce the asteroid's orbital energy (Sanchez and McInnes 2012b).

A very important component on the methodology used to compute the available asteroid resources in Fig. 18.5 and Fig. 18.6 is the NEO orbital and size distributions. A possible concern about the accuracy of these models may for example be raised with the assumption on the orbital distribution being independent of the asteroid size (Bottke et al. 2002). Non-gravitational perturbations affect objects of different sizes in various ways, which implies that the different asteroid sources (e.g., the ν6 resonance, the 3:1 resonance, etc) may be supplying different asteroid size distributions, since non-gravitational perturbations are the main mechanisms that feed the asteroid sources. It is however surprising the resemblance of the expected fireball impact frequencies computed using the NEO orbital and size distributions models described here (and also a linearised impact model such as (Opik 1951)) and the fireball impact frequencies recorded by satellite surveys (Brown et al. 2002). As described in Brown et al. (2002), the Earth is estimated to be struck with approximately 5 ktons of energy as a yearly average. The result computed assuming the NEO orbital and size distributions as described here is 4.2 ktons. Similarly the monthly average compares as 0.3 ktons in (Brown et al. 2002) and 0.2 ktons as computed here, while the decade average is shown to be about 50 ktons in (Brown et al. 2002) and 78 ktons computed here. Thus, despite possible sources of inaccuracies, the results shown in the chapter should represent good order of magnitude estimates of the accessible asteroid material.

18.5.1 Available Resources

Asteroids have a very diverse composition and, thus, the possible uses of the different accessible objects (i.e., as shown in Fig. 18.6) will always depend on the particular object characteristics. Any available space resource however can be envisaged to be put to good use, since for example, lacking any better option, an asteroid could be simply used as an "anchor" for gravity gradient stabilisation or its bulk material (regolith and unprocessed material) as radiation shielding, which would reduce the hardening requirements to survive in space. Nevertheless, one can envisage much more disruptive uses for some particular resources, such as water and volatiles, which could be used to sustain human life in space, as well as a propellant, semiconductors that could be used to build in-situ solar cells and metals used for space structures.

Providing estimates on the amount of resources for different valuable materials can only be based on statistical estimates from spectroscopic surveys and meteoroid recoveries. While spectroscopic surveys of Near Earth Asteroids show a very wide diversity of spectral classes (Bus and Binzel 2002), only meteoroid recoveries provide an accurate account of the materials available in space (Lewis and Hutson 1993). The latter however have a clear body strength bias (i.e., weaker objects ablate in the atmosphere) and are quickly weathered if not recover soon after atmospheric entry.

It is not the purpose of this chapter to provide an accurate census of object types and resources availability, primarily because current knowledge of asteroid composition and their availability on different orbits is inaccurate. On the other hand, a few examples of possible resource availability at different Δv thresholds may cast some more light on the usefulness and possibilities of the exploitation of asteroid resources. The following Table 18.1 is provided only as an example of the amount of material that could be found in different sized objects. Table 18.1 assumes that water should be extracted from hydrated carbonaceous asteroids, while metals and platinum group metals (PGM) would be extracted from M-class objects. This however does not imply that these resources are only available in these types of objects. Quite on the contrary, other asteroid classes, such as for example S-class (Nakamura et al. 2011), may contain more interesting combination of volatiles, metals and semiconductors (Lewis and Hutson 1993), albeit volatiles and metals would likely be found at lower abundances than for the examples described on Table 18.1. On the other hand, some asteroids may offer lower abundances of a given resource but on a much more useful form. For example ordinary LL chondrites contains only ~1 to 5% of Fe-Ni metal but in the form of pulverized regolith that could potentially be easily separated from non-metallic regolith by means of an electromagnetic rake (Kargel 1994).

Table 18.1 Example of possible resources on different asteroid classes

Resource						Asteroid size	
	Ast. Class	Population	Density	Resource mass fraction	10-m	100-m	500-m
Water	Hydrated C-class	10%*	1300 kg/m^3 **	8% ***	54,000 l	54x10^6 l	68x10^9 l
Metals	M-class	5%*	5300 kg/m^3 **	88% ***	2x10^3 t	2x10^6 t	3x10^8 t
PGMs	M-class	5%*	5300 kg/m^3 **	35 ppm ***	97 kg	97 t	12x10^3 t

* (Bus and Binzel 2002);** (Chesley et al. 2002);***(Ross 2001)

One can imagine many interesting scenarios for the utilisation of the resources shown to be accessible at low Δv-budgets. As shown in Fig. 18.6, the largest accessible asteroid at a 100 m/s Δv-budget is estimated to be a 22-m diameter object. Such an object could supply from 10^7 kg to $4\text{x}10^7$ kg of asteroid material, depending on composition and density. As previously mentioned, the particular uses of the resources would depend on the asteroid composition: if the object turns to be a hydrated carbonaceous asteroid, a million litres of water could possibly be extracted (considering, as in Table 18.1, an asteroid of density 1300 kg/m^3 (Chesley et al. 2002) and 8% (Ross 2001) of its weight in water). However, if this object is an M-class asteroid (density 5300 kg/m^3 (Chesley et al. 2002)), of order thirty thousand tonnes of metal could potentially be extracted and even a tonne of Platinum Group Metals (PGM) (88% of metal assumed and 35ppm of PGM (Ross 2001)). The latter resource could easily reach a value of $50M in Earth's commodity markets (Crandall 2009).

Note from the examples in Table 18.1 that hydrated C-class and M-class objects only cover 15% of the asteroid population, implying that this hypothesized 22 m object has only a 15% probability of being one of these two classes. As previously noted, this does not mean that there is an 85% chance that the asteroid would be of no use, or that the 85% remaining population has less resource interest. S-class asteroids, accounting approximately for 40% of the NEA objects, could very well become the most interesting targets to exploit since they likely contain relatively good abundances of semiconductors, metals and even volatiles such as oxygen, including perhaps water. S-class may also contain PGM, albeit at a lower abundance than in M-class. One could possibly imagine the utilisation of a 22 m S-class object to provide between 1 to 4 thousand tonnes of iron to be used as a structural support for a 1.5 km^2 silicon solar array surface (build from the same object), which would generate a minimum of 1 GW of power (O'Leary et al. 1979; Mauk 2003). The capability to exploit small S-class objects may perhaps allow solar power satellites to become commercially viable. Water could also be possibly extracted from S-class objects. Assuming 0.15% of water (O'Leary et al. 1979), this would account for 30 thousand litres of water, which if we assume that a human being requires about 3 litres of water a day, from which at least 90% is recovered through recycling, there should still be enough water to sustain a crew of 25 people for 10 years, and perhaps no resupply of water would be required for the crew responsible to build the solar power orbital plant.

Nevertheless, ideally water should be extracted from asteroids with higher water content, as for example hydrated C-class (see Table 18.1). These are assumed here to likely contain around 8% of water mass fraction, although higher abundances are possible in hydrated C-class objects (~20%), or even ~50% if near earth comets are exploited instead (Lewis and Hutson 1993). Many other volatiles, apart from water, can also be extracted and used for life support or rocket propellant (Lewis 1996). The capability to extract and use water from asteroids would certainly be a disrupting technology allowing much more affordable human access to space, since no resupply of water and oxygen would be required, and also interplanetary travel, if spacecraft are launched without the propellant required to reach their final destinations. Water could be extracted from asteroids by direct sublimation of its native ice, if this is available, or by processing hydrated minerals and clays. Thus, most likely, this commodity in particular may represent a very important resource for exploitation. If water is mined and finally transported to LEO, which would require the addition of 3.3 km/s to the Δv cost estimated in this chapter (cost estimates were only for a weakly bound Earth orbit), the total cost of transportation will still be of order 3 times less than that required to transport water from the Earth's surface. Of course, in order for the transport of asteroid material to LEO to be preferable over the more traditional Earth transport, the cost of mining and transporting the resources back to Earth should be lower than the two-thirds saving on transportation costs. This figure greatly improves if the propellant is transported to the Earth-Moon Lagrangian points and used to fuel interplanetary missions (Lo and Ross 2001). For this latter scenario the energy cost leverage would initially be of 1 to 1000 (i.e., the energy required to access to 1 litre of water from Earth allows access to a 1000 litres in the near Earth orbital space), and progressively reduce as the resource is consumed and larger Δv budgets are required to access the following ore. The utilisation of an orbital *fuel depot* at the L_2, for example, would imply launch mass savings of at least a factor of two for missions to Mars.

As seen in Fig. 18.6, and discussed so far, a myriad of small objects awaits to be cherry-picked and exploited. Retrieval of very small objects suffers however of an important disadvantage with respect exploiting larger objects. Due to their small size, and thus extremely faint luminosity as seen from Earth, they are very difficult to discover and characterise. Thus, asteroid surveys may struggle not only to complete a substantial census of the population of small asteroids, but also to obtain the accuracy required on their ephemeris to encounter an asteroid in deep space during a hypothetical rendezvous mission. Yet, even for those objects sufficiently characterised to be encountered in deep space, we may still lack the compositional data required to tag the object as "of interest", since presumably a mission will not be launched to a random asteroid, but to an object of potential compositional interest.

18.6 Conclusions

While the capture of near-Earth objects for future resource utilisation may seem a distant prospect, we have demonstrated that a significant population of low energy bodies exist, and can be harvested with modest energy costs. Importantly, such objects can be accessed across a spectrum of energy costs, unlike lunar resources utilisation which incurs a minimum (and significant) energy cost. It appears that current and emerging propulsion technologies can now enable these resources to be accessed. This presents an exciting range of possibilities for the capture of small objects for science, or for utilisation in support for future human space exploration, as advocated recently commercial space ventures (Elvis 2012).

Acknowledgements. We would like to thank William Bottke for kindly providing us with the Near Earth Asteroid distribution data. We would also like to acknowledge the reviewers, John Lewis and Haym Benaroya, for their valuable comments on this manuscript. The work reported was supported by European Research Council grant 227571 (VISIONSPACE).

References

Bottke, W.F., Morbidelli, A., Jedicke, R., Petit, J.-M., Levison, H.F., Michel, P., Metcalfe, T.S.: Debiased Orbital and Absolute Magnitude Distribution of the Near-Earth Objects. Icarus 156, 399–433 (2002)

Bowell, E., Hapke, B., Domingue, D., Lumme, K., Peltoniemi, J., Harris, A.W.: Application of Photometric Models to Asteroids. In: Binzel, R.P., Gehrels, R.P., Matthews, M.S. (eds.) Asteroids II. Univ. of Arizona Press, Tucson (1989)

Brophy, J.: The Dawn Ion Propulsion System. Space Science Reviews 163, 251–261 (2011)

Brophy, J., et al.: Asteroid Retrieval Feasibility Study. Keck Institute for Space Studies, Califonia Institute of Technology, Jet Propulsion Laboratory, Pasadena, California (2012)

Brown, P., Spalding, R.E., Revelle, D.O., Ragliaferri, E., Worden, S.P.: The Flux of Small Near-Earth Objects Colliding With the Earth. Nature 420, 294–296 (2002)

Bus, S.J., Binzel, R.P.: Phase II of the Small Main-Belt Asteroid Spectroscopic Survey. Icarus 158, 146–177 (2002)

Chapman, C.R.: The Hazard of Near-Earth Asteroid Impacts on Earth. Earth and Planetary Science Letters 2, 1–15 (2004)

Chesley, S.R., Chodas, P.W., Milani, A., Yeomans, D.K.: Quantifying the Risk Posed by Potential Earth Impacts. Icarus 159, 423–432 (2002)

Crandall, W.B.: Abundant Planet: Enabling Profitable Asteroid Mining. In: Planet, A. (ed.) Redwood City, CA 94062: Abundant Planet 501(C)3 Organization (2009)

Elvis, M.: Let's Mine Asteroids — For Science and Profit. Nature 485, 549 (2012)

Kargel, J.S.: Metalliferous Asteroids as Potential Sources of Precious Metals. J. Geophys. Res. 99, 21129–21141 (1994)

Kawaguchi, J., Fujiwara, A., Uesugi, T.: Hayabusa-Its Technology and Science Accomplishment Summary and Hayabusa-2. Acta Astronautica 62, 639–647 (2008)

Langmaier, J., Elliott, J., Clark, K., Pappalardo, R., Reh, K., Spilker, T.: Assessment of Alternative Europa Mission Architectures. JPL Publication 08-01. NASA (2008)

Lewis, J.S.: Mining the Sky: Untold Riches from Asteroids, Comets and Planets. Helix Books/Perseus Books, Reading, Massachusetts (1996)

Lewis, J.S., Hutson, M.L.: Asteroidal Resource Opportunities Suggested By Meteorite Data. In: Lewis, J.S., Matthews, M.S., Guerrieri, M.L. (eds.) Resources of Near-Earth Space. University of Arizona Press, Tucson (1993)

Lo, M.W., Ross, S.D.: The Lunar L1 Gateway: Portal to the Stars and Beyond. In: AIAA Space 2001 Conference, Albuquerque, New Mexico (2001)

Mainzer, A., Grav, T., Bauer, J., Masiero, J., Mcmillan, R.S., Cutri, R.M., Walker, R., Wright, E., Eisenhardt, P., Tholen, D.J., Spahr, T., Jedicke, R., Denneau, L., Debaun, E., Elsbury, D., Gautier, T., Gomillion, S., Hand, E., Mo, W., Watkins, J., Wilkins, A., Bryngelson, G.L., Del Pino Molina, A., Desai, S., Gómez Camus, M., Hidalgo, S.L., Konstantopoulos, I., Larsen, J.A., Maleszewski, C., Malkan, M.A., Mauduit, J.C., Mullan, B.L., Olszewski, E.W., Pforr, J., Saro, A., Scotti, J.V., Wasserman, L.H.: NEOWISE Observations of Near-Earth Objects: Preliminary Results. The Astrophysical Journal 743, 156 (2011)

Mauk, M.G.: Silicon Solar Cells: Physical Metallurgy Principles. Journal of the Minerals, Metals and Materials Society 55, 38–42 (2003)

Nakamura, T., Noguchi, T., Tanaka, M., Zolensky, M.E., Kimura, M., Tsuchiyama, A., Nakato, A., Ogami, T., Ishida, H., Uesugi, M., Yada, T., Shirai, K., Fujimura, A., Okazaki, R., Sandford, S.A., Ishibashi, Y., Abe, M., Okada, T., Ueno, M., Mukai, T., Yoshikawa, M., Kawaguchi, J.: Itokawa Dust Particles: A Direct Link Between S-Type Asteroids and Ordinary Chondrites. Science 333, 1113–1116 (2011)

Nichols, C.R.: Volatile Products From Carbonaceous Asteroids. In: Lewis, J.S., Matthews, M.S., Guerrieri, M.L. (eds.) Resources of Near-Earth Space. University of Arizona Press, Tucson (1993)

O'Leary, B., Gaffey, M.J., Ross, D.J., Salkeld, R.: Retrieval of Asteroidal Materials. Space Resources and Settlements (1979)

Opik, E.J.: Collision Probabilities With the Planets and the Distribution of Interplanetary Matter. Proceedings of the Royal Irish Academy. Section A: Mathematical and Physical Sciences 54, 165–199 (1951)

Rayman, M.D., Fraschetti, T.C., Raymond, C.A., Russell, C.T.: Dawn: A Mission in Development for Exploration of Main Belt Asteroids Vesta and Ceres. Acta Astronautica 58, 605–616 (2006)

Review of U.S. Human Spaceflight Plans Committee 2009. HSF Final Report: Seeking a Human Spaceflight Program Worthy of a Great Nation. NaASA (2009)

Ross, S.D.: Near-Earth Asteroid Mining. Department of Control and Dynamical Systems, Pasadena (2001)

Ross, S.D.: The Interplanetary Transport Network. American Scientist 94, 230 (2006)

Sanchez, J.P., García, D., Alessi, E.M., Mcinnes, C.R.: Gravitational Capture Opportunities for Asteroid Retrieval Missions. In: 63rd International Astronautical Congress. International Astronautical Federation, Naples (2012)

Sanchez, J.P., Mcinnes, C.R.: On the Ballistic Capture of Asteroids for Resource Utilization. In: 62nd International Astronautical Congress. IAF Cape Town, SA (2011)

Sanchez, J.P., Mcinnes, C.R.: Assessment on the Feasibility of Future Shepherding of Asteroid Resources. Acta Astronautica 73, 49–66 (2012a)

Sanchez, J.P., Mcinnes, C.R.: Synergistic Approach of Asteroid Exploitation and Planetary Protection. Advances in Space Research 49, 667–685 (2012b)

Stokes, G.H., Yeomans, D.K., Bottke, W.F., Chesley, S.R., Evans, J.B., Gold, R.E., Harris, A.W., Jewitt, D., Kelso, T.S., Mcmillan, R.S., Spahr, T.B., Worden, P.: Study to Determine the Feasibility of Extending the Search for Near-Earth Objects to Smaller Limiting Diameters. Report of the Near-Earth Object Science Definition Team. National Aeronautics and Space Administration, Washington, D.C. (2003)

Stuart, J.S.: Observational Constraints on the Number, Albedos, Size, and Impact Hazards of the Near-Earth Asteroids. Massachusetts Institute of Technology (2003)

Tsiolkovsky, K.E.: The Exploration of Cosmic Space by Means of Reaction Devices. Scientific Review (1903)

Chapter 19
Asteroid Capture

Didier Massonnet

Centre National d'Etudes Spatiales, France

19.1 Introduction

The scenario where a celestial body threatens to impact the Earth is very popular and many solutions have been proposed to mitigate the associated risk. Some of them can be considered "soft" and consist in landing on or rendez-vous with the incoming asteroid several years in advance of the collision and then altering its trajectory by various means, such as painting it in order to alter its albedo and therefore the radiation pressure or by pushing it gently using a gravitational tractor (Lu and Love 2005). The mechanical device described below could also be used to change the trajectory. All these scenarios need a lot of time, even discounting mission preparation. If the incoming body is not a near-earth object but, for instance, a comet nucleus, more than ten years are likely to be necessary to reach it. Some "hard" solutions do not require a rendez-vous, but an interception, possibly with a large relative velocity. They can be implemented on a much shorter time-frame. Detonating a nuclear device could disperse the incoming body as the gravitational binding energy of an asteroid is surprisingly low. The difficulty might then be the timing: the explosion should take place before the impact but late enough to create a strong absorption of its energy (mainly X rays), which would turn a percentage of the mass of the asteroid into hot gases generating a shock wave that could disperse the asteroid and additionally creating a mean impulse which would alter the trajectory of the centre of gravity of the cloud of debris. In (Massonnet and Meyssignac 2006), a two step procedure is proposed where a smaller asteroid is first captured and "parked" on a L1-L2 Earth-Sun Lagrange orbit, and then sent into a trajectory impacting the incoming threat.

19.2 The Concept

The concept involves detecting an easy-to-access asteroid, sufficiently light to allow altering its trajectory in three phases: first to capture it into an Earth-bound orbit, second to monitor and correct this orbit as a parking place, third to leave this orbit for a trajectory impacting any oncoming, threatening celestial body. For this last part the piloting law remains to be studied in detail. The main challenge is the low level of acceleration that can be applied by the propulsion mean proposed in (Massonnet and Meyssignac 2006), which should nevertheless always exceed any trajectory instability that may develop. This use of a captured asteroid is the simplest one as the tools used for capturing the asteroid are the same as for handling it further. For using the asteroid as a source of extraterrestrial material such as oxygen only the two first steps are required, but they must be followed by an entirely independent extraction process. If it appears that adequate small size asteroids are currently transitioning between Earth-Sun Lagrange points, the first step may be omitted. The capture can be regarded as a "life insurance" and be undertaken without any actual threat, which would result in cutting considerably our reaction time should a threat materialize quickly. In (Massonnet and Meyssignac 2006), a good candidate for capture (SG344) as well as a time frame for the capture mission (2027-2029), were identified.

19.3 How to Push the Asteroid ?

Here we review which options are available to create the few tens of meters per second we allocate for changing the trajectory on a "reasonable" time frame. By reasonable, we mean similar to the time required to cruise form the initial position to the parking position. This could last a couple of years but certainly less than ten years, even if the operation is conducted before any threat is materialized.

We first discard solar sails or other devices based on the change of the albedo of the asteroid. These methods might only find an application for altering an earthbound trajectory over many years. if we consider an asteroid with a diameter of 10 m and with the density of liquid water, a favourable case of 10000 kg per square meter of cross section, submitted to 1500 W/m^2, also a favourable case as it is the power radiated by the Sun at the level of earth orbit, then the acceleration will never exceed a typical 10^{-9} m/s^2. The asteroid gains 0.3 m/s every 10 years, changing its semi-major axis by some 10^{-5} in relative terms, and in turn its period by about the same order of magnitude. The asteroid would then typically be offset by a few minutes a year on its trajectory, where a typical 200 second offset may suffice to make an earthbound asteroid on a highly elliptical orbit miss its target. However, a capture may require tens of m/s rather that a fraction of one m/s.

Any other method of pushing the asteroid relies on the well known rocket equation which states that the logarithm of the mass ratio (the final mass plus the propellant mass divided be the final mass) equals the velocity ratio (the final velocity of the dry mass divided by the ejection velocity of the propellant). In short $V_F=V_E \ln(M_0/M_F)$ where V_F is the final velocity, V_E is the ejection velocity, M_0 is the initial mass (propellant mass and final mass) and M_F the final mass. In the case of rockets the final mass is the sum of the structural mass and of the payload mass. If the rocket consists of several stages, the rocket equation applies to each stage.

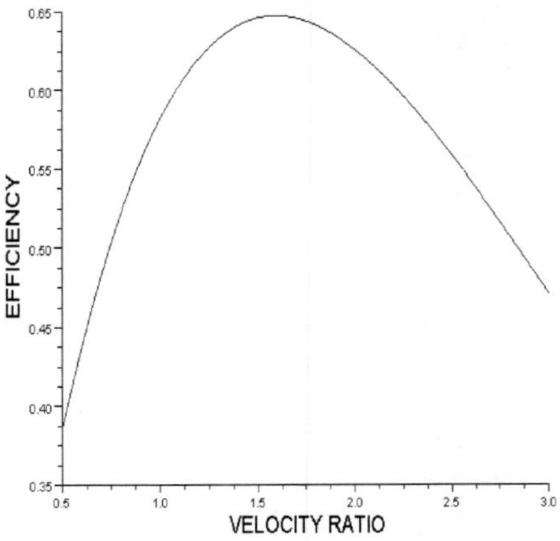

Fig. 19.1 Efficiency of the rocket equation as a function of the ratio of the achieved velocity on the ejection velocity. The efficiency is itself the ratio of the total energy spent in the fuel on the final kinetic energy of the spacecraft.

An interesting point to consider is the efficiency of the rocket propulsion, that is the ratio of the final energy (half the final mass times its final velocity squared) and of the expended energy (half the propellant mass times its ejection velocity squared). Figure 19.1 plots the efficiency as a function of the velocity ratio. There is clearly an optimum which corresponds to a velocity ratio of about 1.6 itself related to a mass ratio of about 5 (meaning that the mass of propellant is 4 times the final mass). At best, the efficiency almost reaches 65%. We observe that the optimum is not very sharp. The equation tells us that, from the point of view of efficiency alone, fireworks rockets designed to reach an altitude of 100 m, or a final velocity of less than 50 m/s, should better be pressurized water rockets with an ejection velocity of 30 m/s rather than gun powder rockets the efficiency of which is 2% in these conditions. In contrast, the best chemical propellants approach ejection velocities of 5 km/s, quite close to the optimum for Earth orbit injection, which requires at least 8 km/s. For very high velocity ratios, the rocket equation can be approximated: the propellant mass divided by the final mass equals the velocity ratio.

If the goal is to give an asteroid a ΔV of 50 m/s, then the most energy-efficient process is to propel mechanically the materials of the asteroid to ejection velocities in the order of a few tens of meter per second. In terms of mass to be brought from the earth, this method is also the best because the propellant is available in situ. If we consider an asteroid with a mass of 50 to 100kT, any scheme of chemical propulsion would require a propellant mass in the order of 1kT. 25 Saturn V lunar rockets, the more potent rocket successfully tested to date, would be required to lift this amount out of earth gravity field. Similarly, nuclear propulsion, albeit standing even farther to energetic optimum, provides plentiful energy. But with ejection velocities of 50 km/s, the required propellant mass would still be in the 50 to 100 T range, not counting the nuclear engine much more complex than the chemical one. We discard here the nuclear technologies based on nuclear blasts because an asteroid, mostly thought to have a granular structure, is not likely to withstand them. These are, however evoked later as a possible use of the asteroid's material. Another important point to consider is the flexibility of the propulsion scheme. If the mission is decided without any prior visit to the target asteroid, and unless it has satellites, its mass will only be known with a large uncertainty, possibly by a factor of two or more. The same factor should be applied to the propellant mass, whether chemical or nuclear. In the case of mechanical propulsion, only the duration of the operation must be optimized after the mass of the asteroid has been precisely assessed once the mission has started to push it.

Not considering cases where a tiny asteroid (less than 1kT) can be captured with a small ΔV, such as in (Brophy et al. 2012), there does not seem to be a practical alternative to mechanical propulsion for pushing small asteroids with this kind of ΔV. If we stick to the energy optimum however, 80% of the mass of the asteroid will be spent and dispersed in space at the end of the propulsion phase. While keeping close to the optimum, it seems wiser to favour the highest ejection velocities available, in order to minimize the loss of mass of the asteroid and the production of space debris, even if we are in orbit around the Sun, not around Earth. For the design of the spacecraft, we set a limit of 100 m/s as the maximum ejection velocity practically achievable with a robust mechanical system.

19.4 A Mechanical Rocket

The parameters of any mission can easily be deduced using the principles of the previous section. For instance, if a 100kT asteroid must be given a 50 m/s impulse in less than a year, we first compute the mass ratio using the rocket equation (with our preferred ejection velocity of 100 m/s) and we find it to be 1.65. We then have to propel 40kT to 100m/s and we will be left with the 60kT remaining after the operation. Now expelling 40kT at 100 m/s in a year timeframe means an average useful power of less than 6.5 kW. For comparison large telecommunication satellites use much larger amounts of power. In addition, placing these satellites on their final, circular geosynchronous orbit requires more ΔV than would sending them toward a near-earth asteroid. The difficulty clearly does not reside here. Expelling 40kT in a year means gathering material on the surface of the asteroid at a rate of 1.2 kg/s. Although any man can shovel this on the long run on earth, this is clearly a challenge in the extremely low gravity on a small asteroid the structure of which is basically unknown. There comes another key design point: the rate of launch sequences. In our example, we can achieve the average of 1.2 kg/s by launching one ton every 15 mn or 120 grams 10 times a second. The more we can launch in one stroke, the less we will have to take care of the selection of the pieces of "unbreakable" rocks, however, it will result in a much larger throwing device. At the other extreme, "sand-sucking" the surface of the asteroid might not generate enough material in total. A critical parameter is therefore the granularity of the asteroid. The granularity may tend to be larger in the smallest asteroids (Delbo et al. 2007, Mueller 2007, Murdoch 2011). It would then be safer to design the mechanical rocket with the capability to launch large pieces with a moderate launch rate. Further studies (Murdoch 2011) and simulations (Murdoch et al. 2012, Richardson et al. 2011, Schwartz et al. 2012), as well as in situ and ground based observations will improve our knowledge on this point. Given the large proportion of the asteroid to be used as propellant, the concept needs most of the mass of the asteroid to be granular in nature, not only superficial layers. Without going into a detailed design, one can show that the orders of magnitude of the mechanical parameters are favourable: Assuming the goal is to launch 100 kg of material at 100 m/s every two minutes or so, we consider a rail-shaped mast equipped with a cable capable of accelerating a basket containing the material over its length. We consider a maximum length of 10 meter, that can be obtained by unfolding once a 5 meter structure, itself compatible with current launchers fairing. The acceleration should therefore be on the order of 500 m/s². The resulting 50 kN force can be given by the cable the cross section of which could be no more than 50 mm² if made of steel. A 20 m cable surrounding the mast with pulleys out of conventional steel would have a mass of only 7 kg even if using steel is very conservative in terms of strength and density. The rest of the mass to be accelerated with the material is essentially the basket, that could take the shape of a metallic mesh. The later must be decelerated at the end of each launch. If we safely assume that the mass of the basket is 20% at most of the one of the ejected material, the deceleration can be at least 5 times stronger than acceleration for the same mechanical strain. The necessary length would also be 20% of the acceleration length. Disregarding the fact that the energy of the accompanying mass could be recovered, we understand that generally speaking the accompanying mass does not change the orders of magnitude set by the material alone. For storing the energy required for a launch (roughly 500kJ) which is to be delivered in 0.2 seconds, using ultra capacitors is a likely option. Finally, the impetus resulting from each launch is uncertain as the mass loaded in the basket might not be precisely known prior to launch. This is not really a problem as the resulting loss of efficiency is negligible, regardless of the chosen option among keeping a constant acceleration, keeping a constant cable tension or keeping a constant energy for each launch. The impetus brought be each launch is, anyway, very low, which removes any urgency in the control and monitoring. The typical ΔV gained by the asteroid following a 100 kg release is always much less than 1 mm/s.

Seemingly simpler systems such as inspired by the medieval trebuchet (Hill 1973), boosted by "slingshot effect", have the drawback of not being able to recover the energy left in the rope after launch. The resulting energy loss is a minor problem compared to the impossibility to dissipate this energy efficiently before the next launch. In addition, these devices are asymmetrical.

The initial rotation of the targeted asteroid might be a problem. In (Massonnet and Meyssignac 2006), we advocated a system capable of bouncing off the asteroid after grasping some material, which would then reach back to the asteroid after ejecting the material in the opposite direction. Such a system is quite insensitive to the rotation of the asteroid and capable of displacement on its surface. The shock of each landing might also be sufficient to overcome a limited strength of the surface and help gathering material. Whatever the concept selected, an asteroid cannot spin with a period of less than two hours (Harris 1996) unless it has a structural strength. Refinements (Holsapple 2001, Holsapple 2004) have lowered this value, which can be further reduced even with a very limited tensile strength (Holsapple 2007). By definition, a large tensile strength is hardly compatible with the mechanical rocket concept, so faster spinning bodies should be discarded as targets. Radar and optical measurements may determine the period of rotation from the ground. Given the size of the body we aim at (typically 40 m in diameter), the equatorial velocity associated with this rotation is less than 2 cm/s. The spin energy is at least three orders of magnitude below the energy involved in the capture. Since the superficial layers of the asteroid are removed, the capture does not necessarily change the spinning rate of the asteroid, but such a change can be programmed in the launch sequences.

19.5 What Is the Interest of Capturing an Asteroid?

In (Massonnet and Meyssignac 2006), our main purpose for capturing an asteroid was the prospect of using it against an incoming, threatening body. We presented it as a "short term" solution as opposed to methods requiring a long preparation such as gravitational tugging (Lu and Love 2005). The actual piloting law used to redirect the captured asteroid into the incoming threat starting from the arbitrary halo orbit achieved after the capture remains to be found. It has to take into account the limitations of the propulsion system in terms a acceleration. Nevertheless it is a purely theoretical work that does not require any further technological development (Meyssignac 2012). Here we focus on other potential uses for a captured asteroid which would all demand an additional technological development.

Mining an asteroid is a quite natural idea. However, it is less attractive after a closer look. As the primordial bricks of the solar system, asteroids are not likely to present the unusual concentration of rare materials that can occur in privileged places on Earth through tectonic forces and the action of underground water. If the amount of, for instance, platinum in an asteroid corresponds to the primeval level of this element in the solar system, which is likely, then its total amount in an asteroid will be extremely low. It is therefore not interesting to capture an asteroid, necessarily small, for the purpose of mining. In addition, high value/low mass element extracted from an asteroid can easily be transported from the initial orbit of the asteroid using conventional means. If the mining requires a human intervention, one can argue that the Sun-Earth L1/L2 Lagrange point is easier to reach for a crew (with a possible one-month one-way trip) than any other point along the earth orbit that could take years to reach. Similarly if the mining requires remote operations, one can argue that the 5 second radio delay of the Lagrange vicinity is better than the few minutes typical delay of interplanetary communications in the vicinity of the sun. These points are nonetheless irrelevant as we cannot tug large asteroids, the most likely targets for mining, if any.

An asteroid capture may contribute to missions aimed at exploring the solar system. If not using nuclear technology, such missions require the launch of a large mass, mainly made of propellants, into trajectories whose first step is the liberation from earth gravity. The idea of producing part of this propellant mass in situ is very attractive. The Moon might be a site for production because the cost of putting Moon-produced material into Earth liberation trajectories (ELT) corresponds to a ΔV of roughly 2.8 km/s against 11.2 km/s in the case of Earth. Similarly, the surface of Mars or the surface of Phobos can be envisioned as in situ sources of propellant or other materials. However, even the moderate ΔV required to lift the material from the Moon into an ELT is a serious limitation to the usefulness of the process as we have shown in (Massonnet and Meyssignac 2006), even disregarding the effort necessary to land the "chemical factory" on the Moon. Therefore the advantage of the process is in the order of a factor two at best compared to lifting everything from earth. The same applies for materials aimed at other purposes, such as mass used for shielding astronauts against cosmic radiations during a long journey.

An asteroid placed on a Lagrange parking orbit is virtually on an ELT. If a proper facility is placed in the vicinity of the asteroid, it could produce a large mass of liquid oxygen within a few years time frame assuming reasonably that a large fraction of the asteroid is made of oxides. This oxygen could then be loaded into the tanks of a planetary mission which would first reach an ELT toward this Lagrange gas station and then come back close to the Earth, still on an ELT, in the proper position for its final (and modest) acceleration to a planetary trajectory. If a chemical propulsion is used for the exploration of Mars or any other planet, up to 80% of the mass of the mission could be liquid oxygen, thus offering a similar gain in heavy lifting requirements. The two-month or so detour to the asteroid does not necessarily apply to the exploration crew which could join the spacecraft when it grazes Earth again on its way back. If necessary, a different crew could manage the refill and enter the earth atmosphere directly on the way back from the asteroid. The cost in ΔV would virtually be zero for the massive interplanetary spacecraft, while the crews will have to reach the liberation velocity for rendezvous, but with a small capsule. In the vicinity of Mars, a similar parking scheme using the Mars-Sun Lagrange point could avoid the cost in ΔV for the largest part of the spacecraft. Although more convincing than mining, this use of a captured asteroid raises a number of hurdles. First, one must be sure that oxygen is available in sufficient proportion in the captured asteroid. The chemical composition of asteroids is not accurately known. Second, the chemical form of this oxygen must be known for a proper design of the chemical facility aimed at turning it into liquid oxygen. Third, the mass of this facility, which must be brought from Earth, must be orders of magnitude lower than its production, making the facility design a challenge.

Obviously the prospects of using a captured asteroid are all the better as the need for transformation from the raw material is minimized. In this category we can think of two applications.

1. First the bulk mass could be used as a radiation shield for deep space manned missions. The cost of adding a large mass to a spacecraft already almost free from Earth gravity might be acceptable as only a small ΔV is required for insertion into a planetary trajectory while grazing the Earth. The Sun-Mars Lagrange points for instance, could be reached using a similar strategy in reverse. We assume the planet would then be reached using unprotected shuttles. This use is simple enough, but if it is not associated to the production of chemical fuel, it might seem a minor improvement in mass not justifying its added complexity.

2. Second the bulk mass could be used together with nuclear propulsion schemes. One metric ton of uranium contains the energy of 7 million tons of O2/H2 fuel within a volume of 50 litre. In theory a nuclear reactor can provide the energy while the bulk material provides the mass either using a power plant (McAndrews et al. 2003) associated with any kind of electromagnetic acceleration (yet to be demonstrated…) or relooking the early concept developed by (Dyson 2002) of using nuclear explosions focussed on a parabolic surface. As strange as it sounds, this use of a captured asteroid appears more likely than any other if we except the interception mission. The bulk material is used to boost the impetus resulting from the explosions themselves by adding neutral mass around the devices. There are no new technological developments as the building of nuclear bombs is unfortunately very mature in several countries. The 30000 km/s average velocity of the post-reaction nuclear material is reduced to much less than 10000 km/s by energy sharing with the bomb casing alone. Placing bulk material around the device increases the energy sharing and further reduces the velocities to less than 1000 km/s, in which case the mass of the bombs is only 1% of the spacecraft mass, most of the rest being made of the asteroid material. Nevertheless the achieved specific impulse is very high. The mission scheme for such a spacecraft would certainly avoid detonations in low Earth orbit and it should thus be operated directly from the Lagrange point. It could however, be used in low orbit around Mars, which unlike Earth has a weak magnetic field and no radio hams on its surface ! Various studies, including some conducted during nuclear tests, suggest that such a vessel could be used for several missions (Dyson 2002).

19.6 Conclusion

Capturing an asteroid is interesting only if a large fraction of its remaining mass is useful, which discards the idea of asteroid mining, among other reasons. Using the asteroid itself as a shield against a threatening celestial body is the most straightforward use, provided the delicate piloting and phasing scheme for interception is demonstrated by appropriate theoretical studies. Using its bulk material as “mass boosting" in association with pulsed nuclear propulsion can be seriously envisioned as few additional technologies are required, and some of them have been tested on the ground. The production of liquid oxygen for ambitious planetary space missions can be envisioned assuming a large fraction of the asteroid mass is made by this element. In this case, the "chemical factory" which must be brought to the asteroid and put to work in the long run is the challenge, even if it is easier on an asteroid than on any other place, including the Moon.

Whatever the use planned for a captured asteroid, it seems paradoxically easier to capture the asteroid than to make use of it. The capture by one or several mechanical rockets requires medium size spacecrafts in terms of mass and energy production capabilities, completed by significant progress in autonomous operations (selecting the material to be launched at a relatively high rate) compared to what has been done so far. In contrast, even the nuclear propulsion scheme discussed above involves launching much larger masses and developing long duration space flights. Capturing asteroids would be a major trump in this direction, but likely a detail in terms of investments.

References

Brophy, J., Culik, F., Friedman, L., et al.: Asteroid Retrieval Feasibility Study. Keck Institute for Space Studies, Pasadena (2012)

Delbo, M., dell'Oro, A., Harris, A.W., Mottola, S., Mueller, M.: Thermal inertia of near-Earth asteroids and implications for the magnitude of the Yarkovsky effect. Icarus 190, 236–249 (2007)

Dyson, G.: Project Orion: The True Story of the Atomic Spaceship. Henry Holt and Company, New York (2002)

Harris, A.W.: The Rotation Rates of Very Small Asteroids: Evidence for 'Rubble Pile' Structure. In: Lunar and Planetary Institute Conference Abstracts, vol. 27, p. 493 (1996)

Hill, D.: Trebuchets. Viator 4, 99–115 (1973)

Holsapple, K.A.: Equilibrium Configurations of Solid Ellipsoidal Cohesionless Bodies. Icarus 154, 432–448 (2001)

Holsapple, K.A.: Equilibrium Figures of Spinning Bodies with Self-Gravity. Icarus 172, 272–303 (2004)

Holsapple, K.A.: Spin limits of Solar System bodies: From the small fast-rotators to 2003 EL61. Icarus 187, 500–509 (2007)

Lu, E.T., Love, S.G.: Gravitational tractor for towing asteroids. Nature 438, 177–178 (2005)

McAndrews, H., Baker, A., Bidault, C., Bond, R., Browning, D., Fearn, D., Jameson, P., Roux, J.-P., Sweet, D., Weston, M.: Future power systems for space exploration. ESA-Report 14565/NL/WK, QinetiQ (2003)

Massonnet, D., Meyssignac, B.: A captured asteroid: Our David's stone for shielding Earth and providing the cheapest extraterrestrial material. Acta Astronautica 59, 77–83 (2006)

Meyssignac, B.: Personal communication (2012)

Mueller, M.: Surface Properties of Asteroids from Mid-Infrared Observations and Thermophysical. Ph.D. Thesis, Freie Universitat, Berlin (2007)

Murdoch, N.: Modelisation of granular material dynamics at the surface of asteroids and under various gravitational conditions (Mars, the Moon). PhD Thesis, Planetary and Space Sciences Research Institute, The Open University and the Côte d'Azur Observatory, The University of Nice, Sophia Antipolis, France (2011)

Murdoch, N., Michel, P., Richardson, D.C., Nordstrom, K., Berardi, C.R., Green, S.F., Losert, W.: Numerical simulations of granular dynamics II: Particle dynamics in a shaken granular material. Icarus 219(1), 231–335 (2012)

Richardson, D.C., Walsh, K.J., Murdoch, N., Michel, P.: Numerical simulations of granular dynamics I: hard sphere discrete element method and tests. Icarus 212, 427–437 (2011)

Schwartz, S., Richardson, D., Michel, P.: An implementation of the soft-sphere discrete element method in a high-performance parallel gravity tree-code. Granular Matter 14, 363–380 (2012)

Chapter 20
Change the Asteroid Trajectory

Alexander A. Bolonkin

C&R Co., New York, USA

20.1 Introduction

The number of asteroids in the Solar System is very large (Friedman and Tantardini 2012). The vast majority are found in a swarm called the asteroid belt, located between the orbits of Mars and Jupiter at an average distance of 2.1 to 3.3 astronomical units (AU) from the Sun. Scientists know of approximately 6,000 large asteroids of a diameter of 1 kilometer or more, and of millions of small asteroids with a diameter of 3 meters or more. Radar observations enable to discern of asteroids by measuring the distribution of echo power in time delay (range) and Doppler frequency. They allow a determination of the asteroid trajectory and spin and the creation of an asteroid image. For delivery asteroid to the Earth author considers theory of three main methods: impact of the space apparatus to asteroid, explosion the conventional explosive on asteroid surface having form of plate and ball, explosion the small nuclear bomb on the asteroids surface, braking asteroid by parachute in Earth atmosphere.

20.2 Kinetic Impact

The impact of a massive object, such as a spacecraft or another near-Earth object, is one possible solution to change the trajectory of the Near Earth asteroid or Object (NEO). Another object (for example, space apparatus) with a high mass close to the Earth could be forced into a collision with an asteroid, knocking it off course.

When the asteroid is still far from the Earth, a mean of deflecting the asteroid to Earth is to directly alter its momentum by colliding a spacecraft with the asteroid.

The European Space Agency is already studying the preliminary design of a space mission able to demonstrate this futuristic technology. The mission, named Don Quijote, is the first real asteroid deflection mission ever designed.

In the case of 99942 Apophis it has been demonstrated by ESA's Advanced Concepts Team that deflection could be achieved by sending a simple spacecraft weighing less than one ton to impact against the asteroid. During a trade-off study one of the leading researchers argued that a strategy called 'kinetic impactor deflection' was more efficient than others.

20.3 Nuclear Bomb for Deflection of Asteroid

Detonating a nuclear explosion above the surface (or on the surface or beneath it) of an NEO would be one option, with the blast vaporizing part of the surface of the object and nudging it off course with the reaction. This is a form of nuclear pulse propulsion. Even if not completely vaporized, the resulting reduction of mass from the blast combined with the radiation blast and rocket exhaust effect from eject could produce positive results.

Another proposed solution is to detonate a series of smaller nuclear bombs alongside the asteroid, far enough away as not to fracture the object. Providing this was done far enough in advance, the relatively small forces from any number of nuclear blasts could be enough to alter the object's trajectory enough to avoid an impact. The 1964 book Islands in Space, calculates that the nuclear megatonnage necessary for several deflection scenarios exists. In 1967, graduate students under Professor Paul Sandorff at the Massachusetts Institute of Technology designed a system using rockets and nuclear explosions to prevent a hypothetical impact on Earth by the asteroid 1566 Icarus. This design study was later published as Project Icarus which served as the inspiration for the 1979 film Meteor.

20.4 Theory of the Asteroids Movement and Changing Trajectory

In Table 20.1 are computed the mass M of the ball asteroid, his energy E for speed V = 16 km/s and explosive power P of asteroids. One ton TNT has 4.184×10^9 J.

The Hiroshima nuclear bomb had power about 15 kilotons of TNT explosive. The small ball asteroid having diameter 10 m has energy in 4 times more for speed 16 km/s.

Table 20.1 Diameter D, mass M of ball asteroid having density 3500 kg/m³, energy E for speed V = 16 km/s and explosive power P of asteroids

D, m	10 m	30 m	100 m	300 m	1 km	3 km	10 km	30 km
M, kg	$1.83\cdot10^6$	$16.5\cdot10^6$	$1.83\cdot10^9$	$16.5\cdot10^9$	$1.83\cdot10^{12}$	$16\cdot10^{12}$	$1.83\cdot10^{15}$	$16.5\cdot10^{15}$
E, J	$2.34\cdot10^{14}$	$21.1\cdot10^{14}$	$2.34\cdot10^{17}$	$21.1\cdot10^{17}$	$2.34\cdot10^{20}$	$21.1\cdot10^{20}$	$2.34\cdot10^{23}$	$21.1\cdot10^{23}$
P, ton	$0.56\cdot10^5$	$5.11\cdot10^5$	$0.56\cdot10^8$	$5.11\cdot10^8$	$0.56\cdot10^{12}$	$5.11\cdot10^{11}$	$0.56\cdot10^{14}$	$5.11\cdot10^{14}$

20.4.1 Equations for Computation of Trajectory in Vacuum Space Near Earth

These equations are following:

$$r=\frac{p}{1+e\cos\beta},\quad p=\frac{c^2}{K},\quad e=\frac{c}{K}\sqrt{H+\frac{K^2}{c^2}},\quad c=v^2r^2\cos^2\nu=const,$$

$$H=2K\frac{M}{R}=const,\quad K=3.98\cdot10^{14}\ \frac{m^3}{s^2},\quad r_a=\frac{p}{1-e},\quad r_p=\frac{p}{1+e},\tag{20.1}$$

$$T=\frac{2\pi}{\sqrt{K}}a^{3/2},\quad a=r_a,\quad b=r_p,\quad b=a\sqrt{1-e^2},$$

where r is radius from Earth center to point in trajectory, m; p is ellipse parament, m; e is ellipse eccentricity, e = 0 for circle trajectory, e < 1 for ellipse, e = 1 for parabola, e > 1 for hyperbola; β is angle from perigee, K is Earth constant, v is speed, m/s; ν is angle between speed and tangent to circle; M = $5.976\cdot10^{24}$ kg is mass of Earth; R = 6378 km is Earth radius; r_a is apogee, m; r_p is perigee, m; b is small semi axis of ellipse, m; a is small semi axis of ellipse, m; T is period of rotation, sec.

20.4.2 Change Asteroid Trajectory by Impact of Space Apparatus

Inelastic head-on collision space apparatus (SA) in the asteroid (As):

$$W=\frac{1}{2}m_1V_1^2+\frac{1}{2}m_2V_2^2,\quad Q=\frac{m_1m_2V_1^2}{2(m_1+m_2)},\quad \eta=\frac{W-Q}{W}=\frac{m_1}{m_1+m_2},\tag{20.2}$$

where W is energy of system, J; Q is heat loss in impact, J; m_1 is mass of space apparatus, kg; m_2 is mass of asteroid, kg; V_1 is speed of SA about center mass of the system asteroid-SA, m/s; V_2 is speed of asteroid about center mass of system asteroid-SA, η is coefficient of efficiency.

Let us place the origin at the center of gravity of an asteroid. The speed of system asteroid-SA will be

$$\Delta V=V\left[\frac{m_1}{m_1+m_2}\left(1-\frac{m_2}{m_1+m_2}\right)\right]^{0.5},\quad \Delta I=(m_1+m_2)\Delta V,\tag{20.3}$$

where ΔV is change of asteroid speed, m/s; V is SA speed relative asteroid, m/s; ΔI is additional impulse of system As+SA.

Example. Let us take the asteroid having diameter 10 m (m_2 = 1830 tons) and SA having mass m_1= 10 tons and speed about asteroid V = 1 km/s. From Eqs. (20.2)-(20.3) we find ΔV = 5.43 m/s, η = 0.00543.

20.4.3 *Change Trajectory by Conventional Plate Explosive Located on the Asteroid Surface*

In this case we get the impulse from the explosive gas. The maximal speed of an explosion gas and asteroid speed received from explosion are

$$V_1 = \sqrt{2q}\,, \quad V_2 = V_1 \frac{m_1}{m_2}\,, \tag{20.4}$$

where V_1 is speed of explosion gas, m/s; q is specific energy of the explosive, J/kg ($q \approx 5.4$ MJ/kg for TNT), V_2 is asteroid speed received from explosion, m/s; m_1 is mass of explosive, kg; m_2 is mass of asteroid, kg.

Example. Let us take the asteroid having diameter 10 m (m_2 = 1830 tons) and explosive having mass m_1= 10 tons and specific energy of the explosive $q \approx 4.2$ MJ/kg. From Eq. (20.4) we find the change of speed of asteroids $V_2 = \Delta V = 15.8$ m/s.

If explosive is not plate (not optimum) and located in one point (ball) on the asteroid surface, the effect from the explosion will be less. Maximum speed is $\pi/4$ = 0.785 from the plate explosion speed: $V_2 = \Delta V = 15.8 \times 0.785 = 12.4$ m/s.

20.4.4 *Nuclear Point Explosion on the Asteroid Surface*

In this case the asteroid gets the impulse from evaporation part of asteroid. The asteroid rest can get the significant speed. If the energy of the nuclear bomb is E, bomb is located on asteroid surface, change the asteroid speed may be estimated by next equations

$$V_1 = \sqrt{\lambda}\,, \quad m_1 = \frac{E}{2\lambda}\,, \quad v = \frac{m_1}{\rho}\,, \quad r^3 = \frac{3v}{2\pi}\,, \quad I = m_1 V_1\,, \quad \Delta V = \frac{I}{m_2 - m_1}\,, \tag{20.5}$$

where V_1 is speed of evaporation gas, m/s; λ is specific energy of the asteroid evaporation, J/kg (heating + melting + heating + evaporation), v is the volume of a sold evaporation mass, m³; ρ is the asteroid density kg/ m³; I is impulse, kg m/s; ΔV is change of the asteroid speed received from nuclear explosion, m/s; m_1 is the asteroid evaporation mass in explosion, kg; m_2 is initial mass of asteroid, kg; r is radius of explosion cavity, m.

For basalt the λ = heating + evaporation = 1191 + 3500 = 4691 kJ/kg, ρ = 3500 kg/ m³. For iron $\lambda \approx 8200$ kJ/kg, ρ = 7900 kg/ m³; for ice $\lambda \approx 3000$ kJ/kg, ρ = 1000 kg/ m³.

Example. Let us take the iron asteroid having diameter 10 m (m_2 = 1830 tons) and energy of a small nuclear bomb is E = 1 kton = $4.2 \cdot 10^{12}$ J. From Eq. (20.4) we find V_1 = 2863 m/s; m_1=256 tons, the change of speed of asteroids $V_2 = \Delta V$ = 460 m/s.

20.4.5 Computation of the Asteroid Trajectory

Equations for computation of the asteroid trajectory are:

$$\dot{r} = \frac{R_0}{R} V \cos\theta,$$
$$\dot{H} = V \sin\theta,$$
$$\dot{V} = -\frac{D + D_P}{m} - g \sin\theta, \tag{20.6}$$
$$\dot{\theta} = \frac{L + L_P}{mV} - \frac{g}{V}\cos\theta + \frac{V\cos\theta}{R} + 2\omega_E \cos\varphi_E,$$

where r is range of ship flight, m; R_0 = 6,378,000 is radius of Earth, m; R is radius of ship flight from Earth's center, m; V is ship speed, m/s; H is ship altitude, m; θ is trajectory angle, radians; D is system drag (asteroid+apparatus), N; D_P is parachute drag, N; m is system mass, kg; g is gravity at altitude H, m/s^2; L is apparatus lift force, N; L_P is parachute lift force, N; ω_E is angle Earth speed; $\varphi_E = 0$ is lesser angle between perpendicular to flight plate and Earth polar axis; t is flight time, sec.

The quantities in Eq. (20.6) compute as:

$$g = g_0\left(\frac{R_0}{R_0 + H}\right)^2, \quad \rho = a_1 e^{(H-10000)/b}, \quad a_1 = 0.414, \quad b = 6719,$$
$$Q = \frac{0.5 \cdot 11040 \cdot 10^4}{R_n^{0.5}}\left(\frac{\rho}{\rho_{SL}}\right)^{0.5}\left(\frac{V}{V_{CO}}\right)^{3.15}, \quad R_n = \sqrt{\frac{S_P}{\pi}},$$
$$T_1 = 100\left(\frac{Q}{\varepsilon C_S} + \left(\frac{T_2}{100}\right)^4\right)^{1/4}, \quad T = T_1 - 273, \tag{20.7}$$
$$D_P = 0.5 C_{DP}\rho a V S_P, \quad L_P \approx (1 \div 4) D_P, \quad L = 2\alpha\rho a V S, \quad D = L/4,$$
$$\Delta V \approx \frac{0.5 C_{DP}\rho a S_p L}{m},$$

where: g_0 = 9.81 m/s^2 is gravity at Earth surface; ρ is air density, kg/m^3; Q is heat flow in 1 m^2/s of parachute, J/s·m^2; R_n (or R_p) is parachute radius, m; S_P (or S_m) is parachute area, m^2; ρ_{SL}= 1.225 kg/m^3 is air density at sea level; V_{CO} = 7950 m/s is circle orbit speed; T_1 is temperature of parachute in stagnation point in Kelvin, K; T is temperature of parachute in stagnation point in centigrade, °C; T_2 is temperature of the standard atmosphere at given altitude, K (T_2 = 253 °K at H = 60 km) ; D_P is parachute drag, N.; L_P is parachute lift force That is control from 0 to 4 D_p, N (the ram-air parachute can produce lift force up 1/3 from its drag); D is ship drag, N; L is ship lift force, N; C_{DP} = 1 is parachute drag coefficient; a = 295 m/s is sound speed at high altitude; $\alpha = 40° = 0.7$ rad is apparatus attack angle. C_S = 5.67 W/($m^2 \cdot K^4$) is coefficient radiation of black body; ε is coefficient of a black ($\varepsilon \approx 0.03 \div 0.99$), ΔV is loss of speed in atmosphere on distance L.

The control is following: if T_1 is more the given safety temperature than the lift force L_P = maximum = $4D_p$. In other case $L_p = 0$. If T_1 is less the given safety temperature than the lift force L_P = negative minimum = $-4D_p$. When the speed is less the sound speed, the control parachute is also used for deliver in given point.

The requested parachute area may be found by equations in lending study at sea level:

$$L_p = C_L \frac{\rho V^2}{2} S_p , \quad D_p = C_D \frac{\rho V^2}{2} S_p , \quad K = \frac{C_L}{C_D}, \quad V_v = \frac{V}{K}, \quad V_v \leq V, \qquad (20.8)$$

where C_L is lift coefficient of parachute, $C_L \approx 2 \div 3$; C_D is drag coefficient of parachute , $C_D \approx 0.5 \div 1.2$; $\rho = 1.225$ kg/m^3 is air density; V is speed system, m/s; S_p is parachute area, m^2; K is ratio C_L/C_D; V_v is vertical speed, m/s.

Example. Let us take the mass of system (asteroid + apparatus) 100 tons = 10^6 N, C_L = 2.5, safety V_v = 20 m/s, K = 4, V = 80 m/s. From Eq. (20.8) we receive the parachute aria is S_p = 100 m^2. The control rectangle parachute is 5.8 x 17.3 m. Result of computation the Eq. (20.7) are presented in Figs. 20.1-20.3.

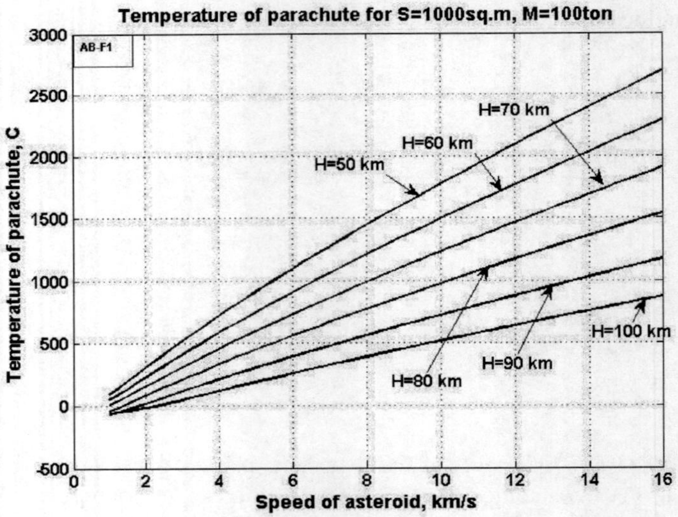

Fig. 20.1 Parachute temperature versus the entrance speed and altitude for mass of asteroid 100 tons and parachute area 1000 m²

Figure 20.1 allows to select the entrance altitude for given safety parachute temperature. For example, asteroid has the entrance speed (perigee speed) 10 km/s, safety parachute temperature is 1000 C. Figure 20.1 gives the perigee altitude 80 km.

For altitude 80 km and distance L = 6378 km the loss is about 150 m/s. The parachute can keep this altitude by lift force to down. In this case the asteroid losses about 2 km/s in during two revolutions around Earth. This allows decreasing the safety altitude up 70 km and increase the speed loss up 1 km on distance L. Control parachute alloys by lift force up and down to decrease speed and to lend the asteroid in need point of Earth surface. The overload in this method is small. One is presented in Fig. 20.3.

20.5 Conclusion

For delivery asteroid to Earth we need in methods for changing the asteroid trajectory and theory for an estimation or computation the impulse which produces these methods. Author develops some methods of this computation. There are: impact of the space apparatus to asteroid, explosion the conventional explosive on asteroid surface having form of plate and ball, explosion the small nuclear bomb on the asteroids surface and braking the asteroid by control parachute in Earth atmosphere.

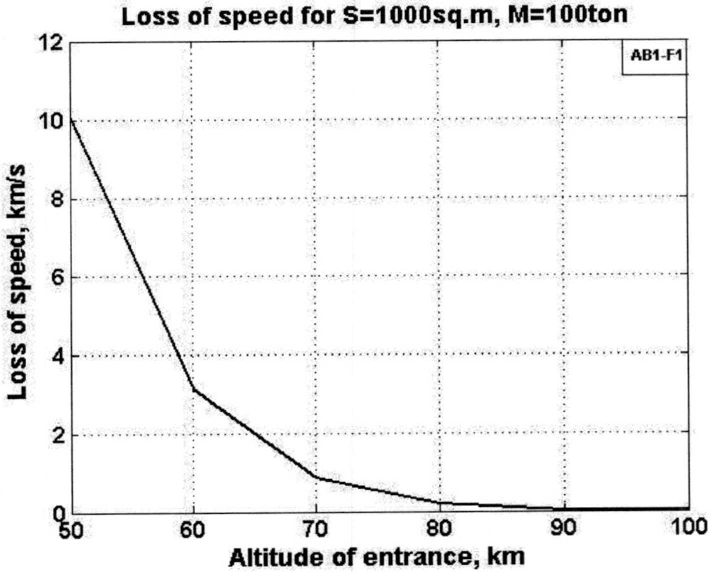

Fig. 20.2 Loss of speed via altitude for distance L = 6378 km (radius of the Earth) for mass of asteroid 100 ton and parachute area 1000 m^2

That method is cheaper than conventional method: flight to asteroid, braking the apparatus to asteroid speed (spending of fuel), braking the asteroid for decreasing of Earth perigee (up to Earth atmosphere)(spending of fuel), non parachute entrée in Earth atmosphere, high heating, destroying of asteroid in atmosphere, non-control flight in atmosphere, powerful impact to Earth surface, possible destructions and earthquake. Delivery of asteroid remains to a plant.

The small control parachute allows multiple using the Earth atmosphere for the braking the asteroid without high heating, deliver the asteroid in given point of Earth and to avoid the asteroid impact to Earth.

The delivery of the metallic asteroid to Earth will be profitable if we dramatic decreases the cost of the space launch (up to 3 – 10 $/kg) as it is offered in (Bolonkin 2005a,b, 2006b, 2011; Bolonkin and Cathcart, 2007). In present time we are spending 200 – 300M of USD for delivery a very small piece of asteroid for scientific purpose. Using the offered method we can deliver the full asteroid (up 3 – 50 tons) to Earth.

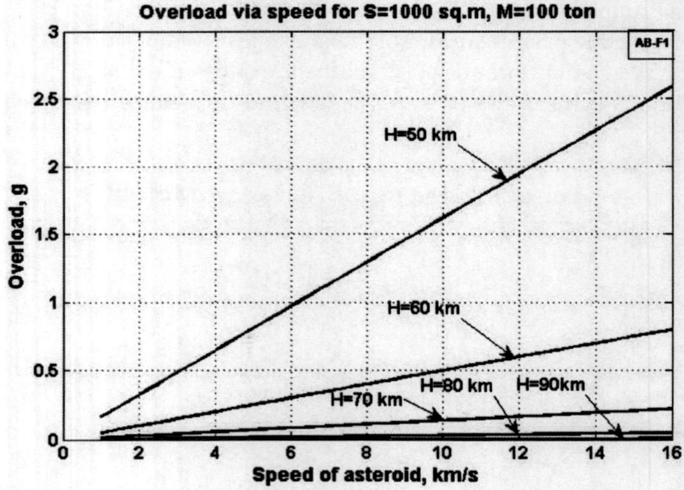

Fig. 20.3 Overload via asteroid speed for different altitudes

If asteroids will contain the very precious metals, their delivery may be profitable. Offered method may be also used for braking of apparatus reentering in the Earth from a space flight. The reader finds useful information about delivery methods also in (Bolonkin 2005a,b, 2006a,b, 2011; Bolonkin and Cathcart 2007).

References

Bolonkin, A.A.: Asteroids as propulsion system of space ship. Journal of The British Interplanetary Society 56, 98–107 (2005a)
Bolonkin, A.A.: Non-Rocket Space Launch and Flight, ch. 11. Elsevier, New York (2005b), http://www.archive.org/details/Non-rocketSpaceLaunchAndFlight, http://www.scribd.com/doc/24056182
Bolonkin, A.A.: A New Method of Atmospheric Reentry for Space Ships. Presented as Paper AIAA- 2006-6985 in Multidisciplinary Analyses and Optimization Conference, September 6-8. Fortsmouth, Virginia (2006a)
Bolonkin, A.A.: New Concepts, Ideas, Innovations in Aerospace. In: Technology and the Human Sciences, 510 p. NOVA, New York (2006b), http://www.scribd.com/doc/24057071, http://www.archive.org/details/NewConceptsIfeasAndInnovationsInAerospaceTechnologyAndHumanSciences
Bolonkin, A.A.: Femtotechnologies and Revolutionary Projects, 538 p. Scribd, USA (2011), http://www.scribd.com/doc/75519828/, http://www.archive.org/details/FemtotechnologiesAndRevolutionaryProjects
Bolonkin, A.A., Cathcart, R.B.: Macro-Projects: Environments and Technologies, 536 p. NOVA, New York (2007), http://www.scribd.com/doc/24057930, http://www.archive.org/details/Macro-projectsEnvironmentsAndTechnologies
Friedman, L., Tantardini, M.: Asteroid Retrieval Feasibility. ESA ESTEC (March 14, 2012), http://www.kiss.caltech.edu/study/asteroid/20120314_ESA_ESTEC.pdf

Chapter 21
Opportunities for Asteroid Retrieval Missions

Daniel García Yárnoz, Joan-Pau Sanchez, and Colin R. McInnes

University of Strathclyde, UK

21.1 Introduction

Recently, significant interest has been devoted to the understanding of minor bodies of the Solar System, including near-Earth and main belt asteroids and comets. NASA, ESA and JAXA have conceived a series of missions to obtain data from such bodies, having in mind that their characterisation not only provides a deeper insight into the Solar System, but also represents a technological challenge for space exploration. Near Earth Objects in particular have also stepped into prominence because of two important aspects: they are among the easiest celestial bodies to reach from the Earth and they may represent a potential impact threat. This increased interest has encouraged the research community to propose further asteroid engineering projects, such as NEO retrieval missions, taking advantage of the synergies with current minor bodies search campaigns and asteroid manipulation technology development initiatives.

Various space macro-engineering projects have as a primary requirement the capture or shepherding of a portion or a full asteroid in useful orbits in the solar system (see Table 21.1).

Early proposals for the space elevator involved the capture of a small body in an orbit close to GEO to serve as counterweight. The size of the counterweight required depends on the radius of the orbit where the asteroid would be placed, with size decreasing exponentially with altitude above GEO, resulting in about a 50 ton asteroid (about 3.3 m diameter) for a counterweight at a circular orbit of radius 100000 km.

The use of captured asteroids has also been proposed for geo-engineering purposes as means of reducing the solar insolation on Earth by generating dust rings or clouds. Depending on the position of the dust cloud, either Earth ring (Pearson et al. 2006), Sun-Earth L_1 (Bewick et al. 2012), or Earth-Moon L_4/L_5 region (Struck, 2007), and the desired reduction in insolation, the asteroid mass requirements and the complexity and cost of the capture transfer vary, but the minimum size for a target asteroid is never below 500 meters diameter. Retrieving objects of this size is probably beyond nowadays technological capabilities. The required mass could be reduced by more than one order of magnitude by the use of space manufactured solar reflectors instead of dust rings or clouds (Pearson et al. 2006), but that would involve a large manufacturing infrastructure on orbit.

Much smaller asteroids can already be of interest for resources exploitation. The in-situ utilisation of resources in space has long been suggested as the means of lowering the cost of space missions, by, for example, providing bulk mass for radiation shielding or distilling rocket propellant for interplanetary transfers (Lewis, 1996). Although the concept of asteroid mining dates back to the very first rocketry pioneers (Tsiolkovsky, 1903), evidences of a renewed interest can be found in the growing body of literature on the topic (Baoyin et al. 2010, Sanchez and McInnes, 2011a, Hasnain et al. 2012), as well as in high profile private enterprise announcements such as by Planetary Resources Inc [http://www.planetaryresources.com/]. A recent asteroid retrieval mission study (Brophy et al. 2012) proposed the capture of a 2-4 m diameter asteroid around the Moon with current technologies, which could serve as test-bed for the development of technologies for in-situ resource utilisation (ISRU). Other proposals (Massonnet and Meyssignac, 2006) suggest a larger asteroid to be used as a NEO shield in combination with resource exploitation. These technologies could become a potentially disruptive innovation for space exploration and utilisation and enable large-scale space ventures that could today be considered far-fetched, such as large space solar power satellites or sustaining communities in space. Fuel depots or permanent space stations that use a small asteroid as a base can be envisaged in the near future. The use of asteroids in cycler orbits to provide structural support and radiation shielding for interplanetary transfers can be considered a more futuristic enterprise also found in literature (Lewis, 1996).

Table 21.1 Macro-engineering projects proposing asteroid capture. To estimate asteroid sizes, given in diameter, an average NEO density of 2.6 gr/cm^3 (Chesley et al. 2002) was used when neccesary.

Project	Target orbit(s)	Size or mass required	Reference
Space Elevator	~GEO	52 x 10^3 kg (> 3.3 m)	(Aravind, 2007)
Geo-engineering: Dust ring LEO	~LEO	2.3 x 10^{12} kg (> 1190 m)	(Pearson et al. 2006)
Geo-engineering: Dust cloud L_1	Sun-Earth L_1	1.9 x 10^{11} kg (> 515 m)	(Bewick et al. 2012)
Geo-engineering: Dust cloud $L_{4/5}$	Earth-Moon L_4/L_5	2.1 x 10^{14} kg (> 5.3 km)	(Struck, 2007)
Tech. demo ISRU/Fuel depot	L_1, L_2, Moon orbit…	> 2 m	(Brophy et al. 2012)
Space station	L_1, L_2, L_4, L_5	> 10 m	
NEO shield	Sun-Earth L_1, L_2	20-40 m	(Massonnet and Meyssignac, 2006)
Cyclers	Earth-Mars resonant orbit	> 100 m	(Lewis, 1996)

In all these studies, the Sun-Earth Lagrangian points repeatedly appear as one of the preferred destinations for captured asteroids. This is relevant as they can also serve as natural gateways to other destinations in the Earth-Moon system through the use of heteroclinic connections (Koon et al. 2000). There are however several main challenges to overcome: the development of techniques to modify asteroid trajectories, the improvement in the minor body census in order to find the most suitable candidates of the appropriate size, and the design of low-cost transfers to the desired final orbits.

Current technologies and methods for deflection of Earth-impacting objects have experienced significant advances, along with increasing knowledge of the asteroid population. While initially devised to mitigate the hazard posed by global impact threats, the current impact risk is largely posed by the population of small undiscovered objects (Shapiro et al. 2010), and thus methods have been discussed to provide subtle orbital changes to these small objects, as opposed to large-scale interventions, e.g. the use of nuclear devices (Kleiman 1968). This latter batch of deflection methods, such as low thrust tugboat (Scheeres and Schweickart 2004), gravity tractor (Edward and Stanley 2005) or small kinetic impactor (Sanchez and Colombo 2012) are moreover based on currently proven space technologies. They may therefore render the apparently ambitious scenario of manipulating asteroid trajectories a likely option for the near future.

With regards the accessibility of asteroid resources, recent work by Sanchez and McInnes (2011a) (see also Chap. 18 in this book) demonstrates that a substantial quantity of resources can indeed be accessed at relatively low energy; on the order of 10^{14} kg of material could potentially be harvested at an energy cost lower than that required to access the resources of the Moon. In their work, the accessibility of asteroid material is estimated by analysing the volume of Keplerian orbital element space from which the Earth can be reached under a given energy threshold by means of a bi-impulsive transfer. This volume of Keplerian element space is then mapped into existing NEO orbital (Bottke et al. 2002) and size distribution models (Mainzer et al. 2011). As discussed in Sanchez and McInnes (2011a) it is perhaps more important that asteroid resources can be accessed across a wide spectrum of energies, and thus, as shown in Chap. 18, current technologies could be adapted to return to the Earth's neighbourhood objects from 10 to 30 meters diameter for scientific exploration and resource utilisation purposes.

Advances in both asteroid deflection technologies and dynamical system theory, which allow new and cheaper means of space transportation, are now enabling radically new mission concepts, including but not limited to asteroid retrieval missions. These envisage a spacecraft reaching a suitable object, attaching itself to the surface and returning it, or a portion of it, to the Earth's orbital neighbourhood. Moving an entire asteroid into an orbit in the vicinity of Earth entails an obvious engineering challenge, but may also allow a much more flexible exploitation phase in the Earth's neighbourhood.

The work presented here aims to provide a feasibility assessment of the latter mission concept by defining a set of preliminary mission opportunities that could be enabled by invariant manifold dynamics. Missions delivering a large quantity of material to the Lagrangian points are of particular interest. The material can be used in a first stage as test bed for ISRU technology demonstration missions and material processing at affordable costs. The science return is also greatly improved, with an asteroid permanently, or for a long duration, available for study and accessible to telescopes, probes and even crewed missions to the Lagrangian points. Finally, it sets the stage for other future endeavours, such as the ones listed in Table 21.1.

21.2 Low Energy Transport Conduits

Current interplanetary spacecraft have masses on the order of 10^3 kg, while a small body of just 10 meters diameter will most likely have a mass of the order of 10^6 kg. Hence, already moving such a small object, or an even larger one, with the same ease that a scientific payload is transported today, would demand propulsion systems orders of magnitudes more powerful and efficient; or alternatively, orbital transfers orders of magnitude less demanding than those to reach other planets in the solar system.

Solar system transport phenomena, such as the rapid orbital transitions experienced by comets Oterma and Gehrels 3, from heliocentric orbits with periapsis outside Jupiter's orbit to apoapsis within Jupiter's orbit, or the Kirkwood gaps in the main asteroid belt, are some manifestations of the sensitivities of multi-body dynamics. The same underlying principles that enable these phenomena allow also excellent opportunities to design surprisingly low energy transfers.

It has for some time been known that the hyperbolic invariant manifold structures associated with periodic orbits around the L_1 and L_2 collinear points of the Three Body Problem provide a general mechanism that controls the aforementioned solar system transport phenomena (Koon et al. 2000). In this analysis, we seek to benefit from these mathematical constructs in order to find *low-cost* trajectories to retrieve asteroid material to the Earth's vicinity. This work assumes the motion of the spacecraft and asteroid under the gravitational influence of Sun and Earth, within the framework of the Circular Restricted Three Body Problem (CR3BP) (Koon et al. 2008). The well known equilibrium points of the system are shown in Fig. 21.1. The mass parameter μ considered in this analysis is $3.0032080443 \times 10^{-6}$, which neglects the mass of the Moon. Note that the usual normalised units are used when citing Jacobi constant values (Koon et al. 2008).

21.2.1 Periodic Orbits and Manifold Structure

In particular, we are interested in the dynamics concerning the Sun-Earth L_1 and L_2 points (see Fig. 21.1), as they are the *gate keepers* for potential ballistic capture of asteroids in the Earth's vicinity.

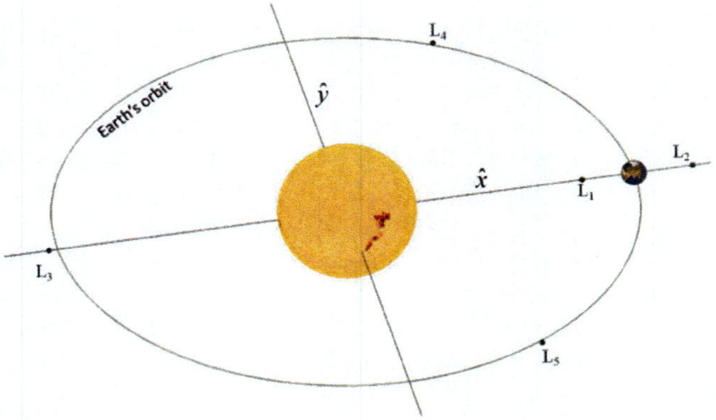

Fig. 21.1 Schematic of the CR3BP and its equilibrium points

During the last half a century there has been an intense effort to catalogue all bounded motion near the libration points of the Circular Restricted Three Body Problem (Howell 2001). The principal families of bounded motion that have been studied are planar and vertical Lyapunov periodic orbits, quasi-periodic Lissajous orbits, and periodic and quasi-periodic halo orbits (Gómez et al. 2000, Koon et al. 2008). Some other families of periodic orbits can be found by exploring bifurcations in the aforementioned main families (Howell 2001).

Theoretically, an asteroid transported into one of these orbits would remain near the libration point for an indefinite time. In practice, however, these orbits are unstable, and an infinitesimal deviation from the periodic orbit will make the asteroid depart asymptotically from the libration point regions. Nevertheless, small correction manoeuvres can be assumed to be able to keep the asteroid within the periodic orbit (Simó et al. 1987, Howell and Pernicka 1993).

The linear behaviour of the motion near the libration points is of the type *centre* x *centre* x *saddle* (Szebehely 1967). All bounded motion near these points arises from the stable *centre* x *centre* behaviour, while the saddle dynamical behaviour ensures that, inherent to any bounded trajectory near the libration points, an infinite number of trajectories exist that asymptotically approach, or depart from, the bounded motion. All these sets of trajectories, both bounded and unbounded motion, associated to a libration point form what is called the invariant manifold structure (Gómez et al. 2005).

There are two classes of invariant manifolds: the central invariant and the hyperbolic invariant. The central invariant manifold is composed of periodic and quasi-periodic orbits near the libration points, while the hyperbolic invariant manifold consists of a stable and an unstable set of trajectories associated with orbits of the central invariant manifold. The unstable manifold is formed by the infinite set of trajectories that exponentially leaves the periodic or quasi-periodic motion to which they are associated. The stable manifold, on the other hand, consists of an infinite number of trajectories exponentially approaching the periodic or quasi-periodic orbit.

It is well known (e.g. Koon et al. (2008)) that the phase space near the equilibrium regions can be divided into four broad classes of motion; bound motion near the equilibrium position (i.e. periodic and quasi-periodic orbits), asymptotic trajectories that approach or depart from the latter, transit trajectories, and, non-transit trajectories (see Fig. 21.2). A transit orbit is a trajectory such that its motion undergoes a rapid transition between orbiting regions. In the Sun-Earth case depicted in Fig. 21.2, for example, the transit trajectory approaches Earth following a heliocentric trajectory, transits through the bottle neck delimited by the halo orbit and becomes temporarily captured at Earth. An important observation from dynamical system theory is that the hyperbolic invariant manifold structure defined by the set of asymptotic trajectories forms a phase space separatrix between transit and non-transit orbits.

It follows from the four categories of motion near the libration points that periodic orbits near the Sun-Earth L_1 and L_2 points can not only be targeted as the final destination of asteroid retrieval missions, but also as natural gateways of low energy trajectories to Earth centred temporarily captured trajectories or transfers to other locations of the cis-lunar space, such as the Earth Moon Lagrangian points.

This work will focus in capture opportunities to periodic orbits near the libration points, which will be enabled by the hyperbolic stable manifold trajectories associated with them. In particular, we will focus on three distinct classes of periodic motion near the Sun-Earth L_1 and L_2 points: Planar and Vertical Lyapunov and Halo Orbits, from now on referred to as a whole as libration point orbits (LPO). This analysis seeks a first insight into the future feasibility of asteroid retrieval missions, and thus we will limit our search to these three classic families of

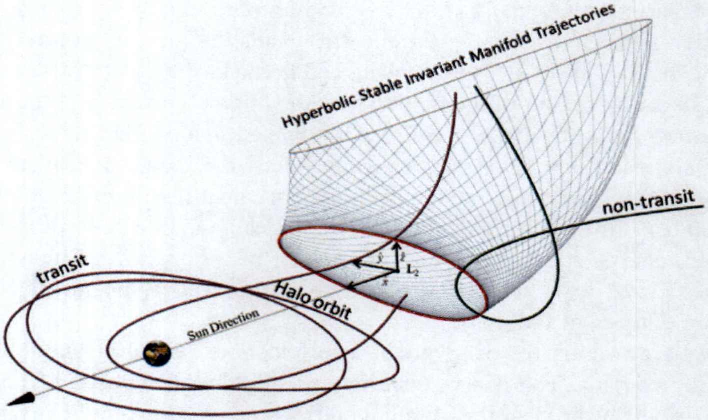

Fig. 21.2 Schematic representation of the four categories of motion near the L_2 point (represented by the set of axes in the figure): periodic motion around L_2 (i.e. halo orbit), hyperbolic invariant manifold structure (i.e. set of stable hyperbolic invariant manifold trajectories), transit trajectory and non transit trajectory.

bounded motion near the Sun-Earth L_1 and L_2 points. Future and more comprehensive searches however should extend to capture trajectories to quasi-periodic orbits, such as Lissajous and quasi-halos. These more exotic families of eventually periodic orbits and their associated sets of asymptotic trajectories will likely increase the range of options for gravitational capture of asteroids.

21.2.1.1 Lyapunov Orbits

As noted earlier, the linear behaviour of the motion near the L_1 and L_2 points is of the type *centre* x *centre* x *saddle*. The *centre* x *centre* part generates a 4-dimensional central invariant manifold around each collinear equilibrium point when all energy levels are considered (Gómez et al. 2005). In a given energy level the central invariant manifold is a 3-dimensional set of periodic and quasi-periodic solutions lying on an invariant torus, together with some chaotic or stochastic regions in between (Gómez et al. 2001). There exist families of periodic orbits with frequencies related to both centers: ω_p and ω_v (Alessi 2010). They are known as *planar Lyapunov family* and *vertical Lyapunov family*, see Fig. 21.3, and their existence is ensured by the Lyapunov center theorem. Halo orbits are 3-dimensional periodic orbits that emerge from the first bifurcation of the planar Lyapunov family.

To generate the entire family of planar and vertical Lyapunov periodic orbits, we start by generating an approximate solution with the associated frequency very close to the libration point (Howell 2001). This initial solution is corrected in the non-linear dynamics by means of a differential correction algorithm (Koon et al. 2008) over a suitable plane section that takes advantage of the known symmetries of these orbits (Zagouras and Markellos 1977). Once one periodic solution has been computed, the complete family can be generated by means of a numerical continuation process that uses the previous solution as initial guess for a periodic orbit on which one of the dimensions on the phase space has been perturbed slightly. By properly choosing the phase space direction that we want to extend; and by repeating the process iteratively, one can build a family of periodic orbits with increasing Jacobi constant, as shown in Fig. 21.3.

21.2.1.2 Halo Orbits

The term halo orbit refers to the orbit's ring shape and its position relative to the secondary mass, which reminds of the ring of light commonly used in religious iconography to denote holiness. The term was coined by Robert Farquhar, who advocated the use of these orbits near the Earth-Moon L_2 point to obtain a continuous communication relay with the far side of the Moon during the Apollo programme (Farquhar 1967).

As previously noted, this type of orbit emerges from a bifurcation in the planar Lyapunov orbits. As the amplitude of planar Lyapunov orbit increases, eventually a critical amplitude is reached where the planar orbits become vertical critical, as defined by Hénon (1973), and new three-dimensional families of periodic orbits bifurcate. Thus, the minimum possible size for halo orbits in the Sun-Earth system is approximately (240 x 660) 10^3 km at L_1 and (250 x 675) 10^3 km at L_2, sizes denoting the maximum excursion from the libration point in the *x* and *y* directions respectively. At the bifurcation point, two symmetric families of halo orbits emerge at each libration point, here referred to as the northern and southern family depending on whether the maximum *z* displacement is achieved in the northern (i.e. $z>0$) or southern (i.e. $z<0$) direction, respectively (see Fig. 21.3).

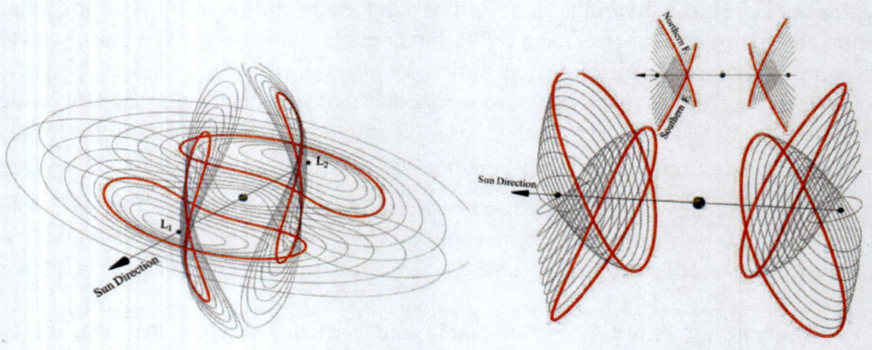

Fig. 21.3 Series of Planar and Vertical Lyapunov orbits (left) and northern and southern halo orbits (right) associated with the Sun-Earth L_1 and L_2 points. Lyapunov orbits are plotted ranging from Jacobi constant 3.0007982727 to 3.0000030032. Halo orbits are plotted ranging from Jacobi constant of 3.0008189806 to 3.0004448196. The thicker red line corresponds to a Jacobi constant of 3.0004448196, which corresponds to half the distance between the energy at equilibrium in L_2 and L_3.

Similarly to planar and vertical Lyapunov, the set of halo orbits, also shown in Fig. 21.3, was computed by means of the continuation of a predictor-corrector process. The initial seed was computed by means of Richardson (1980) third order approximation of a halo orbit. A differential corrector procedure is used to trim Richardson's prediction and obtain the smallest halo possible (Zagouras and Markellos 1977, Koon et al. 2008). We then continue the process by feeding the next iteration with a prediction of a slightly larger displacement in *z*. Iteratively repeating this process provides a series of halo orbits with increasing energy, or decreasing Jacobi constant.

The process can only be continued until a Jacobi constant not far below 3.0004. At this point the direction of the continuation should be changed to the *x* direction, or a more sophisticated processes of continuation on which the direction is modified at each iteration should be used (Ceriotti and McInnes 2012). For this analysis however we chose to stay on the range of halo orbits that can be continued using only the *z* direction to ensure that each halo orbit is defined by a single Jacobi constant. If halo families are continued beyond that point, they become degenerate in energy since a particular Jacobi constant defines more than one halo orbit.

21.3 Asteroid Retrieval Missions

In the past few years, several space missions have already attempted to return samples from the asteroid population (e.g. Hayabusa (Kawaguchi et al. 2008)) and others are planned for the near future [http://www.nasa.gov/topics/solarsystem/features/osiris-rex.html (last accessed 02/05/12)]. As shown by Sanchez and McInnes (2011a) (see also Chap. 18 in this book), given the low transport cost expected for the most accessible objects, it is also possible to envisage the possibility to return to Earth entire small objects with current or near-term technology. The main challenge resides on the difficulties inherent in the detection of these small objects. Thus, for example, only 1 out of every million objects with diameter between 5 to 10 meters is currently known and this ratio is unlikely to change significantly in the coming years (Veres et al. 2009).

In this section then, we will focus our attention to the surveyed population of asteroids in search of the most accessible candidates for near-term asteroid retrieval missions by means of invariant hyperbolic stable manifolds trajectories.

For this purpose, a systematic search of capture candidates among catalogued NEOs was carried out, selecting the L_1 and L_2 regions as the target destination for the captured material. This gives a grasp and better understanding of the possibilities of capturing entire NEOs or portions of them in a useful orbit, and demonstrates a method that can be applied to newly discovered small bodies in the future when detection technologies improve.

21.3.1 Invariant Manifold Trajectories to L_1 and L_2

The design of the transfer from the asteroid orbit to the L_1 and L_2 LPO consists of a ballistic arc, with two impulsive burns at the start and end, intersecting a hyperbolic stable invariant manifold asymptotically approaching the desired periodic orbits. These results consider only the inbound leg of a full capture mission.

Planar Lyapunov, vertical Lyapunov, and halo orbits around L_1 and L_2 generated with the methods described in previous sections were considered as target orbits. The invariant stable manifold trajectories leading to each of these LPO, computed by perturbing the periodic solutions on the stable eigenvector direction (Koon et al. 2008) by a magnitude of 10^{-6}, in normalized units, were propagated backwards in the Circular Restricted 3-Body Problem until they reached a fixed section in the Sun-Earth rotating frame. We refer to this propagation time as the invariant manifold transfer time. The section was arbitrarily selected as the one forming an angle of $\pm\pi/8$ with the Sun-Earth line ($\pi/8$ for the L_2 orbits, see Fig. 21.4, the symmetrical section at $-\pi/8$ for those targeting L_1). This corresponds roughly to a distance to Earth of the order of 0.4 AU, where the gravitational influence of the planet is considered small. No additional perturbations were considered in the backward propagation.

In this analysis, Earth is assumed to be in a circular orbit 1 AU away from the Sun. This simplification allows the orbital elements of the manifold trajectories (and in particular at the selected section) to be independent of the insertion time into the final orbit. The only exception is the longitude of the perihelion, i.e. the sum of the right ascension of the ascending node and the argument of perihelion, which varies with the insertion time with respect to a reference time with the following relation:

$$(\Omega+\omega)=(\Omega_{REF}+\omega_{REF})+\frac{2\pi}{T}(t-t_{REF}) \tag{21.1}$$

where Ω_{REF} and ω_{REF} are the right ascension of the ascending node and the argument of perihelion at the $\pm\pi/8$ section for an insertion into a target orbit at reference time t_{REF}, and T is the period of the Earth. This variation along the Earth orbit has direct implications in the phasing costs and influences the optimal point for final insertion.

For orbits with non-zero inclination, the argument of perihelion of the manifolds is also independent of the insertion time and Eq. (21.1) indicates a variation in Ω. However, in the case of planar Lyapunov with zero inclination, Ω is not defined and an arbitrary value of zero can be selected, resulting in the equation representing a change in argument of perihelion.

The transfer between the NEO orbit and the manifold is then calculated as a heliocentric Lambert arc of a restricted two body problem with two impulsive burns, one to depart from the NEO, the final one for insertion into the manifold, with the insertion constrained to take place before or at the $\pm\pi/8$ section.

The benefit of such an approach is that the asteroid is asymptotically captured into a bound orbit around a collinear Lagrangian point, with no need for a final insertion burn at arrival. All burns are performed far from Earth, so no large gravity losses need to be taken into account. Furthermore, this provides additional time for corrections, as the dynamics in the manifold are "slow" when compared to a traditional hyperbolic approach.

Finally, this type of trajectory is then easily extendable to a low-thrust trajectory if the burns required are small.

The shape of the manifolds in the $r-\dot{r}$ phase space (with r being the radial distance from the Sun) at the intersection with the $\pm\pi/8$ section is shown in Fig. 21.5 for a particular Jacobi constant. For an orbit with exactly the energy of L_1 or L_2, the intersection is a single point; while for lower Jacobi constants, the shape of the intersection is a closed loop. The intersection corresponding to the bifurcation between planar and halo orbits is also plotted. A few capture candidate asteroids have been included in the plot (+ markers) at their intersection with the $\pi/8$ plane near their next closest approach to Earth. In a planar case, this would already provide a good measure of the distance of the asteroid to the manifolds. However, as we are considering the 3D problem, information on the z component or the inclination would also be necessary.

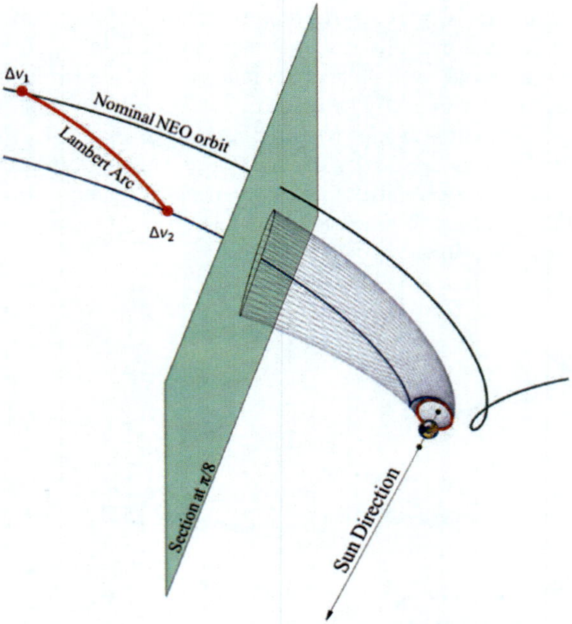

Fig. 21.4 Schematic representation of a transfer to L_2

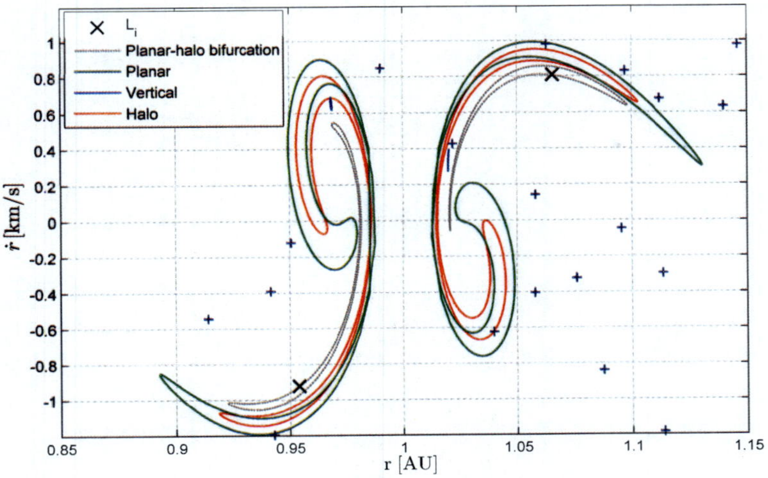

Fig. 21.5 Shape of the manifolds in the $r - \dot{r}$ phase space for a Jacobi constant of 3.0004448196. The manifolds are represented at their intersection with a plane forming a $\pm\pi/8$ angle with the Sun-Earth line in the rotating frame. Manifolds on the left correspond to L_1, on the right to L_2. Candidate NEOs are indicated with a + marker.

Figure 21.6 provides a more useful representation of the manifolds in terms of perihelion radius, aphelion radius and inclination for the two collinear points. The point of bifurcation between the planar Lyapunov and halo orbits, when they start growing in inclination, can easily be identified. Halo orbits extend a smaller range in aphelion and perihelion radius when compared to planar Lyapunov. Vertical Lyapunov orbits have even smaller excursions in radius from a central point, as can already be seen in the smaller loops of vertical Lyapunov in Fig.21.5, but on the other hand they extend to much lower values of the Jacobi constant and cover a wider range of inclinations.

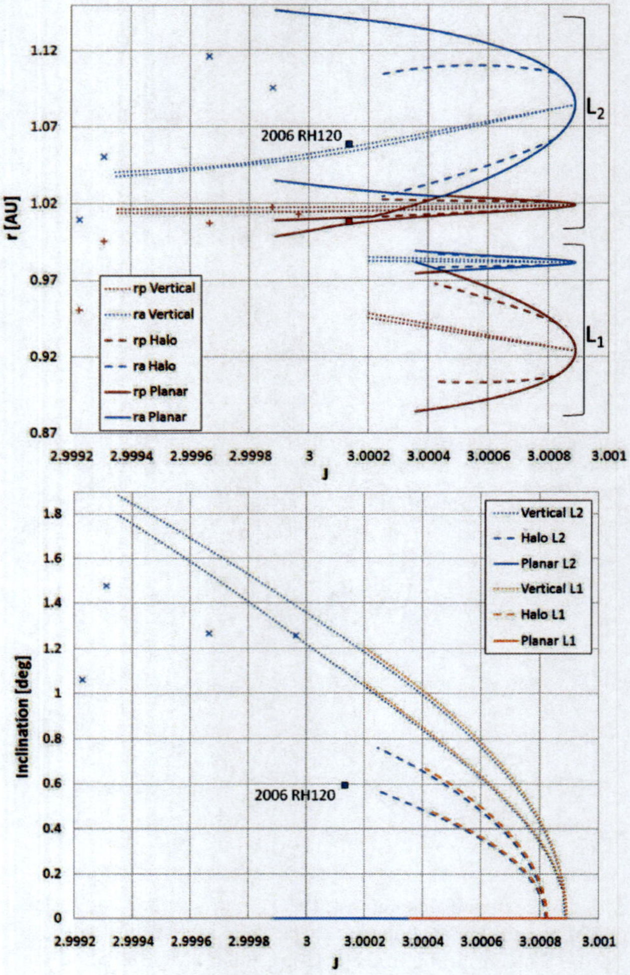

Fig. 21.6 Minimum and maximum perihelion and aphelion radius (top) and inclination (bottom) of the manifolds leading to planar Lyapunov, vertical Lyapunov and halo orbits around L_1 and L_2.

Several asteroids are also plotted with small markers in the graphs. Their Jacobi constant J is approximated by the Tisserand parameter as defined in Eq. (21.2):

$$J \approx \frac{1}{a} + 2\sqrt{a\left(1-e^2\right)}\cos i \qquad (21.2)$$

where a, e and i are the semi-major axis (in AU), eccentricity and inclination of the asteroid orbit.

This illustrates the proximity to the manifolds of a number of NEOs. In particular, asteroid 2006 RH120 has been highlighted, due to its proximity to the L_2 manifolds. From these graphs and ignoring any phasing issues, it can already be identified as a good capture candidate, as its perihelion and aphelion radius is close to or within the range of all the three types of manifolds, and its inclination lies also close to the halo orbit manifolds. The manifold orbital elements appear to be a good filter to prune the list of NEOs to be captured.

21.3.2 Asteroid Catalogue Pruning

For the calculation of capture opportunities, the NEO sample used for the analysis is JPL's Small Bodies Database [http://ssd.jpl.nasa.gov/sbdb.cgi (last accessed 27/07/12)], downloaded as of 27th of July of 2012, containing 9142 small bodies. This database represents the catalogued NEOs up to that date, and as such it is a biased population, most importantly in size, as already noted. A large number of asteroids of the most ideal size for capture have not yet been detected, as current detection methods favour larger asteroids. Secondly, there is an additional detection bias related to the type of orbits, with preference for Amors and Apollos in detriment to Atens or Atiras, as object in Aten/Atira orbits spend more time in the exclusion zone due to the Sun.

Even with this reduced list, it is a computationally expensive problem and preliminary pruning becomes necessary. Previous work by Sanchez et al. (2012) showed that the number of known asteroids that could be captured from a hyperbolic approach with a total Δv less than 400 m/s is of the order of 10. Although their hyperbolic capture approach, which roughly estimates Δv-cost for capture as the Δv necessary to reduce the asteroid two-body energy to zero, ensuring a temporal capture, is inherently different than the manifold capture presented in this work, the number of bodies that could be captured in manifold orbits at low cost is expected to be of the same order.

Without loss of generality, it is possible to immediately discard NEOs with semi-major axis (and thus energy) far from the Earth's, as well as NEOs in highly inclined orbits. However, a more systematic filter needed to be devised.

As a first approximation of the expected total cost in terms of Δv, a bi-impulsive cost prediction with both burns assumed at aphelion and perihelion was implemented. Either of the two burns is also responsible for correcting the inclination. The Δv required to modify the semi-major axis can be expressed as:

$$\Delta v_a = \sqrt{\mu_s\left(\frac{2}{r}-\frac{1}{a_f}\right)} - \sqrt{\mu_s\left(\frac{2}{r}-\frac{1}{a_0}\right)} \tag{21.3}$$

where μ_S is the Sun's gravitational constant, a_0 and a_f are the initial and final semimajor axis before and after the burn, and r is the distance to the Sun at which the burn is made (perihelion or aphelion distance). On the other hand the Δv required to modify the inclination is given by:

$$\Delta v_i = 2\sqrt{\frac{\mu_s}{a_0}r^*}\, sin(\Delta i/2) \tag{21.4}$$

where Δi is the required inclination change, and r^* corresponds to the ratio of perihelion and aphelion distance if the burn is performed at aphelion, or its inverse if performed at perihelion. These formulas only take into consideration the shape and inclination of the orbits, ignoring the rest of the orbital elements: right ascension of the ascending node and argument of pericentre. It is then implicitly assumed that the line of nodes coincides with the line of apsis and the inclination change can be performed at pericentre or apocentre. This can result in an underestimation of the plane change for some cases.

The total cost is then calculated as:

$$\Delta v_t = \sqrt{\Delta v_{a1}^{\ 2} + \Delta v_{i1}^{\ 2}} + \sqrt{\Delta v_{a2}^{\ 2} + \Delta v_{i2}^{\ 2}} \tag{21.5}$$

with one burn performed at each of the apsis, and one of the two inclination change Δv assumed zero.

The estimated transfer Δv corresponds thus to the minimum of four cases: aphelion burn modifying perihelion and inclination followed by a perihelion burn modifying aphelion, perihelion burn modifying aphelion and inclination followed by an aphelion burn modifying perihelion, and the equivalent ones in which the inclination change is done in the second burn.

It is important to note that these formulas are only first order approximations intended for the pruning of the database, and they will not be used to calculate the final transfers. In particular, the plane change is only valid for small changes in inclination and large deviations from the values provided by the filter are expected to be observed for high inclinations. Nevertheless, we are interested in low cost transfers which imply a small plane change, so the approximation is acceptable for filtering purposes.

For simplicity, the target manifold final perihelion, aphelion and inclination values are selected as ranges or bands obtained from Fig. 21.6. For example, planar Lyapunov orbits at L_2 have range of $\{r_p, r_a, i\} \in \{1.00\text{-}1.02, 1.02\text{-}1.15, 0\}$, or $\{1.01\text{-}1.02, 1.025\text{-}1.11, 059\text{-}0.78\}$ for halo orbits at L_2. Note that the inclination range for halos was given as the range corresponding to the highest energy. This is due to the fact that most candidate asteroids have higher energies that the manifolds, and the lowest cost is assumed to take place where the energy difference is minimum. In the case of vertical Lyapunov orbits, due to the narrow ranges and strong dependency with J, polynomial fits for $\{r_p, r_a, i\}$ as a function of J were used.

With this filter, it is then possible to calculate the regions of a three-dimensional orbital element space (in semi-major axis, eccentricity and inclination) than can be captured under a certain Δv threshold. These regions are plotted in Fig. 21.7 for transfers to LPOs around L_2 with a Δv of 500 m/s, and any asteroid with orbital elements inside them could in principle be captured at that

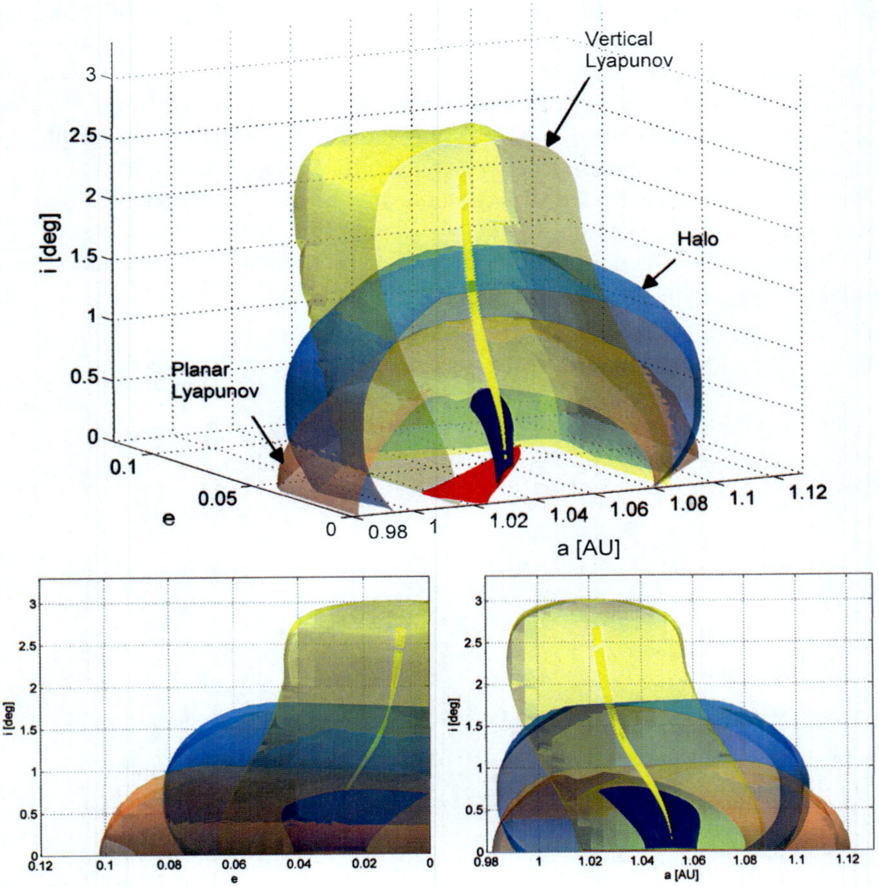

Fig. 21.7 Regions in the orbital element space with total estimated cost for capture into an LPO around L_2 below 500 m/s. The manifolds corresponding to the LPOs are plotted in solid colours.

cost or lower. The figure shows a three-dimensional view of the surfaces that delimit the regions for planar Lyapunov, vertical Lyapunov and Halo, as well as two-dimensional projections in the *a*-*i* and *e*-*i* planes. There is a significant overlap between the regions of different LPO target orbits; therefore, it is expected that several asteroids would allow low-cost captures to more than one family of LPO. A similar plot can be generated for the case of L_1. Figure 21.8 presents the regions for L_1 and L_2 compared to the definitions of the 4 families of NEOs. Objects from all four families seem to be adequate candidates for asteroid retrieval missions, particularly the ones closed to the Apollo-Amor and Aten-Atira divides.

The filter approximation provides in general a lower bound Δv estimate, as it ignores any phasing issues, and assumes the burns can be performed at apocentre or pericentre. Moreover, there is no guarantee, and in fact it is quite unlikely, that a combination of the extremes of the ranges of $\{r_p, r_a, i\}$ used in the filter correspond to proper manifold trajectories. Finally, the plane change does not include a modification in right ascension of the ascending node. Although the final Ω can be tuned by modifying the phasing with the Earth, this is not completely free as the final insertion will take place around a natural close approach of the asteroid with the planet. The combination of this constrained phasing and the plane change will also incur in additional costs. North and south halo obits provide two opportunities with opposite Ω for each transfer, which should result in two different costs, while the filter provides a single value.

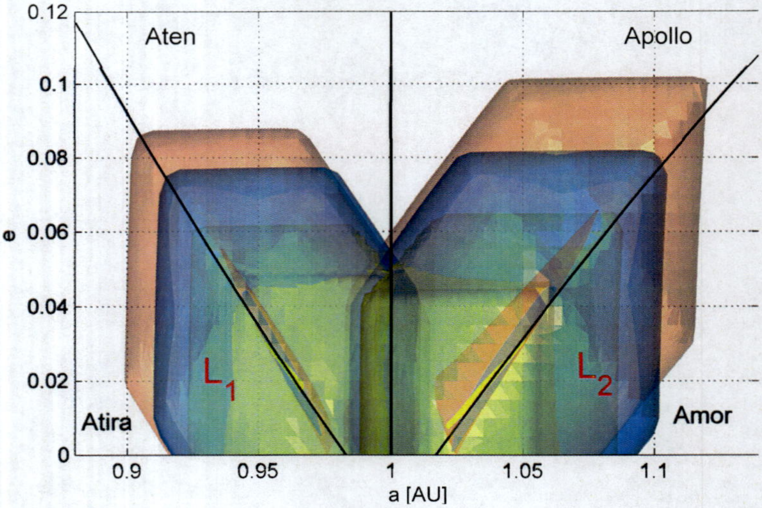

Fig. 21.8 Semi-major axis and eccentricity map of the capturable regions for L_1 and L_2. The boundaries of the main 4 families of NEO objects are also indicated. The manifold orbital elements are enclosed in the capturable regions and closely follow the Apollo-Amor and Aten-Atira divides.

For a few cases, with high initial inclination and associated plane change cost, the filter can over-estimate the Δv. As the inclination increases, solutions splitting the large plane change into the two burns can potentially result in a lower cost. In cases where the filter favours solutions with larger burns at pericentre, it can also incur in higher costs estimation for the plane change than the optimal solution.

21.3.3 Capture Transfers and Mass Estimates

For each of the filtered NEOs with estimated Δv below 1 km/s, feasible capture transfers with arrival date in the interval 2016-2100 were obtained. The NEO orbital elements in JPL's database are only considered valid until their next close encounter with Earth.

The problem can thus be defined with 5 variables: the Lambert arc transfer time, the manifold transfer time, the insertion date at the final target orbit, the Jacobi constant of this target orbit, and a fifth discrete variable determining the point in the target orbit where the insertion takes place. The manifolds are discretised in terms of their Jacobi constant and their insertion point. Five hundred insertion points are considered for each LPO, which propagated backwards translate into 500 sets of orbital elements at the $\pm\pi/8$ section.

The Lambert transfers between the asteroid initial orbit and the manifolds were optimised using EPIC, a global optimisation method that uses a stochastic search blended with an automatic solution space decomposition technique and can handle both continuous and discrete variables (see (Vasile and Locatelli, 2009) for details). Single objective optimisations with total transfer Δv as the cost function were carried out. Trajectories obtained with EPIC were locally optimised with MATLAB's built-in function fmincon. Lambert arcs with up to 3 complete revolutions before insertion into the manifold were considered. For cases with at least one complete revolution, the two possible solutions of the Lambert problem were optimised. This implies that 7 full problem optimisations needed to be run for each NEO.

Figure 21.9 plots the results of the optimisation for the L_2 case together with the filter estimates. It can be observed that the filter provides in general a good approximation of the total cost to be expected. It is a useful tool to select candidates and prioritise lists of asteroids for optimisation, and to quickly predict if any newly discovered asteroid is expected to have low capture costs. Dotted lines have been added to the plot as indicators of the ideal cost of just the inclination change at a circular orbit at 1 AU. Predicted and optimised results are expected to fall above or close to these lines. The filter does however provide a quick and much more accurate estimate of the costs taking into consideration the shape of the original orbit as well as the inclination. Asteroids with capture costs smaller than 500 m/s are indentified in the plots.

Table 21.2 shows the asteroids with costs lower than this selected threshold of 500 m/s. Twelve asteroids of the whole NEO catalogue can be captured at this cost, ten of them around L_2 plus two Atens around L_1.

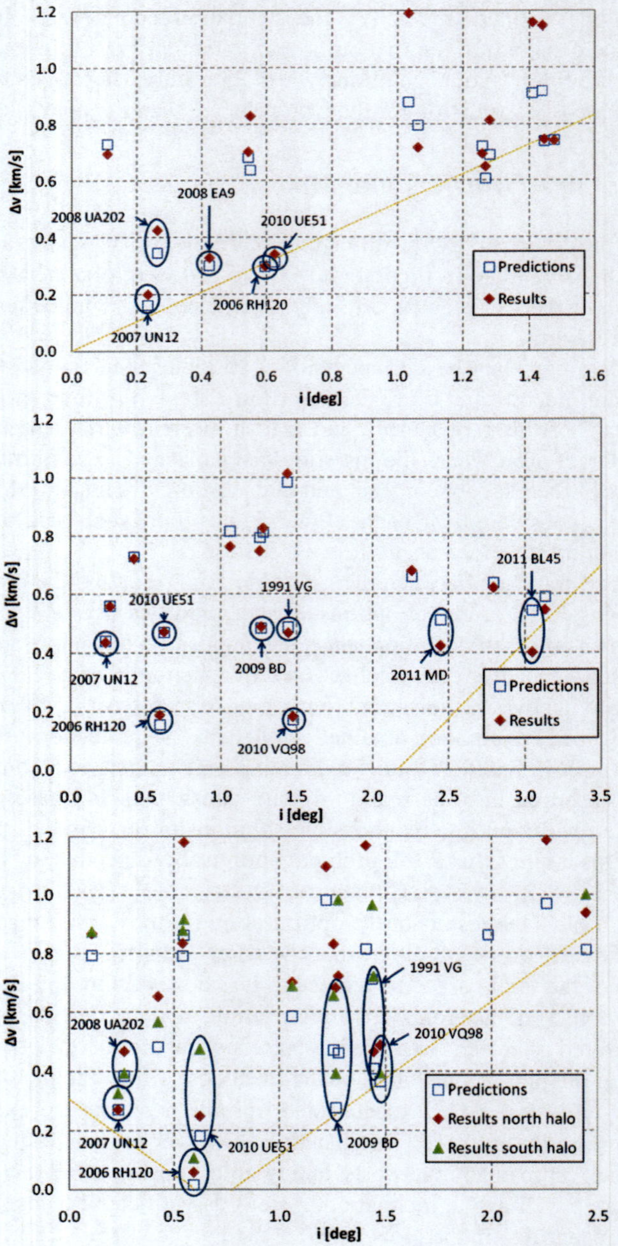

Fig. 21.9 Filter cost estimates and results of the optimisation for planar Lyapunov (top), vertical Lyapunov (middle) and halo orbits (bottom) around L_2. Dotted lines indicate the cost of changing just the inclination.

Table 21.2 NEO characteristics for transfer trajectories with Δv below 500 m/s. The type of transfer is indicated by a 1 or 2 indicating L_1 or L_2 plus the letter P for planar Lyapunov, V for vertical Lyapunov, and Hn or Hs for north and south halo.

	a [AU]	e	*i* [deg]	MOID [AU]	Diameter [m]	Type	Δv [m/s]
2006 RH120	1.033	0.024	0.595	0.0171	2.3- 7.4	2Hs 2Hn 2V 2P	58 107 187 298
2010 VQ98	1.023	0.027	1.476	0.0048	4.3-13.6	2V 2Hn 2Hs	181 393 487
2007 UN12	1.054	0.060	0.235	0.0011	3.4-10.6	2P 2Hs 2Hn 2V	199 271 327 434
2010 UE51	1.055	0.060	0.624	0.0084	4.1-12.9	2Hs 2P 2V 2Hn	249 340 470 474
2008 EA9	1.059	0.080	0.424	0.0014	5.6-16.9	2P	328
2011 UD21	0.980	0.030	1.062	0.0043	3.8-12.0	1Hs 1V 1Hn	356 421 436
2009 BD	1.062	0.052	1.267	0.0053	4.2-13.4	2Hn 2V	392 487
2008 UA202	1.033	0.069	0.264	$2.5 \cdot 10^{-4}$	2.4- 7.7	2Hn 2P 2Hs	393 425 467
2011 BL45	1.033	0.069	3.049	0.0040	6.9-22.0	2V	400
2011 MD	1.056	0.037	2.446	0.0018	4.6-14.4	2V	422
2000 SG344	0.978	0.067	0.111	$8.3 \cdot 10^{-4}$	20.7-65.5	1P 1Hs 1Hn	443 449 468
1991 VG	1.027	0.049	1.445	0.0037	3.9-12.5	2Hs 2V	465 466

The table provides the orbital elements, minimum orbit intersection distance according to the JPL Small Bodies Database, and an estimate of the size of the object. This estimate is calculated with the following relation (Chesley et al. 2002):

$$D = 1329\,km \times 10^{-H/5}\, p_v^{-1/2} \tag{21.6}$$

where the absolute magnitude H is provided in the JPL database, and the albedo p_v is assumed to range from 0.05 (dark) to 0.50 (very bright icy object).

As expected, planar Lyapunov orbits are optimal for lower inclination NEOs, while NEOs with higher inclination favour transfers to vertical Lyapunov. Figure 21.10 shows an example trajectories in a co-rotating frame where the Sun-Earth line is fixed for a transfer of asteroid 2006 RH120 to a south halo orbit around L_2. Close-ups of the final parts of the trajectory are plotted in a three-dimensional view in order to appreciate the shape of the final orbit and manifolds.

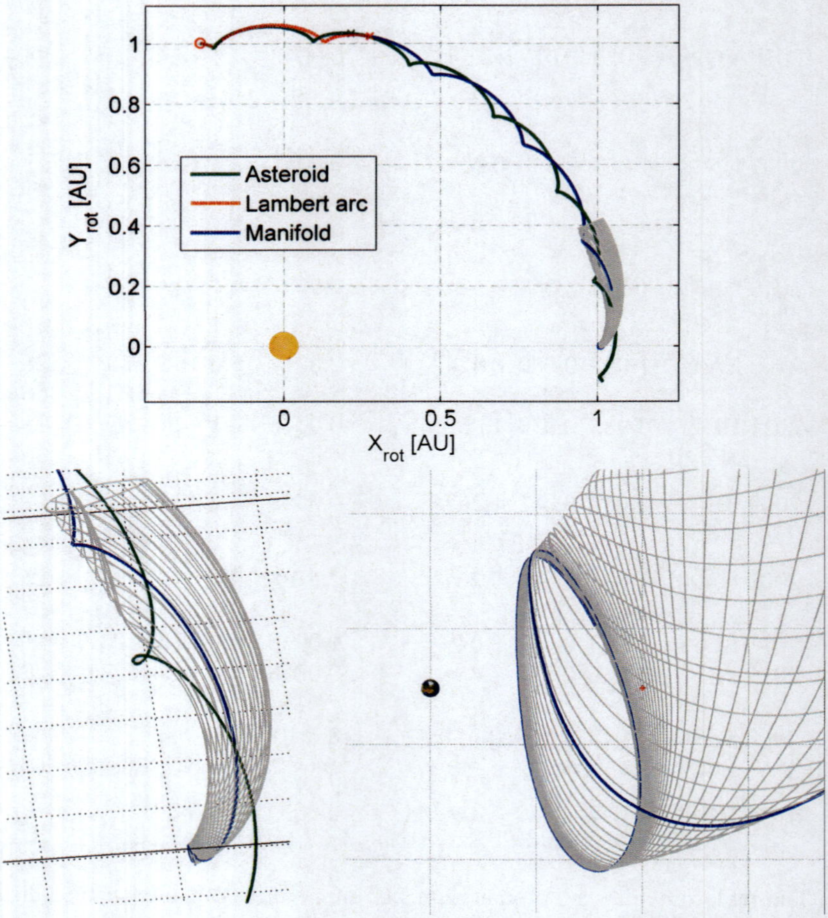

Fig. 21.10 Capture trajectory for asteroid 2006 RH120 to a south halo orbit. Sun and Earth are plotted 10 times their size.

Table 21.3 presents the best trajectory for each type of transfer for L2 and L1. The cheapest transfer, below 60 m/s, corresponds to a trajectory inserting asteroid 2006 RH120 into a halo orbit. Solutions to planar and vertical Lyapunov were also found for 2006 RH120 at higher costs. This agrees well with the interpretation of Fig. 21.6. The pruning method was also predicting that this transfer would be the cheapest, with a minimum estimated Δv of 15 m/s. It is worthwhile to emphasise that the total Δv comprises both burns at departure from the asteroid and insertion into the manifold, but it does not include any navigation costs or corrections. The NEO orbit may intersect the manifold directly, and in that case the transfer to the target orbit can be done with a single burn, as in this particular case.

The total duration of the transfers range from 3 to 7.5 years. For the longer transfers it is possible to find faster solutions with less revolutions in the Lambert arc at a small Δv penalty.

Table 21.3 Capture trajectories for the lowest cost transfers to each type of LPO

		Date [yy/mm/dd]			Jacobi constant manifold	Total Durat. [yr]	Δv [m/s]	
		Asteroid departure	Manifold insertion	L_i arrival			Dep	Ins
2006 RH120	2Hs	21/02/01	21/02/01	28/08/05	3.000421	7.51	58	0
2006 RH120	2Hn	23/05/11	24/02/20	28/08/31	3.000548	5.31	52	55
2010 VQ98	2V	35/02/14	35/09/01	39/11/15	3.000016	4.75	177	4
2007 UN12	2P	13/10/22	13/10/22	21/02/19	3.000069	7.33	199	0
2011 UD21	1Hs	37/11/20	38/07/03	42/07/19	3.000411	4.66	149	207
2011 UD21	1V	36/07/20	38/11/16	41/06/21	3.000667	4.92	226	196
2011 UD21	1Hn	39/10/24	40/06/15	43/08/30	3.000504	3.85	210	226
2000 SG344	1P	24/02/11	25/03/11	27/06/18	3.000357	3.35	195	248

21.3.3.1 Retrieved Mass Estimates

The results presented in the previous section could be used to calculate a limit in the mass that can be captured with current space technology. In order to obtain a first estimate of the size of the asteroid that could be retrieved, we can consider a basic system mass budget exercise. Assuming a spacecraft of 5500 kg dry mass and 8100 kg of propellant at the NEO (as proposed in the Keck study report for asteroid retrieval (Brophy et al. 2012)), it is possible to estimate the total asteroid mass that can be transferred. A full system budget would require a larger fuel mass to deliver the spacecraft to the target, and thus an analysis of the outbound leg, but that is beyond the scope of this work.

Results are presented for each trajectory on Table 21.4 for two different propulsion systems. The total mass for a high thrust engine of specific impulse

300 s ranges from 44 to about 400 tons, which represents 3 to 30 times the wet mass of the spacecraft at arrival to the NEO. The trajectories presented assume impulsive burns, so in principle they are not suitable for low-thrust transfers. However, due to their low Δv and long time of flight, transformation of these trajectories to low-thrust is in principle feasible, and will be considered in future work. If a similar cost trajectory could be flown with a low-thrust engine of higher specific impulse (3000 s) the asteroid retrieved mass would be over ten times that of the high-thrust case, up to an impressive 4000 tons in the case of a hypothetical transfer from the orbit of 2006 RH120 to a halo orbit. That is beyond the maximum estimated size of this particular asteroid. Even if losses of 600% were assumed in the transformation from high to low thrust (and this could well be the case given the mass of the object intended for transfer) the estimated diameter would still be greater than the 7.4 m maximum expected size of asteroid 2006 RH120.

Table 21.4 Retrieved mass estimates with current space technology

		Total Δv [m/s]	*Isp* = 300s Mass [kg]	Ø [m]	*Isp* = 3000s Mass [kg]	Ø [m]
2006 RH120	2Hs	58	398144	6.64	4067256	14.40
2006 RH120	2Hn	107	213657	5.39	2222273	11.77
2010 VQ98	2V	181	121879	4.47	1304330	9.86
2007 UN12	2P	199	110313	4.33	1188630	9.56
2011 UD21	1Hs	356	57441	3.48	659549	7.85
2011 UD21	1V	422	47017	3.26	555160	7.42
2011 UD21	1Hn	436	45236	3.21	537325	7.34
2000 SG344	1P	443	44380	3.19	528741	7.30

For the average NEO density and assuming spherical bodies, the equivalent diameter of the asteroid that can be captured is also included in the table. This shows that reasonably sized boulders of 3-7 m diameter, or small asteroids of that size, could be captured with this method. The capture of entire bodies of larger size is still challenging, but the derived size of a few of the candidates fall actually within this range. With the higher specific impulse, the only NEO in the table that fails to meet the capturable range shown in Table 21.4 is 2000 SG344, with a derived size in the range of 20 to 65 meters.

21.3.4 Overview of the Selected Candidates

The capture candidates are all of small size (perhaps with the exception of 2000 SG344), which is ideal for a technology demonstrator retrieval mission. In fact, seven of them fit the small-Earth approachers (SEA) definition by Brasser and Wiegert (2008). They showed, focusing on object 1991 VG, that the orbit evolution of these type of objects is dominated by close encounters with Earth, with a chaotic variation in the semi-major axis over long periods of time. A direct consequence of this is that reliable capture transfers can only be designed with accuracy over one synodic period, before the next encounter with Earth changes the orbital elements significantly. One could argue that finely tuning these encounters could also be used to shepherd these objects into trajectories that have a lower cost to be inserted into a manifold (Sanchez and McInnes 2011b).

The candidates NEOs in Table 21.2 are well-known, and there has been speculation about the origin of a few of them, including the possibility that they were man-made objects (spent upper stages), lunar ejecta after an impact (Tancredi 1997, Chodas and Chesley 2001, Brasser and Wiegert 2008, Kwiatkowski et al. 2009), or even an alien probe (Steel 1995). In particular object 2006 RH120 has been thoroughtly studied (Kwiatkowski et al. 2009, Granvik et al. 2011), as it was a temporarily captured orbiter that was considered the "second moon of the Earth" until it finally escaped the Earth in July 2007. Granvik shows that the orbital elements of 2006 RH120 changed from being an asteroid of the Atens family pre-capture, to an Apollo post-capture, having followed what we refer to in previous sections as a transit orbit inside Earth's Hill sphere. An additional object in the list, 2007 UN12, is also pointed out by Granvik as a possible candidate to become a TCO (Temporarily Captured Orbiter).

Regarding their accessibility, a recent series of papers (Adamo et al. 2010, Barbee et al. 2010, Hopkins et al. 2010) considered up to 7 of the above objects as possible destinations for the first manned mission to a NEO (and the other 5 were not discovered at the time). They proposed human missions during the same close approaches as the capture opportunities calculated. However, the arrival dates at the asteroids are later than the required departure date for the capture, so their outbound legs could not apply to our proposed capture trajectories. An additional study by Landau and Strange (2011) presents crewed mission trajectories to over 50 asteroids. It shows that a mission to 6 of the considered asteroids is possible with a low-thrust Δv budget between 1.7 and 4.3 km/s. The costs presented are for a return mission of a spacecraft with a dry mass of 36 tons (including habitat) in less than 270 days. A longer robotic mission with a final mass at the NEO of 13,600 kg and a manifold capture as the one proposed would result in much lower fuel costs as the thrust-to-mass ratio increases. NASA also publishes the Near-Earth Object Human Space Flight Accessible Target Study (NHATS) list (Abell et al. 2012), which will be continuously updated and identifies potential candidate objects for human missions to asteroids. The NEOs are ranked according to the number of feasible return trajectories to that object found by an automated search within certain constaints. Eleven of our 12 capturable objects appear in the top 25 of NASA's NHATS list as of September 2012, seven of them in the top 10. This seems to indicate that the objects found by our pruning and optimisation are indeed easily accessible, even if the outbound part of the trajectory was not considered in our calculations.

21.3.5 Method Limitations

One of the first objections that can be raised to the approach presented involves some of the simplifications in the model. The main simplifying assumptions are placing the Earth in a circular orbit, assuming Keplerian propagation for the NEOs orbital elements until the next close encounter with Earth, and not including other types of perturbations, in particular the Moon third body perturbation. While the influence of the first two assumptions should be relatively small, and the trajectories obtained can be used as first guesses for a local optimisation with a more complex model with full Earth and NEOs ephemerides, not including the Moon as a perturbing body can have a much greater influence. Granvik (2011) shows that the Moon plays an important role in the capture of TCO, and so the trajectories of the manifolds would be also affected by it. The lunar third body perturbation can also strongly influence the stability of LPOs, in particular large planar Lyapunov orbits, and it could render some of them unsuitable for target orbits. However, the general behaviour and the type of NEOs that can be captured are not expected to change. Other perturbations, such as the changes in the orbit of small bodies affected by solar radiation pressure are of little importance within the timescales considered.

Other capture possibilities, e.g. by means of a single or double lunar swingby, have not been studied and are outside of the scope of this chapter, but they may provide even cheaper asteroid retrieval opportunities.

21.4 Conclusions

The possibility of capturing a small NEO or a segment from a larger object would be of great scientific and technological interest in the coming decades. It is a logical stepping stone towards more ambitious scenarios of asteroid exploration and exploitation, and possibly the easiest feasible attempt for humans to modify the Solar System environment outside of Earth, or attempting any large scale macro-engineering project.

This analysis has shown that the retrieval of a full asteroid is well within today's technological capabilities, and that there exists a series of objects that can be easily captured into libration point orbits. Taking advantage of this, the utilisation of asteroid resources may be a viable means of providing substantial mass in Earth orbit for future space ventures. Despite the largely incomplete survey of very small objects, the current known population of asteroids provides a good starting platform to begin with the search for easily capturable objects. With this goal, a robust methodology for systematic pruning of a NEO database and optimisation of capture trajectories through the hyperbolic invariant stable manifold into different types of LPO around L_1 and L_2 has been implemented and tested. Twelve possible candidates for affordable full asteroid retrieval missions have been identified among known NEOs with capture opportunities during the next 30 years with a cost below 500 m/s. Transfers to the libration points region have been calculated for all these targets. These transfers enable the capture of bodies within 3-7 meters diameter with low propellant costs.

The proposed method can be easily automated to prune the NEO database on a regular basis, as the number of objects in orbits of interest is expected to grow asymptotically with the new efforts in asteroid detection. Any new occurrence of a low-cost candidate asteroid can be optimised to obtain the next available phasing and transfer opportunities and the optimal target LPO.

Moreover Sun-Earth LPOs can also be considered as natural gateways to the Earth system. Thus, the problem to transfer an asteroid to an Earth or Moon centred orbit can be decoupled into the initial phase of inserting the asteroid into a stable invariant manifold and then providing the very small manoeuvres required to continue the transit into the Earth system. While a method to find optimal Sun-Earth LPO capture trajectories and possible targets has been defined, the transit trajectories can potentially allow the asteroid to move to the Earth-Moon L_1/L_2 or other locations within cis-lunar space taking advantage of heteroclinic connections between collinear points.

This analysis has also shown the costs of accessing the capture material at the Sun-Earth collinear equilibrium points. Given the costs associated with reaching the Sun-Earth LPOs, one can imagine the scientific and exploitation advantages of bringing asteroids close to Earth, as oppose to reaching them on their unperturbed heliocentric orbits.

Acknowledgements. We would like to thank Elisa Maria Alessi for her valuable comments and inputs to this work. We would also like to acknowledge the use of the Faculty of Engineering High Performance Computer Facility, University of Strathclyde. The work reported was supported by European Research Council grant 227571 (VISIONSPACE).

References

Abell, P.A., Barbee, B.W., Mink, R.G., Adamo, D.R., Alberding, C.M., Mazanek, D.D., et al.: The Near-Earth Object Human Space Flight Accessible Targets Study (NHATS) List of Near-Earth Asteroids: Identifying Potential Targets For Future Exploration. NASA (2012)

Adamo, D.R., Giorgini, J.D., Abell, P.A., Landis, R.R.: Asteroid Destinations Accessible for Human Exploration: A Preliminary Survey in Mid-2009. Journal of Spacecraft and Rockets 47(6), 994–1002 (2010)

Alessi, E.M.: The Role and Usage of Libration Point Orbits in the Earth - Moon System [PhD Thesis], Universitat de Barcelona, Barcelona (2010)

Aravind, P.K.: The physics of the space elevator. American Journal of Physics 75(2), 125–130 (2007)

Baoyin, H.-X., Chen, Y., Li, J.-F.: Capturing near earth objects. Research in Astronomy and Astrophysics 10(6), 587–598 (2010)

Barbee, B.W., Espositoy, T., Pinon III, E., Hur-Diaz, S., Mink, R.G., Adamo, D.R.: A Comprehensive Ongoing Survey of the Near-Earth Asteroid Population for Human Mission Accessibility. In: AIAA Guidance, Navigation, and Control Conference, Toronto, Ontario, Canada (2010)

Bewick, R., Sanchez, J.P., McInnes, C.R.: The feasibility of using an L1 positioned dust cloud as a method of space-based geoengineering. Advances in Space Research 49(7), 1212–1228 (2012)

Bottke, W.F., Morbidelli, A., Jedicke, R., Petit, J.-M., Levison, H.F., Michel, P., et al.: Debiased Orbital and Absolute Magnitude Distribution of the Near-Earth Objects. Icarus 156(2), 399–433 (2002)

Brasser, R., Wiegert, P.: Asteroids on Earth-like orbits and their origin. Monthly Notices of the Royal Astronomical Society 386, 2031–2038 (2008)

Brophy, J., Culick, F., Friedman, L., Allen, C., Baughman, D., Bellerose, J., et al.: Asteroid Retrieval Feasibility Study, Keck Institute for Space Studies, Califonia Institute of Technology, JPL, Pasadena, California (2012)

Ceriotti, M., McInnes, C.: Natural and sail-displaced doubly-symmetric Lagrange point orbits for polar coverage. Celestial Mechanics and Dynamical Astronomy 114, 151–180 (2012)

Chesley, S.R., Chodas, P.W., Milani, A., Valsecchi, G.B., Yeomans, D.K.: Quantifying the Risk Posed by Potential Earth Impacts. Icarus 159, 423–432 (2002)

Chodas, P.W., Chesley, S.R.: 2000 SG344: The Story of a Potential Earth Impactor. Bulletin of the American Astronomical Society 33, 1196 (2001)

Edward, T.L., Stanley, G.L.: Gravitational Tractor for Towing Asteroids. Nature 438, 177–178 (2005)

Farquhar, R.W.: Station-keeping in the vicinity of collinear libration points with an application to a Lunar communications problem. In: Space Flight Mechanics. Science and Technology Series, vol. 11, pp. 519–535. American Astronautical Society, New York (1967)

Gómez, G., Jorba, A., Simó, C., Masdemont, J.: Dynamics and Mission Design Near Libration Points: Advanced Methods for Collinear Points, vol. 3. World Scientific Publishing, Singapore (2001)

Gómez, G., Llibre, J., Martínez, R., Simó, C.: Dynamics and Mission Design Near Libration Point Orbits—Fundamentals: The Case of Collinear Libration Points, vol. 1. World Scientific Publishing, Singapore (2000)

Gómez, G., Marcote, M., Mondelo, J.M.: The invariant manifold structure of the spatial Hill's problem. Dynamical Systems 20, 115–147 (2005)

Granvik, M., Vaubaillon, J., Jedicke, R.: The population of natural Earth satellites. Icarus 218, 262–277 (2011)

Hasnain, Z., Lamb, C., Ross, S.D.: Capturing near-Earth asteroids around Earth. Acta Astronautica 81, 523–531 (2012)

Hénon, M.: Vertical Stability of Periodic Orbits in the Restricted Problem, I. Equal Masses. Astronomy and Astrophysics 28, 415–426 (1973)

Hopkins, J., Dissel, A., Jones, M., Russell, J., Gaza, R.: Plymouth Rock: An Early Human Mission to Near Earth Asteroids Using Orion Spacecraft. Lockheed Martin Corporation (2010)

Howell, K.C.: Families of Orbits in the Vicinity of Collinear Libration Points. Journal of the Astronautical Sciences 49(1), 107–125 (2001)

Howell, K.C., Pernicka, H.J.: Stationkeeping Method for Libration Point Trajectories. Journal of Guidance, Control and Dynamics 16(1), 151–159 (1993)

Kawaguchi, J., Fujiwara, A., Uesugi, T.: Hayabusa-Its technology and science accomplishment summary and Hayabusa-2. Acta Astronautica 62, 639–647 (2008)

Kleiman, L.A.: Project Icarus: an MIT Student Project in Systems Engineering. The MIT Press, Cambridge (1968)

Koon, W.S., Lo, M.W., Marsden, J.E., Ross, S.D.: Dynamical systems, the three-body problem and space mission design. Marsden Books (2008) ISBN 978-0-615-24095-4

Koon, W.S., Lo, M.W., Marsden, J.E., Ross, S.D.: Heteroclinic Connections Between Periodic Orbits and Resonance Transitions in Celestial Mechanics. Chaos 10(2), 427–469 (2000)
Kwiatkowski, T., Kryszczynska, A., Polinska, M., Buckley, D.A.H., O'Donoghue, D., Charles, P.A., et al.: Photometry of 2006 RH120: an asteroid temporary captured into a geocentric orbit. Astronomy and Astrophysics 495(3), 967–974 (2009)
Landau, D., Strange, N.: Near-Earth Asteroids Accesible to Human Exploration with High-Power Electric Propulsion. In: AAS/AIAA Astrodynamics Specialist Conference, Girdwood, Alaska (2011)
Lewis, J.S.: Mining the sky: untold riches from asteroids. Helix Books/Perseus Books, Reading, Massachusetts (1996)
Mainzer, A., Grav, T., Bauer, J., Masiero, J., McMillan, R.S., Cutri, R.M., et al.: NEOWISE Observations of Near-Earth Objects: Preliminary Results. The Astrophysical Journal 743(2), 156–172 (2011)
Massonnet, D., Meyssignac, B.: A captured asteroid: Our David's stone for shielding earth and providing the cheapest extraterrestrial material. Acta Astronautica 59, 77–83 (2006)
Pearson, J., Oldson, J., Levin, E.: Earth rings for planetary environment control. Acta Astronautica 58, 44–57 (2006)
Richardson, D.L.: Halo orbit formulation for the ISEE-3 mission. Journal of Guidance, Control and Dynamics 3(6), 543–548 (1980)
Sanchez, J.P., Colombo, C.: Impact Hazard Protection Efficiency by a Small Kinetic Impactor. Journal of Spacecraft and Rockets (2012) (in press)
Sanchez, J.P., García-Yárnoz, D., McInnes, C.R.: Near-Earth Asteroid Resource Accessibility and Future Capture Missions Opportunities. In: Global Space Exploration Conference 2012. IAF, Washington DC (2012)
Sanchez, J.P., McInnes, C.R.: Asteroid Resource Map for Near-Earth Space. Journal of Spacecraft and Rockets 48(1), 153–165 (2011a)
Sanchez, J.P., McInnes, C.R.: On the Ballistic Capture of Asteroids for Resource Utilization. In: 62nd International Astronautical Congress. IAF, Cape Town (2011b)
Scheeres, D.J., Schweickart, R.L.: The Mechanics of Moving Asteroids. In: Planetary Defense Conference. AIAA, Orange County, California (2004)
Shapiro, I.I., A'Hearn, M., Vilas, F., et al.: Defending Planet Earth: Near-Earth Object Surveys and Hazard Mitigation Strategies, National Research Council, Washington DC (2010)
Simó, C., Gómez, G., Llibre, J., Martínez, R., Rodríguez, J.: On the optimal station keeping control of halo orbits. Acta Astronautica 15, 391–397 (1987)
Steel, D.: SETA and 1991 VG. The Observatory 115, 78–83 (1995)
Struck, C.: The feasibility of shading the greenhouse with dust clouds at the stable lunar Lagrange points. Journal of the British Interplanetary Society 60, 82–89 (2007)
Szebehely, V.: Theory of orbits. Academic Press, New York (1967)
Tancredi: An asteroid in a Earth-like orbit. Celestial Mechanics and Dynamical Astronomy 69, 119–132 (1997)
Tsiolkovsky, K.E.: The Exploration of Cosmic Space by Means of Reaction Devices. The Scientific Review 5 (1903)
Vasile, M., Locatelli, M.: A hybrid multiagent approach for global trajectory optimization. Journal of Global Optimization 44(4), 461–479 (2009)
Veres, P., Jedicke, R., Wainscoat, R., Granvik, M., Chesley, S., Abe, S., et al.: Detection of Earth-impacting asteroids with the next generation all-sky surveys. Icarus 203, 472–485 (2009)
Zagouras, C., Markellos, V.V.: Axisymmetric Periodic Orbits of Restricted Problem in Three Dimensions. Astronomy and Astrophysics 59, 79–89 (1977)

Chapter 22
Shaped Metal Earth-Delivery Systems

Richard B. Cathcart[1], Alexander A. Bolonkin[2], Viorel Badescu[3], and Dorin Stanciu[3]

[1] Geographos, Burbank, CA, USA
[2] C & R Co., New York, USA
[3] Polytechnic University of Bucharest, Romania

22.1 Introduction

Michel Verne (1861-1925), son of the famous storyteller Jules Verne (1828-1905), rewrote and posthumously published *The Chase of the Golden Meteor* (1908). An entirely gold metal asteroid of enormous size, the public announcement of its "out-of-the-blue" coming to rest on the Earth's surface provokes an epic worldwide financial crisis. In 1941, Georges Prosper Remi (1907-1983), also widely known by his penname "Herge", published a newspaper serialized children's tale featuring his internationally beloved child-adventurer, the boy TINTIN, performing in "The Shooting Star", and considering the effects on fiercely rival capitalist financiers competing over possession of the strange solid body containing a fictitious mineral, "phostlite", which exhibits previously unknown, environmental-active globally influential properties.

Nowadays, in our non-fictional real-world commercial contexts, cosmochemists seek to examine fallen solid bodies as well as those seen coming to rest on our Earth's land or dredged from bodies of freshwater and seawater, to increase their laboratory-derived factual understanding of such strange extraterrestrial materials (MacPherson and Thiemens 2011). Along with others, cosmochemists presume prudently that extra-Earth raw materials, possibly from our Solar System's Asteroid Belt, when apprehended as a natural potential human infrastructure, will someday become valued collector's items. In a sense, the Asteroid Belt performs the same sociological function for moderns that Philolaus's rendition of the Pythagorean system of cosmology, which included a fabled "Counter-Earth", did for the ancients before the Asteroid Belt was observed and mapped (Burch 1954).

In the past, wealthy and imaginative hobbyists (techno-philanthropists) financed the blossoming of science and technology, including astronomy and ballooning. Moneyed 21st Century investors have funded Planetary Resources, Inc. based in Seattle, State of Washington, USA (Efrati 2012; Elvis, 2012). Planetary Resources Inc. (see http://www. planetaryresources.com.) intends to finance new means to both characterize and examine targeted object-asteroids close to Earth's solar orbit (Granvik et al. 2012). The company may be the international financial community's first major response to worldwide Green "sustainability" propagandized insistence that only natural Earth-biosphere energy and materials flows serve a basic-style human civilization. The concept of human Earth-biosphere dependence stems from 19th Century science, specifically the Earth Science view of our planet as a unified energetic and biogeochemical system. Alfred Lotka (1880-1949) called our homeland "…the world engine" (Lotka 1924). Meantime, some ambitious budding astromineralogists during the 21st Century continue to gather and document systematically mineralogical facts remotely by using asteroid-focused astronomical spectroscopy and powerful radars (Henning 2010).

When cosmic material and energy was aggregating to form a Solar System planet, the early Earth surface was seeded with precious gold by asteroid bombardment (Willbold et al. 2011). Although no asteroid has ever been discovered mostly consisting of socially (over-)valued gold, the 2 km-diameter, 4 x10^9 m^3 near-Earth Asteroid 1986 DA is known to be composed primarily of un-rusted iron and ~8% nickel, both are industrially valuable metals (Kargel 1994). Class M meteorites, typically composed of iron, nickel, cobalt and platinum-group metals, could become resources valuable to a 21st Century asteroid-mining based human civilization in the Earth-biosphere during our species' Anthropocene. In this regard, it is interesting to note that the temporary and economically costly International Space Station orbiting Earth already has all the essential features that architecture's futurists often propound to soon characterize Earth's ever more "sentient cities" (Shepard 2011): artifacts and systems that humans normally interact with on a daily basis store and process information about individuals and are activated by our movements and normal survival and business transactions. Future spaceship and human space colonies will always be aimed at the biological survival of their human occupants, an ethic that only recently has come to dominate the consciousness of the world-public (Hendricks and Mergeay 2007). To transition from one state of like-mindedness, today's Earth-centric macro-engineers must begin to refer to the total use of asteroids such as 1986 DA, which may have a 2012 USD market value of 87.2 trillion (Clark 2012) as if all of it were brought to the Earth-surface, coupled or inter-operable with widely dispersed industrial artifactual systems. If, instead, mankind attempted to concentrate metals dug from the Earth's average bedrock, then "…the estimated [production] costs exceed current prices by orders of magnitude" (Steen and Borg 2002). About 6.6% of all found, recovered and analyzed small-sized object-meteorites (aerosiderites class) consist only of iron. Vast desert-sited solar-powered automated metal processing and fabrication factories are a distinct possibility.

As we will demonstrate herein, even Green marketers, before 2050, may come to tout and sell asteroid-derived stuff that can be advertised as truly "Green" [olivine] (Cruikshank and Hartmann 1984; Schuiling 2006; Olsson et al. 2012). Therefore, macro-engineering, as it is currently being developed by organized wealthy entrepreneurs such as Planetary Resources Inc.'s leaders, is likely to foster a world-public zest (a possibly new acronym, ZEST, for "zero emission stuff transport") for timely and profitable investment outcomes in infrastructure produced with a production plan for uniformity (units of product can be traded on the globalized commodities market), where Earth-surface delivered, cheaply space-shaped metal iron ores as well as other raw materials sell strictly on the basis of bid price.

Allegedly if the world's people enjoyed identical *per capita* metal stock levels like persons living in the Earth's few fully industrially-developed eco-system states, using all technologies currently available, the volume of "...global in-use metal stocks required would be 3-9 times those existing at present" (Gerst and Graedel 2008). Approximately 45% of global iron entering use is dedicated to construction, ~24% is devoted to transportation equipment, and ~20% is formed into industrial machinery (Wang et al. 2007). [As of mid-2012, China is the largest consumer of iron ore.] The energy use per freight metric ton-kilometer (i.e., J/ton-km) for various transport modes is remarkable: aircraft transport is ~29 times more costly than by river or sea-going barge; potentially, ballistic delivery of shaped asteroid-derived iron ore to arid lands such as Australia or the Sahara could have energy use that is nearly zero. Possibly, the Gobi Desert will be more convenient for iron deliveries destined for China. In effect, our proposed outer space-made iron aerospace cargo vehicles are the equivalent of unusually formed ingots that are structurally sound enough not to splinter during ground impact. (Ballistic deliveries, as a major primary operation, or useful secondary side-effect, could excavate capacious surface crust voids in remote land regions, fillable with recoverable "urban ores", and ultimately sustainable geosynthetic fabric-lined landfill cells. Tapped groundwater rock formations must never be penetrated unnecessarily, of course. Splinters of iron won't be wasted because they can be gathered by magnets.)

Within a century there likely will be a significantly different world-economy, one that may be forced by pressing geophysical and social circumstances to deal quickly with abrupt global climate change, more limited Earth-crust availability of some metallic and energy resources (Driscoll 2007; Rauch 2009) and Earth-biosphere degradation instigated by a prosperous human population's worldwide natural increase. In other words, people may encounter some difficult infra-structure (Earth-biosphere) and ultra-structure (outer space) choices and forward-looking macroproject plans must be initiated soon to meet these impending macro-problems; Valero et al. (2011) have even defined Earth's ultimate saturation limit under a scheme excluding all extra-terrestrial resource utilization. Their final end-state planetary environment implies a mandatory modified social apprehension of the late-20st Century predictive slang-like geoscience term "Anthropocene" (Vince 2011). The word technology also has many diverse definitions—so many, in fact, that its popular conceptualization has become deemed usefully "hazardous" (Marx 2010; Machado 2006). The dynamics of radical technical innovation/invention to our world's economy seems to be one of creative destruction: as first proposed comprehensively by Joseph Alois Schumpeter (1883-1950), often referred to as "Schrumpter's Gale", the creative destruction is the key mechanism for global economic development (Thurner et al. 2010). Self-correction is inherent to capitalistic democracy, as is a capacity to fail economically and socially during an unexpected "Schrumpter's Gale" civilizational event. It may be that "Cliodynamics", as it is perfected, will come to support "Schrumpter's Gale" theory more completely than is presently possible (Spinney 2012). Herein, we seek to meld the past natural destruction to the Earth-surface caused by meteorite and asteroid impacts with tolerable induced near-future physical destruction, instituted mainly to promote the advancement of an expanding human, mostly Earth-biosphere lodged civilization. Since the mantra of efficiency and the worldwide prevailing human standard of living [coined in 1902] can vary markedly, the growing global "Middle Class" of humanity allows for greater political awareness, expanding that increasingly stable group's desire for accountable and representative governmental bodies at all levels of any society, and even a demand for free-markets in goods and service (Ali and Dadush 2012).

Taking into account the above considerations, the goal of this chapter is to investigate the physical interaction of iron asteroids with human society and to propose some methods for putting the valuable mineral asteroids resources in the service of our terrestrial needs. These methods consist of safely bringing the asteroids (or pieces of them) to the Earth, so that the removal of potential asteroid threat on functioning human societies is transformed into a useful economic action.

22.2 Generalities

22.2.1 Brief Information about Asteroids

The vast majority of asteroids are found in a ring-like swarm called the Asteroid Belt, located between the orbits of Mars and Jupiter at an average distance of 2.1 to 3.3 astronomical units (AU) from the Sun. Scientists know of approximately 6,000 large asteroids of a diameter of 1 kilometer or more, and of millions of small asteroids with a diameter of 3 meters or more. The largest asteroid, Ceres, is 785 km in diameter. Others range all the way down to dust mote size. The Martian moons (already seen close up) may be asteroids, captured by Mars.

22.2.2 Asteroid Belt

The mass of all the objects of the asteroid belt is estimated to be about 2.8-3.2×10^{21} kg, or about 4 percent of the mass of the Moon. Of this, Ceres comprises 0.95×10^{21} kg, a third of the total. The number of asteroids increases rapidly as their individual masses decrease.

22.2.3 Near-Earth Asteroids

Near-Earth asteroids, or NEAs, are asteroids that have Solar System orbits that pass close to that of Earth's. The approximately 1,000 asteroids that actually cross the Earth's orbital path are known as "Earth-crossers" and are considered to be threatening to our Earth-biosphere's integrity. These are objects of 50 meters or more in diameter in a near-Earth orbit without the tail or coma of a comet. As of May 2012, 8,880 near-Earth asteroids are known to science, ranging in size from one meter up to ~32 kilometers (1036 Ganymede). The composition of near-Earth asteroids is comparable to that of asteroids from the Asteroid Belt, reflecting a variety of asteroid spectral types.

NEAs survive in their orbits for millions of years; eventually they are eliminated by planetary perturbations which cause NEAs to be ejected from our Solar System or sent on a collision with the Sun or another celestial body, such as a planet. With orbital lifetimes short compared to the Solar System's age, additionally asteroids must be constantly moved naturally into near-Earth orbits to explain the newly-observed asteroids found to be present. The accepted origin of these asteroids is that voluminous Asteroid Belt debris is gravitationally pushed into the inner Solar System through orbital resonances with Jupiter. The interaction with Jupiter through the resonance perturbs the asteroid's orbit and it then comes into the inner Solar System. The asteroid belt has gaps, known as Kirkwood gaps, where these resonances occur as the asteroids in these resonances have been moved onto other orbits. New asteroids migrate into these resonances, due to the effect first recognized by a Polish civil engineer during 1888, Ivan O. Yarkovsky (1844-1902) that provides a continuing supply of near-Earth asteroids (Beekman 2006). Earth does have a small population of close-by asteroids that orbit our planet temporarily and it is estimated that 0.1% of all known meteors eventually impacting the Earth were before such impacts temporarily-captured orbiters (Granvik et al. 2012).

22.2.4 Near Earth Objects (NEOs)

A near-Earth object (NEO) is a Solar System object whose orbit brings it into close proximity with the Earth. All NEOs have an apsis distance less than 1.3 AU. They include a few thousand near-Earth asteroids (NEAs), near-Earth comets, a number of solar-orbiting spacecraft, and meteoroids large enough to be tracked in nearby outer space before striking the Earth. NEOs have become of increased interest since the 1980s because of increased awareness of the potential danger some of the asteroids or comets pose to the Earth and its biota, and active mitigations are being currently researched. In a well coordinated joint effort, the USA, the countries forming the extant European Union and other interested nations are systematically scanning the sky for NEOs, especially those endangering human civilization. The NASA has a USA Congressional mandate to catalogue all NEOs that are one kilometer in diameter. Potentially hazardous objects (PHOs) are currently defined based on parameters that measure the outer space solid asteroid-object's potential to move in threatening close approaches to our inhabited Earth-biosphere—truly, the human predicament in the Anthropocene. Mostly objects with an Earth minimum orbit intersection distance of 0.05 AU and corresponds to a 150 m diameter are considered PHOs.

Some NEOs are of high interest to technology-minded, wealthy capitalist individuals as well as other investors because they can be physically explored with lower mission velocity even than the Moon, due to their combination of low velocity with respect to Earth (ΔV) and almost absent gravity, so they may present interesting scientific opportunities both for direct geochemical and astronomical investigation, and as potentially economical sources of extraterrestrial materials for human exploitation. This makes them an attractive target for mineral exploration. By 2008, two near-Earth objects have been visited by spacecraft: 433 Eros by NASA's Near Earth Asteroid Rendezvous probe, and 25143 Itokawa by the JAXA Hayabusa Mission. The Japanese space agency JAXA—Japan Aerospace Exploration Agency, founded 2003--plans to land a spacecraft onto an asteroid in 2018. The spacecraft, Hayabusa 2, will be launched during 2014 with a goal of landing on the targeted asteroid, 1999 JU3, by mid-2018 before arriving at Earth for re-entry in 2020. (While comets are named to honor their discoverers, asteroids are named for scientists, geographic places, celebrities and all manner of other criteria.)

22.2.5 Near-Earth Meteoroids

Near-Earth meteoroids are smaller near-Earth asteroids having an estimated diameter less than 50 meters. They are listed as asteroids on most asteroid tables. The JPL Small-Body Database lists 1,349 near Earth asteroids with an absolute magnitude (H) dimmer than 25 (roughly 50 meters in diameter). The smallest known near-Earth meteoroid is 2008 TS_{26} with an estimated size of one meter.

22.3 Sociometeoritics

The formalized study of the physical and psychical interactions of meteorites/asteroids with living persons housed by Earth's biosphere and various thriving societies encompassed by our world's human civilization is professionalized as "Sociometeoritics". Stone tool-making African predecessors of modern humans, about 1.07 million years ago, experienced the violent central Ghana creation of the 10.5 km-diameter Bosumtwi Crater and early Americans, if any were present perhaps 50,000 years ago, could have witnessed, perhaps only momentarily, the awesome instant formation, Earth-surface "weathering" of a disastrous kind, geomorphological process of Meteor Crater in Arizona, USA, by the impact of an iron object arriving suddenly from outer space. More and more the world-public has become interested, even somewhat entranced, by the fascinating potential for Earth-surface asteroid impacts and asteroid mineral utilization by various industries, some supported by national governments as well as multi-national techno-philanthropist operators (Lee 2012; Marriner et al. 2010). So much so, in fact, that the infamous 25 October 2009 Latvia meteorite hoax was perpetrated by Tele2, a Sweden-based telecommunications company as a globe-wide gimmick to gain public notoriety, a case of media stunt-making rights trumping legitimate meteorite investigators for the fickle world-public's attention and money. Perhaps this hoax was a copycat project since Chinese artist Cai Guo-Qiang (born 1957) had earlier pioneered this kind of faux meteorite land impact with his 7 July 1990 outdoor artwork at Pourrieres, Aix-en-Provence, France *45.5 Meteorite Craters Made by Humans on Their 45.5 Hundred-Million-Year-Old Planet: Project for Extraterrestrials No. 3* (Friis-Hansen et al. 2012) during the exhibition "Chine demain pour hier".

Since there are functioning nuclear reactors orbiting Earth in man-made satellites, the fall of any large chunk of space junk elicits grand-scale public attention that is little different from other extreme events, both experienced (earthquakes, volcanic eruptions, and Solar Cycle 24 magnetic storms) and imagined (global warfare's postulated Nuclear Winter). Civilization's existing nearly globalized populace now demands that geoscientists, space scientists and macro-engineers increasingly undertake complex predictions—that is, macro-project decision-making under conditions of uncertainty that requires an increasingly multi-disciplinary, yet simplified approach to the projection of extended human actions and future Solar System products or byproducts (Reimold and Gibson 2010).

Circa 1970 the biochemist Albert Lehninger (1917-1986) observed that there are very high concentrations of calcium and phosphate in human body fluids—so much he asserted that each of us could imitate Lot's Wife, petrifying if natural inhibitors in our bodies did not actually prevent us from turning into stone. *The Monolith Monsters* is a good example of the 1950s era science fiction-monster movie. In that popular feature film, small black crystals—fragments of an fictitious asteroid strewn widely and anciently over a hot and dry southern California desert impact zone—when wetted transform into gigantic towering rocky obelisks, unintelligible by down-to-earth geoscientists except in the crystals' apparent seeking of global domination, accomplished by parading relentlessly over a mostly uninhabited localized arid landscape. Their growth and movement is supposed by script-writers and movie-goers to function by the subduction of silicon atoms from the surrounding desert's dune sand. For hapless humans, who unknowingly stray too close physically to the immobile small crystals, the result is that they turn to stone too. Natural salt, and salt-laden water, seems to inhibit crystal growth and it is hinted these materials even destroy the crystals eventually. But the answer to that question is left unresolved as it is not part of the exciting cinematic finale; only seemingly temporary immobilization and collapse of the rock towers is demonstrated. We wonder if *The Monolith Monsters* screenplay, authored by Robert M. Fresco (b. 1930) and Jack Arnold (1916-1992) may have stimulated Albert Lehninger's provocative geobiochemical (cryptobiological?) assessment more than a decade later. Their screenplay postulated a horrific rock-making bioinvasion existential threat that has not yet been experienced by humans, unless the ancient Greek's myth of the Gorgons set a precedent based on some long-forgotten factual event-process. B-movie scripts swamped with pseudoscience have impressed the world-public, and presage humanity's now prevalent fears about anything incoming from outer space that enshrouds our Earth.

Fig. 22.1 The *in situ* Hoba iron meteorite. (Copyright free Photo courtesy of J.M.H., 2012.) Notice its flatness, almost wing-like natural shape. Valued only as a chunk of iron scrap metal (average open-market commodity price as of May 2012), Hoba is worth approximately USD 6,000 yet the cost of moving it to an industrial site would probably be exorbitant.

People wandering Earth's glaciated wild-lands harvested iron from a Greenland meteorite (Buchwald 1992) and numerous pure green-glass shards, glass formed from Saharan sand by an aerial bolide explosion above the desert dunes was incorporated as centerpieces in magisterial Egyptian Pharaoh's regalia (Wright 1999); settled people in Canada and South Africa continue to abstract valuable metals—chiefly iron—from very ancient terrestrial impact craters by digging deeply into the Earth-crust (Grieve 1994). Virtually unharmed by human action, the >60 metric ton Hoba meteorite (Fig. 22.1) in Namibia (19^0 35' 32.9" South latitude by 17^0 56' 1.2" East longitude) is the largest known single piece of pure meteoric iron resting on land; it has a flat shape on both major surfaces ("top" and "bottom") and, possibly by skipping through Earth's atmosphere during its entry, it appears to have fallen to its terminal velocity. It apparently, and evidently, blasted no crater in the landscape upon final impact, possibly 80,000 years ago

When material is removed by quarrying, natural landforms are destroyed but when mine wastes are mounded they accumulate to form unnatural positive landforms. Such anthropic landforms are part of our present-day definition of the Anthropocene. Mining Nature-deposited impact/explosion detritus such as iron often is easier, less dangerous and more profitable, than from geographically isolated excavations such as the abandoned Canadian lead and zinc mine at Little Cornwallis Island, active from 1981 until 2004, or the ongoing rhenium extraction effort at an active volcano vent on Iturup Island, Russian Federation (Jones 2000). Very few, if any, mines are without contaminated liquids and solid waste (Chen and Graedel 2012; Sen and Peucker-Ehrenbrink 2012; Reck and Graedel 2012;

Bian et al. 2012) and piled, environmentally exposed gangue of all kinds and mutability affects vast stretches of sub-aerial landscape on the ~15% of Earth's crust imprinted by Homo sapiens' metal-use activities (Kennelly 2011; Hudson-Edwards et al. 2011). Aside from space-weathering, asteroids do not have strong internal alterative "geologic processes" and, therefore, the effects of mining in outer space cannot affect Earth's aquatic ecosystems (the ocean, lakes and rivers) since, herein, only a product (such as, for example, agglomerated and aerodynamically formed pure iron called the lifting-reentry configured "Mega-ASSET" (Powell and Hengeveld 1983), is to be brought to Earth for use in civilization's possibly solar-powered factories located nearly everywhere (Schmidt et al. 2012), by Green-styled establishment of outer space-oriented enterprises.

But, what if human civilization's technology progresses to a new, advanced technology state allowing humans, certainly by mid-21st Century, to look almost exclusively to the sky and beyond for metallic wealth (Muller et al. 2006) made to artificially fall from outer space at near nil post-2012 USD transportation cost for further exploitation? The term exploitation refers to a wide variety of human activities of either internationally legally-sanctioned ecosystem-state or private operators necessary to put Solar System asteroids to use, from prospection to removal of the resource from its initial location to, finally, its deposition upon the accessible Earth-surface.

The macro-engineer Samuel Florman (born 1925), in his techno-thriller epic *The Aftermath* (2001), novelized the post-impact Earth-biosphere recovery process headed by macro-engineers, preserved by being aboard a sturdy ship on the safest side of our planet, to rebuild a crippled civilization after a comet's surprising catastrophic collision with humanity's Solar System homeland. Surprisingly, the astronomer Samuel Herrick (1912-1974) put forth his original idea in 1971 that outer space rock 1620 Geographos, a 5 km by 2 km Paleolithic arrowhead-shaped S-type asteroid (Fig. 22.2), could be used to abruptly excavate an inter-oceanic sea-level canal along the water-eroded valley course of the Atrato River in north-western Columbia sometime after 1994 but before the turn of the 30th Century (Gehrels 1980). [1620 Geographos will closely pass Earth on 23 August 2051 and on 27 August 2119.] Herrick's undeniably unique single-use macro-engineering concept was destined to become outmoded by technology's progress, especially should canals like the Panama Canal in the Americas and the Suez Canal in Eurasia (Finkl et al. 2012) become unwanted or unneeded infra-structures. Since the 6-12 May 2010 closing of some southern Europe airports due to the ash cloud spreading aerially from the erupting Iceland volcano Eyjafjallajokull, it is known that volcanic dust ejections cause havoc with schedules and fixed route international air services. The reflective dust cloud suddenly raised by any impact of an intact 1620 Geographos (Fig. 22.2) would, probably like Mount Pinatubo in 1991, cool the Earth by ~0.5 K within a year following its violent touchdown. (In fact, it would replicate to some degree a "Method and Apparatus for Cooling a Planet", USA Patent Application Publication 2011/0005422 A1 issued to Stephen Trimberger or a horrible global or regional post-conflict Nuclear Winter.) However, Samuel Herrick's macro-project plan is not really an outrageous consideration as at that time, *circa* 1969, the public-empowered Atlantic-Pacific Interoceanic Canal Study Commission was reviewing the possibility for the employment of 250 synchronized nuclear fission device explosions—which if used yielding overall

~120 Megatonnes—to create a series of connected craters producing a massive continuous crustal wave of industrial earthmoving from seacoast to seacoast. Furthermore, two macro-engineers have offered the idea that a wide trans-Panama sea-level seaway must be dug in future to cause "...the restoration of the ancient circum[-]global equatorial [ocean seawater] current...using conventional and nuclear civil engineering methods" in order to control anticipated impending Earthly atmospheric heat fluxes that could obtain during abrupt climate change event-processes (Stevens and Ragheb 2010). For sure, humanity is likely to encounter entirely new ecosystem combinations everywhere by *circa* 2100 (Williams et al. 2007). Such macro-project thinking is unhelpful because it is equivalent to proposing that the potentially huge natural rock landslide from the slope of the active volcano Cubre Vieja, located on the island of La Palma in the Canaries be suddenly triggered artificially through impact of human-directed multiple meteorites, or a single very large asteroid, tried as a substitute for the laborious and costly application of earthmoving machines to that expensive task (McGuire 2005). Hydraulic rock fracture is the means of accomplishment and extra-terrestrial rock-caused blast is the immediate stimulation mechanism, anthropogenic seismicity. The tsunamis generated would, at least, equal those foreseen for a hypothesized Atlantic Ocean impact of the 0.8 x 10^9 m^3 Type M asteroid 1950DA, thought to be rich in the platinum group metals, on 16 March 2880 (Ward and Asphaug 2003; Abadie et al. 2012).

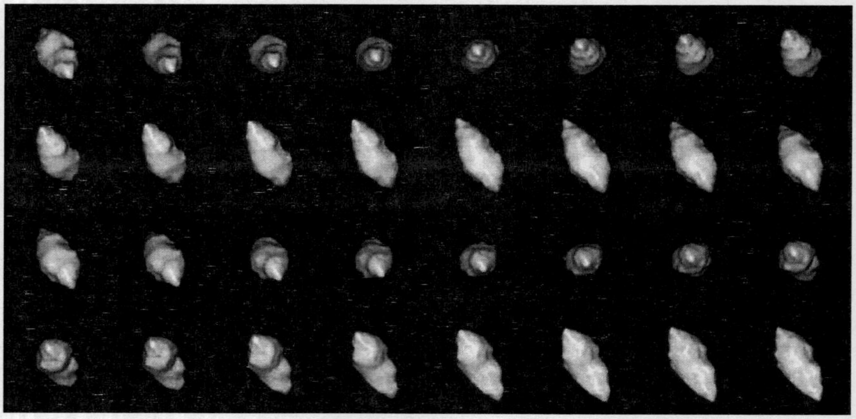

Fig. 22.2 NASA image of 1620 Geographos (copyright free). Foreseen by Samuel Herrick as an instrument not of wanton, doom-like societal anthropogeomorphological destruction but rather of an instantly created negative landform, 1620 Geographos still remains available for other uses. Someday it might be mined. Could it possibly be used to excavate a Kra Canal?

Then, of course, there is the frighteningly awesome possibility for a macro-engineered program of deliberate weaponization of asteroids and/or meteorites as well as manufactured metallic rods dropped from some orbiting outer space platform. Meteors have been postulated as dramatic death-dealing weapons by the popular B-cinema films, as in *The Monolith Monsters*, but also even earlier as, for instance, in the moving picture *This Island Earth* (1955). Accurate geosynchronous Earth-satellite GPS signals make it feasible for modern war-planners and war-gamers (von Hilgers and Benjamin 2012) to ponder the applied use of scored tungsten rods fabricated somewhere within the Earth-biosphere and projected spaceward, released into a place-to-place sub-orbital path and then shed their containers (ground-to-ground or submarine-launched missile warheads) in the terminal phase of re-entry flight, showering a selected fixed or moving military target with a destructive metallic storm approximating an instantaneous hail of vertically descending explosive 50-caliber bullets, or worse (Preston et al. 2002). We do not intend to enlarge humanity's negative cultural actionable choices herein (Hupy and Koehler 2012). Certainly, such self-imposed destruction done by raining lethal masses could never be termed an "act of God", even though some judicial experts have defined "act of God" explicitly as an "unusual [material] volume" that pushes a natural Earthly event to the status of an acknowledged act of God (Fraley 2010; Pierazzo and Artemieva 2012).

22.4 Delivery Methods

In this section two delivery methods are proposed. The first is based on cutting the metal-rich NEAs into many pieces and processing these fragments into shaped metallic flight bodies (Thompson and Peebles 1999; Reed et al. 2011) through difficult extra-terrestrial mining and fabrication operations. Further, these finished selected bodies are sent towards the Earth, and in the dense layers of the Earth's atmosphere they are air broken by deployment of attached AB carbon fiber parachutes. The second method proposed is to capture the NEAs in outer space (somewhere along their natural Solar System trajectories) with the aid of some appropriate-to-the-task smart apparatus launched from Earth. After interception, the kinetic energy of the proposed apparatus is used to fracture the asteroid, so that the whole delivery system (artificially comminuted asteroid and intact smart apparatus) changes the trajectory and enters into the upper layer of the Earth's atmosphere. Below the Karman Line, the aggregate delivery system is slowed in the same manner, by opened AB carbon fiber parachutes.

22.4.1 Mega-ASSET Based Method of Delivery

22.4.1.1 Mining Asteroids

Our civilization, which inhabits only a single, but nevertheless singular, terrestrial-type Solar System planet, soon must cope with the asteroid threat to the human species' native land. The daily observable occurrence of innumerable meteor smoke trails in Earth's upper stratosphere (~90-100 km, in the immediate vicinity of the Karman Line) is a powerful reminder to humans that we are being pelted by solid objects originating some place elsewhere than our planetary biosphere. The infall of ablating micrometeorites that exude ~2400 tonnes of sulfur dioxide annually into Earth's air has some measurable, if rather modest, climate change effect (Court and Sephton 2011).

Probably there are ~10^4 objects bigger than 10 cm in diameter impacting Earth yearly. On 31 January 2012, the 32 km-long non-PHA asteroid Eros made its closest flyby of our planet in 37 years and it was distinctly visible to astronomers, both amateur and professional, using even small low-resolution backyard telescopes. It was Harvey Harlow Nininger (1887-1986) who first established, in 1942, that asteroid impactors hitting Earth could cause mass extinctions of life; the astronomical odds that people alive today will face a major asteroid impact are still, despite the near ubiquity of supercomputers, incalculable and the global social effect of a Torino Impact Hazard Scale warning remains unfathomable (Binzel 2000). [1942 was the year the V-2 rockets first reached outer space so that humans could then and subsequently claim to be on their way to becoming an outer spacefaring species after passing upwards and downwards through the Karman Line.]

Metal-rich asteroids can be used as mineral resources to supply Earth-based needs, allowing the conservation and preservation of scenic landscapes and unmolested wildlife. But the first exploitation of an asteroid is never likely to be a financially profitable venture so that means that Earth's defense from PHOs may be the stimulus that induces humans to forego the tried and true efforts of traditional terrestrial metal mining in usually remote regions of the planet and progressively project our species outwards from the presumed safety of our Earth-biosphere. Encouraging an extensive wild-catting commercial mining operation to piggyback on a United Nations Organization planetary defense operation could accomplish those twin risky tasks synergistically. How the mining of chosen and well-placed near-Earth resources is done is best left to the expertise of the technical and geopolitical elites contributing informative and fascinating chapters to this textbook. Commercial-minded capitalist elites, such as the founders of Planetary Resources Inc. are not to be ignored for sure. For our purpose in this chapter, we assumed the presence of many aggregated containers filled with natural iron and smelted iron—that is, melted or fused ore from which impurities have been extracted—simply await transfer to the Earth-biosphere's surface, whether land or ocean.

22.4.1.2 Description of Tested ASSET Reentry Vehicles

About the same time as the USA's Plowshare Program—the planned peaceful use of nuclear explosives for large-scale geographical alterations of the Earth-surface—was forming, 1960, the USA began its ASSET Program. ASSET signified "Aerothermodynamic Elastic Structural Systems Environmental Tests". ASSET gliders, such as the tested ASV ("Aerothermodynamic Structural Vehicle"), were intended to be winged controlled-reentry vehicles based on a dynamic gliding principal. They featured a bi-conic form with an extremely reduced wing that offered a lift/drag equal to 1.0-1.5. and these vehicles visually resembled the USA's 104,000 kg Space Shuttle. ASSET vehicles, however, had no vertical stabilizers (Fig. 22.3).

Fig. 22.3 ASV-3, a height of 1.79 m, wing span of 1.53 m, a gross mass of 540 kg, and a volume of 0.56 m^3. Six vehicles were launched from 1963 until 1965. (Copyright Free Image: courtesy of Encyclopedia Astronautica; http://www.astronautix.com/craft/ asset.htm downloaded 12/28/11)

22.4.1.3 Proposed Mega-ASSET Vehicle

Based on the ASSET vehicles, one of the authors, in a brief popular magazine essay, first proposed a "Mega-ASSET" machine macro-project entailing the forced descent (fall) of controllable "Directed Meteorite Excavators" (DME), not retarded by any kind of deployed parachute device or rocket blast, for mineral delivery at the Sahara and Central Australia as anthropogenic aerial agents of rapid strip-mining (Cathcart 1981). Central Australia is famous for its five explored large impact structures—Gosse's Bluff, Henbury, Boxhole, Kelly West and Amelia Creek and numerous surface recoveries of small meteorite fragments from shattered meteorites have occurred mainly since about 1971. Interestingly, the Henbury geomorphic feature was initiated by an iron object about 4,200 years ago and it is possible that the Arrernt [aboriginal] people of Central Australia witnessed its impact firsthand (Gammage 2011). From 1986 until the present in northwest Africa—especially in Libya, Algeria and Morocco—uncounted small meteorites, some metal and some stone, have been found and sold commercially, especially at the trading cities of Erfoud and Rissani. Collecting fragments alleged to have fallen from the sky has even become an advertisement-promoted foreign tourist activity. Many of the fragmental finds were gathered from the Saharan regs, spacious and almost featureless flat regions of that vast northern Africa desert. The temporary world-public craze was fostered by e-Bay sales made feasible by the Internet. These scientifically undocumented meteorites are sometimes sold at the major annual international meteorite showcases in Tucson, Arizona and Denver, Colorado in the USA and at Ensisheim in the FRG. Ultimately, the crude

1981 Cathcart DMEs, early conceptual forerunners of the proposed "Mega-ASSET" machines proposed in this chapter, were then considered to be little more than slightly ablative nosecones; the material source for prospective DMEs would likely be the near-Earth asteroids, so DMEs would be shaped by mankind's machines in the realm of outer space surrounding our planet. In the absence of any braking device below the Karman line, the use of perfected DMEs—i.e., 21st Century "Mega-ASSET" machines—must be coordinated globally, perhaps through the auspices of the United Nations Organization, since the terminal blast/potential excavation of one DME might be mistaken by the uninformed and inexpert members of the world-public for a mini-nuclear warhead's detonation (Chyba et al. 1998). However, there is another possibility: the use of AB carbon fiber parachutes (see Sect. 22.4.1.4) could make the terminal impact far less devastating to the landscape directly affected by such Cathcart-postulated DME landings. Such extraordinary, man-caused impacts of extra-terrestrial materials would be a new form of weathering, additional to those usual forms already existing (Hall et al. 2012); as macro-engineers we can foresee landscapes classified in terms of how they can be shaped or gouged by artificial meteor impacts. Could a future Kra Canal be excavated rapidly in Thailand by a well-controlled DME (Cathcart 2008)? Such impacts would, of course, be far less than Samuel Herrick's 2.4 million megaton extravaganza of pure earthly destruction. Hazardous materials would never be delivered in this manner (Cidell 2012). Asteroid mining and material shipments to the other celestial bodies, as well as anywhere in this Solar System, is a far-ranging vision since the delivery systems proposed herein could be used at other planetary places with the potential to be terraformed (Mars etc.). It is worth noting that the NASA Curiosity, arriving at Mars on 5 August 2012, ejected 75 kg of Earth-mined tungsten weights that allowed the atmosphere entry capsule to perform the first planetary lifting-body entry.

It is our intent to *further* adapt the Mega-ASSET configuration, but with a significant change. Instead of pre-figuring an inhabited aerodynamic aerospace vehicle such as the X-20 Dyna-Soar, tested 1958-1963, upgraded "Mega-ASSET" machine-vehicle designs could be transformed into more robust, completely solid flight bodies that are maneuverable remotely (Hallion 1983). Lashing a cargo of asteroid-mined iron, or other metals, would be no problem simply because the modernized ASSET-type vehicle *is* the cargo, and it cannot shift position or morph its exterior shape unexpectedly. Or, alternatively, the cargo could be pelletized iron ore derived from a mined asteroid, even possibly more-valuable-than-iron platinum group metals safely secured inside the sturdy thick ironwalled "Mega-ASSET" machine-styled cargo vehicle.

21st Century upgraded "Mega-ASSET" machines might be shaped by compressing beneficiated iron ore in automated outer space factories that would then ablate somewhat during atmospheric entry (Blanchard 1972), possibly fertilizing the less productive Tropic Zone ocean with iron flakes digestible by phytoplankton. The USA's self-powered, non-ballistic, Hypersonic Technology Vehicle-2 test on 11 August 2011 had to be aborted, even after the vehicle had demonstrated stable aerodynamically controlled flight, because, at about Mach 20, the vehicle's skin commenced to peel, thereby inducing a strong official concern about its continued autonomous flight safety and operations system. This is a strong indicator that bluff-tip solid bodies composed of iron will possibly fair such stressful, short-duration flight circumstances much better. (In the USA, it was Harry Julian Allen (1910-1977) who is credited with devising the blunt missile nosecone as a superior means for high-velocity vehicle atmosphere entry.) Our post-ASSET-3 vehicles, the "Mega-ASSET" machines, will be nowhere near as complex in terms of both structure and controllability as the 4,990 kg, 8.9 m-long X-37B Orbital Test Vehicle used by the US Air Force and even less complicated than the original 540 kg ASSET-3.

22.4.1.4 Deceleration System of MegaASSET-AB Carbon Fiber Parachute

Upon entrance to the Earth-atmosphere, large meteoroids have initial velocities ranging from 11-73 km/s and can produce strong shock waves while ablating, fragmenting and decelerating. These cause a kind of noise pollution (sonic booms). The now honorably retired Space Shuttle fleet, all of which nowadays rest in museums, was said by some of its pilots to have the fly-ability of a "brick". It served the USA and some foreign clients from 1981 until 2011. Almost immediately after its atmospheric entry it was often tracked by seismologists due to the measurable ground shaking it instigated (Kanamori et al. 1992).

For most natural metals the lower limitation for hypervelocity impact on the land is on the order of only ~3 km/s, hence the need for the AB carbon fiber parachute device attachment to the "Mega-ASSET" machine since greater speed would result in splattering. According to Martyn John Fogg (b. 1960), famed for his pioneering scientific work in penning our world's first macro-engineering textbook devoted to Terraforming, a science-fiction author, Peter F. Hamilton (b. 1960), in one of the episodes of his ***The Night's Dawn Trilogy*** published from 1996 until 1999, elucidated a scheme to form and bring nickel-iron ingots into atmospheric entry stage. However, Hamilton added an interesting twist to this feat—by foaming the ingots with induced gas the result was a kind of anthropogenic pumice far less dense than, say, seawater. Thus, after splashing down in an Earth ocean, such floatable non-toxic bodies might be rounded-up by roving tugboats and towed/pushed to some nearby seaport. As Fogg told the authors: "The idea is certainly more elegant than smashing solid masses of iron into land surfaces…" (Fogg 2012). As it turns out, closed-cell iron foam is a metal cellular structure with a large volume fraction of gas-filled pores and a porosity typically of from 75% to 95%. This type of material is not, to the best of our knowledge, found in natural meteors. Closed-cell metal foam was first announced in 1926 by M.A. DeMeller in his French Patent 615,147. In outer space, manufacturing of foamed iron Mega-ASSET machines would entail the use of CO and CO_2 foaming gases, resulting in a massive gray-colored ingot-vehicle with a porosity of 55% that could indisputably survive Earth-atmosphere entry temperatures up to 1543 K (Murakami et al. 2007). And, upon landing, it would float like a well-crafted boat or ship, just as Hamilton foresaw during the late-20th Century.

The AB carbon fiber parachutes were first proposed by Alexander Alexanderovich Bolonkin in the first years of the 21st Century before the USA's Space Shuttle fleet had been retired from active service (Bolonkin 2006). His strong air-resisting parachute device determined by Bolonkin to slow the descent of heavy objects from outer space must, ideally, be attached to balanced solids with a maximum measured mass. Interestingly, the USA's NASA has, since 1960, postulated and tested developed exo-atmospherically deployable inflatable aerodynamic deceleration technology—blunt nosecones, in fact—appropriately designed to greatly increase drag on entry vehicles (or objects) at hypersonic speeds (Mach Number >5) that are reliably capable of surviving high frictional heating during mitigated Earth-atmosphere entry. IRVE-3 (The Inflatable Reentry Vehicle Experiment) test flight conducted on 23 July 2012 was part of the NASA macro-project called HIAD, standing for Hypersonic Inflatable Aerodynamic Decelerator. Roger Bacon demonstrated the first high-performance carbon fibers in 1958 and published his findings two years later. Nowadays, NASA is testing the properties of many strong, heat-resistant fibrous materials: for example, Kapton, Kevlar 29, Kevlar 49, Nomex (type 430), PBO Zylon, Spectra 2000, Technora, Upilex-25S, Vectran (HT) and M5. The strong carbon fibers utilized mimic asbestos and fiber-glass in that they do not ignite. Further testing and wise selection will be necessary to realize a proper material of the AB carbon fiber parachute system. All parachutes rely on tethers to steady the device and permit it to function as it was meant to function. However, some experts have proposed that a single 20 km-long trailing carbon fiber tether, a mere one mm diameter, with a total area of ~20 m^2 be employed to decelerate spacecraft from low-Earth orbit (Krischke et al. 1995); the AB carbon fiber parachute could easily be blossomed after the worst period of the rapid heating caused by rapid descent is past during the atmospheric entry process.

Using the AB carbon fiber parachute permits the safe and controlled landing of any harvested Hoba-type bodies of iron ore that are deemed to be of the correct mass matched to the parachute attached. The long-tether and carbon fiber AB parachute remote-operation smart package will be locked to the rear end of the iron object, where it will be maximally shielded from entry heating and erosion, being delivered regularly to adept and proficient ground-based receiving crews living in two of the Earth-biosphere's largest deserts—the Sahara and Central Australia. Appropriate industrial-scale processing and fabricating facilities, as well as efficient ground/rail/air transportation, must certainly be present at, or very near, both suggested remote desert landing zones to make the entire outer space asteroid mining enterprise economical, not to say profitable as a purely trajectory-based industrial operation. Managers of the author's proposed delivery systems might follow the flights and landings of the ingot-like cargo vehicles by using globular electronic virtual globes (Cathcart 1997; Brovelli and Zamboni 2012). Blessed with enormous subsurface oil, natural gas and fresh groundwater deposits, the Sahara is the fastest changing, most dynamic and wealthiest region of Africa—so much so, in fact, that nowadays it is a major intrinsic component of international commerce.

22.4.1.5 Technology of Delivery

Let's imagine for example that Asteroid 1986 DA has been thoroughly examined for possible contaminants by qualified prospectors in order to completely prevent any significant bio-invasion of the Earth by unwanted anti-biotic organisms. Many reputable astro-biologists currently regard the entire known Universe as an enormous cryogenic habitat for active and dormant microbial life-forms, some of which might be dangerous to the long-term maintenance of human civilization in the Earth-biosphere (Wickramasinghe 2004). Therefore any celestial object-asteroids selected for Earth delivery have to be thoroughly examined for possible contaminants either by qualified prospectors or by some automated devices launched by rockets, in order to completely prevent any significant bio-invasion of the Earth by unwanted organisms. In the first steps of asteroid delivery actions, our choice is to use the qualified prospectors in order to both have an on-the-spot human decision and to acquire or gain the human experience instead of transferring the judgment to some software blindly written on the Earth.

Let's suppose that after inspection and decontamination (if necessary) the asteroid was fractured into a billion equally sized chunks of material, each piece would be valued at 2012 USD 25,000. However, if we then imagine that fragment as being composed of 30,000 kg of only meteoric iron (density = 7.8 g/cm^3), then that segment would have a volume of ~38.5 m^3, or slightly more than 68.7 times the bulk of the ASSET-3 test nosecones from the early-1960s. Since the USA's Space Shuttle weighed 104,000 kg, our AB carbon fiber parachute, as currently designed and calculated, is more than adequate to accomplish our stated Earth-surface delivery task. In other words, our "Mega-ASSET" smart machines are eminently feasible in virtually all phases of the macro-project as outlined herein. We encourage others to pursue its completer definition and final realization as soon as possible in order to foster the capitalistic development of Asteroid Belt mining businesses on behalf of Homo sapiens.

22.4.2 *Kinetic Method of Asteroid Capture and Delivery*

22.4.2.1 Description

The smart apparatus for delivery of processed, shaped asteroids to the Earth contains the rocket, computer, devices for definition of asteroid composition (for example, the laser spectrometer), radio receiver/translator, capture net, a useful long cable and a mechanical energy accumulator, heat-resistance control rectangular parachute and so on.

22.4.2.2 Work of Smart Delivery Apparatus

The smart delivery apparatus works the following way. Most asteroids captured by our planet are moving in the elliptic orbits having a focus at the Earth (Fig. 22.4a). Also, the smart delivery apparatus, in most cases, will also have an elliptic orbit.

The elliptic orbit has the perigee–the nearest point to focus (Earth) and apogee–the most far point from its focus (Earth). The asteroid-objects speed is maximum in the perigee, minimum in the apogee. An object-asteroid captured by Earth commonly has a velocity between 8 km/s and 11 km/s. If its speed is less than 8 km/s, it will likely fall uncontrollably to the Earth's surface. If the asteroid-objects speed exceeds 11 km/s, then the asteroid will probably fly off to the far reaches of outer space.

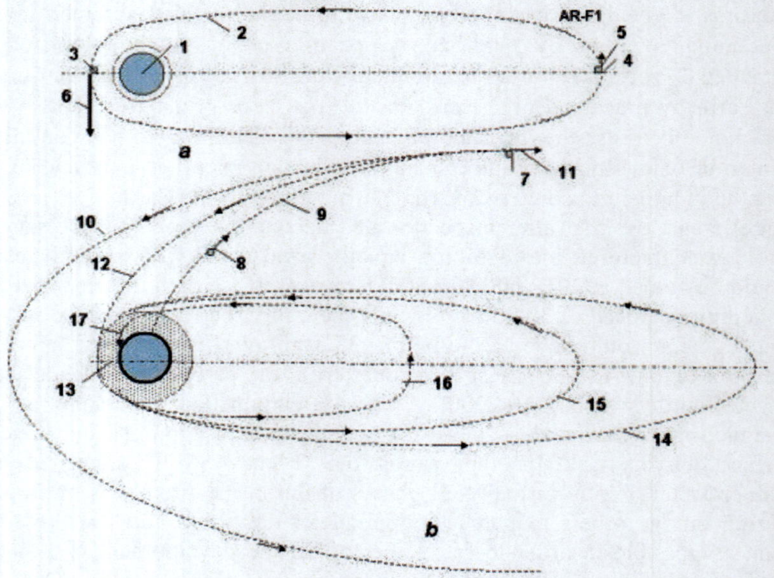

Fig. 22.4 Cheapest method, perhaps, for regular delivery of asteroid-objects to the Earth. (***a***). Elliptic trajectory of any asteroid captured by Earth. (***b***) Capture and delivery of asteroid. 1 – Earth; 2 – elliptic trajectory of asteroid; 3 – perigee; 4 - apogee; 5 – asteroid speed in apogee; 6 - asteroid in perigee; 7 – asteroid and the point of meeting the asteroid and delivery apparatus (DA); 8 – delivery apparatus; 9 – trajectory of delivery apparatus; 10 – initial trajectory of asteroids; 11 – speed of DA; 12 – asteroid/DA trajectory after its braking by DA and connection; 13 – Earth atmosphere; 14 - asteroid/DA trajectory after the first parachute braking in Earth atmosphere; 15 - asteroid/DA trajectory after the second parachute braking in Earth atmosphere; 16 - asteroid/DA trajectory after it's third parachute braking in Earth atmosphere; 17 – lending of asteroid/DA by control parachute.

The speeds in apogee and perigee are connected by the following mathematical relation:

$$r_a V_a = r_p V_p \tag{22.1}$$

where r_a, r_p are radius of apogee and perigee, respectively, and V_a, V_p are speeds in apogee and perigee, respectively. To decrease the perigee (for the object-asteroid), the minimum impulse (minimum of fuel consumption) will be in the apogee; to increase the apogee (for meeting the capture/delivery smart apparatus) the minimum impulse will be induced in the perigee.

If the altitude of smart asteroid/apparatus system (AA) is less than 100 km above the r Earth's surface, the Earth atmosphere decelerates the asteroid by collision with its component gases. The apogee decreases (Fig. 22.4b). After some few grazing contacts of AA with our planet's atmosphere, its trajectory becomes a circle, and the smart asteroid/apparatus package enters the denser layers of our planet's atmosphere. After the launch, the smart delivery apparatus 8 (Fig. 22.4b) must meet the suitable asteroid 7. At the correct/desired meeting point in outer space the apparatus has a speed 11 opposed to the asteroid. The authors propose using the kinetic energy of the smart apparatus itself for breaking the asteroid's velocity as well as for charging the flywheel energy accumulator powering the smart apparatus. The smart apparatus 22 (Fig. 22.5a), by extension of a net 21, then captures the asteroid 20.

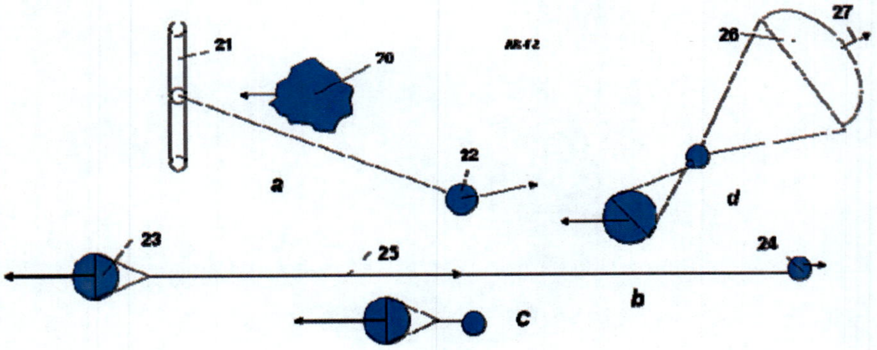

Fig. 22.5 Capturing of asteroid by Delivery Apparatus (DA). (***a***) Capture of asteroid; (***b***) braking of asteroid by kinetic energy of Delivery Apparatus and charging a flywheel energy storage; (c) – final connection DA and asteroid; (***d***) Lending version of asteroid/DA with control lifting parachute for flight in Earth atmosphere. Notations: 20 – asteroid; 21 – capture net; 22 – delivery apparatus; 23 – asteroid into the capture net; 24 – delivery apparatus in position after braking and charging of mechanical energy storage; 25 – brake cable connecting the asteroid to delivery/drive apparatus; 26 – control parachute; 27 - lift/drag force of parachute.

After asteroid-object capture the smart apparatus unwinds the cable 25 and, thus, decreases the asteroid's speed to something eventually for correct for its entry into the Earth's upper atmosphere (Fig. 22.5b). One also decreases its own speed until its value equals the asteroid speed. If the kinetic energy of system AA is very large, the smart apparatus initiates the use of a rocket engine. Further the cable is reeled out/in (Fig. 22.5c), depending on need, and delivery of the smart apparatus is used for object-asteroid trajectory corrections. After entry into Earth's atmosphere, the smart apparatus package opens the controllable lifting/braking parachute 26 (Fig. 22.5d). *This parachute is named an AB carbon fiber parachute.* ["AB" stands for Alexander Bolonkin, its inventor.] It slows the system in the Earth's thin upper atmosphere, decreases the apogee of elliptic orbits (Fig. 22.4b) up to circular orbit (speed is less than 8 km/s). If the smart apparatus package temperature goes beyond its safety value, the smart apparatus package in response then increases the ratio lift/drag of control parachute and causes lifting in the upper atmosphere, where the nose flow is less (air resistance markedly reduced). As a result, the object-asteroid and smart delivery apparatus package cannot overheat and the controlling AB carbon fiber parachute delivers the asteroid-object at a given Earth-surface place. The AB parachute is small in both material bulk and deployed diameter because the lift parachute has a smaller vertical speed and the landing speed of the integrated smart system is allowed to be high by comparison to a man-used parachute (Fig. 22.6).

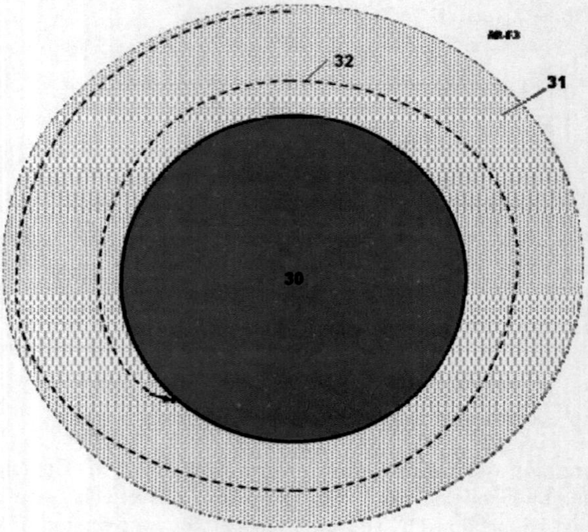

Fig. 22.6 Lending of system with limited heating: Asteroid/DA on Earth surface. *Notations*: 30 – Earth; 31 – Earth atmosphere; 32 – landing trajectory.

The parachute surface is opened from the asteroid-objects backside so that it can emit heat radiation steadily and efficiently to the enveloping Earth-atmosphere. The temperature of the AB carbon fiber parachute may be as much as 1000-1300° C. Without any problem, the carbon fiber is able to keep its functionality up to a temperature of 1500-2000° C. The offered delivery method and system has the following advantages:

1) The system uses for the induced slowing of the asteroid and apparatus the harvested kinetic energy of the asteroid-object and the smart apparatus. In all likelihood that may save a large amount of expensive rocket fuel.

2) The system uses kinetic energy to charge an energy-storage system (this storage/accumulator may be mechanic, electric or chemical and so on). That definitely permits a large amount of energy to gradually accumulate during a long flight time.
3) It is alleged that the offered method of decreasing the high speed of asteroid-objects by a series of entering and re-entering grazing movement into the atmosphere, with serial decreasing of apogee up to circular orbit. That also saves costly rocket fuel and does not need high-quality nose guard protection (note that the heat-shield protection for the manned Apollo Mission spaceships was ~40% of its overall take-off weight).
4) The system has a special cable and brake mechanism for it alone. The flywheel retains/saves energy because outer space is a natural hard vacuum and the gravitational force therein is nil.
5) The system has a control AB fiber carbon parachute with high ratio lift/drag, which allows its operators to avoid too much heating, delivers the asteroid-object to a pre-selected given geographical place and avoids a shock of system touchdown/landing on the Earth-surface.
6) The delivery apparatus may be quickly re-used. Foamed iron bodies, for instance, can be melted to become denser iron ingots used for the common industrial purposes humans practice.

22.4.2.3 Theory of Asteroid Capture, Movement in Outer Space

The speed change of the smart space apparatus or of the packaged asteroid/apparatus system, due to the ignited rocket engine is:

$$\Delta V = -V_g \ln \frac{M_f}{M} \tag{22.2}$$

where ΔV is change of speed, M_f and M represent the final and the initial mass of the system, respectively, and V_g is the discharge velocity of exhaust gas from the rocket engine (for solid fuel $V_g \approx 2500 - 2800$ m/s, for liquid fuel it is either $V_g \approx 3000 - 3200$ m/s (kerosene + O_2), or $V_g \approx 4000$ m/s (hydrogen + O_2)).

In the omnipresent near-Earth vacuum of outer space, the package's trajectory is computed as follows. The radius R from Earth center of mass to a point on the trajectory is given by:

$$R = \frac{p}{1 + e\cos\beta} \tag{22.2'}$$

where β is the angle from perigee, while p and e are the ellipse parameter and eccentricity, respectively ($e = 0$ for circle trajectory, $e < 1$ for ellipse, $e = 1$ for parabola, $e > 1$ for hyperbola). Note that the apogee r_a and the perigee r_p of elliptic trajectory are given by:

$$r_a = \frac{p}{1-e} \ , \ r_p = \frac{p}{1+e} \tag{22.3, 4}$$

The ellipse parameter and eccentricity are computed as:

$$p = \frac{c^2}{K} \quad , \ e = \frac{c}{K}\sqrt{H + \frac{K^2}{c^2}} \tag{22.5, 6}$$

where c and H are expressed by:

$$c = v^2 r^2 \cos^2 \nu = const., \ H = 2K\frac{M_E}{R_0} = const. \tag{22.7, 8}$$

In the above relations, K is the Earth constant (K=3.98 x 10^{14} m^3/s^2), v is the speed, ν represents the angle between the speed and the tangent of the circle, M_E=5.976x10^{24} kg is the mass of the Earth and R_0=6378000 m is Earth radius. The period of rotation is computed as:

$$T = \frac{2\pi}{\sqrt{K}} a^{3/2} \tag{22.9}$$

where:

$$a = r_a \ , \quad b = r_p \ , \quad b = a\sqrt{1-e^2} \tag{22.10}$$

After connection, the velocity of asteroid-connection apparatus ensemble, V, is computed as

$$V = \frac{m_1 V_1 + m_2 V_2}{m_1 + m_2} \tag{22.11}$$

where m_1, V_1 and m_2, V_2 are the masses and velocities of asteroid and connection apparatus, respectively. Denoting by s the length of the cable, the force F can be computed from the following relation:

$$Fs = \frac{m_1 V_1^2}{2} + \frac{m_2 V_2^2}{2} - \frac{(m_1 + m_2)V^2}{2} \tag{22.12}$$

22.5 Economical Efficiency of Asteroid-Object Delivery to Earth-Surface

Only about 10% of known asteroids contain metal. In many cases it is molybdenum and cobalt. Some asteroids, like meteorites, are composed of iron, nickel and various types of stony rock. In composition, they are close to that of the familiar Solar System terrestrial planets. The other main component—nickel-ferrous iron,, which is a solid solution of nickel in iron, and, in any solution, the nickel content in the blend is different-from 6-7% to 30-50%. Occasionally, non-nickel iron objects exist. Sometimes there are significant amounts of iron sulfides Other minerals are also found in small quantities. It was possible to identify a total of about 150 minerals, and it is clear that the number of minerals in the asteroids and meteorites examined is small in comparison with an abundance in the rocks of the Earth, where more than 1,000 known minerals exist, with more being discovered every now and then.

The capture and delivery of a big iron asteroid to Earth requires a huge amount of energy (costly rocket fuel sent to outer space from petroleum refineries in the Earth's biosphere). The delivery of one kilogram of asteroid, by current technology, needs 1 – 5 kg of additional fuel and the launch of one kilogram of the smart delivery apparatus/fuel costs approximately 2012 USD 30-100 thousands. The AD 2012 real cost of purchased metals mined and processed exclusively on Earth is shown in Table 22.1.

Table 22.1 Average cost of metals on 16 May, 2012. 1 lb=0.453 kg

Metal	Price $/Lb	Metal	Price $/Lb	Metal	Price $/Lb
Iron ore	0.063	Nickel	7.69	Silver	27.2
Iron scrap	0.124	Magnesium	1,44	Palladium	592
Molybdenum	13.8	Copper	3.5	Platinum	1433
Cobalt	14	Aluminum	0.9	Gold	1539

The profitable exploitation of outer space metallic resources is possible only after a dramatic decrease of the delivery cost to humans living socially in the ever-changing Earth-biosphere. Our proposal is associated with our strong desire to instigate a maximum decreasing of the current rocket launch costs. One means is offered by Alexander Bolonkin (2011), in an entirely different but relevant outer space exploration and exploitation context, which allows reduction of the rocket launch costs by 3 ÷ 10 $/kg.

22.6 Theory, Computation and Estimation of the Delivery Systems

The entry to Earth atmosphere and the braking by AB carbon fiber parachute are the common actions of both delivery methods. Some theoretic considerations about these processes are presented below.

22.6.1 Theory of Entry to Earth Atmosphere

The distance of ingot-ship's flight, $\dot{r}$, is computed with the relation

$$\dot{r} = \frac{R_0}{R} V \cos\theta \tag{22.13}$$

where R is the radius of ship flight from the Earth's center of mass, V represents the Mega-ASSET vehicle's speed and θ is the angle of its trajectory. The change rate of ship altitude $\dot{H}$, is expressed by:

$$\dot{H} = V \sin\theta \tag{22.14}$$

The lift force L and the drag force D of the delivery system are computed by:

$$L = 2\alpha\rho a V S\,, \quad D = L/4 \qquad (22.15,\ 16)$$

In the above relations $\alpha = 40^\circ = 0.7$ rad is the apparatus attack angle, $a = 295$ m/s is sound's speed through air at high altitude and ρ [units: kg/m^3] denotes the air density at the given altitude H [units :m] which is computed as:

$$\rho = 0.414 e^{(H-10000)/6719} \tag{22.17}$$

On the other hand, the drag force D_p and the lift force L_p of the AB carbon fiber parachute can be expressed as:

$$D_P = 0.5 C_{DP} \rho a V S_P\,, \quad L_P = 4 D_p \qquad (22.18,\ 19)$$

where $C_{DP} = 1$ is parachute drag coefficient and S_p represents the parachute's deployed area. Note that the lift force of the AB carbon fiber parachute is controlled from 0 to 4 D_p (the ram-air parachute can produce lift force up 1/3 from its drag).

The acceleration during object-asteroid atmospheric entry can be computed as:

$$\dot{V} = -\frac{D + D_P}{m} - g \sin\theta \tag{22.20}$$

where g is the acceleration gravity at the given altitude H. Considering g_0=9.81 m/s^2, the variation of g with H is computed as:

$$g = g_0 \left(\frac{R_0}{R_0 + H} \right)^2 \tag{22.21}$$

Further, the angular velocity is expressed by:

$$\dot{\theta} = \frac{L + L_P}{mV} - \frac{g}{V} \cos\theta + \frac{V \cos\theta}{R} + 2\omega_E \cos\varphi_E \tag{22.22}$$

In the above relation, ω_E is the angle of the Earth's speed and φ_E is the smallest angle between a plane perpendicular to flight and Earth polar axis.

Due to intense and unrelenting air friction, the AB carbon fiber parachute is rapidly heated. The specific heat flux [units: W/m^2] generated by air friction phenomena may be expressed by:

$$\dot{q} = \frac{0.5 \cdot 11040 \cdot 10^4}{R_n^{0.5}} \left(\frac{\rho}{\rho_{SL}} \right)^{0.5} \left(\frac{V}{V_{CO}} \right)^{3.15} \tag{22.23}$$

In Eq. (22.23), ρ_{SL}= 1.225 kg/m^3 is the air density at Earth's sea level datum, V_{CO} = 7,950 m/s is circle orbit speed and:

$$R_n = \sqrt{\frac{S_P}{\pi}} \tag{22.24}$$

The cooling of the parachute fiber is mainly due to radiation. Therefore, the stagnation point temperature of the AB carbon fiber parachute surface, T_1, can be expressed by:

$$T_1 = 100 \left(\frac{\dot{q}}{\varepsilon C_S} + \left(\frac{T_2}{100} \right)^4 \right)^{1/4} \tag{22.25}$$

where T_2 represents the temperature of the Standard Atmosphere at a given altitude H, C_S = 5.67 W/(m^2·K^4) is coefficient of black-body radiation and ε is the emissivity of the employed AB carbon fiber parachute's surface. Control is as follows: if T_1 is higher than a given temperature, then the lift force L_P = maximum = $4D_p$. In other case L_p = 0. When the speed is less than the speed of sound, the control parachute is also used for deliveries designated in given absolute geographical location terms (latitude and longitude as furnished constantly by GPS attachment.

22.6.2 Theory of AB Carbon Fiber Parachute Calculus

Let's express now the lift and drag forces of the parachute as defined by Eqs. (22.18) and (22.19) as functions of the lift and drag co-efficients, C_L and C_D, respectively. At sea level, the corresponding relations are:

$$L_p = C_L \frac{\rho_{SL} V^2}{2} S_p , \; D_p = C_D \frac{\rho_{SL} V^2}{2} S_p \tag{22.26, 27}$$

Usually, C_L=2÷3 and C_D=0.5÷1.2. The vertical speed V_v and the absolute speed V of the systems are linked by the relation:

$$V_v = \frac{V}{K_C} \quad (22.28)$$

where, obviously, $V_v \le V$ and:

$$K_C = \frac{C_L}{C_D} \quad (22.29)$$

Let us consider the system's mass (smart delivery system and parachute) m=100 tons , C_L = 2.5, V_v = 20 m/s and K_C = 4 (V = 80 m/s). From Eqs. (22.26)- (22.29) we find that the parachute area is S_p = 100 m^2. The control rectangle parachute is 5.8 m x 17.3 m.

22.7 Some Environmental Impacts of Asteroid-Object Delivery Methods

Earth-surface rocks with low resistance (e.g., rhyolite and tuff) are fractured by lightning while those with high mechanical resistance (e.g., granite) are not usually shattered: lightning does break rocks and bedrock (Wakasa et al. 2012). Until properly-done full-scale outdoor experiments are conducted, it is unknown whether natural iron asteroids, descending via AB carbon fiber parachutes, or as shaped aerodynamic "Mega-ASSET" smart ingot-machine gliders will attract destructive lightning. Civil fireworks displays pollute the air with chemicals rapidly dispersed in the air by rapid combustion and cause booming noises. Some of the chemicals burnt to create brilliant color aerosol clouds in the sky (barium, strontium, copper, radium and sodium, for instance) are bad for humans to breathe (Steinhauser and Musilek 2009). These human-mined chemicals are, of course, additional to Nature's Earth-normal infalling aerial complement: satellite observations suggest that 100-300 metric tons of cosmic dust enters the Earth-atmosphere daily. Metals injected into the atmosphere everyday from evaporating dust particles colliding with the Earth at high speed do cause climate change—as, for example, by affecting the ozone chemistry in the stratosphere and fertilization of the upper layers of ocean seawater with iron. Intentional anthropogenic heating of the planet's stratosphere via injection of aerosols consisting of sulfate, titanium, limestone and soot has been proposed as a climate control activity possibly conducted in the immediate future by the macro-engineering elite (Solomon et al. 2011). In sum, metal importations from extra-Earth places in our Solar System may well change the contents and dynamics of our Earth's atmosphere.

22.8 Conclusions

Sociometeoritics is studying the physical and psychical interactions of meteorites/asteroids with living persons and various efficiently organized human societies. During the modern-day history of human civilization, the world-public has become interested by the fascinating potential for 21st Century Earth-surface asteroid impacts and by asteroid utilization by various industries working for the first time in the outer spaces of our Solar System. In order to make available the economic potential of object-asteroids, in this chapter two delivery methods were proposed.

The first method is a macro-project encompassing extraterrestrial mining, remotely controlled procedure for Earth delivery and the final slowing of delivered material in the lower atmosphere layer (troposphere). So, through the extraterrestrial mining, the iron ore is extracted from the targeted asteroid and is compressed in automated outer space factories to obtain the "Mega-ASSET" smart machines (solid bodies shaped in aerodynamic form) which are sent to Earth. In the dense atmosphere layers the "Mega-ASSET" machines are safely brought to the Earth's surface by the aerially flight-effective AB carbon fiber parachute system. Foamed iron may be used to allow maritime deliveries of Mega-ASSET vehicles.

The second method uses an electronic smart delivery apparatus to capture the entire targeted NEA on its own trajectory and to break the ensemble, the package as it were, by using the kinetic energy of the rocket-launched smart apparatus. This energy is used also for charging the apparatus's energy storage device and batteries. The small AB carbon fiber parachute allows multiple encounters with the Earth atmosphere for breaking the harvest/processed/shaped asteroid without high front-end (nose) "ingot" heating, delivering the selected asteroid-object on time at a given Earth-surface place and to avoid any destructive asteroid impact onto the Earth. The AB carbon fiber parachute system was originally devised, in part, to safely bring the USA's Space Shuttles to their main servicing bases located in Florida and California but also their alternate landing sites distributed worldwide. The delivery of the metallic asteroid to Earth will be profitable if we dramatically decrease the cost of the space launch (up to USD 10 per kg), as it is offered in the essay of Bolonkin (2011). At the present time, AD 2012, America is spending USD 20 –300 millions for delivery of a very small piece of asteroid for scientific laboratory study purposes. Using this chapter's offered method, we can easily and safely deliver an asteroid weighing about 50 metric tonnes to the Earth-surface.

Earth's gravity challenged humans and inspired many artists throughout recorded history. The often recounted ancient myth of Sisyphus strenuously shoving a boulder uphill only later to see it roll uncontrollably downhill where it then rested somewhat menacingly, ready to be pushed uphill again by the tortured Sisyphus, is a myth of humanity's frustration with its homeland environment. If humans begin to mine highly mineralized valuable asteroids in this Solar System, then the castle situated atop the large gray boulder hovering over the Earth-surface in Rene Magritte's painting *La Chateau des Pyrenees* could come to truly symbolize Homo sapiens' 21st Century relationship with Nature and Macro-engineering, both extra-terrestrial, outer space and Earth-biosphere. We fancy Magritte's painting as an ideal evocation for the AB carbon fiber parachute concept we have documented herein.

References

Abadie, S.M., Harris, J.C., Grilli, S.T.T., Fabre, R.: Numerical modeling of tsunami waves generated by the flank collapse of the Cumbre Vieja Volcano (La Palma, Canary Islands): tsunami source and near field effects. Journal of Geophysical Research 117, C05030 (2012)

Ali, S., Dadush, U.: The Global Middle Class Is Bigger Than We Thought. Foreign Policy (May 16, 2012), http://www.foreignpolicy.com/articles/2012/05/16/the_global_middle_class_is_bigger_than_we_thought?pri...

Beekman, G.I.O.: Yarkovsky and the Discovery of 'His" Effect. Journal for the History of Astronomy 37(126, Pt. 1), 71–86 (2006)

Bian, Z., Miao, X., Lei, S., Chen, S., Wang, W., Struthers, S.: The Challenges of Reusing Mining and Mineral-Processing Wastes. Science 337, 702–703 (2012)

Binzel, R.P.: The Torino Impact Hazard Scale. Planetary and Space Science 48, 297–303 (2000)

Blanchard, M.B.: Artificial Meteor Ablation Studies: Iron Oxides. Journal of Geophysical Research 77, 2442–2455 (1972)

Bolonkin, A.A.: A New Method of Atmospheric Reentry for Space Ships. Paper AIAA-2006-6985 at the Multidisciplinary Analyses and Optimization Conference, Portsmouth, Virginia, USA, December 6-8 (2006); Bolonkin, A.A.: New Concepts, Ideas, Innovations in Aerospace, Technology and the Human Sciences, ch. 8, 510 p. NOVA (2006), http://www.scribd.com/doc/24057071, http://www.archive.org/details/NewConceptsIfeasAndInnovationsInAerospaceTechnologyAndHumanSciences

Bolonkin, A.A.: Air Catapult Transportation. Scribd, NY (2011), http://www.scribd.com/doc/79396121/, http://www.archive.org/details/AirCatapultTransport

Brovelli, M.A., Zamboni, G.: Virtual globes for 4D environmental analysis. Applied Geomatics 4, 163–172 (2012)

Buchwald, V.F.: On the use of iron by the Eskimos in Greenland. Materials Characterization 26, 139–176 (1992)

Burch, G.B.: The Counter-Earth. Osirus 11, 267–294 (1954)

Cathcart, R.B.: Meteorite Mining. Future Life #27, 13 (1981)

Cathcart, R.B.: Seeing is Believing: Planetographic Data Display on a Spherical TV. Journal of the British Interplanetary Society 50, 103–104 (1997)

Cathcart, R.B.: Kra Canal (Thailand) excavation by nuclear-powered dredges. International Journal of Global Environmental Issues 8, 248–255 (2008)

Chen, W.-Q., Graedel, T.E.: Anthropogenic Cycles of the Elements: A Critical Review. Environmental Science & Technology (in press, 2013)

Cidell, J.: Just passing through: the risky modilities of hazardous materials transport. Social Geography 7, 13–22 (2012)

Chyba, C.F., van der Vink, G.E., Hennet, C.B.: Monitoring the Comprehensive Test Ban Treaty: Possible ambiguities due to meteorite impacts. Geophysical Research Letters 25, 191–194 (1998)

Clark, S.: New Moon. New Scientist 214(2861), 48–51 (2012)

Court, R.W., Sephton, M.A.: The contribution of sulphur dioxide from ablating micrometeorites to the atmospheres of Earth and Mars. Geochimica et Cosmochimica Acta 75, 1704–1717 (2011)

Cruikshank, D.P., Hartmann, W.K.: The Meteorite-Asteroid Connection: Two Olivine-Rich Asteroids. Science 223, 281–283 (1984)

Driscoll, R.: From Projectile Points to Microprocessors—The Influence of Some Industrial Minerals. USGS Circular 1314, 26 p. (2007)
Efrati, A.: Start-Up Sees New Frontier In Mining: Asteroids in Space. The Wall Street Journal CCLIX: B1 (April 24, 2012)
Elvis, C.: Let's mine asteroids—for science and profit. Nature 485, 549 (2012)
Finkl, C.W., Pelinovsky, E., Cathcart, R.B.: A Review of Potential Tsunami Impacts to the Suez Canal. Journal of Coastal Research 28, 745–759 (2012)
Fogg, M.J.: Personal communication to RBC (September 8, 2012)
Fraley, J.M.: Re-examining Acts of God. Pace Environmental Law Review 27, 669–690 (2010)
Friis-Hansen, D., Zaya, O., Takashi, S.: Cai Guo-Qiang, 160 pages. Phaidon Press Limited, London (2012)
Gammage, B.: The Biggest Estate on Earth: How Aborigines Made Australia, 434 pages. Allen & Unwin, Sydney (2011)
Gehrels, T.: Asteroids, pp. 222–226. University of Arizona Press, Phoenix (1980)
Gerst, M.D., Graedel, T.E.: In-Use Stocks of Metals: Status and Implications. Environmental Science & Technology 42, 7038–7045 (2008)
Granvik, M., Vaubaillon, J., Jedricke, R.: The Population of natural Earth satellites. Icarus 218, 262–277 (2012)
Grieve, R.A.F.: The Economic Potential of Terrestrial Impact Craters. International Geology Review 36, 105–151 (1994)
Hall, K., Thom, C., Sumner, P.: On the persistence of 'weathering'. Geomorphology 148-150, 1–10 (2012)
Hallion, R.P.: The Path to Space Shuttle—The Evolution of Lifting Reentry Technology. Journal of the British Interplanetary Society 36, 523–541 (1983)
Hendricks, L., Mergeay, M.: From the deep sea to the stars: human life support through minimal communities. Current Opinion in Microbiology 10, 231–237 (2007)
Henning, T.: Astromineralogy, 2nd edn., 329 p. Springer, The Netherlands (2010)
Hudson-Edwards, K.A., Jamieson, H.E., Lottermoser, B.G.: Mine Wastes: Past, Present, Future. Elements 7, 375–380 (2011)
Hupy, J.P., Koehler, T.: Modern warfare as a significant form of zoogeomorphic disturbance upon the landscape. Geomorphology 157-158, 169–182 (2012)
Jones, N.: Outrageous Fortune. New Scientist 167, 24–26 (2000)
Kanamori, H., Mori, J., Sturtevant, B., Anderson, D.L., Heaton, T.: Seismic excitation by space shuttles. Shock Waves 2, 89–96 (1992)
Kargel, J.S.: Metalliferous asteroids as potential sources of precious metals. Journal of Geophysical Research 99, 21129–21141 (1994)
Kennelly, P.: Landscape Volumetrics and Vizualizations of the Butte Mining District, Montana. Environmental & Engineering Geoscience XVII, 213–226 (2011)
Krischke, M., Lorenzini, E.C., Sabath, D.: A hypersonic parachute for low-temperature reentry. Acta Astronautica 36, 271–278 (1995)
Lee, R.: Law and Regulation of Commercial Mining of Minerals in Outer Space, 403 pages. Springer, The Netherlands (2012)
Lotka, A.: Elements of Physical Biology (Republished as Elements of Mathematical Biology), p. 23. Dover, New York (1924)
Machado, I.: Impact or explosion? Technological culture and the ballistic metaphor. Sign Systems Studies 34(1), 245–260 (2006)
MacPherson, G.J., Thiemens, M.H.: Cosmochemistry: Understanding the Solar System through analysis of extraterrestrial materials. Proceedings of the National Academy of Sciences 108, 19130–19134 (2011)
Marriner, N., Morhange, C., Skrimshire, S.: Geoscience meets the four horsemen? Tracking the rise of neocatastrophism. Global and Planetary Change 74, 43–48 (2010)

Marx, L.: Technology: The Emergence of a Hazardous Concept. Technology and Culture 51, 561–577 (2010)

McGuire, B.: Surviving Armageddon: Solutions for a Threatened Planet, p. 132. University of California Press, Oxford (2005)

Muller, D.B., Wang, T., Duval, B., Graedel, T.E.: Exploring the engine of anthropogenic iron cycles. Proceedings of the National Academy of Sciences 103, 16111–16116 (2006)

Murakami, T., Ohara, K., Narushima, T., Ouchi, C.: Development of a New Method for Manufacturing Iron Foam Using Gases Generated by Reduction of Iron Oxide. Materials Transactions 48, 2937–2944 (2007)

Olsson, J., Bovet, N., Makovicky, E., Bechgaard, K., Balogh, Z., Stipp, S.L.S.: Olivine reactivity with CO2 and H2O on a microscale: Implications for carbon sequestration. Geochimica et Cosmochimica Acta 77, 86–97 (2012)

Pierazzo, E., Artemieva, N.: Local and Global Environmental Effects of Impacts on Earth. Elements 8, 55–60 (2012)

Powell, J.W., Hengeveld, E.: Asset and Prime—Gliding re-entry Test Vehicles. Journal of the British Interplanetary Society 36, 369–376 (1983)

Preston, B., Johnson, D.J., Edwards, S.J.A., Miller, M., Shipbaugh, C.: Space Weapons Earth Wars. Project Air Force RAND, Santa Monica CA, pp. 40–45 (2002)

Rauch, J.N.: Global mapping of Al, Cu, Fe, and Zn in-use stocks and in-ground resources. Proceedings of the National Academy of Sciences 106, 18920–18925 (2009)

Reck, B.K., Graedel, T.E.: Challenges in Metal Recycling. Science 337, 690–695 (2012)

Reed, D.R., Lister, D., Yeager, C.: Wingless Flight: The Lifting Body Story, 262 pages (2011), `http://WWW.MilitaryBooks.co.uk:London`

Reimold, W.U., Gibson, R.L.: Meteorite Impact. The Danger from Space and South Africa's Mega-Impact, The Vredefort Structure, 326 pages. Springer, Utrecht (2010)

Schmidt, T.S., nine other authors: Geologic processes influence the effects of mining on aquatic ecosystems. Ecological Applications 22, 870–879 (2012)

Schuiling, R.D., Krijgsman, P.: Enhanced Weathering: An Effective and Cheap Tool to Sequester CO2. Climatic Change 74, 349–354 (2006)

Sen, I.S., Peucker-Ehrenbrink, B.: Anthropogenic Disturbance of Element Cycles at the Earth's Surface. Environmental Science & Technology (in press, 2013)

Shepard, M. (ed.): Sentient City: Ubiquitous Computing, Architecture, and the Future of Urban Space, 229 pages. MIT Press, Cambridge (2011)

Solomon, S., Danieal, J.S., Neely III, R.R., Vernier, J.-P., Dutton, E.G., Thomason, L.W.: The persistently variable "background" stratospheric aerosol layer and global climate change. Science 333, 866–870 (2011)

Spinney, L.: Human Cycles: History as Science. Nature 488, 24–26 (2012)

Steen, B., Borg, G.: An estimation of the cost of sustainable production of metal concentrates from the Earth's crust. Ecological Economics 42, 401–413 (2002)

Steinhauser, G., Musilek, A.: Do pyrotechnics contain radium? Environmental Research Letters 4, 1–6 (2009)

Stevens, B., Ragheb, M.: 2010 1st International Nuclear & Renewable Energy Conference (INREC), March 21-24, pp. 1–10 (2010)

Thompson, M.O., Peebles, C.: Flying Without Wings, 254 pages. Smithsonian, Washington DC (1999)

Thurner, S., Klimek, P., Hanel, R.: Schumpeterian Economic Dynamics as a Quantifiable Minimum Model of Evolution. New Journal of Physics 12, 075029 (2010)

Valero, A., Agudelo, A., Valero, A.: The crepuscular planet. A model for the exhausted atmosphere and hydrosphere. Energy 36, 3745–3753 (2011)

Vince, G.: An Epoch Debate. Science 334, 32–37 (2011)

Von Hilgers, P., Benjamin, R.: War Games: A History of War on Paper, 235 pages. MIT Press, Cambridge (2012)

Wakasa, S.A., Nishimura, S., Shimizu, H., Matsukura, Y.: Does lightning destroy rocks?: Results from a laboratory lightning experiment using an impulse high-current generator. Geomorphology 161-162, 110–114 (2012)
Wickramasinghe, C.: The Universe: A cryogenic habitat for microbial life. Cryobiology 48, 113–125 (2004)
Wang, T., Muller, D.B., Graedel, T.E.: Forging the Anthropogenic Iron Cycle. Environmental Science & Technology 41, 5120–5129 (2007)
Ward, S.N., Asphaug, E.: Asteroid impact tsunami of 2880 March 16. Geophysical Journal International 153, F6–F10 (2003)
Willbold, M., Elliot, T., Moorbath, S.: The tungsten isotopic composition of the Earth's mantle before terminal bombardment. Nature 477, 195–198 (2011)
Williams, J.W., Jackson, S.T., Kutzbach, J.E.: Projected distribution of novel and disappearing climates by 2100 AD. Proceedings of the National Academy of Sciences 104, 5738–5742 (2007)
Wright, G.: The riddle of the sands. New Scientist 163, 42–45 (1999)

Chapter 23
Artificial Gravitation on Asteroids

Alexander A. Bolonkin

C&R Co., New York, USA

23.1 Introduction

People have dreamed about a flying freely without any apparatus for many centuries. Physicist knows only two methods for creating repulsive force: magnetism and electrostatics. Magnetism is well studied and the use of superconductive magnets for levitating a train has been widely discussed in scientific journals, but repulsive magnets have only a short-range force. They work well for ground trains but are bad for air flight. Electrostatic flight needs powerful electric fields and powerful electric charges. The asteroids' electric field is very weak and cannot be used for levitation. The main innovations presented in this chapter are methods for creating powerful static electrical fields over surface and powerful, stable electrical charges of small size which allow levitation (flight) of people and vehicles over asteroid surface. The author also shows how this method can be utilized on an asteroid surface for creating artificial gravity.

Magnetic levitation has been widely discussed in the literature for a long time. However, there are few scientific works related to electrostatic levitation. Electrostatic charges have a high voltage and can create corona discharges, breakthrough and relaxation. The asteroids electrostatic field is very weak and useless for flight. That is why many innovators think that electrostatic forces cannot be useful for levitation.

The author's first innovations in this field which changed this situation were offered in (Bolonkin 1982), and some practical applications were given in (Bolonkin 1983). The idea was published in Bolonkin (1990 p. 79). In the following presented work, these ideas and innovations are researched in more detail. Some projects are also presented to allow estimation of the parameters of the offered flight systems.

Currently there is only one method of creating artificial gravity on spacecrafts - rotation. Rotation method cannot be applied to asteroids (the centrifugal force is directed from the center).

23.2 Brief Description of Electro-Gravity Innovation

It is known that like electric charges repel, and unlike electric charges attract (Fig. 23.1a,b,c). A large electric charge (for example, positive) located at altitude induces the opposite (negative) electric charge at the asteroid surface (Fig. 23.1d,e,f,g). Between the upper and lower charges there is an electric field. If a small negative electric charge is placed in this electric field, this charge will be repelled (or attracted) from the like charges (on the asteroid surface) and attracted to the upper charge (Fig. 23.1d). That is the electrostatic lift/gravity force. The lift force is mainly determined by the asteroid charges because the small charges are conventionally located near the asteroid surface. As shown below, these small charges can be connected to a man or an apparatus and have enough force to lift and supports them or attracted them to asteroid surface.

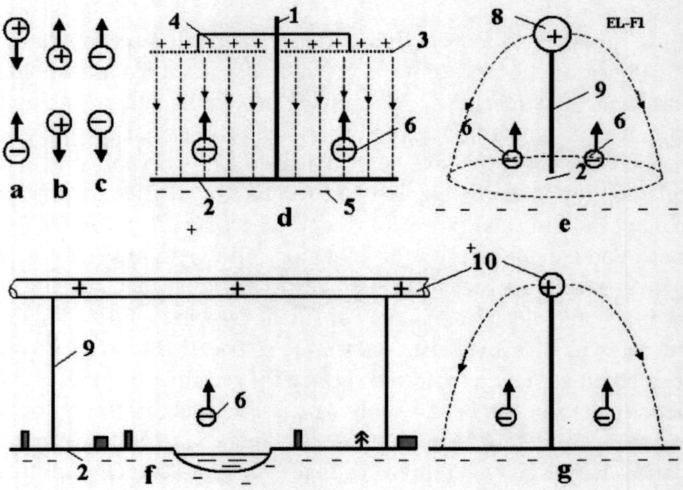

Fig. 23.1 Explanation of electrostatic levitation: a) Attraction of unlike charges; b,c) repulsion of like charges; d) Creation of the homogeneous electric field; e) Electrical field from a large spherical charge ; f,g) Electrical field from a tube (side and front views). Notations are: 1, 9 – column, 2 – asteroid surface charged by induction, 3 – net, 4 – upper charges, 5 – lower charges, 6 – levitation apparatus, 8 – charged fluid balloon, 9 – column, 10 – charged tube.

The upper charge may be located on a column as shown in Fig. 23.1d,e,f,g or a tethered fluid balloon (Fig. 23.1e), or fluid tube, or a tube suspended on columns (Fig. 23.1f,g). In particular, the charges may be at two identically charged plates, used for a non-contact transportation on asteroid (Fig. 23.2a).

An asteroid definitely needs artificial gravity. Any slight carelessness in space can result in the cosmonaut, instruments or devices drifting away from the asteroid. Science knows only two methods of producing artificial gravity and attractive forces: rotation and magnetism. Both methods are bad. The rotation creates artificial gravity only inside the asteroid. Observation of space from a rotating asteroid is very difficult. The magnetic force is only effective over a very short distance. The magnets stick together and a person has to expend a large effort to move (it is the same as when you are moving on a floor smeared with glue).

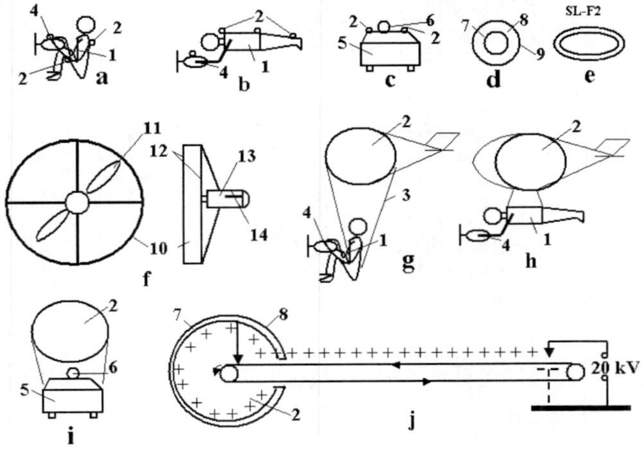

Fig. 23.2 Levitation apparatus: a,b) Single levitated man (mass up to 100 kg) using small highly charged balls 2. a) Sitting position; b) Reclining position; c) Small charged ball for levitating vehicle; d) Small highly charged ball; e) Small highly charged cylindrical belt; f) Small engine for asteroids with atmosphere (forward and side views); g) Single levitated man (mass up to 100 kg) using a big non-highly charged ball which doesn't have an ionized zone (sitting position); h) The same man in a reclining position; i) Large charged ball to levitate a vehicle which doesn't have an ionized zone; j) Installation for charging a ball using a Van de Graaff electrostatic generator (double generator potentially reaches 12 MV) in horizontal position. Notations: 1 – man; 2 – charged lifting ball; 4 – handheld air engine; 5 – vehicle; 6 – engine (turbo-rocket or other); 7 – conducting layer; 8 – insulator (dielectric); 9 – strong cover from from artificial fibers or whiskers; 10 – lagging; 11 – atmospheric propeller; 12 – preventive nets; 13 – engine; 14 – control knobs.

If then is a charge inside the asteroid and small unlike charges attached to object elsewhere, then will fall back to the asteroid if they are dropped. If you charge the asteroid and cosmonauts with unlike electric charges, the cosmonauts will return to the asteroid during any walking and jumping. Artificial gravity on asteroids is possible (Fig. 23.3). The author acknowledges that this method has problems. For example, we need a high electrical intensity if we want to use small charged balls. This problem, and others problems, are discussed below.

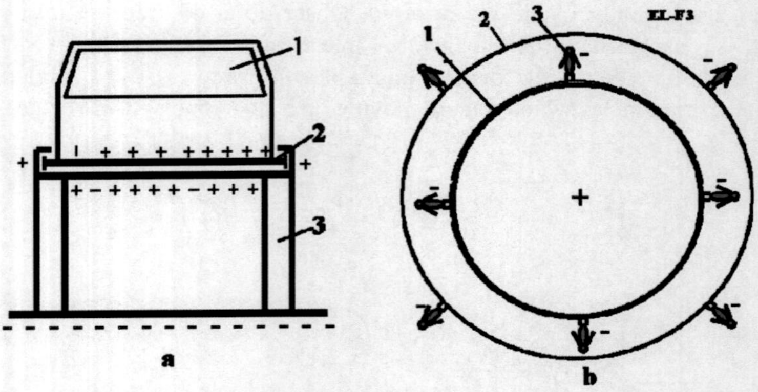

Fig. 23.3 Artificial gravity on asteroids. a) Levitated transportation; b) Artificial gravity on asteroid. Notation: a) 1 – vehicle; 2 – charged plates; 3 – insulated column; b) 1 – charged asteroid; 2 – sphere; 3 – man

23.3 Theory of an Electrostatic Lift Force and Results of Computations

23.3.1 Brief Information about Electric Charges, Electrical Fields, and Electric Corona

Electric charge creates an electrical field. Every point this field has a vector of magnitude called electrical intensity, *E*, measured in Volt/meter. If the unlike charges (or non-insulated electrodes under voltage) are located on asteroids atmosphere, the electrical intensity is less than E_c = (3 to 4)×10^6 V/m, the discharge current will be very small. If $E > E_c = 3\times10^6$ V/m and we have a closed-loop high-voltage circuit (for non-insulated electrodes), electric current appears. The current increases following an exponential law when the voltage is increased. In a homogeneous electric field (as between plates), the increasing voltage makes a spark (flashover, breakthrough, lighting). A non-homogeneous electric field (as between a sphere and plate or an open sphere) makes an electric *corona*. Electrons break away from the metal *negative* electrode and ionize the environment. Positive ions hit the non-insulated positive electrode and knock out electrons. These unlike ions can cause a particle blockade (discharge) of the main charge. The efficiency of ionization by positive ions is much less than for electrons of the same energy. Most ionization occurs as a result of secondary electrons released at the negative electrode by positive ion bombardment. These electrons produce ionization as they move from the strong field at the electrode out into the weak field. This, however, leaves a positive-ion space charge, which slows down the incoming ions. That has the effect of diminishing the secondary electron yield. Because the positive ion mobility is low, there is a time lag before the high field conditions can be restored. For this reason the discharge is somewhat unstable.

The environment on asteroids contains a small amount of free electrons. These electrons can also create an electric corona around the *positive non-insulated electrode*, but under higher voltage than the negative electrode. The effect here is to enable the free electrons to ionize by collision in the high field surrounding this electrode. One electron can produce an avalanche in such a field, because each ionization event releases an additional electron, which can then could further ionization. To sustain the discharge, it is necessary to collect the positive ions and to produce the primary electrons far enough from the positive electrode to permit the avalanche to develop. The positive ions are collected at the negative electrode, and it is their low mobility that limits the current in the discharge. The primary electrons are thought to be produced by photo-ionization.

The particular characteristics of the discharge are determined by the shape of the electrodes, the polarity, the size of gap (ball), and the atmosphere (if any) and its pressure. In high voltage electric lines the corona discharge that surrounds a high-potential power transmission line represents power loss and limits the maximum potential which can be used.

The offered method is very different from the conventional cases described in textbooks (Shortley and Williams 1996). The charges are *isolated* using an insulator (dielectric). They cannot emit electrons in the environment. There is not a closed circuit. This method is nearer to single polarity electrets, when like charges are inserted into an insulator (Kestelman et al. 2000). Electrets have typical surface charges of about $\sigma = 10^{-8}$ C/cm^2, PETP up to 1.4×10^{-7} C/cm^2 (Kestelman et al. 2000, p.17), and TSD with plasticized PVB up to 1.5×10^{-5} C/cm^2 (Kestelman et al. 2000, p. 253). This means the electrical intensity near their surface reaches ($E = 2\pi k\sigma$, $k = 9\times10^9$) 6×10^6 V/m, 80×10^6 V/m, and 8500×10^6 V/m respectively. The charges are not blocked and the discharge (half-life time) continues from 100 days up to several years.

On natural space bodies like Earth and asteroids, radioactivity and cosmic rays create about 1.5–10.4 ions into 1 cm^3 every second (Kikoin 1976, p. 1004). These ions gradually recombine back into conventional molecules.

In a vacuum the discharge mechanism is different. In non-insulated negative metal electrodes, the electrons may be extracted from the conducting electrode by the strong electric field. The critical surface electric intensity, E_o, is about 100×10^6 V/m at the non-insulated negative electrode. This intensity is about 1000 times more at the positive electrode because the ions are very difficult to tear away from the solid material. Conducting sharp edges increase the electric intensity. That is why it is better to charge the asteroid surface with positive charges. A very sharp spike allows the electrical energy of the charged ball to be discharged.

23.3.2 Size of Corona (Ionized Sphere) and Safety of a Ball of Electric Intensity

The size of the corona may be found as a spherical area by using the next relationships:

$$E \ge E_c, \quad \frac{kq}{R_c^2} \ge E_c, \quad q = \frac{E_a a^2}{k}, \quad R_c \le \sqrt{\frac{kE_a a^2}{kE_c}}, \quad \bar{R}_c = \frac{R_c}{a} \le \sqrt{\frac{E_a}{E_c}} \tag{23.1}$$

where E – electrical intensity of the charge, [V/m]; E_c – electrical intensity at the beginning of the corona, [V/m], $E_c \approx 3\times10^6$; E_a – electrical intensity at the ball surface, [V/m]; a – ball radius, m; R_c – radius of corona, [m]; $k = 9\times10^9$.

To find the safe electrical intensity, E_a, for a negatively charged ball in an insulated cover from the point of rupture (spark) into a neutral environment the following equation can be used:

$$\begin{gathered} U \le U_i, \quad U = \frac{kq}{\varepsilon}\left(\frac{1}{a} - \frac{1}{a+\delta}\right), \quad U_i = \varepsilon E_i \delta, \quad \frac{kq}{\varepsilon}\left(\frac{1}{a} - \frac{1}{a+\delta}\right) \le \varepsilon E_i \delta, \quad q = \frac{a^2 E_a}{k}, \\ \bar{\delta} = \frac{\delta}{a} \ge \frac{E_a}{\varepsilon E_i} - 1, \quad for \quad \bar{\delta} = 0, \quad E_a \le \varepsilon E_i \end{gathered} \tag{23.2}$$

where U – ball voltage, [V]; U_i – safe voltage of ball insulator, [V]; E_i – safe electrical intensity of ball insulator, [V/m]; δ – thickness of the ball cover, [m]; ε – dielectric constant. In Eqs. (23.1) and (23.2) the last formulas are the final result.

Example: The ball is covered by Mylar with E_i = 160 MV/m, ε = 3 (see Table 23.1). Then $E_a = 3\times160 = 480$ MV/m, and the relative radius of the ionized sphere [Eq. (23.1)] is $(480/3)^{0.5} = 12.6$. If $a = 0.05$ m, the real radius is $R_c = 12.6\times0.05 = 0.63$ m.

23.3.3 For a Cylindrical Cable or Belt

The radius of the corona (ionized cylinder) can be found using the same method:

$$E \ge E_c, \quad E = \frac{2k\tau}{R_c}, \quad \tau = \frac{aE_a}{2k}, \quad \frac{aE_a}{R_c} \ge E_c, \quad R_c \le a\frac{E_a}{E_c}, \quad \bar{R}_c = \frac{R_c}{a} \le \frac{E_a}{E_c} \tag{23.3}$$

where τ is the linear charge, [C/m]. To find using the same method [Eq. (23.2)] the safe intensity, E_a, for a negatively charged cable (belt, tube) in an insulated cover from point of rupture into a neutral environment the following equation can be used

$$\begin{gathered} U \le U_i, \quad U = 2k\tau \ln\left(\frac{a+\delta}{a}\right), \quad U_i = \varepsilon E_i \delta, \quad 2k\tau \ln\left(\frac{a+\delta}{a}\right) \le \varepsilon E_i \delta, \quad \tau \le \frac{\varepsilon E_i a\delta / a}{2k \ln(1+\delta/a)}, \\ for \quad \frac{\delta}{a} \to 0, \quad \bar{\tau} = \frac{\tau}{a} \le \frac{\varepsilon E_i}{2k}, \quad E_a = k\frac{2\tau}{a}, \quad \frac{\tau}{a} = \frac{E_a}{2k}, \quad \frac{E_a}{2k} \le \frac{\varepsilon E_i}{2k}, \quad E_a \le \varepsilon E_i \end{gathered} \tag{23.4}$$

23.3.4 Discharging by Corona

Below, the author makes computations to show how the milliards (10^9 1/m^3) of charged particles influence the main charge. If 1 m^3 of environment contains *d* like-charged (electrons or ions) particles and the charge density is constant, the charge, *q*, of a sphere with radius *r* is

$$q = \frac{4}{3}\pi r^3 ed \tag{23.5}$$

where $e = 1.6\times10^{-19}$; C – charge of one particle (electron or single charged ion); *d* –particle density. On the other side, the main charge, q_o, will be partially blocked until the intensity at radius *r* becomes E_c. As a result the equation is

$$q_0 - q = \frac{E_c}{k} r^2, \quad \frac{4}{3}\pi edr^3 + \frac{E_c}{k} r^2 - q_0 = 0 \tag{23.6}$$

where $k = 9\times10^9$. Equation (23.6) has only one real root. The results of this computation are presented in Figs. 23.4 and 23.5, which show that a large density only decreases the main charge. But only experiments can show what causes this discharge to take place.

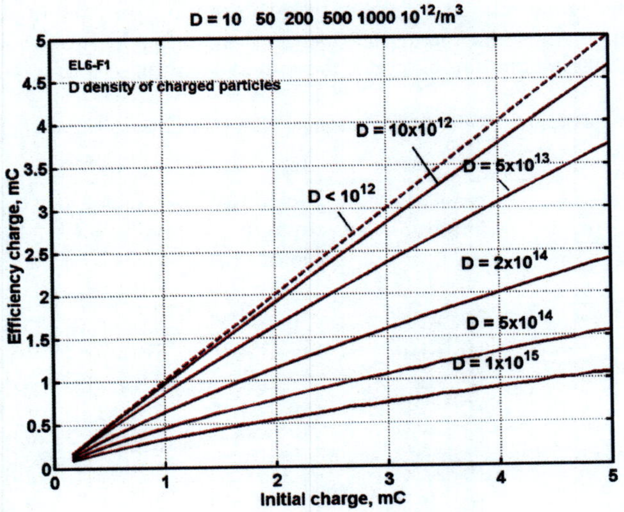

Fig. 23.4 Efficiency charge versus the main charge and density of charged particles in the environment (ionized zone)

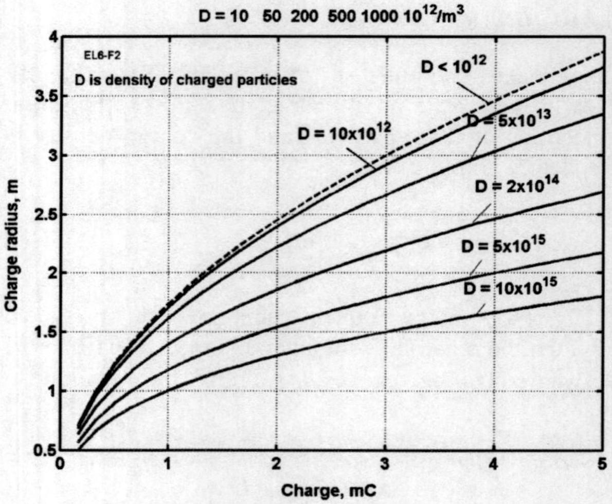

Fig. 23.5 Critical radius of main charge versus the main charge and density of charged particles in the environment (ionized zone)

23.3.5 Some Data about the Ball Material

The properties of electrical insulation vary depending on the impurities in the material, temperature, thickness, etc. and they are different for the same material with a different dieletric. For example, the resistivity of fused quartz is 10^{15} Ohm.cm

Table 23.1 Properties of various good insulators (Encyclopedia 2000, vol. 6, p. 104, p. 229, p. 231)

Insulator	Resistivity Ohm-m	Dielectric strength MV/m. E_i	Dielectric constant, ε	Tensile strength kg/mm^2, $\sigma\times10^7$N/m^2
Lexan	10^{17}–10^{19}	320–640	3	5.5
Kapton H	10^{19}–10^{20}	120–320	3	15.2
Kel-F	10^{17}–10^{19}	80–240	2–3	3.45
Mylar	10^{15}–10^{16}	160–640	3	13.8
Parylene	10^{17}–10^{20}	240–400	2–3	6.9
Polyethylene	10^{18}–5×10^{18}	40–680*	2	2.8–4.1
Poly (tetra-fluoraethylene)	10^{15}–5×10^{19}	40–280**	2	2.8–3.5
Air (1 atm, 1 mm gap)	-	4	1	0
Vacuum (1.3×10^{-3} Pa, 1 mm gap)	-	80–120	1	0

* For room temperature 500–700 MV/m.

** 400–500 MV/m.

for T = 20 °C, the resistivity of the quartz fused (from crystal) reaches about 10^{24} Ohm.cm for $T \approx 20$ °C (see Kikoin 1976, p. 231 and p. 329, their fig. 20.2).Properties of some materials are shown in Table 23.1.

For small balls, the tensile stress is important for reducing the weight (because like charges tear the ball). The author believes that an artificial fiber having a maximum tensile stress at 500–620 kg/mm^2 (fiber) or whiskers up to 2000 kg/mm^2 is better. These fibers can also be used to strengthen balls insulated by a dielectric (for example, as an additional cover).

23.3.6 The Half-Life of the Charge

(1) *Spherical ball.* Let us take a very complex condition; where the unlike charges are separated *only* by an insulator (charged spherical condenser):

$$Ri - U = 0,\quad U = \delta E,\quad E = \frac{kq}{\delta^2},\quad R = \rho\frac{\delta}{4\pi a^2},\quad U = \frac{q}{C},\quad R\frac{dq}{dt} + \frac{a}{C} = 0,$$
$$\frac{dq}{q} = \frac{dt}{RC},\quad C = \frac{a}{k},\quad q = q_0 \exp\left(-\frac{4\pi ak}{\rho\delta}t\right),\quad \frac{q}{q_0} = \frac{1}{2},\tag{23.7}$$
$$-\frac{4\pi ak}{\rho\delta}t_h = \ln\frac{1}{2} = -0.693 \approx -0.7,\quad t_h = 0.693\frac{\rho\delta}{4\pi ka}$$

where: t_h – half-life time, [sec]; R – insulator resistance, [Ohm]; i – current, [A]; U – voltage, [V]; δ – thickness of insulator, [m]; E – electrical intensity, [V/m]; q – charge, [C]; t - time, [seconds]; ρ – specific resistance of insulator, [Ohm-meter, Ωm]; a – internal radius of the ball, [m]; C – capacity of the ball, [C]; $k = 9\times10^9$.

Example: Let us take typical data: $\rho = 10^{19}$ Ω-m, $k = 9\times10^9$, $\delta/a = 0.2$, then $t_h = 1.24\times10^6$ seconds = 144 days.

(2) *Half-life of cylindrical tube*. The computation is same as for tubes (1 m charged cylindrical condenser):

$$q = q_0 \exp\left(-\frac{1}{RC}t\right),\quad C = \frac{1}{k\ln(1+\delta/a)},\quad R = \frac{\rho\delta}{2\pi a},\quad -0.693 = -\frac{1}{RC}t_h,$$
$$t_h = \frac{0.693\rho\delta}{2\pi ka\ln(1+\delta/a)},\quad for\ \ \frac{\delta}{a}\to 0,\quad t_h \approx 0.7\frac{\rho}{2\pi k}.\tag{23.8}$$

23.3.7 Rupture (breakthrough) of Insulator

The breakthrough of a ball can only occur when the charge contacts an unlike charge or conducting material. The voltage between the charges must be less than $U = \delta U_r$, where U_r is the breakthrough voltage for a given insulator and δ is the thickness of the insulator. For a good insulator this is up to $U_r \approx 700$ million V/m, for thin mica it as up to $U_r = 1000$ million V/m.

23.3.8 Levitation between Flat Net and Ground Surface

This is the simplest for both utilization and computation. The top of the column contains the insulated metallic net with a high voltage (it may be a direct current electricity line). This induces the opposite charge in the asteroid and powers the static electric field. The man (vehicle) has charged balls or a ball with like charges to the asteroid's charge. These balls repel from the asteroid's surface (charges) and support the man (vehicle).

The lifting force, L, and radius, a, of a small lifting ball with charge q can be computed by the equations:

$$L = qE_0, \quad E_a = k\frac{q}{a^2}, \quad a = \sqrt{k\frac{q}{E_a}} = \sqrt{\frac{kMg}{E_a E_0}}, \quad U = E_0 h, \tag{23.9}$$

where E_o – electrical intensity between the net and the asteroid's surface [V/m]; E_a – electrical intensity at the ball's surface from the internal ball charge [V/m]; a – internal radius of ball [m]; M – mass of the flight vehicle (man, car); g – asteroid's gravity; U – voltage between the net and asteroid [V]; h – altitude of the net [m].

We can change the single ball to some small highly charged balls or a belt with an ionized zone. The flying vehicles can be protected from contact with the top net by a dielectric (insulator) safety net located below the top net.

23.3.9 Electrostatic Levitation of a Non-contact Transportation

Two identically charged closed plates of area S have a repelling force L:

$$L = 2\pi k \sigma_c^2 S, \tag{23.10}$$

where σ_c is surface charge density [C/m^2].

For example, two 1 m^2 plates with identical charge $\sigma_c = 2\times10^{-4}$ C/m^2 will have a specific lift force of $L = 2260$ N/m^2 = 226 kgf/m^2. Conventional electrets have $\sigma_c = 10^{-4} - 1.4\times10^{-3}$ C/m^2 charge and can be used for a non-contact transportation.

23.3.10 Top Tube Transportation

The parameters of a tube transportation can be calculated by the following equations:

$$\begin{gathered} \tau = \frac{aE_a}{2k}, \quad E_0 = \frac{4k\tau}{h}, \quad \frac{E_0}{E_a} = \frac{a}{h}, \quad C_1 \approx \frac{1}{2k\ln(2h/a)}, \quad U = \frac{\tau}{C_1}, \quad W = \frac{\tau^2}{2C_1}, \\ for \quad \tau = const \quad F_h = \frac{\partial W}{\partial h} = \frac{2k\tau^2}{h}, \quad F_a = \frac{\partial W}{\partial a} = -\frac{2k\tau^2}{a} \end{gathered} \tag{23.11}$$

where: τ – the linear charge of 1 meter of tube [C/m]; a – radius of tube cross section [m]; E_a – electric intensity at tube surface [V/m]; E_0 – electrical intensity at the asteroid's surface at a point under the tube [V/m], for other points $E = E_0\cos^3\alpha$ where α is the angle between a vertical line from the tube center and a line to a given point (electric lines are perpendicular to the asteroid's surface, there is no lateral acceleration); h – altitude [m]; $k = 9\times10^9$ – coefficient [Nm^2/C^2]; C_1 – capacity of 1 meter of tube [C/m] (see Kalashnikov 1985, p. 64); U – voltage [V]; W – electrical energy of 1 meter of tube [J/m]; F_h – electric force of 1 meter of tube between the tube and the asteroid's surface [N/m]; F_a – radial tensile force of 1 meter of tube [N/m].

The thickness and mass of the top tube (with a thin cover) are given by

$$F_a = -2\sigma\delta, \quad \delta = \frac{k\tau^2}{a\sigma}, \quad M_1 = 2\pi\gamma a\delta \tag{23.12}$$

where σ – safe tensile stress of tube cover [N/m^2]; δ – thickness of the tube cover [m]; M_1 – mass of 1 m of tube cover [kg/m]; γ – density of tube cover [kg/m^3].

Assume the case of Earth. The lift force of the tube as a balloon filled by helium can be computed by the equation

$$F_L = (\rho - \rho_g)\pi a^2 \bar{\rho}(h) g \tag{23.13}$$

where ρ = 1.225 kg/m^3 – air density; ρ_g – filling gas density (for helium ρ_g = 0.1785 kg/m^3); a – radius of tube [m]; $\bar{\rho}(h)$ – relative air density at altitude, for h = 0 km the $\bar{\rho}(h) = 1$, for 1 km $\bar{\rho}(h) = 0.908$. Note that E_c decreases in proportional to the atmospheric density. Unfortunately, the attractive electric force F_h in many cases is more than the air lift force F_L. See the example computation in project 2.

23.3.11 Spherical Main Ball on Mast and Fluid Balloon

The parameters of charges of the main ball and spherical balloon can be calculated using the following equations

$$E_a = k\frac{q}{a^2}, \quad E_0 = k\frac{2q}{h^2}, \quad \frac{E_0}{E_a} = 2\left(\frac{a}{h}\right)^2, \quad C = \left[k\left(\frac{1}{a} - \frac{1}{2h-a}\right)\right]^{-1} \approx \frac{a}{2k},$$
$$U \approx \frac{kq}{a}, \quad W = \frac{q^2}{2C}, \; for \quad q = const, \quad F_h = \frac{\partial W}{\partial h} = -\frac{kq^2}{4h^2} \tag{23.14}$$

where q – the charge of fluid balloon (sphere) [C]; a – radius of fluid balloon [m]; E_a – electrical intensity at the balloon's surface [V/m]; E_o – electrical intensity at the asteroid's surface at a point under the balloon [V/m], for other points $E = E_0\cos^3\alpha$ where α is the angle between a vertical line from the balloon center and a line to a given point (electric lines are perpendicular to the asteroid's surface, there is no lateral acceleration); h – altitude [m]; $k = 9\times10^9$ – coefficient [Nm^2/C^2]; C – capacity of balloon [C]; U – voltage [V]; W – electrical energy of the balloon [J]; F_h – electrical force between the balloon and the asteroid's surface [N].

The thickness and mass of the top fluid balloon as a spherical capacitor 1 m (with a thin cover) are

$$F_a = \frac{\partial W}{\partial a}, \quad W = \frac{q^2}{2C}, \quad c = \frac{a}{k}, \quad for \quad q = \frac{a^2 E_a}{k} = const, \quad F_a = -\frac{kq^2}{2a^2} = -\frac{(aE_a)^2}{2k},$$
$$p = -\frac{F_a}{4\pi a^2} = \frac{E_a^2}{8\pi k}, \quad \pi a^2 p = 2\pi a \delta\sigma, \quad \delta = \frac{ap}{2\sigma} = \frac{aE_a^2}{8\pi k}, \quad M_b = 4\pi a^2 \gamma\delta = \frac{a^3 E_a^2 \delta}{4k\sigma}, \tag{23.15}$$

where σ – safe tensile stress of the balloon cover [N/m^2]; δ – thickness of the balloon cover [m]; M_b – mass of the balloon cover [kg]; γ – density of the balloon cover [kg/m^3]; p – internal pressure under like charges [N/m^2].

Assume the case of Earth. The lift force of the air balloon filled by helium can be computed using the equation

$$F_L = \frac{4}{3}(\rho - \rho_g)\pi a^3 \bar{\rho}(h) g \tag{23.16}$$

where ρ = 1.225 kg/m^3 – air density; ρ_g – filling gas density (for helium ρ_g = 0.1785 kg/m^3); a – radius of balloon [m]; $\bar{\rho}(h)$ – relative air density at altitude, for h = 0 km the $\bar{\rho}(h) = 1$, for 1 km $\bar{\rho}(h) = 0.908$. In many cases the attracted electric force F_h is more than the air lift force F_L. See the computation in Project 3.

23.3.12 Small Spherical Lifting Balls

Assume the electrical intensity of the main top charge is significantly more then the lifting charges. The parameters of large spherical balls with thin covers can be computed using the equations above. The parameters of small balls with thick covers can be computed using the following equations

$$L = nqE_0, \quad q = CU, \quad C = \frac{r}{k}, \quad U = aE_a, \quad q = \frac{a^2 E_a}{k}, \quad L = n\frac{a^2 E_a E_0}{k} \tag{23.17}$$

where L – total lift force [N]; n – number of balls; E_a – electrical intensity at the ball surface from electrical charge of the ball q [C]; a – internal radius of the ball [m].

The thickness and mass of the ball (with a thick cover) are

$$p_b = \frac{E_a^2}{8\pi k}, \quad \sigma = \frac{p_b}{(R/a)^2 - 1} = \frac{E_a^2}{8\pi k\left[(R/a)^2 - 1\right]},$$
$$(R/a)^2 - 1 = \frac{E_a^2}{8\pi k\sigma}, \quad M_b = \frac{4}{3}\pi\gamma a^3\left(\frac{R^3}{a^3} - 1\right) \tag{23.18}$$

where $R = a + \delta$ – external radius of ball [m]; σ – safe tensile stress of the ball material [N/m^2]; δ – thickness of the ball cover [m]; M_b – mass of the ball cover [kg]; γ – density of the ball cover [kg/m^3].

Results of this computation are in Figs. 23.6 to 23.9. Notice that the lifting balls have a large ratio of lift force/ball mass, about 10,000–20,000.

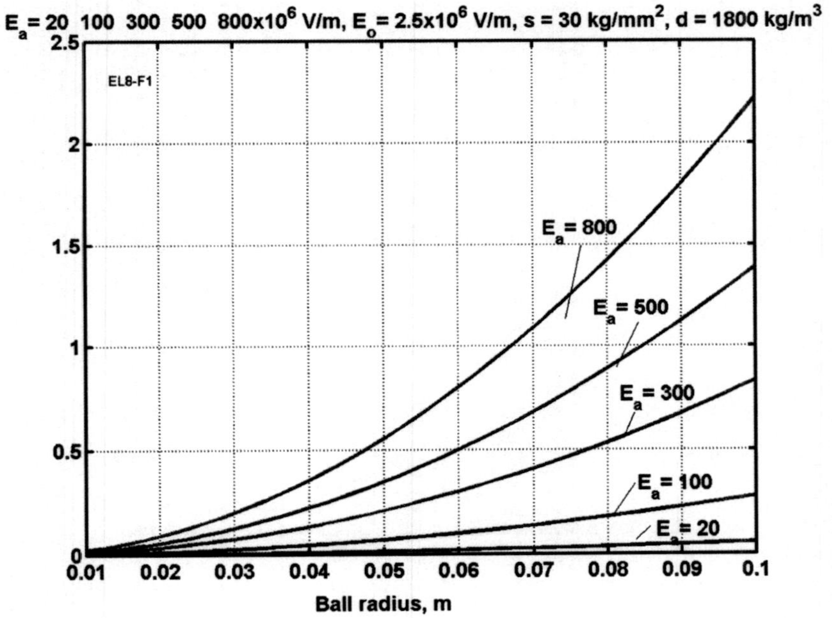

Fig. 23.6 Electrostatic lift force (kN) of small lifting ball versus radius of ball for the electrical intensity of the ball's surface $E_a = (20–800)\times10^6$ V/m, general electrical intensity $E_o = 2{,}5 \times10^6$ V/m, safe tensile stress of the ball cover 30 kg/mm^2, specific density of the ball cover 1800 kg/m^3

23.3.13 Long (a << l) Cylindrical Lifting Belt

The maximum charge and mass of a 1 meter long cylindrical lifting belt may be computed using the following equations

$$L = E_0 \tau l, \quad 2\sigma\delta = F_a, \quad F_a = \frac{2k\tau^2}{a}, \quad \tau = \sqrt{\frac{a\delta\sigma}{k}}, \quad q = \tau l,$$
$$M_1 = 2\pi\gamma a\delta, \quad M = M_1 l, \quad E_a = k\frac{2\tau}{a} \tag{23.19}$$

where τ – charge of 1 meter [C/m]; a – internal radius of the belt cross-section area [m]; δ – thickness of the belt [m]; σ – safe tensile stress of the belt cover [N/m^2]; q – charge of the belt [C]; l – length of the belt [m]; M – mass of the belt [kg]; E_a – electrical intensity of the belt surface [V/m]; F_a – electrostatic force in the tube [N] (see Eq. (23.11)); γ – density of belt cover [kg/m^3]. Computations are presented in Fig. 23.10 and 23.11. See also example computation in project 3.

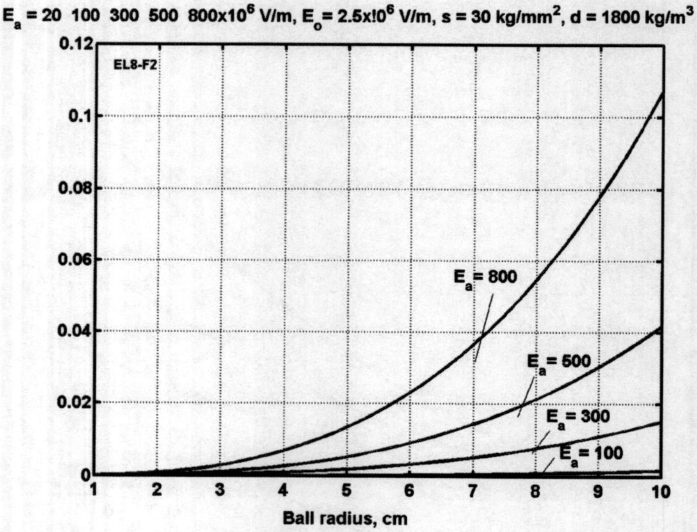

Fig. 23.7 Mass (kg) of small lifting ball versus radius of ball for the electrical intensity of the ball's surface E_a = (100–800)×10^6 V/m, general electrical intensity E_o = 2.5 ×10^6 V/m, safe tensile stress of the ball cover 30 kg/mm^2, specific density of the ball cover 1800 kg/m^3

23.3.14 Aerodynamics of the Levitated Vehicles

On asteroids with natural or artificial atmosphere, the drag, D, (required thrust, T) and required power of the levitated person, car and vehicles can be computed by the following equations

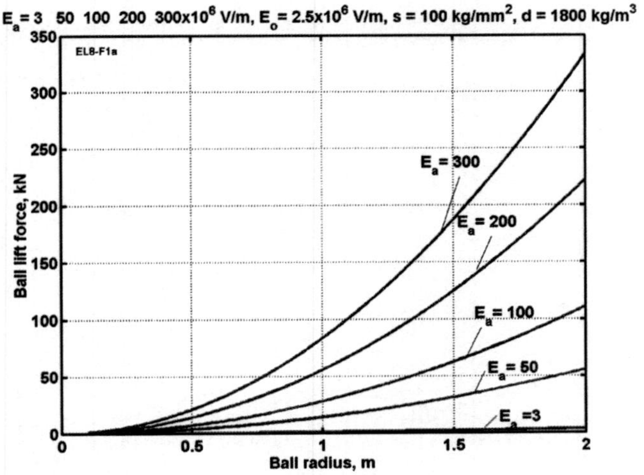

Fig. 23.8 Electrostatic lift force (kN) of the lifting ball versus radius of ball for the electrical intensity of the ball's surface $E_a = (3\text{–}300)\times10^6$ V/m, general electrical intensity $E_o = 2.5 \times10^6$ V/m, safe tensile stress of the ball cover 100 kg/mm^2, specific density of the ball cover 1800 kg/m^3

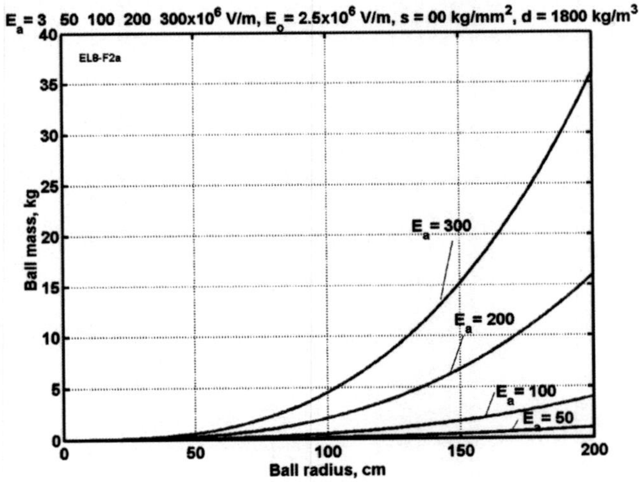

Fig. 23.9 Mass (kg) of the lifting ball versus radius of ball for the electrical intensity of the ball's surface $E_a = (3\text{–}300)\times10^6$ V/m, general electrical intensity $E_o = 2.5 \times10^6$ V/m, safe tensile stress of the ball cover 100 kg/mm^2, specific density of the ball cover 1800 kg/m^3

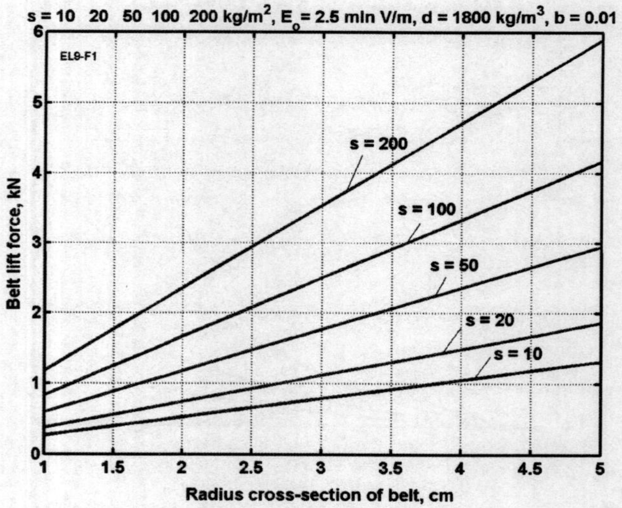

Fig. 23.10 Electrostatic lift force (kN) of a 1 m small lifting belt via radius of ball cross-section area for general electrical intensity $E_o = 2.5 \times 10^6$ V/m, safe tensile stress of the ball cover 10–200 kg/mm^2, $\sigma = (10–200) \times 10^7$ N/m^2, density of ball cover $\gamma = 1800$ kg/m^3

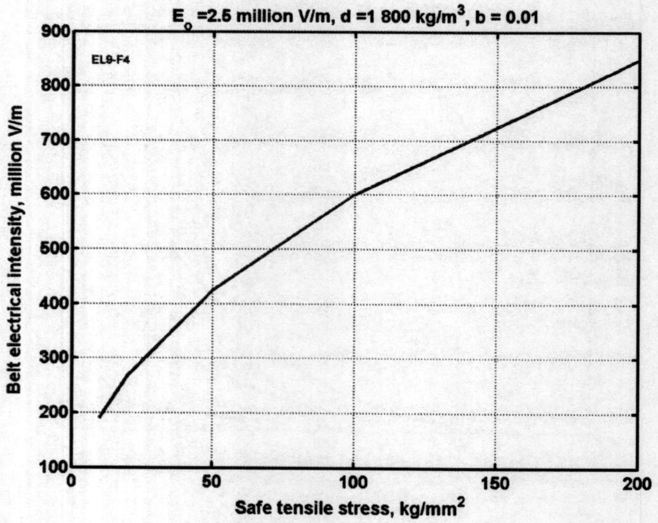

Fig. 23.11 Belt electrical intensity via safe tensile stress for general electrical intensity $E_o = 2.5 \times 10^6$ V/m, safe tensile stress of the ball cover 10–200 kg/mm^2, $\sigma = (10–200) \times 10^7$ N/m^2; γ – specific density of ball cover 1800 kg/m^3, relative thickness of belt cover, $\delta = 0.01 \times a$

$$T = D = C_D \frac{\rho V^2}{2} S, \quad W = \frac{VD}{\eta}, \quad a = \frac{T}{M}, \tag{23.20}$$

where C_D – aerodynamic drag coefficient, for a sitting person $C_D \approx 0.5$, for a lying man $C_D \approx 0.3$, for a car $C_D \approx 0.25$, for a sphere $C_D \approx 0.1 - 0.2$ (depending on the size and speed), for a dirigible $C_D \approx 0.06 - 0.1$; ρ atmosphere density; V – speed [m/s]; S – vehicle cross section area [m^2]; W – required power [W]; η – propeller coefficient of efficiency, $\eta = 0.7 - 0.8$.

For example, on Earth a flying person ($S = 0.3$ m^2) has D = 5.5 N for speed $V =$ 10 m/s (36 km/hour). He only needs a small motor, W = 0.073 kW.

23.3.15 Control and Stability

Control is accomplished using the direction (and magnitude) of motor thrust (and variable torque) and the charging and discharging of lifting charges. The levitated vehicle will be stable in a vertical position if its center of gravity is lower than the center of levitation (lift) force. The dipole moment of the particular vehicle's design can give additional stability. Note that electric lines are vertical at the asteroid's surface (Figs. 23.1d,e,f,g), which means that the lift force is vertical.

23.3.16 Flight in Thunderstorms

On asteroids where thunderstorms are produced, an electric field of about 300,000 to 1,000,000 V/m is generated. This field can be used for levitation.

23.3.17 Charging

In the author's opinion, the easiest method of charging and maintaining the charge is by using a Van de Graaff electrostatic generator (Fig. 23.2j). Any other high voltage generation devices can also be used.

23.3.18 Safety

It is not known exactly how static fields of electrical intensity affect a human body. People in an electric field of about 300,000 to 1,000,000 V/m during a thunderstorm or under high voltage electrical lines feel normal. The inside space of a conventional car with a metal body (or conductive paint) does not have an electric field. People can wear clothes armored by conductive filaments as a defense against the electric field.

23.3.19 Charged Ball as an Accumulator of Energy

The energy required to charge a ball (accumulate at in the ball) can be calculated by the following equations

$$W=\frac{1}{2}\frac{q^2}{C},\quad C=\frac{a}{R},\quad E_a=k\frac{q}{a^2},\quad q=\frac{a^2E_a}{k},\quad W=\frac{1}{2}\frac{a^3E_a^2}{k} \tag{23.21}$$

where W is energy [J]. The ball mass at safe stress levels with repelling charges can be calculated using Eq. (23.17):

$$\left[\left(\frac{R}{a}\right)^2-1\right]=\frac{E_a^2}{8\pi k\sigma},\quad \left(\frac{R}{a}\right)=\sqrt{\frac{E_a^2}{8\pi k\sigma}+1},$$
$$M=\frac{4\pi\gamma a^3}{3}\left[\left(\frac{R}{a}\right)^3-1\right],\quad \frac{W}{M}=\frac{3E_a^2}{8\pi\gamma k\left[(R/a)^3-1\right]} \tag{23.22}$$

where M – mass [kg]; σ – safe tensile stress of the ball cover [N/m^2]; γ – specific density of the ball cover [kg/m^3]; $R = a + \delta$ – external radius of the ball [m].

The accumulated relative energy for σ = 200 kg/mm^2 may be close to conventional powder and last a lot langer than electrical energy in a typical condenser. This electrical energy can be reclaimed (by using a sharp spike) or used for launching or accelerating space vehicles if we take two like charges (balls) and allow them to repel each other. This method of transforming electrical energy into thrust may be more useful than the thrust from a conventional electric space engine because one can create a big thrust by utilizing asteroids.

23.4 Projects

Let us estimate the main parameters for some offered applications. Most people understand the magnitudes and properties of applications better than theoretical reasoning and equations. The suggested application parameters are not optimal, but our purpose is to show the method can be utilized by current technology.

23.4.1 Levitation Transportation (Fig. 23.1d)

The height of the top net is 20 m. The electrical intensity is $E_o = 2.5\times10^6$ V < $E_c = (3–4)\times10^6$ V. The voltage between the top net and the asteroid is $U = 50\times10^6$ V. The width of each side of the road is 20 m. We first find the size of the lifting ball for the man (100 kg), small vehicle (1000 kg), or large vehicle (10,000 kg). Here R_c is the radius of the ionized zone [m]:

1) Flying man on Earth (mass M = 100 kg, ε = 3, E_i = 200×10^6 V/m, g ≈10 m/s^2)

$$E_a \le \varepsilon E_i = 3\times 200\times 10^6, \quad a = \sqrt{\frac{kMg}{E_0 E_a}} = \sqrt{\frac{9\times 10^9\times 100\times 10}{2{,}5\times 10^6\times 6\times 10^8}} \approx 0.08\, m, \quad R_c = \sqrt{\frac{E_a}{E_c}} \approx 1\, m$$

Notice that the radius of a single ball supporting the man is only 8 cm, or the man can use two balls a = 5–6 cm., R_c=0.75 m (or even more smaller balls). If the man uses a 1 m cylindrical belt, the radius of the belt cross-section area is 1.1 cm, σ = 100 kg/mm^2, E_a = 600×10^6 V/m (Figs. 23.10 and 23.11). The belt may be more comfortable for some people.

2) With the same calculation you can find that a small vehicle of mass M = 1000 kg will be levitated using a single charged ball a = 23 cm, R_c = 3.2 m (or two balls with a = 16 cm. R_c = 2.3 m).

3) A large vehicle of mass M = 10,000 kg will be levitated using a single charged ball a = 70 cm, R_c = 10 m (or two balls with a = 0.5 m. R_c = 7 m).

23.4.2 Levitating Tube Transportation

Assume the levitation tube transportation has the design of Fig.23.1f,g where the top net is changed to a tube. Take the data E_o = 2.5×10^6 V < E_c = (3–4)×10^6 V, E_a = 2×10^8 V/m, h = 20 m. This means the electrical intensity, E_o, at ground level is the same as in the previous case. The required radius, a, of the top tube is

$$\frac{a}{h} = \frac{E_0}{2E_a} = 0.00625, \quad a = 0.00624h = 0.125\, m, \quad R_c = a\frac{E_a}{E_0} = 10\, m$$

The diameter of the top tube is 0.25 m, the top ionized zone has a radius of 10 m.

23.4.3 Charged Ball Located on a High Mast or Tower

Assume there is a mast (tower) 500 m high with a ball of radius a = 32 m at its top charged up to E_a = 3×10^8 V/m. The charge is

$$q = \frac{a^2 E_a}{k} = 34\, C, \quad E_0 = k\frac{2q}{h^2} = 2.45\times 10^6\, V/m$$

This electrical intensity at ground level means that within a radius of approximately 1 km, people, small vehicles and other loads can levitate.

23.4.4 *Levitation in Low Clouds*

On asteroids with natural or artificial clouds, the electrical intensity at ground level is about $E_o = 3\times10^5 - 10^6$ V/m. A person can take more (or more highly charged) balls and levitate.

23.4.5 *Artificial Gravity on Asteroids*

Assume the asteroid is a sphere with an inner radius at a = 10 m and external radius of 13 m. We can create the electrical intensity $E_o = 2.5\times10^6$ V/m without an ionized zone. The electrical charge is $q = a^2E_o/k = 2.8\times10^{-2}$ C. For a man weighing 100 kg on Earth (g = 10 m/s^2, force F = 1000 N), it is sufficient to have a charge of $q = F/E_o = 4\times10^{-4}$ C and small ball with a = 0.1 m and $E_a = qk/a^2 = 3.6\times10^8$ V/m. At the asteroid's surface, the artificial gravity will be $(10/13)^2 = 0.6$ = 60% of g (Bolonkin 2005a).

23.4.6 *Charged Ball as an Accumulator of Energy and Rocket Engine*

The computations show the relative *W/M* energy calculated from safe tensile stress does not depend on E_a. A ball cover with a tensile stress of σ = 200 kg/mm^2 reaches 2.2 MJ/kg. This is close to the energy of conventional powder (3 MJ/kg). If whiskers or nanotubes are used the relative electrical storage energy will be close to than of liquid rocket fuel.

Two like charged balls repel one another and can give significant acceleration for a space vehicle, VTOL aircraft, or weapon.

23.5 Discussion

Electrostatic levitation can be used for transportation on asteroids.The offered method needs development and testing. The experimental procedure it is not expensive. We just need a ball with a thin internal conducting layer, a dielectric cover, and high voltage charging equipment. This experiment can be carried out in any high voltage electric laboratory. The proposed levitation theory is based on proven electrostatic theory. There may be problems may be with discharging, blockage of the charge by the ionized zone, breakdown, and half-life of the discharge, but careful choice of suitable electrical materials and electric intensity may be also to solve them. Most of these problems do not occur in a vacuum.

Another problem is the affects of the strong electrostatic field on a living organism. Only experiments using animals can solve this. In any case, there are protection methods – conducting clothes or vehicle is (from metal or conducting paint) which offer a defense against the electric field.

Other related ideas from the author are shown in Bolonkin (2005a,b,c,d, 2006, 2007a,b, 2008, 2010; Bolonkin and Cathcart, 2007).

References

Bolonkin, A.A.: Installation for creating open electrostatic field. Patent applications #3467270/21 116676, USSR Patent office (July 9, 1982)

Bolonkin, A.A.: (Electrostatic) method for tensing of films. Patent application #3646689/10 138085, USSR Patent office (September 28, 1983)

Bolonkin, A.A.: Aviation, Motor, and Space Design, Collection Emerging Technology in the Soviet Union, pp. 32–80. Delphic Ass., USA (1990)

Bolonkin, A.A.: Electrostatic Utilization of Asteroids for Space Flight, AiAA-2005-4032. 41st Joint Propulsion Conferences, Tucson, Arizona, USA, July 10-13 (2005a)

Bolonkin, A.A.: Electrostatic Solar Wind Propulsion System, AIAA-2005-4225. 41st Joint Propulsion Conferences, Tucson, Arizona, USA, July 10-13 (2005b)

Bolonkin, A.A.: Kinetic Anti-Gravitator, AIAA-2005-4505. 41st Joint Propulsion Conferences, Tucson, Arizona, USA, July 10-13 (2005c)

Bolonkin, A.A.: Sling Rotary Space Launcher, AIAA-2005-4035. 41st Joint Propulsion Conferences, Tucson, Arizona, USA, July 10-13 (2005d)

Bolonkin, A.A.: Non-Rocket Space Launch and Flight, 488 p. Elsevier, London (2006), http://www.archive.org/details/Non-rocketSpaceLaunchAndFlight, http://www.scribd.com/doc/24056182

Bolonkin, A.A.: New Concepts, Ideas, and Innovations in Aerospace, Technology and Human Life. NOVA Publishers (2007a), http://www.scribd.com/doc/24057071, http://www.archive.org/details/NewConceptsIfeasAndInnovationsInAerospaceTechnologyAndHumanSciences

Bolonkin, A.A.: AB Levitation and Energy Storage. This work presented as paper AIAA-2007-4613 to 38th AIAA Plasma Dynamics and Lasers Conference in conjunction with the16th International Conference on MHD Energy Conversion, Miami, USA, June 25-27 (2007b)

Bolonkin, A.A.: New Technologies and Revolutionary Projects, Lambert, 324 p. (2008), http://www.scribd.com/doc/32744477, http://www.archive.org/details/NewTechnologiesAndRevolutionaryProjects

Bolonkin, A.A.: Life. Science. Future (Biography notes, researches and innovations), 208 p. Publish America (2010), http://www.scribd.com/doc/48229884, http://www.lulu.com, http://www.archive.org/details/Life.Science.Future.biographyNotesResearchesAndInnovations

Bolonkin, A.A., Cathcart, R.B.: Macro-Projects: Environments and Technologies, 536 p. NOVA Publishers (2007), http://www.scribd.com/doc/24057930, http://www.archive.org/details/Macro-projects EnvironmentsAndTechnologies

Encyclopedia, McGraw-Hill Encyclopedia of Science & Technology (2000)

Kalashnikov, C.K.: Electricity, Nauka, Moscow (1985) (in Russian)

Kestelman, V.N., Pinchuk, L.S., Goldale, V.A.: Electrets in Engineering, Fundamentals and Applications. Kluwer Academic Publishers (2000)

Kikoin, I.K. (ed.): Tables of physical values, Directory, Atomisdat, Moscow (1976) (in Russian)

Koshkin, N.I., Shirkevich, M.G.: Directory of Elementary Physics, Nauka, Moscow (1982) (in Russian)

Shortley, G., Williams, D.: Elements of Physics, 5th edn. Prentice-Hall, Inc., New Jersey (1996)

Chapter 24
Making Asteroids Habitable

Alexander A. Bolonkin

C&R Co., New York, USA

24.1 Introduction

The real development of outer space (permanent human life in space) requires two conditions: all-sufficient space settlement and artificial life conditions close to those prevailing currently on the Earth. (Such a goal extends what is already being attempted in the Earth-biosphere—for example at the 1st Advanced Architecture Contest, “Self-Sufficient Housing”, sponsored by the Institute for Advanced Architecture of Catalonia, Spain, during 2006.) The first condition demands production of all main components needed for human life: food, oxidizer, and energy within the outer space and Solar System body colony. The second requisite condition is a large surface settlement having useful plants, attractive flowers, splashing water pools, walking and sport areas, etc. All these conditions may be realized within large 'greenhouses' that will produce food, oxidizer and “the good life” conditions. Human life on asteroids will be more comfortable if it uses A.A. Bolonkin’s macro-project proposal - staying in outer space without special spacesuit (Bolonkin 2006a, p. 335) (mass of current spacesuit reaches 180 kg). The idea of this chapter follows the approach proposed in Bolonkin (2006b, 2007a). The current life conditions on asteroids are far from comfortable. For example, the asteroids do not have any atmosphere and there are deadly space radiation and meteor bombardments. Future humans living on asteroids must be more comfortable for humans to explore and properly exploit these distant places.

24.2 ‘Evergreen’ Inflated Domes

Possibly the first true architectural attempt at constructing effective artificial life-support systems on the climatically harsh asteroid will be the building of greenhouses. Greenhouses are maintained nearly automatically by heating, cooling, irrigation, nutrition and plant disease management equipment. Humans share commonalities in their responses to natural environmental stresses that are stimulated by night cold, day heat, absent atmosphere, so on. Darkness everywhere inflicts the same personal visual discomfort and disorientation as cosmonauts/astronauts experience during their space-walks—that of being adrift in featureless space. With special clothing and shelters, humans can adapt successfully to the well-land marked asteroid.

Incontrovertibly, living on the asteroid is difficult, even when tempered by strong conventional protective buildings.

Our macro-engineering concept of inexpensive-to-construct-and-operate "Evergreen" inflated surface domes is supported by computations, making our macro-project speculation more than a daydream. Innovations are needed, and wanted, to realize such structures upon the asteroid of our unique but continuously changing life.

24.3 Description and Innovations of Dome

24.3.1 Dome

Our basic design for the asteroid people-housing "Evergreen" dome is presented in Fig. 24.1, which includes the thin inflated double film dome. The innovations are listed here: (1) the construction is air-inflatable; (2) each dome is fabricated with very thin, transparent film (thickness is 0.2 to 0.4 mm) without rigid supports; (3) the enclosing film is a two-layered structural element with air between the layers to provide insulation; (4) the construction form is that of a hemisphere, or in the instance of a roadway/railway a half-tube, and part of the film has control transparency and a thin aluminum layer about 1 μm or less that functions as the gigantic collector of incident solar radiation (heat). Surplus heat collected may be used to generate electricity or furnish mechanical energy; and (5) the dome is equipped with sunlight controlling louvers [also known as, "jalousie", a blind or shutter having adjustable slats to regulate the passage of air and sunlight] with one side thinly coated with reflective polished aluminum of about 1 μm thickness. Real-time control of the sunlight's entrance into the dome and nighttime heat's exit is governed by the shingle-like louvers or a controlled transparency of the dome film.

Variant 1 of artificial inflatable Dome for asteroid is shown in Fig. 24.1. Dome has top thin double film 4 covered given area and single under ground layer 6. The space between layers 4 - 6 is about 3 meters and it is filled by air. The support cables 5 connect the top and underground layers and Dome looks as a big air-inflated beach sunbathing or swimming mattress. The Dome includes hermetic sections connected by corridors 2 and hermetic lock chambers 3. Topmost film controls the dome's transparency (and reflectivity). That allows people to closely control temperature affecting those inside the dome. Topmost film also is of a double-thickness. When a meteorite pushes hole in the topmost double film, the lowermost layer closes the hole and puts temporary obstacles in the way of the escaping air. Dome has a fruitful soil layer, irrigation system, and cooling system 9 for supporting a selected given humidity. That is, a closed-biosphere with a closed life-cycle that regularly produces an oxidizer as well as sufficient food for people and their pets, even including some species of farm animals. Simultaneously, it is the beautiful and restful Earth-like place of abode. The offered design has a minimum specific mass, about 7-12 kg/m^2 (air - 3 kg, film - 1 kg, soil - 3 - 8 kg). Mass of an example area of 10×10 m is about 1 metric ton (oftentimes spelt "tones").

Figure 24.2 illustrates the second thin transparent dome cover we envision. The Dome has double film: semispherical layer (low pressure about 0.01 - 0.1 atmosphere, atm.) and lower layer (high 1 atm. pressure). The hemispherical inflated textile shell—technical "textiles" can be woven (weaving is an interlacement of warp and weft) or non-woven (homogenous films)—embodies the innovations listed: (1) the film is very thin, approximately 0.1 to 0.3 mm. A film this thin has never before been used in a major building; (2) the film has two strong nets, with a mesh of about 0.1 × 0.1 m and $a = 1 \times 1$ m, the threads are about 0.3 mm for a small mesh and about 1 mm for a big mesh.

The net prevents the watertight and airtight film covering from being damaged by micrometeorites; the film incorporates a tiny electrically-conductive wire net with a mesh of about 0.001 x 0.001 m and a line width of about 100 μm and a thickness near 1 μm. The wire net can inform the "Evergreen" dome supervisors (human or automated equipment) concerning the place and size of film damage (tears, rips, punctures, gashes); the film is twin-layered with the gap — $c = 1$ m and $b = 2$ m—between the layer covering (Fig. 24.3). This multi-layered covering is the main means for heat insulation and anti-puncture safety of a single layer because piercing won't cause a loss of shape since the film's second layer is unaffected by holing; the airspace in the dome's twin-layer covering can be partitioned, either hermetically or not; and part of the covering may have a very thin shiny aluminum coating that is about 1 μm for reflection of non-useful or undesirable impinging solar radiation.

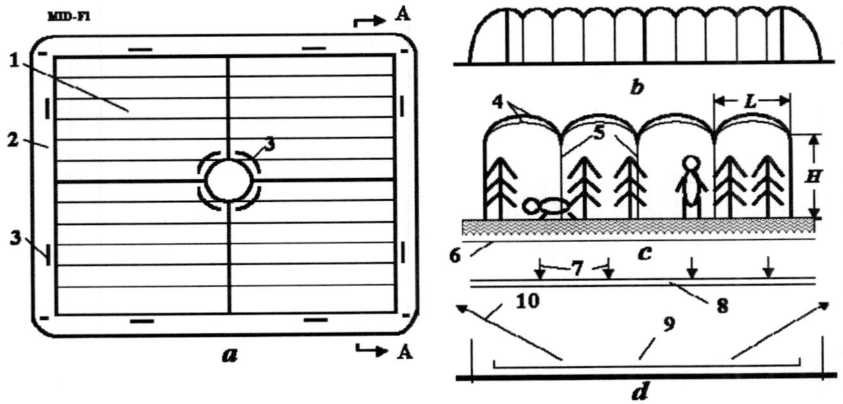

Fig. 24.1 Variant 1 of artificial inflatable Dome for asteroid. (a) top view of dome; (b) cross-section AA area of dome; (c) inside of the Dome; (d) Cooling system. Notations: 1 - internal section of Dome; 2 - passages; 3 - doors; 4 - transparence thin double film ("textiles") with control transparency; 5 - support cables; 6 - lower underground film; 7 - solar light; 8 - protection film; 9 - cooling tubes; 10 - radiation of cooling tubes

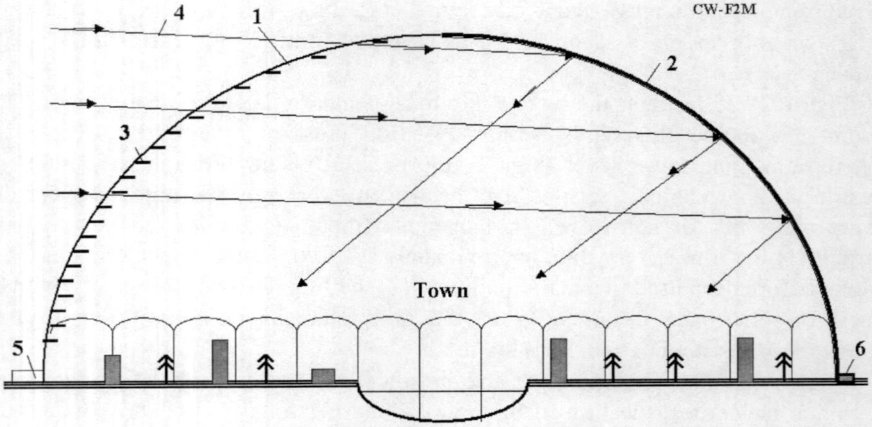

Fig. 24.2 Variant 2 of artificial inflatable Dome for larger asteroids. Notations: 1 - transparent thin double film ("textiles"); 2 - reflected cover of hemisphere; 3 - control louvers (jalousie); 4 - solar beams (light); 5 - enter (dock chamber); 6 - water extractor from air. The lower section has air pressure about 1 atm. The top section has air pressure of 0.01 - 0.1 atm.

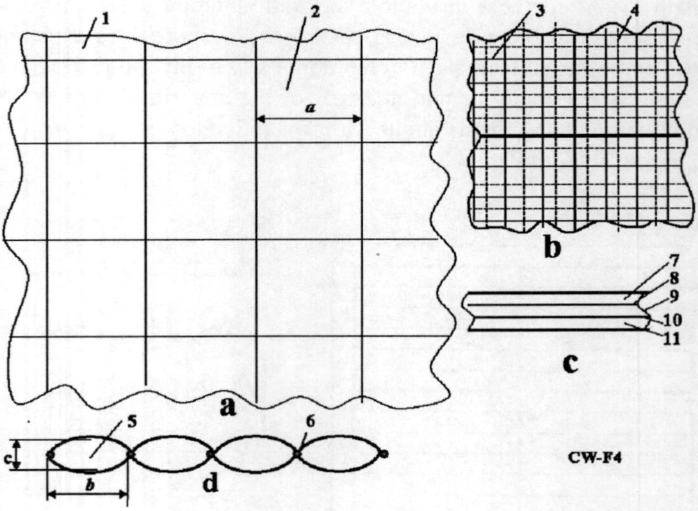

Fig. 24.3 Design of membrane covering. *Notations*: (a) Big fragment of cover with control clarity (reflectivity, carrying capacity) and heat conductivity; (b) Small fragment of cover; (c) Cross-section of cover (film) having 5 layers; (d) Longitudinal cross-section of cover for cold and hot regions; 1 - cover; 2 -mesh; 3 - small mesh; 4 - thin electric net; 5 - cell of cover; 6 - tubes;: 7 - transparant dielectric layer, 8 - conducting layer (about 1 - 3 μm), 9 - liquid crystal layer (about 10 - 100 μ), 10 - conducting layer, and 11 - transparant dielectric layer. Common thickness is 0.1 - 0.5 mm. Control voltage is 5 - 10 V.

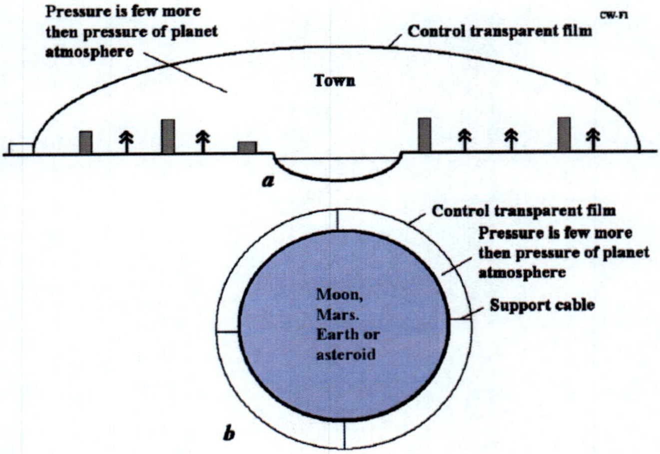

Fig. 24.4 (*a*) Inflatable film dome over a single town; (*b*) Inflatable film dome covering the whole surface of the asteroid

Offered inflatable Dome can cover a big region (town) and create beautiful Earth-like conditions on the asteroid (Fig. 24.4a). In future, the "Evergreen" dome can cover the whole surface of the asteroid (Fig. 24.4b).

24.3.2 Location, Illumination and Defending Human Settlements from Solar Wind and Space Radiation

The asteroids have rotation. If we want to have conventional Earth artificial day and natural solar lighting, the settlement must have a magnetic control mirror suspended at high altitude in given (stationary) place (Fig. 24.5). For building this mirror (reflector) may use idea and theory of magnetic levitation developed in Bolonkin (2007b). If reflector is made with variable focus, as in Bolonkin (2006b, p. 306, Fig. 16.3), then it may well be employed as a concentrator of sunlight and be harnessed for energy during "night" (Earth-time).

The second important feature of the offered installation is defense of the settlement from solar wind and all cosmic radiation. It is known that the Earth's magnetic field is a natural defense for living animals, plants and humans against high-energy particles, such as protons, of the solar wind. The artificial magnetic field near asteroid settlement is hundreds of times stronger than the Earth's magnetic field. It will help to defend delicate humans. The polar location of the planned settlement also decreases the intensity of the solar wind. People can move to an underground cosmic radiation protective shelter, a dugout or bunker, during periods of high Sun activity (solar flashes, coronal mass ejections).

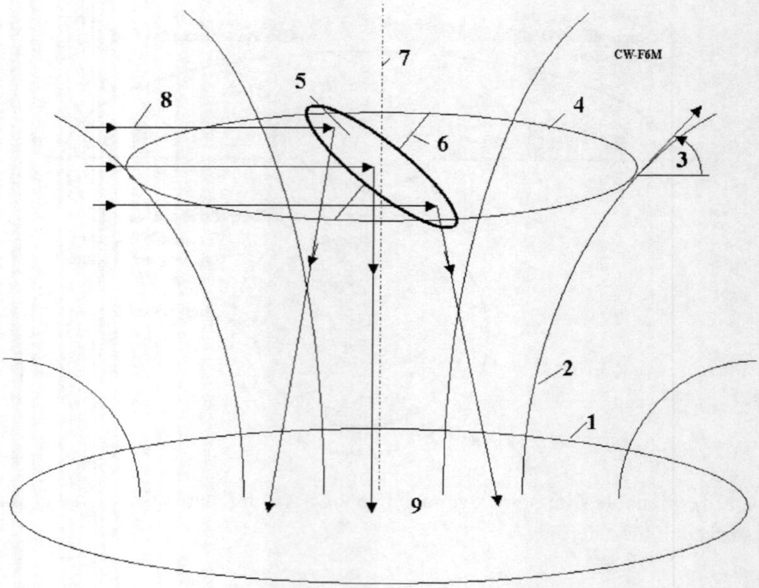

Fig. 24.5 Magnetic control mirror is suspended at high altitude over human asteroid settlement. *Notations:* 1 -superconductivity ground ring; 2 - magnetic lines of ground superconductivity ring; 3 -angle (α) between magnetic line of the superconductivity ground ring and horizontal plate (see Eq. (24.6)); top superconductivity ring for supporting the mirror (reflector) 5; 6 - axis of control reflector (which allows turning of mirror); 7 - vertical axis of the top superconductivity ring; 8 - solar light; 9 – human settlement.

The theory and computation of this installation is in theoretical section, below. The mass of the full reflector (rings, mirror, head screens is about 70 - 80 kg; if the reflector is used also as powerful energy source, then the mass can reach 100 - 120 kg. Note: for lifting, the reflector does not need a rocket. The magnetic force increases near ground (see Eq. (24.3)). This force lifts the reflector to the altitude that is required by its usage. The reflector also will be structurally stable because it is located in magnetic hole of a more powerful ground ring magnet.

The artificial magnetic field may be used, too, for free flying of men and vehicles, as it is described in Bolonkin (2007a,b). If an asteroid does not have enough gravity, then electrostatic artificial gravity may be used (Bolonkin 2006b, Ch. 15).

The magnetic force lifts the reflector to needed altitude. Centrifugal force and cables can also to keep the reflector at a rotated asteroid.

Figure 24.6 illustrates a light-weight, possibly portable house, using the same basic construction materials as the dwelling/workplace.

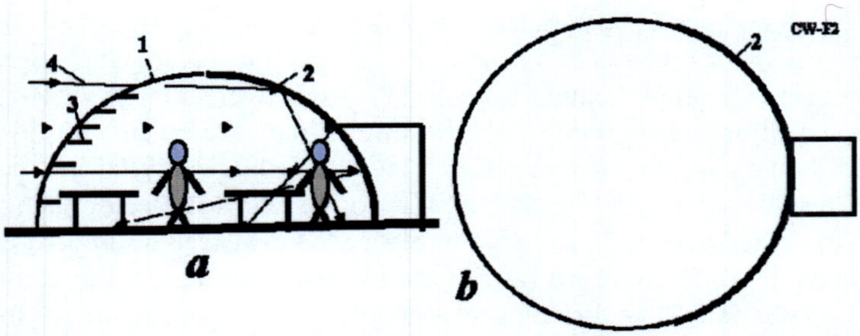

Fig. 24.6 Inflatable film house for asteroid. Notation: (a) Cross-section area; (b) Top view. The other notations are same with Fig. 24.2.

24.3.3 Description and Innovations of Closed-Loop Water Cycle

Irrigation is the artificial application of freshwater to the potentially fertile soil, usually for assisting the growth of crops of food and fiber. In crop production, it is mainly used to replace absent rainfall during periods of drought, but also to protect plants against sudden frosts. Additionally, irrigation suppresses weed growth in rice-fields. In contrast, agriculture that relies only on direct rainfall is sometimes referred to as dry land farming or as rain-fed farming. Irrigation techniques are generally studied in conjunction with drainage techniques, which are the natural or artificial removal of surface and sub-surface water from a selected land region. Irrigation gives high-stability harvests which are ~ 3 – 5 times more than conventional agriculture would provide humanity.

Fig. 24.7 Film dome over agriculture region or a city. *Notations*: 1 - area, 2 - thin film cover with a control heat conductivity and clarity, 3 – control support cable and tubes for rain water (height is 50 – 300 m), 4 - exits and ventilators, 5 - semi-cylindrical border section.

Our idea is a closed-loop water cycle in dome covering a local region of an asteroid by a thin film with controlled heat conductivity and optionally-controlled clarity (reflectivity, albedo, carrying capacity of solar spectrum)(Fig. 24.7). The film is located at an altitude of ~50 – 300 m. The film is supported at this altitude by a small additional air pressure produced by ventilators sitting on the ground. The film is connected to asteroid's surface by tethering cables. The cover may require double-layer film. We can control the heat conductivity of the dome cover by pumping in air between two layers of the dome film cover and change the solar heating due to sunlight heating by control of the cover's clarity. That allows selecting for different conditions (solar heating) in the covered area and by pumping air into dome. Envisioned is a cheap film having liquid crystal and conducting layers. The clarity is controlled by application of selected electric voltage. These layers, by selective control, can pass or blockade the available sunlight (or parts of solar spectrum) and pass or blockade the asteroid's radiation. The incoming and outgoing radiations have different wavelengths. That makes control of them separately feasible and, therefore, possible to manage the heating or cooling of the asteroid's surface under this film. In our closed water system the rain (or at least condensation) will occur at night when the temperature is low. In our case, the ground heats *only* the air in the dome (as in a hotbed). We have, then, a literal greenhouse effect. That means that many cold asteroids may absorb more solar energy and became a temperate climate or sub-tropic climate (under the dome, as far as plants are concerned).

The building of a film dome is easy. We spread out the film over asteroid's surface, turn on the pumping and the film is raised by air pressure to the needed altitude limited by the support cables.

24.3.4 Brief Data on Cover Film

Our dome filmic cover has 5 layers (Fig. 24.3c): transparent dielectric layer, conducting layer (about 1 - 3 μm), liquid crystal layer (about 10 - 100 μm), conducting layer (for example, SnO_2), and transparent dielectric layer. Common thickness is 0.1 - 0.5 mm. Control voltage is 5 - 10 V. This film may be produced by industry relatively cheaply.

24.3.4.1 Liquid Crystals (LC)

Liquid crystals (LC) are substances that exhibit a phase of matter that has properties between those of a conventional liquid, and those of a solid crystal. Liquid crystals find general employment in liquid crystal displays (LCD), which rely on the optical properties of certain liquid crystalline molecules in the presence or absence of an electric field. On command, the electric field can be used to make a pixel switch between clear or dark. Color LCD systems use the same technique, with color filters used to generate red, green, and blue pixels. Similar principles can be used to make other liquid crystal-based optical devices. Liquid crystal in fluid form is used to detect electrically generated hotspots for failure analysis in the semiconductor industry. Liquid crystal memory units with extensive capacity were used in the USA's Space Shuttle navigation equipment. Worth noting also is the fact that many common fluids are, in fact, liquid crystals. Soap, for instance, is a liquid crystal, and forms a variety of LC phases depending on its concentration in water.

The conventional control clarity (transparency) film reflected all superfluous energy to outer space. If the film has solar cells then it may convert the once superfluous solar energy into harnessed electricity.

24.3.4.2 Transparency

In optics, transparency is the material property of passing natural and artificial light through any material. Though transparency usually refers to visible light in common usage, it may correctly be used to refer to any type of radiation. Examples of transparent materials are air and some other gases, liquids such as water, most non-tinted glasses, and plastics such as Perspex and Pyrex. The degree of material transparency varies according to the wavelength of the light. From electrodynamics it results that only a vacuum is really transparent in the strictest meaning, any matter has a certain absorption for electromagnetic waves. There are transparent glass walls that can be made opaque by the application of an electric charge, a technology known as electrochromics. Certain crystals are transparent because there are straight-lines through the crystal structure. Light passes almost unobstructed along these lines.

24.3.4.3 Electrochromism

Electrochromism is the phenomenon displayed by some chemical species of reversibly changing color when a burst of electric charge is applied. One good example of an electrochromic material is polyaniline which can be formed either by the electrochemical or chemical oxidation of aniline. If an electrode is immersed in hydrochloric acid which contains a small concentration of aniline, then a film of polyaniline can be grown on the electrode. Depending on the redox state, polyaniline can either be pale yellow or dark green/black. Other electrochromic materials that have found technological application include the viologens and polyoxotungstates. Other electrochromic materials include tungsten oxide (WO_3), which is the main chemical used in the production of electrochromic windows or smart windows.

As the color change is persistent and energy need only be applied to effect a change, electrochromic materials are used to control the amount of light and heat allowed to pass through windows ("smart windows"), and has also been applied in the automobile industry to automatically tint rear-view mirrors in various lighting conditions. Viologen is used in conjunction with titanium dioxide (TiO_2) in the creation of small digital displays. It is hoped that these will replace LCDs as the viologen (which is typically dark blue) has a high contrast to the bright color of the titanium white, therefore providing a high visibility of the display.

24.3.5 Visit Asteroids without a Space Suit

Current spacesuit designs are very complex and expensive “machines for living”. They must, at minimum, unfailingly support human life for some period of time. However, the spacesuit makes a cosmonaut/astronaut barely mobile, slow moving, prevents exertive hard work, creates bodily discomfort such as pain or irritations, disallows meals in outer space, has no toilet, etc. Mass of current spacesuits is up 180 kg. Cosmonauts/Astronauts—these should be combined into "Spationauts" as the 20th Century descriptions were derived from Cold War superpower competition—must have spaceship or special outer space home habitat located not far from where they can undress for eating, toilet, and sleep as well as rest.

Why do humans need the special spacesuit in outer space, or on asteroids? There is only one reason – we need an oxygen atmosphere for breathing, respiration. Human evolution in the Earth-biosphere has created lungs that aerate our blood with oxygen and delete the carbonic acid. However, in a particularly harsh environment, we can do it more easily by artificial apparatus. For example, surgeons when they perform surgery on heart or lungs connect the patient to the apparatus "Heart-lung machine", temporarily stopping the patient's respiration and hear-beat. In Bolonkin (2006b, p. 335), it is suggested that a method exists by donating some human blood, with the use of painless suture needles, is possible and that the blood can then be passed through artificial "lungs", just as is done in hospitals today.

We can design a small device that will aerate people's blood with oxygen infusion and delete the carbonic acid. To make offshoots from main lungs arteries to this device, we would turn on/off the artificial breathing at anytime and to be in vacuum. On asteroids we can be in conventional spacesuit defending the wearer from harmful solar light. Some type of girdle-like total body wrapping is required to keep persons on asteroids from expanding explosively.

This idea may be checked with animal experiments in the Earth. We use the current "Heart-Lung" medical apparatus and put an animal under bell glass and remove the air inside the bell jar. We can add into the blood all appropriate nutrition and, thusly, be without normal eating food for a long period of time; it is widely known that many humans in comas have lived fairly comfortably for many years entirely with artificial nourishment provided by drip injection. The life possible on asteroids without spacesuit will be easier, comfortable and entirely safe.

24.4 Computations of Inflatable Dome

1. Specific mass of inflatable Dome. The mass (and relative mass) of film is summary of top double layer and support cable (Fig. 24.1):

$$M = \frac{pS\gamma}{\sigma} H + \frac{\pi pS\gamma}{2\sigma} L \quad or \quad \overline{M} = \frac{M}{S} = \frac{p\gamma}{\sigma}(H + \frac{\pi}{2} L) \qquad (24.1)$$

where M is film and cable mass, kg; p is air pressure, N/m^2; S is cover area, m^2; γ is specific mass of film and support cables, kg/m^3; σ is safety tensile stress of film and the support cable, N/m^2; H is height of Dome, m; L is distance between support cable, m.

The needed thickness of film δ is

$$\delta = \frac{\pi p L}{2\sigma} \tag{24.2}$$

Example: Let us take $p = 10^5$ N/m^2; $\sigma = 10^9$ N/m^2 = 100 kgf/mm^2; $\gamma = 1800$ kg/m^3; $H = 3$ m; $L = 2$ m. Then $\overline{M} = 1.1 \quad kg/m^2, \quad \delta = 0.314 \quad mm.$

2. *Magnetic stationary solar space reflector.* Magnetic intensity from ground ring

$$B \approx \mu_0 \frac{iS}{2\pi H^3}, \quad S = \pi R^2, \tag{24.3}$$

where B is magnetic intensity, T; $\mu_0 = 4\pi \times 10^{-7}$ is magnetic constant; i is electric currency, A; S is area of ground ring, m^2; R is ground ring radius, m; $H >> R$ is altitude of reflector, m.

Example: for R = 1000m, H = 1000 m; i = 10^5 A, magnetic intensity is $B = 6.3\times10^{-5}$ T.

The mass of electronically superconducting wire is

$$M_R = 2\pi R s \gamma_w \tag{24.4}$$

where s is cross-section area of wire, m^2; γ_w is specific mass of wire, kg/m^3.

For density of electric currency $j = 10^5$ A/mm^2 and $\gamma_w = 8000$ kg/m^3 the mass density of ground wire is about 50 kg. The mass of thin film heat screens, which defend the wire from solar heat radiation, is about 20 kg (Bolonkin 2007b).

The mass of solar thin film reflector is

$$m_r = k_1 \pi r^2 \delta_r \gamma_r \tag{24.5}$$

where r is reflector radius, m; δ_r is thickness of reflector film, m; γ_r is mass density of reflector, k_1 is coefficient of reflector mass increasing from additional support parts (for example, from inflatable ring). For r = 20 m, δ_r = 5 μm, γ_r = 1800 kg/m^3, k_1 = 1.2 the reflector mass is 13.6 kg.

The mass of top ring is

$$m = \frac{m_r}{\left(B j_r \cos\alpha / g_m \gamma_w\right) - k_2} \tag{24.6}$$

where j_r is density of electric currency, A/m^2; g_m is gravity of the asteroid, m/s^2; α is angle between magnetic line and asteroid surface (Fig. 24.5); $k_2 > 1$ is coefficient of top ring mass increasing from heat radiation screens. The mass of top ring is remarkably quite small (less than 0.5 kg).

The energy emitted by a body may be computed by employment of the Stefan-Boltzmann law:

$$E = \varepsilon \sigma_S T^4, \text{ [W/m}^2\text{]}, \tag{24.7}$$

where ε is the coefficient of body blackness (ε =0.03 ÷ 0.99 for real bodies), σ_s = 5.67×10^{-8} Stefan-Boltzmann constant.

The common daily average Sun-furnished heat (energy) at Earth's orbit is calculated by equation

$$Q = 86400 c\, qt, \tag{24.8}$$

where c is daily average heat flow coefficient, $c \approx 0.5$; t is relative daily light time, 86400 = 24 × 60 × 60 is number of seconds of an Earth-day, q = 1400 W/m^2 is a rough estimate of solar energy flux density at Earth's orbit.

The heat loss flow per 1 m^2 of dome film cover by convection and heat conduction is (Bolonkin and Cathcart 2007b):

$$q = k(t_1 - t_2), \quad \text{where} \quad k = \frac{1}{1/\alpha_1 + \sum_i \delta_i / \lambda_i + 1/\alpha_2} \tag{24.9}$$

where k is heat transfer coefficient, $t_{1,2}$ are temperatures of the initial and final multi-layers of the heat insulators, $\alpha_{1,2}$ are convection heat transfer coefficients of the initial and final multi-layers of heat insulators (α = 30 ÷ 100 W/(m^{2}K)), δ_i are thickness of insulator layers; λ_i are coefficients of heat transfer of insulator layers (see Table 24.1), $t_{1,2}$ are temperatures of initial and final layers, $^\circ$C.

The radiation energy flux per 1 m^2 of the service area is computed by relationships:

$$q = C_r \left[\left(\frac{T_1}{100} \right)^4 - \left(\frac{T_2}{100} \right)^4 \right], \quad \text{where} \quad C_r = \frac{c_s}{1/\varepsilon_1 + 1/\varepsilon_2 - 1}, \quad c_s = 5.67 \tag{24.10}$$

where C_r is general radiation coefficient, $\varepsilon_{1,2}$ are the emissivities of the plates (see Table 24.2); $T_{1,2}$ are the temperatures of the plates, K.

The radiation energy flux across a set of the heat screens is computed by equation

$$q = 0.5 \frac{C'_r}{C_r} q_r \tag{24.11}$$

where C'_r is computed by Eq. (24.10) between plate and reflector.

The data of some construction materials is found in Naschekin (1969, p.331) and Table 24.1.

As the reader will see, the air layer is the very best heat insulator. We do not limit the used material's thickness δ.

Table 24.1 Blackness (Naschekin 1969, p. 465)

Material,	Blackness, ε	Material	Blackness, ε	Material	Blackness, ε
Bright Aluminum $t = 50 \div 500$ °C	0.04 - 0.06	Baked brick $t = 20$ °C	0.88 - 0.93	Glass $t = 20 \div 100$ °C	0.91 - 0.94

As the astute reader will notice, the shiny aluminum louver coating is excellent mean jalousie offsetting radiation losses from the dome.

The general radiation flux Q is computes by Eq. (24.10). Equations (24.7)-(24.11) allow computation of the heat balance and comparison of incoming heat (gain) and outgoing heat (loss).

The heat from combusted fuel is found by equation

$$Q = c_t m/\eta, \tag{24.12}$$

where c_t is heat rate of fuel [J/kg]; $c_t = 40$ MJ/kg for liquid oil fuel; m is fuel mass, kg; η is efficiency of heater, $\eta = 0.5$ - 0.8.

The thickness of the dome envelope, its sheltering shell of film, is computed by formulas (from equation for tensile strength):

$$\delta_1 = \frac{Rp}{2\sigma}, \quad \delta_2 = \frac{Rp}{\sigma} \tag{24.13}$$

where δ_1 is the film thickness for a spherical dome, m; δ_2 is the film thickness for a cylindrical dome, m; R is radius of dome, m; p is additional pressure into the dome, N/m^2; σ is safety tensile stress of film, N/m^2.

For example, compute the film thickness for dome having radius R =100 m, additional air pressure p = 0.01 atm at top section in Fig. 24.2 (p = 1000 N/m^2), safety tensile stress σ= 50 kg/mm^2 (σ= 5×10^8 N/m^2), cylindrical dome:

$$\delta = \frac{100 \times 1000}{5 \times 10^8} = 0.0002\, m = 0.2\, mm \tag{24.14}$$

The dynamic pressure from surface wind is

$$p_w = \frac{\rho V^2}{2} \tag{24.15}$$

where ρ is atmospheric density, kg/m^3; V is wind speed, m/s.

If the asteroid has a long nighttime, the heat loss protection can reduce the head losses as we can utilize inflated dome covers with additional layers and additional heat screens. One heat screen decreases heat losses by 2, two screens can decrease heat flow by 3 times, three by 4 times, and so on. If the inflatable domes have a multi-layer structure, the heat transfer decreases proportional to the total thickness of its enveloping film layers. The data of some construction materials is found in Tables 24.2-24.4.

Table 24.2 Heat Transfer Data (Naschekin 1969, p. 331)

Material	Density, kg/m^3	Thermal conductivity, λ, W/m °C	Heat capacity, kJ/kg. °C
Concrete	2300	1.279	1.13
Baked brick	1800	0.758	0.879
Ice	920	2.25	2.26
Snow	560	0.465	2.09
Glass	2500	0.744	0.67
Steel	7900	45	0.461
Air	1.225	0.0244	1

Table 24.3 Emittance (Naschekin 1969, p. 465)

Material	Temperature, T °C	Emittance, ε
Bright Aluminum	50 ÷ 500 °C	0.04 - 0.06
Bright copper	20 ÷ 350 °C	0.02
Steel	50 °C	0.56
Asbestos board	20 °C	0.96
Glass	20 ÷ 100 °C	0.91 - 0.94
Baked brick	20 °C	0.88 - 0.93
Tree	20 °C	0.8 - 0.9
Black vanish	40 ÷ 100 °C	0.96 – 0.98
Tin	20 °C	0.28

24.4.1 *Irrigation without Water. Closed-Loop Water Cycle*

A reader can derive the equations below from well-known physical laws. Therefore, no detailed explanations of these are furnished here.

1. *Amount of water in atmosphere*. Amount of water depends entirely upon temperature and humidity. For relative humidity 100%, the maximum partial pressure of water vapor is shown in Table 24.4.

Table 24.4 Maximum partial pressure of water vapor in atmosphere for given air temperature

t, C	-10	0	10	20	30	40	50	60	70	80	90	100
p,kPa	0.287	0.611	1.22	2.33	4.27	7.33	12.3	19.9	30.9	49.7	70.1	101

2. The amount of water in 1 m^3 of air may be computed by equation

$$m_W = 0.00625\ [p(t_2)h - p(t_1)] \tag{24.16}$$

where m_W is mass of water, kg in 1 m^3 of air; *p(t)* is vapor (steam) pressure from Table 24.4, relative $h = 0 \div 1$ is relative humidity. The computation of Eq. (24.16) is presented in Fig. 24.8. Typical relative humidity of atmosphere air is 0.5 - 1.

24.4.1.1 Computation of Closed-Loop Water Cycle

Assume the maximum safe temperature is achieved in the daytime. When dome reaches the maximum (or given) temperature, then the control system fills the space 5 with air (Fig. 24.4) between double–layers of the film cover. That protects the inside part of the dome from further heating by hot air outside the dome. The control system decreases also the solar radiation input, increasing reflectivity of the liquid crystal layer of the film cover. In short, we can then support a constant temperature inside our imagined filmic dome.

The *heating* of the dome in the daytime may be computed by equations:

$$q(t)=q_0\sin(\pi t/t_d),\quad dQ=q(t)dt,\quad Q=\int_0^{t_d}dQ,\quad Q(0)=0,\quad M_w=\int_0^{t_d}adT,$$
$$dT=\frac{dQ}{C_{p1}\rho_1\delta_1+C_{p2}\rho_2H+rHa},\quad a=10^{-5}(5{,}28T+2),\quad T=\int_0^{t_d}dT,\quad T(0)=T_{\min}, \tag{24.17}$$

where q is heat flow, J/m^2 s; q_o is maximum Sun energy flux in noon daily time, $q_o \approx 100 \div 1000$, J/ m^{2}s; t is time, s; t_d is daily (Sun) time, s; Q is heat, J; T is temperature in dome (air, soil), °C; C_{p1} is heat capacity of soil, $C_{p1} \approx 1000$ J/kg; $C_{p2} \approx 1000$ J/kg is heat capacity of air; $\delta_1 \approx 0.1$ m is thickness of heating soil; $\rho_1 \approx 1000$ kg/m^3 is density of the soil; $\rho_2 \approx 1.225$ kg/m^3 is density of the air; H is thickness of air (height of cover), $H \approx 50 \div 300$ m; r = 2,260,000 J/kg is evaporation heat, a is coefficient of evaporation; M_w is mass of evaporation water, kg/m^3; T_{min} is minimum temperature into dome after night, °C.

The convective (conductive) *cooling* of dome at night time may be computed as below

$$q_t=k(T_{\min}-T(t)),\quad \text{where}\quad k=\frac{1}{1/\alpha_1+\sum_i\delta_i/\lambda_i+1/\alpha_2}$$
$$dQ=[q_t(t)+q_r(t)]dt,\quad Q=\int_0^{t_d}dQ,\quad Q(0)=0,\quad M_w=\int_0^{t_d}adT,$$
$$dT=\frac{dQ}{C_{p1}\rho_1\delta_1+C_{p2}\rho_2H+rHa},\quad a=10^{-5}(5{,}28T+2),\quad T=\int_0^{t_d}dT,\quad T(0)=T_{\min}, \tag{24.18,19}$$

where q_t is heat flow through the dome cover by convective heat transfer, J/m^{2}s or W/m^2; see the other notation in Eq. (24.7). We take $\delta = 0$ in nighttime (through active control of the film).

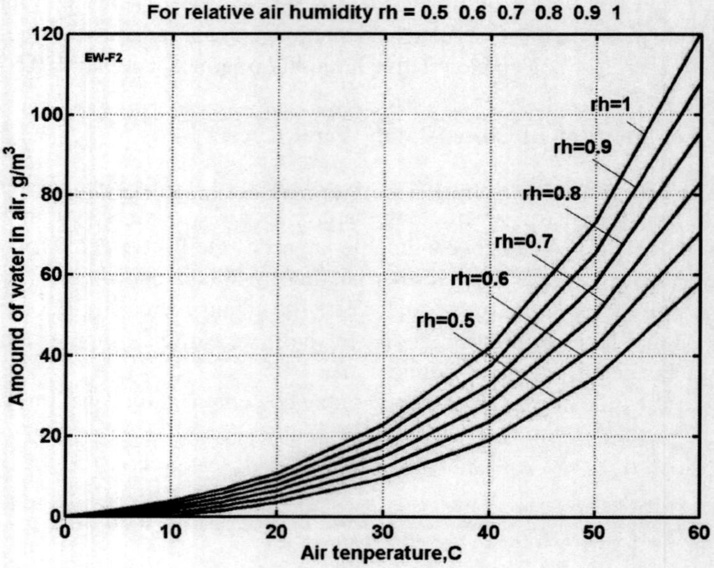

Fig. 24.8 Amount of water in 1 m^3 of air versus air temperature and relative humidity (rh). $t_1 = 0\,^{\circ}$C

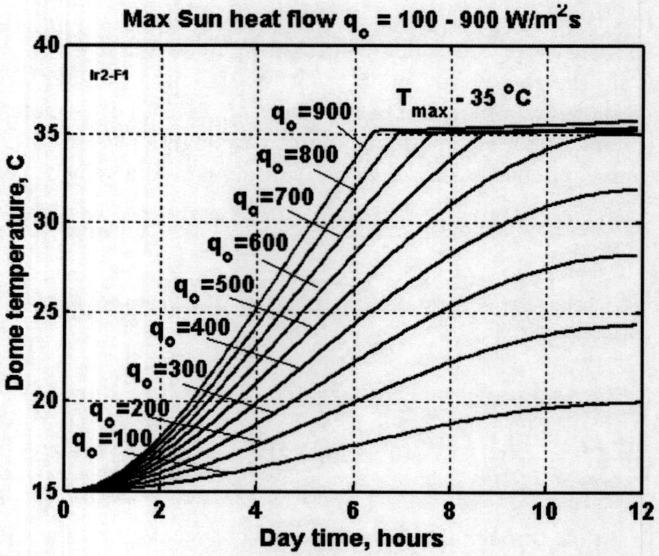

Fig. 24.9 Heating of the dome by solar radiation from the night temperature of 15° C to 35° C via daily maximum solar radiation (W/m^2) for varying daily time. Height of dome film cover equals H = 135 m. The control temperature system limits the maximum internal dome temperature to 35° C. Compare with Fig. 24.6.

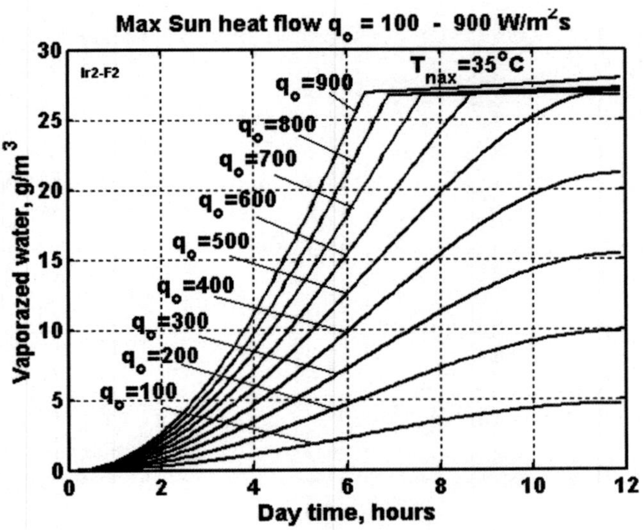

Fig. 24.10 Water vaporization for 100% humidity of the air for different maximum solar radiation (W/m^2) levels delivered over varying daily time. Height of dome film cover equals H = 135 m. The temperature control system limits the maximum internal dome temperature to 35^0 C.

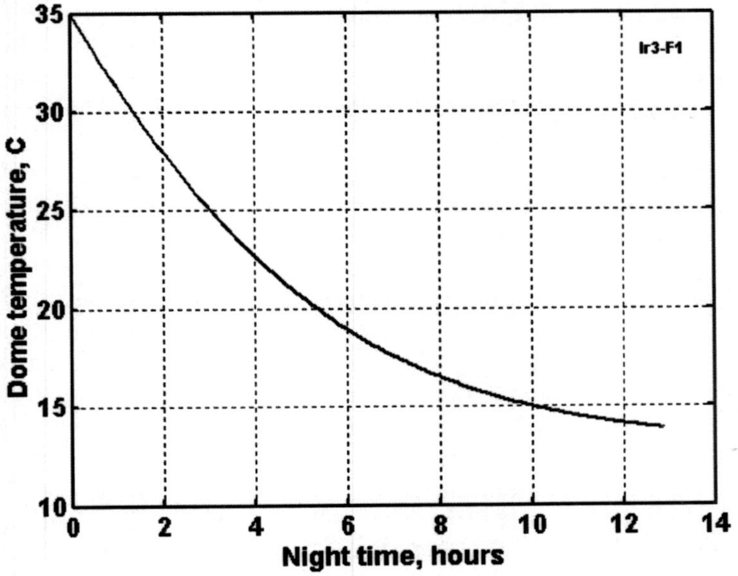

Fig. 24.11 Cooling of the Dome via nighttime for initial daily temperature 35 ^{o} C and the night outer temperature 13 ^{o} C

Let us take the following parameters: $H = 135$ m, $\alpha = 70$, $\delta = 1$ m between cover layers (see #5 in Fig. 24.4), $\lambda = 0.0244$ for air. Result of computation for given parameter are presented in Figs. 24.9-24.11.

For dome cover height $H = 135$ m the night precipitation (maximum) is $0.027 \times 135 = 3.67$ kg or 3.67 mm/day (Fig. 24.10). The annual precipitation is 1336.6 mm (maximum). If it is not enough, we can increase the height of dome cover. The globally-averaged annual precipitation is about 1000 mm.

Our dome design, most assuredly, is not optimal but rather it is selected for realistic parameters.

24.5 Macro-projects

The dome shelter innovations outlined here can be practically applied to many types of asteroids. We suggest initial macro-projects could be small (10 m diameter) houses (Fig. 24.6) followed by an "Evergreen" dome covering a land area 200 m × 1000 m, with irrigated vegetation, homes, open-air swimming pools, playground, "under the stars style" concert hall.

The house and "Evergreen" dome have several innovations: magnetic suspended Sun reflector, double transparent insulating film, controllable jalousies coated with reflective aluminum (or film with transparency control properties and/or structures) and an electronic cable mesh inherent to the film for dome safety/integrity monitoring purposes. By undertaking to construct a half-sphere house, we can acquire experience in such constructions and explore more complex constructions. By computation, a 10 m diameter home has a useful floor area of 78.5 m^2, airy interior volume of 262 m^3 covered by an envelope with an exterior area of 157 m^2. Its film enclosure material would have a thickness of 0.0003 m with a total mass of about 100 kg.

A city-enclosing "Evergreen" dome of 200 m × 1000 m (Fig. 24.2, with spherical end caps) could have calculated characteristics: useful area = 2.3×10^5 m^2, useful volume 17.8×10^6 m^3, exterior dome area of 3.75×10^5 m^2, comprised of a film of 0.0003 m thickness and about 200 tones. If the "Evergreen" dome were formed with concrete 0.25 m thick, the mass of the city-size envelope would be 200×10^3 tones, which is a thousand times heavier. Also, just for comparison, if we made a gigantic "Evergreen" dome with stiff glass, thousands of tones of steel, glass would be necessary and such materials would be very costly to transport hundreds or thousands of kilometers into outer space to the asteroid where they would be assembled by highly-paid but risk-taking construction workers. Our film is flexible and plastically deformable. It can be relatively cheap in terms of manufacturing cost. The single greatest boon to "Evergreen" dome construction is the protected cultivation of plants under a protective dome that efficiently captures energy from the available and technically harnessed sunlight.

24.6 Discussion

As with any innovative macro-project proposal, the reader will naturally have many questions. We offer brief answers to the two most obvious questions our readers are likely to ponder.

(1) *Cover damage*

The envelope contains a rip-stopping cable mesh so that the film cannot be damaged greatly. Its structured cross-section of double layering governs the escape of air inside the living realm. Electronic signals alert supervising personnel of all ruptures and permit a speedy repair effort by well-trained responsive emergency personnel. The topmost cover has a strong double film.

(2) *What is the design-life of the dome film covering?*

Depending on the kinds of materials used, it may be as much a decade (up 30 years). In all or in part, the durable cover can be replaced periodically as a precautionary measure by its owners.

24.7 Conclusions

Utilization of "Evergreen" domes can foster the fuller economic development of the asteroid, increasing the effective area of territory dominated by humans. "Evergreen" domes may find soon application on asteroids since a vertical variant, inflatable space towers (Bolonkin 2006a), is to become available for launching spacecraft inexpensively into Earth orbit or on to long-duration interplanetary outer spaceflights. The reader finds further information in Bolonkin (2008, 2009, 2010), Bolonkin and Cathcart (2007a,b).

References

Bolonkin, A.A.: Non-Rocket Space Launch and Flight, 488 p. Elsevier, London (2006a), http://www.archive.org/details/Non-rocketSpace LaunchAndFlight, http://www.scribd.com/doc/24056182

Bolonkin, A.A.: Control of Regional and Global Weather (2006b), http://arxiv.org/ftp/physics/papers/0701/0701097.pdf

Bolonkin, A.A.: New Concepts, Ideas, and Innovations in Aerospace, Technology and Human Life. NOVA, Hauppauge (2007a), http://www.scribd.com/doc/24057071, http://www.archive.org/details/NewConceptsIfeasAnd InnovationsInAerospaceTechnologyAndHumanSciences

Bolonkin, A.A.:AB Levitation and Energy Storage. This work was presented as paper AIAA-2007-4613 to 38th AIAA Plasma Dynamics and Lasers Conference in conjunction with the16th International Conference on MHD Energy Conversion, Miami, USA, June 25-27 (2007b)

Bolonkin, A.A.: New Technologies and Revolutionary Projects, Lambert, 324 p. (2008), http://www.scribd.com/doc/32744477, http://www.archive.org/details/NewTechnologiesAnd RevolutionaryProjects

Bolonkin, A.A.: Man in Outer Space without a Special Space Suit. American Journal of Engineering and Applied Science 2(4), 573–579 (2009), http://www.scribd.com/doc/24050793/, http://www.archive.org/details/LiveOfHumanityInOuterSpaceWithoutSpaceSuite

Bolonkin, A.A.: Life. Science. Future (Biography notes, researches and innovations), 208 p. Publish America (2010), http://www.scribd.com/doc/48229884, http://www.lulu.com, http://www.archive.org/details/Life.Science.Future.biographyNotesResearchesAndInnovations

Bolonkin, A.A., Cathcart, R.B.: Inflatable 'Evergreen' dome settlements for Earth's Polar Regions. Clean Technologies and Environmental Policy 9(2), 125–132 (2007a)

Bolonkin, A.A., Cathcart, R.B.: Macro-Projects: Environments and Technologies, NOVA, Hauppauge, 536 p. (2007b), http://www.scribd.com/doc/24057930, http://www.archive.org/details/Macro-projectsEnvironmentsAndTechnologies

Naschekin, V.V.: Technical thermodynamic and heat transmission. Public House High Universities, Moscow (1969) (in Russian)

Chapter 25
Usage of Asteroid Resources for Space-Based Geoengineering

Russell Bewick, Joan-Pau Sanchez, and Colin R. McInnes

University of Strathclyde, Glasgow, UK

25.1 Introduction

Currently, climate change is a significant threat to our way of life, with global mean temperatures predicted to increase by 1.1-6.4°C by the end of the century (IPCC 2007). This increase is driven by multiple factors, with the main contributors being the increasing concentrations of Greenhouse Gases (GHG), mainly CO_2, CH_4 and N_2O, in the atmosphere, which is altering the Earth's current energy balance and therefore the present climate. The current consensus within the scientific community is that the dominant factor in the changing climate of the Earth is the anthropogenic emission of GHG's, with the probability of this being true termed "very likely" (90% probability) by the IPCC (IPCC 2007). Whilst the main effort within the global community should be to control climate change by reducing our emissions of GHG's, it is prudent to investigate other methods of managing the climate system. The field of deliberately manipulating the Earth's climate is called geoengineering, or climate engineering.

25.1.1 Geoengineering Methods

Several proposals for possible geoengineering methods have been made. These can generally be placed into two categories; Solar Radiation Management (SRM) and carbon sequestration (Shepherd 2009). Solar radiation management focuses on the reduction of the amount of sunlight being absorbed by the Earth's atmosphere by one of two main methods; increasing the Earth's albedo to reflect more incoming sunlight away from the surface or by reducing the level of sunlight reaching the surface, primarily by scattering incoming solar radiation before it reaches the surface. Alternatively, carbon capture techniques aim to deal with the fundamental cause of global warming, by either directly taking CO_2 out of the atmosphere or indirectly, by inducing an increase in the ability of current carbon sinks to take CO_2 out of the atmosphere. In general, solar radiation management techniques are expected to be fast acting, once implemented, whereas carbon sequestration will take many years to significantly reduce CO_2 concentrations in the atmosphere.

Due to the nature of the two methods, carbon sequestration, in its direct form, is inherently safer than SRM techniques since the root cause is treated, with the expectation of there being fewer side effects, though this is dependent upon the safe storage of the captured CO_2. Additionally, SRM methods cannot mitigate all the effects of increased atmospheric CO_2 concentrations, primarily the acidification of the oceans, which is likely to harm the populations of algae and other photosynthetic organisms that provide 99% of the organic matter used in the oceans food chains (Raven 2005).

A report into geoengineering conducted by the Royal Society in 2009 (Shepherd 2009) examines the feasibility of, the then latest, schemes based on the criteria of effectiveness, affordability, timeliness and safety. The ratings applied to three geoengineering method can be seen in Table 25.1. In general the report appears to show that there is no perfect solution, with the schemes that appear most effective suffering in other criteria such as affordability. For example, the reflectivity of roofs and roads in an urban environment can be increased, thus reflecting more sunlight out of the atmosphere. This is a relatively cheap method, with added benefits of reducing the energy required to cool buildings in warm climates. However, the overall effectivity of the method is low, as noted by Jacobson and Ten Hoeve (2011).

One of the most widely known methods of geoengineering suggests ejecting large quantities of sulphur particles into the upper atmosphere, with the effect that sunlight is reflected out of the atmosphere before reaching the surface. This method is inspired by observations of volcanic eruptions which eject large numbers of particles into the upper atmosphere. After such events a significant global cooling is observed. There are also many negative aspects to this geoengineering method however, as noted by Robock et al. (2009), most importantly the ozone layer will be negatively affected, whilst the key benefits are the relatively low cost in comparison to other geoengineering methods and that it can likely be deployed with existing technologies. Due to these factors the sulphuric aerosol concept rates highly in all categories except safety.

Ocean fertilisation is a currently popular carbon capture technique which aims to increase the marine uptake of CO_2 by the injection of iron particles into the upper layers of the ocean. This will then encourage the growth of algae, which uses CO_2 during photosynthesis. Though promising in principle, this scheme has yet to be conclusively proven to be effective as the increased uptake of CO_2 is only likely to be a fraction of the total emitted annually by anthropogenic sources (Shepherd 2009). Other downsides are that side effects are likely, with reduced carbon uptake in other regions of the ocean and some areas becoming starved of oxygen, creating 'dead zones' (Shepherd 2009), likely increasing the acidification of the ocean due to the increased level of carbon sequestration (Raven 2005). Due to these reasons this method rates lowly with the exception of affordability, since iron particles can be easily produce.

Table 25.1 Rating applied by the Royal Society to three geoengineering methods (Shepherd 2009), rating from 1-5

Method	Effectiveness	Affordability	Timeliness	Safety
Sulphuric Aerosol	4	4	4	2
Space Reflector	4	1.5	1	3
Ocean fertilisation	2	3	1.5	1

25.1.2 Space-Based Geoengineering

One of the most effective solutions discussed in the Royal Society report is the use of space-based solar reflectors to reduce incident solar insolation. Whilst this technique is not rated by the Royal Society as being affordable or timely (Shepherd 2009), it does have a significant advantage over other schemes; neither the physical properties of the Earth's surface or atmosphere needs to be changed. This is a significant advantage in comparison to the stratospheric aerosol or ocean fertilisation techniques which are likely to have noticeable side effects.

It has been estimated that in order to offset the effects of global warming caused by a doubling of the CO_2 concentration in the atmosphere (compared to pre-industrial levels and corresponding to an increase in global temperature of approximately 2°C) solar insolation must be reduced by 1.7% (Govindasamy and Caldeira 2000). Similarly for a quadrupling of CO_2 the required insolation change is 3.6% (Govindasamy et al. 2003).

There have been several proposals to date regarding the reduction of solar insolation using space-based geoengineering (SBGE) methods, the key characteristics of which can be seen in Table 25.2. The methods either utilise a cloud of dust (Pearson et al. 2006, Struck 2007, Bewick et al. 2012a, Bewick et al. 2012b, Bewick et al. 2013, Teller 1997) or solid reflectors or refractors (Pearson et al. 2006, McInnes 2012, Early 1989, Mautner 1991, Angel 2006) to reduce the level of solar insolation. Typically, the methods that require the least mass are those that use solid reflectors or refractors, with the mass for the dust cloud methods being orders of magnitude higher. This is mostly due to the increased level of control that can be placed upon the solid reflectors or refractors, hence they can be stationed in optimum positions with active control to extend their lifetimes. In contrast dust clouds cannot be controlled and can only be placed with suitable initial conditions, with subsequent replenishment necessary due to the orbital decay or perturbation of the particle orbits. The dominant perturbation on these dust grains, and one that can be easily accounted for, is solar radiation pressure, described later. Other perturbations that are more complex to model include the Lorentz force due to a charge on the dust grains and the gravity of other solar system bodies.

Conflicting with this, though, is the consideration of the engineering complexity of the system. Whilst dust clouds are a relatively crude method, the material can be readily found in the populations of near Earth asteroids, whereas solid reflectors or refractors must either be manufactured terrestrially, then launched into position, or manufactured in-situ. Clearly, taking this into account, the low rating for affordability and timeliness indicated in the Royal Society report (which only considered solid reflectors and refractors) can be understood.

Table 25.2 Masses of the proposed space-based geoengineering schemes

Position	Method	Insolation Change [%]	Mass [kg]	Ref.
Earth Orbit	Dust Ring	1.6	2.3×10^{12}	(Pearson et al. 2006)
		1.7	2×10^{12}	(Bewick et al. 2013)
	Reflector	1.6	5.0×10^{9}	(Pearson et al. 2006)
Earth-Moon L_4/L_5	Dust Cloud	1.4	2.1×10^{14}	(Struck 2007)
Sun-Earth L_1	Reflector	1.8	2.6×10^{11}	(McInnes 2010)
	Refractor	1.8	2.0×10^{10}	(Angel 2006)
	Dust Cloud	1.7	1.9×10^{10} kg/yr	(Bewick et al. 2012a)
	Gravitationally anchored dust cloud	1.7	1.3×10^{17}	(Bewick et al. 2012b)

However, though the masses of the existing dust cloud methods are higher than the solid device methods, they are still promising avenues of research for SBGE due to the lower engineering challenges involved. In this chapter new and existing proposals for SBGE will be discussed.

25.2 Usage Scenarios

The methods suggested by McInnes (2010) and Angel (2006) suggest the placement of large swarms of reflective or refractive devices at the first Lagrange point, L_1, in between the Earth and the Sun. This point is unstable and thus the devices will require an active control mechanism, increasing the complexity of the system. The main challenge with these scenarios is the manufacture and placement of the reflective or refractive devices. To terrestrially manufacture and then place the mass of material required to the L_1 point would require many thousands of launch vehicles, more than has ever been launched in the entire space age. Angel describes a system of mass drivers, launching spacecraft from Earth to the L_1 point, to overcome this, though this is a hypothetical scenario, as the technology to build such a device cannot be anticipated in the near or mid-term. A more likely scenario, suggested by McInnes, is the in-situ manufacture of the devices at L_1 from captured near-Earth objects (NEOs). Similarly this method cannot be seen as a near-term option, though recent advancements in 3D printing technology and the interest in capturing NEOs for resource utilisation (Brophy et al. 2012), suggest that this method could be seen as a mid-term possibility.

A novel method proposed by Pearson et al. (2006) is to place a ring of dust or reflecting satellites in Earth orbit. Though the masses of these two methods are comparatively low there are clearly possible side effects including an increased danger to Earth orbiting satellites and to the Earth, as the dust ring method requires two shepherding satellites, each of which would be 1km in size. As well as having a highly uneven insolation reduction pattern on Earth, discussed further in Sec. 2.3, the ring will have the additional side effect of increasing reflected light onto the night side of the Earth under certain conditions. For these reasons this method is not seen as the most optimal space-based geoengineering solution.

An additional factor that affects the relative mass of the different methods is the amount of time that the reflectors spend in between the Earth and the Sun. For example the method proposed by Struck (2007) to place clouds of dust at the L_4/L_5 Lagrange libration points of the Earth-Moon system has a clear benefit, as these points are passively stable. However, as these points effectively orbit around the Earth they are only occasionally in a position to reduce solar insolation. This leads to a large mass requirement, in the region of 10^{14}kg. Furthermore, the movement of the clouds will create a flickering effect. On most occasions there will be no change in insolation whilst at those times when the cloud is present the insolation change will be much greater than the average 1.7% reduction required to offset a 2°C increase in global temperature.

25.2.1 Unstable Dust Cloud at the L_1 Position

The method proposed by Struck, with the L_4/L_5 dust cloud in the Earth-Moon system, has the main disadvantage of only occasionally being in a position to intercept solar radiation destined for Earth. Considering this, a better position for the reduction of solar insolation is the first Lagrange point, L_1, in between the Earth and Sun, since a constant insolation reduction can be achieved. However, this position has the disadvantage of being an unstable equilibrium, therefore any dust grains released at this position will drift away over time.

To implement a dust cloud method here a scenario is imagined whereby an asteroid can be captured and positioned in the vicinity of the L_1 point. It is discussed later that sufficient material can be captured with a velocity less than is necessary to exploit lunar resources, 2.37km s^{-1}. Following capture, material can be mined from the asteroid, and processed if necessary, before being ejected from the surface. This material will then form a dust cloud which will drift away over time (see Fig. 25.1). The continual ejection of material will then create a steady state distribution of dust flowing away from the asteroid. It is assumed that the material ejected from the asteroid can be used to stabilise the asteroid at the L_1 point, though not when a significant thrust is required. As a first approximation of the cloud steady state cloud, the process shall be modelled here by considering an initially static, homogeneous, spherical cloud of dust that is then propagated over numerous time steps. By adding at each time step a given amount of dust, also homogenously distributed within the same initial spherical volume, a steady state distribution of flowing mass can be computed.

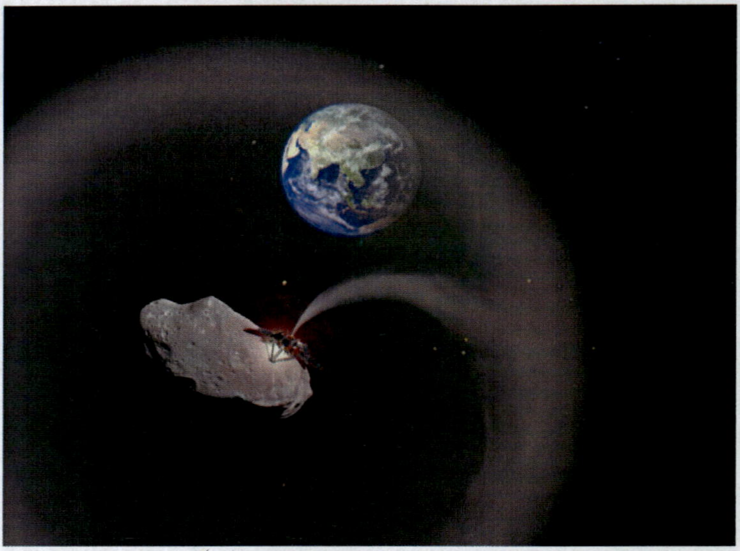

Fig. 25.1 Impression of an L_1 positioned dust cloud for space-based geoengineering

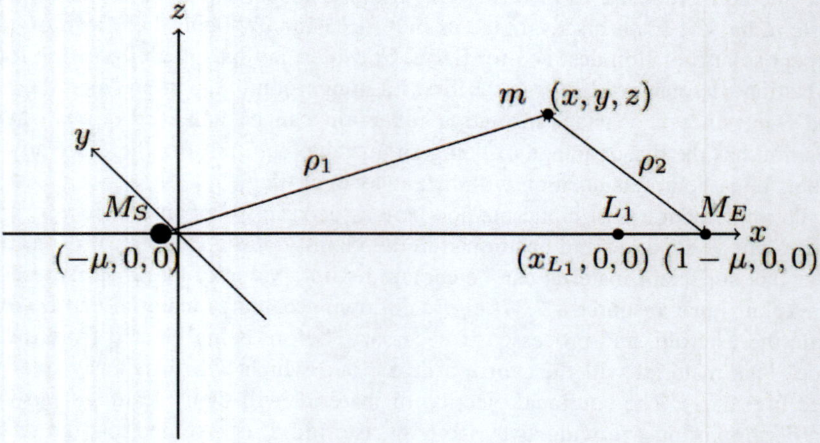

Fig. 25.2 Dimensions in the Circular Restricted Three-body Problem

A determination of the lifetime of dust grains around the L_1 point will be determined using the circular restricted three-body problem (CR3BP), the equations of motion of which, expressed in dimensionless units, are:

$$\begin{aligned} \ddot{x} - 2\dot{y} &= \frac{\partial U}{\partial x} \\ \ddot{y} + 2\dot{x} &= \frac{\partial U}{\partial y} \\ \ddot{z} &= \frac{\partial U}{\partial z} \end{aligned} \tag{25.1}$$

where x and y are the position in the rotational plane of the Earth's orbit and z is the out-of-plane position, shown in Fig. 25.2, whilst U is the non-dimensional potential function, defined as:

$$U=\frac{1}{2}(x^2+y^2)+\frac{1-\mu}{\rho_1}+\frac{\mu}{\rho_2} \tag{25.2}$$

The dimensionless equations of motion are found from the conventional dynamics by setting the distance between the Sun and Earth to 1, i.e. dividing by 1AU, and the period to 2π, under the assumption that the unit of mass is the total system mass. This gives μ as the ratio of the mass of the Earth to the combined mass of the Earth and Sun, $\mu = m_E/(m_S + m_E)$, and the parameters $\rho_{1,2}$ are the non-dimensional distances between the dust particle and the Sun and Earth respectively, defined by Eq. (25.3) and Eq. (25.4). In dimensionless co-ordinates the Sun and Earth are positioned at $M_S(-\mu, 0{,}0)$ and $M_E(1-\mu, 0{,}0)$ respectively. Hence:

$$\begin{aligned} \rho_1 &= \sqrt{(x+\mu)^2 + y^2 + z^2} \\ \rho_2 &= \sqrt{(x+\mu-1)^2 + y^2 + z^2} \end{aligned} \tag{25.3,4}$$

The equilibrium, or libration points, are located where the combined gravitational force of the Sun and Earth on a particle is equal to the centripetal force required for it to orbit in a fixed position relative to the two primary bodies. These positions are at the stationary points of the potential function, Eq. (25.2). In particular, the equilibrium points required for this geoengineering method must lie along the Sun-Earth line and must therefore be along the x-axis, hence y=z=0 with $\dot{x} = \dot{y} = \dot{z} = 0$. For the Sun-Earth system (μ=3x10^{-6}) the L_1 point is located approximately 1.5x10^6 km sunwards of the Earth.

25.2.1.1 Solar Radiation Pressure

The material used to populate the dust cloud should be in the micrometre length scale, since larger dust grains would have a lower surface area to mass ratio, hence they would be less mass efficient in reducing solar insolation. Small dust grains would be perturbed by a variety of forces including solar radiation pressure (SRP) and the Lorentz force. It has been shown by Bewick et al. (2012a) that, above 0.1 μm, SRP is the dominant perturbation of the dust grains, therefore, for simplicity only dust grains above this size are considered. Solar radiation pressure is the transfer of momentum between solar photons and the dust grains of the cloud. The effect of SRP can be quantified using the 'lightness' parameter, β, which is the ratio of the force due to SRP and solar gravity (de Pater and Lissauer 2001);

$$\beta = \frac{F_{rad}}{F_g} \approx 570 \frac{Q}{\rho_m R_{gr}} \tag{25.5}$$

where ρ_m is the mass density of the grain, Q is the momentum transfer coupling between the solar photons and the dust grain and R_{gr} is the radius of the grain, expressed in μm.

The value of Q can vary from 0, for a completely transmissive particle, to 2 for a completely reflective particle For relatively large radius particles, $R_{gr} > 1\mu m$, the value of Q varies little, with a value of approximately 1 (Wilck and Mann 2006), but as the size decreases the interaction between the solar photons and the dust grains becomes more complex. The β-value for a range of particle radii is calculated by Wilck and Mann (2006) using Mie theory for different composition models. The β-value peaks with a value of approximately 0.9 at a radius of 0.2μm, before decreasing to 0.1 for a radius of 0.01 μm.

Since SRP has an inverse square relationship with heliocentric distance, its effect is to reduce the effective gravitational force of the Sun. Hence, the mass parameter, μ, for the CR3BP is now;

$$\mu = \frac{m_E}{m_E + (1 - \beta) m_S} \tag{25.6}$$

Due to the increase in the value of μ with increasing β the L_1 equilibrium point is found to shift towards the Sun. For particles with $\beta>0$, placed at the conventional L_1 point, a shorter instability timescale will be observed due to the displacement from the new equilibrium position. Additionally, as β increases the gradient of the potential function, U, around the new equilibrium position is found to decrease. Thus, dust grains placed here will experience greater stability as β increases.

To propagate the dust cloud over time a transition matrix is used. This uses the linearised dynamics around L_1 to generate a matrix with which a dust cloud can be propagated collectively over time. This is in contrast to the conventional method which is to apply the equations of motion to each initial position in turn and propagate until the final time. The method for the generation of the transition matrix can be found in (Bewick et al. 2012a). Using the transition matrix greatly decreases the time taken to simulate the motion of the cloud, though at the expense of some accuracy. However, it was found that, within the time frame of interest, the accuracy of the transition matrix was sufficient (Bewick et al. 2012a).

25.2.1.2 Solar Radiation Model

A numerical model was used to determine the shading effect of the dust cloud on the Earth. In this model the surfaces of the Sun and Earth are divided into segments, with each segment on the Sun being connected to each segment on the Earth by a path. Along each path the Beer-Lambert law is applied to calculate the attenuation of the solar flux due to the presence of the cloud. The Beer-Lambert law is expressed as (Ingle and Crouch 1988);

$$I = I_0 e^{-\int_0^L \rho(l)\sigma(l)dl} \tag{25.7}$$

where $\rho(l)$ and $\sigma(l)$ are the number density and cross-sectional area of the dust grains at position l along the path and I and I_0 are the final and initial flux respectively. Using Eq. (25.7), the average reduction in solar insolation over the Earth can be calculated by averaging over all paths. A full description of the solar radiation model can be found in (Bewick et al. 2012a).

25.2.1.3 Results

The key quantifiable parameter for this method of geoengineering is the cloud mass necessary to create the required level of solar insolation reduction. This shall be presented in terms of the mass-per-year of asteroid material required. This is calculated using the SRM mentioned previously by optimising the initial number density of the spherical, homogeneous cloud that is propagated using the transition matrix. The results shall be found for dust clouds with sizes ranging from 1,000-12,000km radius, placed at the classical Lagrange point and the new displaced equilibrium point created for a grain β-value 0.751, corresponding to a grain size of 0.1µm. In terms of terrestrial particles this grain size corresponds to condensed gas particles, and is the limit above which solar radiation pressure can be considered the dominant force. For this grain size the distance by which the equilibrium point is displaced sunwards of the conventional L_1 point is 875,000km.

The mass required to reduce solar insolation by 1.7% can be seen in Fig. 25.3 for the cloud released at the classical L_1 and displaced equilibrium positions. It can clearly be seen that the cloud released from the displaced equilibrium position requires significantly less mass than if it would be released from the classical L_1 point. This is due to the greater lifetime of the dust grains when released at the displaced equilibrium position.

For the optimum cloud radius of 4,000km the mass requirement is 7.60×10^{10} kg yr^{-1} for the cloud released at the classical L_1 point. In comparison to the method proposed by Struck (2007) this is a mass saving of several orders of magnitude. The result for the steady state solution for a cloud ejected at the displaced equilibrium position has an optimum mass requirement of 1.87×10^{10} kg yr^{-1}. As a useful comparison, the asteroid Apophis, with a mass of approximately 2×10^{10}kg (Binzel et al. 2009), can supply enough material to sustain this method for several years.

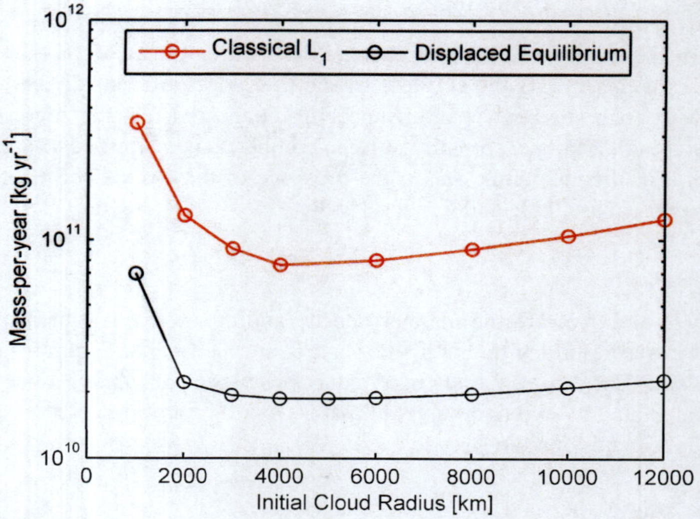

Fig. 25.3 Mass required to reduce solar insolation by 1.7% for the static cloud ejected in the vicinity of the L_1 position

25.2.2 *Gravitationally Anchored Dust Cloud at the L_1 Point*

The concept of an L_1 positioned dust cloud for geoengineering, described previously does not take into account the mass of the asteroid from which the dust cloud is generated. This is a logical assumption for small asteroids. However, there are many asteroids in the population of near Earth objects that have a considerable mass. This mass can be accounted for by using the circular restricted four-body problem (CR4BP), including the masses of the Sun, Earth and asteroid, and the resultant effect on the gravitational potential in the vicinity of the L_1 point can be observed. These dynamics can be seen to generate a zero-velocity curve which bounds the asteroid, within which dust grains ejected from the asteroid will remain trapped, if ejected below the escape velocity. This will gravitationally anchor a dust cloud, in the vicinity of the L_1 point, which can be then used for geoengineering, thus negating the instability of the classical L_1 point.

In the CR4BP the equations of motion are the same as for the CR3BP with the dimensionless potential function now expressed as;

$$U = \frac{1}{2}(x^2 + y^2) + \frac{1-\mu}{\rho_1} + \frac{\mu}{\rho_2} + \frac{\gamma}{\rho_3} \tag{25.8}$$

As well as the parameters defined in the previous section, the parameter γ is the mass fraction of the asteroid in relation to the mass of the three-body system, $\gamma = m_A/(m_s + m_E)$, and ρ_3 is the distance between a particle and the asteroid;

$$\rho_3 = \sqrt{(x - x_3)^2 + y^2 + z^2} \tag{25.9}$$

When the gravitational potential of a body placed at the classical L_1 point is considered, two new collinear equilibrium positions appear. These can be found by substituting the potential function of the CR4BP, Eq. (25.9), into the equations of motion defined in Eq. (25.1). The location of the new equilibrium positions are along the x-axis and can be then found by setting y=z=0 and $\ddot{x} = \dot{y} = 0$, thus resulting in the following equation:

$$x: x - \frac{1-\mu}{(x+\mu)^2} + \frac{\mu}{(x+\mu-1)^2} \pm \frac{\gamma}{(x-x_3)^2} = 0 \tag{25.10}$$

The new collinear equilibrium positions are located on either side of the classical L_1 position, as shown in Fig. 25.4. These new equilibria, like the conventional L_1 position, are unstable, but will bound the asteroid, thus approximating the size of the dust cloud. As for the static cloud, small dust grain sizes will be used and therefore the effects of SRP must be included by the addition of the factor β to the mass parameter, μ. The effect of a non-zero value for β can be seen in Fig. 25.4.

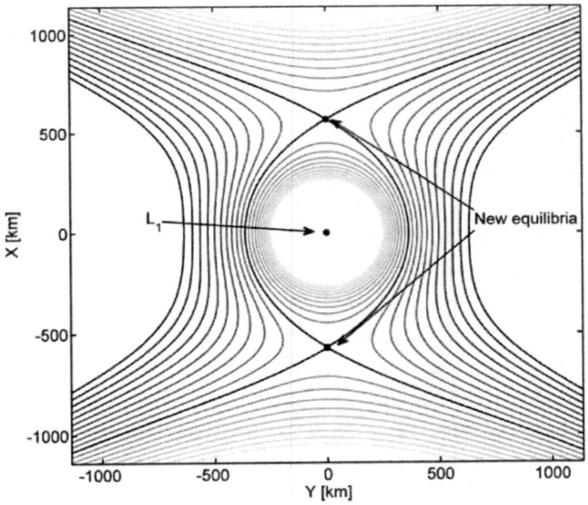

Fig. 25.4 Contour plot showing the variation in the effective potential of the four-body problem for a body of mass 1x10^{15}kg placed at the conventional L_1 point for β=0, with bold lines showing the contours with the Jacobi constant of the equilibrium points. Also shown are the new equilibrium positions along the x-axis.

The increase in β can be seen to shrink the area encompassed by the zero-velocity curve significantly (Fig. 25.5).

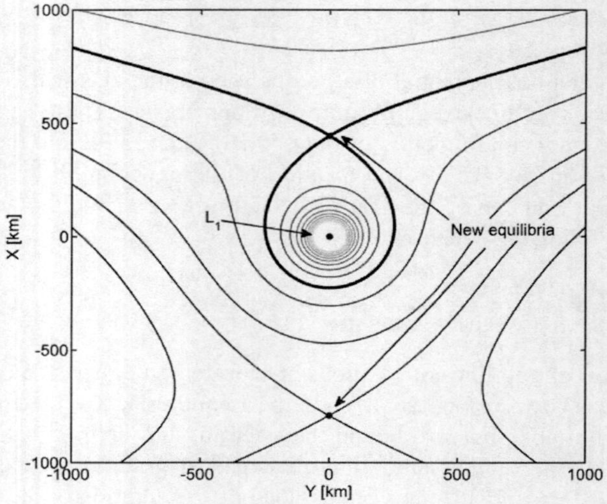

Fig. 25.5 Contour plot showing the variation in the effective potential of the four-body problem for a body of mass 1x10^{15}kg placed at the conventional L_1 point for β=0.005, with bold lines showing the contours with the Jacobi constant of the equilibrium points. Also shown are the new equilibrium positions along the *x*-axis.

The speed of a particle in the artificial 4-body system can be described by the Jacobi integral as;

$$V^2 = 2U - C \tag{25.11}$$

where V is the particle speed and C is the Jacobi constant. Since kinetic energy can only be strictly positive, it follows from Eq. (25.11) that the particle can only move within a region delimited by a zero velocity curve (i.e., when the right hand side of Eq. (25.11) vanishes). This constraint can then be used to investigate the size of the region around the third body, the asteroid at L_1, where a particle can become trapped if the energy, or Jacobi constant, of the particle is not large enough for escape. Clearly, the maximum enclosed volume will be found for a zero velocity surface with a Jacobi constant equal to that of one of the new equilibrium points in the CR4BP. From Eq. (25.11), the Jacobi constant for each of the new equilibrium positions can be found and then used to find the volume enclosed by the associated zero-velocity curve. Note that when β>0 and the anchoring asteroid is located in the classic L_1 position (i.e., when SRP is neglected), as in Fig. 25.5, only the smallest Jacobi constant guarantees a closed volume. Instead if the asteroid is located at the equilibrium position required for a given value of β, the encompassed volume increases, as in Fig. 25.4. The acceleration required to achieve the displacement associated with a grain size of 32μm is 9x10^{-7} m/s^2. This is a small value, however, due to the mass of asteroid required for this scheme, the thrust is of order 10^{11}N, which is clearly unfeasible with current technology. The thrust required to stabilise the asteroid at the classical L_1 is also high. For example to position a body at a distance of 1km away from the L_1 point would require an acceleration of $3.6\text{x}10^{-10}$ m/s^2. This thrust level is still significant and beyond current technologies, though should it be feasible to capture the largest near-Earth objects it will likely be achievable.

25.2.2.1 Results

The potential of this zero-velocity curve method for space-based geoengineering is investigated by estimating the size of the dust cloud anchored by a range of asteroid sizes above $1\text{x}10^{13}$kg, and then applying the solar radiation model described previously. The masses of the asteroids were determined by applying Bowell's relation (Bowell and Hapke 1989), which estimates the mass from the absolute magnitude, assuming an average density of 2,600 kg m^{-3} and albedo of 0.154. The results of the insolation reduction calculation, under the assumption of a value for β of 0.005, corresponding to a grain radius of 32μm, can be seen in Fig. 25.6 for asteroids positioned at the classical L_1 point and at the displaced equilibrium position of dust with β of 0.005.

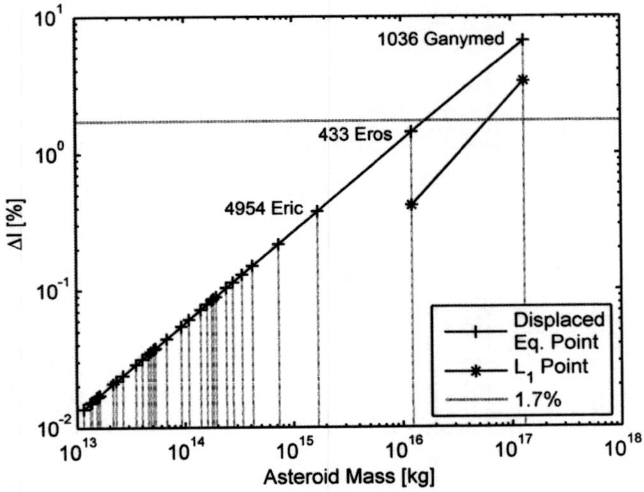

Fig. 25.6 Maximum insolation change available for the masses of the asteroids on the Pareto front situated at the displaced equilibrium position and the classical L_1 point

Initially, in order to estimate the maximum achievable insolation reduction, it is assumed that all light passing through the zero-velocity curve is blocked, by simply assuming a very high attenuation factor in Eq. (25.7). These results, in

Fig. 25.6, show a linear trend on a log-log plot with the maximum insolation reduction of 6.58% being achieved for the largest asteroid Ganymed at the β-displaced equilibrium position, with a maximum insolation reduction of 3.3% when the asteroid is assumed fixed at the classical L_1 position. This result meets the required 1.7% reduction in solar insolation required to offset a doubling of CO_2 in the atmosphere (Govindasamy and Caldeira 2000). The maximum change in solar insolation reduces significantly for the next largest asteroid, Eros, to 1.42% and 0.42% for the β-displaced equilibrium and classical L_1 positions respectively. These values do not meet the required 1.7% insolation reduction noted previously. However, these results, and some of the insolation reductions achieved by a few of the following large known NEO, could still be significant enough to be considered as part of a portfolio of geoengineering schemes. The smallest asteroids considered, with a mass in the region of 10^{13}-10^{14}kg are capable of offsetting the 1-2 W/m^2 variation in solar insolation experienced over a solar cycle (Willson and Hudson 1991) though are not capable of contributing significantly to a space based geoengineering scheme.

From this analysis, it can be seen that placing an asteroid at the β-displaced equilibrium position is clearly favourable for reducing solar insolation. However, as was described previously the thrust required to stabilise the asteroid at this position is orders of magnitude greater than to maintain the asteroid at the classical L_1 point. Therefore, the potential for insolation reduction at the conventional L_1 position must be maximised. Since the size of the zero-velocity curve at the classical L_1 point quickly reduces as β increases, the use of dust grains much larger than the 32μm previously investigated will be required. Figure 25.7 shows the variation in maximum insolation reduction for the asteroid Eros for increased grain radii. As well as for the mean Sun-Earth distance, the maximum insolation reduction was also calculated for the separation of the Earth and Sun at aphelion and perihelion. The results show that as the grain size increases the maximum insolation reduction will increase significantly before levelling off for the largest grain radii assessed, at a level similar to the scenario where the asteroid is positioned at the displaced equilibrium position. This is because as β decreases the zero velocity curve shape will resemble the equilibrium position shape and volume, as seen in Fig. 25.2. As the maximum reduction at the displaced equilibrium position is less than the desired 1.7%, the maximum reduction for the L_1 position will similarly be less than the desired amount. Nevertheless it can be shown that the geoengineering potential is close to the amount necessary and can therefore be considered as an option to provide a large portion of any insolation reduction required.

The negative aspect of the increase in grain size is the reduction in efficiency associated with a decrease in area-to-mass ratio. Therefore, a greater mass of dust grains is required to block the same level of sunlight. The fraction of the asteroid that must be used, assuming no loss of particles, to achieve the maximum insolation reduction for the largest grains would require a mass greater than that of the asteroid. Therefore only a smaller insolation reduction will be achievable.

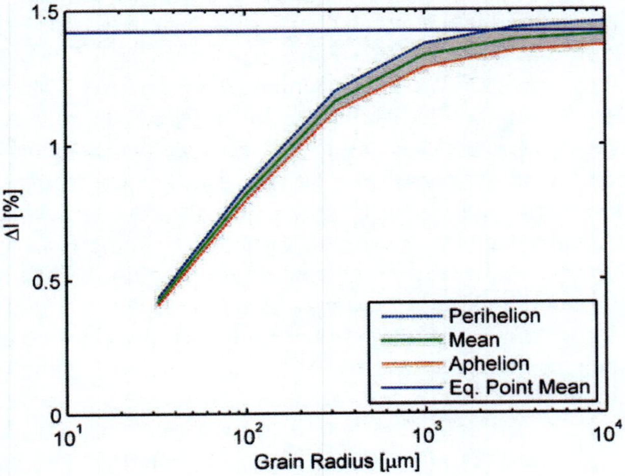

Fig. 25.7 Maximum insolation reduction for the asteroid Eros at the classical L_1 for grain sizes of 32μm and above. Also shown is the maximum insolation reduction at the displaced equilibrium position.

25.2.3 Earth Ring

As discussed previously, the concept of an Earth ring for geoengineering has been suggested by Pearson et al. (2006). In this concept a circular ring of dust grains is to be placed in the equatorial plane. It is assumed that above a grain size of 1μm there is no perturbation of the dust grains orbits due to solar radiation pressure (SRP), hence the orbits are assumed stable and no replenishment is necessary. However, a more detailed study of the effects of SRP and the Earth oblateness J_2 effect, discussed fully by Bewick et al. (2013) concludes that this lower bound on the grain size is not feasible. By including the factors of SRP and the J_2 effect, a set of stable, eccentric orbits with Sun-pointing apogee, termed heliotropic orbits, were found to exist, in the equatorial plane, for dust grain sizes above 6.5μm. Applying the same principles to the circular ring suggested by Pearson reveals that only above a grain radius of 13μm are stable orbits found. By trading off the minimum grain size possible and the fraction of the orbit spent in between the Earth and Sun, a semi-major axis of 9,316km with an eccentricity of 0.1 was chosen for the optimal stable orbit. The dust grain orbits will librate around this 'feeder orbit'. For these heliotropic orbits atmospheric drag was assumed to cause the orbital decay of a dust grain below an altitude of 2,000km. This places bounds on the semi-major axis and eccentricity for the stable heliotropic orbits since the altitude of perigee cannot fall below this altitude.

Despite the close proximity to Earth, the most optimal source for the dust grains required for the Earth ring is from captured near Earth objects, due to the high velocity required to reach orbit from the surface of the Earth. The asteroid is envisaged to be captured into a circular, equatorial generator orbit with an orbital radius of 10,250. It follows that 10,250km is also the radius of the apogee for the feeder orbit. The dust grains will be continuously extracted from the asteroid, milled to achieve a certain radii distribution and collected during one orbit of the asteroid. Whenever the asteroid passes directly between the Earth and Sun it ejects the collected dust with the correct Δv, using a mass driver, to inject it into the feeder orbit. From this feeder orbit the grains will then evolve and spread due to their different area-to-mass ratios and thus form the dust ring. Because the period of the generator orbit and the feeder orbit are different, grains will be distributed at all positions in the orbit. After a maximum of one year the dust grains will have spread over all positions in the ring. From a study of the 3D evolution of the dust grains the maximum angular spread away from the equatorial plane was assumed to be 0.2° (Colombo et al. 2012).

Three log-normal dust grain distributions were used, to analyse the feasibility of the Earth ring method. The distributions can be seen in Fig. 25.8 and are a representation of the milled dust size distribution for an optimistic technology capability (D1), a case achievable with existing terrestrial milling machinery (D2) and a pessimistic case of low milling efficiency (D3). The narrower distribution has a lower mean radius, though the spread of the dust grains within the orbit is reduced (Bewick et al. 2013), and vice-versa for the broader distribution. For this geoengineering method a simpler solar radiation model was used to calculate the reduction in insolation, with the incoming solar photons assumed to be plane parallel, due to the close proximity to the Earth negating any non-point source characteristics of the Sun. Again the Beer-Lambert law was used to calculate the attenuation. The

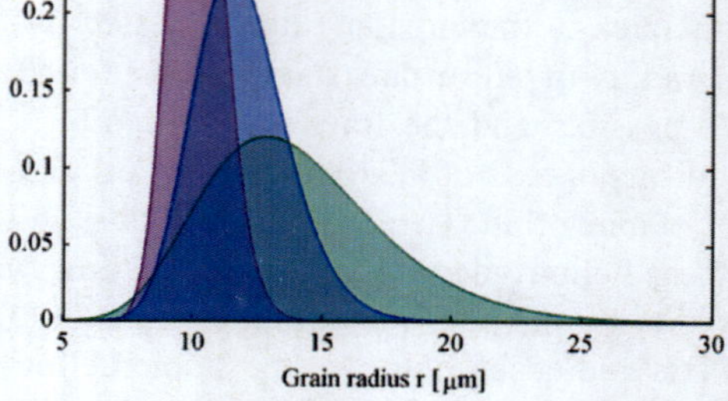

Fig. 25.8 Probability density functions for the three distributions of grain radii considered for the Earth ring

insolation reduction was calculated for different tilt angles of the equatorial plane away from the ecliptic, to simulate the changing angle caused by the orbit of the Earth around the Sun. This varying tilt leads to a changing aspect angle of the ring as seen from the Sun, leading to variable shading over the course of a year.

The mass of dust required to lead to an average insolation reduction of 1.7% over the course of a year can be seen in Fig. 25.9 for the different distributions. It can be seen that the D1 distribution, with the smaller mean value, requires the least amount of mass, though the difference in value between the three distributions is not significant. The mass shown is a lower estimate of the mass required since the reflection of solar radiation by the ring onto the Earth and thermal re-radiation from the ring must be accounted for. This amounts to approximately 40% (Pearson et al. 2006, Bewick et al. 2013), giving a final mass requirement of $1\text{x}10^{12}$kg, which is marginally smaller than the mass already suggested of $2.3\text{x}10^{12}$kg (Pearson et al. 2006). Finally, Fig. 25.10 shows a representation of where the insolation over the Earth's surface will be greatest. Thus, such a ring of dust would shade primarily the equatorial and low tropical regions. In these latitudes the tropics receive a greater insolation reduction than the equator since only when the ring is edge-on to the Sun, during the spring and autumn equinox, will the equatorial plane of the Earth be edge-on to the Sun. During other times the precession of the tilt of the Earth's rotational axis will cause the equatorial plane to be angled such that higher latitudes will be shaded.

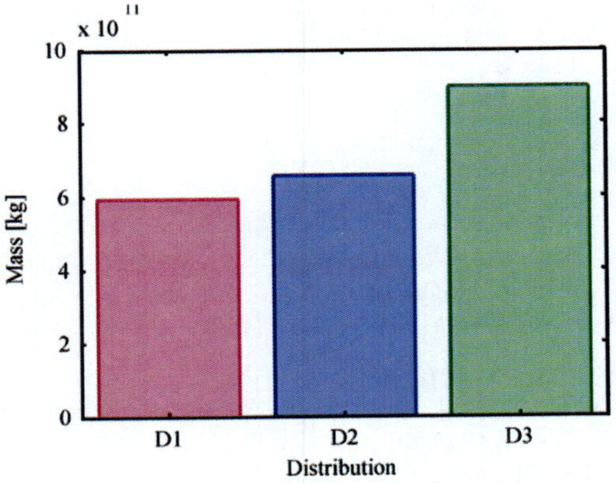

Fig. 25.9 Mass required to give an average 1.7% insolation change for three dust distributions

One of the negative aspects related to this method of space-based geoengineering is the increased risk to spacecraft due to space debris. For the Earth ring model it is predicted that very few dust grains will not be stable and will de-orbit quickly due to atmospheric drag, meaning that there will not be a significantly increased risk of space debris. However, other smaller perturbations can act over a longer period of time to distort the ring. These other perturbations include the third body effects of the gravity of the Sun and Moon, the charging of dust grains leading to the Lorentz force and grain collisions within the ring. These perturbations will likely increase the risk of space debris. As well as the Earth ring, the dust cloud positioned at the L_1 point will likely lead to an increased risk to spacecraft, though for the L_1 cloud case the grain sizes are likely too small to pose a significant risk, but will increase the degradation rate of external elements such as solar panels.

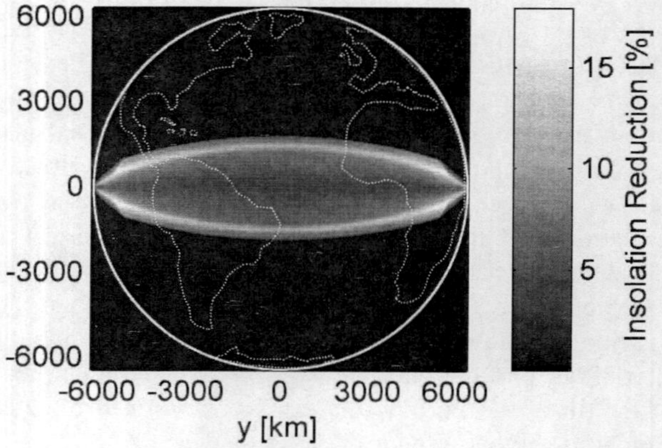

Fig. 25.10 Insolation change over the Earth's surface over the course of a year, as seen from the Sun

25.3 Availability of Resources

The mass needed by the space-based geoengineering methods described so far is a significant requirement, with the smallest mass being in the region of 10^{10}kg yr^{-1}, which is comparable to the mass of concrete in the Three Gorges dam in China. Therefore, consideration must be taken regarding the availability of asteroid resources that can fulfil this requirement.

It is assumed here that the accessibility of asteroid/comet material from the Sun-Earth L_1 point is, as a first approximation, similar to the asteroid accessibility from weakly-bound Earth orbits. The method of capture will depend on the size of asteroid, for example small asteroids can potentially be captured with conventional technologies, however, larger asteroids will require other methods, for example ejecting material from the surface at high velocities. The approximated amount of material accessible at an energy lower than that required to exploit the Moon can be shown to be of order 10^{14} kg (as seen in Chap. 18). This estimation results from summing up the mass of all objects, described by a Near-Earth object population distribution, that can reach a weakly-bound Earth orbit (i.e., Earth parabolic orbit) with a total Δv budget lower than the Moon's escape velocity of 2.37km s^{-1} (Sanchez and McInnes 2010, Sanchez and McInnes 2011). In particular, for the value presented here, a three impulse transfer model was used to assess the Δv cost of the transfer, not including the velocity to reach the asteroid (Sanchez and McInnes 2011). Thus, this result suggests that the Sun-Earth L_1 dust cloud could be theoretically sustained for 3,000 years by depleting all the asteroid/comet material that is energetically more accessible than the surface of the Moon. On the other hand, the two largest dust cloud methods, the gravitationally anchored dust cloud and Struck's Earth-Moon L_4/L_5 dust cloud, could not be supplied, even by depleting the same asteroid material. Note also that, in order to provide asteroid material for the planetary dust ring in low Earth orbit, presented here and also suggested by Pearson (Pearson et al. 2006), would require an extra 3.3km s^{-1} to transport the asteroidal material to the required position at low to medium Earth orbit.

Work undertaken by Sanchez and McInnes (Sanchez and McInnes 2011), as well as briefly in Chap. 18, describes how to estimate the Δv cost to access individual objects as a function of object size. A simplified version of the *asteroid resource map* represented in Chap. 18 is presented here in Fig. 25.11. This figure shows the average available resources by using the first, tenth, hundredth and thousandth largest accessible asteroid or cometary object in near Earth space. The figure also represents the 90% confidence region, which accounts for the statistical uncertainty of the Near-Earth object population distribution. This particular figure has been computed with the accessibility provided by the three impulse transfer model and NEO distribution model as described in (Sanchez and McInnes 2011).

Figure 25.11 suggests that it is likely to find one large object able to sustain the L_1 dust cloud concept for at least 150 years with a Δv lower than that required to exploit the Moon. The minimum size object to sustain the cloud for 1 year is found to be accessible with a Δv of order 1km s^{-1}, while 3km s^{-1} would be needed to provide 1 large object able to sustain the concept for 1,000 years. This, of course, assumes that all the material on the asteroid is milled to fine dust and expelled. These are general feasibility considerations on the availability of the material, which strongly suggests the benefits of using dust sourced in-situ over previous published dust-cloud concepts.

To assess the feasibility of the largest dust cloud methods, the gravitationally anchored dust cloud and Struck's dust cloud in the Earth-Moon system, a crude approximation of the energy needed to capture the 250 largest near-Earth objects was determined. These correspond to all asteroids above a mass of $1x10^{13}$kg. In this process the mass of the objects was estimated using Bowell's relation, discussed previously. Then, the minimum Δv of the Lambert-arc connecting the asteroid and the Earth is optimised. A global optimisation procedure is used to select the optimal Lambert arc conditions to transfer the asteroid to an Earth-like orbit. The design parameters of the optimisation are the true anomalies at both departure and arrival. A global optimisation method is used that blends a stochastic search with an automatic solution space decomposition technique (Sanchez and McInnes 2010, Sanchez and McInnes 2011). This is however a very rough approximation of the costs of transporting material to an Earth bound orbit, since as seen in Chaps. 19 and 21, the judicious use of the Earth gravitational perturbation can substantially reduce these costs.

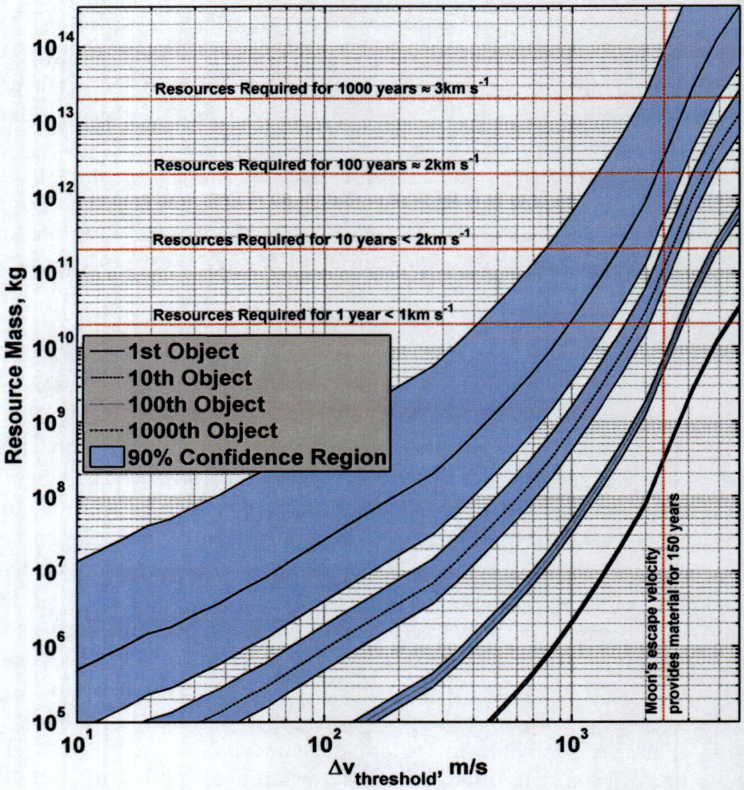

Fig. 25.11 Expected resources for each statistical accessible object

The impulse required to capture the 250 largest objects can be seen in Fig. 25.12. The figure shows a Pareto front of asteroid mass, m_A, against transfer impulse (i.e., $I = m_A \Delta v$), and represents a measure of incremental engineering effort to capture different known NEO objects. This Pareto front provides the list of the, a priori, most efficient asteroids to be captured with masses up to the largest known near Earth asteroid mass, approximately $1.3\text{x}10^{17}$kg for the asteroid 1036 Ganymed. The impulse obtained from the Lambert arc method is used here as a sorting parameter only, since we envisage continuous low thrust used for capture. The Δv required to capture all of these objects to the classical L_1 point is greater than the velocity required to exploit lunar resources. Therefore, for the Earth-Moon cloud suggested by Struck, where only the dust grains are required, utilising lunar resources would be optimal. For the gravitationally anchored dust cloud it will still be necessary to capture the a near-Earth object, since to maintain the position of the dust cloud the asteroid position would need to be maintained in the correct position.

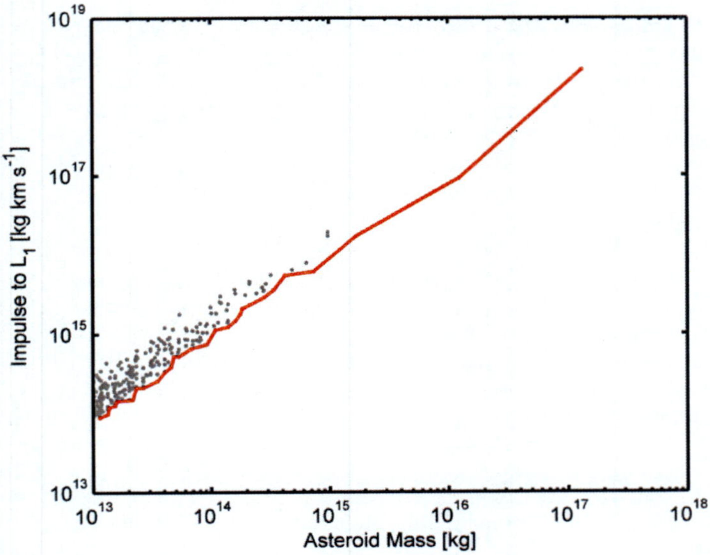

Fig. 25.12 Impulse required for capture to the L_1 point for the population of near Earth asteroids with masses above 1×10^{13}kg, with a Pareto front showing the optimum bodies for capture

25.4 Conclusion

In this chapter a selection of methods of space-based geoengineering, that utilise large clouds of asteroid derived dust grains, have been discussed. The masses required to achieve a 1.7% solar insolation reduction, to offset the temperature effects of a doubling of CO_2 in the atmosphere, range from 2×10^{10}kg yr^{-1} to 10^{17}kg. The lowest mass is for the cloud released for the vicinity of the L_1 point and allowed to drift, thus requiring constant replenishment. This mass is of the same order of magnitude as the solid reflector/refractor methods proposed by McInnes (2010) and Angel (2006), but requires much less manufacturing and launchers to initiate the scheme. Therefore, this represents a significant improvement in the affordability and timeliness of SBGE. The improved Earth ring concept requires in the region of 10^{12}kg of asteroid material, though requires no replenishment. Not considering the negative aspects of this scheme, discussed in (Bewick et al. 2013), this method of SBGE will be optimal should geoengineering be required for many decades. The concept of a gravitationally anchoring a dust cloud in the vicinity of the L_1 point will require the capture of the largest near-Earth objects and represents a significant requirement that is beyond current technology. Though the use of a gravitationally anchored dust cloud is beneficial, the extra engineering complexity outweighs this advantage. It has been shown that the mass required for the unstable L_1 cloud and the Earth ring concept, though it will require the capture of many objects, can be found in the populations of near-Earth objects and is more accessible, in terms of velocity required to capture, than lunar material. For the reasons discussed in this chapter SBGE can be seen as a promising concept for a mid-term scenario for geoengineering to reduce the effects of climate change.

References

Angel, R.: Feasibility of cooling the Earth with a cloud of small spacecraft near the inner Lagrange point (L1). Proceedings of the National Academy of Sciences 103, 17184–17189 (2006)

Bewick, R., Sanchez, J.P., McInnes, C.R.: The feasibility of using an L1 positioned dust cloud as a method of space-based geoengineering. Advances in Space Research 49, 1212–1228 (2012a)

Bewick, R., Sanchez, J.P., McInnes, C.R.: Gravitationally bound geoengineering dust shade at the inner Lagrange point. Advances in Space Research 50, 1405–1410 (2012b)

Bewick, R., Lücking, C., Colombo, C., Sanchez, J.P., McInnes, C.R.: Heliotropic Dust Rings for Earth Climate Engineering. Advances in Space Research 51(7), 1132–1144 (2013)

Binzel, R.P., Rivkin, A.S., Thomas, C.A., Vernazza, P., Burbine, T.H., DeMeo, F.E., Bus, S.J., Tokunaga, A.T., Birlan, M.: Spectral properties and composition of potentially hazardous Asteroid (99942) Apophis. Icarus 200, 480–485 (2009)

Bowell, E., Hapke, B., Dominique, D., Lumme, K., Peltoniemi, J.I., Harris, A.W.: Application of photometric models to asteroids. In: Asteroids II. University of Arizona Press (1989)

Brophy, J., Culick, F., Friedman, L.: Asteroid retrieval feasibility study. Keck Institute for Space Studies (2012)

Colombo, C., Lücking, C., McInnes, C.R.: Orbital Dynamics of High Area-to-Mass Ratio Spacecraft with J2 and Solar Radiation Pressure for Novel Earth Observation and Communication Services. Acta Astronautica 81, 137–150 (2012)

Early, J.T.: Space-based Solar Shield to Offset Greenhouse Effect. Journal of the British Interplanetary Society 42, 567–569 (1989)

Govindasamy, B., Caldeira, K.: Geoengineering Earth's radiation balance to mitigate CO2-induced climate change. Geophysical Research Letters 27, 2141–2144 (2000)

Govindasamy, B., Caldeira, K., Duffy, P.B.: Geoengineering Earth's radiation balance to mitigate climate change from a quadrupling of CO2. Global and Planetary Change 37, 157–168 (2003)

Ingle, J.D.J., Crouch, S.R.: Spectrochemical Analysis. Prentice Hall (1988)

IPCC, Contribution of Working Groups I, II and III to the Fourth Assessment Report of the Intergovernmental Panel on Climate Change, IPCC, Geneva (2007)

Jacobson, M.Z., Ten Hoeve, J.E.: Effects of Urban Surfaces and White Roofs on Global and Regional Climate. Journal of Climate 25, 1028–1044 (2012)

Mautner, M.: A Space-based Solar Screen against Climatic Warming. Journal of the British Interplanetary Society 44, 135–138 (1991)

McInnes, C.R.: Space-based geoengineering: challenges and requirements. Proceedings of the Institution of Mechanical Engineers, Part C: Journal of Mechanical Engineering Science 224, 571–580 (2010)

de Pater, I., Lissauer, J.J.: Planetary Sciences. Cambridge University Press (2001)

Pearson, J., Oldson, J., Levin, E.: Earth rings for planetary environment control. Acta Astronautica 58, 44–57 (2006)

Raven, J., Caldeira, K., Elderfield, H., et al.: Ocean acidification due to increasing atmospheric carbon dioxide. Royal Society, Science Policy Section (2005)

Robock, A., Marquardt, A., Kravitz, B., Stenchikov, G.: Benefits, risks, and costs of stratospheric geoengineering. Geophysical Research Letters 36, L19703 (2009)

Sanchez, J.P., McInnes, C.R.: Asteroid Resource Map for Near-Earth Space. Journal of Spacecraft and Rockets 48, 153–165 (2011)

Sanchez, J.P., McInnes, C.R.: Accessibility of the resources of near Earth space using multi-impulse transfers. In: 2010 AIAA/AAS Astrodynamics Specialist Conference, Toronto (2010)

Shepherd, J., Caldeira, K., Cox, P., Haigh, J.: Geoengineering the climate. Report of Royal Society working group of geo-engineering (2009)

Struck, C.: The feasibility of shading the greenhouse with dust clouds at the stable lunar Lagrange points. Journal of the British Interplanetary Society 60, 82–89 (2007)

Teller, E., Wood, L., Hyde, R.: Global warming and ice ages. I. Prospects for physics-based modulation of global change. In: Proc. 22nd Int. Seminar on Planetary Emergencies, Erice, Italy, August 19-24 (1997)

Wilck, M., Mann, I.: Radiation pressure forces on "typical" interplanetary dust grains. Planetary and Space Science 44, 493–499 (1996)

Willson, R.C., Hudson, H.S.: The Sun's luminosity over a complete solar cycle. Nature 351, 42–44 (1991)

Chapter 26
Using Asteroids for Launch/Landing, Change of Trajectory and Acceleration of Space Ships

Alexander A. Bolonkin

C&R Co., New York, USA

26.1 Introduction

At the present time, rockets are used to carry people and payloads into space (Bolonkin 2006). Other than rockets, methods used to reach space speed include the space elevator (Bolonkin 2006, 2007a), tethers (Bolonkin and Cathcart 2007; Bolonkin 2008), the electromagnetic system (Bolonkin 2007b, 2010), and the tube rocket (Bolonkin 2002a). The space elevator for big planets is not technically feasible at the present time; it would require substantial costs for development. In particular, the space elevator concept requires extremely strong nanotubes. Tethers are very complex and would require two artificial bodies. Electromagnetic systems are also complex and expensive. The author has previously discussed several other non-rocket launch methods that are potentially low cost, but which require much additional research. These include cable launchers (Bolonkin 2006), circle launchers (Bolonkin 2002a-f, 2006, 2007a,b, 2008, 2010; Bolonkin and Cathcart 2007) and inflatable towers (Bolonkin 2002g).

There are many asteroids in the Solar System. The vast majority are found in a swarm called the asteroid belt, located between the orbits of Mars and Jupiter at an average distance of 2.1 to 3.3 astronomical units (AU) from the Sun. Scientists know of approximately 6,000 large asteroids of a diameter of 1 kilometer or more, and of millions of small asteroids with a diameter of 3 meters or more. Ceres, Pallas, and Vesta are the three largest asteroids, with diameters of 785, 610 and 450 km, respectively. Others range all the way down to meteorite size. In 1991 the Galileo probe provided the first close-up view of the asteroid Caspra. There are many small asteroids, meteorites, and comets outside the asteroid belt. For example, scientists know of 1,000 asteroids of diameter larger than one kilometer located near the Earth. Every day 1 ton meteorites with mass of over 8 kg fall on the Earth. The orbits of big asteroids are well known. The small asteroids (from 1 kg) may be also located and their trajectory can be determined by radio and optical devices at a distance of hundreds of kilometers.

Radar observations enable to discern of asteroids by measuring the distribution of echo power in time delay (range) and Doppler frequency. They allow a determination of the asteroid trajectory and spin and the creation of an asteroid image.

There are also the asteroids located at the stable Lagrange points of the Earth–Moon system. These bodies orbit with the same speed as Jupiter, and might be very useful for propelling spacecraft further out into the solar system. It seems likely that the kinetic and rotational energy of asteroids will eventually find application in space flight.

Most asteroids consist of carbon-rich minerals, while most meteorites are composed of stony-iron.

The present ideas (Bolonkin 1965, 2002a-j, 2003a,b, 2005a,b, 2011; Weekly News 1998) is to utilize the kinetic energy of asteroids to change the trajectory and speed of space ships (probes). Any space bodies more than 10% of a ship's mass may be used, but here mainly bodies with a diameter of 2 meters (6 feet) or larger are considered. In this case the mass (20–100 tons) of the asteroid is some 10 times more than the mass of probe (1 ton, 2200 lb) and the probe mass can be disregarded.

26.2 Asteroid Space Elevator. Optimal Cable for Any Space Elevator

This section proposes a new method and transportation system to fly into space to the asteroids. This transportation system uses a mechanical energy transfer and requires only minimal energy so that it provides a "Free Trip" into space. It uses the rotational and kinetic energy of planets, asteroids, meteorites, comet heads, moons, satellites and other natural space bodies (Bolonkin 1965, 2002a-j, 2003a,b, 2005a,b, 2011; Weekly News 1998) .

This chapter contains the theory and results of computations for three projects. The projects use artificial materials like nanotubes and whiskers that have a ratio of tensile strength to density equal to 4 million meters. In the future, nanotubes will be produced that can reach a specific stress up 100 million meters and will significantly improve the parameters of suggested projects.

The concept of the space elevator first appeared in 1895 when a Russian scientist, Konstantin Tsiolkovsky, considered a tower that reached a geosynchronous orbit. The tower was to be built from the ground up to an altitude of 35,800 kilometers (geostationary orbit). Comments from Nikola Tesla suggest that he may have also conceived such a tower. His notes were sent behind the Iron Curtain after his death.

Tsiolkovsky's tower would be able to launch objects into orbit without a rocket. Since the elevator would attain orbital velocity as it rode up the cable, an object released at the tower's top would also have the orbital velocity necessary to remain in geosynchronous orbit.

Building from the ground up, however, proved an impossible task; there was no material in existence with enough compressive strength to support its own weight under such conditions. It took until 1957 for another Russian scientist, Yuri N. Artsutanov, to conceive of a more feasible scheme for building a space tower. Artsutanov suggested using a geosynchronous satellite as the base from which to construct the tower. By using a counterweight, a cable would be lowered from geosynchronous orbit to the surface of Earth while the counterweight was extended from the satellite away from Earth, keeping the center of mass of the cable motionless relative to Earth. Artsutanov published his idea in the Sunday supplement of *Komsomolskaya Pravda* in 1960. He also proposed tapering the cable thickness so that the tension in the cable was constant–this gives a thin cable at ground level, thickening up towards GEO (http://www.liftport.com/files/Artsutanov_Pravda_SE.pdf).

26.2.1 Short Description of the Space Elevator

The space elevator is a cable installation which connects the planet/asteroid's surface/mast to an over geostationary planet orbit (For the Earth GEO (Geostationary Earth Orbit) = 37.786 km in altitude). Schematic diagram of Space elevator for asteroid is shown in Fig. 26.1, the space elevator for Earth/planet/asteroid is shown in Fig. 26.2.

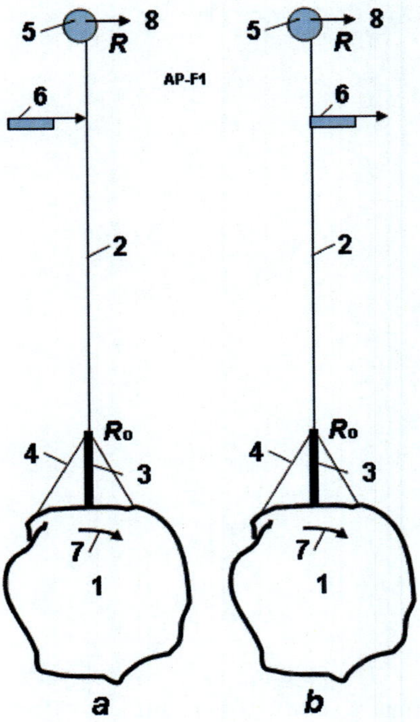

Fig. 26.1 Space elevator of asteroid. *Notations*: 1 - asteroids, 2 – elevator cable; 3 – tower or mast; 4 – bracing; 5 – counterweight; 6 – space ship or apparatus; 7 - angle speed of asteroid; 8 – speed of cable end; R_o is maximum radios (from center of asteroid gravity to top of mast).

The GEO is a 24-hour orbit and stays over the same point above the equator as the Earth rotates on its axis. The installation center of mass is at or above this altitude. The space elevator has a counterweight, which allows it to have its center of gravity at or above GEO, and climbers. Once sent far enough, climbers would be accelerated further by the planet's rotation. A space elevator, also known as a space bridge, is one of the technology concepts that are aimed at improving access to space. Also called a geosynchronous orbital tether, it is one kind of skyhook.

The elevator would have to be built of a material that could endure tremendous stress while also being light-weight, cost-effective, and manufacturable. A considerable number of other novel engineering problems would also have to be solved to make a Earth space elevator practical. Today's technology does not meet these requirements for Earth's elevator.

There are a variety of space elevator designs. Almost every design includes a base station, a cable, climbers, and a counterweight. The base station designs typically fall into two categories: mobile and stationary. Stationary platforms are generally positioned in high-altitude locations.

Fig. 26.2 Space elevator

The building of a space elevator on Earth has two main problems: tire need for material with a very high tensile stress/specific density ratio, and the very large cost of Earth installation. But a Earth space elevator could be made relatively economically if a cable with a density similar to graphite, with a tensile strength of around 65–120 GPa could be produced in bulk at a reasonable price. By comparison, the strongest steels are no more than 5 GPa (1 GPa = 100 kg/mm^2 = 0.1 ton/mm^2), but steel is heavy. The much lighter material kevlar has a tensile strength of 2.6 – 4.1 GPa, while quartz fiber can reach upwards of 20 GPa; the tensile strength of diamond filaments would theoretically be minimally higher.

Carbon nanotubes have a theoretical tensile strength and density that lie well within the desired range for space elevator structures, but the technology to manufacture bulk quantities and fabricate them into a cable has not yet been developed. Theoretically carbon nanotubes can have tensile strengths beyond 120 GPa. Even the strongest fiber made of nanotubes is likely to have notably less strength than its components (30–60 GPa). Further research on purity and different types of nanotubes will hopefully improve this value.

Note that at present (year 2012), carbon nanotubes have an approximate cost of $10÷50/gram, and 20 million grams would be necessary to form even a seed elevator. This price is decreasing rapidly, and large-scale production would reduce it further.

Climbers cover a wide range of designs. On elevator designs whose cables are planar ribbons, some have proposed to use pairs of rollers to hold the cable with friction. Other climber designs involve magnetic levitation (unlikely due to the bulky track required on the cable).

Power is a significant obstacle for climbers. Some solutions have involved nuclear power, laser or microwave power beaming. They are very complex, or expensive, or have very low efficiency. The primary power methods (laser and microwave power beaming) have significant problems with both efficiency and heat dissipation on both sides. Bolonkin (2006) offers a cable transport system which is more realistic at the present time.

There have been two methods proposed for dealing with the counterweight needed: a heavy object, such as a captured asteroid, positioned just beyond geosynchronous orbit; and extending the cable itself well beyond geosynchronous orbit. The latter idea has gained more support, it is simpler and the long cable located out of GEO (up to 144,000 km) may be used for launching payload to asteroids.

A space elevator could also be constructed on asteroids or other planets. A Mars space elevator could be much shorter than one on Earth. Exotic materials might not be required to construct such an elevator. A lunar tather would need to be very long—more than twice the length of an Earth elevator. It could also be made of existing engineering materials.

There are a lot of problems in the development and design of a Earth space elevator: corrosion of cable, meteoroids, micrometeorites and space debris, Earth's weather, Earth's satellites, failure modes and safety issues, sabotage, vibrational harmonics, the event of failure, breaking of the cable, elevator pods, Van Allen Belts (radiation region), political issues, economics problems, etc. Many of them are absent in asteroid space elevator.

Many problems of Earth's space elevator is absent for the asteroid space elevator because the asteroid gravity is closed to zero. In particular, the artificial fiber can be used for the asteroid space elevator.

The centrifugal space launcher (Bolonkin 2006, Ch.10) is very suitable for non-rotational asteroids.

26.2.2 General Theory of Space Elevator

1. **General case of a planet**. Let us take a small segment (dR) of cable. Equilibrium (forces) of this segment is

$$dF_1 = dF_2 + dF_3\,, \quad dF_1 = \sigma dA\,, \quad dF_2 = gdm\,, \quad dF_3 = V^2 dm / R\,, \qquad (26.1)$$

where dF_1 is stress force, N; dF_2 is gravity force, N; dF_3 is centrifugal force, N; σ is safety stress, N/m^2; A is gross section cable area, m^2; g is planet gravity, m/s^2; m is mass kg; V is speed, m/s^2; R is radius, m. The values in Eq. (26.1) equal:

$$dm = \gamma A dR\,, \quad g = g_0\left(\frac{R_0}{R}\right)^2, \quad V = \omega R\,, \qquad (26.2)$$

where γ is cable density, kg/m^3; g_0 is planet gravity at $R = R_o$, m/s^2; R_o is planet radius, m; ω is planet angle speed, rad/s.

If we substitute Eq. (26.2) in Eq. (26.1), we obtain the differential equation

$$\frac{1}{A}dA = \frac{\gamma g_0}{\sigma}\left[\left(\frac{R_0}{R}\right)^2 - \frac{\omega^2 R}{g_0}\right]dR \qquad (26.3)$$

Solution to Eq. (26.3) is

$$a(R) = \frac{A}{A_0} = \exp\left[\frac{\gamma g_0 B(R)}{\sigma}\right]$$

$$B(r) = R_0^2\left\{\left(\frac{1}{R_0} - \frac{1}{R}\right) - \frac{\omega^2}{2g_0}\left[\left(\frac{R}{R_0}\right)^2 - 1\right]\right\} \qquad (26.4a,b)$$

or

$$a(R) = \frac{A}{A_0} = \exp\left[\frac{\gamma B(R)}{\sigma}\right]$$

$$B(r) = R_0^2\left\{g_0\left(\frac{1}{R_0} - \frac{1}{R}\right) - \frac{\omega^2}{2}\left[\left(\frac{R}{R_0}\right)^2 - 1\right]\right\} \qquad (26.4'a,b)$$

where a is the relative cable area, $B(r)$ is the work of lifting 1 kg mass. These Eqs. (26.4'a,b) have been reported by author in Bolonkin (2006 p.13, Eq. (1.4)).

The mass of the cable M and a volume v can be calculated by equations

$$v\,(R) = A_0\int_{R_0}^{R} a dR\,, \quad M(R) = \gamma_0 v(R) \qquad (26.5)$$

For asteroid $g_o \approx 0$ and Eq. (26.4'a) is

$$a(R) = \frac{A(R)}{A_0} = \exp\left[-\frac{\gamma}{2\sigma}\omega^2\left(R^2 - R_0^2\right)\right] \tag{26.6}$$

4. Lift force (L) of the asteroid space elevator is

$$L(R) = A(R)\,\sigma \tag{26.7}$$

The fibers suitable foe asteroid space elevator are presented in Table 26.1.

Table 26.1 Some data of fibers. The data are from best cases, and have been established for giving a rough figure.

Material	Strength (MPa)	Density (g/cm³)	Specific Strength (kN·m/kg)	Breaking length on Earth (km)
Brass	580	8.55	67.8	6.91
Aluminium	600	2.80	214	21.8
Stainless Steel	2000	7.86	254	25.9
Titanium	1300	4.51	288	29.4
Bainite	2500	7.87	321	32.4
Scifer steel wire	5500	7.87	706	71.2
carbon-epoxy composite	1240	1.58	785	80.0
spider silk	1400	1.31	1069	109
Silicon carbide	3440	3.16	1088	110
Glass Fiber	3400	2.60	1307	133
Basalt fiber	4840	2.70	1790	183
1 μm iron whiskers	14000	7.87	1800	183
Vectran	2900	1.40	2071	211
Carbon fiber (AS4)	4300	1.75	2457	250
Kevlar	3620	1.44	2514	256
Dyneema (UHMWPE)	3600	0.97	3711	378
Zylon	5800	1.54	3766	384
Carbon nanotube	62000	.037-1.34	46268-N/A	4716-N/A
Colossal carbon tube	6900	.116	59483	6066

Note that multiwalled carbon nanotubes have the highest tensile strength of any material yet measured, with labs producing them at a tensile strength of 63 Gpa, still well below their theoretical limit of 300 GPa. The first nanotube ropes (20 mm long) whose tensile strength was published (in 2000) had a strength of 3.6 GPa, still well below their theoretical limit. The density is different depending on the manufacturing method, and the lowest value is 0.037 or 0.55(solid).

26.3 Cable Method for Using the Asteroids

26.3.1 Connection Method

The connection method includes the following main steps (Bolonkin 2002b):

(a) Finding an asteroid using a locator or telescope (or looking in catalog) and determining its main parameters (location, mass, speed, direction, rotation); selecting the appropriate asteroid; computing the required position of the ship with respect to the asteroid.
(b) Correcting the ship's trajectory to obtain the required position; convergence of the ship with the asteroid.
(c) Connecting the space apparatus (ship, station, and probe) to the space body (planet, asteroid, moon, satellite, meteorite, etc.) by a net, anchor, and a light strong rope (cable), when the ship is at the minimum distance from the asteroid.
(d) Obtaining the necessary position for the apparatus by moving around the space body and changing the length of the connection rope.
(e) Disconnecting the space apparatus from the space body; spooling the cable.

The equipment required to change a probe (spacecraft) trajectory includes:

(a) A light strong cable (rope).
(b) A device to measure the trajectory of the spacecraft with relative to the space body.
(c) A device for spacecraft guidance and control.
(d) A device for the connection, delivery, control, and disconnection and spooling of the rope.

26.3.2 Description of Utilization

The following describes the general facilities and process for an asteroid with a small gravitational force to change the trajectory and speed of a space apparatus.

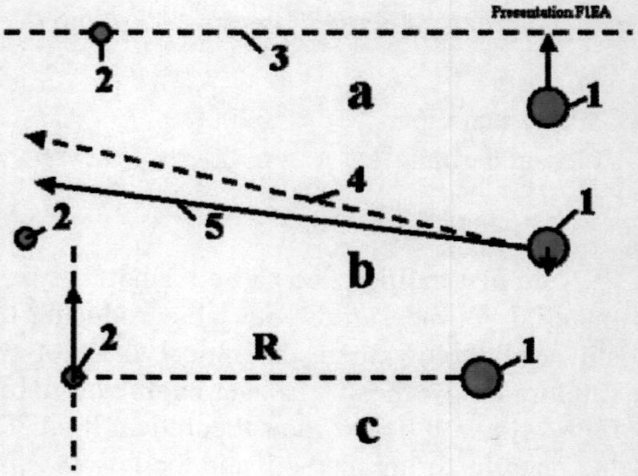

Fig. 26.3 Preparing for employment of the asteroid. Notations: 1 – space ship, 2 – asteroid, 3 – plane of maneuver. 4 – old ship direction, 5 – corrected ship direction. a) Reaching the plane of maneuver; b) Correcting the flight direction and reaching the requested radius; c) Connection to the asteroid.

Figures 26.3a,b,c show the preparations for using an asteroid to change the trajectory of the space apparatus; for example, the asteroid 2, which is moving in the same direction as the apparatus (perpendicular to the sketch, Fig. 26.3a). The ship wants to make a maneuver (change direction or speed) in plane 3 (perpendicular to the sketch), and the position of the apparatus is corrected and moved into plane 3. It is assumed that the asteroid is more massive than the apparatus.

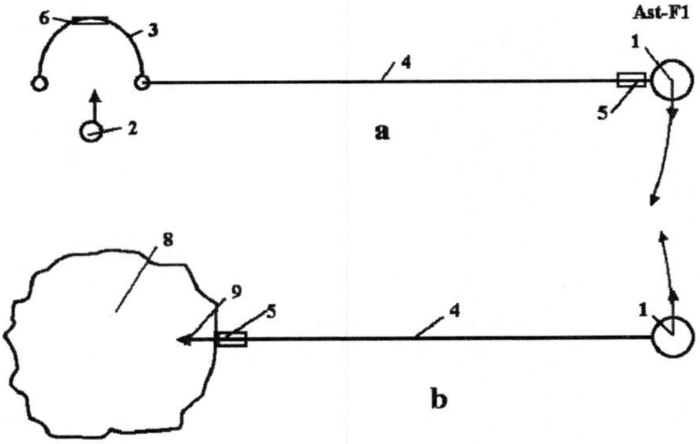

Fig. 26.4 a) Catching a small asteroid using net; b) Connection to a big asteroid using an anchor and cable. Notation: 1 – space ship, 2, 8 – asteroid, 3 – net with inflatable ring, 4 – cable (rope), 5 – load cabin, 6 – valve, 9 – anchor.

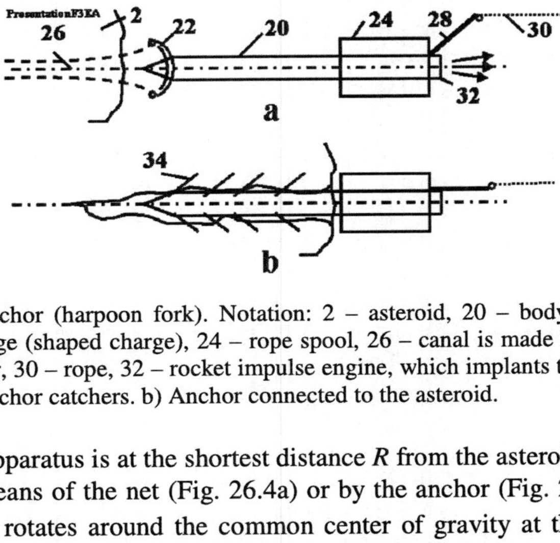

Fig. 26.5 a) Anchor (harpoon fork). Notation: 2 – asteroid, 20 – body of anchor; 22 – cumulative charge (shaped charge), 24 – rope spool, 26 – canal is made by shaped charge, 28 – rope keeper, 30 – rope, 32 – rocket impulse engine, which implants the anchor into the asteroid, 34 – anchor catchers. b) Anchor connected to the asteroid.

When the apparatus is at the shortest distance R from the asteroid, it connects to the asteroid means of the net (Fig. 26.4a) or by the anchor (Fig. 26.5b) and rope. The apparatus rotates around the common center of gravity at the angle φ with angular speed ω and linear speed ΔV (Fig. 26.6). The cardioids of additional speed and direction of the apparatus are shown in Fig. 26.6 (right side). The maximum additional velocity is $\Delta V = 2V_a$, where V_a is the relative asteroid velocity when the coordinate center is located in the apparatus. Figure 26.6b shows the case where the asteroid moves in the opposite direction to the apparatus with velocity ΔV.

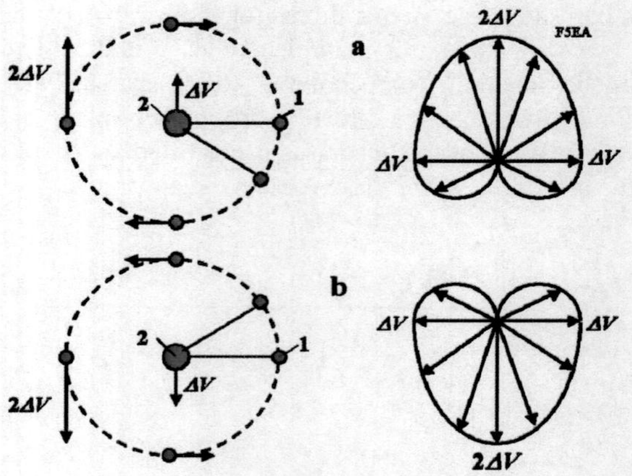

Fig. 26.6 Using the kinetic energy of an asteroid to change the space ship trajectory (speed and direction). On the right are cardioids of the additional velocity and its direction. The ship can get this velocity from the asteroid. Notations: 1 – space ship, 2 – asteroid, ΔV – difference between velocities of space ship and asteroid. a) Case when the asteroid has the same direction as the ship; b) Case when the asteroid has the opposite direction to the ship a

Figure 26.4a shows how a net can be used to catch a small asteroid. The net is positioned in the trajectory of the small asteroid, supported in an open position by the inflatable ring and connected to the space apparatus by the rope. The net catches the asteroid and transfers its kinetic energy to the space apparatus. The space apparatus changes its trajectory and speed and then disconnects from the asteroid and spools the cable. If the asteroid is large, the astronaut team can use the asteroid anchor (Figs. 26.4b, 26.5).

The astronauts use the launcher (a gun or a rocket engine) to fire the anchor (harpoon fork) into the asteroid. The anchor is connected to the rope and spool. The anchor is implanted into the asteroid and connects the space apparatus to the asteroid. The anchor contains the rope spool and a disconnect mechanism (Fig. 26.5). The space apparatus contains a spool for the rope, motor, gear transmission, brake, and controller. The apparatus may also have a container for delivering a load to the asteroid and back (Fig. 26.4b). One possible design of the space anchor is shown in Fig. 26.5. The anchor has a body, a rope, a cumulative charge (shared charge), the rocket impulse (explosive) engine, the rope spool and the rope keeper. When the anchor strikes the asteroid surface the cumulative charge burns a deep hole in the asteroid and the rocket-impulse engine hammers the anchor body into the asteroid. The anchor body pegs the catchers into the walls of the hole and the anchor's strength keeps it attached to the asteroid. When the apparatus is to be disconnected from the asteroid, a signal is given to the disconnect mechanism.

If the asteroid is rotated with angular speed ω (Fig. 26.7), its rotational energy can be used for increasing the velocity and changing the trajectory of the space apparatus. The rotational asteroid spools the rope on its body. The length of the rope is decreased, but the apparatus speed is increased (see a momentum theory in physics).

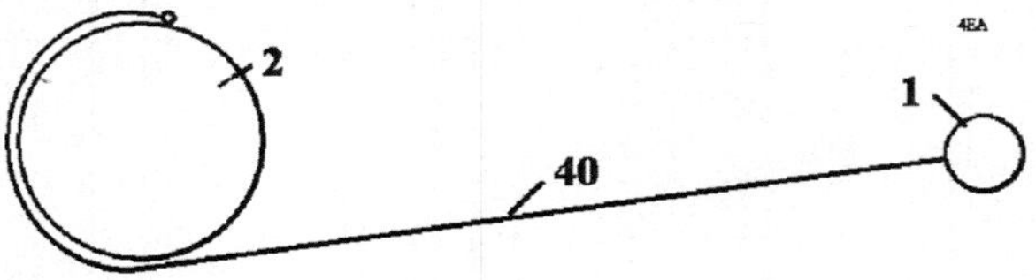

Fig. 26.7 Using the rotational energy of a rotating asteroid. Notations: 1 – space ship, 2 – asteroid, 40 – connection cable.

The ship can change the length of the cable. When the radius decreases, the linear speed of the apparatus increases; conversely, when the radius increases the apparatus speed decreases. The apparatus can obtain energy from the asteroid by increasing the length of the rope.

The computations and estimations show the possibility of making this method a reality in a short period of time (see projects than follows).

26.3.3 Theory of Cable Method and Computation

The differential equation can be found from the equilibrium of a small cable part under centrifugal force. This equation for optimal (equal stress from centrifugal force) cable is

$$da/A = (\omega^2\gamma/\sigma)RdR$$

Its solution is the cable equal stress

$$a(R)=A/A_o=\exp(V^2/2k) = \exp(\omega^2R^2/2k)\,. \qquad (26.8)$$

where a – relative cross-section area of cable; A – cross-section area of cable [m^2]; A_0 – initial (near probe) cross-section area of cable [m^2]; V – speed of probe or space ship about asteroid [m/s]; $k = \sigma/\gamma$ – ratio of cable tensile stress to density [Nm/kg]; $K = k/10^7$ – coefficient; R – radius from the common gravity center: asteroid + probe [m]; ω - angular speed of a probe around asteroid [rad/s]; σ – tensile strength [N/m^2]; γ – density of cable [kg/m^3];

Mass W [kg] of cable is

$$W = A_0\gamma\int_0^R a(r)dr = \frac{F_0}{k}\int_0^R e^{\omega^2 r^2/2k}dr \qquad (26.9)$$

where r – variable [m]; F_o – force from the probe [N]

Relative cable mass $W_r = W/W_s$ is

$$W_r = B(1+B), \quad B = \frac{n}{k}\int_0^{v^2/ng}\exp\left[\left(\frac{ng}{V}\right)^2\frac{r^2}{2k}\right]dr \qquad (26.10)$$

where the integration interval is [0, V^2/ng], $n = F_o/g$ is the overload, V is the circle apparatus speed around the common center of gravity, W_s is ship (probe) mass, g is Earth gravitation [m/s^2], $g = 9.81$ m/s^2 .

4. Circular velocity of ship around asteroid

$$R = V^2/gn\ ,\ V = (gnR)^{0.5}\ . \tag{26.11}$$

Computations are represented in Figs. 26.8-26.10.

5. Relative mass of cable with constant cross-section area for small speed

$$W_r = W/W_s = \gamma V^2/\sigma g = V^2/kg\ . \tag{26.12}$$

6. Finding the additional velocity the space vehicle has received from the asteroid.

Let us take the coordinate axis along the positive direction of the asteroid speed and write the momentum and energy laws of the asteroid-apparatus system for this axis

$$m_1V_1 + m_2V_2 = m_1u_1 + m_2u_2\ ,$$
$$0.5m_1V_1^2 + 0.5m_2V_2^2 = 0.5m_1u_1^2 + 05m_2u_2^2 + A, \tag{26.13,14}$$

where m_1, V_1 are the mass and speed of the asteroid respectively before connection to apparatus; m_2 , V_2 are the mass and speed of the apparatus respectively before connection to asteroid; u_1 is speed of the asteroid after disconnection from the apparatus; u_2 is speed of the apparatus after disconnection from the asteroid; A is energy (work) applied by the apparatus to change the length of the connection cable.

Let us locate beginning of the axis of at the apparatus (this means $V_2 = 0$) and apply the variable $V = V_1$ - asteroid speed around apparatus; $u = u_2$ – the additional apparatus speed; *and* $m = m_2/m_1$ – the relative apparatus mass.

Substitute u_1 from Eq. (26.13) into Eq. (26.14), we receive the quadratic equation about u

$$(m+1)u^2 - 2Vu + 2A/m_1m = 0\ . \tag{26.15}$$

Solution of this equation is

$$u = \{V \pm [V^2 + 2A(m+1)/mm_1]^{0.5}\}/(m+1)\ . \tag{26.16}$$

Investigating this equation, if $A = 0$ (the apparatus does not change the length of the connection cable) and the asteroid mass is large ($m \approx 0$), the maximum additional speed of the apparatus is $u = 2V$ in the asteroid direction and $V = 0$ in the opposite direction. If $A \neq 0$, the maximum work (energy) apparatus can receive from the asteroid by increasing the connection cable length, is less than

$$A \le mm_1V^2/2(m + 1)\ . \tag{26.17}$$

If the apparatus expends internal energy (decreases the length of the connection cable), the additional apparatus speed is limited only by the safe cable strength and apparatus overload. The apparatus does not lose mass to increase its speed.

If apparatus is disconnected in a direction with an angle of φ to the asteroid speed direction, the additional apparatus speed is

$$\Delta V = V(1 + \cos \varphi). \tag{26.18}$$

where V is initial speed of the asteroid around the space ship [m/s] (coordinate center is located at the space apparatus); ΔV is additional speed received by the ship from the asteroid [m/s]; φ is the angle between the old velocity vector of the asteroid and the new velocity vector of the apparatus.

The additional kinetic energy of the apparatus is then

$$E_k = 0.5m_2(\Delta V)^2. \tag{26.19}$$

The known formulas below may be useful:

$$V = \omega R, \quad V_3 R_3 = V_2 R_2. \quad V_1 = R_o\left(\frac{g_0}{R}\right)^{0.5},$$
$$V_2 = \sqrt{2}V_1, \quad R_g = \left(\frac{g_0 R_0^2}{\omega^2}\right)^{1/3}, \tag{26.20}$$

where V_1 is the circular speed around the Earth, V_2 is the escape speed, R_o is the Earth's radius, and R_g is the radius of Earth's geosynchronous orbit.

26.3.4 Project

The capability to change the trajectory and speed of a space vehicle using an asteroid is shown in Fig. 26.6. The space ship could obtain a maximum additional speed equal to twice the speed difference between the space vehicle and the asteroid (speed of the asteroid around the space ship). If the length of the connection cable is changed, the speed of the space ship could change by more than double the speed difference. If the asteroid is rotating, the space ship can also obtain an additional speed increase from the rotation. The additional speed from one asteroid is also limited (for a manned ship) by the mass of the cable. For an additional speed of 1,000 m/s and $K = 0.2$, the mass of cable would equal 5% of the mass of the space apparatus. For an additional speed of 2,000 m/s, the mass of cable would equal 23% of the mass of the space apparatus. To travel to an asteroid, a connection device may be mounted onto the transport cable. The cable may be used many times.

The reader finds the reports closed to this topics in Bolonkin (1965, 2002a-j, 2003a,b, 2005amb, 2011), Weekly News (1998), data for computation in Naschekin (1969), Galasso (1989), Kroschwitz (1990), Palmer (1991), Directory (1995), Cosmo and Lorenzini (1997), Anonymous (1996-1997, 2001), Dresselhous (2000), Smitherman (2000), Ziegler and Cartmell (2001).

The results of computation for different cases are shown in Figs. 26.8-26.13.

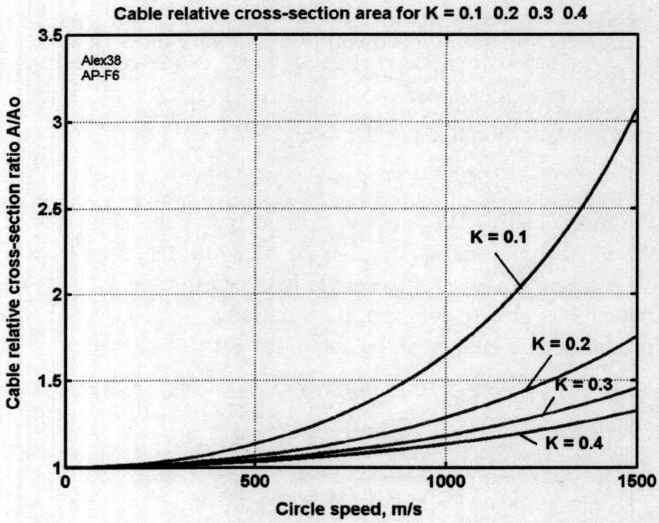

Fig. 26.8 Asteroid cable relative ratio via circle speed; coefficient $K = 0.1 - 0.4$

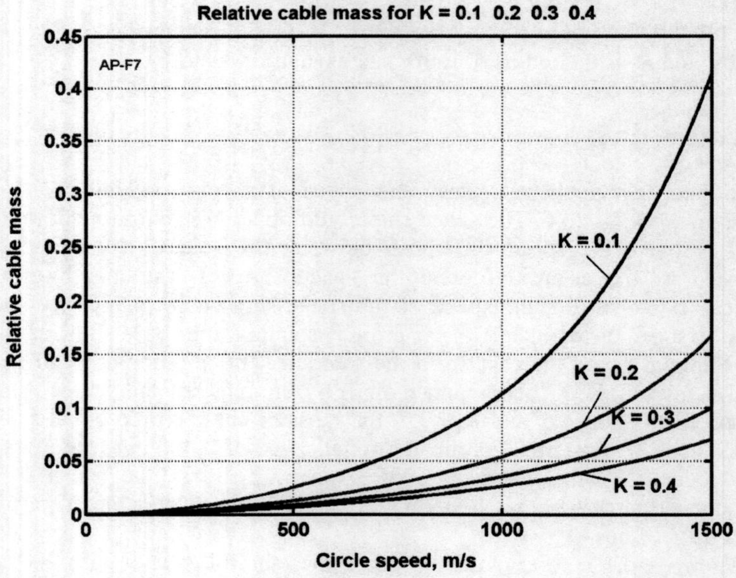

Fig. 26.9 Relative cable asteroid mass via circle speed in m/s; coefficient $K = 0.1 - 0.4$

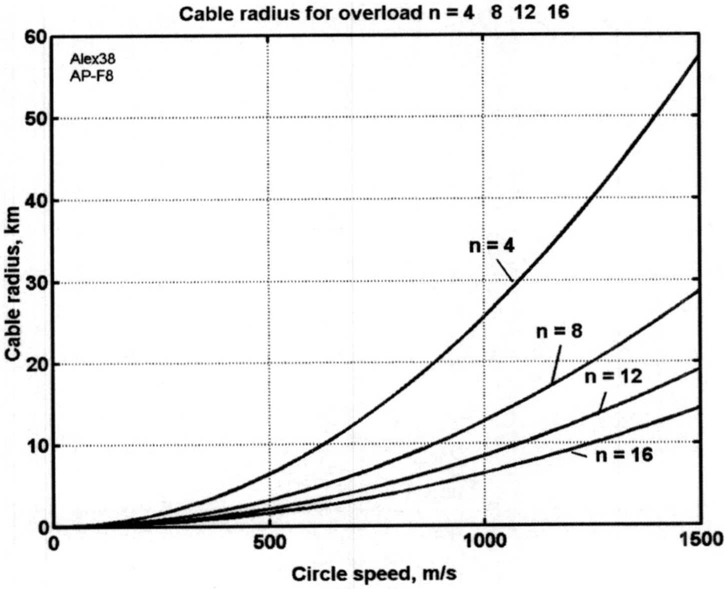

Fig. 26.10 Cable radius in km via circle speed in m/s and overload $n = 4$–16

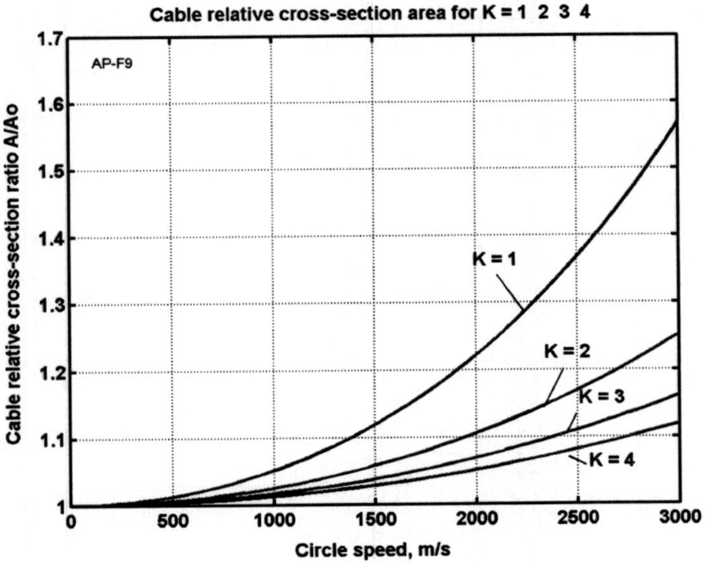

Fig. 26.11 Asteroid cable relative ratio via circle speed and coefficient $K = 1$– 4

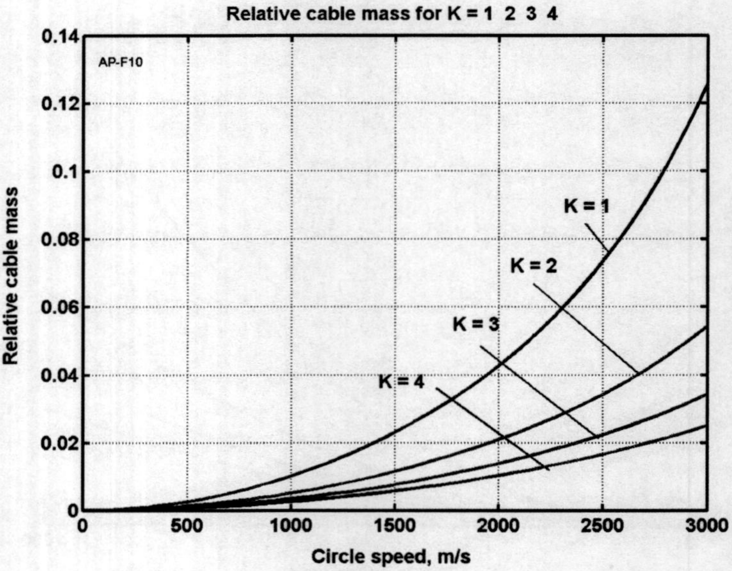

Fig. 26.12 Relative cable asteroid mass via circle speed in m/s and coefficient $K = 1–4$

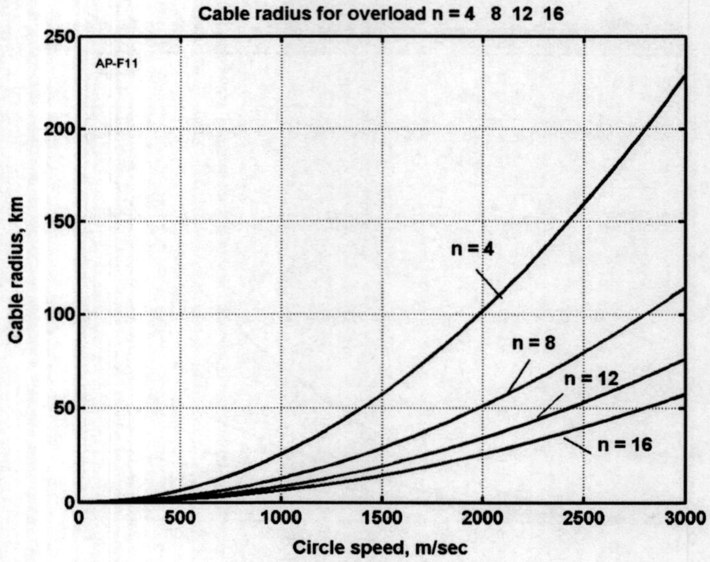

Fig. 26.13 Cable radius in km via circle speed in m/s and overload n

26.3.5 Discussion about Cable Method

If the change in the ship's speed is less than 1000 m/s, the conventional widely produced fiber (safe $K = 0.1$) can be used. The cable mass is about 8% of the ship's mass. After disconnection the cable will be spooled and can be used again. The reader can make estimations for other cases. Radio or optical devices can locate asteroids at distance of thousands of kilometers. Their speed, direction of flight and mass can be computed. The ship (probe) can make small corrections to its own trajectory to obtain the required position relative to the asteroid. All big asteroids with a diameter of more than 1 kilometer are listed in astronautic catalogs and their trajectories are well known. One thousand of them are located near the Earth. For those, we can compute in advance the intercept parameters. At the present time, long-range space apparatus uses the gravity of a planet to change its trajectory. However, the solar system has only nine planets, and they are located very far from one another. The employment of asteroids increases this possibility a million times over.

Estimation of the Probability of Meeting a Small Asteroid. It is known that every day about a ton of meteorites with a mass greater than 8 kg fall into the Earth's atmosphere. The Earth's surface area is about 512 million km^2. If the average mass of meteorites is 10 kg, then the Earth encounters 100 meteorites per day or one meteorite a day for every 5 millions km^2. If a space probe has a mass around 100 kg, a 10 kg meteorite has enough mass for it to be employed to change the direction and speed of the space probe. Ground locators can detect a 1 kg space mass at distances up to thousands of km. If the space ship can detect over a range of 1000 km, it means it can see a space body with an area of one million km^2, or about one meteorite in every 5 days. If one meteorite in ten is suitable for employment, it means every 50 days the space apparatus will meet an eligible meteorite near the Earth. The likelihood is ten times greater in the asteroid belt between Mars and Jupiter. For 6,000 big asteroids, we can compute the intercept parameters now. This number is expected to increase as more small asteroids are registered.

Note that the kinetic energy of space bodies may be used if the space body has a different speed or direction. It is difficult to use a tether system (for example, the last stage of a rocket and the Shuttle ship) because they have the same speed and direction.

Cable. If the required change of speed is less than 1,000 m/s, then cable from current artificial fibers can be used.

26.3.6 Conclusion

The availability of both current and new materials makes the suggested propulsion system and projects more realistic for a long trip to outer space with a minimum expenditure of energy (Bolonkin 2006).

26.4 Electrostatic Utilization of Asteroids for Space Flight

This section offers an electrostatic method for changing the trajectory of space probes. The method uses electrostatic force and the kinetic or rotational energy of asteroids to increase or decrease ship/ probe speed by 1000 m/s or more and to achieve any new direction in outer space. The flight possibilities of spaceships and probes are thereby increased by a factor of millions.

26.4.1 Introduction

The method includes the following main steps (Fig. 26.14):

(a) Finding an asteroids using a locator or telescope (or looking in a catalog) an asteroid and determining its main parameters (location, mass, speed, direction, rotation); selecting the appropriate asteroid; computing the required position of the ship with respect to the asteroid.

(b) Correcting the ship's trajectory to obtain the required position; convergence of the ship with the asteroid.

(c) Charging the asteroid and space apparatus ball using a charge gun.

(d) Obtaining the necessary apparatus position and speed for the apparatus by flying it around the space body and changing the charge of the apparatus and asteroids.

(e) Discharging the space apparatus and the asteroid.

The equipment requires for changing a probe (spacecraft) trajectory includes:

(a) A charging gun.

(b) Devices for finding and measuring the asteroids, and computing the trajectory of the spacecraft relative of the space body.

(c) Devices for spacecraft guidance and control.

(d) A device for discharging of the apparatus and asteroid (see Fig. 26.14).

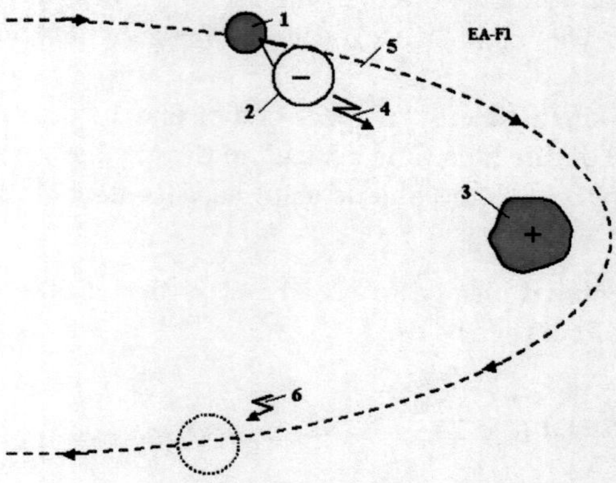

Fig. 26.14 Method of electrostatic maneuvers of the space apparatus. Notations: 1 – space apparatus, 2 – charged ball, 3 – asteroid, 4 – charged gun, 5 – new apparatus trajectory, 6 – discharging the apparatus and asteroid.

26.4.2 Description of Asteroid Utilization

The following describes the general facilities and process for an asteroid to change the trajectory and speed of a space probe.

Figure 26.15a,b,c,d show the preparations for using an asteroid for changing the trajectory of the space apparatus', for example, the asteroid 2, which moves near the space apparatus. The ship needs to make a maneuver (change direction or speed) in plane 3 (perpendicular to the sketch), and the position of the apparatus is corrected and it is moved into the required plane 3 using small rocket impulses. It is assumed that the asteroid has more mass than the artificial apparatus, and the space body speed is close to that of the apparatus (the difference can be as much as 1000 m/s).

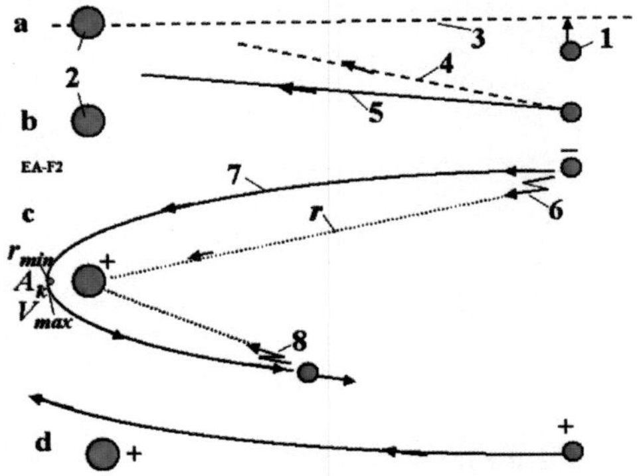

Fig. 26.15 Maneuvers of electrostatic space apparatus. a – preparing for aneuver, correcting the plane of maneuver; b – correcting the apparatus trajectory in the maneuver plane; c – charging the apparatus and asteroids, changing apparatus trajectory and velocity, discharging apparatus and asteroid; d – the case of like charges (which repel). *Notations:* 1 – space apparatus (ball); 2 – asteroid; 3 – plane of maneuver; 4 - initial apparatus trajectory; 5 – primary correction of the trajectory in the maneuver plane; 6 – charge impulse from apparatus to asteroid using charge gun; 7 – new apparatus trajectory; 8 – discharge impulse from apparatus (return part of charge energy) using sharp edge.

In the early computed point of the trajectory the space apparatus sends the asteroid a charge which takes root on asteroid surface, electrifying it. The apparatus also is charged (with unlike or like charges). If the apparatus and asteroid have unlike charges, they will be attracted one to another (Fig. 26.15c). If they have like charges, they will repel such other (Fig. 26.15d). The electrostatic force changes the apparatus trajectory and speed. The trajectory gains a new direction with a new velocity. When the trajectory has the selected direction and speed, the apparatus is then discharged and leaves the asteroid. The charge energy returns to the apparatus. If efficiency equals 1, this energy will be less when the new apparatus speed is more than before, equal when the speeds are same, and more when the apparatus speed is less than the previous speed.

Charging is done using a special charge gun, which accelerates charged particles (electrons or ions) to the required speed, and discharging is done using the edge of a pike. The return energy is an electric current through the pike's edge. The apparatus can also utilize the asteroid's speed and asteroid kinetic energy by mechanical connection as described herein.

The charging equipment may be used to land the apparatus on an asteroid surface and to launch the apparatus from an asteroid (see point 5 in the Theory and Computation section). The apparatus ball is usually charged with negative charges because negative charges (electrons) are emitted from an open surface which does not have a high resistance electric cover, when the electrical intensity (in a vacuum) is over 100 MV/m.

26.4.3 Theory of Electrostatic Method and Computation

1. **The electrostatic force** between charged bodies is

$$F = k\frac{Q_1 Q_2}{r^2}, \quad if \quad Q = Q_1 = Q_2, \quad F = k\frac{Q^2}{r^2}, \quad Q = \frac{a^2 E}{k}, \qquad (26.11)$$

where F is the force [N]; k is a coefficient, $k = 9\times10^9$; Q is the charge [C]; r is the distance between the center of charges [m]; a is radius of the charged ball apparatus [m]; E is the electrical intensity at the ball's surface [volts/m] (it may be up 100–200 MV/m in the vacuum and negative charge, and is more for positive charge).

2. Computation of space apparatus trajectories. Assume the asteroid mass is many times greater than the apparatus mass and start the origin of the coordinate system at the asteroid's center of gravity. The initial apparatus position is shown in Fig. 26.16.

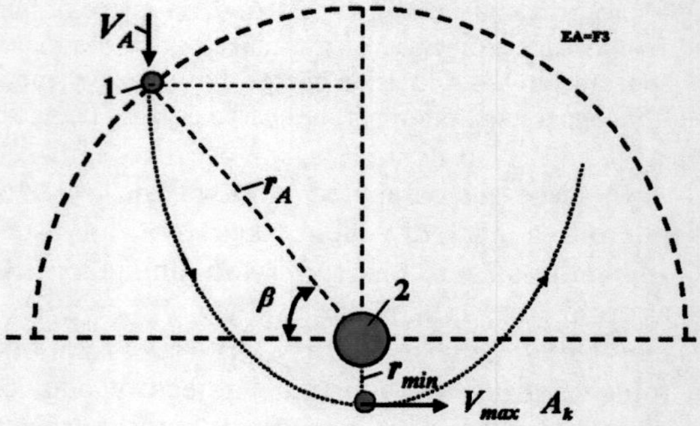

Fig. 26.16 Interaction between the charged ball apparatus and the charged asteroid. Notations are: 1 – space apparatus and electrically charged ball (balloon); 2 – asteroid; r_A – initial radius, V_A – initial apparatus speed.

We assume the charging happens instantly and the charges are constant up to point of discharge.

The equations for computation of the hyperbolic ($e > 1$) apparatus trajectory are (asteroid mass >> apparatus mass):

$$K=\frac{kqQ}{m},\quad H=V_A^2\mp\frac{2K}{r_A},\quad c=r_AV_A\cos\beta,\quad e=\frac{c}{K}\sqrt{H+\frac{K^2}{c^2}},\quad \alpha=2\arctan\sqrt{e^2-1},\ e>1,$$
$$\gamma=\pi-\alpha,\quad V=\sqrt{H+\frac{2K}{r}},\quad p=\frac{c^2}{K},\quad r=\frac{p}{1+e\cos\varphi},\quad T=\frac{2\pi}{\sqrt{K}}\left(\frac{p}{1-e^2}\right)^{3/2},\quad r_{\min}=\frac{p}{1+e},\qquad (26.22)$$
$$V_{\max}=\sqrt{H+\frac{2K}{r_{\min}}},\quad A_k=\frac{V_{\max}^2}{p}.$$

Here: K – coefficient [constant]; m – apparatus mass [kg]; q – electrical charge of the space apparatus [C]; Q – electrical charge of the asteroid [C], which is conventionally $q = Q$; H – coefficient of apparatus energy (kinetic and electric potential) around the asteroid; V_A – relative speed of the asteroid about spaceship [m/s] at the initial radius r_A (moment of charge impulse); $V(r)$ – spaceship speed [m/s], c – momentum constant, e – eccentricity of the apparatus trajectory ($e > 1$ for a hyperbolic trajectory, $e = 1$ for parabolic trajectory, $0 < e < 1$ for an elliptical trajectory, $e = 0$ for a circle), β – the angle between V_A and perpendicular to r_A at the moment of charging (see Fig. 26.16), α – angle between the asymptotes of the hyperbolic trajectory, γ – final deviation of the hyperbolic trajectory from the initial direction [radians]; p – parameter of the hyperbolic trajectory [m]; r – variable radius-vector at the trajectory point [m]; r_{min} – minimum distance of the charge apparatus center from the asteroid [m]; V_{max} – maximum apparatus speed relative to the asteroid [m/s]; A_k – maximum centrifugal acceleration of apparatus [m/s^2].

The constant (H) may be found from the initial position (r_A) and initial speed of the apparatus (V_A) (see the second formula in Eq. (26.22)). Sign "–" is used for attraction, sign "+" is used for repelling charges.

Note than the trajectory may be circular or elliptical and the apparatus can orbit for a long time around the asteroid.

Readers can find initial formulas for gravity fields in space mechanics and physics text-books and modify them for electrical charged bodies (electric fields). Formulas in Eq. (26.22) are used for deviation of the charged apparatus by the electrically charged asteroid.

The reader finds the reports closed to this topics in Bolonkin (1965, 2002a-j, 2003a,b, 2005a,b, 2011), Weekly News (1998) , data for computation in Naschekin (1969), Galasso (1989), Kroschwitz (1990), Palmer (1991), Directory (1995), Cosmo and Lorenzini (1997), Anonymous (1996-1997, 2001), Dresselhous (2000), Smitherman (2000), Ziegler and Cartmell (2001).

Computations of a typical parameter of trajectory versus parameter r_A for space apparatus of mass 100 kg are presented in Figs. 26.17 – 26.20.

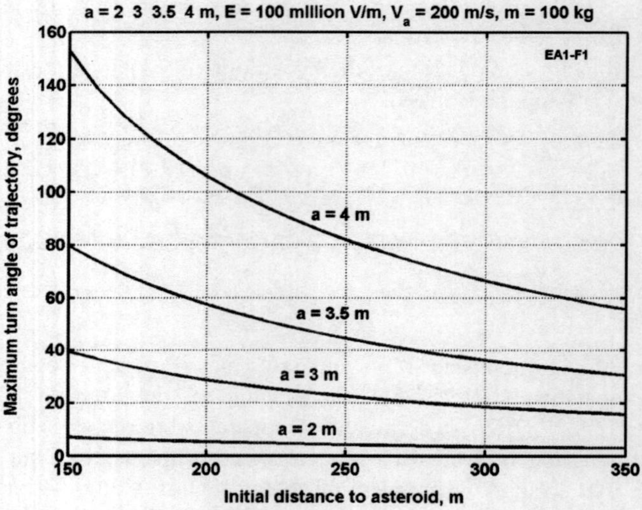

Fig. 26.17 Maximum turn angle of trajectory, γ, versus the initial charge distance to asteroid and radius of the charged ball for electrical intensity of 100 million volts/meter, initial apparatus speed 200 m/s, and apparatus mass 100 kg, $\beta = \pi/4$

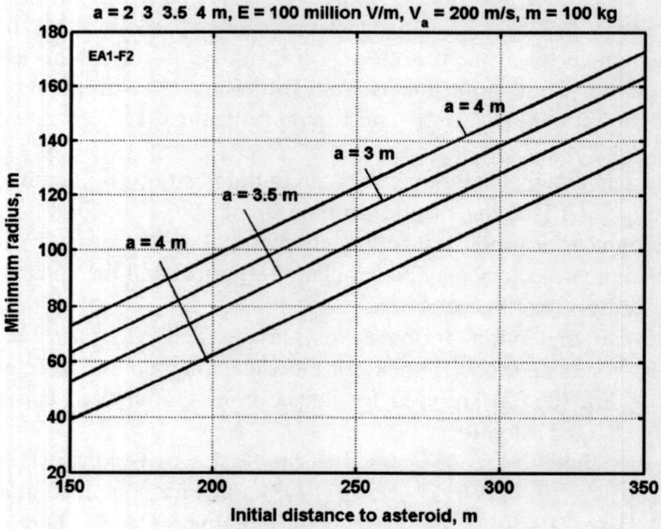

Fig. 26.18 Minimum radius of trajectory versus the initial charge distance to the asteroid and radius of the charged ball for electrical intensity 100 million volts/meter, initial apparatus speed 200 m/s, and apparatus mass 100 kg, $\beta = \pi/4$

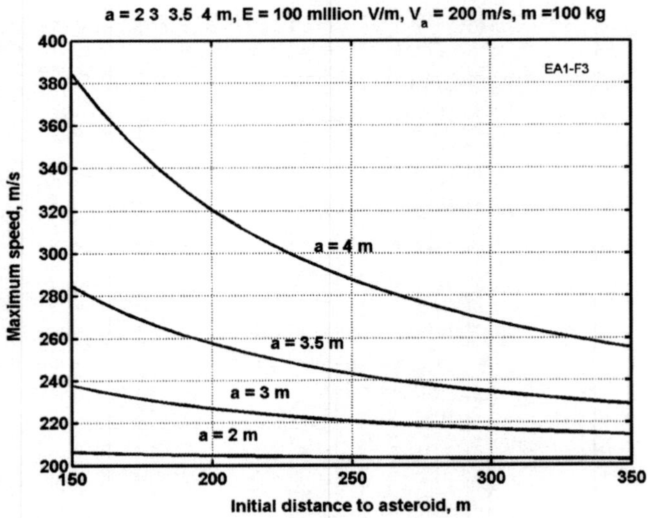

Fig. 26.19 Maximum speed of the space apparatus versus the initial charge distance to the asteroid and radius of the charged ball for electrical intensity 100 million volts/meter, initial apparatus speed 200 m/s, and apparatus mass 100 kg, $\beta = \pi/4$

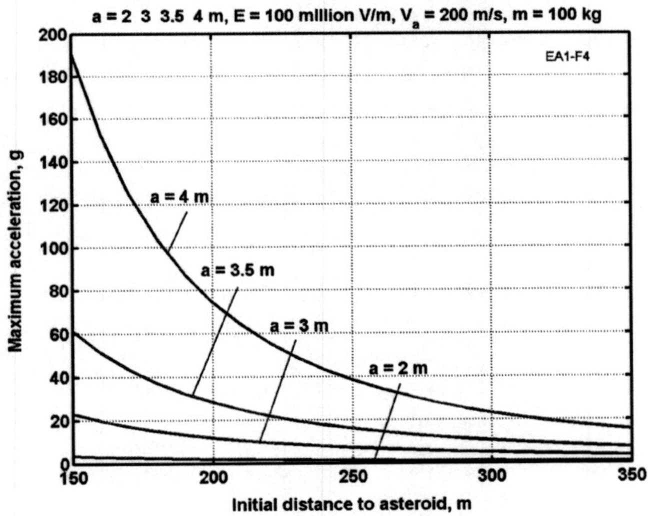

Fig. 26.20 Maximum acceleration (in *g*) of space apparatus versus the initial charge distance to the asteroid and radius of the charged ball for electrical intensity 100 million volts/meter, initial apparatus speed 200 m/s, and apparatus mass 100 kg, $\beta = \pi/4$

3. Initial expenditure of electrical energy needed to charge of the ball. The ball must be charged with electrical energy of high voltage (millions of volts). Let us estimate the minimum energy, when charged the device has 100% efficiency. This energy is equals to the work of moving of the ball charge to infinity. This may be computed using the equation

$$W=\frac{Q^2}{2C},\quad Q=\frac{a^2E}{k},\quad C=\frac{a}{k},\quad W=\frac{a^3E^2}{2k}, \tag{26.23}$$

where W is ball charge energy [J]; C is ball capacitance [F]; Q is ball charge [C].

The result of this computation is presented (same) in Fig. 26.18. As you can see this energy is not enormous – it is about 1 – 10 kWh for a ball radius of a = 2–4 m and electrical intensity of 25–100 MV/m. This energy (equal, or less, or more) may be returned when the ball is discharged by emitting of the charge into space using a sharp edge.

The energy (work) requirement for the separation of unlike charges (apparatus – asteroid, $q = Q$) can be computed by the following equations (for efficiency is equal to 1):

$$W=\int_a^{r_A} k\frac{Q^2}{r^2}dr = kQ^2\left(\frac{1}{a}-\frac{1}{r_A}\right),\quad \Delta W=\frac{mV_A^2}{2}-\frac{mV_f^2}{2},\quad \Delta W = kQ^2\left(\frac{1}{r_A}-\frac{1}{r_f}\right) \tag{26.23'}$$

where ΔW is increment in energy [J]; V_f is apparatus speed on the moment of discharge [m/s]; r_f – is the distance to the asteroid at the moment of discharge [m]. Charging the asteroid requires less energy because it is located at a limited distance from the space apparatus. The increment ΔW may be either positive, zero, or negative.

4. **The ball stress, cover thickness and ball mass.** The ball has tensile stress from the like electric charges. The equations are same as Eqs. 13.9-13.12 of Bolonkin (2006).

5. **Landing on and launching the space apparatus from asteroids**

If the apparatus and the asteroid have like charges, they will repel each other. This can be used to brake the apparatus for landing on the asteroid or for launching the apparatus from the asteroid surface. The change in apparatus speed can be computed using the following equations:

$$Q=\frac{a^2E}{k},\quad \frac{mV^2}{2}=k\frac{Q^2}{a}=\frac{a^3E^2}{k},\quad V=\sqrt{\frac{2a^3E^2}{km}} \tag{26.24}$$

where V is the initial (for braking) or final (for launching) speed of the apparatus [m/s]. Computations of braking (landing) and launch speed are presented in Fig. 26.21. The launch speed can be quite high, at up to 1600 m/s or more.

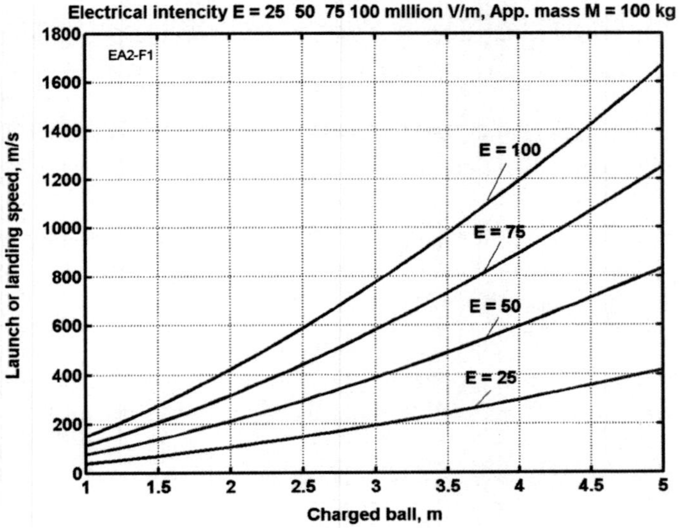

Fig. 26.21 Launch and landing (braking) speed of apparatus versus the charged ball radius and the electrical intensity (in million of volts/meter)

26.4.4 Project

The reader finds the reports closed to this topics in Bolonkin (1965, 2002a-j, 2003a,b, 2005a,b, 2011), Weekly News (1998).

As a project we can take the space apparatus of the mass 100 kg. All its parameters and maneuver capabilities can be estimated from the figures above. Data for computation may be found in Naschekin (1969), Galasso (1989), Kroschwitz (1990), Palmer (1991), Directory (1995), Cosmo and Lorenzini (1997), Anonymous (1996-1997, 2001), Dresselhous (2000), Smitherman (2000), Ziegler and Cartmell (2001).

26.4.5 Discussion

1. Estimation of probability of meeting a small asteroid. This problem was considered in Chap. 11 of Bolonkin (2006). Note that the kinetic energy of space bodies may be used through mechanical connections if the asteroid has a different speed or direction. However, electrostatic apparatus can use asteroids that have the same direction and speed for its acceleration if the apparatus has electrical energy (through charging and discharging the asteroid).
2. **Charging the ball.** This problem was considered in Chap. 13 of Bolonkin (2006), point 9.
3. **Blockading of the ball charge.** Blockading of the charge on the ball and the asteroid by unlike solar wind particles may be a problem with this method. The

charge on the ball attracts unlike particles and repels like particles and opposite-charge particles accumulate near the ball and block its charge. As the result the efficient area near the ball is less than electrostatic theory predicts. The area has a radius of about 7–25 km in Earth orbit. The electrostatic forces can be reduced. The method of computation for a neutral (efficient) area is offered by author elsewhere (Bolonkin 2010)(see also Chap. 13 of Bolonkin (2006)). This problem is not important for suggested method by following reasons:

1. The apparatus trajectory is usually located inside the neutral sphere.
2. The maneuvers are usually made far from the Sun where the solar wind intensity (density of space plasma) is very small.
3. The maneuvers are made quickly.

Possible drawbacks of the method may include the mass of the charge gun, electrical energy storage, and the mass of ball because the apparatus has big centrifugal acceleration.

References

Anonymous, Space technology Application. International Forum, Albuquerque, MN, parts, 1–3 (1996-1997)

Anonymous, Newsletter. Chim.& Rng (October 8, 2001)

Bolonkin, A.A.: Theory of Flight Apparatus with Control Radial Force. In: Ostoslavsky, I.V. (ed.) Collection Researches of Flight Dynamic, pp. 79–118. Mashinostroenie, Moscow (1965) (in Russian)

Bolonkin, A.A.: Hypersonic Gas-Rocket Launch System, AIAA-2002-3927. 38th AIAA/ASME/SAE/ASEE Joint Propulsion Conference and Exhibit, Indianapolis, IN, USA, July 7-10 (2002a)

Bolonkin, A.A.: Inexpensive Cable Space Launcher of High Capability, IAC-02-V.P.07. 53rd International Astronautical Congress, The World Space Congress, Houston, Texas, USA, October 10-19 (2002b)

Bolonkin, A.A.: Non-Rocket Missile Rope Launcher, IAC-02-IAA.S.P.14. 53rd International Astronautical Congress, The World Space Congress, Houston, Texas, USA, October 10-19 (2002c)

Bolonkin, A.A.: Hypersonic Launch System of Capability up 500 tons per day and Delivery Cost $1 per lb. IAC-02-S.P.15. 53rd International Astronautical Congress, The World Space Congress, Houston, Texas, USA, October 10-19 (2002d)

Bolonkin, A.A.: Employment Asteroids for Movement of Space Ship and Probes. IAC-02-S.6.04. 53rd International Astronautical Congress, The World Space Congress, Houston, Texas, USA, October 10-19 (2002e)

Bolonkin, A.A.: Non-Rocket Space Rope Launcher for People, IAC-02-V.P.06. 53rd International Astronautical Congress, The World Space Congress, Houston, Texas, USA, October 10-19 (2002f)

Bolonkin, A.A.: Optimal Inflatable Space Towers of High Height. COSPAR 02-A-02228. 34th Scientific Assembly of the Committee on Space Research (COSPAR), The World Space Congress, Houston, Texas, USA, October 10-19 (2002g)

Bolonkin, A.A.: Non-Rocket Earth-Moon Transport System, COSPAR-02 B0.3-F3.3-0032-02, 02-A-02226. 34th Scientific Assembly of the Committee on Space Research (COSPAR). The World Space Congress, Houston, Texas, USA, October 10-19 (2002h)

Bolonkin, A.A. Non-Rocket Earth-Mars Transport System, COSPAR 02-A-02224. 34th Scientific Assembly of the Committee on Space Research (COSPAR). The World Space Congress, Houston, Texas, USA, October 10-19 (2002)

Bolonkin, A.A.: Transport System for delivery Tourists at Altitude 140 km. IAC-02-IAA.1.3.03. 53rd International Astronautical Congress. The World Space Congress, Houston, Texas, USA, October 10-19 (2002j)

Bolonkin, A.A.: Non-Rocket Transport System for Space Travel. Journal of British Interplanetary Society 56, 231–249 (2003a)

Bolonkin, A.A.: Asteroids as Propulsion Systems of Space Ship. Journal of British Interplanetary Society 56, 98–107 (2003b)

Bolonkin, A.A.: Electrostatic Solar Wind Propulsion System, AIAA-2005-3653. 41st Propulsion Conference, Tucson, Arizona, USA (2005a)

Bolonkin, A.A.: Electrostatic Utilization of Asteroids, AIAA-2005-4032. 41st Propulsion Conference, Tucson, Arizona, USA (2005b)

Bolonkin, A.A.: Non-Rocket Space Launch and Flight, 488 p. Elsevier, London (2006), http://www.archive.org/details/Non-rocketSpaceLaunchAndFlight, http://www.scribd.com/doc/24056182

Bolonkin, A.A.: New Concepts, Ideas, and Innovations in Aerospace, Technology and Human Life. NOVA, New York (2007a), http://www.scribd.com/doc/24057071, http://www.archive.org/details/NewConceptsIfeasAndInnovationsInAerospaceTechnologyAndHumanSciences

Bolonkin, A.A.: AB Levitation and Energy Storage. Work presented as paper AIAA-2007-4613 to 38th AIAA Plasma Dynamics and Lasers Conference in Conjunction with the 16th International Conference on MHD Energy Conversion, Miami, USA, June 25-27 (2007b)

Bolonkin, A.A.: New Technologies and Revolutionary Projects, 324 p. Lambert, New York (2008), http://www.scribd.com/doc/32744477, http://www.archive.org/details/NewTechnologiesAndRevolutionaryProjects

Bolonkin, A.A.: Life. Science. Future (Biography notes, researches and innovations), 208 p. Publish America, New York (2010), http://www.scribd.com/doc/48229884, http://www.lulu.com, http://www.archive.org/details/Life.Science.Future.biographyNotesResearchesAndInnovations

Bolonkin, A.A.: Universe, Human Immortality and Future Human Evaluation. Elsevier, New York (2011), http://www.scribd.com/doc/75519828/, http://www.archive.org/details/UniverseHumanImmortalityAndFutureHumanEvaluation

Bolonkin, A.A., Cathcart, R.B.: Macro-Projects: Environments and Technologies, 536 p. NOVA, New York (2007), http://www.scribd.com/doc/24057930, http://www.archive.org/details/Macro-projectsEnvironmentsAndTechnologies

Cosmo, M.L., Lorenzini, E.C. (eds.): Tethers in Space Handbook, 3rd edn. Smithsonian Astronomic Observatory, New York (1997)

Directory, Carbon and High Performance Fibers. Springer, New York (1995)

Dresselhous, M.S.: Carbon Nanotubes. Springer, New York (2000)

Galasso, F.S.: Advanced Fibers and Composite. Gordon and Branch Scientific Publisher, New York (1989)

Kroschwitz, J.I. (ed.): Concise Encyclopedia of Polymer Science and Engineering, New York (1990)

Naschekin, V.V.: Technical thermodynamic and heat transmission. Public House High Universities, Moscow (1969) (in Russian)

Palmer, M.R.: A Revolution in Access to Space Through Spinoff of SDI Technology. IEEE Transactions on Magnetic 27, 11–20 (1991)

Smitherman Jr., D.V.: Space Elevators, NASA/CP-2000-210429 (2000)

Weekly News, Asteroids as Engine of Space Ships (Suggestion of American Scientist Alexander Bolonkin), Israel (April 28, 1998) (in Russian)

Ziegler, S.E., Cartmell, M.P.: Using Motorized Tethers for Payload Orbital Transfer. Journal of Space and Rockets 38, 904–913 (2001)

Chapter 27
Observation of Asteroids for Searching Extraterrestrial Artifacts

Csaba Kecskes

Comptech Ltd, Budapest, Hungary

27.1 Introduction

Since the beginning of modern astronomy all evidence indicates that mankind's place in the Universe is not central or exceptional. A major step was made into this direction recently when the availability of very high resolution spectrographs made it possible to detect giant and subgiant planets around Sun-like or smaller stars (Mayor 1995) and it was found that planetary systems are not rare. If mankind is not unique then one may ask obviously: where are the others? This problem is traditionally named as the Fermi paradox, although there are just some vague recollections that Enrico Fermi once discussed it with his colleagues (Finney 1985). There are hundreds of hypothetical answers to this question (for a not too long summary see Webb (2002) but none of them is supported by any evidence.

Probably the two most popular answers are the following:

- UFO (Unidentified Flying Objects) theories: extraterrestrials visit the Earth frequently, they have bases in the Solar system, etc.
- SETI (Search for Extraterrestrial Intelligence): interstellar travel never happened and never will happen, the extraterrestrials can be contacted only by radio (or visible light, neutrino etc.) messages (Drake 1992).

UFO theories are not popular among scientists; this might be understandable, considering the "anecdotal evidences" (sometimes with quite incredible details) usually presented by their supporters. But replacing the basic assumption of the UFO theorists (i.e. extraterrestrials visited the Earth in the past or they are visiting it frequently nowadays) with a weaker assumption (i.e. extraterrestrials visited the Solar system in the past or they might be present in the Solar system even now) one gets into an area where testable (verifiable/falsifiable by repeated observations) theories can be created. For example in the 1970s it was suggested that extraterrestrials may inhabit some asteroids (Papagiannis 1978) and their presence could be detected by the excess infrared radiation caused by their activities (Papagiannis 1985). The IRAS (Neugebauer 1984) was the first spaceborne instrument which was able to map the Solar system in infrared radiation but no asteroids were found with significant excess radiation.

The extreme pessimism of the SETI theorists is also not very convincing, especially if one takes into consideration that most Sun-like stars in the Galactic Habitable Zone (Balazs 1988) are older than the Sun by approximately one billion years (Lineweaver 2004); therefore, there should be many technical civilisations which had at their disposal many millions of years to develop a method of interstellar spaceflight.

A possible solution for the Fermi paradox is to assume that technical civilisations evolve in such a way that Earth-like planets (and, eventually, Sun-like stars) are becoming unimportant for them. As John Allen Ball wrote: "more likely an Earthlike planet is to ETI what an empty eggshell is to a bird" (Ball 1985). In Kecskes (1998) a step-by-step evolutionary model of the technical civilisations was suggested which is summarised in Table 27.1.

According to the above model, level 3 civilisations regularly visit planetary systems where they use asteroids or other small objects in order to refuel, repair or even rebuild their spaceships. Because they are adapted (technologically and biologically) to a low gravity environment, planets and large moons are not suitable for them. Because they use stellar energy when not travelling these small objects cannot be too far from a star. Therefore the main asteroid belt of the Solar system is a suitable place for them. The materials what they need are probably light metals (Al, Mg) for structural components and organics and water for replenishing the life support system. If they use a fusion based propulsion system then water also can be used for deuterium and tritium production.

There is an opinion that a closed-cycle biological life support system will always be too heavy to be used on an economically designed interstellar (or even interplanetary) spaceship; instead of using it, the astronauts should be turned into cyborgs (Clynes 1960). This could be possible with future advances of nanotechnology (Freitas 2002). But the model presented in Table 27.1 basically depends on the gradual development of the spaceships: interplanetary spaceships should be developed before the interstellar spaceships, and before the design of the advanced interstellar spaceships (which can extract the necessary materials from the interstellar medium), "basic" interstellar spaceships should be developed, which should be operated by using materials available around a star. Cyborg astronauts may not adapt biologically to the low gravity environment as it is stated at the level 2 in Table 27.1, but their technology will be adapted to low gravity environment because extracting materials from a small celestial object is cheaper than raising things from the "gravity well" of a planet.

As an "all in one" solution the optimal place for the repairing and refueling operations of an interstellar spaceship can be a C class asteroid with significant amounts of hydrated minerals and/or water ice (Rivkin 2002). Even if the extraterrestrials have advanced nanotechnological fabrication units and they are using diamondoid-type materials (Freitas 2011) a C class asteroid is a good choice because of it's carbon content. Of course, we cannot expect that there are extraterrestrials right now in the Solar system but the traces of their visitation (furnaces, slag heaps, damaged spaceship parts which were not worth to be repaired) can be visible even after millions of years because the erosion of the asteroid surfaces is quite slow (Clark 2002).

Table 27.1 Evolutionary levels of technical civilisations (abridged version of Table 2 of Kecskes (1998))

General description	Typical long-range transportation method	Material resources used	Energy resources used	Biological properties
level 1: civilisation on the surface of a planet	airplane, ship	ores from the crust of the planet, organics from the biosphere	naturally conserved stellar energy (coal, oil, uranium)	adapted to planetary conditions (high air pressure, strong gravity, natural food resources)
level 2: civilisation in the interplanetary space	interplanetary spaceship with small acceleration	asteroids, extinct comet nuclei	direct sunshine (solar cells, solar furnaces)	adapted to space habitats (low air pressure, near-zero gravity, closed biological life support system)
level 3: civilisation with interstellar travel capability	interstellar spaceship (with a limited range)	like level 2 and stocked materials during travels	like level 2 and artificially conserved forms (deuterium, He3, antimatter?) during travels	like level 2 and very long lifetime (or hibernation capability?), small-size closed life support system
level 4: civilisation in the interstellar space	interstellar spaceship (with unlimited range)	interstellar dust, controlled fusion products?	fusion of interstellar hydrogen, exotic resources?	like level 3, life support system integrated in their body?

Even if one thinks that the model presented in Table 27.1 is too speculative, the "Search for Extraterrestrial Artifacts" (SETA) can be a worthwhile proposal from the viewpoint of other theories, too. In the 1960s it was suggested that extraterrestrials may have sent spaceprobes into the Solar system in order to observe the Earth (Bracewell 1960). An astronomical search was conducted in the most probable places (e.g. in the Earth-Moon Lagrangian points, see Freitas (1983)) but no evidence of artificial objects was found (Valdes 1983). It must be noted here that the term "SETA" was coined by R.A. Freitas Jr.; this was first used in Freitas (1983).

According to the above evolutionary model it is improbable that advanced civilisations would send intentionally spaceprobes into the Solar system. On level 3 they are not interested in Earth-like planets. An asteroid belt around a nearby star can be detected even with instruments on our technological level; by using more advanced technology, even a mineralogical characterisation is possible with remote observations. This is far more faster (and cheaper) than sending an interstellar spaceprobe to another star and waiting for the results.

But there is a possibility of unintentional sending, too: a level 2 civilisation with large scale industrial activities in the asteroid belt around its star may produce enormous amounts of garbage and "throwing it out" into the interstellar space may be cheaper than reprocessing. If a level 2 civilisation becomes extinct or passes onto level 3 then some of the artificial objects left in their asteroid belt may be ejected into the interstellar space via orbit evolution. If there are many such sources, then the random arrival of an artificial object into the Solar system might be not very improbable. It was suggested that the remnants of such artificial objects should be searched on the Moon (Arkhipov 1998).

27.2 Search Method

The problem of finding artificial objects in the asteroid belt is twofold:

- how to find the proper asteroids
- how to identify the artificial objects

The second question can be discussed more easily because there is an ongoing space mission with one of its objectives defined as photographing artificial objects on the surface of the Moon. This is the Lunar Reconnaissance Orbiter (LRO) of the NASA (Chin 2007). The artificial objects are the remnants of the previous Soviet Moon landing missions (Luna, Lunokhod) and American missions (Surveyor, Apollo). It was a major success of the LRO mission when the exact location of the Lunokhod-1 rover was determined (Abdrakhimov 2011).

But the resolution of the LRO pictures is not very high compared to the sizes of the man-made objects on the Moon, i.e. approximately 50 cm/pixel on the best pictures. On such pictures the artificial objects appear only as 10...50 pixel blobs. This is enough if one knows what, and approximately where, should be searched, but certainly not enough for proving (or at least hinting) that a previously unseen object is artificial. For that purpose at least a 10 cm/pixel resolution would be necessary if one assumes that the typical sizes of the objects made by extraterrestrials are similar to the sizes of the man-made objects. This is a reasonable assumption if one takes into consideration that in similar environments (Earth-like planets) probably similar beings evolve, and the typical sizes of man-made objects follow from the sizes of humans. Table 27.2 shows the most important parameters of the LRO high resolution camera and some other high resolution cameras which were used on spacecrafts for photographing asteroids or comets.

If one asks "which camera should be used on an LRO-like spacecraft for achieving a 10 cm/pixel resolution", then there is only one adequate choice in Table 27.2, the HRI of the Deep Impact mission. This indicates that the searching of previously unseen artificial objects on the surface of celestial bodies is possible within the current limits (including financial limits) of space technology, but the manufacturing and testing of an adequate imaging system requires a significant investment. Of course, lander type missions can make even some mm/pixel resolution photographs with rather small cameras after landing, but sending a lander onto every "larger than a km" asteroid (or into every "larger than a meter" crater of the Moon) is certainly a financial impossibility.

Table 27.2 Cameras used on spacecrafts

Mission name	Camera name and type	Focal length (mm)	Aperture or f ratio	CCD size	Expected resolution
NASA LRO (Chin 2007)	Narrow Angle Camera (NAC), Cassegrain reflector	700	195 mm	5000 pixels linear	0.5 m/pixel at 50 km
NASA Galileo (Russell 1992)	Solid State Imager (SSI), Cassegrain reflector	1500	f/8.5	800x800 pixels	100 m/pixel at 10000 km
NASA NEAR (Cheng 2002)	Multispectral Imager (MSI), refractive optics	167	18.6 cm^2	537x244 pixels	10 m/pixel at 100 km
ESA Rosetta (Keller 2007)	Narrow Angle Camera (NAC), 3-mirror asymmetric (anastigmatic) telescope	717	90 mm	2048x2048 pixels	20 m/pixel at 1000 km
NASA Deep Space 1 (Rayman 2000)	MICAS/VISCCD imager, reflector with asymmetric mirrors	677	100 mm	1024×1024 pixels	50 m/pixel at 4000 km
JAXA Hayabusa (Ishiguro 2009)	AMICA, refractor telescope	120	15 mm	1000x1024 pixels	1 m/pixel at 10 km
NASA Stardust (Brownlee 2003)	Imaging and Navigation Camera, Petzval-type refractor	200	f/3.5	1024x1024 pixels	12 m/pixel at 150 km
NASA Deep Impact (A'Hearn 2005)	High Resolution Imager (HRI), Cassegrain reflector	10500	300 mm (f/35)	1024x1024 pixels	1.4 m/pixel at 700 km (actual resolution was worse because of focus misalignment)
NASA DAWN (Russell 2007)	Framing camera, refractive optics	150	f/7.9	1024x1024 pixels	20 m/pixel at 200 km

The first question (how to find asteroids which have artificial objects on their surface) is much more difficult. An asteroid used for repairing and refueling an interstellar spaceship by extraterrestrials can be named as a "repair station asteroid". The probable properties of such a "repair station asteroid" are the following:

- It is a C type asteroid; therefore the semimajor axis of it's orbit is probably between 2.3 and 3.5 AU (Nelson 1993);
- The eccentricity of it's orbit is probably less than 0.06. Level 3 civilisations use solar (stellar) energy when they stay around a star and any technological process can be operated more easily if the energy input does not vary too much;
- Its diameter is probably greater than 1 km. Even a much smaller asteroid probably contains enough materials for the repairing and refueling operations but strip mining is easier than shaft mining and on a larger surface the solar plants, furnaces, etc, can be placed more easily.
- The inclination of it's orbit is probably greater than 10 degrees. If an interstellar spaceship arrives from an arbitrary direction to the Solar system, then its first orbit around the Sun will be probably a high inclination orbit. If they have only little fuel left (at the end of an interstellar travel it is certainly not unusual) then they will choose an asteroid which can be reached with a minimum amount of delta-v from their initial orbit that is an asteroid with a high inclination orbit.

The numbers in the above criteria are only "rough estimates" but enough for some calculations. A recent yearbook of the minor planets (Shor 2010) contains data for 231,665 numbered asteroids. The Internet site of the IAU Minor Planet Center (www.minorplanetcenter.net) displays that there are 599,955 known main belt asteroids in May 2012. Using the interactive database search facility of this website, the author made the following search based on the above listed criteria:

- absolute magnitude < 17 (a rough approximation of the "diameter > 1 km" criterion)
- 2.3 AU < semimajor axis < 3.5 AU
- eccentricity < 0.06
- inclination > 10 degrees

The result: 9039 objects match search criteria (in May 2012), 4885 of them are numbered asteroids. This is certainly a too big number for a one-by-one examination. Spectral characterisation cannot help too much because of the following causes:

- In the region 2.3 AU < semimajor axis < 3.5 AU most of the asteroids are in the C class and most of them contain water (Rivkin 2002); excluding the "not C type" and "dry C type" ones may halve the above numbers but that is still too much;

- Asteroids are faint objects, observing asteroid spectras requires the expensive time of a large telescope. The number of asteroids with a measured spectrum was approximately 3000 in the year 2001 (Bus 2002) and this number is growing far more slowly than the number of asteroids with known orbits.
- If an asteroid is used as a "repair station" by extraterrestrials, then probably only a smaller part of it's surface is used for mining and slag depositing; in this case the asteroid spectrum will not change significantly. Space weathering tends to suppress even this minor changes.

But one must take into consideration that the search for artificial objects does not mean that a single specific asteroid must be found. The concept of "mankind is not unique" works for the extraterrestrials, too. If there were some extraterrestrials who visited the Solar system, then there were others, too. At this point it is obvious to ask: how many visitations were there? This question leads to the usual great questions about the extraterrestrials: "how many Earth-like planets are in the Milky Way", "what is the probability of the evolution of a technical civilisation on an Earth-like planet" etc. These questions sometimes are discussed in the context of the Drake Equation (Drake 1992) but the suggested numbers are only guesses because of the lack of observational data.

If one accepts the evolutionary model of technical civilisations described in Table 27.1, then other similar questions may arise: "what is the probability (and how much time is necessary) for a level 1 civilisation to reach level 2 and then level 3", "how many interstellar travels are made by level 3 civilisations before reaching level 4" etc. There are many possible hypothetical answers to these questions, but because of the lack of observational data it does not worth to discuss them. One thing is certain: if we don't try to search something then we will find nothing.

Considering the large number of potential targets the obvious method for a SETA mission in the asteroid belt is a multiple asteroid flybys mission. Using this method a single spacecraft can visit many asteroids. It can be estimated that the average distance between the known asteroids in the main belt is around 10 million kilometers. A spaceprobe in theory could approach one in almost every week while orbiting in the main belt. This is not possible in reality because of the delta-v limitation, but with a careful trajectory design a 100...200 m/sec delta-v can be enough between two flybys (this estimate is based on the trajectory calculations of the once planned Vesta mission, see Harvey (2007). In this case a spacecraft capable of a 1...2 km/sec total speed change while flying in the main belt can visit 5...20 asteroids. Another advantage of the flyby method is that with appropriate timing an asteroid with a high inclination orbit can be approached when it flies near the ecliptic plane, therefore the spacecraft need not use lots of fuel for changing it's orbit inclination.

The concept of "multiple asteroid flybys" is not a new idea; in the 1980s the leaders of the Soviet space program planned such a mission named Vesta (Harvey 2007). The Vesta mission would have been an international project, the CNES (France) providing the spaceprobes (two was planned) and the Interkosmos (Soviet Union) providing the launches and penetrators to be released at the most important targets. In order to reduce the relative speed during the flybys Mars swingbys were planned for the spaceprobes. Trajectories were planned for launches in 1994 and 1996 (Veverka 1989), the 1994 trajectories are summarized in Table 27.3.

Table 27.3 Planned trajectories for the Vesta mission

Trajectory 1	Trajectory 2
launch from Earth	launch from Earth
Mars gravity assist	Mars gravity assist
flyby: 2335 James (speed: 15km/s)	flyby: 1204 Renzia (speed: 4.3 km/s)
Mars gravity assist	Mars gravity assist
flyby: 109 Felicitas (speed: 6.3 km/s)	flyby: 435 Ella (speed: 3.5 km/s)
flyby: 739 Mandeville (speed: 7 km/s)	flyby: P/Tempel 1 (short period comet, speed: 7.1 km/s)
flyby: 4 Vesta (penetrator is released, speed: 3.3 km/s)	flyby: 46 Hestia (penetrator is released, speed: 3.6 km/s)

Because of the fall of the Soviet Union the above plans were not realised, but some asteroid flybys were performed as subordinate missions in space missions where the main targets usually were not asteroids. These flybys are summarized in Table 27.4.

Table 27.4 Asteroid flybys

Asteroid name, mission name, year	Relative velocity	Closest approach	Best photographic resolution achieved
951 Gaspra, NASA Galileo, 1991	8 km/s	1600 km	54 m/pixel
243 Ida (+ Dactyl), NASA Galileo, 1993	12 km/s	2390 km	25 m/pixel
253 Mathilde, NASA NEAR, 1997	10 km/s	1210 km	160 m/pixel
9969 Braille, NASA Deep Space 1, 1999	15 km/s	28 km	200 m/pixel (instrument problem)
5535 Annefrank, NASA Stardust, 2002	7.4 km/s	3100 km	185 m/pixel
2867 Steins, ESA Rosetta, 2008	8.6 km/s	800 km	80 m/pixel (instrument problem with NAC, best images from WAC)
21 Lutetia, ESA Rosetta, 2010	15 km/s	3160 km	60 m/pixel

The following lessons can be drawn from Table 27.4 and from the descriptions of the missions:

- With careful trajectory design a very close (< 30 km) flyby distance is possible, an onboard (autonomous) navigation system can help a lot.
- The geometry of the flyby must be selected properly, a high phase angle approach (worst case: flying over the dark side of the asteroid) leds to imaging problems.
- The imaging system (or other vital subsystems of the spacecraft) may fail temporarily (or get into "safe mode") during a flyby because the operations in this mode differ significantly from the cruising mode operations. The first flyby target of the spacecraft should be a not very important target and the problems occuring during the first flyby must be carefully analysed and corrected.
- A second, wide angle camera with separate mechanical, optical and electronic components is necessary. If the first (narrow angle, high resolution) camera fails (temporarily or permanently) then the second camera still can provide useful pictures from the viewpoint of "astrogeology" (for geological research purposes images with 5...20 m/pixel resolution of an asteroid with a diameter > 1 km are quite satisfactory). A wide angle camera collects much more light than a narrow angle camera therefore a spectrometer can be added to it easily. The wide angle camera also can replace the star tracker / navigation camera if that fails.
- The cameras should be mounted on a mobile platform. With such a platform the movement of the asteroid can be tracked more precisely and the other vital parts of the spacecraft (high gain antenna, solar panels) can be kept in an optimal position during the whole flyby.
- An onboard tracking system (analysing the images of the cameras during flybys and correcting the orientation of the camera platform if necessary) may be very useful.

As it can be seen from Table 27.4 the asteroid flybys hitherto were not used for obtaining high resolution images. This was caused by various reasons (relatively small cameras, large flyby distances, instrument failures, unfavorable approach geometries) but apart from these there is a difficult problem here. Lets imagine an asteroid flyby with a 50 km minimum distance and with a 10 km/s relative speed. In this case the time available for making high resolution images is approximately 10 seconds. If the narrow angle camera can make an image in every 0.05 seconds (an optimistic figure for an f/35 camera and the speed of the electrons is ignored here) then there will be 200 images, each covering a 10000 m^2 rectangle if the CCD size is 1024x1024 pixels and the resolution is 10 cm/pixel. The total area covered by these images is 2 km^2 or somewhat smaller if we assume more realistically that there will be overlaps and gaps between the pictures because of the imprecision in the camera positioning.

In the case of a small asteroid with 1...2 km diameter the visible area of the asteroid is 1...5 km^2 during the flyby, therefore the difference between the areas "visible during the flyby" and "photographed with high resolution" is not very big. But in the case of a bigger asteroid (for example if the diameter of the asteroid is 10 km and the flyby geometry is close to the optimal, then the visible surface will be over 100 km^2) the difference is enormous. This is a very big discrepancy and within the current limits of technology (including the expected financial limits) this problem cannot be solved but mitigated. For mitigation, the followings can be suggested:

- Try to reduce the flyby speeds, the planned trajectories for the Vesta mission are good examples.
- Try to use bigger CCDs (like the 2048x2048 pixels CCD of the Rosetta spacecraft's narrow angle camera) if the proper sensitivity can be provided.
- If the CCD read time is a problem then multiple CCDs can be used with a rotating mirror, but this would increase costs.
- Theoretically it is possible to characterize the surface of the asteroid using lower resolution images and to use this information for pointing the narrow angle camera to the "interesting" parts of the surface during the flyby. This requires a significant onboard image analyzing capability and an onboard tracking system.

27.3 Conclusion

For the purpose of SETA (trying to find artificial objects on the surface of main belt asteroids) many "multiple asteroid flyby" type missions would be necessary. Such missions would be also useful for "astrogeological research" purposes (studying the geology and the mineralogy of the asteroids). The propulsion system of the spacecraft used in such a mission must provide at least 1...2 km/s total delta-v while flying in the main asteroid belt, with this amount of delta-v the spacecraft can fly by 5...20 asteroids. The science payload of the spacecraft must contain a narrow angle camera (capable of making images with 10 cm/pixel resolution at 50 km distance), a wide angle camera and a spectrometer (this can be integrated with the wide angle camera). The cameras should be mounted on a mobile platform. The computers of the spacecraft must provide a significant data processing capability (onboard navigation, tracking and possibly image analysis).

References

Abdrakhimov, A.M., Basilevsky, A.T., Head, J.W., Robinson, M.S.: Luna 17/Lunokhod 1 and Luna 21/Lunokhod 2 Landing Sites as Seen by the Lunokhod and LRO Cameras. In: The 42nd Lunar and Planetary Science Conference, The Woodlands, Texas, March 7-11 (2011)

Arkhipov, A.V.: Earth-Moon System as a Collector of Alien Artefacts. Journal of the British Interplanetary Society 51, 181–184 (1998)

A'Hearn, M.F., et al.: Deep Impact: A Large-Scale Active Experiment on a Cometary Nucleus. Space Science Reviews 117, 1–21 (2005)

Balazs, B.S.: The Galactic belt of intelligent life. In: Marx, G. (ed.) Bioastronomy - The Next Steps, pp. 61–66. Kluwer Academic Publishers, Dordrecht (1988)

Ball, J.A.: Extraterrestrial intelligence - Where is everybody? In: Papagiannis, M.D. (ed.) The Search for Extraterrestrial Life: Recent Developments, pp. 483–486. D. Reidel Publishing Co., Dordrecht (1985)

Bracewell, R.N.: Communication from superior galactic communities. Nature 186, 670–671 (1960)

Brownlee, D.E., et al.: Stardust: Comet and interstellar dust sample return mission. Journal of Geophysical Research 108(E10), 1–15 (2003)

Bus, S.J., Vilas, F., Barucci, M.A.: Visible-Wavelength Spectroscopy of Asteroids. In: Bottke, W.F., Cellino, A., Paolicchi, P., Binzel, R.P. (eds.) Asteroids III, pp. 169–182. University of Arizona Press (2002)

Cheng, A.F.: Near Earth Asteroid Rendezvous: Mission Summary. In: Bottke, W.F., Cellino, A., Paolicchi, P., Binzel, R.P. (eds.) Asteroids III, pp. 351–366. University of Arizona Press (2002)

Chin, G., et al.: Lunar Reconnaissance Orbiter Overview: The Instrument Suite and Mission. Space Science Reviews 129, 391–419 (2007)

Clark, B.E., Hapke, B., Pieters, C., Britt, D.: Asteroid Space Weathering and Regolith Evolution. In: Bottke, W.F., Cellino, A., Paolicchi, P., Binzel, R.P. (eds.) Asteroids III, pp. 585–599. University of Arizona Press (2002)

Clynes, M.E., Kline, N.S.: Cyborgs and space. Astronautics, 26–27 (September 1960)

Drake, F., Sobel, D.: Is Anyone Out There? Delacorte, New York (1992)

Finney, B.R., Jones, E.M.: Interstellar Migration and the Human Experience. University of California Press, Berkeley (1985)

Freitas, R.A.: The search for extraterrestrial artifacts (SETA). Journal of the British Interplanetary Society 36, 501–506 (1983)

Freitas, R.A.: The future of nanofabrication and molecular scale devices in nanomedicine. Studies in Health Technology and Informatics 80, 45–59 (2002)

Freitas, R.A.: Diamondoid Mechanosynthesis for Tip-Based Nanofabrication. In: Tseng, A. (ed.) Tip-Based Nanofabrication: Fundamentals and Applications, pp. 387–400. Springer, Heidelberg (2011)

Harvey, B.: Russian planetary exploration: history, development, legacy, prospects. Springer Praxis Books, Heidelberg (2007)

Ishiguro, M., et al.: The Hayabusa Spacecraft Asteroid Multi-band Imaging Camera (AMICA). Icarus 207, 714–731 (2009)

Kecskes, C.: The Possibility of Finding Traces of Extraterrestrial Intelligence on Asteroids. Journal of the British Interplanetary Society 51, 175–180 (1998)

Keller, H.U., et al.: OSIRIS – The Scientific Camera System Onboard Rosetta. Space Science Reviews 128(1-4), 433–506 (2007)

Lineweaver, C.H., Fenner, Y., Gibson, B.K.: The Galactic Habitable Zone and the Age Distribution of Complex Life in the Milky Way. Science 303, 59–62 (2004)

Mayor, M., Queloz, D.: A Jupiter-mass companion to a solar-type star. Nature 378, 355–359 (1995)

Nelson, M.L., Britt, D.T., Lebofsky, L.A.: Review of Asteroid Compositions. In: Lewis, J.S., Matthews, M.S., Guerrieri, M.L. (eds.) Resources of Near-Earth Space, pp. 493–522. University of Arizona Press (1993)

Neugebauer, G., et al.: The Infrared Astronomical Satellite (IRAS) Mission. Astrophysical Journal, Part 2 - Letters to the Editor 278, L1–L6 (1984)

Papagiannis, M.D.: Are we all alone or could they be in the asteroid belt? Quarterly Journal of the Royal Astronomical Society 19, 236–251 (1978)

Papagiannis, M.D.: An Infrared Search in our Solar System as part of a more flexible Search Strategy. In: Papagiannis, M.D. (ed.) The Search for Extraterrestrial Life: Recent Developments, pp. 505–511. D. Reidel Publishing Co., Dordrecht (1985)

Rayman, M.D., et al.: Results from the Deep Space 1 Technology Validation Mission. Acta Astronautica 47, 475–488 (2000)

Rivkin, A.S., Howell, E.S., Vilas, F., Lebofsky, L.A.: Hydrated Minerals on Asteroids: The Astronomical Record. In: Bottke, W.F., Cellino, A., Paolicchi, P., Binzel, R.P. (eds.) Asteroids III, pp. 235–253. University of Arizona Press (2002)

Russell, C.T. (ed.): The Galileo Mission. Kluwer Academic Press, Boston (1992)

Russell, C.T., et al.: Dawn Mission to Vesta and Ceres. Earth Moon and Planets 101, 65–91 (2007)

Shor, V.A. (ed.): Ephemerides of Minor Planets. Institute of Applied Astronomy, St. Petersburg (2010)

Valdes, F., Freitas, R.A.: A search for objects near the Earth-Moon Lagrangian points. Icarus 53, 453–457 (1983)

Veverka, J., Langevin, Y., Farquhar, R., Fulchignoni, M.: Spaceraft Exploration of Asteroids, the 1988 Perspective. In: Binzel, R.P., Gehrels, T., Matthews, M.S. (eds.) Asteroids II, pp. 970–995. University of Arizona Press (1989)

Webb, S.: If the Universe is Teeming with Aliens ... Where is Everybody?: Fifty Solutions to the Fermi Paradox and the Problem of Extraterrestrial Life. Springer, Heidelberg (2002)

Weissman, P.R., Bottke, W.F., Levison, H.F.: Evolution of Comets into Asteroids. In: Bottke, W.F., Cellino, A., Paolicchi, P., Binzel, R.P. (eds.) Asteroids III, pp. 669–686. University of Arizona Press (2002)

Chapter 28
The Case for Asteroids: Commercial Concerns and Considerations

Mike H. Ryan and Ida Kutschera

Bellarmine University, Louisville, KY, USA

28.1 Introduction

The idea that one could mine the asteroids has long been a topic for science fiction writers including such notable authors as Robert A. Heinlein and Ben Bova. Science fiction has been the preeminent place where the access to extra-terrestrial resources has been envisioned for decades. The lure of space-based resources has been a consistent focal point as a metaphor for moving toward a new frontier or simply a reflection of humanity's desire to expand beyond the earth. However, it has been the nonfiction proponents of space resource utilization that generate the most interest in some respects (Lewis and Lewis 1987; Lewis 1996).

Intelligent life, once liberated by the resources of space, is the greatest resource in the solar system. The material and energy resources of the solar system allow humankind an infinite future: we can not only break the surly bonds of Earth but break free of the Sun and escape its fate (Lewis 1996, p. 256).

Discussions of asteroid mining and descriptions of the nonfiction possibilities of extra-terrestrial resources became increasingly popular in the last twenty years as humans have tentatively entered Near-Earth space (Belfiore 2012). However, these visions of the future have not yet achieved the substantiality that their proponents would prefer.

28.2 Using Space-Based Resources

The importance of the philosophy that underlies efforts to mine the asteroids cannot be over emphasized. Asteroid mining represents the triumph in the minds of many of the "pie makers", those who provide a means for new resources, over the "pie slicers", those seeking to control and allocate current resources. The underlying behavior that relates to resources is often about taking that which is available and parsing it out in ever smaller pieces. Scarcity provides a means to justify increased prices or restricted availability. Entire societies have built up rules for the allocation of scarce resources including everything from water to minerals. Given the vast amount of resources accessible on earth both now and as a byproduct of new and improved technologies, scarcity is more a matter of a willingness to invest in increasingly costly processes rather than the availability of actual resources

(Miller 2012). Once we get beyond the Earth, the availability of a great number of resources is practically limitless; iron ore and electricity from solar power being but two examples. As a consequence, taking advantage of space-based solar power systems, mining asteroids or alternatively looking to other planets for resources may mean the end of a significant amount of resource allocation based on scarcity. Resources are virtually unlimited in space for there is always another asteroid waiting to be mined. "Of the 585,081 known asteroids between the sun and Jupiter, 562,224 are in the main belt between Mars and Jupiter" (Geggel and Peek 2012, p. 60). Therefore, expansive sources for raw materials exist if it is practical to get to and from their locations in space. Once humanity moves beyond the parameters of the Earth's orbit, we could enter a different phase for resource utilization for the entire planet. At that point, many resources should become almost unlimited and inexhaustible in terms of time scales that can be envisioned.

28.3 Asteroid Mining: Wants, Needs and Capability

Part of the attraction of asteroid mining as stated previously is that it puts forth an economic and business argument as part of the rationale to do more things in space. Raw materials can be obtained and a rational business model developed supporting that activity. The business questions then are "do we need to do this" and if so, "can it be done at a cost competitive with other alternatives." The first question from a business perspective is whether or not we require the materials that might be obtained from extra-terrestrial sources. Given the materials readily available from comparatively closer and less expensive locations the answer is probably, *not yet*. But the potential for using space-based resources still represents a critical topic for conversation and exploration. Many significant and interesting business opportunities get deferred until a need either develops or circumstances catch up to the available opportunity. Scarcity of metals and minerals is one rationale for proposing space mining operations. Unless some material that is not abundant on earth becomes crucial to commerce or industry, resource analysis suggests that off-planet sources may not be needed for a considerable period of time (Miller 2012). Scarcity, however, can be as much about strategic access as availability. For example, several nations have had concerns about access to rare earth metals. The major sources of supply are controlled primarily by China. The political implications of limited access to materials needed for high technology products have encouraged other nations to embark on searches for other sources even if those sources entailed higher costs. Similar scenarios might be imagined that would make strategic use of asteroid resources. The higher costs would be justified by the strategic advantage of controlling a source of critical supply. Investment makes sense in either case only when the overall cost of the operation can be justified for a business or strategic need.

The second part of an asteroid mining discussion is about "can we do this" with current or foreseeable technology. The difficulties of getting people and materials to and from low-Earth orbit (LEO) alone make it somewhat intimidating at present to talk about engaging in projects well beyond LEO. The technical problems involved with asteroid mining appear to be daunting when combined with the problem that potential resource targets are extremely distant from Earth and that humans have not ventured back to the moon in more than 40 years. Of course, this assumes that humans on location are necessary as an integral part of any future mining scenario. Many approaches to gathering space-based materials may not require a direct human presence. For example, human operators in space may not be required with improvements in robotics and automation. If so, many of the issues related to having people on long duration space missions including asteroid mining ventures could be reduced or eliminated. This would not fulfill the desires of those seeking to expand human presence beyond earth. However, reducing the extent of human exposure might improve the prospects for initial mission successes. It could also improve business possibilities by shrinking the venture to a more manageable and marketable size.

Managing an asteroid mining venture, even using remote technology, poses significant limitations and would likely require an extensive infrastructure to support sustained activity. A hybrid solution that combines human and automated activity solves some problems but creates others. Most notably are those issues related to keeping people safe, secure and happy, particularly if human operators and support personnel are moved closer to the mining and extremely distant from their home - earth. These issues are comparatively straightforward but addressing them will increase the costs associated with any asteroid mining project by several orders of magnitude (Ryan and Kutschera 2012). Having humans in space is never easy. Human presence in space is costly but would provide a level of flexibility that technology alone might not provide. Tradeoffs exist with any venture. Initial decisions when contemplating something as expensive and difficult as asteroid mining will impact the prospect for success and revenue or profit at the other end of the venture.

28.4 Reality versus Possibility

Current space capabilities pragmatically position asteroid mining ventures in between fiction and reality. Some suggest that the time frame is only a decade away while others imply that the time frame is closer to twenty plus years. Most of the differences in time perspectives are functions of how soon activities to develop needed infrastructure are undertaken and how many investment or research dollars get applied to the problems. The sooner fundamental questions of technology are addressed, the sooner a funded mission may be expected. But the relative speed of advancement is also a function of the financial investments made in expanding known mining and space technologies, understanding the keys to asteroid composition and developing new systems needed for infrastructure, propulsion, software, and other key technologies. Smaller investments lengthen the time needed to address critical technologies and push asteroid activities further into the future. Clearly, the possibility of asteroid mining exists, and the simple truth is that its probability in the near term is lowered by the number of nontrivial problems to overcome.

Still, serious minded investors are beginning to talk about the potential that asteroids have both for resource utilization and the generation of new wealth as Planetary Resources, a startup space venture, suggests is possible and may be feasible (Greenwald 2012). Many practical problems facing asteroid mining have clear and unambiguous solutions. The majority of technical problems for such ventures with few exceptions revolve around cost and not practicality. The question is not whether asteroids can be mined, but whether or not the financial return warrants the extensive investment necessary for such mining operations, making the overall concept of mining asteroids problematic.

28.5 The Practical Problems

Just getting to and from the mining locations are significant issues. The convergence of automated technology, new mechanisms for space propulsion, and increasing recognition that artificial intelligence in the form of self-automated machines is possible increases the probability that an organization could successfully mine one or more asteroids. One might well imagine two competing scenarios for asteroid mining ventures. The first involves extensive robotic and semi-autonomous vehicles. The second involves the use of humans or human operators either on-site or via remote connections. Different people may prefer one approach versus the other. Considering the trade-offs between remote mining and human presence is important to the discussion. The absence of human beings on location suggests that some sort of artificial intelligence would be critical to mining success. However, current technology does not provide a level of artificial intelligence as flexible or as resourceful as a human being. Therefore, in the absence of significant improvements in artificial intelligence, it is likely that people will be involved at some point during the mining process. Providing for those people will consequently become a critical management concern. Previous research has demonstrated the factors that need to be considered when people are involved in space-based business ventures (Kutschera and Ryan 2010; McPhee and Charles 2009). Long-duration missions space missions for exploration or mining will be no different in that regard.

It can be argued that successful operations can be sustained on long-duration space voyages with appropriate management oversight if adequate considerations are made for the realities of operating in space. The success of the International Space Station clearly demonstrates that the scenario of people working in space is possible but will impart some unique challenges if extended to business operations.

28.6 Problems of Time and Distance

One common description of space is that it is big. That this phrase is an obvious understatement in the extreme is apparent to anyone who has ever considered what it takes to travel from point to point in space. The time it would take to travel from Earth to Mars and its environs is measured in months with the best propulsion systems currently imagined. Economical transfers from Earth orbit to the vicinity of the asteroid belt quickly become multi-year undertakings with present technology. Asteroids do frequently pass close enough to Earth to make their examination, geological exploration, and possible exploitation by mining a less time-consuming task. Such opportunities could represent an ideal test environment for asteroid mining technology with a measure of reduced risk. While potentially very helpful at exploring mining processes, such approaches would not be the norm. It is far more likely that industrial groups would seek to maximize their returns by setting up asteroid ventures in a manner that would allow for ongoing mining and material transport. It might be expected that a significant amount of these ventures could be accomplished using remote mining equipment. The lag time between communications both to and from the mining site could create serious operational issues. For example, going only as far out as Mars can create lag times in communications ranging between seven minutes and over twenty minutes. Traveling beyond Mars increases the lag times and communications may require multiple relay stations because of the relative positions of the earth and the mission vehicles in the asteroid belt. A number of possible scenarios to improve communication and reduce the inherent difficulties of lagged instructions among remote devices exist, and an obvious answer would be to put trained operators closer to the point of extraction. The difficulties that both time and distance impart to the venture significantly change the nature of the project and add an entire host of management problems the moment that human beings enter the operational equation.

Even as communication technology improves, the need for more capability will expand as the range of deep space operations including asteroid mining grows. An ongoing supply stream to a remote point in space and a return stream of mined materials back to the vicinity of earth could be established. The movement of people to and from remote mining locations may be more complex. The logistic and operational parameters required to maintain humans in space for prolonged periods of time are greatly complicated by the distance from Earth. This problem has historical analogs. Whaling expeditions of the 19th century frequently spent two to four years harvesting whales. The whaling ships made infrequent trips to port while at sea as such diversions interfered with their ability to turn a profit. Whaling ships were largely self-sufficient and able to operate independently during their long voyages. Their crews were generally capable of dealing with the boredom and routine inherent in such trips (Villiers 1973). It is therefore conceivable that multi-year asteroid mining ventures could be designed so as to enhance their likelihood of success. The issues of 19th century whalers regarding weather, sea conditions, provisions, and fresh water, however, pale in the face of attempting to live and work in space.

28.7 A Question of Command-and-Control

Space missions have largely been controlled from the ground from the early days of space activity. A mission control would be established and staffed by well-trained support personnel to provide ongoing monitoring and troubleshooting for each space mission. NASA's Mission Control is correctly complimented on the role it played in orchestrating successful missions. For example, the Apollo 13 mission is a clear and relatively unambiguous demonstration of how a well-run supporting infrastructure can make a difference between success and failure or even life and death. Apollo 13 had about two hours to make a decision for a quick return to Earth after the liquid oxygen tank explosion. Problems such as carbon dioxide removal from the lunar excursion module, water utilization, and power usage for reentry were evaluated, solutions developed and implemented by the astronauts with the assistance of mission control. It is probable that the mission failure would have led to a loss of the crew without quick and direct action. Direct command and control support from Earth was crucial to the ability of the mission to overcome a catastrophic situation. This would not be as practical for long duration missions where the increased distance from earth-based mission control would produce lengthy delays in getting information to and from the mission vehicles (Ryan and Luthy 2003). The importance of operational self-sufficiency will increase the more distant the operation is from its communication and operational support. Earth-based advice could be quite critical to the success of distant missions and the lagged communications impart some serious questions in a time sensitive crisis.

Self-sufficiency suggests that many, if not most, critical command-and-control decisions would be resident within the mission itself. This also affects situations of operational crisis, medical emergency or any other circumstance where a decision is required and the outcome determined under the time required for complete communications from Earth, assuming that complete roundtrip communications are possible at that time. Asteroid mining in the context of operational self-sufficiency may therefore become more analogous to colonization. And people living further out in space would not have the built-in advantages of living off the land that historical settler examples provide. Issues of survival and well-being become acutely critical when missions need to carry oxygen, create water, and protect crew and equipment from a truly hostile environment without the benefit of the Earth's magnetic field or its atmosphere.

28.8 Issues of Self-sufficiency

The problems of self-sufficient enterprises are pretty straightforward and there are clear examples of colonies disappearing, such as the lost Roanoke colony, North Carolina, USA, and exploration vessels failing to return, such as the 1845 John Franklin northwest passage arctic expedition. Space missions literally have to be prepared for every eventuality without timely assistance from Earth. Most importantly, redundancy takes on an entirely new dimension when the likelihood of getting replacement parts, trained personnel, or other essential life-sustaining elements is so remote as to be virtually impractical. A primer on the possibilities of Moon colonization would describe the problems of food, water, solar radiation, isolation, communication, medical care, morale, and an ever widening range of similar issues (Benaroya 2010). A lunar settlement principle would be, if you do not take it with you or think about it before you go, it is likely to be too late to do anything once you arrive. And the distance from earth to lunar base sites in lunar base scenarios is comparatively short relative to ventures further into the solar system.

Medical considerations alone present formidable problems for asteroid mining. Knowing that you cannot ship people home for more complicated medical procedures requires determining how extensive the medical equipment and training should be on extended multi-year ventures into space. The questions include how many physicians, and with what skill sets, should be assigned to a mission, and how do you divide your medical resources to provide appropriate redundancy in the event of a mission-threatening crisis?

Genuine difficulties in providing medical care in space include, but are not limited to: (a) resource constraints resulting from the boundaries of the mission design and architecture (volume, mass, power) and dictating that only the most critical medical equipment can be stored on board the space vehicles and delivered to the space habitats; (b) lack of trained medical professionals among the crew members; (c) limited pre-flight crew training time, necessitating the restriction of the training to only medical knowledge, techniques, and procedures that address the medical situations that are most likely to occur or that are most critical; (d) the probability that the crew members on the vehicle or in the habitation module may have to respond to emergency medical conditions without real-time support from Earth; (e) limited shelf-life of medical therapeutics and supplies; and (f) the possibility of encountering unpredicted illnesses and ailments that may be unique to the space exploration environment (Risin 2009, p. 241).

The prospect of surgery within a zero gravity environment alone requires some very ingenious technology and skills. For example, NASA is experimenting with the Aqueous Immersion Surgical System (AISS) developed by a team of researchers from the University of Louisville and Carnegie Mellon University to develop technology to make "astro surgery" possible (University of Louisville 2012).

In the weightless atmosphere of deep space, the absence of gravity will make it nearly impossible to control the escape of blood and bodily fluids during surgery…This lack of control would both compromise the health of the patient as well as contaminate the spacecraft cabin (Pantalos 2012).

Worst-case scenarios do not even begin to approach the level of detail that will be required to ensure self-sufficiency for missions of multi-year duration (Ryan and Kutschera 2007). A superbly equipped medical bay is worthless to the people who might need it if some catastrophe such as a power surge renders it useless which some might suggest is a worst case situation. It does not take much foresight to envision circumstances that would make having multiple vessels on a single mission a reasonable backup procedure. This approach has worked quite well on numerous explorations and similarly risky undertakings including Columbus's initial voyages to the Western Hemisphere with the Nina, Pinta and Santa Maria. And even then, future space exploration vessels could be expected to have multiple backups including well-equipped machine shops suitable for major repair work or even capable of building new systems as part of redundant capabilities. One answer to the inevitable problem of not being able to carry every possible spare would be found in the adaption of various types of three dimensional (3-D) printing systems that are beginning to appear (Betancourt 2012). Making parts as needed would not only reduce the mass necessary for transporting but ensure that a spare would be available. Transporting sufficient feedstock for printers and CAD files for parts to be printed could support long-duration space activities without massive numbers of parts being carried for a "just in case" moment. According to one example, a "Mars bound ship would have roughly 20 metric tons of machined parts. To ensure one spare for every part, the crew would need an extra 20 tons. But they need only two tons of feedstock to print their own spares" (Betancourt 2012, p. 29). Even new parts for some unforeseen mission requirement could be designed and constructed on board the spacecraft. It is also conceivable that once at a mining location local ores could be adapted for use by 3-D printers to create all manner of items from tools to larger modularized habitats. That type of capability would be a critical component for the level of self-sufficiency needed for operations far from Earth-based assistance and supply (Roach 2010).

28.9 Property Rights

Selling a property, including an asteroid, requires ownership. Without property rights, those seeking to mine asteroids face an impossible situation. The current state of extraterrestrial law suggests that ownership by private entities operating in space is not possible. The language that common or shared heritage imparts to private ownership of space resources is equally problematic. Companies seeking to exploit extraterrestrial resources are likely to run into major legal hurdles until private rights in space are clarified. Private claims on space-based property have simply not been able to stand. The 1967 Outer Space Treaty is very clear on that: "Outer Space, including the moon and other celestial bodies, is not subject to national appropriation by claims of sovereignty, by means of use or occupation, or by any other means" (Szoka and Dunstan 2012). Some suggest that nations such as the United States might allow private entities to operate under a national charter for purposes of securing space resources. Such space privateers could well be regarded as icons of industry by the nation whose charter they carry and as pirates by those in other countries who do not recognize their right to obtain, transport and profit from resources obtained in space. Fortunately, Article II of the Outer Space Treaty provides that parties to the treaty will bear responsibility for their national activities and their nationals regardless of who carries out the activity. Nations who launch things into space are required to ensure that their citizens abide by the treaty's provisions (Szoka and Dunstan 2012). Private sector privateers therefore need not apply for permission to use space-based resources from their national governments as such authorizations are disallowed under the treaty.

However, some interesting exceptions for rights to objects and vehicles launched into, or constructed in space by private firms are recognized by the treaty. Orbital slots are pretty much recognized as de facto property in space. Satellites are regularly bought and sold, and orbital rights are subject to negotiation. And, materials extracted from the moon have established that portions of a celestial body can be subject to ownership if removed from a celestial body. This alone suggests that the possibility for mining materials from asteroids is likely to have a substantial foundation at some point in international law. What exact form commercial rights for asteroid mining may take depends on how the various nations are able to agree. One possibility might be similar to exclusive seabed mining rights where licenses are granted and recognized multilaterally for a limited area, for a limited period of time. Without such agreements it would be very unlikely that any effort to take advantage of space-based resources could secure sufficient capital to get started. Long before private companies travel into space in search of wealth, many more legal areas will have to be negotiated and written into agreements. The situation may spin out of control similar to how allocation of deep ocean resources threatens the peace of nations unless great care is taken in advance of proposed mining efforts.

The argument that those that have the ability to exploit the resources should be allowed to do so has little weight with nations who see those resources as being part of their concern also. This is evident from the actions of Russia and Canada at the North Pole and those of the various players with interests in the Pacific Ocean seafloor. Such situations could lead to confrontations between different groups who either refuse to or see no need to recognize the property rights of those currently working in a particular location. Activities to secure and exploit resources in space could easily end up mirroring the increasing conflicts and disagreements over resources here on earth. It would be beneficial to all concerned to avoid this situation lest a significant barrier be placed on any future commercial space mining operations.

28.10 Amelioration of Risk

Risk from the perspective of asteroid mining missions falls into two major categories: operational risk and organizational risk. The first category is the most obvious which involves the actual mission itself. A distinct tendency exists to focus on the problems and constraints an asteroid mission engenders in terms of operational risk. This is not surprising given the difficulties involved in virtually all space travel. Getting to and from space is inherently dangerous; the United States lost two space shuttles and crews on routine missions. And nothing involving space is ever truly routine, at least not yet. Russia has lost cosmonauts and had several major mission failures as well. Harnessing the power needed to put objects into orbit is dangerous and will likely remain so for the foreseeable future. That space related ventures are and will be dangerous is a simple reflection of the space environment.

Operational risk has generally taken precedence in as much as programs or activities of the scale of an asteroid mission have generally fallen to governments. Financial constraints, resource constraints, and organizational constraints of all sorts are important and do not represent impenetrable barriers to undertaking the activity. Governments simply do not have the same limitation faced by private individuals and organizations. The organizational risks assumed by both individuals and private firms seeking to engage in space-based activities are both significant and challenging. Offsetting such risk will require a reinvention of insurance and related financial instruments that will parallel those that rose from commercial needs for vessels at sea (Bernstein 1996). Not the least of those burdens is the extreme financial cost associated with all past, present and projected future operations. In simple terms, it takes a very great deal of money to do things in space. These costs are compounded by the lack of infrastructure and the scarcity of proven business models that suggest operational activities in space can be profitable.

Virtually all private sector, profit-driven, independently-owned commercial enterprises focused on space or space-based resources involve a large amount of guesswork in the absence of known, profitable, historical analogs. The essential elements required to develop a thoughtful financial and organizational plan are simply not present. The questions related to profitability, sustainability, economies of scale and scope are sufficiently difficult to answer as to make the enterprise more speculative than might be preferred without such elements, regardless of the detail provided in operational plans.

The difficulty of working in space, far from relief or rescue, is complicated by the nature of space itself. It is possible to compare prospective asteroid mining operations to similar large scale innovative commercial enterprises of the 19th century (Cadbury 2003). Risk traveled hand in hand with innovation and opportunity. Entrepreneurs traveled to the far corners of the earth in search of financial opportunities and sought to exploit tangible resources of the planet and pushed technology to its limits. In their effort to prepare for any eventuality, the lists of their preparations in the historical records of the time demonstrate the complexity of their undertakings. Hundreds of commercial vessels were lost in the 19th century in spite of extensive preparation, with over 150 vessels lost on transatlantic routes alone. Those losses did not deter or slow down the development of new avenues for commercial enterprise but rather shaped how future excursions were staged (e.g., avoiding the North Atlantic in winter). It is unlikely that similar obstacles will shake the confidence of those who view space as the next commercial frontier regardless of the risks.

28.11 The Business Model

It is impossible to talk about the use of asteroids as a resource without considering the business models that such utilization might require. A business model generally represents the manner and mechanisms that a business uses to make money. An enterprise is virtually guaranteed to be unsuccessful if its circumstances do not allow it to make money. As previously described, the practical impediments for using asteroids as resources suggest that the overall operational risks are quite high. Increased investment risks are generally associated with expectations of higher levels of return. The rationale for this is that if one assumes that investing a given level of money in a venture having comparatively little risk provides a specific financial return, then to assume greater risk requires a greater return, respectively. The inherent risks are extremely high regardless of the startup capital required to undertake an asteroid mining venture. These risks are exacerbated if the operation becomes functionally limited to mining only a relatively small number of asteroids. An asteroid mining venture must successfully locate, process, and return the resources from asteroids to earth. The concentration of materials expected to provide the greatest value is unlikely to be evenly distributed among the virtually limitless piles of rock floating throughout the solar system. Ensuring valuable materials are effectively captured and returned to earth implies a business model requiring multiple asteroid captures and processing, and suggests an ongoing activity that captures and processes possibly hundreds of asteroids. The structure of such an operation and the logistics involved might be straight forward, while the actual implementation of making such an operation possible is much more problematic.

An example may be if an initial asteroid capture mission sampled several asteroids but discovered only a few small pockets of valuable material; that would make it difficult for investors to justify putting capital into future operations. Equally daunting, only a comparatively small number of prospective investors would be interested if the timeframe for an asteroid mining venture became too long. Asteroid ventures would be viewed as quite speculative and likely unsuitable investments for the majority of individuals or firms. Asteroid ventures may become the speculative investment of the very rich or the very desperate depending on the level of human or robotic activity. Businesses related to asteroid-based activities would likely be viewed as yet another form of financial speculation until the actual process of locating, processing, and returning the processed materials to earth is proven. Therefore, those who view the eventual use of asteroids as a mechanism to improve life on planet Earth should develop sufficiently detailed business models as to bolster confidence and credibility in such ventures. Creating a "go for broke" mentality may undermine serious efforts needed to make asteroid mining ventures both attractive and successful over the long-term given the inherent costs and risks associated with all space ventures.

Mining for resources on earth is by its very nature risky and those seeking quick and easy fortunes are much more likely to lose their money than to see any financial rewards. Historical precedents for mining ventures are very easy to find. The people who often made the most money were those who either sold mining supplies, provisions or other necessities. Comparatively few individual miners made money, and great mining fortunes were made by only a few. It is likely that more money was made selling picks and shovels, wheelbarrows, food and sundries, and by providing means for turning mined metals into goods and services. Such analogies do lose some comparative value once the sheer size and complexity of asteroid mining is considered. And it is likely that substantive fortunes will be made on some other dimension of the space venture process, including non-mining activities, if history is an indication. Still, the importance of creating a sustainable business model that puts revenue generation on a long-term path cannot be over emphasized.

28.12 Conclusion

...resources from space change all the rules. They offer us boundless increasing-sum game, with wealth beyond our wildest Earth-bound dreams and opportunities for travel that boggle the mind (Lewis 1996, p. 235).

Space-based ventures have always held the promise of unlimited resources. It is no longer a question of whether those resources will be exploited but when they will be exploited. Estimates based on current technology suggest such ventures will begin to become practical in a minimum of ten to fifteen years. The range of possible scenarios for how such ventures might develop is far too extensive at this junction, and many essential elements requisite to their success are clear. First and foremost, it is highly probable that efficient ventures will include a significant amount of robotic possibly even autonomous equipment. It is equally probable that even with great improvements in technology and the possibility of autonomous robotic machines, humans would be dispatched fairly early in the venture process due to their inherent flexibility. The price of that flexibility will be complex environmental systems needed for the support and well-being of humans in space. The questions of ownership, transportation and ancillary activities will be addressed and answers implemented. Undoubtedly, new problems and issues will be discovered that were not even considered initially. Such is the nature of new opportunities and new enterprises. What should not surprise anyone is how poorly the predictions of the future will prove to be (Davidson 1983). That too, along with all the missed and unforeseen opportunities, is part of the pattern for new ventures. Asteroid mining, along with human expansion into the solar system, is not a question of "if" but "when."

In April (2012 - authors' insert), Planetary Resources, a newly formed private space company, announced that it would begin mining asteroids for water in 2020 (Geggel and Peek 2012, p. 61).

References

Belfiore, M.: How to Mine an Asteroid. Popular Mechanics, 50–55 (August 2012)
Benaroya, H.: Turning Dust to Gold: Building a Future on the Moon and Mars. Praxis Publishing Ltd., Chichester (2010)

Bernstein, P.: Against the Gods: The Remarkable Story of Risk. John Wiley & Sons, Inc., New York (1996)

Betancourt, M.: Printed in Space. Air & Space Smithsonian 27, 26–31 (2012)

Cadbury, D.: Dreams of Iron and Steel. Forth Estate. An Imprint of HarperCollins Publishers, New York (2003)

Davidson, F.P.: Macro: A Clear Vision of How Science and Technology Will Shape Our Future. William Morrow and Company, Inc., New York (1983)

Geggel, L., Peek, K.: Space Metal: What do scientists know about mining's final frontier? Popular Science 281, 60–61 (2012)

Greenwald, T.: The X Man. Wired Magazine, 88–96 (July 2012)

Kutschera, I., Ryan, M.H.: The Future Role of Human Resource Management in Non-Terrestrial Settlements: Some Preliminary Thoughts. In: Benaroya, H. (ed.) Lunar Settlements, pp. 87–99. CRC Press, Boca Raton (2010)

Lewis, J.S., Lewis, R.A.: Space Resources: Breaking the Bonds of Earth. Columbia University Press, New York (1987)

Lewis, R.A.: Mining the Sky: untold riches from the asteroids, comets and planets. Helix Books, Addison-Wesley Publishing Company, Inc., Reading, Massachusetts, USA (1996)

McPhee, J.C., Charles, J.B. (eds.): Human Health and Performance Risks of Space Exploration Missions. National Aeronautics and Space Administration (NASA), Lyndon B. Johnson Space Center, Houston, TX (2009)

Miller, J.M.: Under Earth: Rocks or a Hard Place? Wall Street Journal, B1 and B6 (June 5, 2012)

Pantalos, G.: As cited in University of Louisville (2012), `http://louisville.edu/medschool/news-archive/uofl-carnegie-mellon-researchers-to-test-zero-gravity-surgical-technology` (accessed October 4, 2012)

Risin, D.: Risk of Inability to Adequately Treat an Ill or Injured Crew Member. In: McPhee, J.C., Charles, J.B. (eds.) Human Health and Performance Risks of Space Exploration Missions, pp. 239–249. National Aeronautics and Space Administration (NASA), Lyndon B. Johnson Space Center, Houston, TX (2009)

Roach, M.: Packing for Mars. WW Norton and Company, New York (2010)

Ryan, M.H., Kutschera, I.: Lunar-based enterprise infrastructure—hidden keys for long-term business success. Space Policy 23, 44–52 (2007)

Ryan, M.H., Kutschera, I.: Fundamental of Modern Lunar Management: Private Sector Considerations. In: Badescu, V. (ed.) Moon Prospective Energy and Material Resources, pp. 703–724. Springer, Heidelberg (2012)

Ryan, M.H., Luthy, M.R.: Management Architecture: Problems Facing Lunar-Based Entrepreneurial Ventures. Journal of Space Mission Architecture 3, 20–38 (2003)

Szoka, B, Dunstan, J., `http://www.wired.com/wiredscience/2012/04/opinion-space-property-rights` (accessed September 22, 2012)

University of Louisville, `http://louisville.edu/medschool/news-archive/uofl-carnegie-mellon-researchers-to-test-zero-gravity-surgical-technology` (accessed October 4, 2012)

Villiers, A.: Men, Ships and the Sea. National Geographic Society, Washington, DC (1973)

Chapter 29
Legal Considerations on Asteroid Exploitation and Deflection

Virgiliu Pop

Romanian Space Agency, Bucharest, Romania

29.1 Exploitation: The Property Status of Asteroids

29.1.1 Introduction

The sky is, and has always been, a mine. Human civilization has been using meteoritic iron ever since the Neolithic period (Tylecote 1992, p.3). And as humankind evolved, it realized it need not wait for stones to fall from above, but it can go up there and mine the sky.

In April 2012, a group of adventurous, wealthy and famous entrepreneurs – among them being Google's Larry Page, Space Adventures' Eric C Anderson, filmmaker James Cameron and space visionary Peter Diamandis were announced as backers of the newly established company Planetary Resources, Inc – a venture which, according to Wired's Adam Mann (2012) "plans to send swarms of robots to space to scout asteroids for precious metals and set up mines to bring resources back to Earth, in the process adding trillions of dollars to the global GDP, helping ensure humanity's prosperity and paving the way for the human settlement of space".

Is the vision of these entrepreneurs anchored in reality? Is there really "gold in'em thar' asteroids"? In a mind-blowing sample of astro-demographic mathematics, planetary scientist John S. Lewis (1996, pp. 195-196) estimates that using asteroidal iron and steel would generate wealth amounting to $7 billion per person; adding in this equation the other ingredients composing the asteroidal belt - such as gold, silver, uranium, etc. - the total would rise to over $100 billion for each person on Earth.

But - should humans be billionaires in asteroidal metals -, how is this wealth going to be appropriated and shared? Who, after all, really own the asteroids? What is the place and significance of property rights in the context of space activities?

The depths of space are doubled, in Kenneth Silber's (1998) view, by an even more forbidding realm: the "uncharted legal territory, and unpredictable politics, of owning property out there". He remarks the "general silence of national and international law on extraterrestrial property rights", this field being in his view "murky at best, downright hostile at worst". Other commentators see the extraterrestrial realms as harboring "a legal vacuum almost as complete as the vacuum of space itself"; they deem the present space law as "rather uneven", its principles "often retreat[ing] into ambiguous generalities or leave gaping holes which must be filled before any serious commercial development can occur" As for domestic standards, they are considered "either absent or so cumbersome as to inhibit exploitation" (Roberts et al 1996). H. Nauges (1979, p.269), at his turn, voices a very genuine concern:-

> "The difficulty is not … that we are starting from a legal vacuum but that there are a number of abstract, imprecise, insufficient and sometimes contradictory legal rules which are likely to be subject to genuinely differing legal interpretations".

In an era where private enterprise is poised to play a major role in opening the high frontier, treaty law norms on this field are indeed scarce and, when they do exist, they are imprecise. They fail to define concepts that are fundamental to the law of landed property in outer space, such as 'celestial body'. This particular definitional issue raises the question of ownership of asteroids: are they celestial bodies from a legal point of view and, hence, outside the sphere of appropriation?

29.1.2 Are the Asteroids "Celestial Bodies" in the Legal Sense?

The 1967 Outer Space Treaty (OST) contains in article II a fundamental principle, outlawing the national appropriation by any means of outer space and celestial bodies. Strict as it may be in this prohibition, the Treaty fails however to define the precise object of its application. This silence has prompted two disputes in the specialized academic circles: the legal definition of outer space, and the legal definition of a celestial body.

The question of "how tall is the sky" has been brought to practice in 1976, on the occasion of the First Meeting of Equatorial Countries in Bogota, whereby eight Equatorial countries claimed their right to exercise national sovereignty over the segments of the geostationary synchronous orbit (GSO) located over their territories. Whereas flawed in its assertion that the GSO "must not be considered part of the outer space" due to "its existence depend[ing] exclusively on its relation to gravitational phenomena generated by the earth", the Declaration of the First Meeting of Equatorial Countries adopted on December 3, 1976 is correct in pointing out that "there is no valid or satisfactory definition of outer space".

The fact that there is, also, no valid or satisfactory definition of what is a "celestial body" in the context of space law has not yet been challenged in practice – yet it raises the theoretical question whether asteroids and comets are immovable land-like territorial extensions that cannot be legally appropriated, or floating movable goods, capable of being captured and reduced into private ownership. The legal treatment applicable to various classes of goods is fundamentally different, material extensions having a separate legal dimension from territorial extensions, and movables from immovables. Pursuant to the non-appropriation principle of Article II of the OST, celestial bodies cannot be appropriated. In practice, should asteroids and comets be considered celestial bodies, they would fall under this prohibition; adversely, if they are not celestial bodies, they may become the object of private property rights.

It is a generally accepted principle of Private International Law that, while the law of the country where the thing is situated regulates the legal regime of the rights over immovables, its importance is diminished regarding the rights over movables; in this second instance, the law of the owner's domicile has an important role to play according to the principle that "chattels follow the person" (North 1979, pp. 483 and 552).

Land is the archetypal spatial extension, as opposed to a material extension. Should one take away the substance of the land, the spatial value still remains. One cannot consume land; it may, at worst, make it unsuitable for use, but it cannot completely destroy it. Should one dig a hole in the ground and take away all the mass, there will still be the space of the land that is left. The same, an orbit may become unsuitable for use by accumulation of debris, but it does not physically disappear. While landed ownership may seems a flat concept, in fact landowners do not own surfaces; they own three-dimensional entities.

Celestial bodies proper, orbits, points in space and outer space proper are spatial extensions. Outer space and orbits are purely spatial extensions, as they do not have any material existence. Unlike incorporeal things, they do exist in three dimensions and, in the absence of the non-appropriation principle, they could be brought under the sway of national territorial jurisdiction. Celestial bodies, outer space and its sub-categories (orbits and point positions) have characteristics analogous to the municipal category of immovables. Other extensions in outer space have characteristics analogous to the municipal category of movables. Such is the case, for instance, with space objects such as artificial satellites. Unlike territorial extensions, national jurisdiction is not prohibited regarding material extensions located in the extraterrestrial realms.

[Some] asteroids and comets could be viewed not as landed extension, but as movables. Land has two unique characteristics which distinguish it from all other commodities, namely – it is immovable, hence it cannot be physically transferred from one person to another, and it is everlasting - the owner of land cannot destroy it in its legal sense, his power being limited to the enjoyment or disposition of rights in or over it (Simpson 1976, pp.5-6). [Some] asteroids and comets, however, do not have these characteristics; with the appropriate technology, they could be moved; and they can be destroyed, i.e. consumed in their totality. Thus, they may qualify as movables.

In the question of the legal definition of outer space, several approaches have been used – such as the "spatialist" and "functionalist" ones. The same approaches could be used in the quest for a legal definition of a celestial body.

Applied to the present topic, a spatialist approach would define celestial bodies as objects over a certain size, while objects under that size would not be celestial bodies. The practical problem - falling under the spell of the "*sorites paradox*" - is to quantify that size, and to reach a consensus over that. Some concepts are vague, lacking sharp boundaries; such is the case for a heap, whereby single grain or two grains of wheat cannot be described as a heap. By adding grains, one must admit the presence of a heap sooner or later, so where does one draw the line? (Hyde 2005) In our case, if we accept that the Moon is a celestial body whereas a piece of dust floating in space is not, where does one draw the line between celestial bodies and space dust? At what dimension a "stone in space" ceases to be legally movable, and becomes legally immovable? Difficult as it is, lawyers have the ability to find the mythical "straw that broke the camel's back". Where there is no natural boundary or one cannot discover it, law can set a conventional boundary - such being the case with the legal age. A spatialist approach seems to be favored by Article I of the OST, providing for the "free access to all areas of celestial bodies." From this, it results that one cannot consider a small space rock as a celestial body insofar as it is not viewed primarily as an area permitting landing on it. The spatialist approach has its merits insofar as it distinguishes between small objects – which are not celestial bodies – and big objects, which are celestial bodies. However, the problem still remains to agree on how small is small.

Another approach – a functional one - would take into account the actual use of the asteroid – i.e. for building a base, or for exploitation of resources, differentiating between objects used in their spatial dimension – these being deemed as celestial bodies - or in their material dimension, these being movable orebodies.

Yet another way of setting apart non-appropriable celestial bodies from appropriable orebodies would be the use of the criterion of actual movability from orbit by human action – or the effective control approach. This would distinguish between immovables – celestial bodies – and movables in outer space literally, according to the actual ability of moving them by human intervention. The fact that everything in space moves by itself is not relevant in this context –land is the immovable *par excellence*, yet the Earth orbits the Sun. According to the control approach, therefore, it is movable what it can actually be moved, and immovable what it cannot be moved. By the action of actually moving it, one makes it appropriable. Change occurs in the moment of actual movement; property would install when moved.

Most of the scholars that have studied the problem of defining celestial bodies belong to the control school. In the 1960's, the members of Working Group III of the International Institute of Space Law concerning the legal status of celestial bodies came with a definition considering celestial bodies in the legal sense as "natural objects in outer space … which cannot be artificially moved from their natural orbits" (Smirnoff 1966, p. 13). As with any theory not enshrined into law, the control approach is not accepted by everybody. In July 1980, testifying in front of the US Senate, NASA General Counsel Neil Hosenball expressed his view that if an asteroid were moved into Earth orbit for exploitation, it would still be a celestial body within the meaning of that term, and would not change its character by its moving (Leich 1980). However, there is merit in considering that the process of actually moving an asteroid/comet would qualify as extraction, the body in cause ceasing to be a resource "in place" and thus by-passing even the general prohibition in article 11.3 of the Moon Treaty.

A number of authors envisage using asteroids as interplanetary vehicles; the legal classification of such objects as movables / space objects would be supported by a functional approach. Ernst Fasan (1984, p.243) examines the case when an asteroid, by way of human intervention, is used as a shell for a space station, losing its natural appearance, together with its legal status of 'celestial body' by becoming a manmade structure, i.e. legally a space object. This so-called "Asteroid Base" would then have to be registered internationally with the Secretary General of the United Nations. Should small asteroids be considered space objects they could be claimed by way of registration in the national registry of space objects referred to by Article VIII of the OST: -

> "A State Party to the Treaty on whose registry an object launched into outer space is carried shall retain jurisdiction and control over such object"

These provisions have been detailed in the 1975 Convention on Registration of Objects Launched into Outer Space, that in Article I(c). defines the term "State of Registry" as "a launching State on whose registry a space object is carried…" and provides both for a national and international registration of space objects. The Registration Convention is very liberal insofar as it gives the State of registry concerned the freedom to determine "[t]he contents of each registry and the conditions under which it is maintained" (Article II.3). This may be interpreted as entitling a State to register small asteroids as space objects on its registry.

Of course, the status of launching state would come with its privileges and responsibilities, one of the latter being its international liability as provided by several Space treaties. In the light of this international liability, the registration of small asteroids as space objects and their private ownership would be in fact of benefit to the possible survivors of accidents provoked by small asteroids. Several questions spring from this possible approach – such as whether compulsory insurance should be legally imposed upon owners - like the one imposed to atomic energy providers. Would this discourage people in claiming asteroids? Were asteroids *res nullia*, nobody would be responsible for damage caused by them; were they *res communis*, then the whole humankind would be responsible for the damage to a private spaceship or other object by such a common asset.

Interesting consequences would also result from using the legal status of icebergs as a paradigm for the legal classification of comets and asteroids. In fact, comets have been often described as the "icebergs of space", offering striking similarities to icebergs (composition, dimension, location in an international area). It is to be noted that icebergs have as well a rather unclear legal status, though their small-scale exploitation has already begun. Like asteroids and comets, icebergs have a spatial dimension but are used mainly in their material dimension, as a floating mineral resource. While article 89 of the UN Convention on the Law of the Sea prohibits the national appropriation of the high seas, we have no knowledge of States having protested appropriation of icebergs. In the same time, we have no knowledge of a formal declaration of ownership over icebergs by the entities using them in their material extension; the principle of extraction seems to apply, given that icebergs have been appropriated either in their entirety and displaced from their initial location, or parts of them have been moved away without claims being laid for the exclusion of others from the exploitation of that particular iceberg.

Given that, as shown *supra*, the notion of celestial bodies is not legally defined, it could be argued that several asteroids might escape the non-appropriation principle because of this lacuna. The issue of defining celestial bodies is extremely intricate, and there is no absolute answer to be given *ex cathedra*. We have only attempted to present the existing theories and some new approaches, but at the end of the day only practice will decide whether [some] asteroids are places or movables.

29.1.3 The Commons Regime – Everybody's and Nobody's

Article I of the Outer Space Treaty proclaims that the extraterrestrial realms – outer space, including the Moon and other celestial bodies – shall be "free for exploration and use by all States without discrimination of any kind, on a basis of equality", and that "there shall be free access to all areas of celestial bodies". The exploration and use of the extraterrestrial realms is declared as being the "province of all mankind". This norm effectively establishes among the States Parties an open access and free use regime on the extraterrestrial realms, making them a public good whose owner is everybody and nobody.

Far from being Orwellian doublethink - that is, simultaneously accepting two mutually contradictory beliefs -, this state of affairs hails from the "bundle of rights" theory of property. Property is an embodiment of three attributes - *jus utendi* (the right to use), *jus fruendi* (the right to enjoy the fruits) and *jus abutendi* (the right to "abuse" or "dispose" of one's own good). The commons regime is built around *jus utendi* (use is permitted, hence "everybody owns the extraterrestrial realms") yet forbids *jus abutendi* (title is denied, hence "nobody owns the extraterrestrial realms"). As to *jus fruendi* – i.e, to collect the "fruits" of the extraterrestrial realms - this will be addressed later in this chapter.

The current incarnation of space law is not strictly an open access regime, containing several regulations. According to Article I of the Outer Space Treaty, the freedom of exploration and use of the extraterrestrial realms pertains to States. Article VI of the same Treaty requires non-governmental entities to obtain authorization from the appropriate State Party in order to carry out activities in the extraterrestrial realms, and to consent to being continually supervised by same. States bear international responsibility for national activities carried out in outer space and on the celestial bodies, whether these are performed by governmental entities or by private enterprise. The regime is therefore a hybrid of *res communis* (whereby ownership pertains to the community as a whole, and every member has the non-exclusive right to use the property) at international level – and *res publica* (whereby property is centrally enclosed and cannot be used without permission from the community) at municipal level, given the need for a nationally issued license.

As explained by French jurist Pothier – quoted by the U.S. Supreme Court in *Geer v. Connecticut* [1896], *res communis* originates from "all those things which God had given to the human race", this community being different from the "positive community of interest, [existing] between several persons who have the ownership of a thing in which each have their particular portion". This original community was -

> "'a negative community,' which resulted from the fact that those things which were common to all belonged no more to one than to the others, and hence no one could prevent another from taking of these common things that portion which he judged necessary in order to subserve his wants. Whilst he was using them, others could not disturb him; but when he had ceased to use them, if they were not things which were consumed by the fact of use, the things immediately re-entered into the negative community, and another could use them".

In a nutshell, *res communis* implies freedom of an actor to use a good; as long as one uses this good, one may not be obstructed by another, yet this latter is free to use the good as soon as the first has ceased to use it.

Under space law, *jus utendi* is not absolute; the right of using the celestial bodies has to be carried out, under the provisions of Article IV of the Outer Space Treaty, "exclusively for peaceful purposes". Article IX of the same pact requires the State Parties to use the principle of co-operation and mutual assistance as a guide in the exploration and use of the outer space and celestial bodies. The State Parties need to conduct all their activities in the extraterrestrial realms "with due regard to the corresponding interests of all other States Parties", according to the same legal norm. The Outer Space Treaty elaborates, up to a certain degree, the mechanism for accommodating the interests of the other members of the public. Thus, Article IX requires States Parties to publicize, "to the greatest extent feasible and practicable" the nature, conduct and locations of their space activities. If an activity by a State Party or one of its nationals is likely to cause "potentially harmful interference" with the space activities of other States Parties, the first State Party is required by the same tenet to "undertake appropriate international consultations" prior to carrying out such activity. In the same time, should a State Party to the Outer Space Treaty rightfully fear that a space activity of another State Party could harmfully affect its own activities, the first State is entitled to request appropriate consultations on this subject. While the Treaty makes no mention of it, it is nonetheless understood that a "first-come, first-served" regime exists.

Whereas several rules are established and ought to be respected by the State Parties, most of these remain at the stage of principle – as recognized by the document's title – "Treaty on Principles Governing the Activities of States in the Exploration and Use of Outer Space, Including the Moon and Other Celestial Bodies". Many open access regimes are self-regulatory, and where the text of the law is silent, custom is bound to develop.

The status quo of the commons regime is not very stable, being challenged on two fronts, as we will show next.

29.1.4 *The Common Heritage of Mankind: From Res Communis to Res Communist*

On the left, *res communis* is bordered by the Common Heritage of Mankind (CHM) regime, whereby users have to share with the community the benefits accrued from the use of the commons.

The post-Sputnik and post-colonial era could not escape the Marxian historic materialism, where the world and its extraterrestrial surroundings are the scene of the class struggle between the antagonistic spacefaring and non-spacefaring nations, developed and developing states, the haves and have-nots. The Marxist ideas received escape velocity through the adoption by a very small number of countries of the "Agreement Governing the Activities of States on the Moon and Other Celestial Bodies". Informally known as the Moon Agreement, its provisions also apply, according to Article 1.1., "to other celestial bodies within the solar system, other than the earth, except in so far as specific legal norms enter into force with respect to any of these celestial bodies". Article 11.1 of this legal document proclaims that the celestial bodies and their natural resources (literally, "the moon and its natural resources", but see article 1.1. above) are "the common heritage of mankind", whereas Article 11.3 contains a facial prohibition of landed property in outer space:-

> "Neither the surface nor the subsurface of the [celestial bodies], nor any part thereof or natural resources in place, shall become property of any State, international intergovernmental or non-governmental organization, national organization or non-governmental entity or of any natural person".

Besides prohibition of title, the Moon Agreement contains another key tenet, namely an egalitarian distribution of benefits. As an attribute of property, *jus fruendi* embodies the right to enjoy the income (fruits) derived from an asset. Under the Moon Agreement regime, this cannot be fully enjoyed, as Article 11.7.d of the said document provides for –

> "[a]n equitable sharing by all States Parties in the benefits derived from [the natural resources of the celestial bodies], whereby the interests and needs of the developing countries, as well as the efforts of those countries which have contributed either directly or indirectly to the exploration of the [celestial bodies], shall be given special consideration".

These provisions are in resonance with the 1848 Manifesto of the Communist Party, whereby Karl Marx and Friedrich Engels (1848) called for the "[a]bolition of property in land and application of all rents of land to public purposes". Lenin (1917) defined Socialism as the "social ownership of the means of production and the distribution of products according to the work of the individual"; in his view, Socialism will "ripen into Communism, whose banner bears the motto: 'From each according to his ability, to each according to his needs'. The tenets of the Moon Agreement hold middle ground between Socialism and Communism, providing for a share of the lunar benefits from each according to his ability, to some according to their work and to some according to their needs.

In the 1979 report of the Independent Commission on International Development Issues, the organism chaired by Willy Brandt (quoted in Vicas, 1980, p.303) considered that "'Global commons' is a neat catchword, but hardly appropriate", because -

> "It connotes villagers in medieval England who have the right to pasture their cattle in the village commons. The space analogy is nations 'pasturing' their satellites in the global commons. The term connotes something of free access to outer space, but none of the distributional aspects of the 'common benefit' or 'common heritage'".

Indeed, whereas *res communis* offers free access, it does not entail a share of the benefits. The villager pasturing his cow in the village commons needed not share the meat and milk with the other villagers, even if originated from the grass grazed from a common pool. In contrast, under the CHM regime, the asteroid miner has to share with all other humans what his equipment extracted from the asteroids.

Critics lament the whole society having moved toward a culture of entitlement over the years, an ethos whose motto is "weaken the strong to strengthen the weak" (Anderson 2002). This culture promotes dependence on government handouts - rating justice by how the government rewards those who don't succeed, rather than the way it encourages those who do, denying thus individual responsibility and creating convenient excuses for failure (Urbahn 2005). Instead, as explained by John Locke (1690 chap.5, sect. 34) more than three centuries ago, the world was given -

> "to the use of the industrious and rational, … not to the fancy or covetousness of the quarrelsome and contentious. He that had as good left for his improvement, as was already taken up, needed not complain, ought not to meddle with what was already improved by another's labour: if he did, it is plain he desired the benefit of another's pains, which he had no right to, and not the ground … whereof there was … more than he knew what to do with, or his industry could reach to".

An egalitarian regime, which not only prohibits the appropriation of asteroids but also calls for an 'equitable' *jus fruendi* offers no incentive to exploit them when the investors have to share the benefits with free riders who believe in a culture of entitlement. Together with Locke, we believe that law ought to provide for the "industrious and rational" and not for the "quarrelsome and contentious".

29.1.5 Homesteading the Final Frontier

What distinguishes the *res communis* from full property rights is its lack of marketability. As explained by Robert P. Merges and Glenn H. Reynolds (1997, p.121), rights can be separated in two broad classes, namely usufruct – "a right to continued use for a limited time" and fee – "a more permanent interest that can be traded, devised, or otherwise transferred". According to the two authors, while the usufruct may prove to be valuable for some purposes in the space environment, a fee interest akin to the "fee simple" of common law would have the advantages of predictability flexibility.

In 2004, the US President's Commission on Implementation of United States Space Exploration Policy recommended that -

> "Congress increase the potential for commercial opportunities related to the national space exploration vision ... by assuring appropriate property rights for those who seek to develop space resources and infrastructure" (Aldridge 2004, Recommendation 5-2).

This *lex ferenda* proposal would represent an important shift from the res communis approach consecrated in the OST.

Private property rights have long been promoted by space advocates as the most appropriate – if not the only way for developing and settling outer space. In the tradition of the American Frontier, whose settlement was encouraged by governmental plans of privatizing the public domain, the proponents of private property in space seek a similar privatization of the extraterrestrial realms. The frontier is inexorably linked with the idea of individualism and private ownership, of transforming *res nullia* and *res publica* into private property. The association between space colonization and the American Frontier is a recurrent theme in much of the pro-space ethos, complementing the earlier view of the American continent as a "New World". At the end of the 19th Century, historian Frederick Jackson Turner (1893) explained American development through "[t]he existence of an area of free land, its continuous recession, and the advance of American settlement westward". According to historian Barbara Tuchman (1976), the frontier and private property rights are deeply connected, in her view -

> "[t]he open frontier, the hardships of homesteading from scratch, the wealth of natural resources, the whole vast challenge of a continent waiting to be exploited, combined to produce a prevailing materialism and an American drive bent as much, if not more, on money, property, and power than was true of the Old World from which we had fled."

In a 2003 testimony before American legislators, Rick Tumlinson, founder of the appropriately named Space Frontier Foundation, pleaded for "the right to own new land in space", asserting the crucial need to "begin putting in place the rights of those who explore and develop such new 'lands' in space to own them" in order for these to live up to their potential of great sources of wealth:-

> "Throughout history, it has been the ability to gain and hold land which has driven [the explorers and developers] forth, and given them the will to carve new human domains out of wilderness. Space is no different. If people are going to invest their wealth and lives in opening the frontier, they should have the right to pass what they have done down to the next generations. When the time is right, the US should stand up and recognize that in space, the same rights to won property exist as on Earth." (Tumlinson 2003, p.16)

Most of the resources of outer space are, in practical terms, unlimited; given their abundance, it is illogical to forbid their private appropriation.

Perhaps the best arguments for the privatization of the extraterrestrial realms have been brought by John Locke, long before the start of the Space Age. In his view – to which we adhere -, privatization enhances the common heritage, provided there is still enough left for the others:-

> "[M]en had a right to appropriate, by their labour ... as much of the things of nature, as he could use: yet this could not be much, nor to the prejudice of others, where the same plenty was still left to those who would use the same industry" (Locke 1690, sect. 37).

Locke did not justify greed, bearing in mind the interests of the fellow humans. Yet, as shown in the previous section, this interest cannot be the entitlement to the work of the other, but the right to use one's "industry" for homesteading the common heritage. In his correct reasoning, he who proceeds at appropriating land by the means of his labour "does not lessen, but increase the common stock of mankind", because enclosing an acre and using it yields more essential products than if the land were left "waste in common". The privatization is seen, thus, as an active administration of the public trust stemming from the common heritage in its original sense:-

> "And therefore he that incloses land, and has a greater plenty of the conveniences of life from ten acres, than he could have from an hundred left to nature, may truly be said to give ninety acres to mankind" (Locke 1690, sect. 37).

29.1.6 Means of Acquiring Ownership over Asteroids

If allowed, how could one establish ownership over an asteroid? On March 3rd, 2000, Greg Nemitz, a space activist and owner of a company named Orbital Development registered a claim for asteroid Eros with the "Archimedes Institute Private Property Rights Registry". When in February 2001 the NEAR Shoemaker spacecraft landed on Eros, Nemitz sent NASA a $20 invoice for parking their spacecraft on his property, setting a low amount as a proof his goal was in fact to bring publicity to the issue of space property rights. Following a lengthy exchange of letters with an uncooperative NASA and other institutions, Nemitz would then engage into an actual lawsuit against NASA lasting for over five years and culminating with Nemitz renouncing to take the case to the Supreme Court, and considering thus the case closed (Nemitz 2005).

We have shown before that claims to extraterrestrial real estate unsubstantiated by physical acts of possession are not valid means of acquiring ownership [Note: For a more elaborate analysis of the subject of extraterrestrial real estate, see Pop, V, 2006. Unreal Estate: The Men who Sold the Moon. Liskeard: Exposure Publishing]. Simply saying "Asteroid Eros is mine", and publicizing this claim without however taking possession of the object of desire has no legal effect.

A mere claim is not tantamount with ownership – or, in plain language, claiming does not mean owning. In the acquisition of possession, two concurrent elements - "the mind" (animus possidendi) and "the body" (corpus possidendi) are required. One is insufficient without another; there must be "both an intention to take the thing and some act of a physical nature giving effect to that intention" (Reid 1996, p.103). The Scottish jurist Stair (1693, II.i.18) has explained this in very illustrative terms: "if any act of the mind were enough, possession would be very large and but imaginary". An imagination as large as the Universe, in the case of the purported extraterrestrial salesmen - who present a very valid animus, but no corpus at all.

While entities such as the "Lunar Embassy" who "sell" plots of "extraterrestrial real estate" base their claims on pure *animus*, several space exploration companies intend to claim ownership of space resources – namely, asteroids - only after having established effective possession by means of robotic prospectors. In 1998, the late US entrepreneur Jim Benson (1998, pp. 46-49) stated his intention to "fly a privately sponsored deep space science mission" and claim ownership of an asteroid by means of effective possession with the aim of creating "an important and historic precedent" in support for space property rights. Can title (should space assets be claimable) be gained this way?

Advocates of telepossession – i.e., property acquisition through telepresence, appeal to the 1989 case of Columbus-America Discovery Group (1989). In that case, a U.S. District Court in Virginia ruled that a salvage firm could claim an undersea wreck it had explored via "telepresence" with an undersea probe equipped with a TV camera. Human presence on the ocean floor was not required to claim the wreck; instead, the court allowed telepossession, defined as:

> "(i) locating the object searched; (ii) real time imaging of the object; (iii) placement or capability to place teleoperated or robotic manipulators on or near the object, capable of manipulating it as directed by human beings exercising control from the surface; (iv) present intent to control … the location of the object."

According to Richard Westfall (2003), establishing telepossession of resources in space should involve three tasks:

- *Telepresence* (visual observation of the site) – telepossessor must be able to show live video pictures of the resource site;
- *Telemetry* (communication with and knowledge of the location of the site) – telepossessor must know where the resource is and be able to communicate with the equipment thereon;
- *Telerobotics* (manipulation of the resources at the site) – telepossessor must demonstrate ability to manipulate the resource site presumably (but not necessarily) through purifying the ore, reducing metals, or manufacturing parts in situ.

An interesting precedent justifying the acquisition of property through telepossession can be found in the Roman law, where a person could acquire ownership or other rights not only by his own direct actions, but also through the actions of slaves under his power (Roby 1902, p. 432). Slaves had a similar legal status as modern robotic space probes; the word "robot" itself was derived from the Czech "robota"- meaning "labor" in the early 20th Century (Zunt 1998). Roman slaves were mere instruments for their master, being nothing more than chattels, objects of property rights and having no independent legal existence as persons. Their master had the property in anything that the slaves acquired. In the absence of a statement to the contrary, a slave owned in common by two or more masters acquired for them in the ratio of their respective shares in him (Roby 1902, pp.452, 53, and 433).

Space lawyer Glenn H. Reynolds agrees that claiming property by landing a robot may well be possible; he nonetheless considers that the size of the claim has to be reasonable – e.g., one can claim "the quarter of a square mile in which your robot rolled around", but not a whole planet (Scripps Howard News Service 1998). Should such actions be taken, we, too, believe that entities performing them may well have a legal standing to make a property claim in the presence of both animus and corpus. Of course, the validity of ownership acquisition through space telepossession is subject to the permissibility of fee simple property rights in the extraterrestrial realms.

29.1.7 Property Status of the Extracted Asteroidal Resources

We promised above to address the issue of *jus fruendi* – i.e, the right to collect the "fruits" of the extraterrestrial realms. As we have shown, the CHM regime calls for the share of the extracted planetary resources with the international community, whereas the supporters of property rights want the enclosure of the lands. Who owns the extracted resources under the current, commons regime?

It is first to be remarked that the legal treatment applied to real estate and substances removed thereon is different. As C. Sweet (1882, p. 259) says, "[w]hile unsevered, minerals form part of the land, and as such are real estate. When severed, they become personal chattels". Whereas the immovables examined above are subject to the *lex situs* of outer space, extracted resources are movables, subject mainly to the *lex domicilii* of the person who caused their removal.

Is the conversion of immovables into movables by way of extraction, allowed in the light of the non-appropriation principle of the Outer Space Treaty? The legal text is silent as to the permissibility of appropriating natural materials in the course of exploration and use of the celestial bodies. In the absence of a specific norm clarifying the ownership of extraterrestrial resources, most scholars draw a clear distinction between the appropriation of outer space and celestial bodies, and the appropriation of materials thereon. Article I of the OST consecrates the freedom of scientific investigation, exploration and use of the extraterrestrial realms, whereas the "residuary rule of presumptive freedom of action" (Lauterpacht 1975, p.220) proclaims that what is not prohibited is permitted.

Indeed, a strong opponent of national appropriation of the high seas, Hugo Grotius did not discourage the appropriation of the resources found there. In support of this view, he found it appropriate to cite the play *Rudens* ("The Rope") by Titus Maccius Plautus: -

> "[W]hen the slave says: - the sea is certainly common to all persons - the fisherman agrees; but when the slave adds: - then what is found in the common sea is common property -, he rightly objects, saying: - But what my net and hooks have taken, is absolutely my own" (Grotius 1608).

A similar logic is found in John Locke (1690, sect. 30)'s Second Treatise of Civil Government, where -

> "by virtue thereof [of the original law of nature], what fish any one catches in the ocean ... is, by the labour that removes it out of that common state nature left it in, made his property who takes the pains about it".

Whereas the Moon Agreement contains a more restrictive regime, as long as the space powers do not ratify the document, their private enterprises are "entitled to acquire and retain space resources" for their own disposition "without limitation on possible profit", as agreed by Stephen Gorove (1985, p. 227).

In December 1993, Russia set an important precedent by commercializing extraterrestrial material retrieved by a Soviet probe. Sotheby's auctioned three small particles of lunar regolith weighing about one carat (200 mg), who sold for US$442,500 - i.e. US$2.2 million per gram (Arthur 1998). As no objections were voiced by third States, it can be stated that, as an attribute of ownership, the right to commercialize extraterrestrial material has entered into customary international law. Indeed, Gyula Gal (1996, p.47) finds worthy of mentioning

> "some facts of international practice", namely that both USA and USSR collected and returned lunar samples – objects that "were appropriated by U.S. and Soviet authorities respectively and have been owned by them without objection from the international community."

It is safe, therefore, to conclude that although fee simple ownership is prohibited, under the Outer Space Treaty regime the conversion of immovables into movables by way of extraction is allowed in the light of the non-appropriation principle, a distinction being made between the appropriation of outer space and celestial bodies and the appropriation of materials thereon. While lacking fee simple ownership over the land thereof, private actors are entitled to "enterprise rights" - that is, to explore and exploit the extraterrestrial natural resources. Ownership of extraterrestrial products vests in those who sponsored their removal, through the labor invested in seizing them and, as an attribute of ownership, the right to extract and commercialize extraterrestrial material is part of customary international law.

29.2 Deflection: The Legality of Planetary Defense

29.2.1 Introduction

The sky is, and has always been, a mine. The asteroids and comets lie up there as a giant mine field and, from time to time, some of them deliver their deadly blow to Earth. A planet where Homo Sapiens is now the dominant species – but it has not always been so. "The dinosaurs" – says Larry Niven (quoted by Chaikin, 2001) – "became extinct because they didn't have a space program. And if we become extinct because we don't have a space program, it'll serve us right!"

Blockbusters such as "Armageddon" and "Deep Impact" made the subject of planetary defense against potentially hazardous objects (PHOs) a part of the popular culture, in the current context of the millennaristic fears surrounding the year 2000 and, more recently, 2012. Together with other global issues such as climate change, planetary defense has also prompted vivid debates in the scientific and policy circles.

The prospective of interacting with asteroids and comets raise, indeed, a number of legal and policy questions. We examined above the issue of appropriation and of benefits share – yet another important topic pertains to security. Is planetary defense a right, or an obligation? Who is entitled - or obliged - to defend the Earth from the PHO menace? Are all deflection technologies legal? Can nuclear explosions be used not only for deflection, but also for exploiting the mineral riches of asteroids? And, last but not least, could asteroids be used for waging war?

29.2.2 Defending the Planet: A Right, Or an Obligation?

Interacting with potentially hazardous objects is, undoubtedly, a space activity. Does space law permit such an interaction – and we speak now not of the means for carrying out such a deed, but of the end? Under Article I of the OST, the extraterrestrial realms-

> "shall be free for ... use by all States without discrimination of any kind, on a basis of equality and in accordance with international law, and there shall be free access to all areas of celestial bodies."

The same article mandates the States Parties to conduct the use of the extraterrestrial realms for the benefit and in the interests of all countries, whereas Article III of the OST requires them to carry out their space activities in the interest of maintaining international security. We believe that the above norms instate a right for the States Parties to the OST to defend the planet from PHOs, and that such actions are indeed beneficial and in the interest of all countries and in that of maintaining global security.

In the same time, in the past, pirates and slave traders were held as being beyond legal protection, being deemed *"hostis humani generi"* – enemies of all mankind. Accordingly, they could be dealt with as seen fit by any nation – under the principle of universal jurisdiction - even if that nation had not been directly attacked. It can be argued that, like the pirates and the slavers before, potentially hazardous objects are *"hostis humani generi"* hence lawfully in the gunsight of any nation. All of the above, coupled with the right of self-defense, demonstrate beyond any doubt that planetary defense is a State right.

Is this right also a legal obligation? Unlike municipal law, international law is consensual, binding only the States who adhered to the norms in cause - apart from the peremptory norms of international law ("*jus cogens*"). The latter bar unlawful acts rather than imposing obligations to act. From a legal point of view, the exploration and use of outer space is a right, but is not a duty. There are no legal norms compelling States to explore space or to deflect asteroids. While to Declan O'Donnell (quoted by Rodriguez, 2004, p.2), a nation capable of deflecting asteroids is liable for an asteroid impact if it acts and any damage is inflicted, if does not act and any damage is inflicted; and if it could have acted and didn't, irrespective of the outcome, we believe that such responsibility is not a legal one, but a moral one.

A moral obligation can become a legal one if enshrined into law. There have been calls for international treaties whereby signatory states would commit finances and technology for planetary defense and would also solve the security issues at stake. In the same time, active members of the society consider themselves bound by moral obligations – such as eradicating poverty and diseases, and defending the planet from PHOs – hence an active volunteer and privately funded network of organizations involved in cataloguing rogue asteroids and raising the awareness – such as the B16 foundation, the Association for Space Explorers, the Secure World Foundation, various Spaceguard efforts, and so on.

We have concluded that planetary defense, as an end, is lawful. What about the means for achieving it?

29.2.3 The Legality of Deflection Strategies

Unlike many rogue entities, potentially hazardous objects are not hiding their deadly intentions, and their moves are quite predictable. Astronomy is, for most of it, an exact science, hence detecting PHOs and calculating their orbits is an activity which is becoming more and more precise.

Like with any enemy, the earlier its intentions are known, the easier it is to avoid a deadly encounter, the gentler the methods for confronting it, and the least controversial. Dealing with an asteroid likely to strike in 20 years is different from dealing with an asteroid expected to strike in 3 months.

Surveys of PHOs as a first line of defense are, undoubtedly, lawful; NASA, as a Government entity of the United States, is actively involved in discovering, verifying, and providing follow-up observations for Earth-crossing asteroids – and such are LINEAR, NEAT, CINEOS, Orbit@home and other similar efforts.

When it comes to the actual deflection strategies following the identification of a rogue object, the legal issues become more complicated. Various deflection strategies have been imagined, with various variables such as rapidity of reaction, performance, cost, etc. Some strategies are aimed at destroying the PHO, while others at altering its orbit so that it misses the earth; some are directly impacting the object, while others are indirect. While all strategies have security implications, some of them make use of systems that are, in practice, weapons – being able to be used both against potentially hazardous objects and against other States. As with many other dual use technologies, under the guise of a lawful aim such as planetary protection, States can develop aggressive means of attacking other countries or their space systems.

The technology of a nuclear explosive device used for peaceful purposes has no essential distinction from the technology of a nuclear explosive device that would be used as a weapon. Given the prohibition in Article IV of the Outer Space Treaty of installing weapons of mass destruction on celestial bodies and in orbit around Earth, as well as the stationing of WMDs in outer space, important security issues are raised by the possible use of peaceful nuclear explosions for planetary defense and the exploitation of extraterrestrial resources, as well as by the presence on the celestial bodies of radioactive materials. Peaceful Nuclear Explosions (PNE) were legitimated, inter alia, by the 1967 Treaty for the Prohibition of Nuclear Weapons in Latin America, whose Article 18.1 allows the Contracting Parties to "carry out explosions of nuclear devices for peaceful purposes - including explosions which involve devices similar to those used in nuclear weapons", and by the 1976 Treaty Between the USA and USSR on Underground Nuclear Explosions for Peaceful Purposes (PNE Treaty). Would PNE be allowed for the deflection of rogue asteroids and for the exploitation of extraterrestrial minerals?

The language of article IV of the Outer Space Treaty - "[t]he use of any equipment or facility necessary for peaceful exploration of the moon and other celestial bodies shall ... not be prohibited" - suggests that the use of absolutely any exploitation equipment, including nuclear explosion devices, is lawful as long as it is used for peaceful purposes. The same conclusion would arise from article 3.4 of the Moon Agreement that, reiterating the quoted paragraph from Article IV of the Outer Space Treaty, omits the term "facility" but extends this provision to the "use" of the moon. However, Article III.2.d. of the PNE Treaty forbids, inter alia, any explosion except in compliance with the provisions of the 1963 Moscow Treaty Banning Nuclear Weapons Tests in the Atmosphere, in Outer Space and Under Water. The Moscow treaty, in the words of Eugene Brooks (1997, p. 246), "flatly bans any nuclear explosion in outer space"; its Article I.1.a. contains the undertaking of States Parties "to prohibit, to prevent and not to carry out any other nuclear explosions … in the atmosphere; beyond its limits, including outer space". John Kish considers that these provisions apply "in outer space … in the eventual atmosphere of celestial bodies" and "to all areas of celestial bodies, including their subsoil", as

> "[a]rticle I (1) (b) of the [Moscow Treaty] prohibits nuclear explosions outside national territory, namely "...in any other environment if such explosion causes radioactive debris to

be present outside the territorial limits of the state under whose jurisdiction or control such explosion is conducted""(Kish 1973, p.185).

Therefore, use of PNE for the deflection of asteroids or the exploitation of extra-terrestrial minerals would be illegal.

Furthermore, the 1996 Comprehensive Nuclear Test-Ban Treaty (CTBT) signed by most states has outlawed the use of PNE. Initially opposed by China, the CTBT provides for a compromise by providing in article VIII for a Review Conference who may consider "the possibility of permitting the conduct of underground nuclear explosions for peaceful purposes", taking into account "any new scientific and technological developments relevant to this Treaty".

Eugene Brooks (1997, p.247) believes that "perhaps the time is ripe for the UN Space Committee to initiate discussions on whether an exception should not be made for nuclear or other drastic means to alter asteroid orbits". He further reckons that the Space Committee "may also review whether nuclear explosions might not be used on celestial bodies for mineral extraction, with provisions for international consultations and agreed safeguards".

As a historical footnote, another security issue was, until 2002, the ABM capability of asteroid deflection systems. The bilateral USA-USSR 1972 Anti Ballistic Missile Treaty (unilaterally denounced by the USA in 2002) prohibited in Art. V the development, testing and deployment of ABM systems or components, including space-based ones. Art. II of the ABM Treaty defined the ABM system as a "system to counter strategic ballistic missiles or their elements in flight trajectory". Asteroid deflection systems such as lasers, nuclear, or railguns, although not directly aimed at countering strategic ballistic missiles, could have been accused of having an ABM "hidden agenda", given their real ABM capabilities. The unilateral deployment of such a deflection system either by the USA or Russia would have entailed the risk of apparent violation of the ABM treaty.

29.2.4 The Deflection Dilemma(s)

In the early 1990s, Carl Sagan and other authors described a challenging situation: "if one can deflect an asteroid away from a collision, one can also deflect an asteroid toward a collision" (Schweickart 2004, p.1) Indeed, using asteroids as kinetic weapons of (mass) destruction has been envisaged by many SF authors, such as R.A Heinlein whose Klendathu bugs wipe out Buenos Aires by crashing an asteroid on Earth, or Stephen Baxter with a man-made extinction scenario in his 1997 novel "Titan". The use of space-based kinetic weapons has also been studied by military strategists, such as those involved in "Project Thor".

The use of asteroids as weapons poses even more theoretical security problems than the ones of employing weapon-like systems for deflecting them - supplemental safeguards may be needed in order to avoid misuse of otherwise peaceful technologies. It is hereby offered that deflection techniques do exist which would solve the Sagan Dilemma – techniques effective enough to deflect a harmful asteroid yet imprecise enough to make unfeasible the modification of the orbit of an otherwise harmless asteroid into a collision course with the Earth. Such is the case with the surface albedo treatment system currently under development at the Scientific Preparatory Academy for Cosmic Explorers (SPACE), a project entailing the use of the Yarkovsky effect by actually painting an asteroid and altering hence its orbit away from a collision course [Note: the Yarkovsky effect is the result of anisotropic heating of a celestial object. The amount of heat absorbed by the object is closely related to its albedo; as uneven heating occurs, the warmer surface areas radiate more thermal energy and a resulting net force acts on the object. For instance, the side of an asteroid that faces the sun will be heated. As the sunlit side of the asteroid rotates away from the sun, the warmer dusk side radiates more energy than on the cooler dawn side. The resulting net force acts in a direction that is determined by the asteroid's spin axis, rotation rate, and orbital period.] As a non-dual use technology, this solves Sagan's dilemma, as it cannot be used in practice to transform an asteroid into a kinetic weapon, answering also to legal issues raised by other, more aggressive techniques (see Ge and Pop, 2011).

Astronaut Rusty Schweickart (2004, p.1) believes nonetheless that the deflection dilemma as expressed by Carl Sagan is virtually non-existent, as "this malicious deflection capability is of so little 'practical' use as a weapon that it is of no real concern"; however, considers he, there is another dilemma, and that is the choice "between doing nothing, thereby suffering the consequences of an impact, or pro-actively deflecting an asteroid which will, in the process of "protecting the Earth", necessarily place otherwise non-threatened people and property at risk".

We consider that there is yet another dilemma, between doing what is lawful and doing what is right in the area of planetary defense. The Latin maxim "*Fiat justitia ruat caelum*" - "Let justice be done though the heavens fall" has found a literal meaning. In the case of a possible impact, should positive law be followed – even if this may mean that nuclear explosions are forbidden in deflecting asteroids – or should one use illegal, yet effective means of protecting Earth from asteroids? We believe that measures should be taken towards the earliest detection of potentially threatening asteroids, that efforts should be taken for their early deflection by gentle means, that the use of nuclear and other dangerous forces be legalized when employed for deflecting asteroids that cannot be deflected by other means and, even if not legal, these should still be used when confronted with such a tremendous menace.

29.3 Instead of Conclusion: Mines and Mines

The sky is, and has always been, a mine – in both senses of the word. The mediaeval cartographers used to put sea serpents in the blank areas of the maps, in order to denote unexplored or dangerous territories. "Here be dragons" and "here be tigers" were phrases that scared away some travelers, yet attracted intrepid explorers ready to slay the fierce dragons and sell their miraculous skin for many a case of gold. "Here be asteroids" is the new "here be dragons" – a place marked on the celestial maps where dangerous, yet priceless astronomical creatures abide.

Throughout the history, the ability of humans to tame the destructive forces of nature has been one of the main factors in the evolution of our species. Deadly wild boars were hunted for their meat; deadly wolves were domesticated and transformed into faithful guardians; deadly rapids were jammed and their force used to create electricity. The asteroids will too be tamed, and transformed from a mine field, into a field of mines; from a force that wipes out civilizations into a resource that builds them.

In order for this to happen, we need a framework helping humankind to realize the above goal. We need a framework that would interest not only governments, but also private actors, into developing the resources of space. That framework is best represented by private property rights, whereby the third meaning of the word "mine" – that of possession – would be the most appropriate. A private asteroid mining industry would not only make the human civilization transcend its terrestrial boundaries – it would also save it from extinction. Private enterprise, if allowed to make a profit from exploiting the asteroids, would catalogue, survey and track them better than a voluntary or a government program; private enterprise would devise means of exploiting and moving the asteroids that would not only extract their riches, but would also be able to alter their orbits. Mining the sky would, then, mean also demining it – but in order for this to happen, humans ought to be able to say the word "mine" in its third meaning – that of property – when it comes to asteroids.

References

Aldridge, E.C. (chairman) A Journey to Inspire, Innovate, and Discover - Report of the President's Commission on Implementation of United States Space Exploration Policy. US Government Printing Office, Washington, DC (2004)

Anderson, D.: Changing a culture of entitlement into a culture of merit. The CPA Journal (November 2002)

Arthur, C.: Jewellery's final frontier. The Independent (May 11, 1998)

Benson, J.W.: Space resources: first come first served. In: Proceedings of the 41st Colloquium on the Law of Outer Space, p. 46 (1988)

Brooks, E.: Dangers from Asteroids and Comets: Relevance of International Law and the Space Treaties, IISL-97-IISL.3.14, 40 Proceedings of the Colloquium on the Law of Outer Space, pp. 234–263 (1997)

Chaikin, A.: Meeting of the Minds - Buzz Aldrin visits Arthur C. Clarke. Space.com (February 27, 2001)

Columbus-America Discovery Group, Columbus-America Discovery Group Inc. v. The Unidentified, Wrecked and Abandoned Sailing Vessel, S.S. Central America, A.M.C. 1955 (E.D. Va. 1989) (1989)

Fasan, E.: Large space structures and celestial bodies. In: Proceedings of the 27th Colloquium on the Law of Outer Space, p. 243 (1984)

Gal, G.: Acquisition of property in the legal regime of celestial bodies. In: Proceedings of the 39th Colloquium on the Law of Outer Space, p. 45 (1996)

Ge, S., Pop, V.: Solution to the Sagan Dilemma via Surface Albedo Modification Technique. In: Annual Technical Symposium 2011, NASA/JSC Gilruth Center, Houston, Texas (May 20, 2011)
Geer v. Connecticut, Supreme Court of the United States. 161 U.S. 519 (1896)
Gorove, S.: Private rights and legal interests in the development of international space law. Space Manufacturing 5, 226 (1985)
Grotius, H.: Mare Liberum. Lodewijk Elzevir, Leiden (1608)
Hyde, D.: Sorites Paradox. In: Zalta, E.N. (ed.) The Stanford Encyclopaedia of Philosophy (2005), http://plato.stanford.edu/archives/fall2005/entries/sorites-paradox/
Kish, J.: The Law of International Spaces. AW Sijthoff, Leiden (1973)
Lauterpacht, H.: International law ('the collected papers'), vol. 2. Cambridge University Press, Cambridge (1975)
Leich, M.N.: Digest of United States Practice in International Law. Department of State, Washington, D.C. (1980)
Lenin, V.I.: The Tasks of the Proletariat in Our Revolution. Priboi Publishers, St. Petersburg (1917)
Lewis, J.S.: Mining the sky. Addison-Wesley, Reading (1966)
Locke, J.: The Second Treatise of Civil Government, London (1690)
Mann, A.: Tech Billionaires Plan Audacious Mission to Mine Asteroids. Wired (2012), http://www.wired.com/wiredscience/2012/04/planetary-resources-asteroid-mining/ (April 23, 2012)
Marx, K., Engels, F.: Manifesto of the Communist Party, London (1848)
Merges, R.P., Reynolds, G.H.: Space resources, common property, and the collective action problem. New York University Environmental Law Journal 6, 107 (1997)
Nauges, H.: Legal aspects. In: Proceedings of the 22nd Colloquium on the Law of Outer Space (1979)
Nemitz, G.W.: Personal communication (November 28, 2005)
North, P.M.: Cheshire and North private international law, 10th edn. Butterworths, London (1979)
Reid, K.G.C.: The law of property in Scotland. The Law Society of Scotland/Butterworths, Edinburgh (1996)
Roberts, L.D., Pace, S., Reynolds, G.H.: Playing the commercial space game: time for a new rule book? Ad Astra (May/June 1996)
Roby, H.J.: Roman private law in the times of Cicero and of the Antonines, vol. I. Cambridge University Press, Cambridge (1902)
Rodriguez, G.: Telepossession Transforms Asteroids into Resources. Space Resources Roundtable VI #6033 (2004)
Schweickart, R.L.: The Real Deflection Dillema. In: 2004 Planetary Defense Conference, Orange County, CA (2004)
Scripps Howard News Service, Own your piece of the Rock. San Francisco Examiner (February 21, 1998)
Silber, K.: A Little Piece of Heaven. Reason Magazine (November 1998)
Simpson, S.R.: Land law and registration. Cambridge University Press, Cambridge (1976)
Smirnoff, M.: Introductory report and summary of discussions - draft resolution on the legal status of celestial bodies. In: Proceedings of the 9th Colloquium on the Law of Outer Space, p. 8 (1966)
Stair, J.: The Institutions of the law of Scotland, Edingburgh (1693)
Sweet, C.: A dictionary of English law. Sweet, London (1882)
Tuchman, B.: On our birthday—America as idea, Newsweek (July 12, 1976)
Tumlinson, R.: Testimony before the Senate Committee on Commerce. Science and Transportation (October 29, 2003)

Turner, F.J.: The significance of the frontier in American history. In: Proceedings of the State Historical Society of Wisconsin (December 14, 1893)
Tylecote, R.F.: A History of Metallurgy, 2nd edn. The Institute of Materials, London (1992)
Urbahn, K.: Putting the torch to a culture of entitlement. Yale Daily News (March 23, 2005)
Vicas, A.G.: The New International Economic Order and the emerging space regime. In: Space Activities and Implications: Where from and Where to at the Threshold of the 80's, p. 293. McGill University Press, Montreal (1980)
Westfall, R.M.: Establishing possession of debris and resources in space. Galactic Mining Industries (2003), http://www.angelfire.com/trek/galactic_mining/rmwgmi.html
Zunt, D.: Who did actually invent the word "robot" and what does it mean? The Karel Čapek website (1998), http://capek.misto.cz/english/robot.html

Author Index

Subject Index

C

D

E

F

G

H

I

J

K

L

R

S